Museo La Specola Florence

ANATOMICAL WAXES

Anatomische Wachsmodelle
Cires anatomiques

With contributions by Monika von Düring & Marta Poggesi
Photographs by Saulo Bambi

Museo di Storia Naturale dell'Università di Firenze,
sezione di zoologia La Specola

TASCHEN

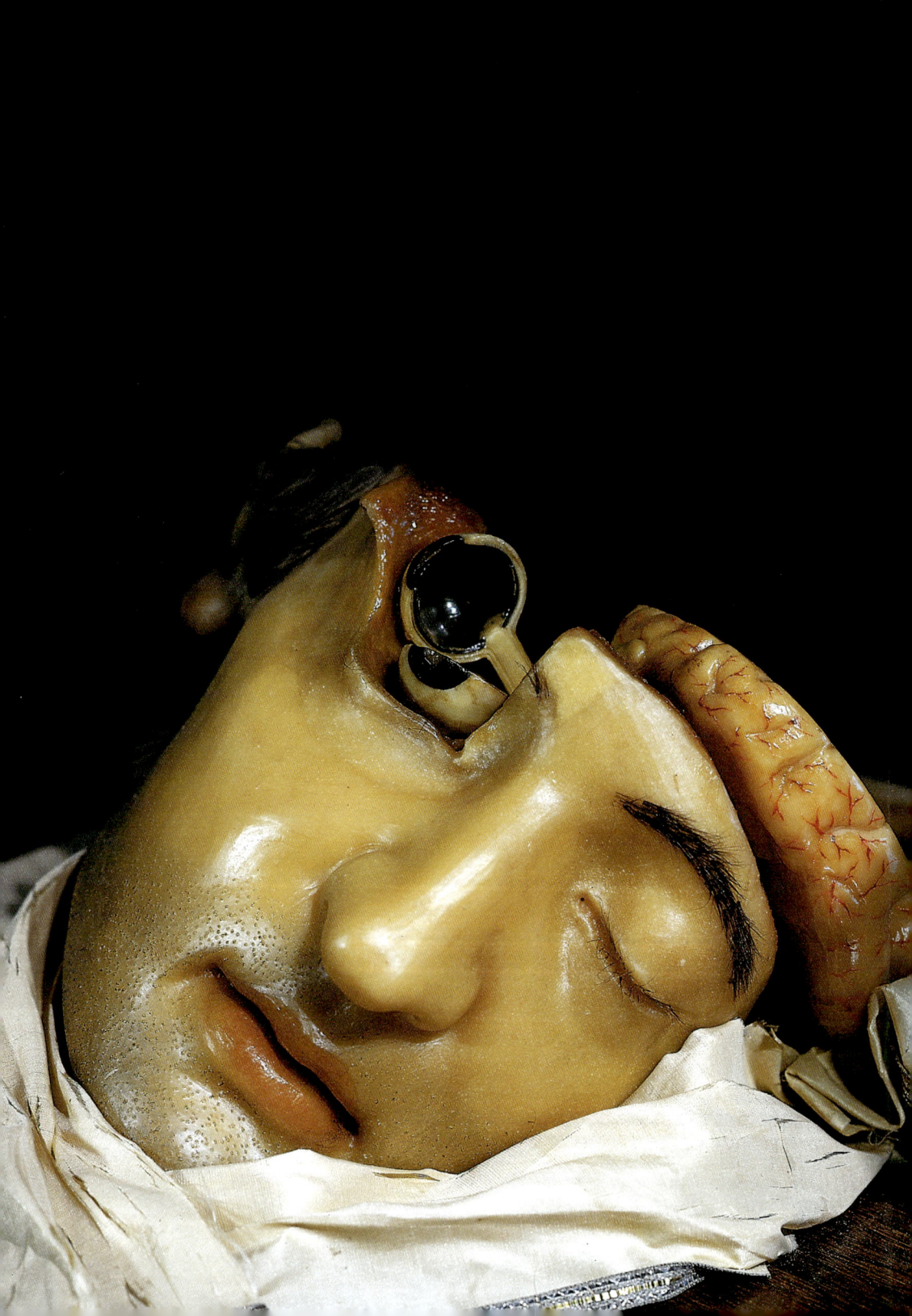

Contents

Anatomical Waxes

with introductory texts by Monika von Düring
60

The Wax Figure Collection at La Specola in Florence

The History of the Collection

When Peter Leopold of Habsburg-Lotharingen (1747–1792), Grand Duke of Tuscany (ill. p. 7), decided in 1771 to bring together all the "scientific" collections from the various galleries in the Grand Duchy, the result was an innovation that was unique not only in Europe but anywhere in the world.

Much earlier than other potentates this enlightened ruler – an enthusiastic student of the natural sciences – had understood the importance of the sciences for the cultural advancement of any society. The first thing he did was to examine the various ways in which the findings of science could be made accessible to all those who were interested.

And indeed the Imperial Regio Museo di Fisica e Storia Naturale (The Imperial-Royal Museum for Physics and Natural History, later widely known as "La Specola", which means "observatory" in Italian) was the first of its kind, in that from the day of its opening on 21 February 1775 it admitted the general public to its collections. It is true that there were separate opening hours for the upper and the lower classes: the latter – "provided they were cleanly clothed" – were allowed to visit between 8.00 and 10.00 in the morning, which then left enough time before the "intelligent and well-educated people" were admitted at 1.00 in the afternoon. But however discriminatory this distinction may seem to us today, nevertheless one can still sense how innovative it was to open the museum's doors to this broader public.

These collections had originally been started by the Medici family; immensely important as patrons and connoisseurs of the arts they had also done much to promote the sciences. Ample evidence of this may be seen in the Accademia del Cimento (1657–1667) which had among its staff such famous scientists as Redi, Magalotti and Galileo's favourite student Viviani. After the death of Giangastone, the last descendant of the family, the Grand Duchy of Tuscany went to Francis III of Habsburg-Lotharingen, who decided to have an inventory made of all the treasures in his residence. This task was entrusted to the physician and natural historian Giovanni Targioni-Tozzetti (1712–1783), who completed the work in just under a year in 1763/64.

When Peter Leopold succeeded his father to the throne in 1765 – after the latter had become Emperor of Austria – he therefore found that the groundwork for a reorganisation of the collection had already been done. The work involved in this major undertaking fell to Felice Fontana (1730–1805; ill. p. 27), by profession a teacher of logic at the University of Pisa, but also an anatomist, physicist, chemist and, above all, an internationally renowned physiologist. He devoted himself to restructuring the buildings with such passion that by the end of 1771 the first items could already be moved into the new rooms. As early as 1771 the Grand Duke had already bought the Palace of the Torrigiani in the via Romana, very close to the Palazzo Pitti. He had also bought a number of neighbouring

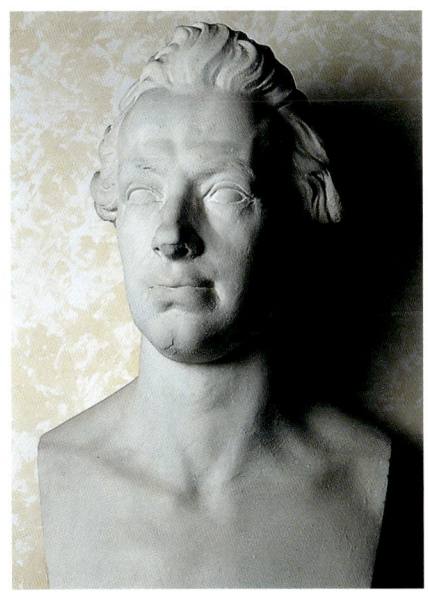

houses and had commissioned the Abbot Felice Fontana of Rovereto to draw up plans for alterations to these buildings in order to create a home for his scientific collections. As director of the new museum, in the early years Fontana travelled throughout Europe acquiring books and collections and establishing contacts in different countries. As a result the museum in Florence became one of the most important museums of its day, not least for its rich scientific library.

Fontana directed the museum until his death. Throughout this time Giovanni Fabbroni was at his side – at first as his assistant (and constant antagonist) and from 1784 onwards as the museum's deputy director, accompanying Fontana on numerous journeys. In 1805 he took over as director but only for one year.

It is interesting to note that the money to construct the museum for physics and natural history (and many other projects instigated by Peter Leopold) was raised by the sale of valuable objects once owned by the Medici family – despite a testament left by the Palatinate Electress Anna Maria Luisa (1667–1743), Giangastone's sister, which decreed that the estate of the Medici family should in its entirety and irrevocably remain in the ownership of the city of Florence.

The core holdings of the museum, which came principally from the Uffizi, consisted of collections (minerals and shells, for instance), natural history curiosities from the era of

Bust of Peter Leopold of Habsburg-Lotharingen
Büste von Peter Leopold von Habsburg-Lothringen
Buste de Pierre-Léopold de Habsbourg-Lorraine

the Medici family, Galileo's instruments and equipment from the Accademia del Cimento as well as four wax figures by the Sicilian sculptor Giulio Gaetano Zumbo.

In 1771 a ceroplastic studio was set up together with other workshops essential to the running of the museum (a carpentry shop, a glaziery and a taxidermy studio): these will be discussed in more detail later in this essay. Keen to have astronomy and meteorology included in the museum as well, in 1780 Peter Leopold commissioned the architect Gaspare Paoletti (who had earlier been involved in the restructuring of the palace) to build the Osservatorio Astronomico (the so-called "little tower"). This was later to be the source of the name "Specola", which means observatory. The work turned into a major project which led to the building's being much larger than originally planned. In fact many experts had advised against it and had advocated building on the Acetri hills instead. In 1789, the year of the completion, a part of the Boboli Park was incorporated as the museum's Botanic Garden (ill. p. 8).

After the death of his brother Josef in 1790, Peter Leopold became Emperor of Austria and put Tuscany into the hands of his second son Ferdinand III (1769–1824), who had neither his father's vision nor his skill in matters of government. But the times were also against him: Napoleonic expansionism forced the Lotharingians to give up Tuscany which, after various complications, went to the Bourbons of Parma. During this period the museum

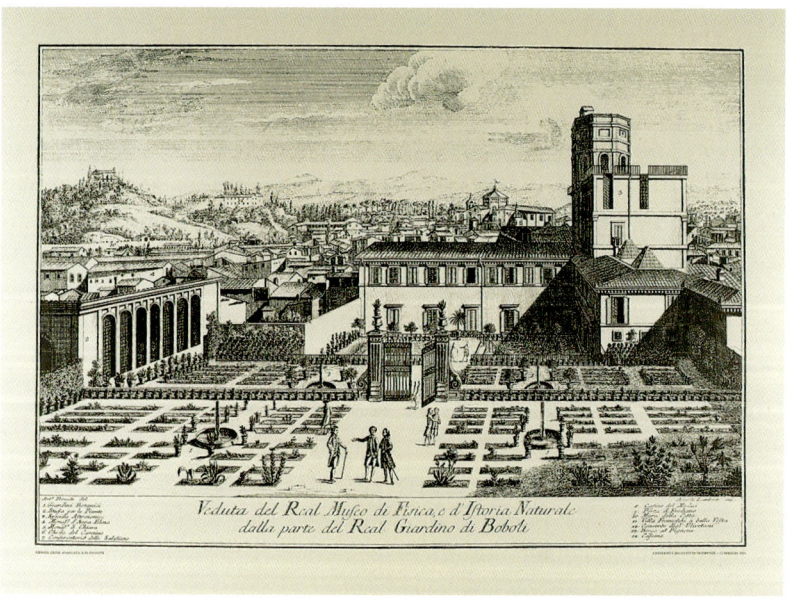

Veduta del Real Museo di Fisica, e d'Istoria Naturale dalla parte del Real Giardino di Boboli

La Specola seen from the Boboli Garden; engraving from the late 17th century
Die Specola vom Boboli-Garten aus gesehen; Kupferstich vom Ende des 17. Jahrhunderts
La Specola vue du jardin Boboli ; gravure de la fin du XVIIᵉ siècle

started to organise the teaching of scientific subjects, which was continued in 1814 after the restoration of the Lotharingians.

Under Ferdinand's rule the museum had to survive a period of unrest and lost a great deal of the scientific importance attached to it in its early years when it enjoyed a reputation as one of the most important centres of learning in Europe according to illustrious foreign authors such as Goethe and Bernoulli. Ferdinand also had major structural alterations made to the palace: in 1820, under the direction of the architect Pasquale Poccianti a corridor was constructed joining La Specola with the *Meridiana* wing of the Palazzo Pitti, so that the Poccianti Corridor now extends the Vasari Corridor – which leads from the Palazzo Vecchio via the Uffizi to the Palazzo Pitti – as far as La Specola.

After his death in 1824, Ferdinand was succeeded by his son Leopold II (1797–1870), affectionately nicknamed "Canapone" by the Florentines on account of his blond hair. It was thanks to him that the teaching and lectures in the sciences were reinstated, with a particular emphasis on the applied sciences relevant to agriculture and the cultivation of reclaimed land. During his time as regent, the Tribuna di Galileo was constructed. Dedicated to the memory of one of the greatest scientists, it was inaugurated in 1841 on the occasion of the third congress of Italian scientists in Florence. The Tribuna is a large room on the first floor of the building and was partially rebuilt by the architect Giuseppe Marelli. Work was started in 1830, the initial plan being simply to add an apse to the existing room, but the scheme was then altered in accordance with the wishes of the Grand Duke: in order to lend more weight to his intended celebration of Galileo and his work, the whole room was to be dedicated to the renowned scientist. As well as a statue, it was to contain all the surviving Galileo memorabilia and the instruments from the Accademia del Cimento. The building of the Tribuna, one of the few examples of late Neo-Classicism in Florence, led to a number of important alterations to the palace, most particularly to the floors below the Tribuna. It also meant that a part of the courtyard was roofed over. The room itself was decorated specifically and exclusively with Tuscan marble and with works by artists – sculptors and painters – who were also exclusively from Tuscany.

After the fall of Leopold II – the last Grand Duke of Tuscany – in 1859 the Istituto di Studi Superiori e di Perfezionamento (Institute of Higher Study and Further Education) was founded. In 1923 this became the University of Florence. The museum buildings became part of the Institute and now contain the University's Department of Physics and Natural Sciences.

While the founding of the Institute was of inestimable importance, it also led to the breaking up of the museum into its various disciplines, which – along with the relevant collections and libraries – had to be gradually relocated to other sites in view of the constantly increasing numbers of students. Even the resistance of many scientists at the time could not hinder this process, with the result that the historical buildings in the via Romana now only contain the zoological collections and the bulk of the anatomical ceroplastics.

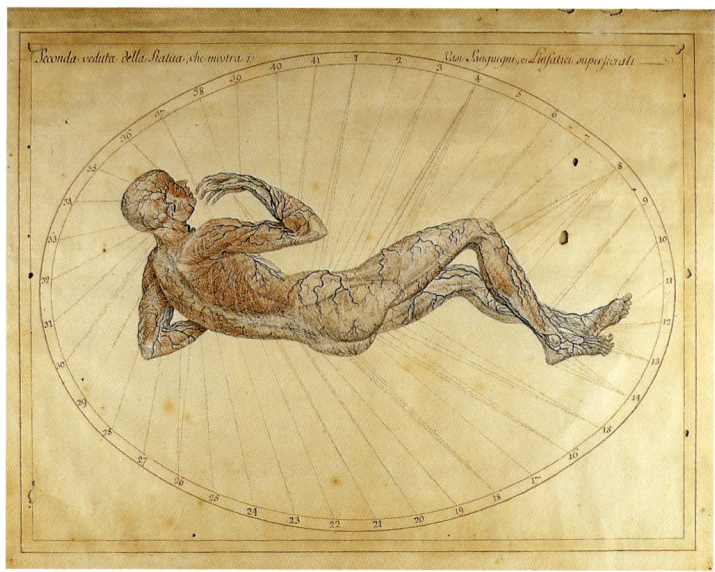

The Early Days of Anatomical Wax Models

Wax has been used to make models since time immemorial; it has always been a popular material for artists, both for aesthetic and for technical reasons, because it is easy to work and allows the finished piece to be cast in metal afterwards. It was used widely for religious motifs such as statues of saints or cribs and votive panels. At the time of the Renaissance and throughout the 17th century, huge numbers of wax figures could be found in churches, particularly in Orsanmichele and in Santissima Annunziata. These included limbs, organs or parts of organs as well as portraits, objects and statues – often even life-sized. In the 17th century, wax models were also increasingly used for scientific purposes.

Although the state – and even more so the Church – did its best to prevent corpses being used for anatomical studies and penalised anyone who attempted to do so, by the mid-15th century the first anatomical drawings and treatises were beginning to emerge, generally drawn by great artists and sculptors such as Leonardo da Vinci, Michelangelo, Raphael and Titian, to name but a few. Renowned anatomists such as Fallopio, Cesalpino and Vesalius used drawings by these masters to illustrate their treatises.

Towards the end of the 17th century, Gaetano Giulio Zumbo, working in Bologna where there was a famous school of anatomy, was the first to make anatomical models using coloured waxes. Zumbo came from Syracuse and his works, which are illustrated here, include two male heads (one now in Florence and one in Paris), a female figure (now unfortunately lost) and some other smaller works. Bologna was also the first place to have a proper school of ceroplastics and it was here that the wax modellers working in the museum in

Original drawings of a
recumbent figure, called
lo spellato

Originalzeichnungen einer
liegenden Figur, genannt
lo spellato

Dessins originaux d'une
figure couchée appelée
lo spellato

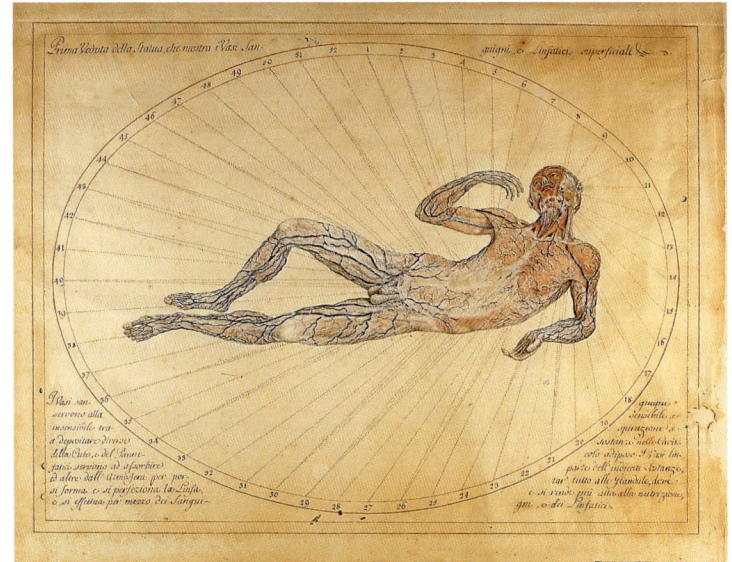

Florence during the second half of the 18th century were trained. The painter, sculptor and architect Ercole Lelli (1702–1766) is credited with founding the Bolognese school and, like Zumbo before him, he too made wax models of muscles and inner organs by studying real corpses. He was able to obtain access to the corpses he needed for this work through the support of Cardinal Prospero Lambertini, who became Pope Benedetto XIV in 1740. As an amateur scientist and a man of learning, the Cardinal encouraged the study of anatomy at the school in Bologna and bought all the models Lelli made; they are magnificent pieces, now preserved in the Istituto d'Anatomia Umana Normale at the University of Bologna.

Among Lelli's assistants were Giovanni Manzolini (1700–1755), who with his wife Anna Morandi (1714–1774) also made numerous models that can likewise still be seen at Bologna University.

At the time, the study of anatomy was progressing throughout Europe but the school of ceroplastics in Florence was in fact a direct offshoot of the school in Bologna. The link was made by the surgeon and childbirth specialist Giuseppe Galletti. Having been much impressed by models made by Lelli and Manzolini, the physician Galetti – together with the wax modeller Giuseppe Ferrini – himself made a number of models in wax and terracotta (now in the Museo Galileo – Istituto e Museo di Storia della Scienza in Florence) in order to demonstrate different birth procedures, both normal and with complications.

The Wax Figures at La Specola:
The School, Works, Techniques

Although the ceroplastic workshop existed for more or less a century – from 1771 until the mid 19th century – the majority of the wax models in the museum, like the commissioned works for other institutions in Florence and elsewhere, were made during the first fifty to sixty years of the workshop's activities.

The founding of the school goes back to Felice Fontana. Supported by Grand Duke Peter Leopold, Fontana devoted himself with great energy and determination to setting up a wax workshop, particularly since – as an anatomist and pathologist – he was personally involved in the creation of the models. At first there was just the wax modeller Giuseppe Ferrini working under Fontana's guidance in the workshop, but subsequently the anatomist Antonio Matteucci and the very young Clemente Susini were also employed. Immensely talented and hugely productive, Susini was later to become the most important and famous of the wax modellers of the Florentine School (ill. p. 14).

We have no precise information today as to where the wax was worked, although it seems most likely that this would have been in studios on the ground floor in the south wing of the palace which is divided into a number of small courtyards and has windows looking out on to the via Romana. Practically none of the original equipment used has survived, but archive records indicate that the following at least were purchased: copper vats in various sizes for melting the wax; modelling tools (ill. p. 31); metal wire in different thicknesses; marble slabs for pressing the wax into thin sheets; scales; tripods for heating substances; slates for making notes and drawings during dissections; crates with handles for transporting the corpses, wooden crates with carrying poles for transporting the wax figures; containers, vases and bottles made from clay or glass for pigments and other substances that were added to the wax. Many of the latter were found with their contents still intact in the museum's storerooms (ill. p. 30).

Furthermore, it is also clear from the archives how many corpses or parts of a corpse were necessary to make a wax model: astoundingly, over two hundred for a single figure. This startlingly large number is a direct consequence of the fact that there was no way of preserving or freezing the corpses. Consequently there was a constant need for fresh subjects if a dissection were to be accurate and thorough. In order to keep an exact record of all the corpses or parts of corpses – which came from the Santa Maria Nuova hospital about two kilometres away – a register was put up at the door where all corpses being admitted or

G. G. Zumbo: Specimen of a head (cf. pp. 41, 299, 302)
Präparat eines Kopfes (vgl. S. 41, 299, 302)
Préparation anatomique d'une tête (cf. p. 41, 299, 302)

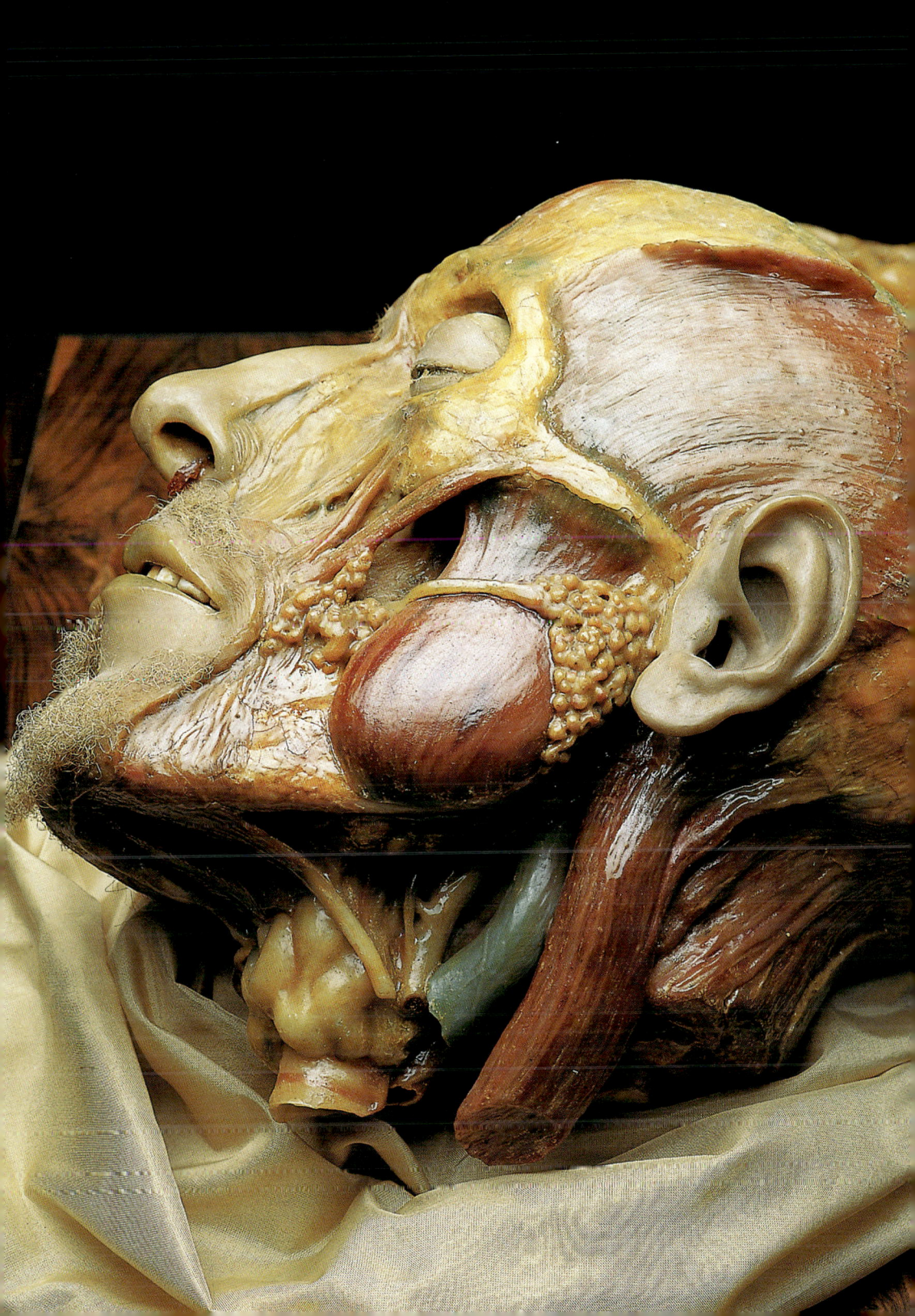

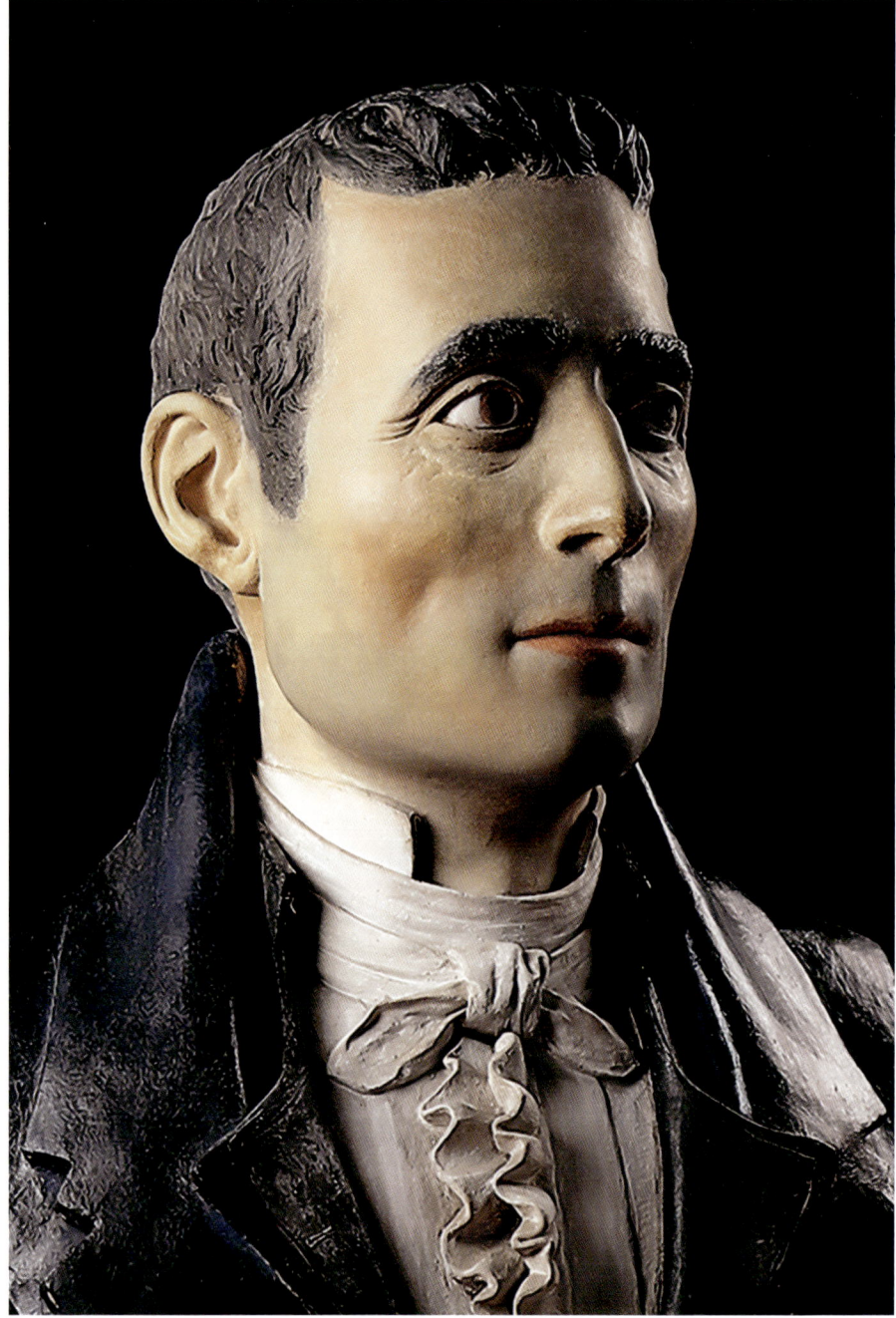

Egisto Tortori: Coloured plaster relief of Clemente Susini
Gipsbüste von Clemente Susini, farbig gefasst
Buste en plâtre polychrome de Clemente Susini

ANATOMICAL WAXES

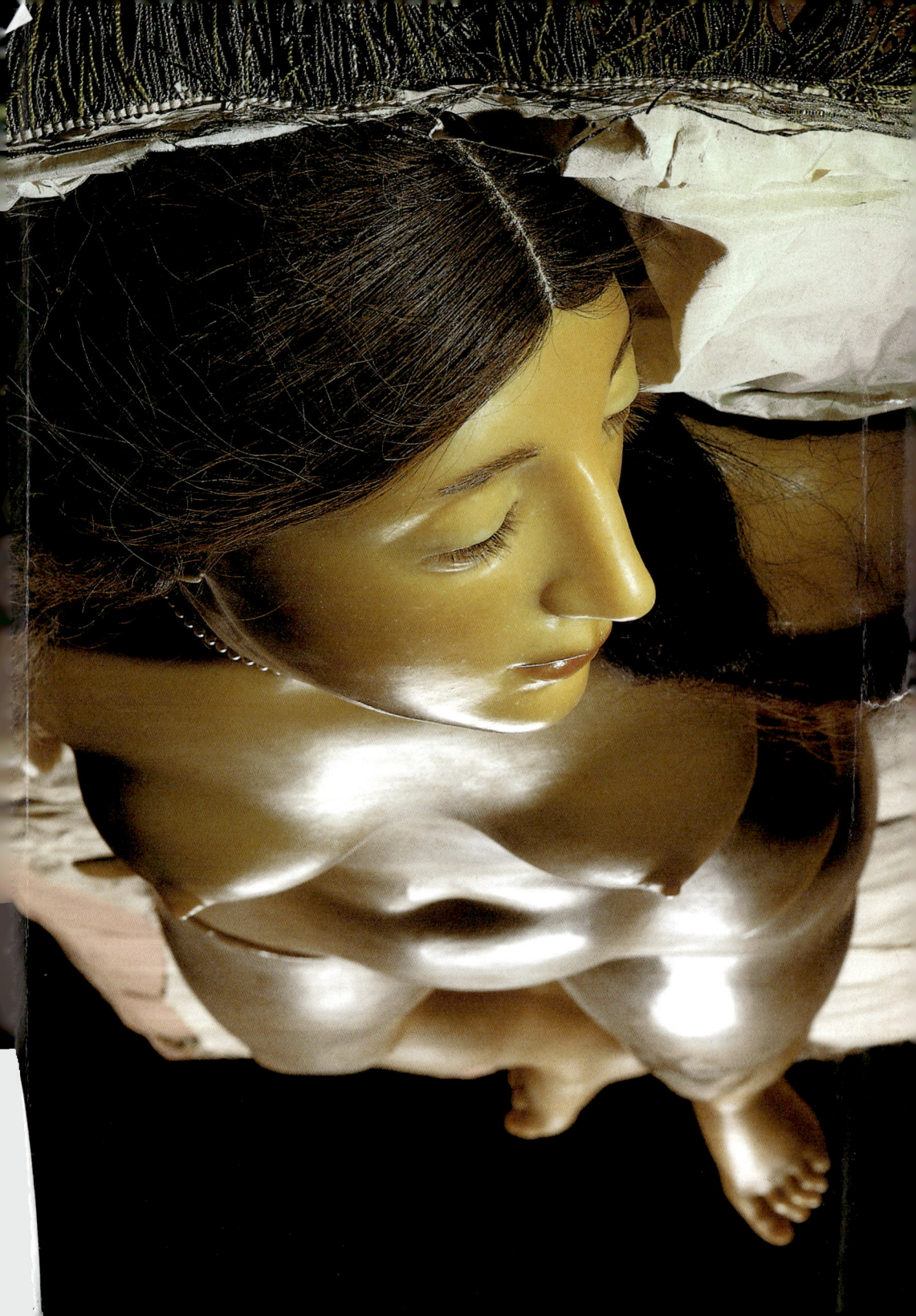

dispatched to the cemetery were listed: "[...] with regard to the corpses from Santa Maria Nuova, of which it is not known how many there are and because one may not have people die purely for the benefit of the wax collection, a register was put up at the entrance door on which all admissions were noted as well as the number of corpses sent to the cemetery." (from a document written in 1793).

Despite the small number of employees the work progressed very rapidly, when one considers that in 1790 – twenty years after the workshop was set up – the models already filled eight rooms, and this did not even include the models that were made specifically for the hospital in Florence and for other institutions in Italy and abroad (cf. ill. p. 46). Most outstanding among the wax figures from the Specola workshop that were soon to be found all over Europe was the collection commissioned in 1781 by the Austrian Emperor Josef, Peter Leopold's older brother, for the military school of medicine in Vienna (called the Josephinum after the Emperor). This collection contains 1,200 items made in 1786 and transported to Vienna in two deliveries on the backs of mules. Other models were destined for Pavia, Cagliari, Bologna, Budapest, Paris (today in Montpellier), Uppsala, London, Leiden and elsewhere. These outside commissions occasionally caused problems since the Grand Duke was determined that they should not hold up the production of works for the museum. Thus in the case of such commissions the stipulation was that the modellers and anatomists – Fontana himself worked on the collection for Vienna – were allowed to use the museum's moulds and tools, but that they had to procure the materials themselves and carry out the work with the assistance of craftsmen not in the employ of the museum.

It was Fontana's ambition to produce as many wax figures as possible in order to create a teaching resource which would in the future obviate the need to exhume corpses for the study of the human anatomy. To this end he included in his collections a number of tempera drawings (ill. pp. 10, 11) showing individual parts with numbers around the drawn figure linked by fine lines to the different organs, which could then be identified using these numbers. The relevant written explanations would then be found in a drawer that was part of every wax figure's display cabinet: by this means the user would, as it were, have access to a complete three-dimensional anatomical treatise (ill. p. 20).

Around 1790 Fontana began another immensely ambitious project which was, however, never realised according to the technical difficulties encountered: the plan was to make a series of painted wooden anatomical models – either life-sized or somewhat larger – which could be taken apart for teaching purposes in order to demonstrate how the organs connect with each other. After a few initial attempts of which only a few examples have survived – one model in the museum, one in Paris and another enormous bust – Fontana had to abandon his plans. Not only was the wood harder to work but it also had a tendency to warp, which meant that all the craftsmen's efforts to make the individual parts fit into each other were rendered useless with time (ill. p. 37). The only wax model in the museum which has the desired qualities is the so-called "Venere medicea": a recumbent female figure which can be dismantled from above, layer by layer, eventually revealing a uterus with a small foetus in it.

By this time the museum had increased its personnel to include a graphic artist (Claudio Valvani) and a woodcarver (Luigi Gelati) as well as various anatomists for carrying out the dissections. Among the most famous of these was Paolo Mascagni (1755–1815) who left behind extremely beautiful anatomical plates. He specialised in the study of the lymphatic system, and it is no coincidence that many of the recumbent statues and the smaller works display this part of the human body in minute detail.

In 1805 Felice Fontana died (ill. p. 17), but the wax workshop continued with its work, and even after the death of Clemente Susini in 1814, other wax sculptors, such as the Calenzuolis (Francesco Calenzuoli and his son Carlo) and Luigi Calamai (1800–1851), took over. Under Calamai the work was concentrated on comparative anatomical models and botany as well as on pathological models that were made for the Santa Maria Nuova hospital, which can still be seen today in the Department of Pathological Anatomy in the University. After Calamai's death, Egisto Tortori (1829–1893) took over. Besides making figures of animals and models for the Department of Pathological Anatomy, Tortori also made a bust of Clemente Susini. When Tortori died, however, no one was appointed to replace him and the workshop was closed down forever.

Of the wax figures now stored in urns at La Specola, there are 513 of human anatomy and 65 comparative figures; 5 made by G. G. Zumbo. There are 26 whole figures (including the half-finished figure of a young boy): 13 standing and 13 recumbent, 18 of which are life-sized (6 standing and 12 recumbent) and 8 approximately 60 cm in height (7 standing and 1 recumbent). Almost all are on display; only 14 of the human figures and parts are kept in storage. There are more than 800 framed drawings and almost 900 explanatory notes, and yet it is clear that there must have been more that have gone missing with time. In recent years both the original drawings and the written explanations have been removed from display in order to prevent them suffering further damage through light and humidity. The original frames now contain copies so that the collection is as useful as it ever was for teaching purposes and also still constitutes a worthwhile display.

The scientific perfection of these works could hardly be surpassed, and although this is mainly appreciated by experts in the field, most visitors are equally impressed by their artistic quality. For however scientifically accurate these exhibits are and whatever their didactic purpose was, they are also artworks in their own right. In addition their aesthetic qualities are heightened still further by the beauty of the wooden display cabinets in which they are kept, despite the fact that many show signs of wear and tear and are badly in need of restoration. The care that originally went into the presentation of this collection is also evident from the equally fragile 18th-century silk drapes and veils – now faded but once bright green.

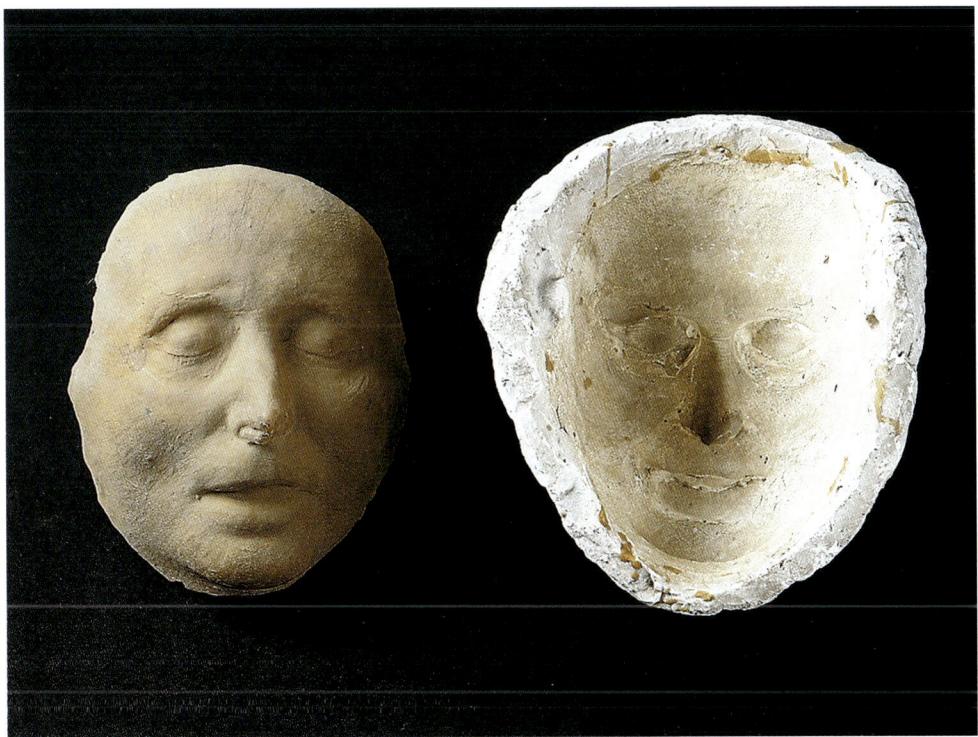

The importance that was placed on aesthetic appearances may be seen again in the collections of comparative anatomy, of human pathology and above all of botany: small ceramic vases were specially made in the workshop of Ginori von Sesto Fiorentino for the plant models.

We have no detailed knowledge of the technical processes involved in the production of wax figures, and all that is known has had to be deduced from various letters and documents in the state archives and in the archives of the Museum for the History of Science in Florence. Besides this, each wax modeller also had his own technique, and – like any other craftsmen or artists – they were not keen to have their methods spread abroad. However, the following is relatively certain: the dissected item would first of all be copied exactly in chalk or low-grade wax. A plaster cast would be taken of this, which could either be very large or made up of a number of sections. It was sometimes possible to make a cast directly from the pieces in question, such as bones for instance. These moulds, which are still stored in the museum (ill. p. 55), subsequently served as a kind of matrix which could also be used

Wax death-mask of Felice Fontana and plaster relief
Totenmaske aus Wachs von Felice Fontana und Gipsabdruck
Masque mortuaire en cire de Felice Fontana et moulage en plâtre

repeatedly for making additional casts of the same item. The hardest, most problematic part of the process was, however, the construction of the final model. This required great precision and knowledge of the substances that had to be added to the wax in order to achieve the desired colour and consistency – it was a matter of both skill and experience. The most crucial part of the process was the slow melting of the wax in a water bath at exactly the right temperature so that it did not discolour. Most commonly the modellers used white wax from Smyrna, Chinese wax or Venetian wax; in order to make it more elastic, turpentine was added as well as paints (thinned with turpentine) and other substances that are recorded in the lists of purchases "needed for the wax models". Before the wax mixture was poured into the mould, the latter would be moistened with lukewarm water and rubbed with soft soap to facilitate the removal of the model. Apart from a few which were solid, most of the wax models were hollow, and in order to stabilise them, they would be stuffed with rags, hemp waste or wood chippings. Figures which were constructed from a number of separate parts generally had a supporting metal frame. After the moulds had been re-moved, each item would be cleaned and finished, that is to say, equipped with the relevant organs, vessels, nerves and so on until it was complete; a final coating of clear varnish lent the whole a suitably glossy appearance. To ensure that the models were completely accurate each stage had to be supervised by the anatomists who were also responsible for deciding on how to place certain organs so that they might be seen to best advantage.

As has already been mentioned, each wax modeller developed his own technique and equipment, which would become increasingly refined, through the invention of new in-struments such as a "drawing iron to turn the wax cylinders" which they used from 1786. Previously this process, which was essential for producing blood vessels, had to be carried out by hand. Towards the close of the century, the wax modellers kept notebooks (ill. p. 21) in which they recorded what they had done each day and which tell us that even in those days it was frequently necessary to carry out restoration work, repairs, improvements and reworkings. In 1793 Susini made a note "about anatomical wax models in need of improve-ment as a result of flaws that should be treated and which are described in the following by me, Clemente Susini, modeller to the Royal Museum". For instance, on "examination of models of the pharyngeal artery and the jugular vein, leading off from the windpipe one sees two foreign muscles, that is to say, muscles which do not belong in the human form". Having noted further anomalies Susini ends with the comment that "these errors are too grave to be left uncorrected and, for the sake of brevity, I should rather not speak of the many others". Almost all the models have survived in good condition, and only a few have become at all discoloured, although here and there a foetus has darkened in colour and we know that some now greenish veins were once a clear "blue-purple".

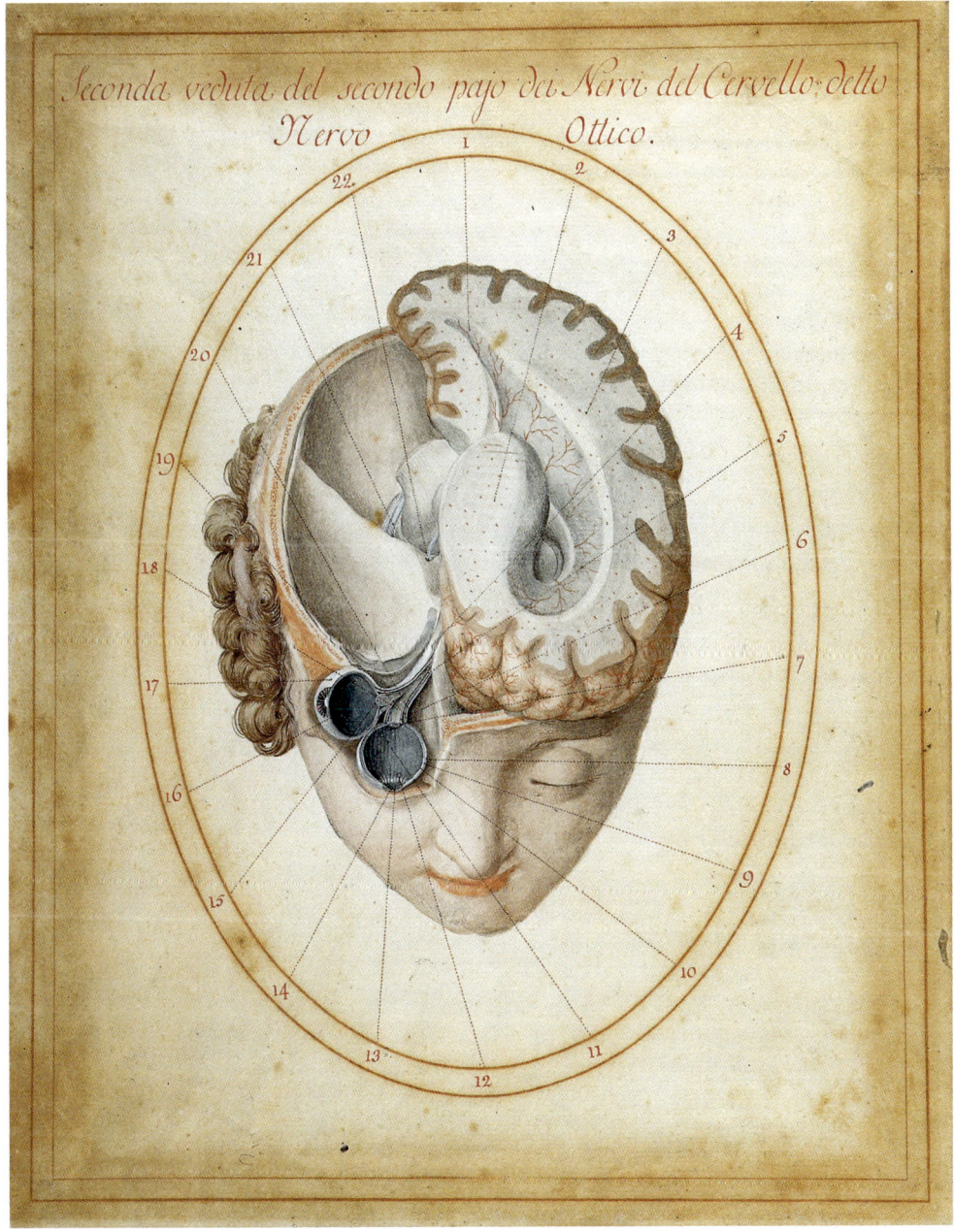

Drawing of a head, front view, 2nd half of the 18th century
Zeichnung eines Kopfes: Frontalansicht, 2. Hälfte des 18. Jahrhunderts
Dessin d'une tête vue de face, 2ᵉ moitié du XVIIIᵉ siècle

The Wax Figures of Gaetano Giulio Zumbo

Gaetano Giulio Zumbo's wax figures deserve a chapter of their own, partly on account of their unusual conception and production but above all because they pre-empted the use of wax for anatomical models by a good century and laid the foundations for everything that was to follow.

Zumbo was born into a noble family in 1656 in Syracuse; the family name was probably Zummo, but this name has now disappeared without trace, and since the debased version "Zumbo" is much better known and widely used it seems best to use it here. We have little information on his life – and what we have is only fragmentary. It is known that he was educated in the Jesuit Institute in his hometown but had to leave following some – otherwise unspecified – "irksome incident". He may well have received his earliest inspiration as an artist from the Classical relics of his homeland, followed later by the pictorial world of Mannerism and the early Baroque. From 1687 to 1691 he lived in Naples and it may well have been there that he created two of the groups still preserved at La Specola: the "teatrini" with the title *Il Trionfo del Tempo* (The Triumph of Time) and *La Peste* (The Plague). According to scholarly experts in the field who have studied Zumbo's work, this may be concluded from the painterly background and the particular arrangement of the figures, both of which point unmistakably to southern Italian precursors. From 1691 to

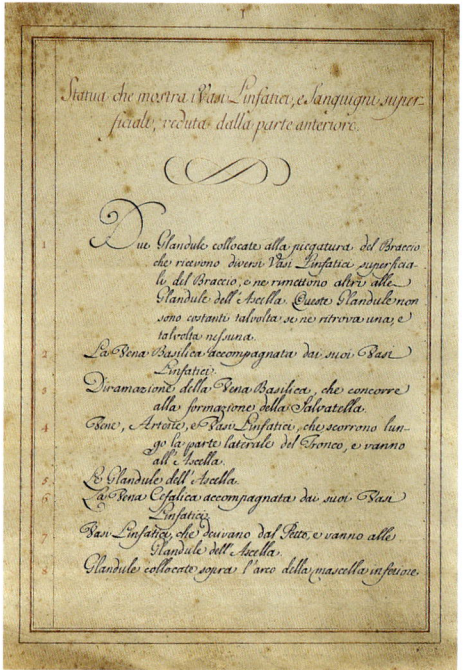

1694 he worked for Cosimo III de' Medici in Florence, where he created another two 'teatrini': *Il Sepolcro* (or *La Vanità della Gloria Umana*) and *Il Morbo Gallico* (or *Sifilide*) which display many Renaissance features. In 1695 he moved to Genoa where it seems he completed *The Anatomy of a Head* on show in the museum there, and he also made the acquaintance of the French surgeon Guillaume Desnoues. They later worked together on a number of anatomical wax models, including the life-sized figure of a woman giving birth, a smaller model of a woman who has died during childbirth, and the burial of Christ – unfortunately none of which has survived. In 1700 Zumbo and the French physician fell out, and Zumbo moved to Marseilles where he created a wax head. His last move was to Paris where he entered the service of Louis XIV, and where he died from a brain haemorrhage in 1701. On his death his estate included a beautiful wax head, most probably the very one that is now in the Museum National d'Histoire naturelle. His grave in the L'Église de Saint Sulpice was destroyed during the Revolution.

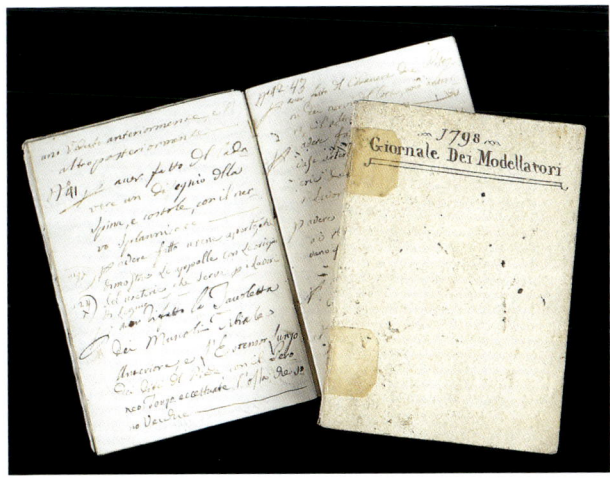

The bulk of this artist's work, largely ignored until fifty years ago, is now at La Specola. The three "teatrini" which are also known as the *Cera della peste* ("plague figures") and *The Anatomy of a Head* were in fact in the Royal gallery (that is to say in the Uffizi) when the Medici were succeeded by the Lotharingians, and were then passed to the Museo de Fisica e Storia Naturale, but never put on display there – perhaps because of the shocking nature of their subject matter. When Peter Leopold set off for Vienna in 1790 he gave the "plague figures" to the court physician Giovanni Giorgio Hasenöhrl, better known as Lagusius. The latter did not even take possession of them but commissioned first Fontana and then Agostino Renzi to sell them for 150 sequins. Renzi was well aware of the importance of these groups and offered them to the new Grand Duke Ferdinand III for the Royal Academy of Fine Arts. When the pieces were rejected as neither "useful nor suitable" he offered them instead to the Duke for the Royal Museum. This proposal was accepted in principle by Giovanni Fabbroni, the deputy director of the museum, and the three works were valued by various experts including Clemente Susini. Since the valuations turned out higher than the price set by Lagusius, the Grand Duke authorised their purchase. Thus Zumbo's works remained in the museum until 1878 when these three dramatic scenes were given for safekeeping to the museum Nazionale del Bargello. From there they went to the Museo di Storia della Scienza. There they were badly damaged during the torrential rainstorms of 1966 but, on the basis of photographic records, were meticulously and miraculously restored over a period of eighteen months by Guglielmo Galli from the Pietre Dure workshop in Florence. *The Anatomy of a Head*, on the

Diary kept by the wax modellers
Tagebuch der Wachsbildner / Cahiers des modeleurs

← Explanations of the drawings for *lo spellato*
Erläuterungen zu den Zeichnungen für *lo spellato*
Explications des dessins réalisés pour *lo spellato*

other hand, never left La Specola since it was rightly regarded as an essential component in the scientific collection of the Institute.

The technique used by Zumbo was considerably different from that used later in Florence. *The Anatomy of a Head* was made using a real human skull, which X-ray examination has shown must have belonged to a man of around 25 years old. By contrast, the head now in Paris and another head (now lost) were made completely from wax, which would seem to indicate advances in the master's technique, perhaps as a result of his collaboration with the French physician Desnoues. The figures in Zumbo's dramatic scenes, executed with the greatest artistry, were almost certainly produced using plaster moulds which in turn had been made from meticulously sculpted clay models. The more or less liquid wax (a mixture of bees' wax, resin, turpentine and pigments) was applied in relatively thin layers in order to achieve the desired consistency and colouration. The scant knowledge we have of Zumbo's technique we owe to the restoration work carried out by Guglielmo Galli after the floods of 1966. This work also revealed evidence of older repairs, for during his work Galli identified pigments that could not have existed in the late 17th century. It also seems that the small self-portrait (ill. p. 53) in the group *Trionfo del Tempo* cannot with any certainty be credited to Zumbo and could well be the work of a contemporary or a later work of Susini's. The *Morbo Gallico*, originally a similar composition, of which now only some fragments remain, was a present from Cosimo III to Filippo Corsini. Unfortunately, this was kept in the cellars of the Palazzo Corsini by the River Arno and was almost completely destroyed during the floods. A few of the figures were found in the gardens, but since there were no photographs of the group it was impossible to reconstruct it. On the occasion of the 200th anniversary of the opening of the museum, which was celebrated with a conference on wax modelling, the Corsini family donated the remains of the work to the museum.

For many years Zumbo's masterpieces were misread as the diseased caprices of an artist revelling in macabre and repellent details. It was not until the mid-20th century that they began to be seen in a more appropriate historical perspective: as realistic documentations of an era when death was omnipresent in the shape of war, famine and major epidemics. This vision of the destruction of human life, of its fragility, of the relentless passage of time, and the constant, specific reminders of death that face us in Zumbo's works are simply typical of the culture and thinking of the 17th century, and put him in the company of other great artists of the time such as Luca Giordano and Mattia Preti.

G. G. Zumbo: *Il Trionfo del Tempo*
The Triumph of Time (detail) / Der Triumph der Zeit (Detail)
Le Triomphe du Temps (détail)

Pages 24–25:
G. G. Zumbo: *Il Morbo Gallico* (or *Sifilide*)
Syphilis / Die Syphilis / Le Morbo Gallico

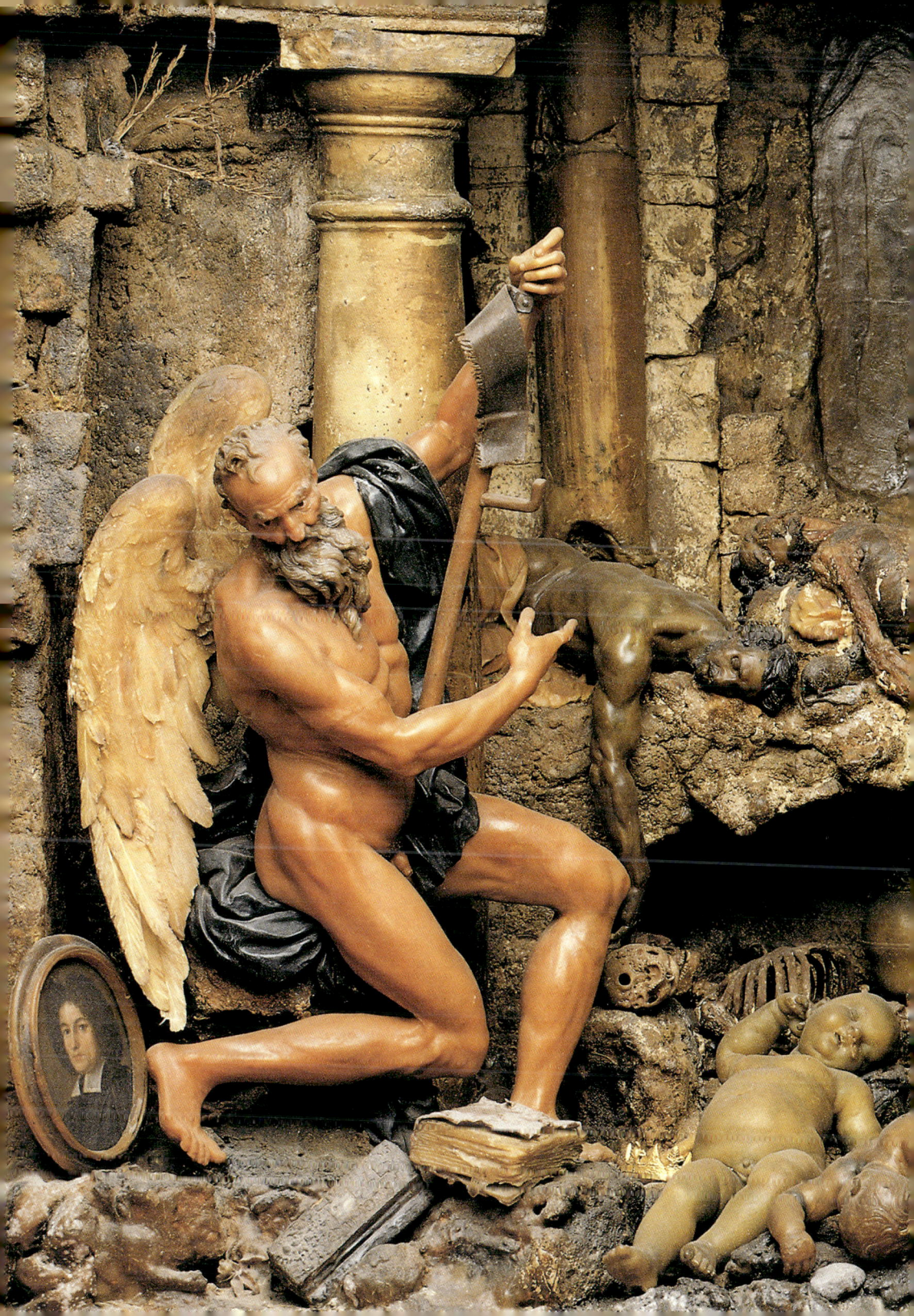

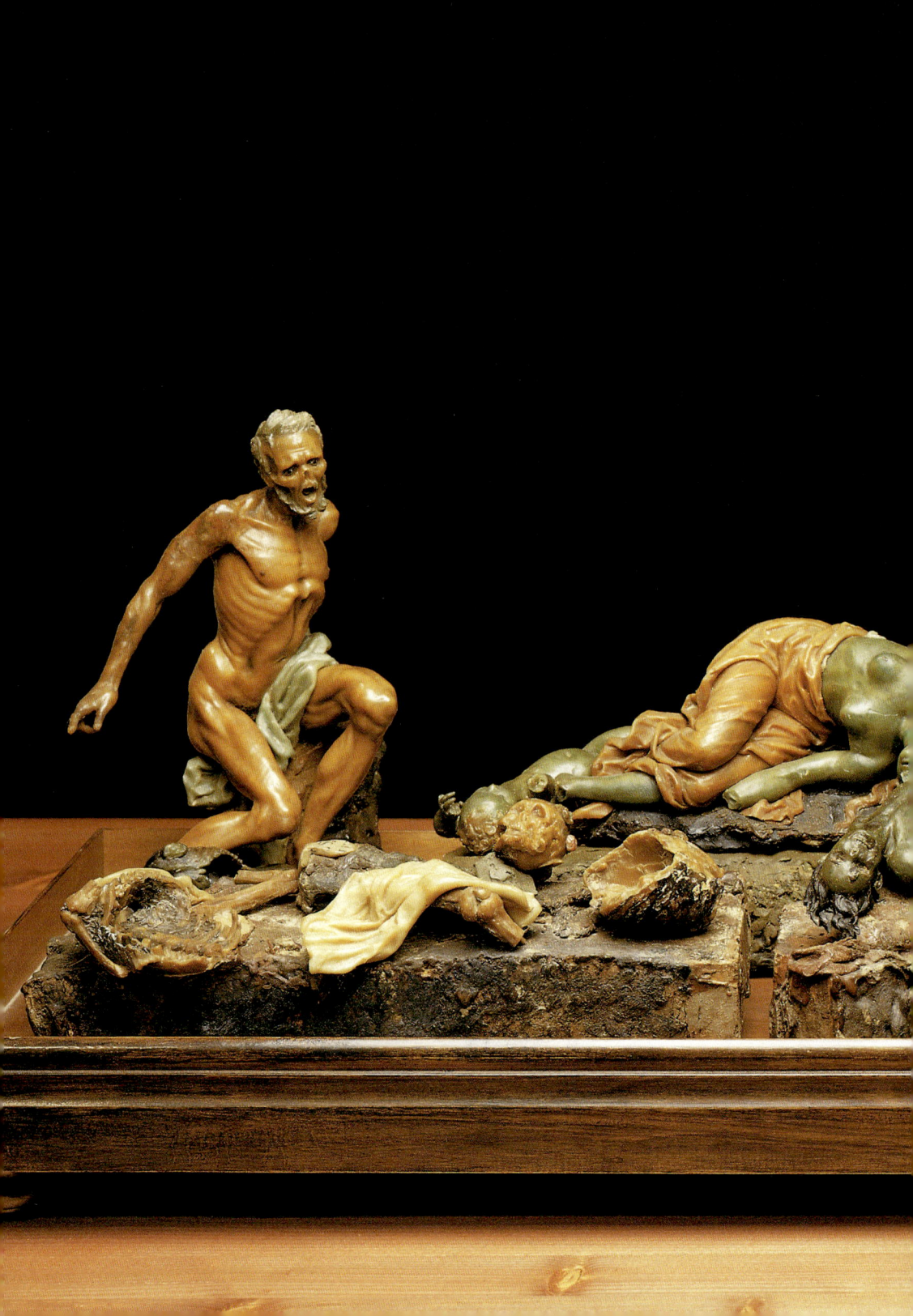

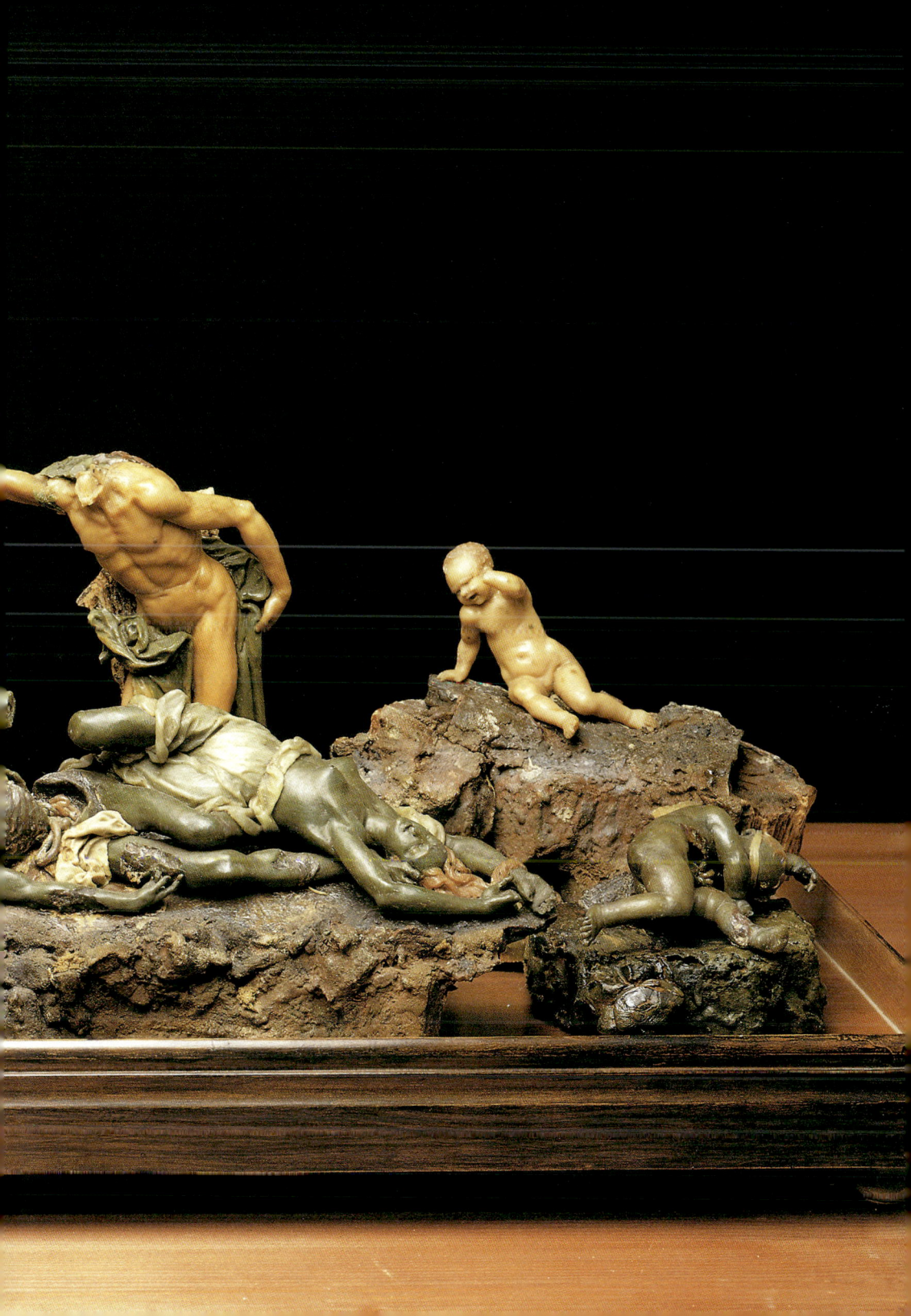

Die Wachsfigurensammlung des Museums La Specola in Florenz

Die Geschichte

Als Peter Leopold von Habsburg-Lothringen (1747–1792), von 1765 bis 1790 Großherzog der Toskana (Abb. S. 7), 1771 beschloss, alle „wissenschaftlichen" Sammlungen der großherzoglichen Galerien in einem Museum zu vereinigen, führte er eine Neuerung ein, die in Europa und in der Welt ihresgleichen suchte. Viel früher als anderen Herrschern war dem aufgeklärten Landesfürsten, selbst ein begeisterter Student der Naturwissenschaften, deren Bedeutung für die kulturelle Entwicklung der Gesellschaft bewusst geworden. Als Erster sann er über Möglichkeiten nach, die Errungenschaften der Naturwissenschaften allen Interessierten zugänglich zu machen.

Und in der Tat war das Imperial Regio Museo di Fisica e Storia Naturale (das Kaiserlich-Königliche Museum für Physik und Naturkunde, das später allgemein „La Specola" – „Sternwarte" im Italienischen – genannt wurde) das weltweit erste seiner Art, das von seiner Eröffnung am 21. Februar 1775 an für das allgemeine Publikum freigegeben wurde. Zwar gab es getrennte Besuchszeiten für Gebildete und für das gemeine Volk; Letzteres hatte – „reinliche Bekleidung vorausgesetzt" – von acht bis zehn Uhr morgens Zutritt, um genügend Zeit verstreichen zu lassen bis zum Einlass „der intelligenten und gelehrten Leute [...] um ein Uhr nachmittags". Auch wenn diese Unterscheidung heute diskriminierend erscheint, kann man doch ermessen, wie innovativ die Öffnung der Bestände für die breite Öffentlichkeit damals gewesen sein muss.

Die Einrichtung der Sammlungen ging auf die Medici zurück, die sich auch um die Förderung der Wissenschaft verdient gemacht hatten, wie das Beispiel der Accademia del Cimento (1657–1667) zeigt, an der zur Zeit Ferdinandos II. de' Medici so bedeutende Wissenschaftler wie Redi, Magalotti und Galileis Lieblingsschüler Viviani lehrten. 1737, nach dem Tod Giangastones, des letzten Abkömmlings der Familie, fiel das Großherzogtum Toskana aufgrund der Bestimmungen des Wiener Friedens von 1735 an den späteren Kaiser Franz I. von Habsburg-Lothringen, der eine Aufstellung aller in seiner Residenz enthaltenen Schätze vornehmen ließ. Von 1763 bis 1764 inventarisierte der Arzt und Naturforscher Giovanni Targioni-Tozzetti (1712–1783) die Bestände.

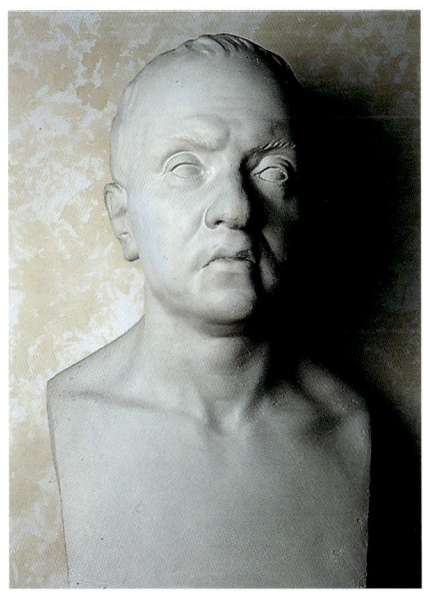

Als Peter Leopold 1765 seinen Vater, der zum Kaiser von Österreich ernannt worden war, als Großherzog der Toskana ablöste, fand er eine gute Arbeitsgrundlage zur Reorganisation der wissenschaftlichen Sammlungen vor. Diese Aufgabe sollte dem Abt Felice Fontana (1730–1805) aus Rovereto (Abb. S. 27) zufallen, seines Zeichens Dozent für Logik an der Universität von Pisa, aber auch Anatom, Physiker, Chemiker und v. a. ein international renommierter Physiologe, der sich der geplanten Neustrukturierung mit solcher Leidenschaft widmete, dass bereits Ende 1771 der erste Teil der Bestände in die neuen Räumlichkeiten verlegt werden konnte. Bereits 1771 hatte Großherzog Peter Leopold den Palast der Torrigiani (zuvor im Besitz der Bini) in der Via Romana in unmittelbarer Nähe des Palazzo Pitti sowie einige angrenzende Häuser erworben und Fontana mit dem Entwurf des Umbaus der Gebäude beauftragt, um in ihnen die wissenschaftlichen Sammlungen unterzubringen.

Als Direktor des neuen Museums bereiste Fontana in den ersten Jahren ganz Europa, um Bücher und Sammlungen zu erwerben und Kontakte zu Gelehrten in vielen Ländern herzustellen. So wurde das Florentiner Museum zu einem der bedeutendsten seiner Zeit, zumal es auch über eine reichhaltige wissenschaftliche Bibliothek verfügte. Fontana leitete das Museum bis zu seinem Tod 1805, wobei ihm Giovanni Fabbroni erst als Assistent und ab 1784 als Vize-Direktor (und permanenter Widersacher) zur Seite stand und ihn auf zahlreichen Reisen begleitete.

Das Geld zur Verwirklichung des Museums für Physik und Naturkunde (und auch vieler anderer Projekte Peter Leopolds) stammte aus dem Verkauf wertvoller Gegenstände aus dem Besitz der Medici; und dies trotz des Testaments der palatinischen Kurfürstin Anna Maria Luisa (1667–1743), der Schwester Giangastones, in dem das Vermögen der Medici unwiderruflich an die Stadt Florenz gefallen war. Zum Grundbestand des Museums, der vornehmlich aus den Uffizien stammte, gehörten Sammlungen (etwa von Mineralien und Muscheln), naturkundliche Kuriositäten aus der Zeit der Medici, die Instrumente Galileis

Bust of Felice Fontana
Büste von Felice Fontana
Buste de Felice Fontana

und die Gerätschaften der Accademia del Cimento sowie vier Wachsfiguren des sizilianischen Bildhauers G. G. Zumbo.

Um den Bestand des Museums um die Gebiete Meteorologie und Astronomie zu erweitern, beauftragte Peter Leopold 1780 den Architekten Gaspare Paoletti (der bereits an der Umstrukturierung des Palastes beteiligt gewesen war) mit dem Bau des Osservatorio Astronomico (dem sog. „Türmchen"), der dem ganzen Komplex später den Namen „Specola" (Sternwarte) geben sollte. Das aufwendige Bauvorhaben, das – trotz der abweichenden Meinung vieler Experten, die einen Neubau auf den Hügeln von Acetri befürwortet hatten – zu einer beträchtlichen Erweiterung des Gebäudes führte, wurde 1789 abgeschlossen. In dasselbe Jahr fällt die Verlegung eines Teils des Boboli-Parks in den Botanischen Garten des Museums (Abb. S. 8).

Nach dem Tod seines Bruders Josef wurde Peter Leopold 1790 Kaiser von Österreich und überließ die Toskana seinem Zweitgeborenen Ferdinand III. (1769–1824), dem die Weitsicht seines Vaters ebenso abging wie dessen Geschick bei den Regierungsgeschäften. Erschwerend kam hinzu, dass die napoleonische Expansion die Lothringer zur Aufgabe der Toskana zwang. 1801 fiel die Toskana schließlich als Königreich Etrurien an Bourbon-Parma. Während dieser Periode richtete das Museum Lehrveranstaltungen in wissenschaftlichen Fächern ein, die auch nach der Restauration der Lothringer 1814 fortgeführt wurden.

Unter Ferdinand III. durchlebte das Museum unruhige Zeiten und verlor seinen Ruf als eines der bedeutendsten Zentren europäischen Wissens, was ihm auch von illustren Besuchern aus dem Ausland wie Goethe und Bernoulli attestiert worden war. Auch Ferdinand III. ließ wichtige Umbauten an dem Palast vornehmen: 1820 wurde unter der Leitung des Architekten Pasquale Poccianti der bereits bestehende Vasari-Korridor, der vom Palazzo Vecchio über die Uffizien zum Palazzo Pitti führt, um den Poccintiani-Korridor bis zur Specola verlängert und so die Specola mit dem Flügel der Meridiana, des Palazzo Pitti, verbunden.

Auf Ferdinand III. folgte 1824 sein Sohn Leopold II. (1797–1870), von den Florentinern seiner blonden Haare wegen wohlwollend „Canapone" genannt. Ihm gebührt das Verdienst, den wissenschaftlichen Studien, insbesondere den anwendungsbezogenen wie etwa der Landwirtschaft, neue Impulse gegeben zu haben. In seine Regentschaft fällt die Realisierung der Tribuna di Galileo, eine Hommage an den großen Forscher, die 1841 anlässlich des dritten Kongresses italienischer Wissenschaftler in Florenz eingeweiht wurde. Bei der Tribuna handelt es sich um einen großen Saal im ersten Stock des Gebäudes, der von dem Architekten Giuseppe Marelli z. T. neu gebaut wurde. Die 1830 begonnenen Arbeiten sahen ursprünglich lediglich eine zusätzliche Apsis in einem bereits existierenden Saal vor. Um jedoch dem Vorhaben, dem berühmten Gelehrten ein würdiges Monument zu errichten, größeren Nachdruck zu verleihen, sollte ihm auf Wunsch des Großherzogs der gesamte Saal gewidmet werden und neben einer Statue alle Andenken Galileis sowie die Instrumente aus der Accademia del Cimento beherbergen. Der Bau der Tribuna, eines der seltenen Beispiele des späten Neoklassizismus in Florenz, führte zu bedeutsamen Änderungen der Palastarchitektur; insbesondere wurden die darunterliegenden Stockwerke umgestaltet und ein Teil des Hofes überdacht. Der Saal wurde mit toskanischem Marmor verkleidet, und für die Ausstattung zeichneten ausschließlich Künstler aus der Toskana – Bildhauer, Maler – verantwortlich.

Nach dem Fall Leopolds II., des letzten Großherzogs der Toskana, wurde 1859 das Istituto di Studi Superiori e di Perfezionamento (Institut für höhere Studien und Weiterbildung) gegründet, aus dem 1923 die königliche Universität von Florenz hervorgehen sollte. Die Gebäude des Museums wurden als Institut für Physik und Naturwissenschaften Teil der Universität.

Trotz der Bedeutung der Einrichtung dieses Instituts markierte es andererseits für das Museum den Beginn seiner Zergliederung in die verschiedenen Disziplinen, die mit der stetig wachsenden Zahl der Studenten mitsamt der dazugehörigen Sammlungen und Bibliotheken ausgelagert werden mussten. Diesen Prozess konnte selbst der Widerstand vieler Naturwissenschaftler jener Epoche nicht aufhalten, sodass im historischen Gebäude in der Via Romana nur die zoologischen Sammlungen und der Großteil der anatomischen Zeroplastiken verblieben.

Die Anfänge der Wachsbildnerei in der Anatomie

Die Verwendung von Wachs zur Modellierung von Figuren geht bis auf die Zeit der Römer zurück. Vor allem in der Kunst war das Material seit jeher sehr beliebt, sowohl aus ästhetischen Gründen als auch aus technischen. Wachs lässt sich einfach verarbeiten und eignet sich hervorragend für die Herstellung von Gussformen, wie etwa für den Bronzeguss. Insbesondere im europäischen Mittelalter wurden Weihgaben oder Heiligenstatuen, Krippen und auch Votivtafeln aus Wachs in großer Stückzahl gefertigt. In der Renaissance und während des gesamten 17. Jahrhunderts konnte man gewaltige Mengen dieser Wachsmodelle in Kirchen, besonders in Orsanmichele und in Santissima Annunziata, bewundern, darunter Gliedmaßen, Organe oder Teile davon ebenso wie Porträts, gegenständliche Darstellungen und Statuen, nicht selten sogar lebensgroß. Im 17. Jahrhundert kamen dann in Wachs nachgebildete anatomische Objekte zu wissenschaftlichen Zwecken auf.

Obwohl der Staat und mehr noch die Kirche das Studium der menschlichen Anatomie am Leichnam nach Kräften behinderten und sanktionierten, tauchten Mitte des 15. Jahrhunderts erstmals anatomische Zeichnungen und Traktate auf, in der Regel aus der Feder so berühmter Maler und Bildhauer wie Leonardo da Vinci, Michelangelo, Raffael und Tizian, mit denen bedeutende Anatomie-Forscher wie Falloppio, Cesalpino oder Vesalius ihre Traktate illustrierten.

Gaetano Giulio Zumbo war gegen Ende des 17. Jahrhunderts der Erste, der in der berühmten Anatomie-Schule von Bologna für die Modellierung anatomischer Präparate farbige Wachse verwendete. Der aus Syrakus stammende Künstler, dessen Werke im Folgenden abgebildet sind, gestaltete zwei Männerköpfe (einer befindet sich heute in Florenz, einer in Paris), eine (leider verlorengegangene) Frauenfigur sowie einige kleinformatige Kompositionen. Ebenfalls in Bologna formierte sich die erste Schule der Zeroplastik, aus der in der zweiten Hälfte des 18. Jahrhunderts auch die Wachsbildner des Florentiner Museums hervorgingen. Als Begründer der Bologneser Schule gilt der Maler, Bildhauer und Architekt Ercole Lelli (1702–1766), der – wie schon zuvor Zumbo – die Modellierung wächserner Muskeln und Eingeweide auf der Basis echter Skelette vornahm. Als Kenner und Liebhaber der Naturwissenschaften förderte Kardinal Prospero Lambertini, der 1740 zum Papst Benedikt XIV. ernannt wurde, die anatomischen Studien der Bologneser Schule und kaufte Lellis sämtliche Modelle, die man noch heute im Istituto di Anatomia Umana Normale der Universität Bologna bewundern kann. Von besonderer Bedeutung war auch seine Unterstützung für die Beschaffung der für diese Arbeit notwendigen Leichname. Zu den Assistenten Lellis gehörte außerdem Giovanni Manzolini (1700–1755), der mithilfe seiner Frau Anna Morandi (1714–1774) zahlreiche Modelle ausführte, die ebenfalls in der Bologneser Universität aufbewahrt werden.

In jener Zeit wurde das Studium der Anatomie zwar in ganz Europa vorangetrieben, doch die zeroplastische Schule von Florenz stammt direkt von derjenigen Bolognas ab. Die Vermittlung besorgte der Chirurg und Geburtshelfer Giuseppe Galletti, der unter dem Eindruck der Werke von Lelli und Manzolini zusammen mit dem Wachsbildner Giuseppe Ferrini eine Reihe von Modellen aus Wachs und Terracotta anfertigte (die sich zur Zeit

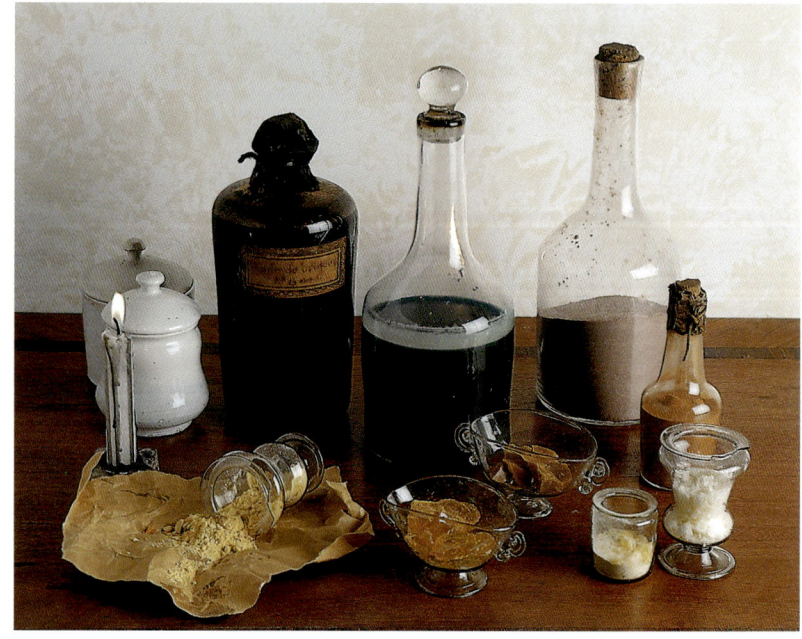

Pigments and other materials used in the production of wax models in their original containers

Farbstoffe und andere bei der Wachsfigurenherstellung erforderliche Materialien in ihren ursprünglichen Behältern

Colorants et autres matières nécessaires à la création de sculptures en cire dans leurs récipients d'origine

im Museo Galileo – Istituto e Museo di Storia della Scienza in Florenz befinden), um verschiedene Geburtsvorgänge – normale und dystokische – zu demonstrieren.

Die anatomischen Wachsfiguren der Specola: Schule, Werke, Techniken

Die zeroplastische Werkstatt in Florenz bestand etwa ein Jahrhundert lang – von 1771 bis in die zweite Hälfte des 19. Jahrhunderts hinein –, aber der überwiegende Teil der im Museum aufbewahrten Wachsbildnisse entstand ebenso wie die Auftragsarbeiten für andere Florentiner Institutionen und Auftraggeber in den ersten fünfzig bis sechzig Jahren ihres Bestehens.

Die Gründung der Specola geht auf Felice Fontana zurück, der sich mit voller Unterstützung Großherzog Peter Leopolds deren Ausbau widmete, zumal er sich als Anatom und Pathologe persönlich an der Erstellung der Modelle beteiligte. Anfangs arbeitete lediglich der Wachsbildner Giuseppe Ferrini unter Fontanas Anleitung in der Werkstatt. In der Folge wurden dann der Anatom Antonio Matteucci und der damals noch sehr junge Clemente Susini eingestellt, der aufgrund seines Geschicks und seiner Produktivität zum bedeutendsten und berühmtesten Zeroplastiker der Florentiner Schule avancierte (Abb. S. 14).

Die Wachsemodelle wurden vermutlich in den Räumen des Erdgeschosses verarbeitet, deren Fenster zur Via Romana im Südflügel des Palastes mit seinen vielen kleinen Höfen

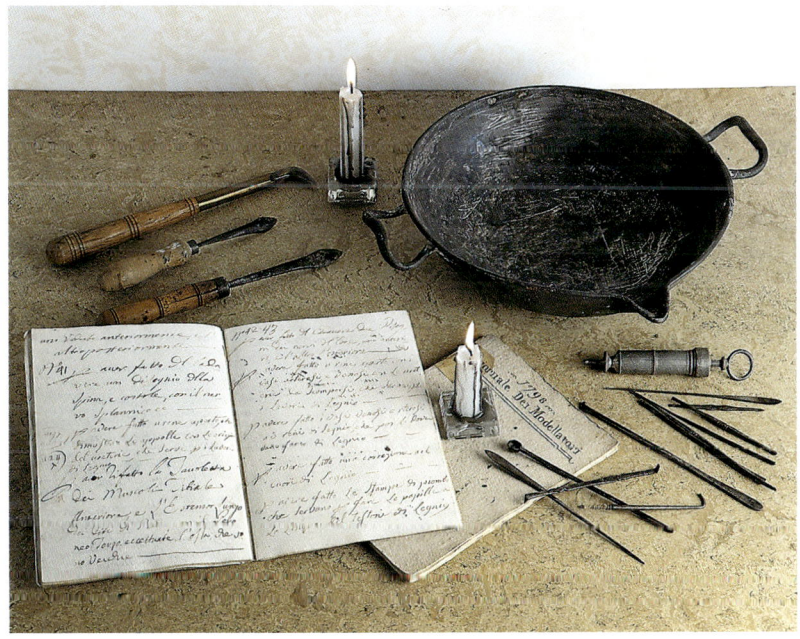

Copper vat with modelling tools

Kupferwanne und Modellierbesteck

Bassine en cuivre et instruments à modeler

hinausgingen. Von den einstigen Gerätschaften ist kaum etwas erhalten, aber aus Archivunterlagen gehen folgende Ankäufe hervor: Kupferwannen verschiedener Größe, in denen das Wachs geschmolzen wurde; Modellierbesteck (Abb. S. 31); Eisendraht in unterschiedlichen Stärken; Marmorplatten zum Pressen des Wachs; Waagen; Dreifüße zum Erhitzen des Materials; Schiefertafeln für Anmerkungen und Zeichnungen während der Sektionen; Kisten mit Handgriffen zum Transport der Leichen, Holzkisten mit Stangen zum Transport der Wachsfiguren; Behälter, Vasen und Flaschen aus Keramik oder Glas für Farbmittel und andere Substanzen, die dem Wachs beigegeben wurden. Von Letzteren wurden in den alten Magazinen des Museums noch etliche mitsamt Inhalt gefunden (Abb. S. 30).

Aus den Archivdokumenten geht ferner hervor, wie viele Leichname oder Leichenteile für die Ausführung eines Modells benötigt wurden: über zweihundert für eine einzige Figur! Die erstaunlich hohe Zahl erklärt sich aus dem Umstand, dass es keine Konservierungsmöglichkeiten gab, sodass für genaue anatomische Sektionen ständig neue Leichname gebraucht wurden. Um alle Leichen oder Leichenteile, die vom (etwa zwei Kilometer entfernten) Krankenhaus Santa Maria Nuova herbeigeschafft wurden, numerisch genau zu erfassen, war an der Eingangstür eigens ein Register angebracht, auf dem alle Eingänge, aber auch alle für den Friedhof bestimmten Abgänge notiert wurden: „[…] bezüglich der Leichname aus Santa Maria Nuova, von denen man nicht weiß, wie viele anfallen, und weil man die Leute nicht passend für die Wachssammlungen sterben lassen kann: Darum wurde an der Tür des Museums ein Register angebracht, in dem die Leichname, die daselbst gebraucht werden und deren Zahl beträchtlich erscheint, festzuhalten sind und ebenso zu vermerken, wie oft und wie viele zum Friedhof geschickt werden." (aus einem Dokument von 1793)

Ungeachtet der geringen Zahl der Präparatoren gingen die Arbeiten recht schnell voran, wenn man bedenkt, dass um 1790, also zwanzig Jahre nach Beginn, die Modelle bereits acht Säle füllten, nicht gerechnet all jene, die für das Florentiner Krankenhaus und für andere Institutionen in Italien und im Ausland angefertigt worden waren (vgl. Abb. S. 46). Unter den über ganz Europa verstreuten Wachsplastiken aus der Specola-Werkstatt ragen die vom österreichischen Kaiser Josef, dem älteren Bruder von Peter Leopold, für die militärische Medizinschule Wien (das nach ihm benannte Josephinum) 1781 in Auftrag gegebenen hervor. Die Sammlung umfasst 1 200 Stücke, die 1786 fertiggestellt und in zwei Lieferungen auf dem Rücken von Maultieren nach Wien gebracht wurden. Weitere Plastiken waren für Pavia, Cagliari, Bologna, Budapest, Paris (heute in Montpellier), Uppsala, London, Leiden und andere Städte bestimmt. Diese externen Aufträge warfen gelegentlich Organisationsprobleme auf, da dem Großherzog daran gelegen war, die Herstellung der für das Museum bestimmten Modelle nicht zu verzögern. Darum wurde die Durchführung dieser Aufträge an die Bedingung geknüpft, dass die Plastiker und Anatomen – im Fall der Wiener Sammlung Fontana selbst – zwar museumseigene Abdrücke und Utensilien benutzen durften, das Material aber selbst besorgen und mithilfe auswärtiger Handwerker verarbeiten mussten.

Fontana hatte den Ehrgeiz, so viele anatomische Wachsmodelle wie nur irgend möglich zu schaffen, um einen Fundus für didaktische Zwecke einzurichten, der die direkte Exhumierung von Leichen für das Studium der Anatomie entbehrlich machen sollte. Hierfür stattete er seine Sammlungen mit einer Reihe von Tempera-Zeichnungen aus (Abb. S. 10, 11), auf denen die einzelnen Teile dargestellt werden; von einem Nummernkranz um die gezeichnete Figur ziehen sich dünne gestrichelte Linien zu den verschiedenen Organen, Muskeln, Knochen usw., die anhand der Ziffern identifiziert werden können. Die Blätter mit den entsprechenden Erläuterungen finden sich in einer Schublade, die an jeder Vitrine angebracht wurde: So erhielt man zu jedem Präparat einen regelrechten Anatomie-Traktat (Abb. S. 20).

Um 1790 nahm Fontana ein weiteres ambitioniertes Projekt in Angriff: Es ging um eine Serie anatomischer Teile aus bemaltem Holz, die – im Originalmaßstab oder in Vergrößerungen – zu didaktischen Zwecken auseinandergenommen werden konnten, um die Beziehungen zwischen den Organen veranschaulichen zu können. Nach einigen Versuchen, von denen nur einige wenige Stücke erhalten sind, darunter zwei ganze Plastiken (eine im Museum, die andere in Paris) und eine enorme Büste, musste Fontana das Vorhaben aufgeben, weil das Holz sich nicht nur schwerer bearbeiten lässt, sondern darüber hinaus als organisches Material nicht formbeständig ist, wodurch mit der Zeit die einzelnen Teile ihre Passung zueinander verlieren (Abb. S. 37). Die einzige Wachsplastik des Museums, die die gewünschten Merkmale aufweist, ist die sogenannte „Venere medicea": eine liegende weibliche Figur, von der mehrere Schichten entfernt werden können, bis der Uterus mit einem kleinen Fötus zum Vorschein kommt.

Das Museum hatte mittlerweile seine Mannschaft um einen Zeichner (Claudio Valvani) und einen Holzschnitzer (Luigi Gelati) erweitert und beschäftigte diverse Anatomen für die Sezierung der Leichname. Zu den berühmtesten zählte Paolo Mascagni (1755–1815), der wunderschöne anatomische Tafeln hinterlassen hat und sich auf das Studium des lymphatischen Systems spezialisiert hatte: Nicht zufällig stellen viele liegende Statuen, aber auch kleinere Präparate diesen Apparat des Organismus minutiös dar.

Auch nach dem Tod Felice Fontanas 1805 (Abb. S. 17) setzte die zeroplastische Werkstatt ihre Arbeit fort, und nach dem Tod seines Nachfolgers Clemente Susinis 1814 übernahmen andere Wachsbildner wie die beiden Calenzuoli (Vater Francesco und Sohn Carlo) oder Luigi Calamai (1800–1851) den Betrieb. Unter Calamai konzentrierte sich die Aktivität auf Modelle der vergleichenden Anatomie und der Botanik sowie auf jene der pathologischen Anatomie, die für das Krankenhaus von Santa Maria Nuova bestimmt waren und heute in der biomedizinischen Abteilung des Museo di Storia Naturale zu sehen sind. Auf Calamais Stelle rückte nach dessen Tod Egisto Tortori (1829–1893) nach, der neben Tierpräparaten und solchen der anatomischen Pathologie auch eine Büste Clemente Susinis modellierte. Nach Tortoris Tod 1893 löste sich die Werkstatt endgültig auf.

Von den Wachsfiguren, die das Museum La Specola gegenwärtig in Urnen verwahrt, befassen sich 513 mit menschlicher und 65 mit vergleichender Anatomie; 5 Exponate stammen von G. G. Zumbo. Es gibt 26 Figuren (einschließlich eines halbfertigen Jünglings): 13 stehende und 13 liegende Exemplare, davon 18 in Originalgröße (6 stehende und 12 liegende), während acht Figuren etwa 60 cm messen (7 stehende und 1 liegende). Mit Ausnahme von 14 Zeroplastiken zur menschlichen Anatomie, die sich im Magazin befinden, sind alle öffentlich ausgestellt. Hinzu kommen mehr als 800 gerahmte Zeichnungen und annähernd 900 erläuternde Blätter, doch es müssen mehr gewesen sein, da im Laufe der Zeit einiges verlorengegangen ist. In den letzten Jahren wurden die Zeichnungen und die Blätter mit den Erläuterungen entfernt, um sie vor weiteren, durch Licht und Feuchtigkeit bedingte Schäden zu bewahren, und durch Farbkopien ersetzt, sodass die Struktur der Sammlung unter didaktischen Gesichtspunkten wie auch unter solchen einer ausstellungsgerechten Präsentation erhalten blieb.

Während Experten der Materie vor allen Dingen die wissenschaftliche Perfektion der Exponate beeindrucken werden, steht für die meisten Besucher in erster Linie die künstlerische Leistung im Vordergrund, denn bei aller wissenschaftlichen Strenge und didaktischem Anspruch sind diese Exponate regelrechte Kunstwerke. Ihre ästhetischen Qualitäten unterstreichen überdies die kunstvoll gearbeiteten Holzvitrinen, in denen sie aufbewahrt werden, auch wenn viele zeitbedingte Abnutzungserscheinungen aufweisen und ständig restauriert werden müssen. Welche Sorgfalt bei der Präsentation dieser Sammlung aufgewendet wurde, bezeugen auch die mittlerweile verblassten, ursprünglich hellgrünen seidenen Draperien und Schleier aus dem 18. Jahrhundert, die sich ebenfalls in einem äußerst prekären Zustand befinden. Welche Aufmerksamkeit der Ästhetik gewidmet wurde, belegen die Sammlungen der vergleichenden Anatomie, der Humanpathologie und v. a. der Botanik, wo für die kleinen Pflanzenmodelle eigens Keramikvasen aus der Werkstatt Ginori von Sesto Fiorentino angefertigt wurden.

Die technischen Verfahren zur Herstellung der Wachsfiguren können nur aus verschiedenen Briefen und Dokumenten aus dem Staatsarchiv und dem des Museums für Wissenschaftsgeschichte in Florenz erschlossen werden. Außerdem hatte jeder Wachsbildner seine eigene Technik, die er – wie alle Handwerker und Künstler – geheim hielt. Von der in der anatomischen Sektion vorbereiteten Vorlage wurde zuerst eine exakte Kopie aus Kreide oder minderwertigem Wachs hergestellt, von der ein Gipsabdruck genommen wurde, der, wenn er sehr groß war, aus verschiedenen Einsatzstücken bestand. Diese Abdrücke (Abb. S. 55), die in einem Depot des Museums aufbewahrt werden, stellten eine Art Matrix dar, die mehrfach für die Reproduktion desselben Modells verwendet werden konnte. Der schwierigste und heikelste Teil war die Konstruktion des definitiven Modells, die große Präzision, Kenntnis der Substanzen, die dem Wachs beigemischt werden mussten, um die gewünschte Farbe und Konsistenz zu erhalten, sowie große Erfahrung und Geschick verlangten. Das Wachs musste bei der richtigen Temperatur langsam im Wasserbad schmelzen, damit es sich nicht verfärbte. Am häufigsten wurden weißes Wachs aus Smyrna, chinesisches Wachs sowie venezianisches Wachs verwendet. Um das Wachs formbarer zu machen, fügte man Terpentin

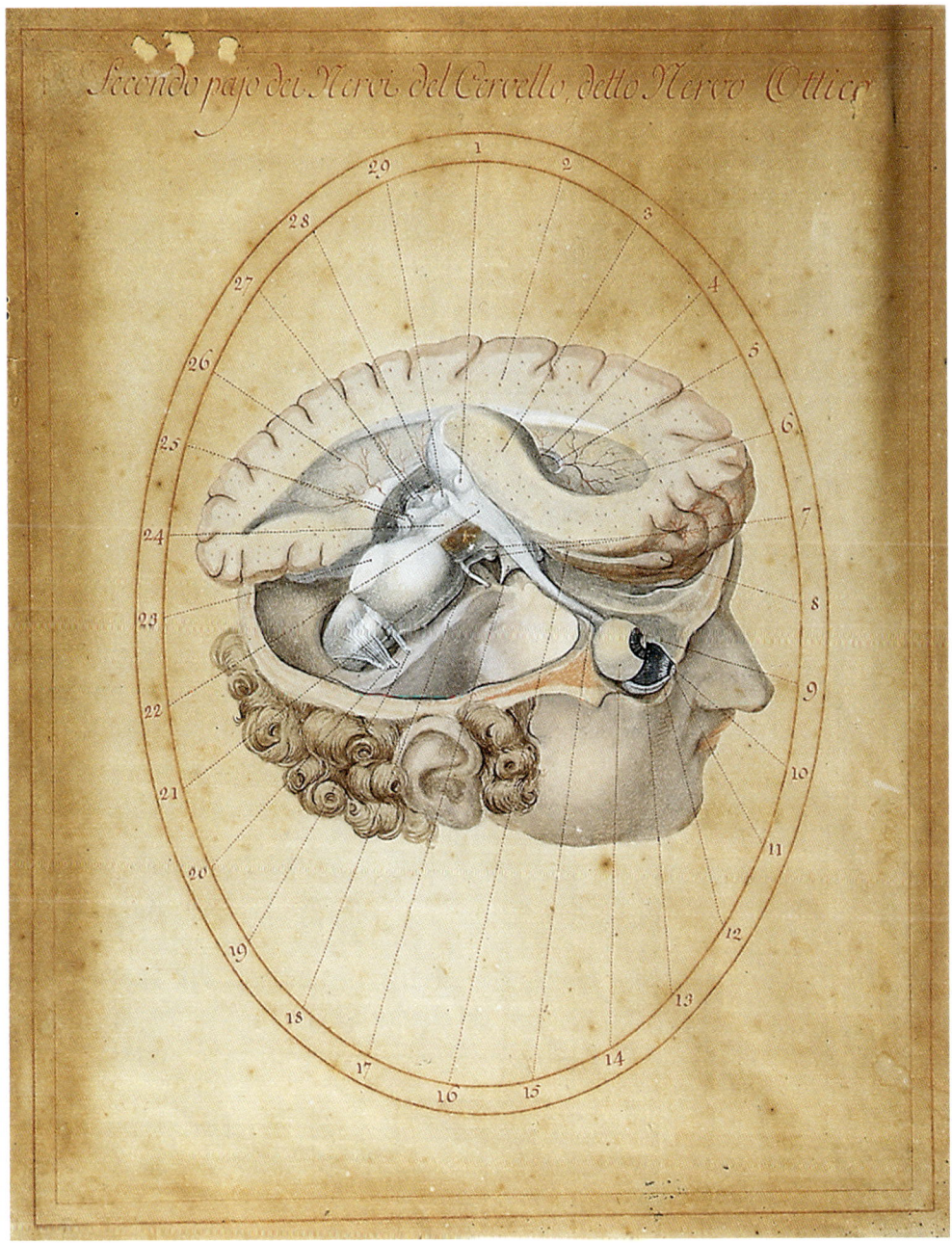

Drawing of a head, lateral view, 2nd half of the 18th century
Zeichnung eines Kopfes: Seitenansicht, 2. Hälfte des 18. Jahrhunderts
Dessin d'une tête vue latérale, 2ᵉ moitié du XVIIIᵉ siècle

hinzu, ferner – ebenfalls mit Terpentin verdünnte – Farbstoffe und andere Substanzen, die in den Einkaufslisten als „für die Wachsarbeiten benötigte Sachen" nachgewiesen sind. Damit sich das Modell später leicht aus der Form lösen ließ, wurde die Gussform mit lauwarmem Wasser angefeuchtet und mit weicher Seife eingerieben, bevor man das Wachsgemisch in den Abdruck goss. Abgesehen von einigen Vollwachspräparaten waren die meisten innen hohl und wurden zur Stabilisierung mit Lappen, Werg oder Holzstücken ausgestopft. Figuren, die aus mehreren Teilstücken bestehen, wurden in der Regel durch eine Stützarmatur aus Metall verstärkt. Nach Entfernung des Abdrucks wurde jede Komponente gesäubert und mit entsprechenden Geräten ausgekehlt sowie zur Vervollständigung mit Organen, Gefäßen, Nervensträngen usw. versehen. Eine letzte Schicht durchsichtigen Lacks verlieh dem Ganzen den nötigen Glanz. Um die Originaltreue des Modells sicherzustellen, mussten all diese Arbeitsgänge ständig von den Anatomen überwacht werden, denen auch die Entscheidung über die jeweils günstigste Position zur Hervorhebung bestimmter Organe oblag.

Jeder Wachsbildner verfügte über eine eigene Technik, die mit der Zeit immer mehr verfeinert wurde, u. a. dank neuer Instrumente wie z. B. einem „Zieheisen zum Drehen der Wachszylinder", das ab 1786 den Plastikern zur Verfügung stand. Früher musste dieser Arbeitsgang, der für die Gestaltung der Blutgefäße unumgänglich war, von Hand bewältigt werden. Um 1790 führten die Wachsbildner einige Jahre lang „Hefte" (Abb. S. 21), in denen sie täglich die ausgeführten Arbeiten eintrugen, und aus denen hervorgeht, dass bereits damals des Häufigeren Reparaturen, Verbesserungen und Überarbeitungen anfielen. 1793 legte Susini eine „Notiz [an] über zu korrigierende anatomische Wachspräparate aufgrund im Folgenden beschriebener, von mir, Clemente Susini, Modellierer des Königlichen Museums, erkannter Fehler, die es verdienen, behoben zu werden." Zum Beispiel: „Überprüfung des Präparats der Schlundader und der Halsschlagader: Man sieht von der Luftröhre zwei fremde Muskeln abgehen, das heißt solche, die nicht zum menschlichen Körper gehören, und einen, der am Schilddrüsenfortsatz ansetzt, der aber weder der Zungen- noch der Schlund- und schon gar nicht der Schilddrüsengriffel zu sein scheint." Susinis Notiz endet mit der Feststellung: „Diese Fehler sind zu gravierend, um nicht berichtigt zu werden und von den vielen anderen schweige ich lieber, um mich kurz zu fassen." Fast alle Modelle sind gut erhalten, lediglich bei einigen sind Verfärbungen aufgetreten: Mancher Fötus ist nachgedunkelt, und von den heute grün schimmernden Venen wissen wir, dass sie einst „blauviolett" waren.

Die Wachsfiguren des Gaetano Giulio Zumbo

Ein besonderes Kapitel gebührt den Wachsfiguren Gaetano Giulio Zumbos, zum einen wegen ihrer ungewöhnlichen Konzeption und Herstellungsmethode, v. a. aber, weil sie der allgemeinen Entwicklung bei der Verwendung von Wachs für anatomische Darstellungen ihrer Zeit um ein Jahrhundert voraus waren. Zumbo wurde 1656 in Syrakus als Sohn einer

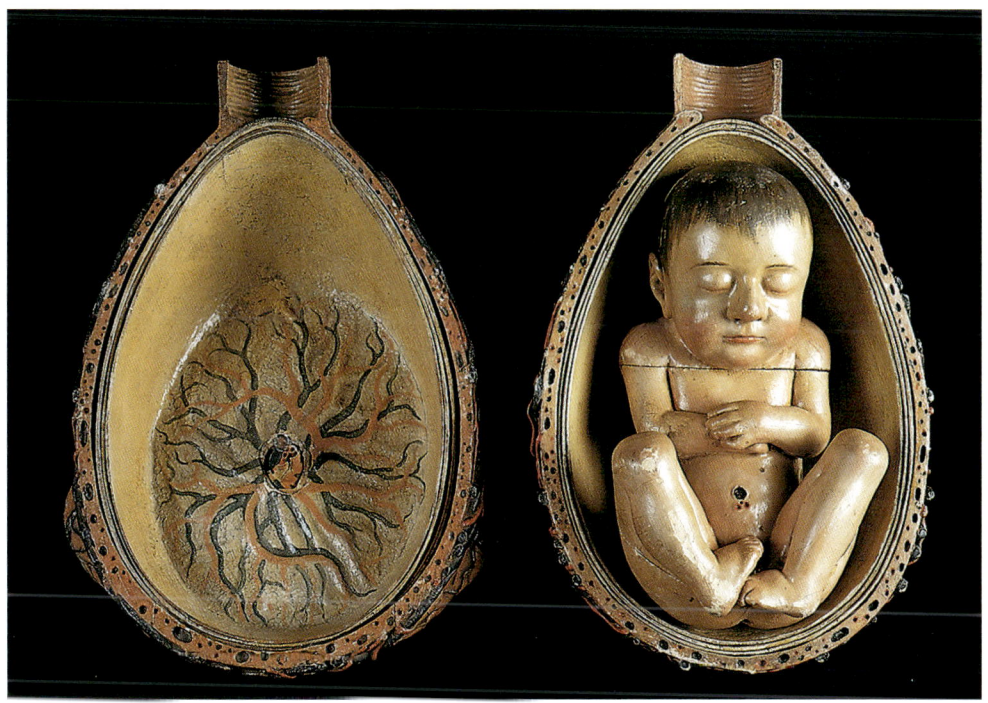

alten Adelsfamilie geboren, deren Nachname vermutlich Zummo war und von der es heute keinerlei Spuren mehr gibt; da die entstellende Version „Zumbo" sich allgemein eingebürgert hat, scheint es angebracht, dabei zu bleiben. Von seinem Leben ist wenig überliefert, und auch das nur bruchstückhaft; man weiß, dass er im Jesuiten-Internat seiner Geburtsstadt erzogen wurde und wegen eines nicht näher benannten „lästigen Unfalls" die Schule aufgeben musste. Die klassischen Zeugnisse seiner Heimat dürften ihm erste künstlerische Anregungen gegeben haben. Später wurde sein künstlerischer Werdegang vom Manierismus und der Frühphase des Barocks beeinflusst. Von 1687 bis 1691 lebte er in Neapel und schuf vermutlich hier zwei der in der Specola aufbewahrten Gruppenszenen, der „teatrini", nämlich *Il Trionfo del Tempo* (Der Triumph der Zeit) und *La Peste*. Seine Urheberschaft schließen die Gelehrten, die sich mit Zumbos Werk befasst haben, aus dem malerischen Hintergrund und der Anordnung der Figuren, die deutlich auf süditalienische Vorbilder verweist. Von 1691 bis 1694 arbeitete Zumbo für Cosimo III. de' Medici in Florenz, wo er zwei weitere „teatrini" mit Bezügen zur Renaissancekunst fertigstellte: *Il Sepolcro* (oder *La Vanità della Gloria Umana*) und *Il Morbo Gallico* (oder *Sifilide*). 1695 unternahm er einige

Uterus with foetus; wood, made of individual sections which can be taken apart
Uterus mit Fötus, Holz, in Einzelteile zerlegbar
Utérus et fœtus ; bois, démontable

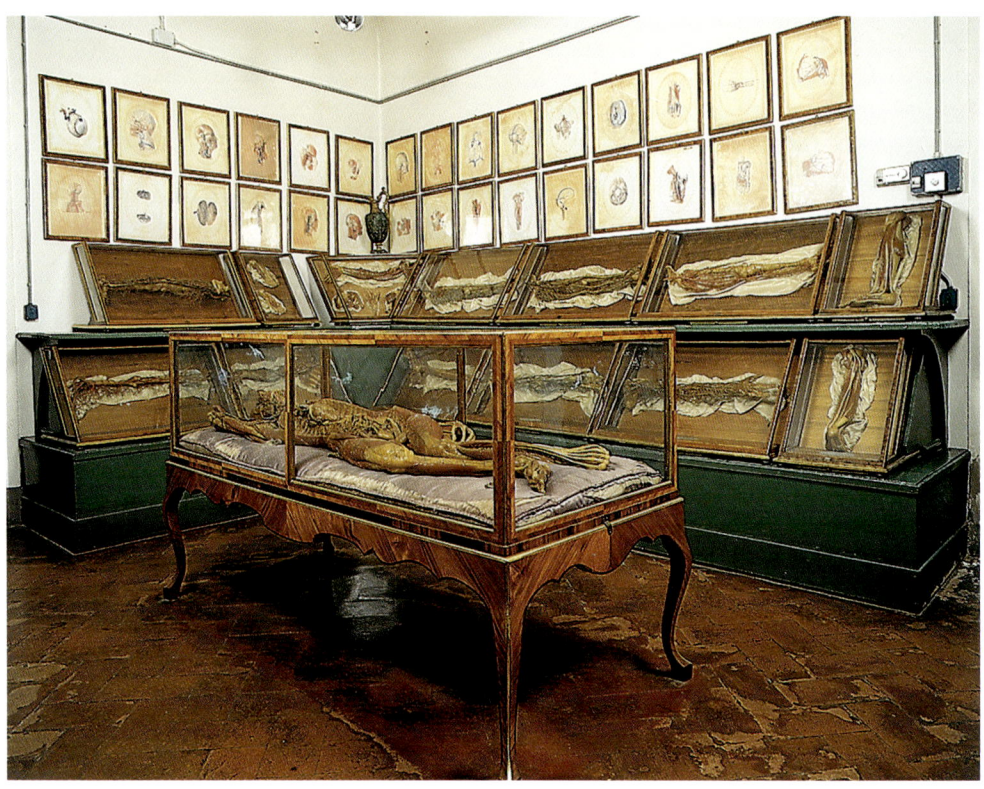

Reisen nach Bologna, wo er mit großem Interesse die gerade im Aufschwung begriffenen anatomischen Studien verfolgte. Ende 1695 zog er nach Genua, wo er vermutlich die im Museum ausgestellte *Anatomie des Kopfes* vollendete. Hier lernte er den französischen Chirurgen Guillaume Desnoues kennen; auf ihre Zusammenarbeit gehen diverse anatomische Wachsmodelle zurück, darunter die lebensgroße Figur einer Gebärenden, eine bei der Geburt gestorbene Frau in kleinerem Format sowie Geburt und Grablegung Christi – Werke, die leider nicht mehr erhalten sind. 1700 entzweite sich Zumbo mit Desnoues und siedelte zuerst nach Marseille über, wo er einen weiteren Wachskopf anfertigte, und schließlich nach Paris, wo er in die Dienste Ludwigs XIV. trat, bis er 1701 an einer Gehirnblutung verstarb. Unter seinen Sachen fand man einen wunderschönen Wachskopf, der höchstwahrscheinlich mit demjenigen identisch ist, der im Museum National d'Historie Naturelle aufbewahrt wird. Sein Grab, das sich in der Kirche von Saint Sulpice befand, wurde während der Revolution zerstört.

View of Room XXVII
Gesamtansicht des Saals XXVII
Vue d'ensemble de la salle XXVII

Ein Großteil der Werke dieses bis noch vor einem halben Jahrhundert weitgehend verkannten Künstlers befindet sich im Museum La Specola. Die drei „teatrini", die auch als *Cera della peste* („Pestfiguren") bekannt sind, und die *Anatomie des Kopfes* befanden sich, als die Medici von den Lothringern abgelöst wurden, in der Königlichen Galerie (also in den Uffizien) und wurden anschließend dem Museo di Fisica e Storia Naturale überantwortet, dort aber, vielleicht wegen der Drastik der dargestellten Szenen, nie ausgestellt. Als Peter Leopold 1790 nach Wien aufbrach, schenkte er die „Pestfiguren" dem Hofarzt Giovanni Giorgio Hasenöhrl, besser bekannt als Lagusius. Dieser beauftragte erst Fontana und dann Agostino Renzi, Superintendent der Königlichen Apotheke, sie für 150 Zechinen zu verkaufen. Letzterer war sich der Bedeutung der Ensembles durchaus bewusst und bot sie dem neuen Großherzog Ferdinand III. an, erst für die Königliche Akademie der Schönen Künste und daraufhin, da diesem die Stücke weder „nützlich noch angemessen" erschienen, für das Königliche Museum. Dieser Vorschlag wurde vom Vizedirektor des Museums Giovanni Fabbroni befürwortet, und die drei Werke wurden von verschiedenen Experten, darunter Clemente Susini, geschätzt. Da der Schätzwert höher war als der von Lagusius geforderte Preis, autorisierte der Großherzog den Ankauf, und so blieben Zumbos Werke im Museum, bis die drei Theaterszenen 1878 dem Museum Nazionale in Bargello zur Aufbewahrung übergeben wurden; von diesem kamen sie zum Museo di Storia della Scienza, um 1974 schließlich wieder zur Specola zurückzukehren. Archivdokumente belegen, dass die Wachsfiguren verschiedentlich restauriert wurden, u. a. von Susini und Tortori. Die Pestdarstellungen, die sich im Museo di Storia della Scienza befanden, wurden während der sintflutartigen Regenfälle im Jahre 1966 schwer beschädigt. In über 18-monatiger Kleinarbeit gelang es Guglielmo Galli aus der Werkstatt der Pietre Dure in Florenz, sie nach vorhandenen Fotografien wiederherzustellen. Die *Anatomie des Kopfes* hingegen hat die Specola nie verlassen, da sie zu Recht als einschlägig für die wissenschaftliche Sammlung des Instituts betrachtet wurde.

Die von Zumbo angewandte Technik unterscheidet sich stark von jener der späteren Florentiner Schule. Die *Anatomie des Kopfes* wurde an einem echten Schädel ausgeführt, der nach dem Befund der Röntgenaufnahmen einem etwa 25-jährigen Mann gehört haben muss. Der in Paris befindliche Kopf und ein weiteres, verschollenes Exemplar sind hingegen ganz aus Wachs modelliert, was auf eine Fortentwicklung seiner Verarbeitungstechnik schließen lässt, die möglicherweise auf die Zusammenarbeit Zumbos mit Desnoues zurückgeht. Die Figuren der „teatrini", deren Gestaltung bis in die kleinsten Details von großer Kunstfertigkeit zeugt, wurden mit an Sicherheit grenzender Wahrscheinlichkeit von Gipsabdrücken hergestellt, denen ihrerseits sorgfältig geformte Tonmodelle zugrunde lagen. Das mehr oder weniger flüssige Wachs (ein Gemisch aus Bienenwachs, Kolophonium, Terpentin und Farbstoffen) wurde in dünnen Schichten aufgetragen, um die jeweils gewünschte Konsistenz und Farbigkeit zu erhalten. Die wenigen überlieferten Notizen über Zumbos Technik verdanken wir den von Guglielmo Galli nach den Wasserschäden von 1966 durchgeführten Restaurierungen, die zudem auch Spuren älterer Reparaturen anhand des Nachweises von Farbstoffen, die es Ende des 17. Jahrhunderts noch nicht gab, zutage förderten. Auch das kleine Selbstporträt (Abb. S. 53) aus dem Ensemble *Il Trionfo del Tempo* kann nicht mit Bestimmtheit Zumbo zugeschrieben werden, es könnte das Werk eines Zeitgenossen oder des späten Susini sein. Der *Morbo Gallico*, von der Anlage her eine vergleichbare Komposition, von der nur noch wenige Bruchstücke existieren, war ein Geschenk Cosimos III. an Filippo Corsini, das in den Kellern des Palazzo Corsini am Arno aufbewahrt wurde und von der Überschwemmung praktisch vollständig zerstört wurde; man fand ein paar Figuren im Garten wieder, aber da es keine Fotografien des Ensembles gab, war eine Rekonstruktion nicht möglich. Anlässlich des 200-jährigen Jubiläums der Museumseröffnung, das mit einem Kongress über Wachsplastik zelebriert wurde, übergab die Familie Corsini die Reste dieses Werks dem Museum.

Lange Zeit hindurch wurden Zumbos Meisterwerke als krankhafte Kaprizen eines sich an der Darstellung makabrer und abscheulicher Details ergötzenden Künstlers erachtet. Erst Mitte dieses Jahrhunderts ging man dazu über, sie in ihrer historischen Perspektive wahrzunehmen: als realistische Dokumente einer Epoche, in der der Tod mit Kriegen, Hungersnöten und großen Epidemien allgegenwärtig war. Diese Vision der Zerstörung, der Hinfälligkeit menschlichen Lebens, des unerbittlichen Vergehens der Zeit, dieses beständige Memento mori, mit dem uns Zumbos Werke in einer Fülle von Details konfrontieren, ist ein typisches Beispiel für die Kultur des 17. Jahrhunderts, wie sie uns auch von anderen großen Künstlern jener Zeit, etwa von Luca Giordano oder Mattia Preti, überliefert wird.

G. G. Zumbo: Specimen of a head (cf. pp. 13, 299, 302)
Präparat eines Kopfes (vgl. S. 13, 299, 302)
Préparation anatomique d'une tête (*cf.* p. 13, 299, 302)

Pages 42–43:
G. G. Zumbo: *Il Sepolcro*
The burial / Das Begräbnis / L'Enterrement

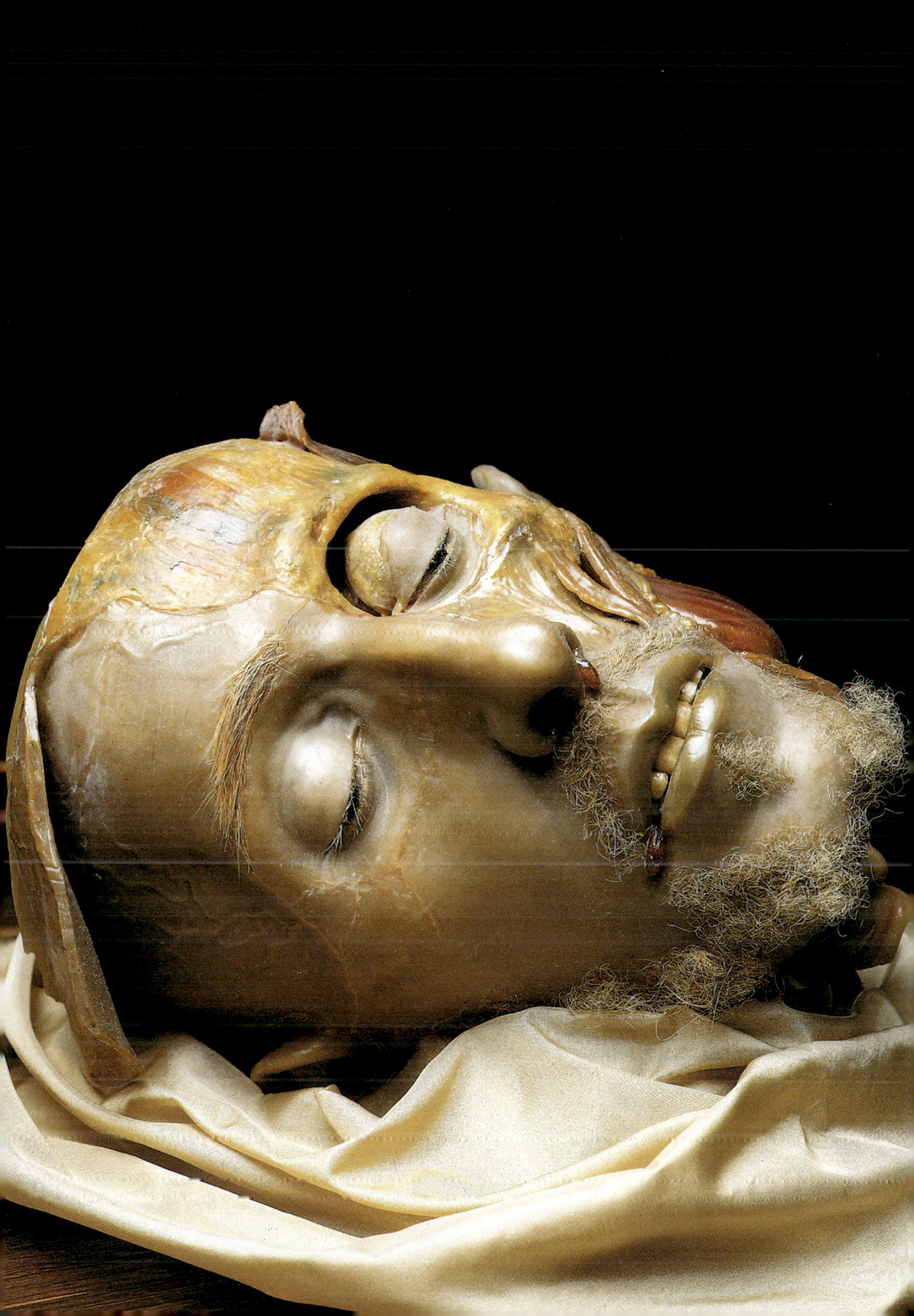

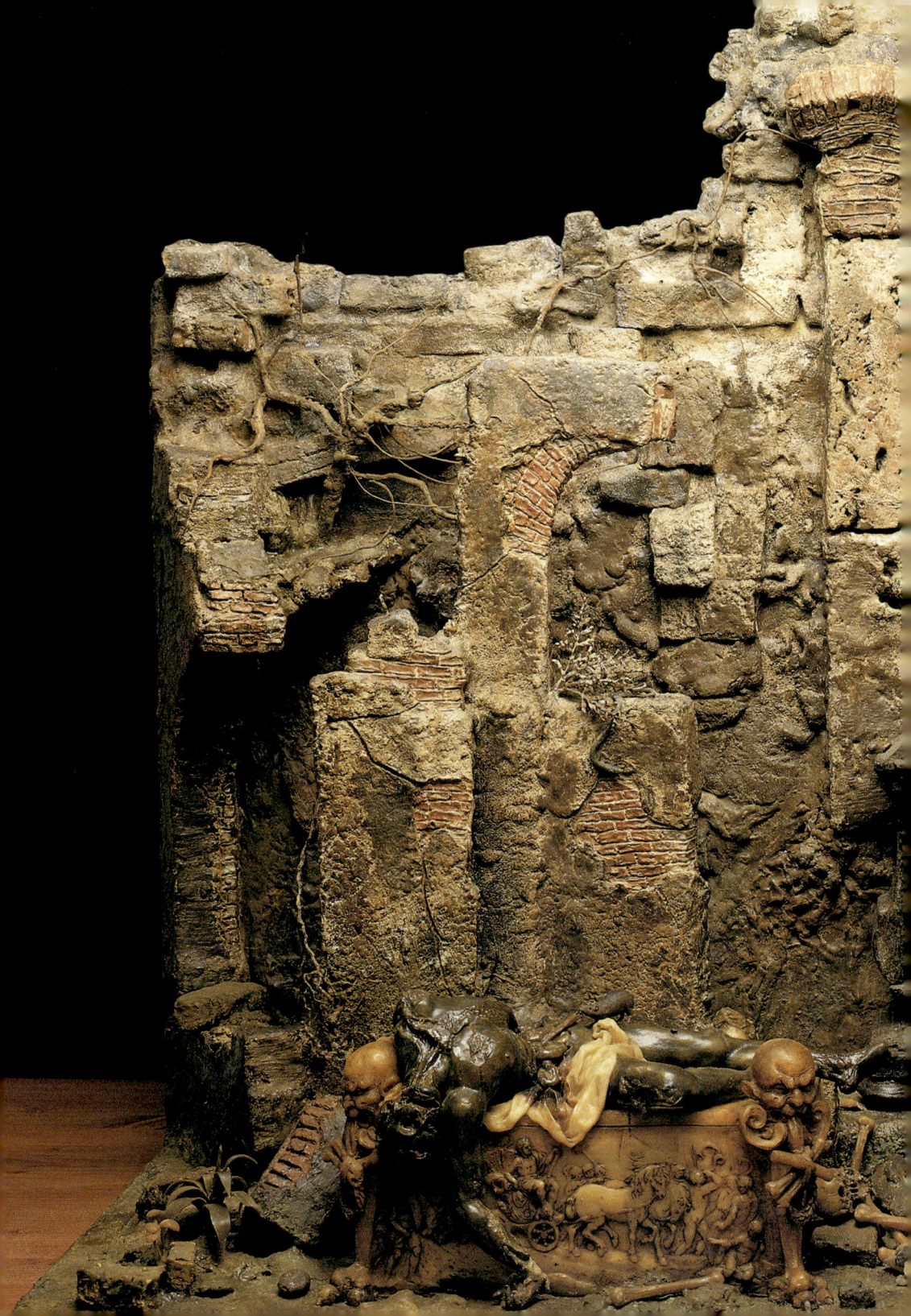

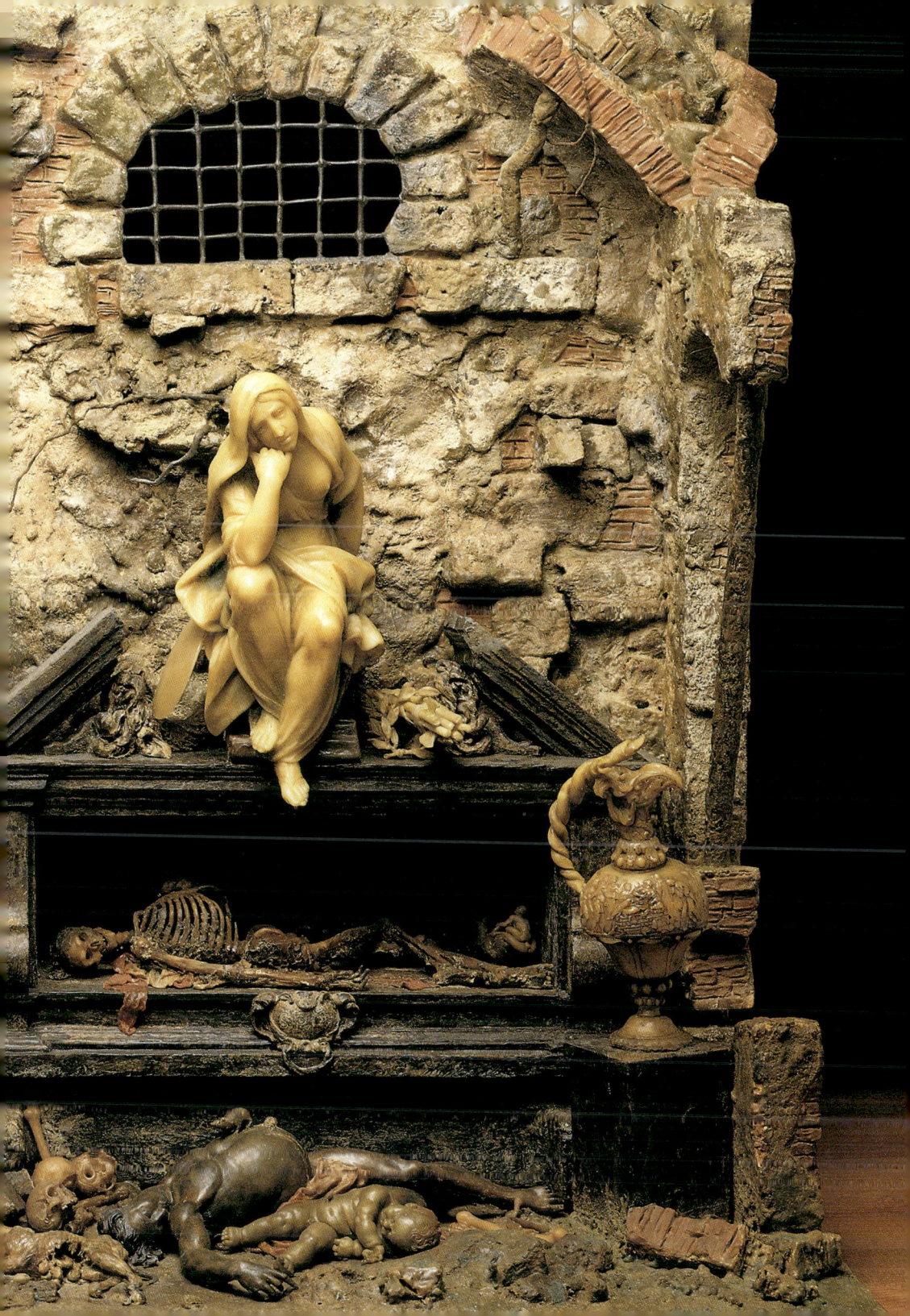

La collection de figures de cire du musée La Specola à Florence

Le musée et son histoire

Lorsqu'en 1771 Pierre-Léopold de Habsbourg-Lorraine (1747–1792), grand-duc de Toscane de 1765 à 1790 (ill. p. 7), décida de réunir en un seul musée toutes les collections « scientifiques » des galeries grand-ducales, il entreprenait quelque chose d'absolument inédit en Europe et dans le monde entier. Homme éclairé, bien en avance sur les autres souverains de son époque, le prince était passionné de sciences naturelles, conscient de leur importance pour le développement culturel de la société. C'est lui qui songea le premier à rendre accessibles à tous les découvertes réalisées dans ce domaine.

Appelé plus tard « La Specola », « observatoire » en italien, l'« Imperial Regio Museo di Fisica e Storia Naturale » (Musée impérial royal de physique et d'histoire naturelle) ouvrit ses portes le 21 février 1775, au grand public, ce qui le rendit unique en son genre dans le monde entier. Certes, les heures d'ouverture étaient différentes pour les hommes cultivés et le commun des mortels : ces derniers dont on exigeait une « tenue propre » ne pouvaient visiter le musée que de huit heures à dix heures du matin afin de laisser suffisamment de temps entre leur passage et l'arrivée « des personnes intelligentes et savantes […] à une heure de l'après-midi ». Mais même si cette distinction nous paraît discriminatoire de nos jours, on peut se figurer que l'accès des collections au grand public représentait jadis une incroyable innovation.

La création de ces collections remontait aux Médicis, grands mécènes et amateurs d'art, qui encouragèrent également les sciences avec la fondation, par exemple, de l'Accademia del Cimento (1657–1667) où enseignaient, à l'époque de Ferdinand II de Médicis, d'illustres savants comme Redi, Magalotti et Viviani, l'élève favori de Galilée. Après la mort de Jean Gaston, le dernier rejeton de la famille en 1737, le grand-duché de Toscane passa, conformément au traité de Vienne, aux mains du futur empereur François Ier de Habsbourg-Lorraine (1745–1765) qui fit dresser la liste des trésors de sa nouvelle résidence. En un peu moins d'un an, de 1763 à 1764, le médecin et naturaliste Giovanni Targioni-Tozzetti (1712–1783) parvint à faire l'inventaire des collections. Lorsqu'en 1765 Pierre-Léopold succéda à son père devenu empereur d'Autriche, il s'aperçut qu'en raison du travail déjà effectué il pouvait aisément entreprendre la réorganisation des collections scientifiques. Cette tâche devait revenir à l'abbé Felice Fontana (1730–1805), originaire de Rovereto (ill. p. 27). Professeur de logique à l'université de Pise, il était également anatomiste, physicien, chimiste et surtout physiologiste de réputation internationale. L'abbé se lança dans cette entreprise avec tant d'ardeur qu'à la fin de 1771, on pouvait déjà transporter la première partie des collections dans les nouvelles salles. La même année, le grand-duc Pierre-Léopold avait fait l'acquisition du palais des Torrigiani (qui appartenait auparavant à la famille Bini), situé dans la via Romana à deux pas du palais Pitti, ainsi que

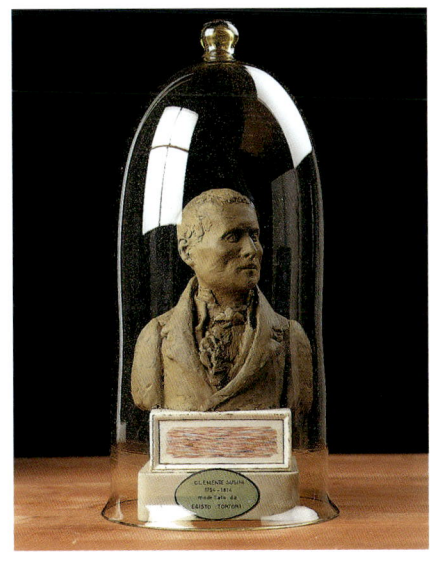

de quelques maisons avoisinantes. Il chargea ensuite Fontana de transformer les bâtiments afin qu'ils puissent abriter les collections scientifiques. Nommé par ailleurs directeur du nouveau musée, l'abbé sillonna l'Europe durant les premières années pour acheter des livres et des collections et nouer des contacts avec les savants d'autres pays. C'est ainsi que le musée florentin devint l'un des plus importants de son époque d'autant plus qu'il possédait une vaste bibliothèque scientifique.

Fontana dirigea le musée jusqu'à sa mort en 1805. Il fut secondé dans son travail par Giovanni Fabbroni qui l'accompagna dans ses nombreux voyages et devint son adversaire permanent en accédant au poste de directeur adjoint en 1784.

L'argent pour la réalisation du Musée de physique et d'histoire naturelle (et de beaucoup d'autres projets de Pierre-Léopold) fut obtenu en vendant des objets précieux appartenant aux Médicis, et ce malgré le testament de la princesse palatine Anna Maria Luisa (1667–1743), la sœur de Jean Gaston, qui avait légué d'une façon irrévocable toute la fortune des Médicis à la Ville de Florence.

Les fonds du musée, provenant essentiellement des Offices, comprenaient des collections diverses (minéraux, coquillages…), des curiosités de sciences naturelles de l'époque des Médicis, les instruments de Galilée, les appareils de l'Accademia del Cimento et quatre figures de cire du sculpteur Giulio Gaetano Zumbo.

Voulant élargir le musée aux domaines de l'astronomie et de la météorologie, Pierre-Léopold demanda en 1780 à l'architecte Gaspare Paoletti (qui avait déjà participé à la

Egisto Tortori: Small clay bust, model for a plaster bust of C. Susini
Kleine Tonbüste, Modell für eine Gipsbüste von C. Susini
Petit buste en argile, modèle d'un buste en plâtre de C. Susini

transformation du palais) de construire l'Osservatorio Astronomico (la « petite tour ») qui devait donner plus tard à l'ensemble du complexe le nom de « Specola » (« observatoire »). Il s'agissait d'un projet ambitieux qui agrandissait considérablement le bâtiment et fut achevé en 1789, malgré les avis divergents de nombreux experts, lesquels auraient préféré bâtir une nouvelle construction sur les collines de l'Acetri. La même année, une partie du parc Boboli fut intégrée au jardin botanique du musée (ill. p. 8).

À la mort de son frère Joseph, Pierre-Léopold lui succéda à la tête de l'Empire en 1790 et céda la Toscane à son fils cadet Ferdinand III (1769–1824) qui était bien loin de posséder la largeur d'esprit de son père et son habileté à gouverner. Pour couronner le tout, la Toscane fut prise par les Français en 1799. Après maintes complications, elle devint en 1801 le royaume d'Étrurie et fut remise aux Bourbon-Parme. Durant ces années, le musée organisa des cours dans les disciplines scientifiques, qui se poursuivirent après la restauration des Lorrains en 1814.

Le musée vécut une période mouvementée sous Ferdinand III. Il perdit son importance scientifique qui l'avait caractérisé à ses débuts, époque où il était considéré comme l'un des plus grands centres du savoir européen, ce qu'attestaient d'ailleurs des visiteurs étrangers

1793 catalogue of the wax models
Katalog der Wachspräparate von 1793
Catalogue des préparations en cire de 1793

aussi illustres que Goethe et Bernoulli. À l'instar de son père, Ferdinand III entreprit des transformations importantes dans le palais : en 1820, un corridor reliant la Specola à l'aile de la Méridienne, un pavillon du palais Pitti, fut construit sous la direction de l'architecte Pasquale Poccianti. Ce nouveau passage prolongeait jusqu'à la Specola le corridor Vasari existant qui conduisait du palais Vecchio au palais Pitti en passant par les Offices.

À la mort de Ferdinand III en 1824, le grand-duché passa aux mains de son fils Léopold II (1797–1870), surnommé gentiment « Canapone » par les Florentins en raison de ses cheveux blonds. C'est à lui que revient le mérite d'avoir donné un nouvel élan aux études scientifiques, en particulier à celles que l'on pouvait appliquer concrètement, comme dans l'agriculture par exemple. C'est également sous sa régence que fut construite la Tribuna di Galileo, hommage au grand chercheur, qui fut inaugurée à Florence, en 1841, à l'occasion du troisième congrès des scientifiques italiens. La Tribuna est une grande salle au premier étage du bâtiment en partie reconstruit par l'architecte Giuseppe Marelli. Les travaux commencés en 1830 ne prévoyaient à l'origine qu'une abside supplémentaire dans une salle déjà existante, mais les plans furent modifiés à la demande du grand-duc. Voulant honorer avec grandeur la mémoire de l'illustre savant, il décida de lui consacrer la salle entière qui abriterait, outre

A page of the register recording receipts for corpses from the hospital
Registerseite mit Vermerk der Eingänge der Leichname aus dem Krankenhaus
Page du registre avec les indications des arrivées des cadavres provenant des hospices

une statue de Galilée, tous les objets que l'on avait conservés de lui ainsi que ses instruments se trouvant à l'Accademia del Cimento. La construction de la Tribuna, l'un des rares témoignages du néoclassicisme à Florence, entraîna d'importantes modifications dans l'architecture du palais : les étages inférieurs furent transformés et la cour fut en partie couverte. On fit venir des peintres et des sculpteurs de la région pour décorer la salle recouverte entièrement de marbre de Toscane.

L'Istituto di Studi Superiori e di Perfezionamento (institut d'études supérieures et de perfectionnement), transformé en 1923 en université royale, fut fondé en 1859 après la chute de Léopold II, le dernier grand-duc de Toscane. Événement déjà important en soi, la fondation de l'Institut a marqué par ailleurs pour le musée le début de son démembrement en différents départements qui, en raison du nombre toujours croissant des étudiants, durent s'installer en d'autres lieux avec leurs collections et leurs bibliothèques. Seul le département de physique et de sciences naturelles du musée demeura au siège de l'Institut. Malgré l'opposition de nombreux scientifiques de l'époque, le bâtiment historique de la via Romana n'abrita plus finalement que les collections zoologiques et la plus grande partie de la céroplastie anatomique.

Les débuts de la céroplastie dans l'anatomie

L'utilisation de la cire pour modeler des figures remonte à l'époque des Romains. De tout temps, la cire fut un matériau très apprécié, dans l'art en particulier, pour des raisons esthétiques et techniques. La cire se laisse aisément travailler, elle est idéale comme moule pour couler le bronze par exemple. C'est surtout dans l'Europe du Moyen Âge que l'on fabriqua en cire un grand nombre d'offrandes, de statues de saints, de crèches et d'ex-voto. À la Renaissance et durant tout le XVIIᵉ siècle, on pouvait admirer d'innombrables modèles en cire dans les églises italiennes, en particulier dans l'Orsanmichele et la Santissima Annunziata. Ces modèles étaient des membres et des organes humains, mais aussi des portraits et des statues bien souvent grandeur nature. Au XVIIᵉ siècle apparurent ensuite des modèles anatomiques reproduits pour les besoins de la science.

Bien que l'État et encore plus l'Église eussent interdit et puni l'étude du corps humain sur les cadavres, les premiers dessins anatomiques firent leur apparition au milieu du XVᵉ siècle. Ils furent réalisés en général par des grands peintres et sculpteurs comme Léonard de Vinci, Michel-Ange, Raphaël et Titien, pour ne citer que les plus célèbres. D'illustres chercheurs, tels Falloppio, Cesalpino ou Vesalius, utilisèrent les dessins de ces artistes pour illustrer leurs traités d'anatomie.

À la fin du XVIIᵉ siècle, Gaetano Giulio Zumbo fut le premier à se servir dans la célèbre école d'anatomie de Bologne de cire colorée pour modeler des préparations en cire. Originaire de Syracuse, l'artiste, dont les œuvres sont reproduites plus loin, réalisa deux têtes d'homme (l'une conservée à Florence, l'autre à Paris), une figure féminine (malheureusement disparue) et quelques petites compositions. C'est à Bologne également que fut fondée la première école de céroplastie dont sortirent, durant la seconde moitié du XVIIIᵉ siècle, les sculpteurs du musée florentin. Considéré comme le fondateur de l'école de Bologne, le peintre, sculpteur et architecte Ercole Lelli (1702–1766) modela, comme Zumbo avant lui, des muscles et des viscères en cire sur de vrais squelettes. Amateur de sciences naturelles, le cardinal Prospero Lambertini, élu pape en 1740 sous le nom de Benoît XIV, encouragea les études anatomiques de l'école de Bologne et acheta tous les modèles de Lelli, de magnifiques exemplaires que l'on peut admirer aujourd'hui encore à l'Istituto di Anatomia Umana Normale de l'université de Bologne. Enfin, et ceci est d'une grande importance, il apporta son soutien aux chercheurs devant se procurer des cadavres pour leurs travaux. Parmi les assistants de Lelli, signalons Giovanni Manzolini (1700–1755) qui, avec l'aide de sa femme Anna Morandi (1714–1774), réalisa de nombreux modèles conservés également à l'université de Bologne.

À cette époque, l'étude de l'anatomie était en plein essor dans toute l'Europe. L'école céroplastique de Florence fut fondée sur le modèle de l'école de Bologne à l'instigation de Giuseppe Galletti. Chirurgien et accoucheur, il s'inspira des œuvres de Lelli et de Manzolini pour créer avec le sculpteur Giuseppe Ferrini toute une série de modèles en cire et en terre cuite (elles se trouvent actuellement au Museo Galileo – Istituto e Museo di Storia della Scienza de Florence) devant illustrer différents types d'accouchements, normaux et dystociques.

Les figures en cire de la Specola : l'école, les œuvres, les techniques

L'atelier de céroplastie à Florence fut actif pendant un siècle environ, de 1771 à la seconde moitié du XIXᵉ siècle, mais la majeure partie des objets en cire conservés au musée, de même que les travaux réalisés pour les institutions florentines et d'autres commanditaires, furent effectués durant les soixante premières années de son existence.

La fondation de la Specola remonte à Felice Fontana qui, pleinement soutenu par le grand-duc Pierre-Léopold, s'y consacra avec une énergie d'autant plus grande qu'il participait lui-même à la création des modèles en tant qu'anatomiste et pathologiste. Au début, le sculpteur Giuseppe Ferrini travailla seul à l'atelier sous la direction de Fontana. Plus tard, on engagea l'anatomiste Antonio Matteucci ainsi que Clemente Susini, un tout jeune homme à l'époque, qui se révéla si habile et si productif qu'il ne tarda pas à devenir le plus célèbre céroplasticien de l'école florentine (ill. p. 14).

On ignore aujourd'hui où on travaillait la cire, probablement dans les salles du rez-de-chaussée dont les fenêtres donnent sur la via Romana. Ces salles sont situées dans l'aile sud du palais, qui est pourvue de nombreuses courettes. La plupart des appareils d'origine ont disparu, mais leur acquisition figure encore dans les archives : bassines en cuivre de plusieurs tailles dans lesquelles on faisait fondre la cire ; instruments à modeler (ill. p. 31) ; fils de fer de grosseur différente ; plaques de marbre servant à presser la cire en fines couches ; balances ; trépieds pour chauffer le matériau ; ardoises pour les notes et les dessins réalisés pendant les dissections ; caisses munies de poignées pour le transport des cadavres, caisses en bois munies de tiges pour le transport des figures de cire ; récipients, vases et bouteilles en céramique ou en verre pour les couleurs et autres substances ajoutées à la cire. On retrouva dans les anciens entrepôts du musée un grand nombre de ces récipients avec leur contenu (ill. p. 30).

Les archives nous indiquent par ailleurs qu'il fallait réunir plus de deux cents cadavres ou parties de cadavres pour fabriquer une seule figure. Ce nombre incroyablement élevé s'explique pour la bonne raison que le formol et la réfrigération étant inconnus, il fallait constamment recourir à des cadavres frais pour les dissections. Afin de recenser exactement les cadavres provenant des hospices de Santa Maria Nuova à deux kilomètres de là, on tenait un registre dans lequel étaient notées toutes les arrivées, mais aussi tous les départs pour le cimetière : « [...] à propos des cadavres de Santa Maria Nuova dont on ignore le nombre et parce qu'on ne peut pas faire mourir les gens pour les besoins des collections de cire, il a été décidé de placer un registre à la porte du musée afin de consigner les cadavres utilisés en ce lieu, dont le nombre semble considérable, et afin de noter la fréquence des envois au cimetière. » (extrait d'un document de 1793).

Malgré le petit nombre de préparateurs, les travaux allaient bon train si l'on considère qu'aux environs de 1790, soit vingt ans après leur commencement, les modèles remplissaient déjà huit salles, sans compter ceux que l'on avait fabriqués pour les hospices de Florence et d'autres institutions en Italie et à l'étranger (*cf.* ill. p. 46). De toutes les sculptures en cire qui sortirent de l'atelier de la Specola pour être envoyées aux quatre coins de l'Europe, les plus remarquables sont celles effectuées pour l'école militaire de médecine à Vienne. Commandée en 1781 par l'empereur d'Autriche Joseph II, le frère aîné de Pierre-Léopold, la collection fut achevée en 1786. Elle comprenait 1 200 pièces qui furent acheminées à dos d'âne jusqu'à Vienne et nécessitèrent deux voyages. D'autres sculptures prirent la route de Pavie, de Cagliari, de Bologne, de Budapest, de Paris (elles se trouvent aujourd'hui à Montpellier), d'Uppsala, de Londres, de Leyde et de bien d'autres villes encore. Ces commandes de l'extérieur soulevaient parfois des problèmes d'organisation car le grand-duc ne tolérait aucun retard dans la fabrication de certains modèles pour le musée. On décida donc que pour ces commandes, les sculpteurs et les anatomistes – Fontana lui-même dans le cas de la collection viennoise – pouvaient certes utiliser les moulages et les instruments du musée, mais qu'ils devaient se procurer eux-mêmes le matériel et se faire assister d'artisans étrangers au musée.

Si Fontana tenait absolument à créer le plus grand nombre possible de modèles anatomiques en cire, c'est parce qu'il voulait constituer une réserve qui rendrait superflue l'exhumation de cadavres pour les cours d'anatomie. C'est également dans ce but qu'il réalisa toute une série de dessins (ill. p. 10, 11) illustrant les différentes parties du corps : les organes, les muscles, les os, etc., étaient reliés par des lignes discontinues à des chiffres qui permettaient leur identification. Les dessins et les explications étaient conservés dans un tiroir du coffre où se trouvait la figure de cire correspondante. Chaque préparation était ainsi accompagnée d'un véritable traité d'anatomie (ill. p. 20).

Vers 1790, Fontana se lança dans un projet encore plus ambitieux qu'il ne put mener à bien à cause des difficultés qu'il rencontra. Il s'agissait d'une série de pièces d'anatomie en bois peint – de taille réelle ou agrandies – que l'on aurait pu démonter afin de montrer les liens entre les différents organes. Après quelques tentatives dont on a gardé seulement un petit nombre d'exemplaires, à savoir deux sculptures entières (une au musée, l'autre à Paris) et un buste gigantesque, Fontana dut abandonner son projet. Non seulement le bois se laissait difficilement travailler, mais étant un matériau organique, il se déformait avec le temps et les différentes pièces ne s'adaptaient plus (ill. p. 37). La seule sculpture du musée présentant les caractéristiques souhaitées fut fabriquée en cire. Il s'agit de la « Venere medicea », une figure féminine allongée dont on peut retirer les éléments superposés jusqu'à l'apparition de l'utérus et d'un petit fœtus.

Le musée avait entre-temps agrandi son équipe. Il avait engagé le dessinateur Claudio Valvani, le sculpteur sur bois Luigi Galati ainsi que plusieurs anatomistes pour la dissection des cadavres. L'un des plus célèbres fut Paolo Mascagni (1755–1815) qui laissa à la postérité de magnifiques planches anatomiques. Il s'était spécialisé dans l'étude du système lymphatique, ce n'est donc pas un hasard si à la Specola cette partie de l'organisme est représentée avec une grande minutie sur de nombreuses statues, mais aussi sur des préparations plus petites.

La mort de Felice Fontana (ill. p. 17) en 1805 n'empêcha pas l'atelier de poursuivre son travail, et à la mort de son successeur Clemente Susini, en 1814, ce furent d'autres sculpteurs, comme les Calenzuoli (le père Francesco et le fils Carlo) ou Luigi Calamai (1800–1851), qui se chargèrent de la direction. Sous Calamai, les activités se concentrèrent sur des modèles d'anatomie comparative et de botanique ainsi que sur les modèles d'anatomie pathologique qui étaient destinés aux hospices de Santa Maria Nuova et que l'on peut voir aujourd'hui au département de biomédecine du Museo di Storia Naturale. Après sa mort, Calamai fut remplacé par Egisto Tortori (1829–1893) qui, à côté de préparations d'animaux et d'anatomie pathologique, modela également un buste de Clemente Susini. À la mort de Tortori, en 1893, l'atelier cessa définitivement ses activités.

De nos jours, le musée La Specola abrite 513 modèles en cire d'anatomie humaine, soixante-cinq d'anatomie comparative et cinq réalisés par G. G. Zumbo. Il possède aussi vingt-six figures (dont un tronc de jeune garçon) : treize debout et treize couchées, dont dix-huit grandeur nature (six debout et douze couchées) et huit mesurant soixante centimètres environ (sept debout et une allongée). À l'exception de quatorze sculptures sur l'anatomie humaine se trouvant dans les entrepôts, toutes sont exposées au public. Il faut également ajouter plus de huit cents dessins encadrés et près de neuf cents feuillets explicatifs dont le nombre était jadis certainement supérieur, mais qui ont été égarés au cours des siècles. Ces dernières années, on a retiré des salles d'exposition les dessins et les feuilles explicatives pour les préserver des dommages causés par la lumière et l'humidité. Ils ont été remplacés par des photocopies en couleur de sorte que la structure de la collection a pu être conservée aussi bien du point de vue de la présentation que du point de vue didactique.

Tandis que les experts en la matière sont impressionnés par la perfection scientifique des objets exposés, la plupart des visiteurs s'intéressent surtout à leur côté artistique, car ils sont, malgré toute leur rigueur scientifique et leur vocation didactique, de véritables œuvres d'art. La beauté de ces objets est mise en valeur par les coffres en bois très travaillés, dans lesquels ils sont conservés. Malheureusement, beaucoup de ces coffres se sont abîmés avec le temps et doivent être souvent restaurés. Datant du XVIIe siècle, les draperies et voiles en soie, à l'origine vert clair, et se trouvant eux aussi dans un bien triste état, montrent quel soin fut toujours apporté à la présentation de la collection.

Ce souci d'esthétique se constate par ailleurs dans les collections d'anatomie comparative, d'anatomie pathologique et en particulier dans les collections de botanique où des vases en céramique furent réalisés spécialement pour les petits modèles de plantes par l'atelier Gionori de Sesto Fiorentino.

Comme les procédés techniques pour la fabrication des figures en cire ne nous ont pas été transmis de façon explicite, nous ne pouvons les reconstituer qu'à partir des diverses lettres et documents des archives nationales et du Musée des sciences de Florence. Pour compliquer le tout, chaque sculpteur avait sa propre technique qu'il se gardait bien de divulguer. Une chose est sûre cependant : une fois les parties du corps disséquées, on en faisait une copie exacte en craie ou en cire de qualité inférieure dont on tirait un moulage en plâtre qui, s'il s'avérait de grande taille, pouvait être constitué de plusieurs pièces. Pour certains organes, comme les os, on pouvait faire directement un moulage en plâtre (ill. p. 55). Ces moulages conservés aujourd'hui encore dans les entrepôts du musée constituaient une sorte de matrice que l'on pouvait utiliser plusieurs fois pour reproduire le même modèle. L'étape la plus complexe et la plus délicate était la construction du modèle définitif. Elle exigeait une grande précision, une connaissance approfondie des substances qu'il fallait mélanger à la cire pour obtenir la couleur et la consistance souhaitées, ainsi que beaucoup d'expérience et de savoir-faire. La cire devait fondre lentement au bain-marie pour que sa couleur ne soit pas altérée. On utilisait le plus souvent de la cire blanche de Smyrne, de la cire chinoise et de la cire de Venise. Pour la rendre plus malléable, on y ajoutait de la térébenthine ainsi que des colorants, dilués eux aussi dans de la térébenthine, et d'autres substances apparaissant sur les listes d'achat sous la mention « matières nécessaires pour les travaux en cire ». Afin que plus tard, le modèle se détache facilement du moule, celui-ci était humidifié à l'eau tiède et frotté avec un savon doux avant que l'on y coule le mélange de cire. Mis à part quelques préparations complètement en cire, la plupart étaient creuses à l'intérieur et remplies de chiffons, d'étoupe ou de morceaux de bois pour une plus grande stabilité. Les statues constituées de plusieurs pièces assemblées les unes aux autres étaient en général renforcées par une armature métallique. Une fois le moule ôté, la figure était soigneusement nettoyée, puis cannelée avec les outils appropriés et, finalement, dotée de ses organes, vaisseaux, nerfs, etc. Une dernière couche de vernis transparent lui conférait le brillant nécessaire. Afin d'être sûr que le modèle soit conforme à l'original, toutes les étapes étaient constamment surveillées par les anatomistes qui décidaient également de la meilleure position pour mettre en valeur tel ou tel organe.

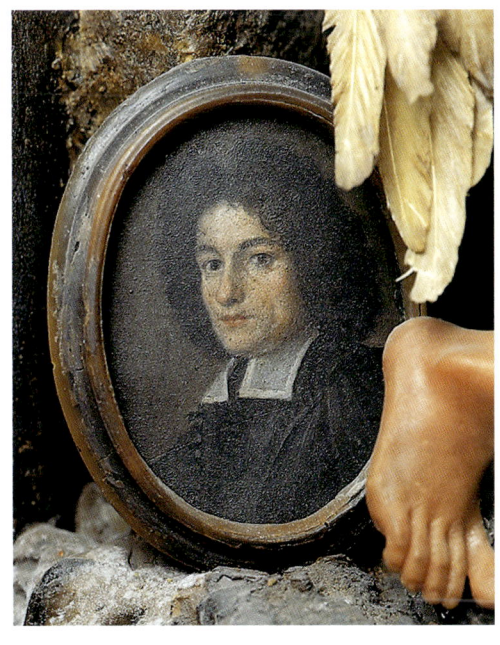

Comme nous l'avons déjà dit, chaque sculpteur possédait sa propre technique et ses méthodes particulières qu'il perfectionnait avec le temps en utilisant de nouveaux instruments, comme cette « filière à tirer les cylindres en cire » mise à leur disposition dès 1786. Avant, cette opération permettant de modeler les veines et les artères devait être faite à la main. Autour de 1790, les sculpteurs tinrent pendant quelques années des « cahiers » (ill. p. 21) dans lesquels ils notaient quotidiennement les travaux effectués, et qui nous permettent de constater qu'à l'époque déjà ils devaient entreprendre des restaurations, des réparations, des améliorations et des corrections. En 1793, Susini rédigea une « [...] note sur les préparations anatomiques en cire à corriger en raison des erreurs constatées par moi, Clemente Susini, modeleur du musée royal, erreurs qui méritent d'être réparées ». Par exemple : « Vérification de la préparation de l'artère pharyngée et de la carotide : on peut voir partant de la trachée-artère deux muscles étrangers n'appartenant pas au corps humain et un muscle rattaché au prolongement de la thyroïde, mais qui ne semble être ni le muscle de la langue, ni du pharynx et encore moins de la thyroïde. » La note de Susini se termine avec la constatation suivante : « Ces erreurs sont trop graves pour ne pas être corrigées. Je passe toutes les autres sous silence car je désire être bref. » Presque tous les modèles se trouvent en bon état de conservation, quelques-uns seulement ont changé de couleur : certains fœtus sont devenus foncés et les veines qui présentent aujourd'hui une teinte verte étaient jadis « bleu-violet ».

G. G. Zumbo: Self-portrait (detail from *The Triumph of Time*)
Selbstporträt (Ausschnitt aus *Der Triumph der Zeit*)
Autoportrait (détail du *Triomphe du Temps*)

Les figures en cire de Gaetano Giulio Zumbo

Les sculptures en cire de Gaetano Giulio Zumbo méritent qu'on leur consacre un cha-
pitre à part, non seulement en raison de leur conception et de leur méthode de fabrication
inhabituelles, mais surtout parce qu'étant en avance d'un siècle sur les autres représentations anatomiques en cire, elles constituent la base de toutes celles qui suivront.

Né en 1656 à Syracuse, Zumbo est issu d'une famille de la vieille noblesse qui s'appelait
probablement Zummo et dont il ne reste plus aucune trace aujourd'hui. Comme le nom de
« Zumbo » est plus connu et est passé dans l'usage, il nous semble opportun de le garder.
Nous savons peu de choses sur sa vie, quelques fragments tout au plus. Élevé chez les Jé-
suites dans sa ville natale, il dut quitter le collège à cause d'un « incident ennuyeux » qui n'est
pas précisé. Les œuvres classiques qu'il put voir dans sa patrie ont certainement eu au début
une influence sur sa formation artistique et culturelle. Plus tard, ce furent les thèmes et les
sujets du maniérisme et du baroque qui exercèrent un ascendant sur lui. Il vécut à Naples de
1687 à 1691 où il réalisa probablement deux des groupes allégoriques appartenant à la Specola, les « teatrini », à savoir *Il Trionfo del Tempo* (Le Triomphe du Temps) et *La Peste*. Les
spécialistes qui se sont penchés sur l'œuvre de Zumbo affirment que le fonds pittoresque
et l'agencement des figures trahissant l'influence des maîtres d'Italie du Sud sont le signe
de sa paternité. De 1691 à 1694, Zumbo travailla à Florence pour Cosme III de Médicis et
acheva d'autres « teatrini » : *Il Sepolcro* (ou *La Vanità della Gloria Umana*) et *Il Morbo Gallico*
(ou *Sifilide*) qui se réfèrent à l'art de la Renaissance. En 1695, il entreprit quelques voyages
pour Bologne où il manifesta beaucoup d'intérêt pour les études anatomiques alors en plein
essor. À la fin de 1695, il s'installa à Gênes. C'est là probablement qu'il acheva *L'Anatomie de
la tête* exposée à la Specola. C'est là aussi qu'il rencontra le chirurgien français Guillaume
Desnoues avec lequel il collabora pour des modèles anatomiques en cire comprenant, entre
autres, une parturiente grandeur nature, une femme morte en couches de format plus petit
ainsi que la naissance et la mise au tombeau du Christ. Malheureusement, ces œuvres n'ont
pas été conservées. En 1700, Zumbo se brouilla avec le médecin français et partit pour Marseille où il effectua une tête en cire, puis pour Paris où il entra au service de Louis XIV. Il
demeura dans cette ville jusqu'à sa mort, en 1701, emporté par une hémorragie cérébrale.
Parmi ses affaires, on trouva une magnifique tête en cire qui est, selon toute probabilité,
celle conservée de nos jours au Musée national d'histoire naturelle. Enterré en l'église de
Saint-Sulpice, sa tombe fut détruite pendant la Révolution.

Une grande partie des œuvres de cet artiste encore largement méconnu il y a un demi-
siècle est conservée au musée La Specola. Les trois « teatrini », connus également sous le
nom de *Cera della peste* (figures de la peste), et *L'Anatomie de la tête* se trouvaient à la Galerie
royale (c'est-à-dire aux Offices) lorsque les Lorrains succédèrent aux Médicis. Ils furent ensuite envoyés au Museo di Fisica e Storia Naturale où ils ne furent jamais exposés, peut-être

à cause du réalisme brutal des scènes. Lorsque Pierre-Léopold partit pour Vienne en 1790, il offrit les « figures de la peste » au médecin de la cour, Giovanni Giorgio Hasenöhrl, plus connu sous le nom de Lagusius. Celui-ci ne les garda pas mais chargea Fontana et, plus tard, Agostino Renzi, surintendant de la Pharmacie royale, de les vendre pour 150 sequins. Conscient de leur valeur, le surintendant les proposa au nouveau grand-duc Ferdinand III, d'abord pour l'Académie royale des beaux-arts où l'on jugea ces figures « ni utiles ni appropriées », puis pour le Musée royal. Cette proposition fut appuyée par Giovanni Fabbroni, directeur adjoint du musée, et les trois œuvres furent estimées par différents experts dont Clemente Susini. Comme les estimations dépassaient le prix réclamé par Lagusius, le grand-duc consentit à leur acquisition. Les « teatrini » restèrent au musée jusqu'en 1978, date à laquelle ils furent remis au Museo Nazionale de Bargello. Après avoir été cédés au Museo di Storia della Scienza, ils revinrent finalement à la Specola en 1974. Comme le prouvent les archives, les sculptures en cire furent plusieurs fois restaurées dans le passé, entre autres par Susini et Tortori. Se trouvant au Museo di Storia della Scienza, celles « de la peste » furent gravement endommagées par les pluies diluviennes de 1966. Très méticuleusement, Guglielmo Galli de l'atelier de Pietre Dure à Florence parvint au bout de dix-huit mois à les remettre en état en s'aidant de photographies. *L'Anatomie de la tête* en revanche n'a jamais quitté la Specola puisqu'elle était à juste titre considérée comme présentant un intérêt particulier pour la collection scientifique de l'institut.

La technique employée par Zumbo se distingue fortement de celle de l'école florentine à venir. *L'Anatomie de la tête* fut exécutée sur un vrai crâne ayant appartenu, si l'on en croit les

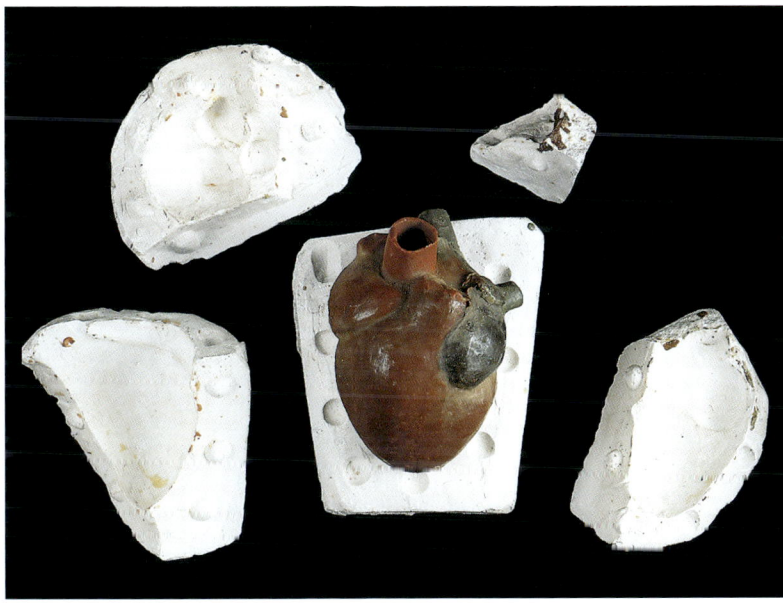

Plaster mould of a heart

Gipsabdruck eines Herzens

Moulage d'un cœur

radios, à un homme âgé de vingt-cinq ans. En revanche, la tête conservée à Paris et une autre disparue aujourd'hui ont été fabriquées entièrement en cire, ce qui permet de conclure à une évolution de sa technique, due probablement à la coopération du céroplasticien sicilien avec Desnoues. Les figures des « teatrini » exécutées avec une grande habileté dans les moindres détails ont été très probablement obtenues à partir de moulages de pièces qui, à leur tour, se basaient très soigneusement sur des modèles en argile. La cire plus ou moins liquide (un mélange de cire d'abeille, de colophane, de térébenthine et de colorants) était appliquée en couches de diverses épaisseurs suivant la consistance et la coloration souhaitées. Si l'on possède aujourd'hui quelques précisions sur la technique de Zumbo, c'est grâce aux travaux de restauration de Guglielmo Galli entrepris après les inondations de 1966, qui révélèrent par ailleurs les traces de réparations plus anciennes du fait de la présence de colorants n'existant pas encore à la fin du XVIIe siècle. Le petit autoportrait (ill. p. 53) du groupe *Il Trionfo del Tempo* ne peut pas être attribué à Zumbo avec certitude, il pourrait être l'œuvre d'un contemporain ou de Susini plus tard. Celui intitulé *Il Morbo Gallico*, à l'origine une composition du même type et dont il n'existe plus que quelques fragments, fut offert à Filippo Corsini par Cosme III. Conservé en permanence dans les caves du palais Corsini sur l'Arno, il fut pratiquement détruit par les inondations. On retrouva quelques figures dans les jardins mais, n'ayant pas de photographie de l'ensemble, une reconstruction s'avéra impossible. À l'occasion du bicentenaire de l'ouverture du musée, qui fut célébré avec un congrès sur la céroplastie, la famille Corsini fit don au musée des restes de cette œuvre.

Pendant longtemps, les chefs-d'œuvre de Zumbo furent considérés comme les caprices malsains d'un artiste se plaisant à reproduire des détails macabres et repoussants. Ce n'est qu'au milieu de notre siècle qu'on les considéra dans leur véritable perspective historique, comme des documents réalistes d'une époque où les guerres, les famines et les grandes épidémies étaient le lot quotidien des hommes. Cette vision de la destruction, de la précarité de l'existence humaine, de la fuite inexorable du temps, ce « memento mori » continuel auquel les œuvres de Zumbo nous confrontent avec une foule de détails, est un exemple typique de la culture du XVIIe siècle, telle qu'elle nous a été également transmise par d'autres grands artistes de l'époque, comme Luca Giordano ou Mattia Preti.

G. G. Zumbo: *Il Trionfo del Tempo*
The Triumph of Time (detail) / Der Triumph der Zeit (Detail)
Le Triomphe du Temps (détail)

Pages 58–59:
G. G. Zumbo: *La Peste*
The plague / Die Pest / La peste

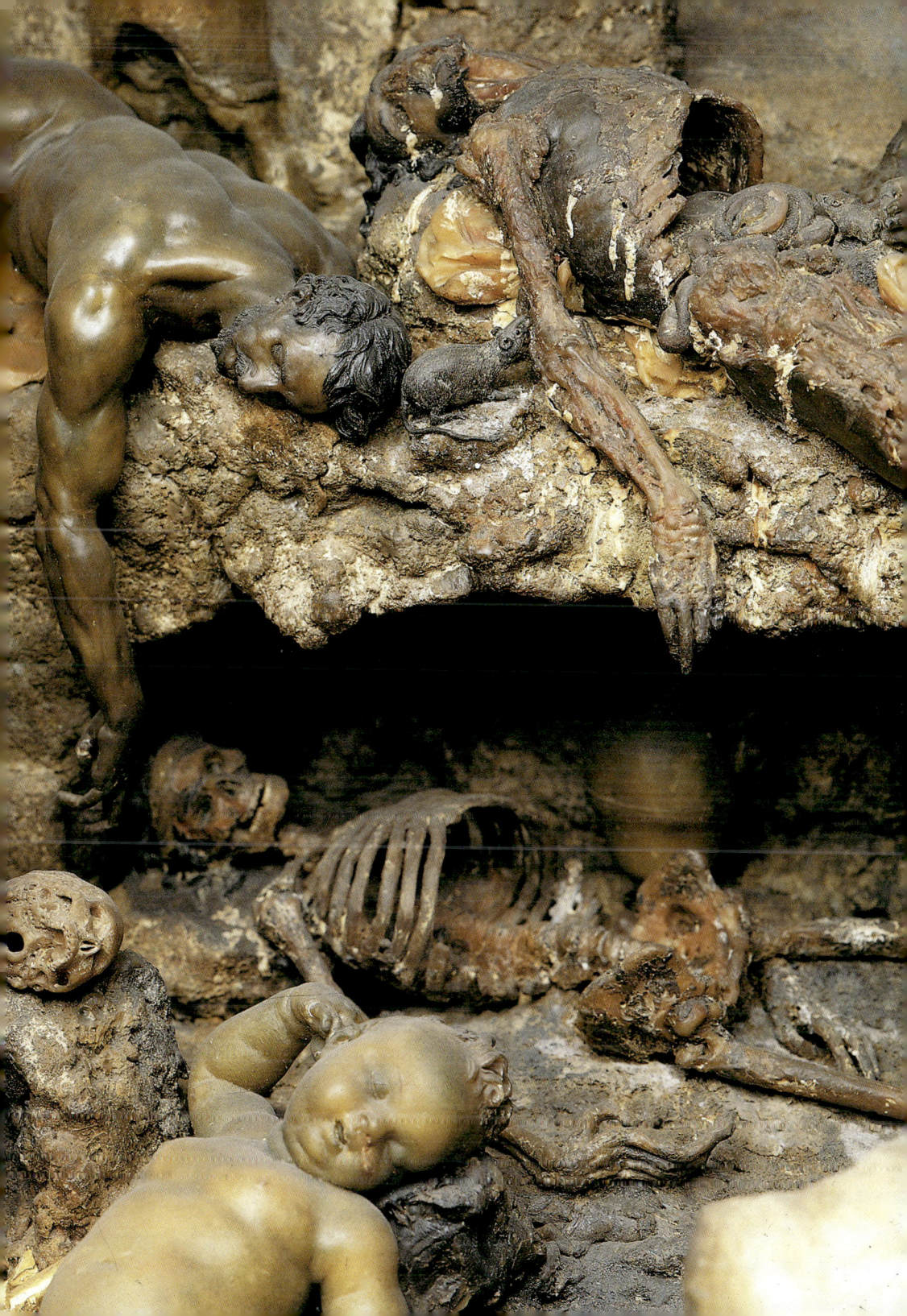

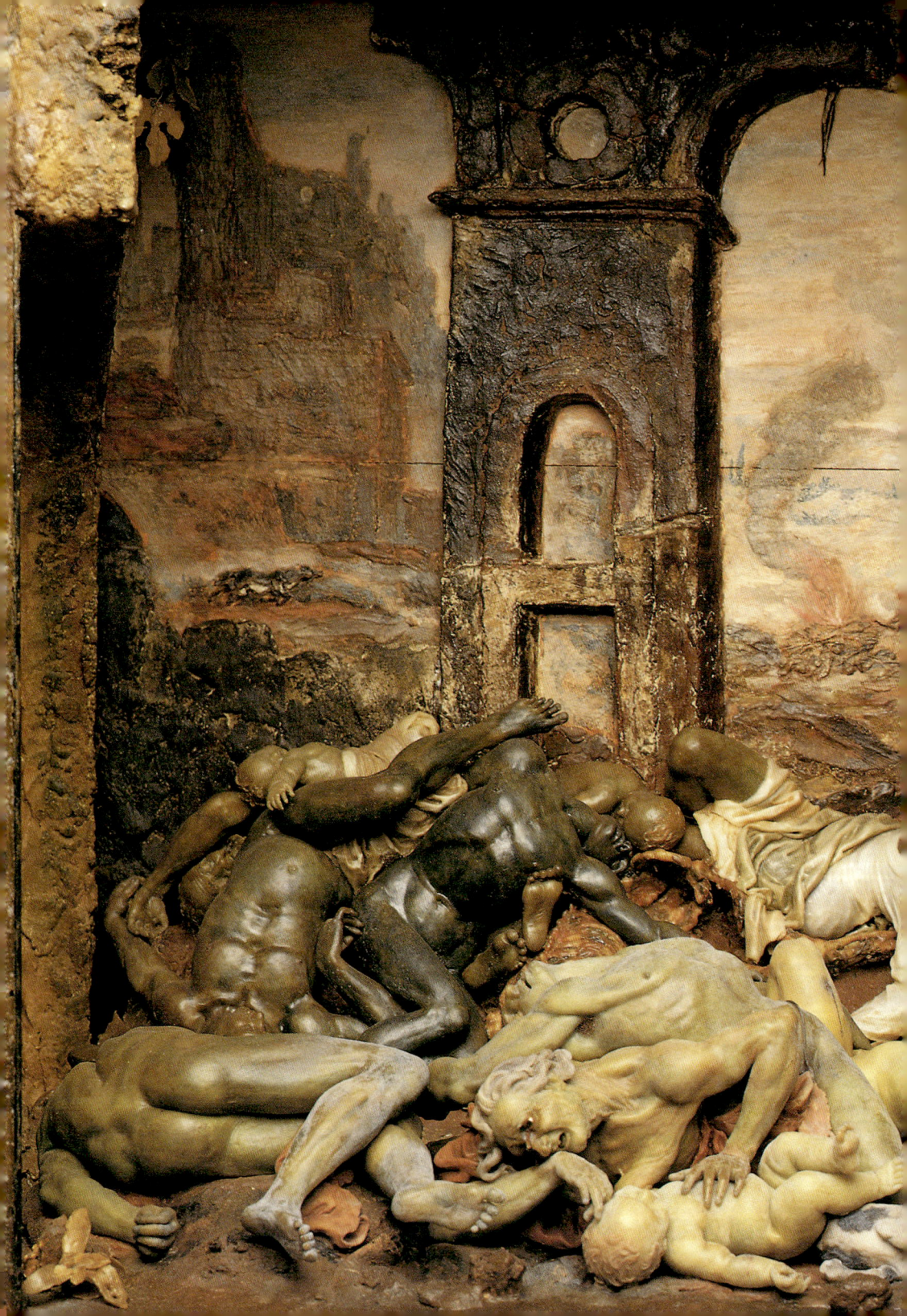

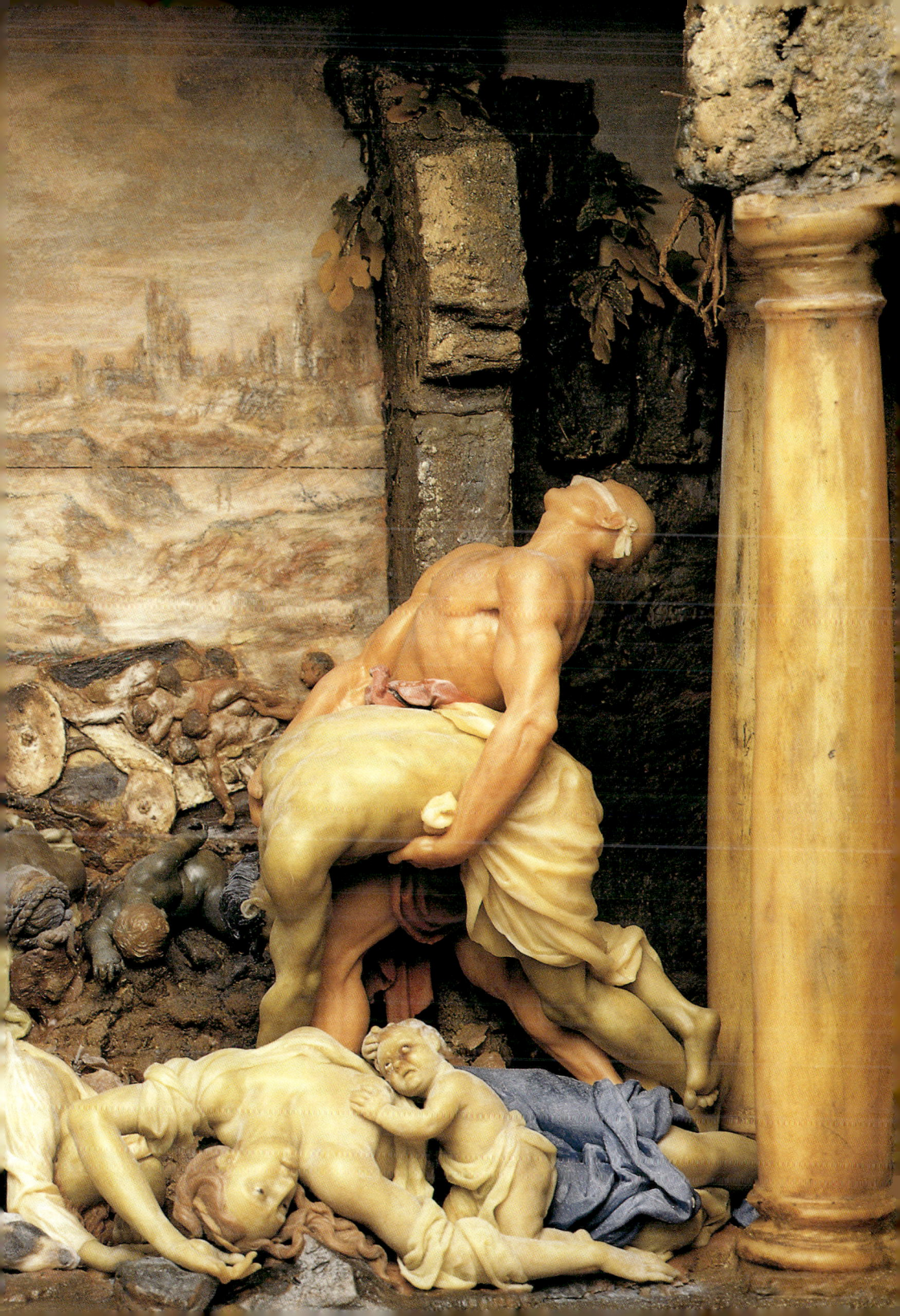

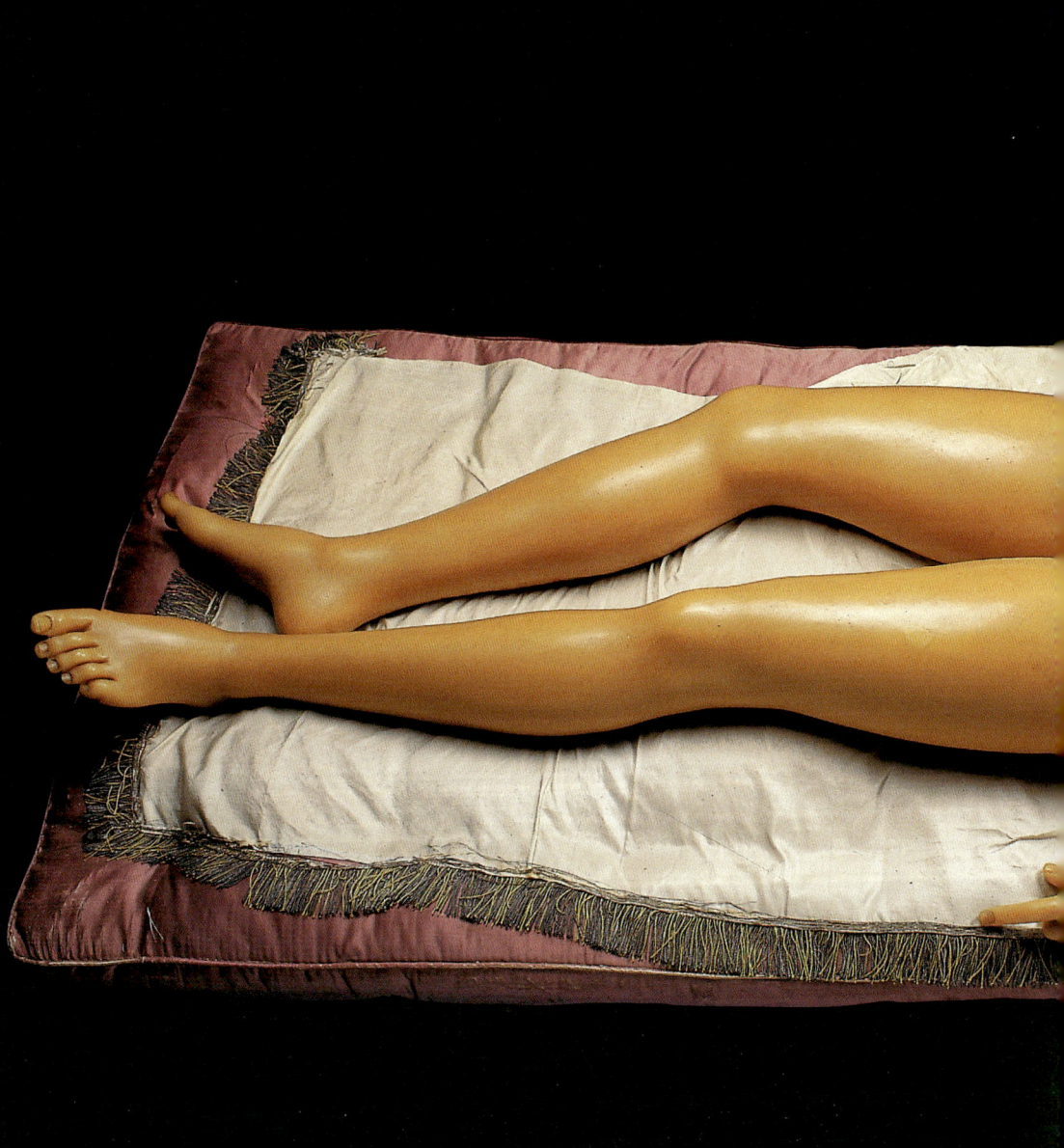

Anatomical Waxes
Anatomische Wachsmodelle
Cires anatomiques

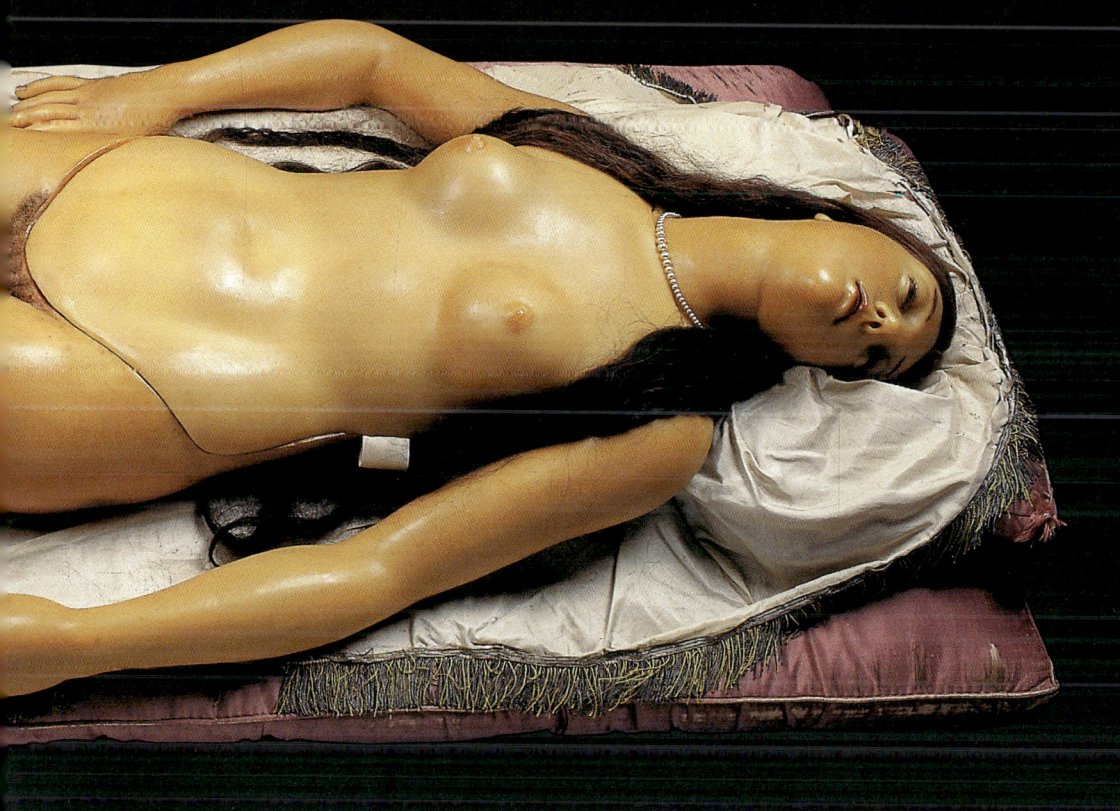

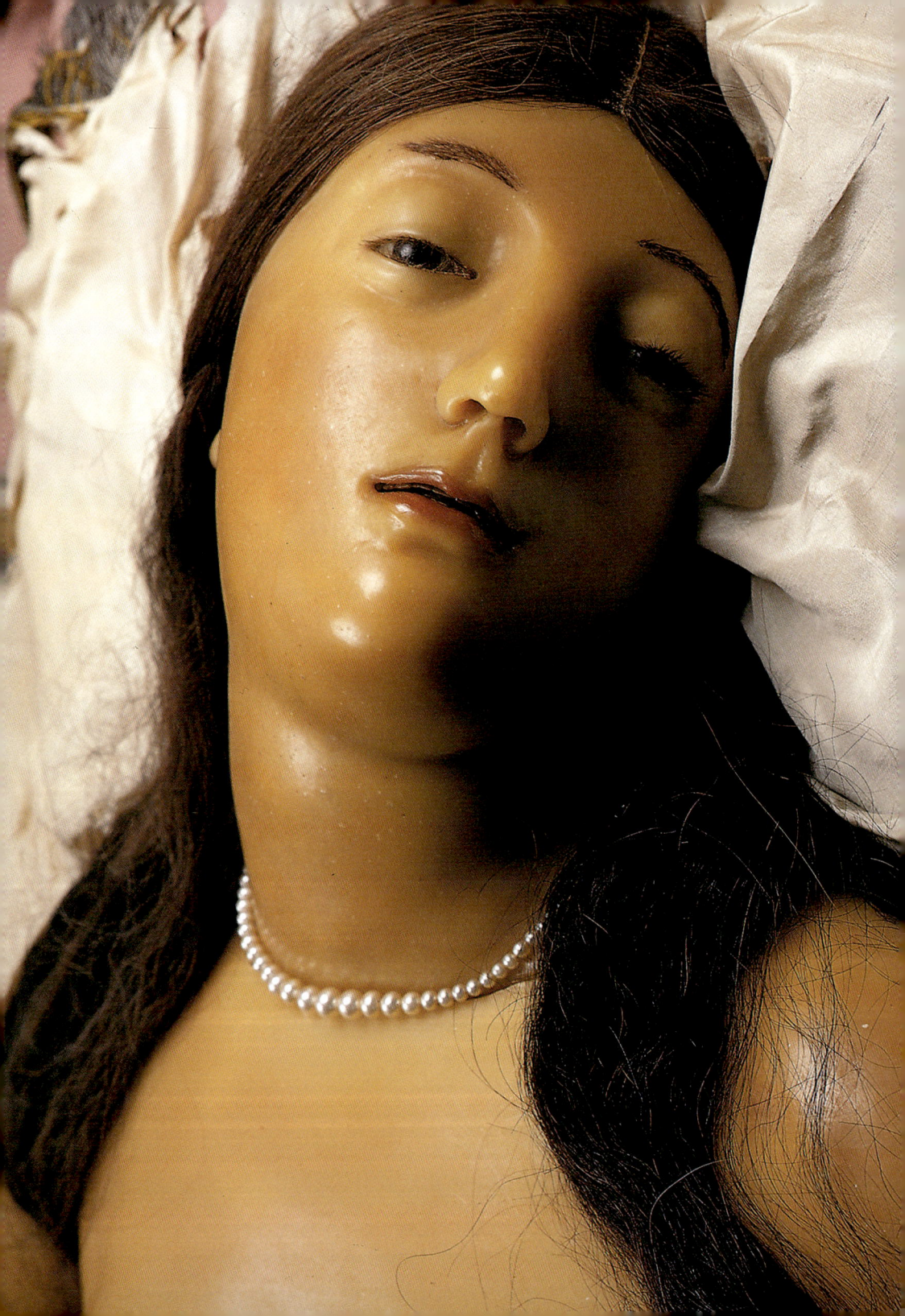

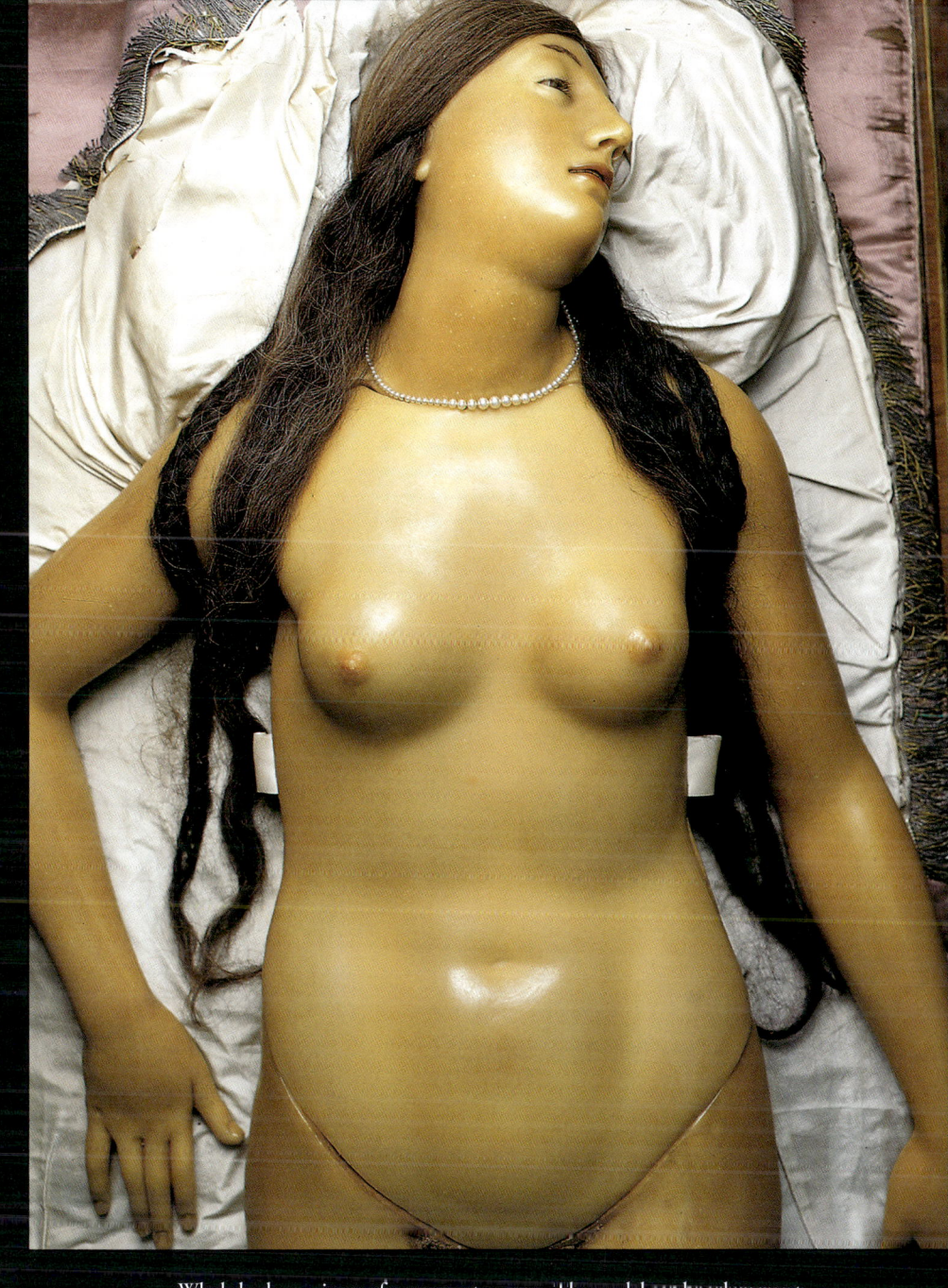

Whole body specimen of a pregnant woman. The model can be taken apart

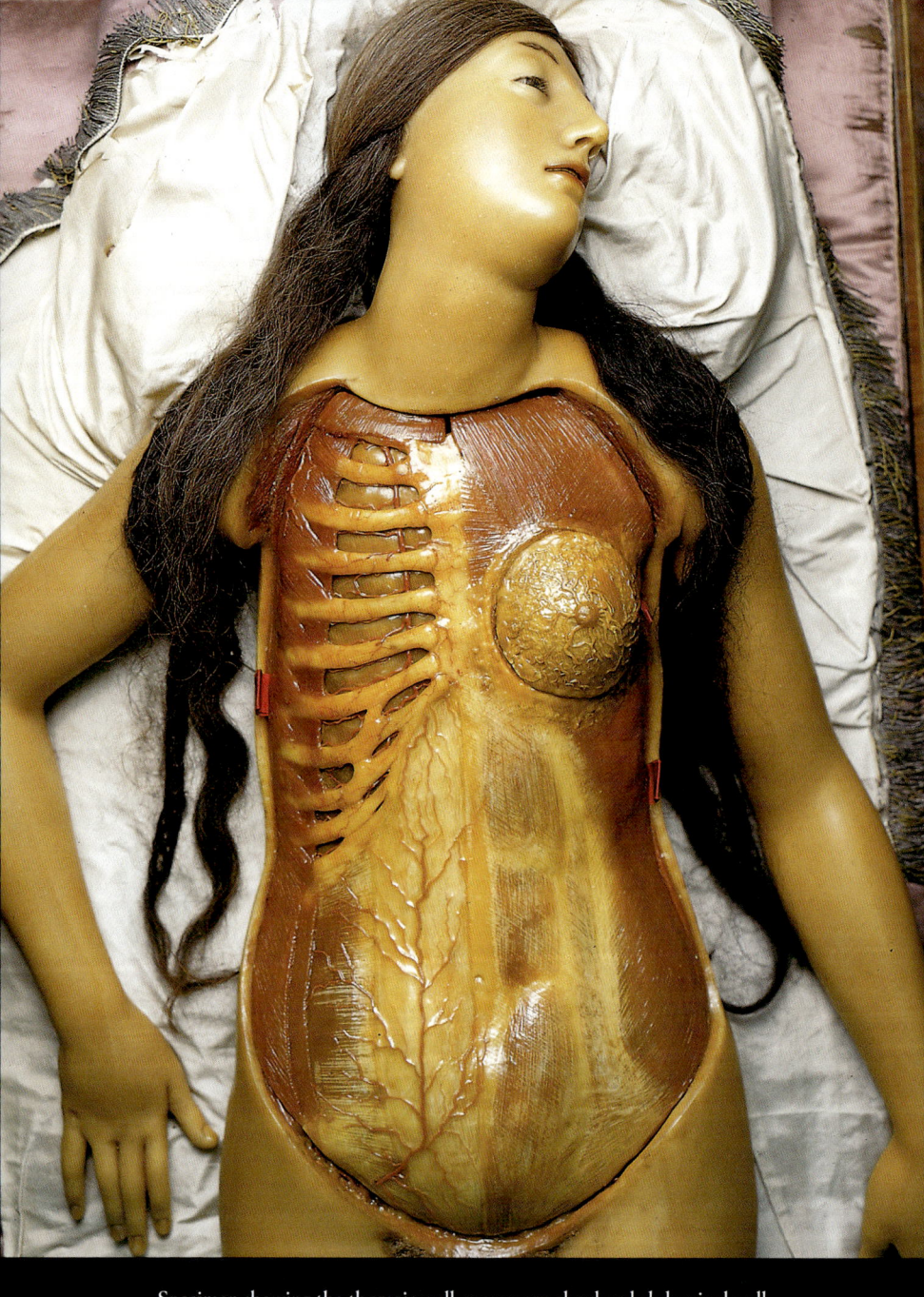

Specimen showing the thoracic wall, mammary gland and abdominal wall

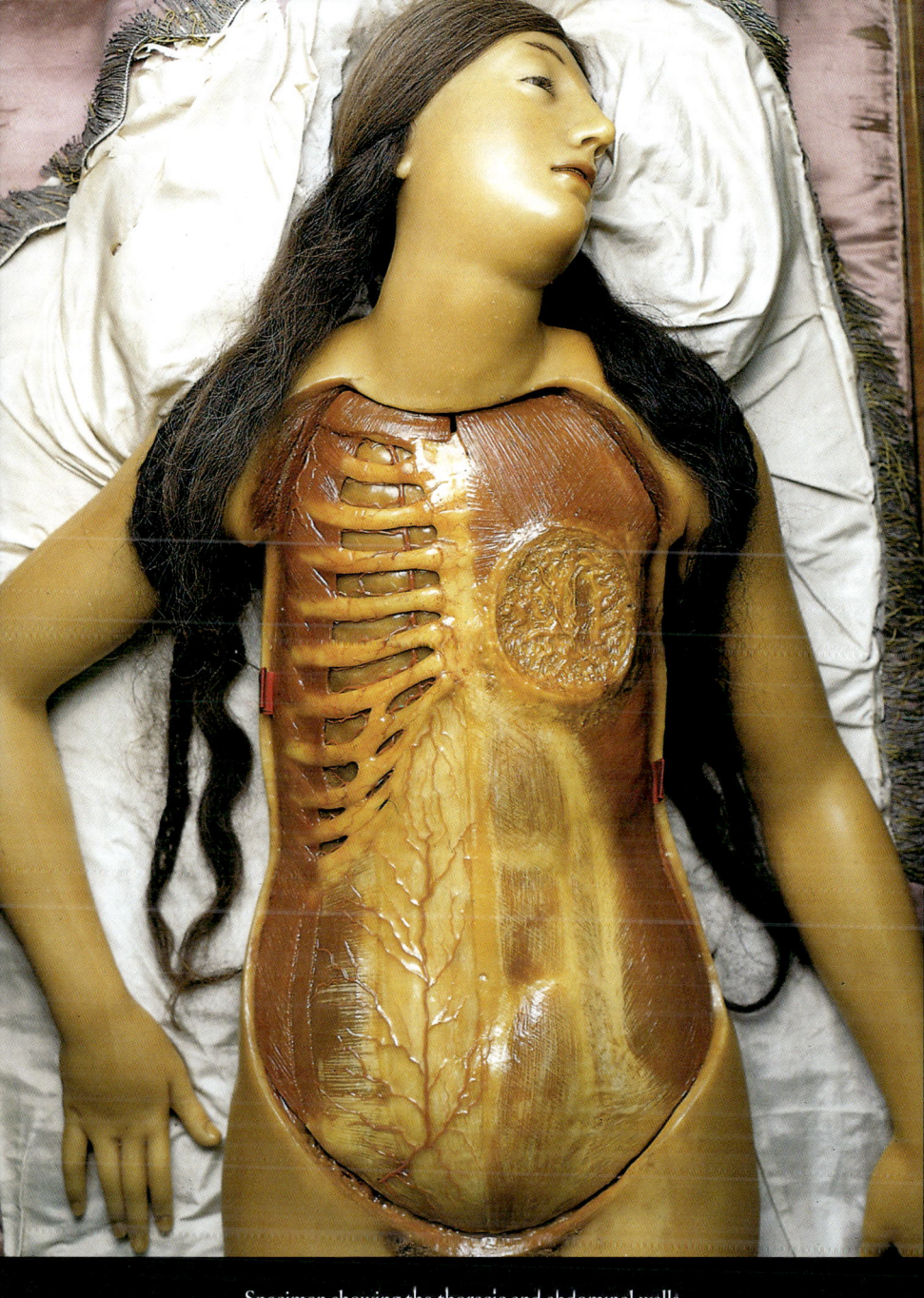

Specimen showing the thoracic and abdominal walls

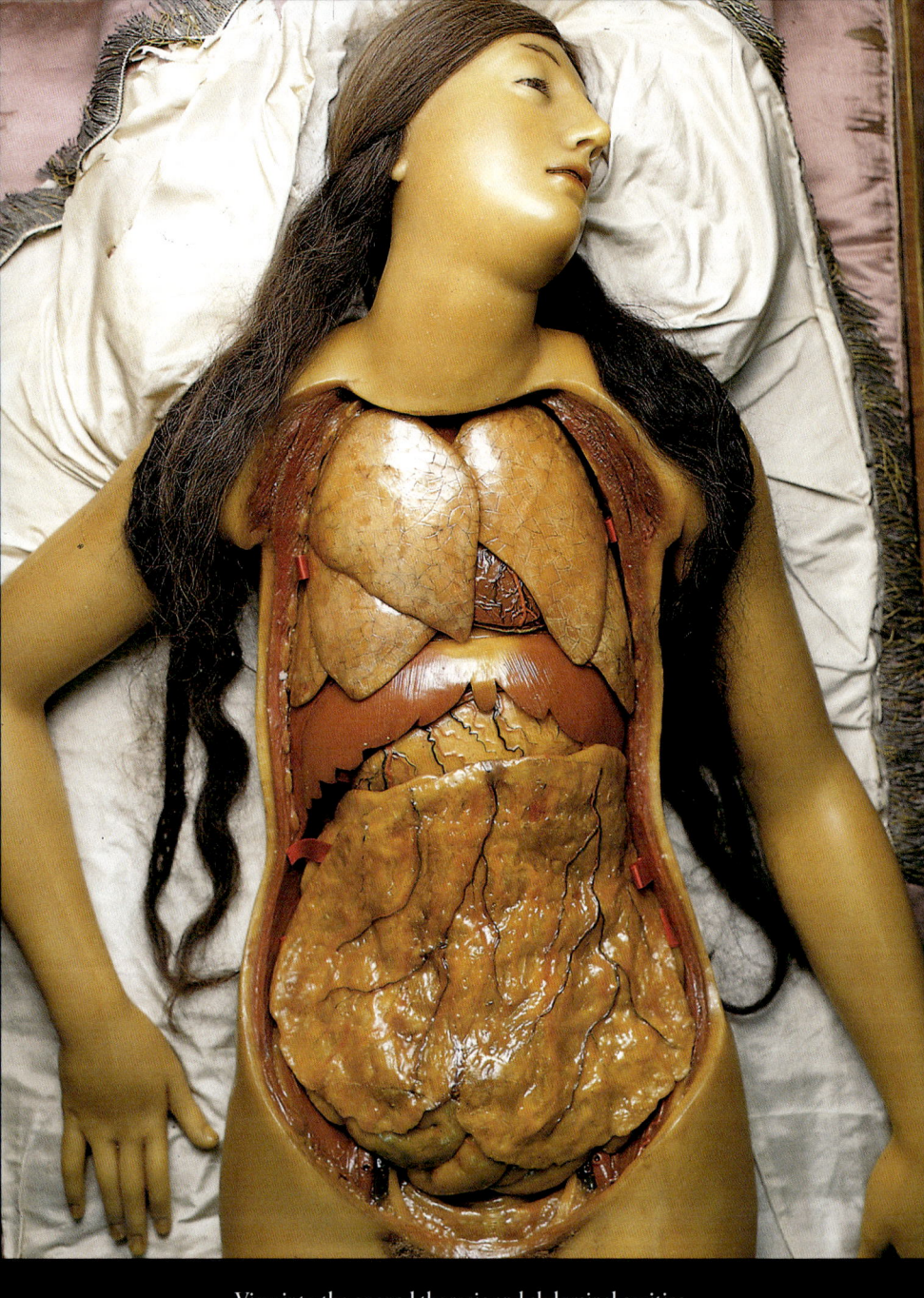

View into the opened thoracic and abdominal cavities.
In the thorax the right and left lungs can be seen, and part of the anterior surface of the heart.
The intestines are hidden behind the greater omentum

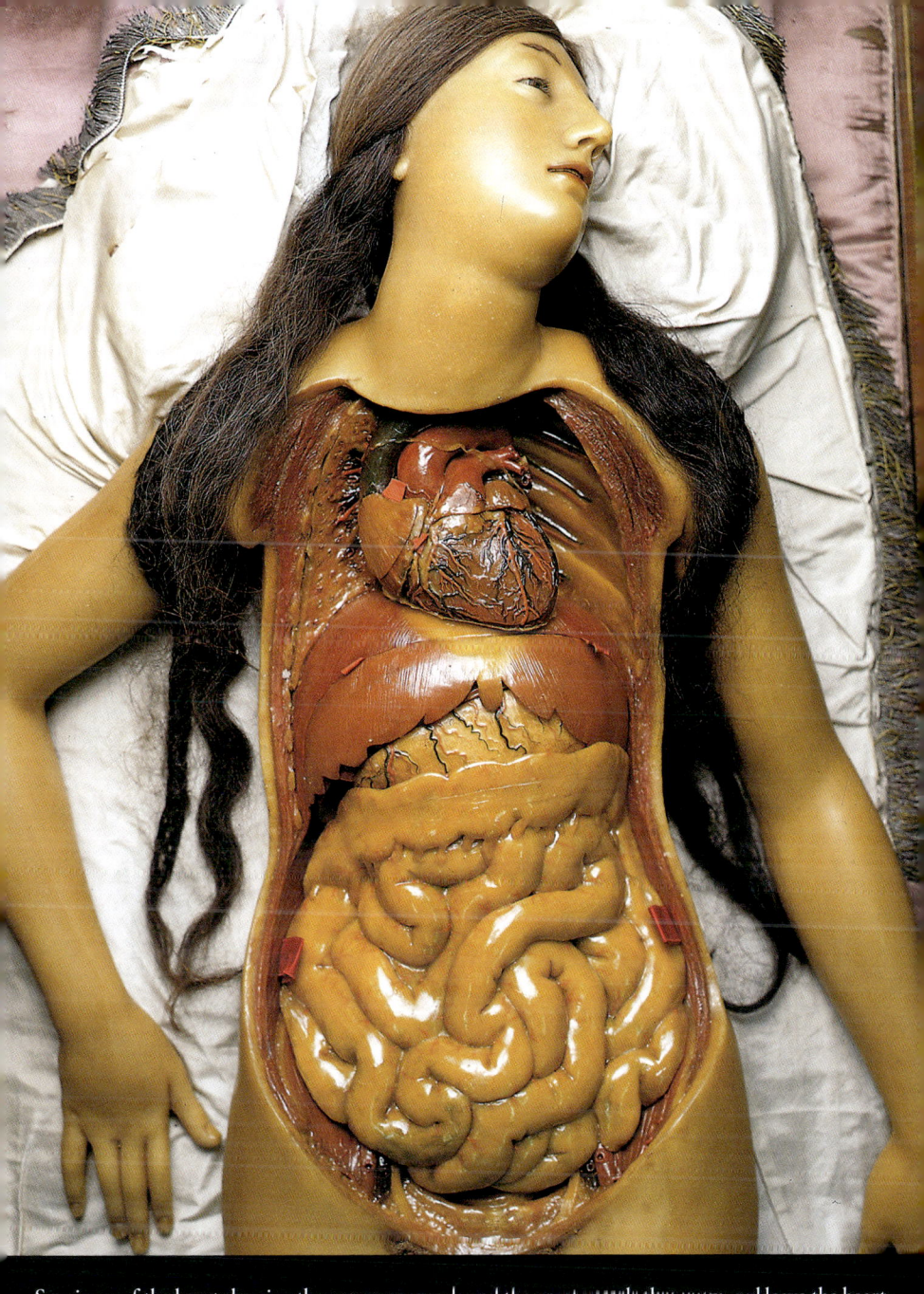

Specimen of the heart showing the coronary vessels and the great vessels that enter and leave the heart.

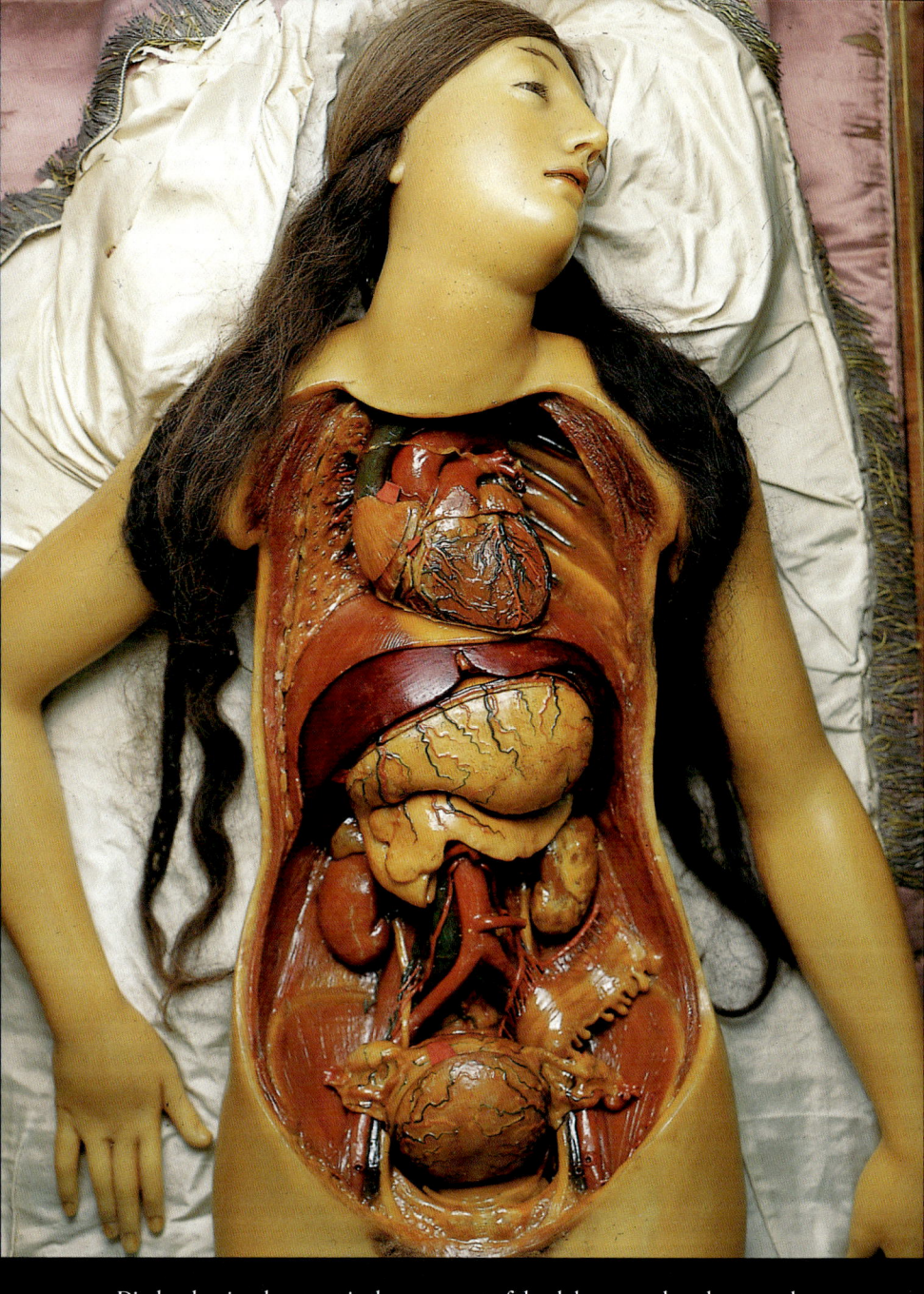

Display showing the organs in the upper part of the abdomen, such as the stomach, liver and duodenum, and also the right and left kidneys, the right adrenal gland, the abdominal aorta and the uterus

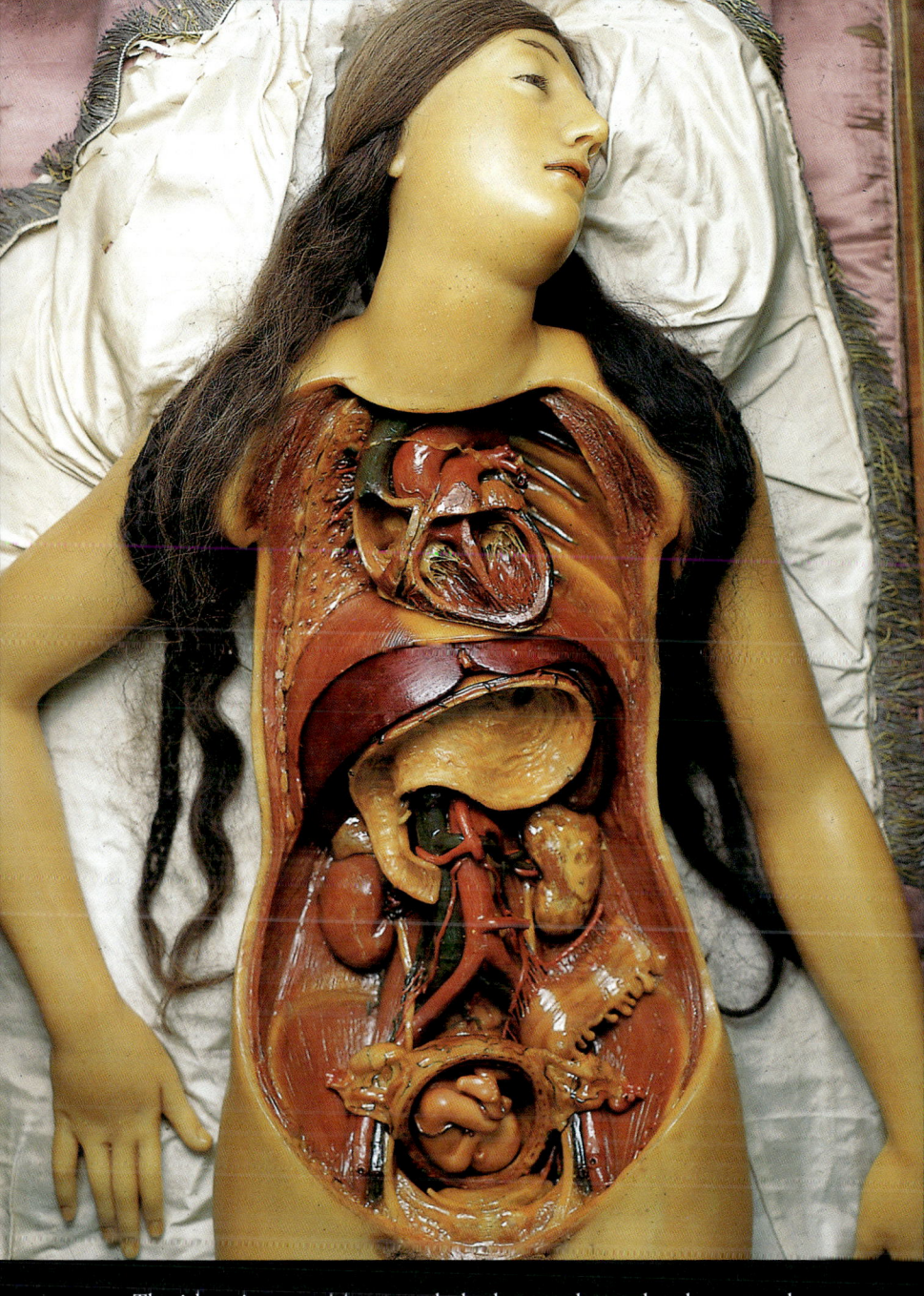

The right atrium, ventricles, stomach, duodenum and uterus have been opened

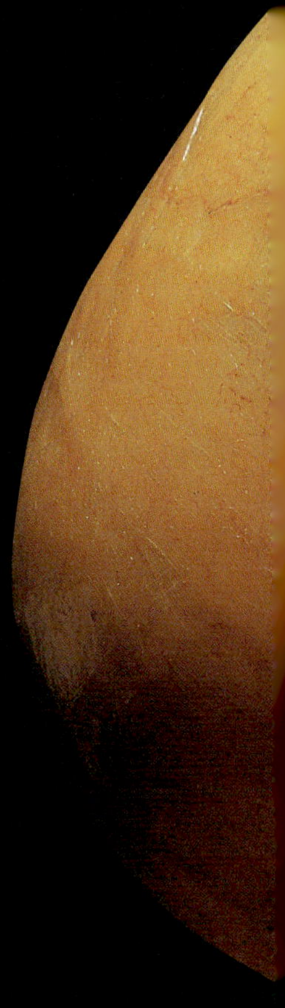

Osteologia et Arthrologia

Bones and Joints
Knochen und Gelenke
Squelette et articulations

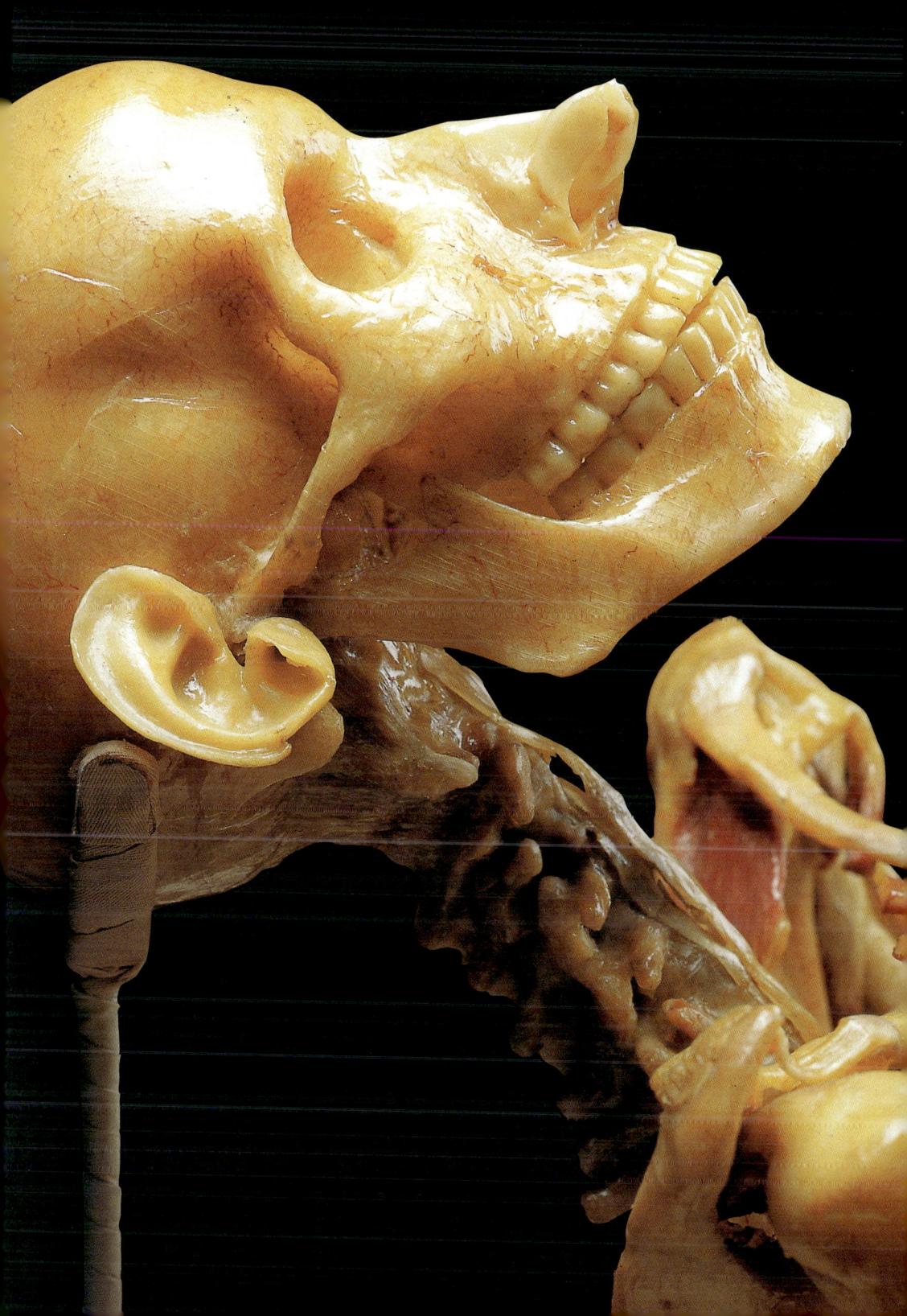

The form of the human body, which is genetically determined, is a particular example of the general plan underlying that of all vertebrate animals. As the supporting element in this plan, the skeleton forms the bony framework of our bodies. The high proportion of crystalline minerals in bone is responsible for the skeleton being the organic system that remains intact long after the death of the individual.

Under certain circumstances, such as the exclusion of air, or even the various forms of embalming to which the body was subjected in some cultures before burial, the bones may even survive for thousands of years. This means that long after death, the age, sex, dietary habits and health or disease of a person can be determined. The preservation of the skeleton after death also explains why it has become a symbol of death and the transitoriness of human life in the religions and cultural history of Europe.

In the first table, examples are given of the structural characteristics of various bones. These consist of a compact outer shell (the cortex) and an inner layer of fine bony trabeculae (the spongiosa), the particular structure and architecture of which is the morphological correlate revealing the organisation of the pressure and tension lines of the bone under normal loading. These patterns remain dynamic throughout life, and alter in response to the functional demands made upon the bone. The red bone marrow lies between the cancellous trabeculae of the spongiosa. It contains the stem cells which during life give rise to the newly developing cells of the blood.

In the following table the many different shapes and rich variation in form of the individual bones of our bodies make up an impressive picture. Differences in size, projections, ridges and crests of whole bones or sections through bones, symmetry or asymmetry, are further clues to the interpretation of the function of the individual components of the entire skeletal system.

Considerable space is taken up by the different aspects of the bones which make up the skull, and which in their entirety form the cranium and facial skeleton. In addition to these, single vertebrae, sections of the vertebral column, the bones of the thoracic cage, the pelvis, and also the bones of the upper and lower limbs, including those of the hands and feet, are depicted.

A few specimens of joints displaying the ligaments which strengthen them and which stabilise and define their movements are also represented. Some of the joints have been opened to reveal the articular surfaces, or particular structures such as, for instance, the menisci of the clavicular and knee joints.

Knochen und Gelenke

Die Gestalt des menschlichen Körpers ist genetisch festgelegt und repräsentiert den allgemeinen Bauplan der Wirbeltiere. Als tragender Bestandteil dieses Bauplans bildet das Skelettsystem das knöcherne Gerüst unserer Körperformen. Der hohe Anteil an kristallisierten Mineralien im Knochen macht das Skelett zu jenem Organsystem, das über den Tod des Menschen hinaus erhalten bleibt. Unter bestimmten Voraussetzungen – wie etwa Luftabschluss, aber auch durch Konservierungsmaßnahmen, die in einigen Kulturen vor der Bestattung des Leichnams durchgeführt wurden – kann der Knochen sogar über Jahrtausende erhalten bleiben. So lassen sich am Skelett noch lange nach dem individuellen Tod, Alter, Geschlecht, Ernährungsgewohnheiten, körperliche Leistung und Erkrankungen nachweisen und ablesen. Die postmortale Erhaltung des Skeletts erklärt auch, warum es in der europäischen Religions- und Kulturgeschichte zum Symbol für den Tod an sich wie auch für die individuelle Vergänglichkeit geworden ist.

Die erste Tafel zeigt Beispiele des prinzipiellen Aufbaus verschiedener Knochen, bestehend aus einer kompakten äußeren Rindenschicht (Corticalis) und einer aus feinen Knochenbälkchen aufgebauten Innenschicht (Spongiosa). Die besondere Strukturierung und Architektur der Spongiosa aus unterschiedlich feinen Knochenbälkchen ist das morphologische Korrelat für die Ausrichtung der mechanischen Druck- und Zugspannungslinien des unter natürlicher Belastung stehenden Knochen. Diese Strukturen sind zeitlebens dynamisch und verändern sich entsprechend der jeweiligen funktionellen Belastungen des Knochens. Zwischen den Spongiosabälkchen befindet sich das rote Kochenmark mit den Blutstammzellen, die zeitlebens Ursprung der Neubildung unserer Blutzellen sind.

Auf den nachfolgenden Tafeln wird die vielfältige Gestalt sowie der Formenreichtum der einzelnen Knochen unseres Körpers eindrucksvoll dokumentiert. Größenunterschiede, Knochenvorsprünge und Knochenleisten an einzelnen Knochen oder Knochenabschnitten, Symmetrie oder auch Asymmetrie sind weitere Kriterien der funktionellen Interpretation einzelner Knochen im Verband des Gesamtsystems.

Einen breiten Raum nehmen die unterschiedlichen Ansichten der Schädelknochen ein, die in ihrer Gesamtheit den Hirn- und Gesichtsschädel bilden. Darüber hinaus werden die einzelnen Wirbel, ganze Wirbelsäulenabschnitte, die Knochen des Brustkorbes, des Beckens sowie der oberen und unteren Extremitäten, einschließlich der Hand- und Fußknochen, dargestellt. Einige Knochen- und Gelenkpräparate sind um die Darstellung ihrer Bänder, die ein Gelenk stärken, stabilisieren und durch ihren Verlauf die Gelenkbewegungen definieren, ergänzt und erweitert. Zum Teil sind die Gelenke eraöffnet, um die Form der miteinander artikulierenden Gelenkflächen oder auch Sondereinrichtungen, wie beispielsweise die Menisken am Schlüsselbeingelenk und Kniegelenk zu demonstrieren.

Squelette et articulations

Le squelette constitue l'armature osseuse du corps, élément de charpente de l'architecture du corps humain, comme pour tous les vertébrés. La richesse en sels minéraux des os rend leur conservation possible après la mort. Des os peuvent ainsi être conservés pendant des milliers d'années, notamment grâce à l'utilisation de techniques de conservation pratiquées dans certaines cultures avant l'inhumation. Le squelette permet de révéler, longtemps après la mort de l'individu, son âge, son sexe, ses habitudes alimentaires, ses capacités physiques, ses maladies... La conservation des os après la mort explique pourquoi le squelette est devenu, dans l'histoire des religions et des cultures européennes, le symbole de la mort et de l'existence éphémère de l'individu.

L'architecture générale des os est démontrée par la première planche qui en présente des exemples. Les os sont constitués d'une couche compacte périphérique, l'os cortical, entourant une zone profonde formée par de fines travées osseuses, l'os spongieux. La disposition et l'architecture spécifique des différentes travées de l'os spongieux sont la traduction morphologique des forces de pression et de traction exercées de façon physiologique. Ces structures évoluent de façon dynamique au cours de la vie et se transforment sous l'action des contraintes fonctionnelles. Dans l'interstice entre les travées d'os spongieux se trouve la moelle osseuse rouge contenant les cellules souches des éléments figurés du sang permettant le renouvellement des globules sanguins durant toute la vie.

La grande variété et la richesse des formes des os du corps apparaissent de façon remarquable sur la deuxième planche. La variation de taille, la présence de tubercules et d'apophyses, la symétrie ou l'asymétrie constituent des critères pour l'interprétation fonctionnelle dans l'ensemble squelettique de pièces osseuses isolées. Différentes vues des os du crâne, articulés pour former la voûte et la base du crâne ainsi que le squelette de la face, occupent une place importante de la collection. Les os de la colonne vertébrale, les os de la cage thoracique, les os du bassin, et les os des membres supérieurs et inférieurs sont ensuite représentés.

Les préparations concernant le squelette et les articulations sont complétées par la représentation de ligaments ; ceux-ci renforcent et stabilisent chaque articulation et guident les mouvements. La capsule articulaire est ouverte dans certains cas pour démontrer la cavité articulaire, la forme des surfaces articulaires, ou la présence de dispositifs intra-articulaires particuliers tels que le disque sterno-claviculaire ou les ménisques du genou.

Skeleton humanum
Human Skeleton

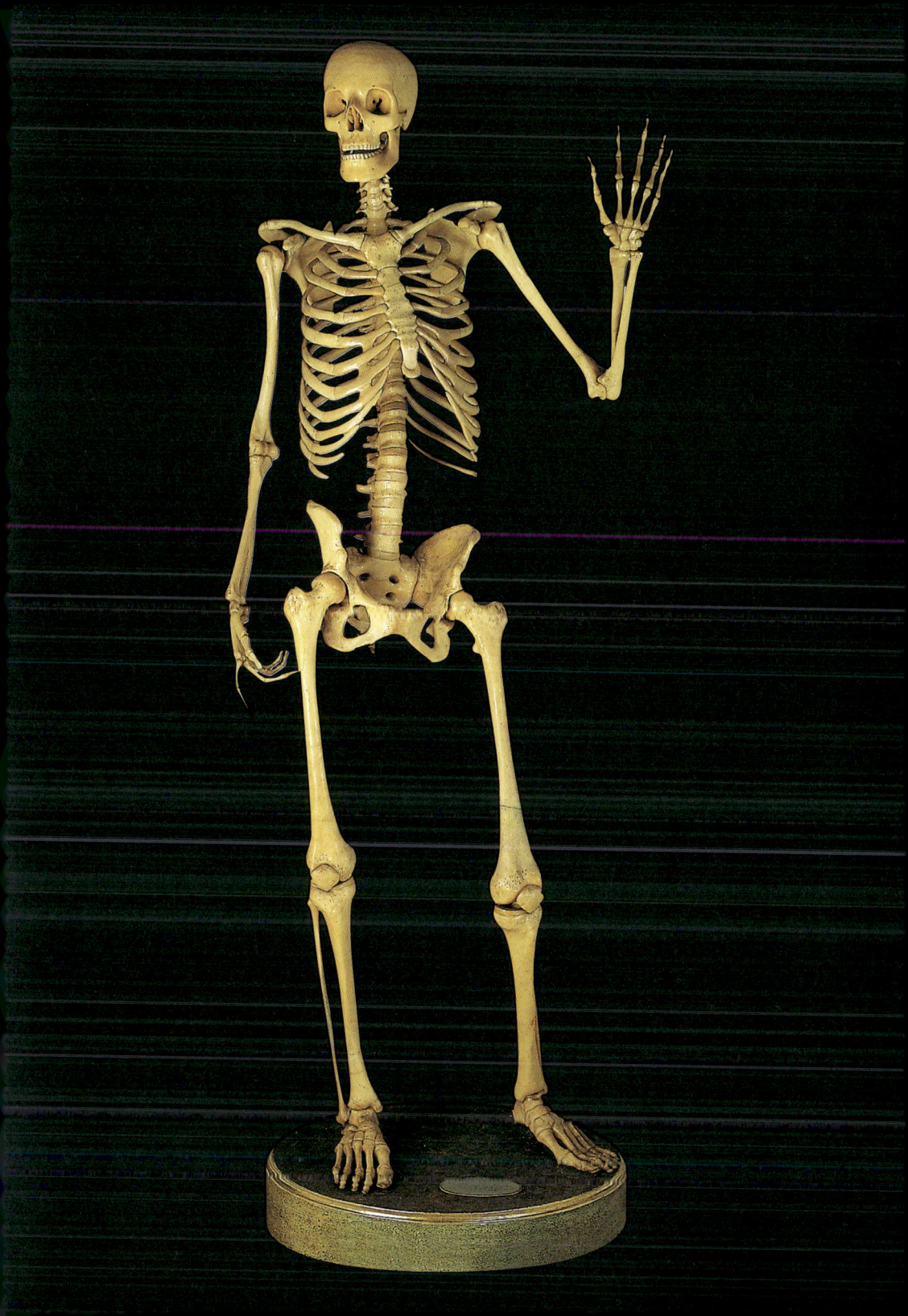

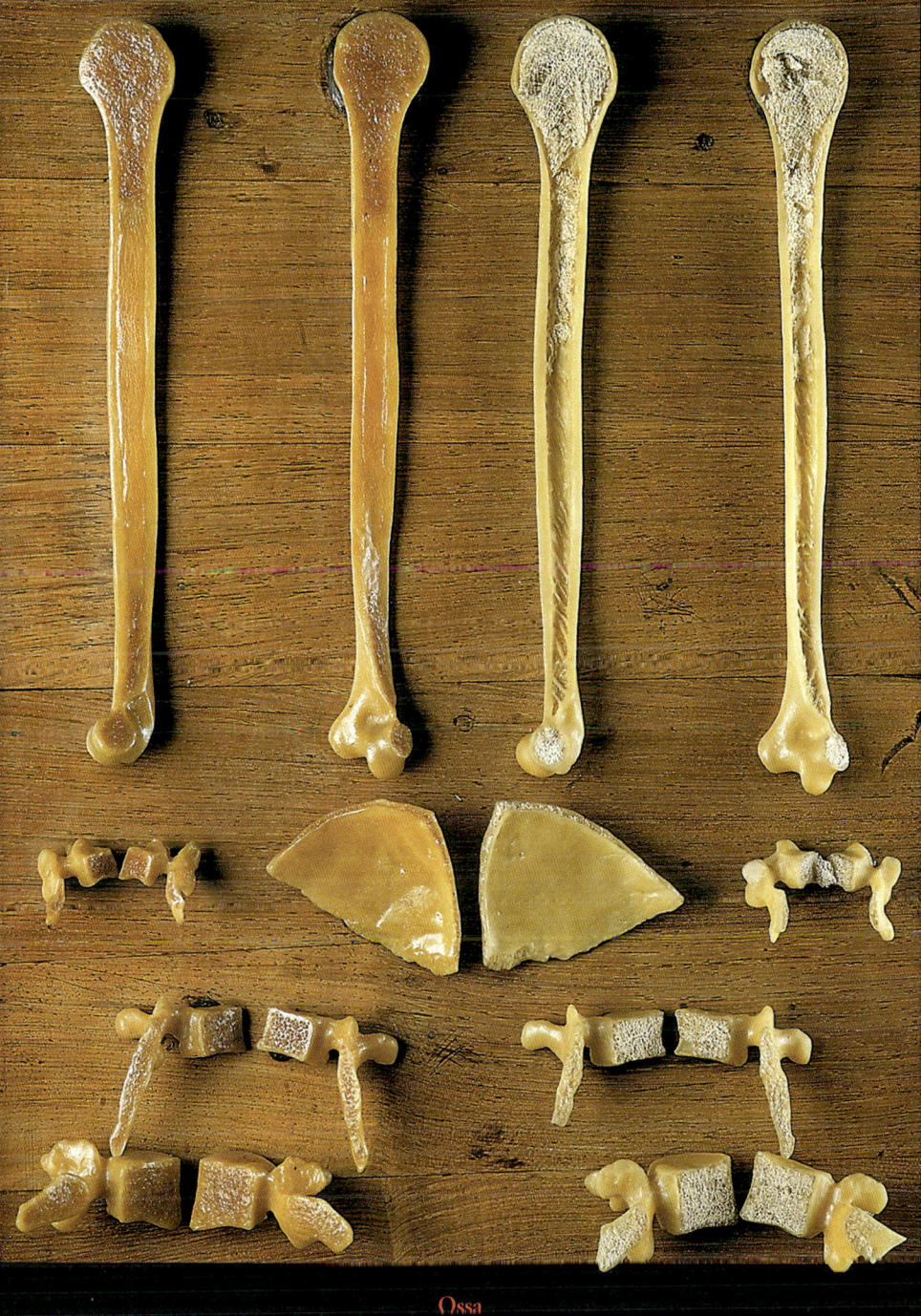

Internal structure of long bones, skull bones and vertebral bodies

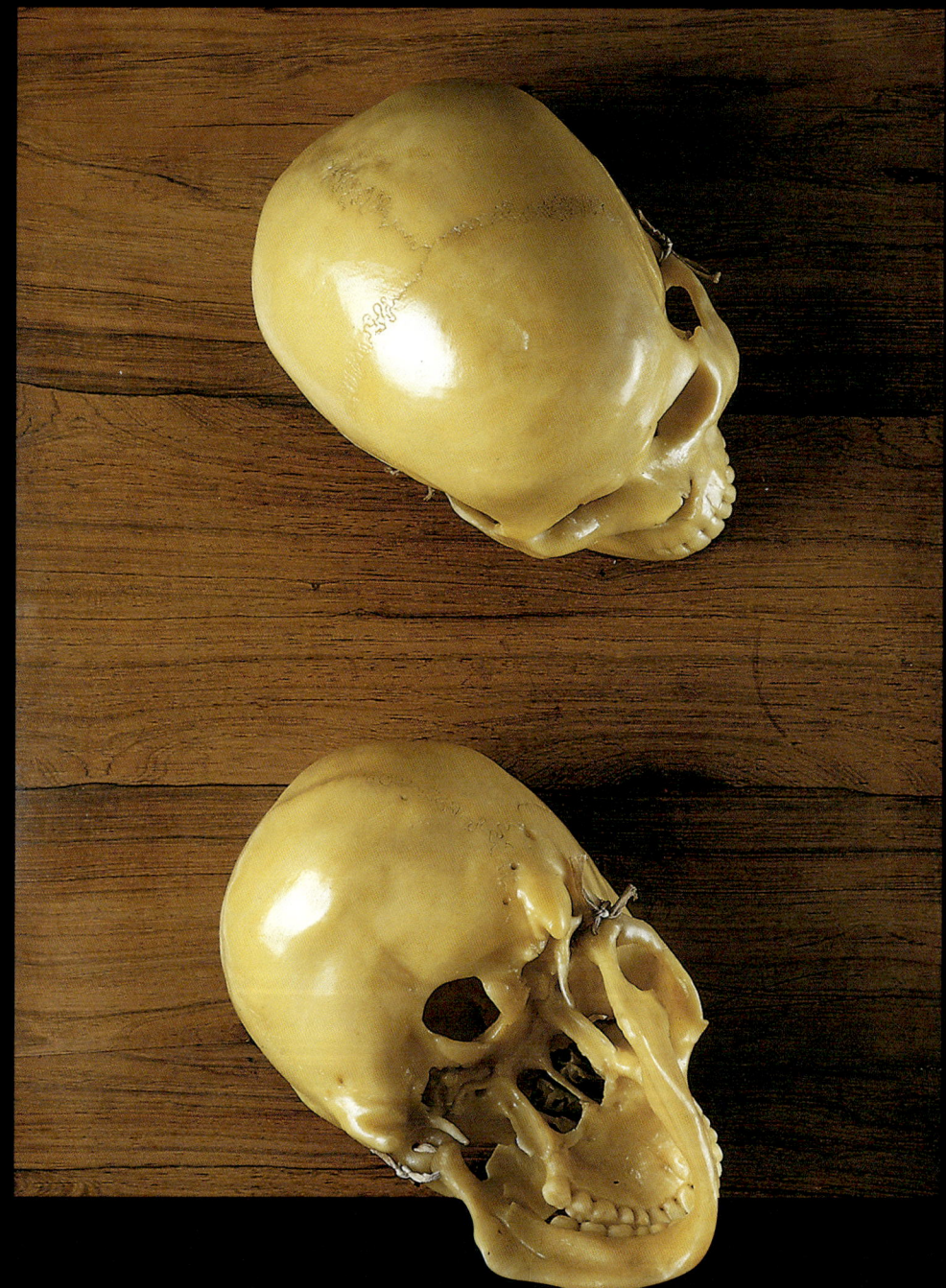

Neurocranium, Viscerocranium

View of the calvaria, facial skeleton and skull base

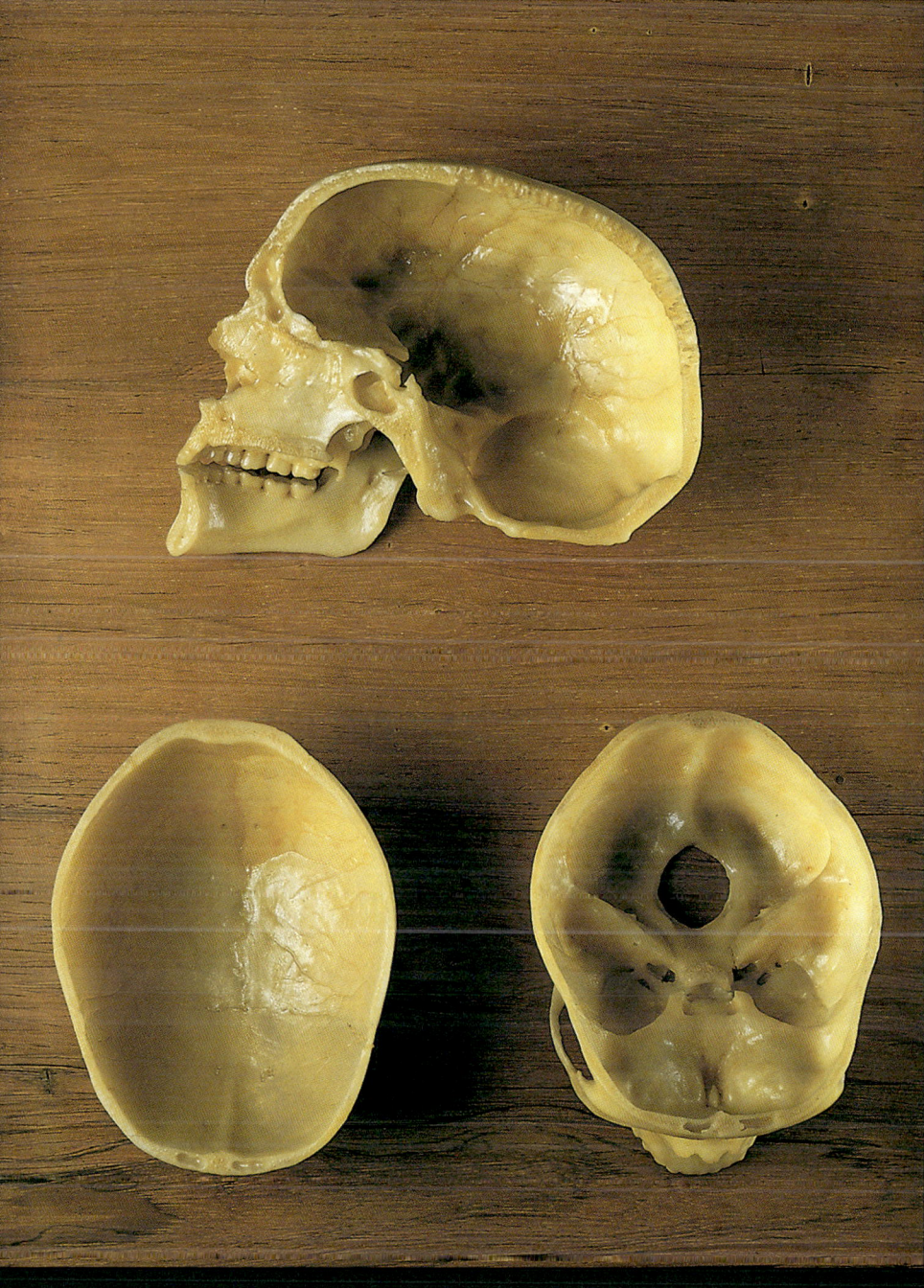

Neurocranium, Viscerocranium

Above: median section through calvaria and facial skeleton; inner view of right half of the skull.
Below: horizontal section with inner view of the skull roof and skull base

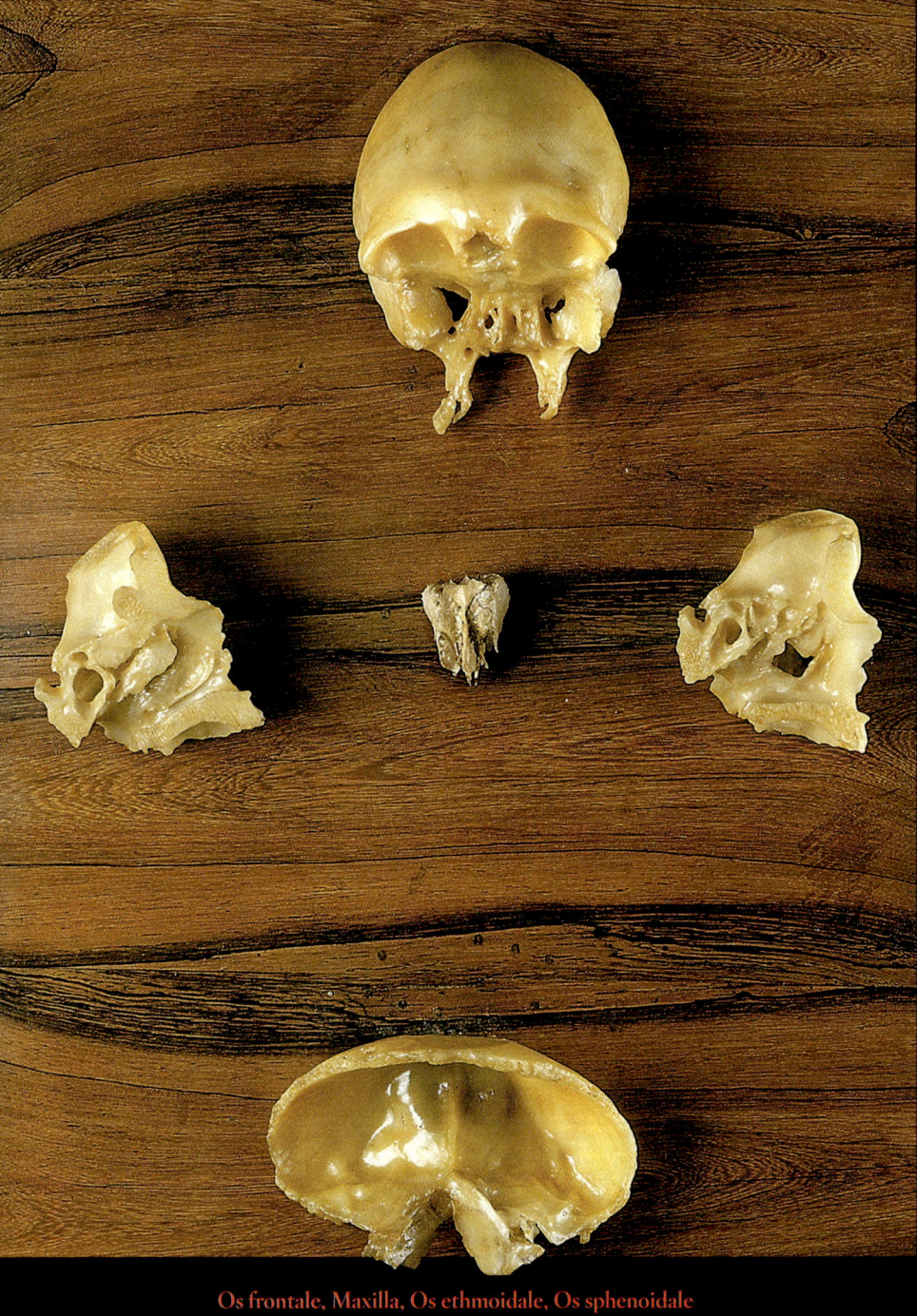

Os frontale, Maxilla, Os ethmoidale, Os sphenoidale

Frontal bone (upper and lower specimens) together with two views of the upper jaw bone (specimen at side).
In the centre is the ethmoid bone

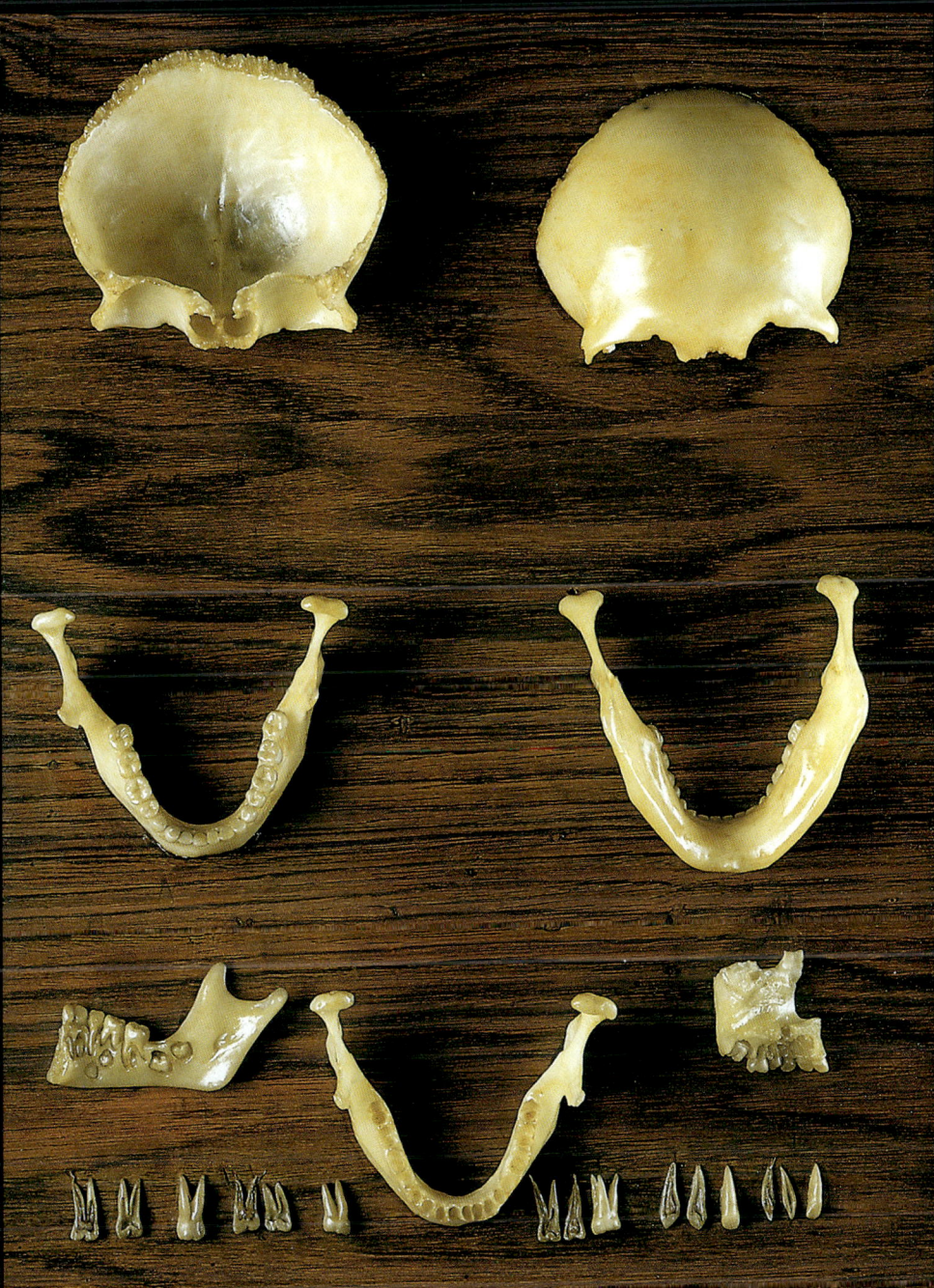

Os frontale, Mandibula, Dentes permanentes, Dentes decidui

Upper specimen: two views of the frontal bone. Lower specimen: lower jaw bone with teeth.
Left: milk teeth with tooth germs

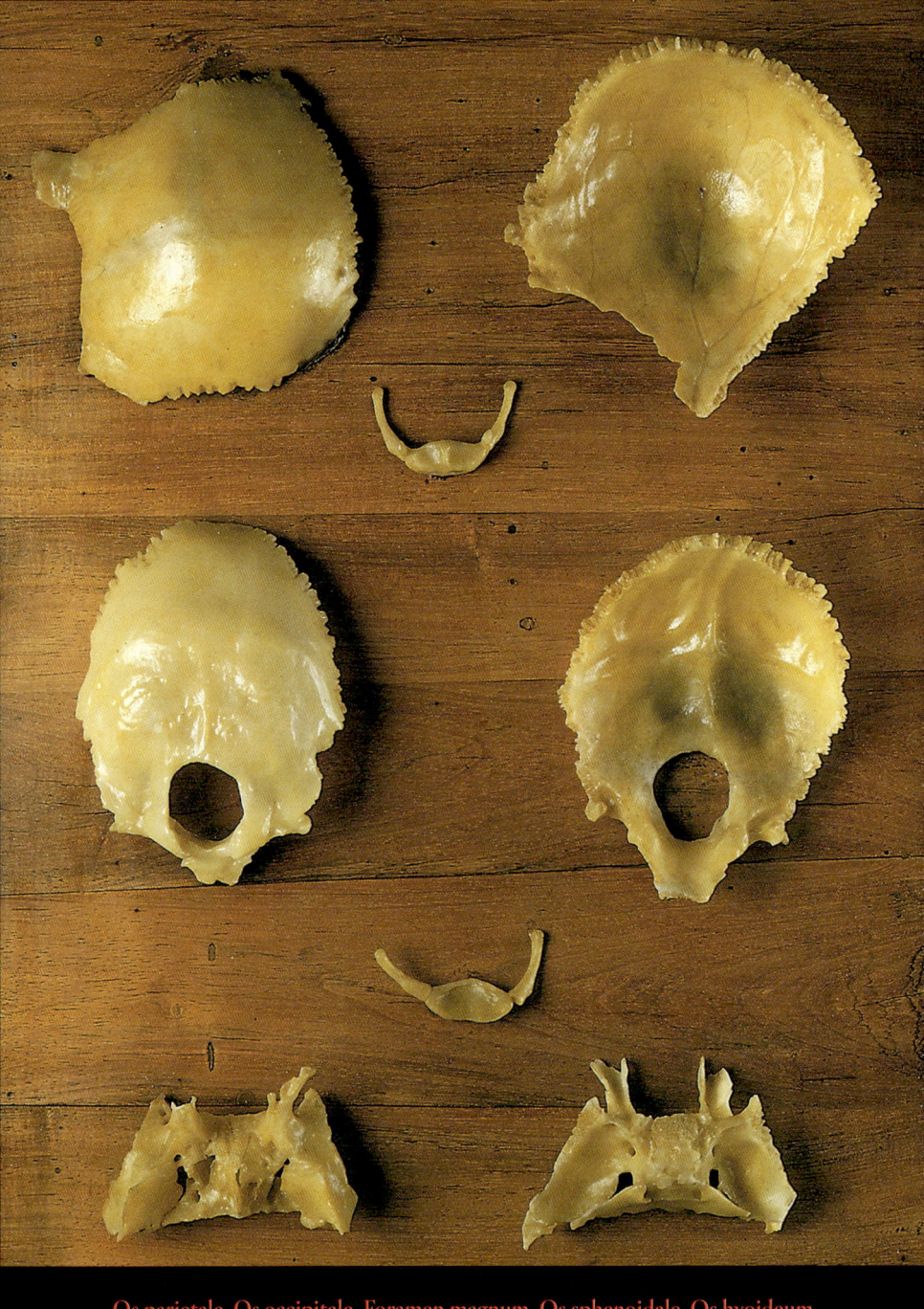

Os parietale, Os occipitale, Foramen magnum, Os sphenoidale, Os hyoideum

Parietal bone, occipital bone with foramen magnum, two views of the sphenoid.
Middle specimen: hyoid bone

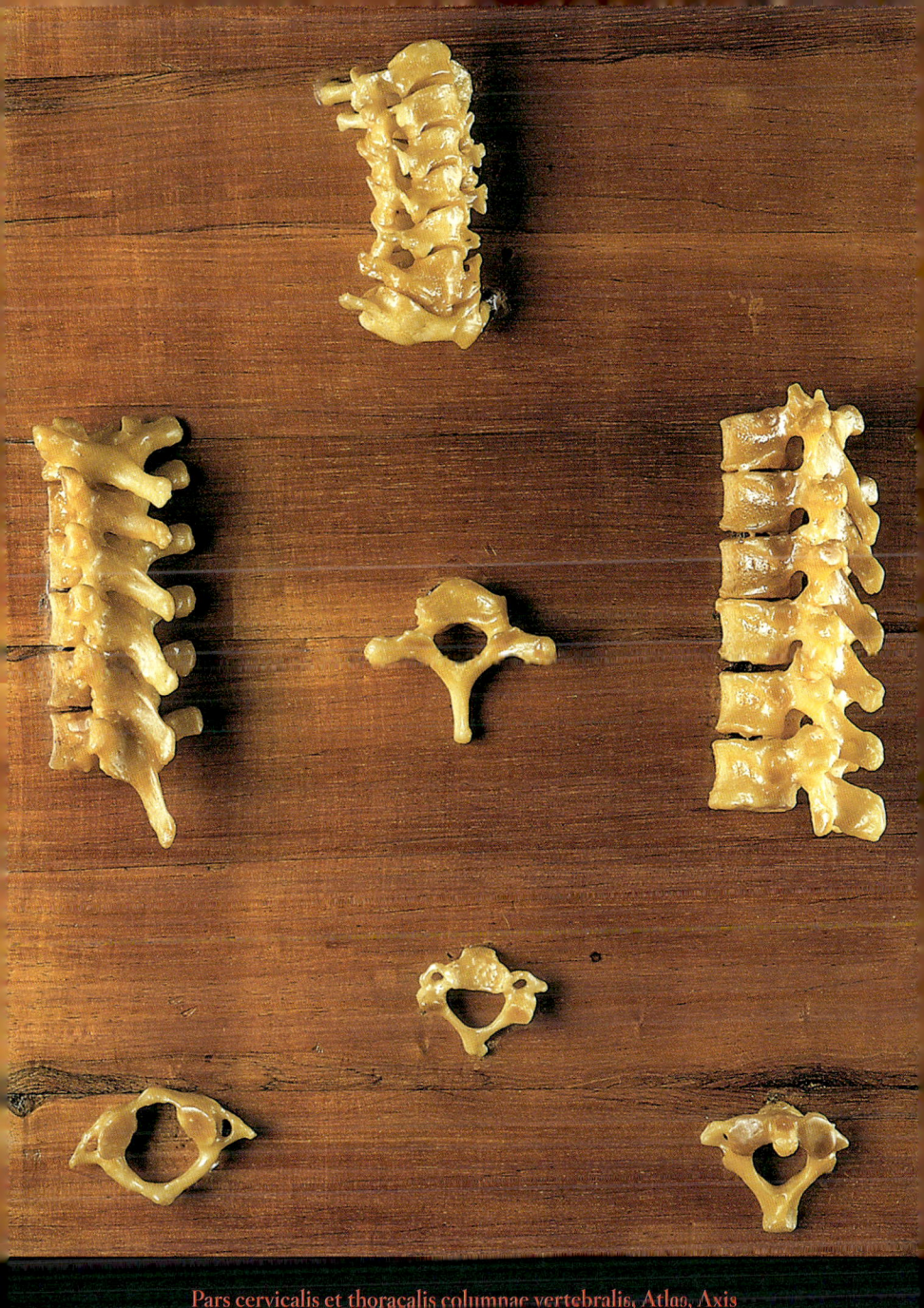

Pars cervicalis et thoracalis columnae vertebralis, Atlas, Axis

Cervical and thoracic vertebral column. Below left: the first vertebra (atlas), which supports the skull

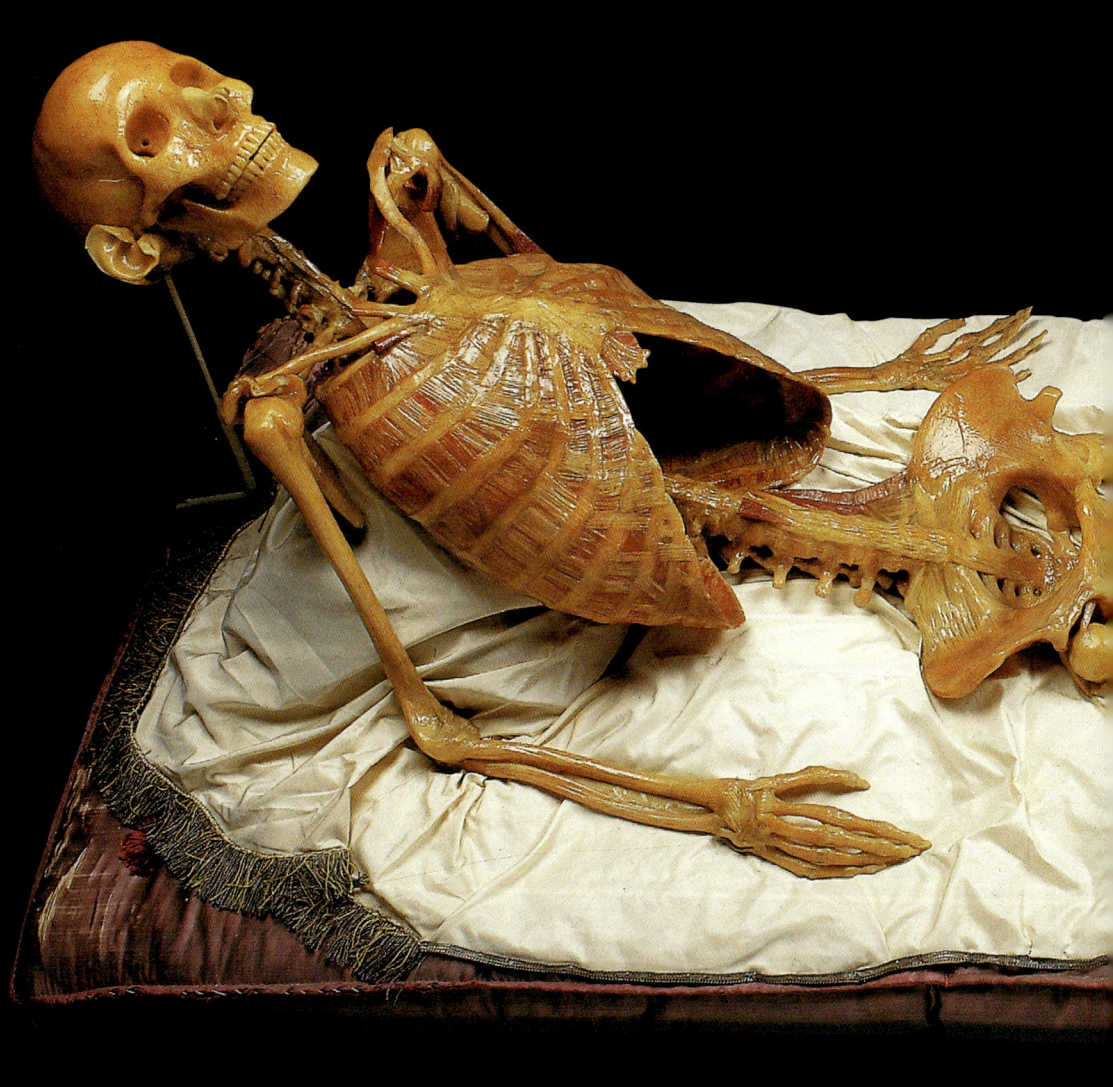

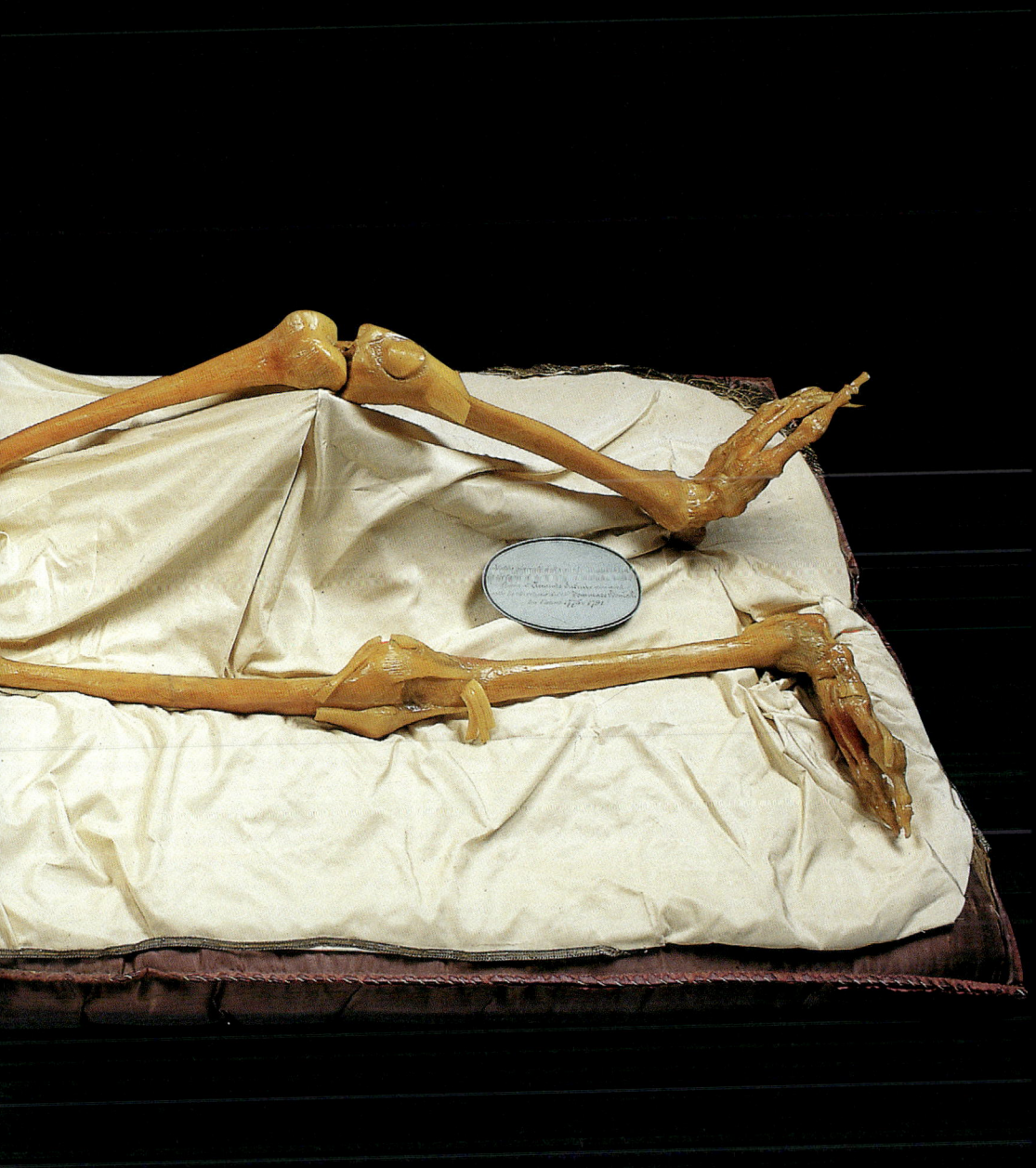

Whole body specimen showing the ligaments, joints and some single muscles and tendons

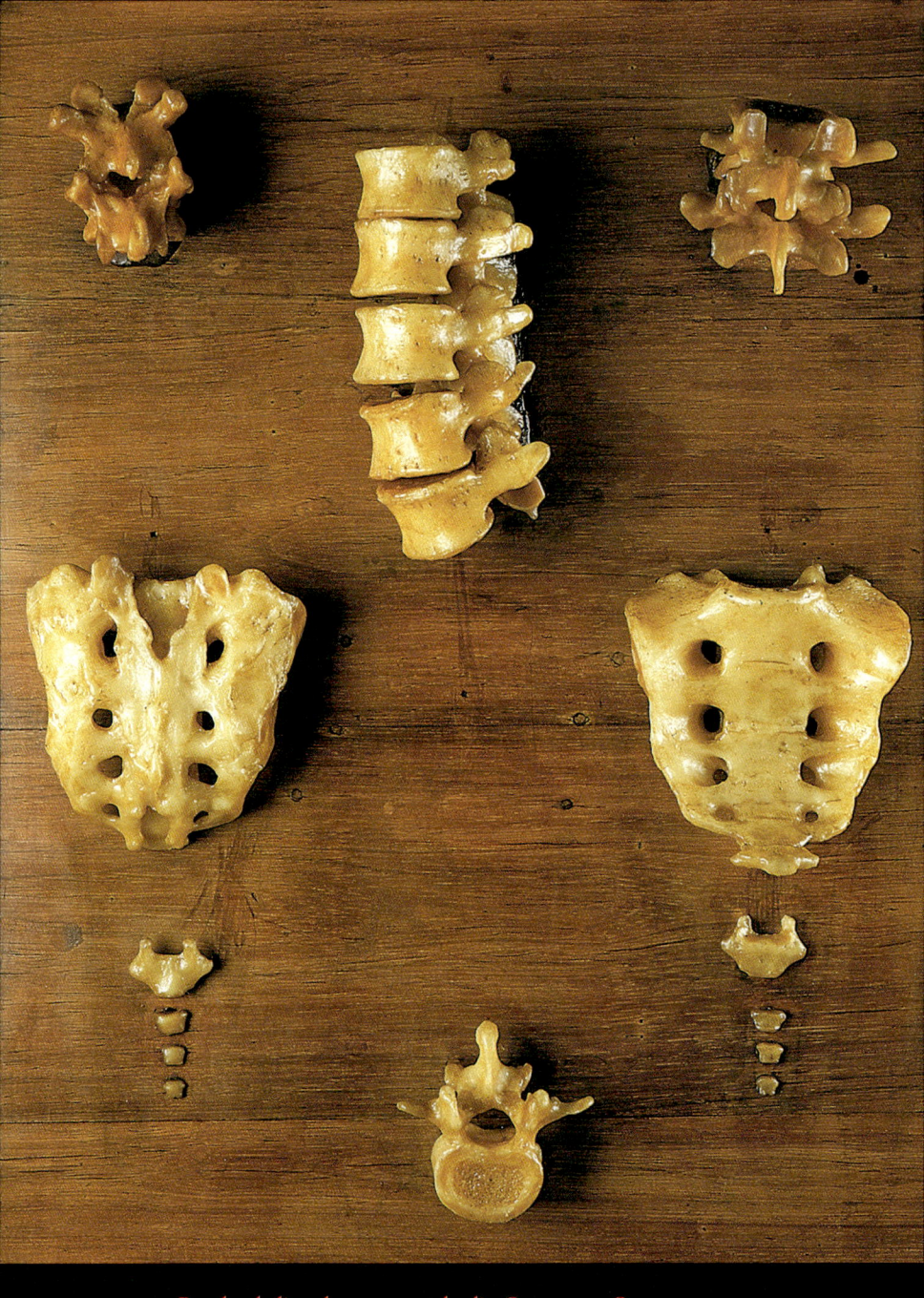

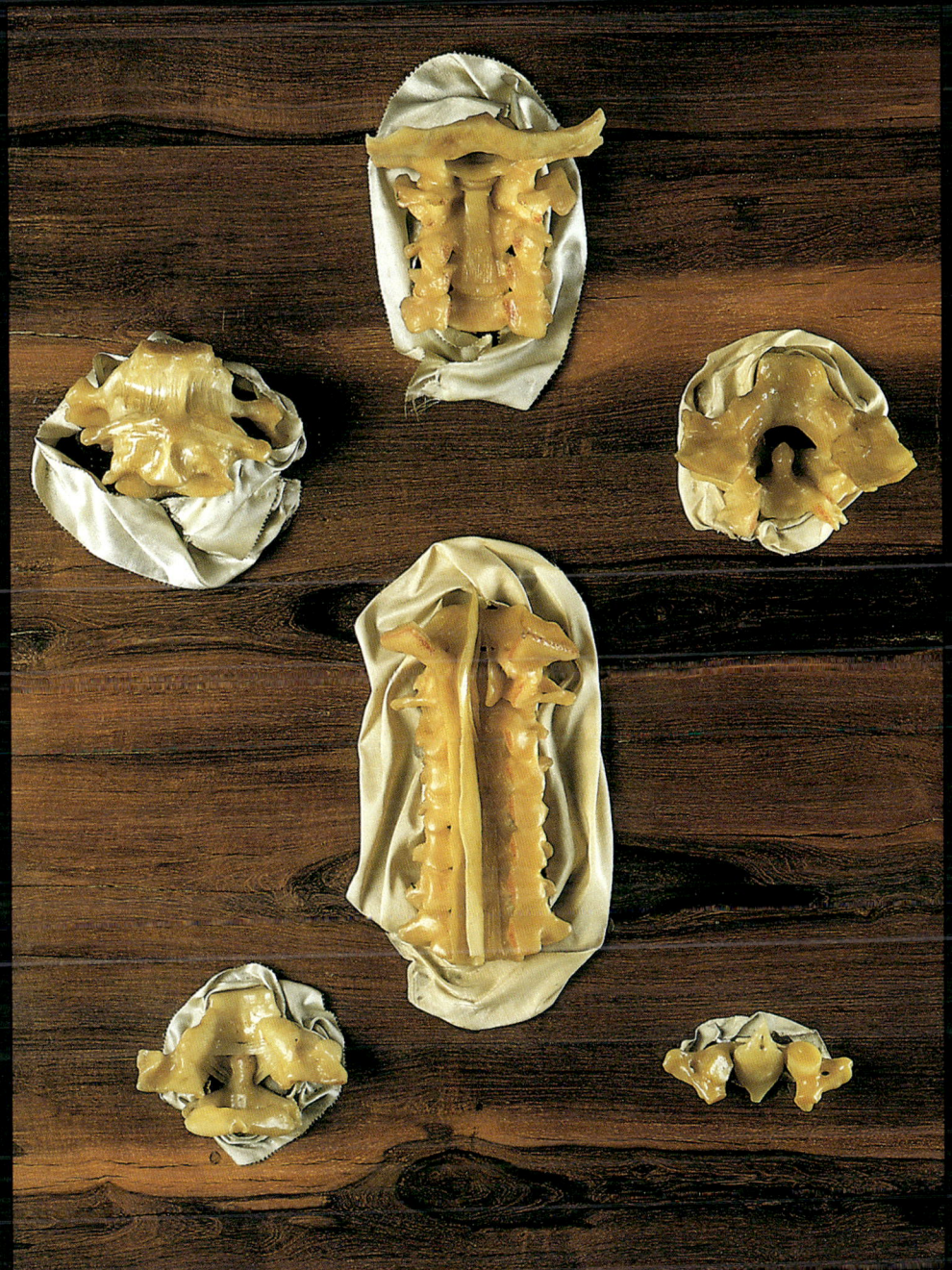

Pars cervicalis columnae vertebralis, Ligamentum longitudinale posterius,
Articulatio atlantooccipitalis

View from behind into the vertebral canal showing ligaments of the upper cervical vertebral column

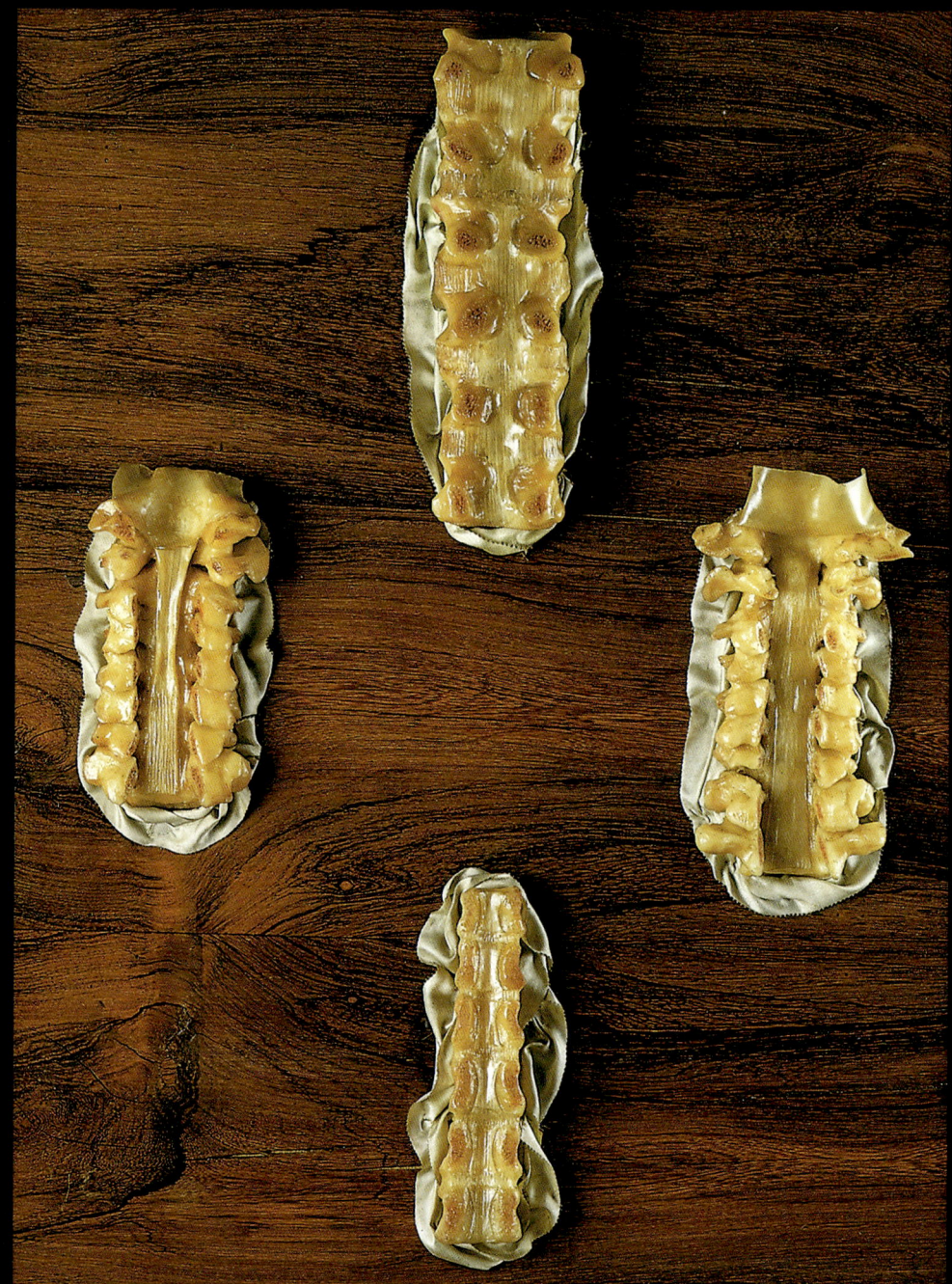

Pars cervicalis columnae vertebralis, Ligamentum longitudinale posterius
Ligaments of the upper cervical vertebral column seen from behind

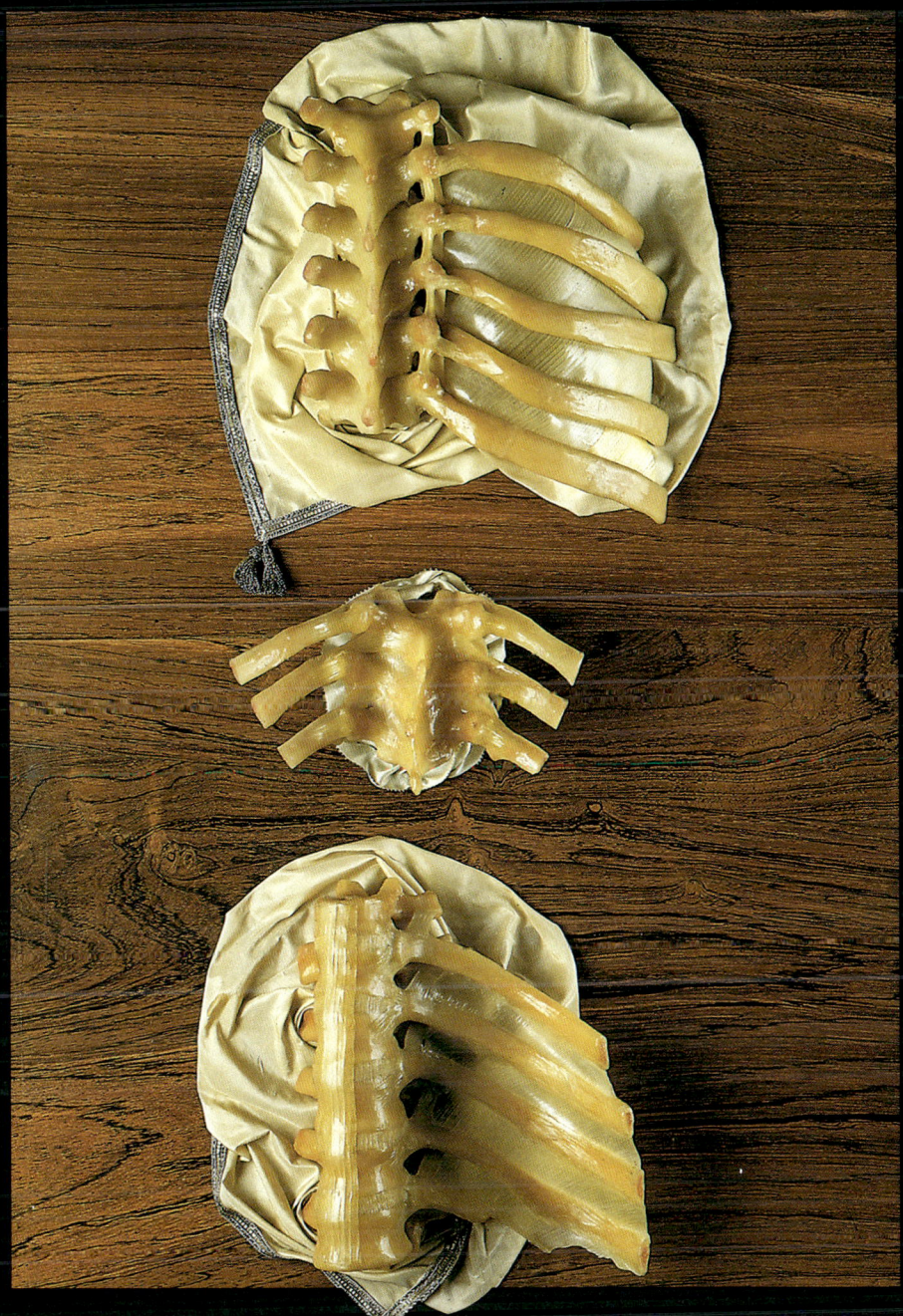

Pars thoracalis columnae vertebralis, Costae

Ligaments of the thoracic vertebral column and of the costovertebral joints, thoracic vertebrae and ribs

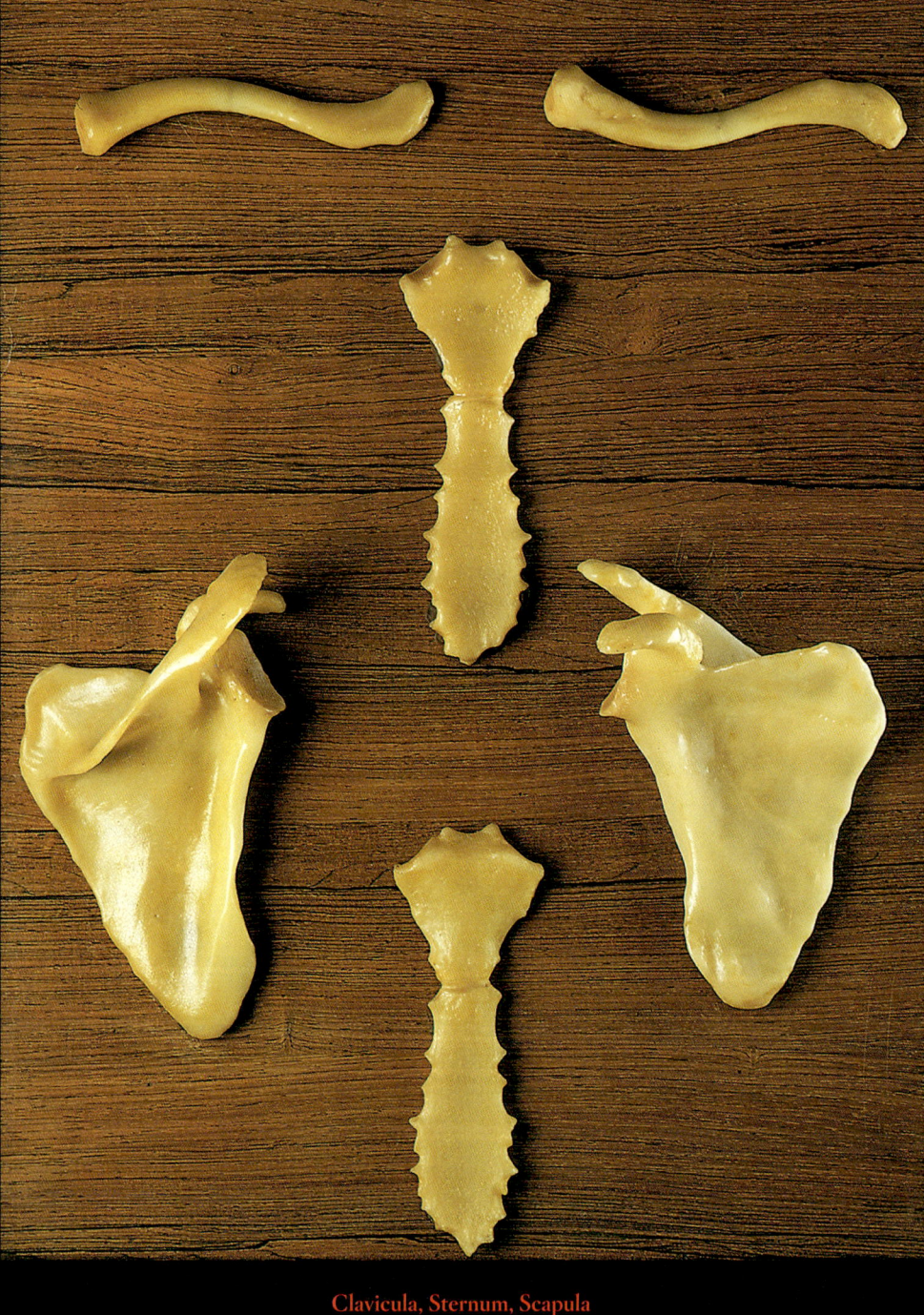

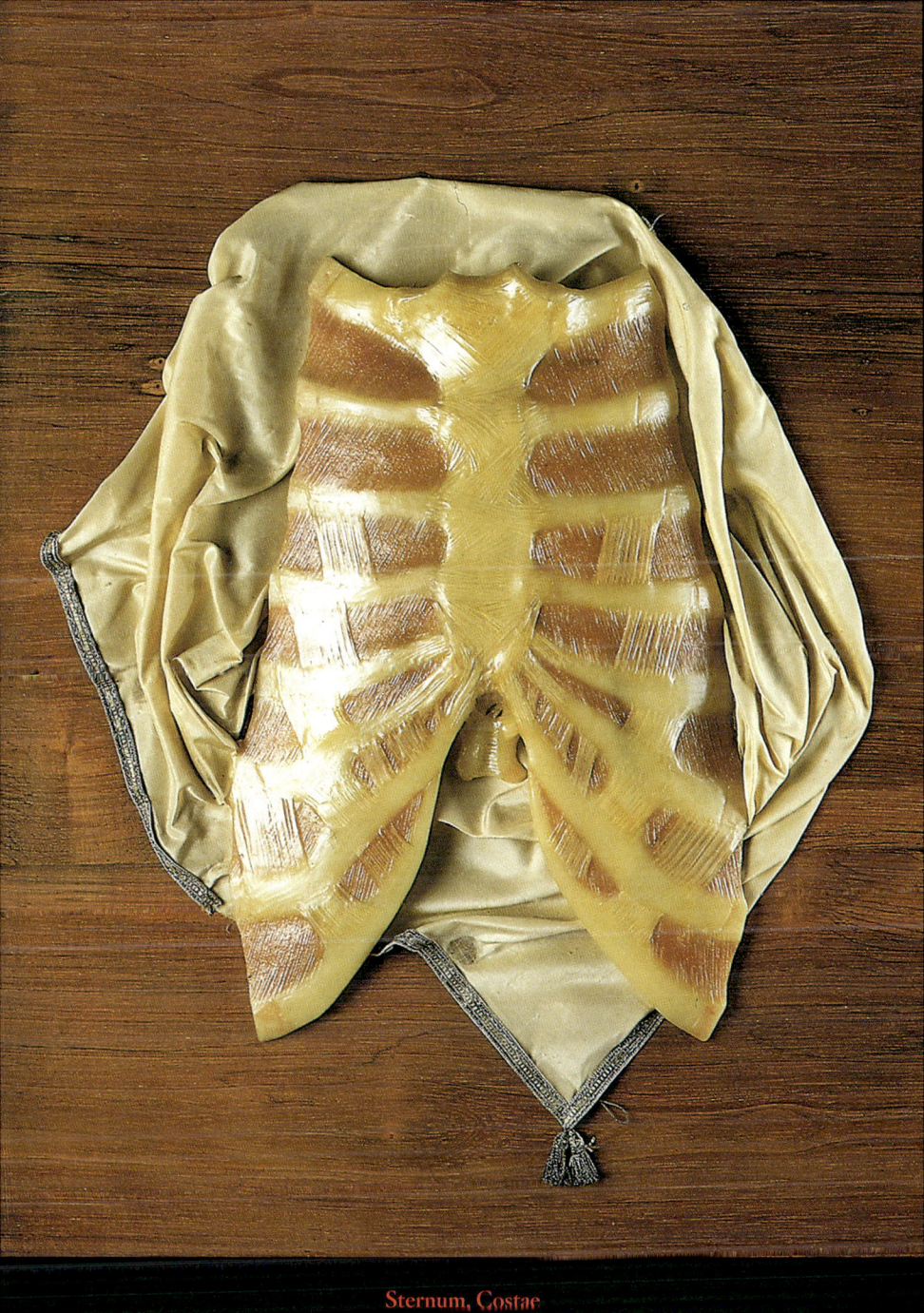

Sternum, Costae

Breastbone with rib attachments, ligaments and muscles, seen from in front

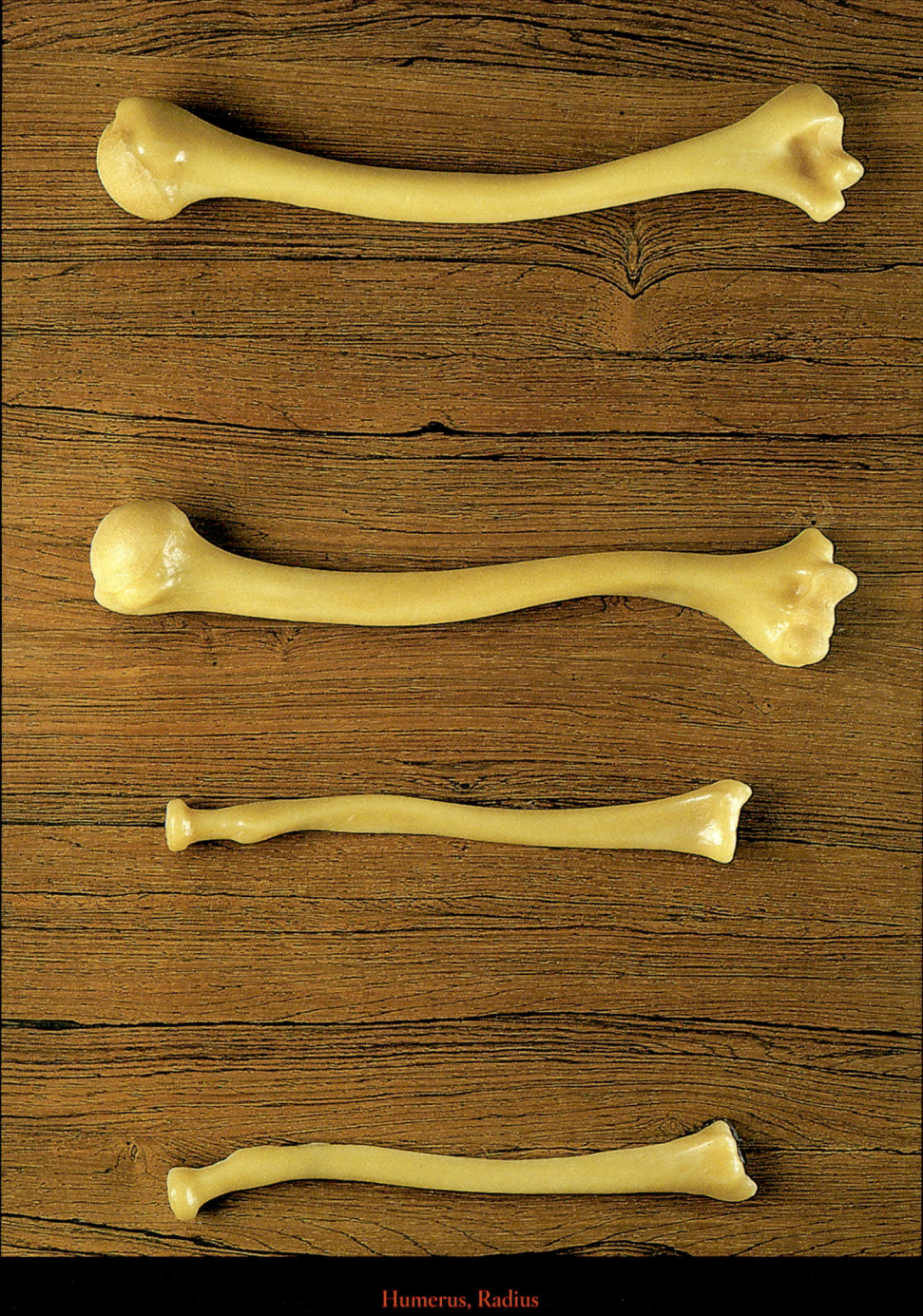

Humerus, Radius

Upper arm and radius

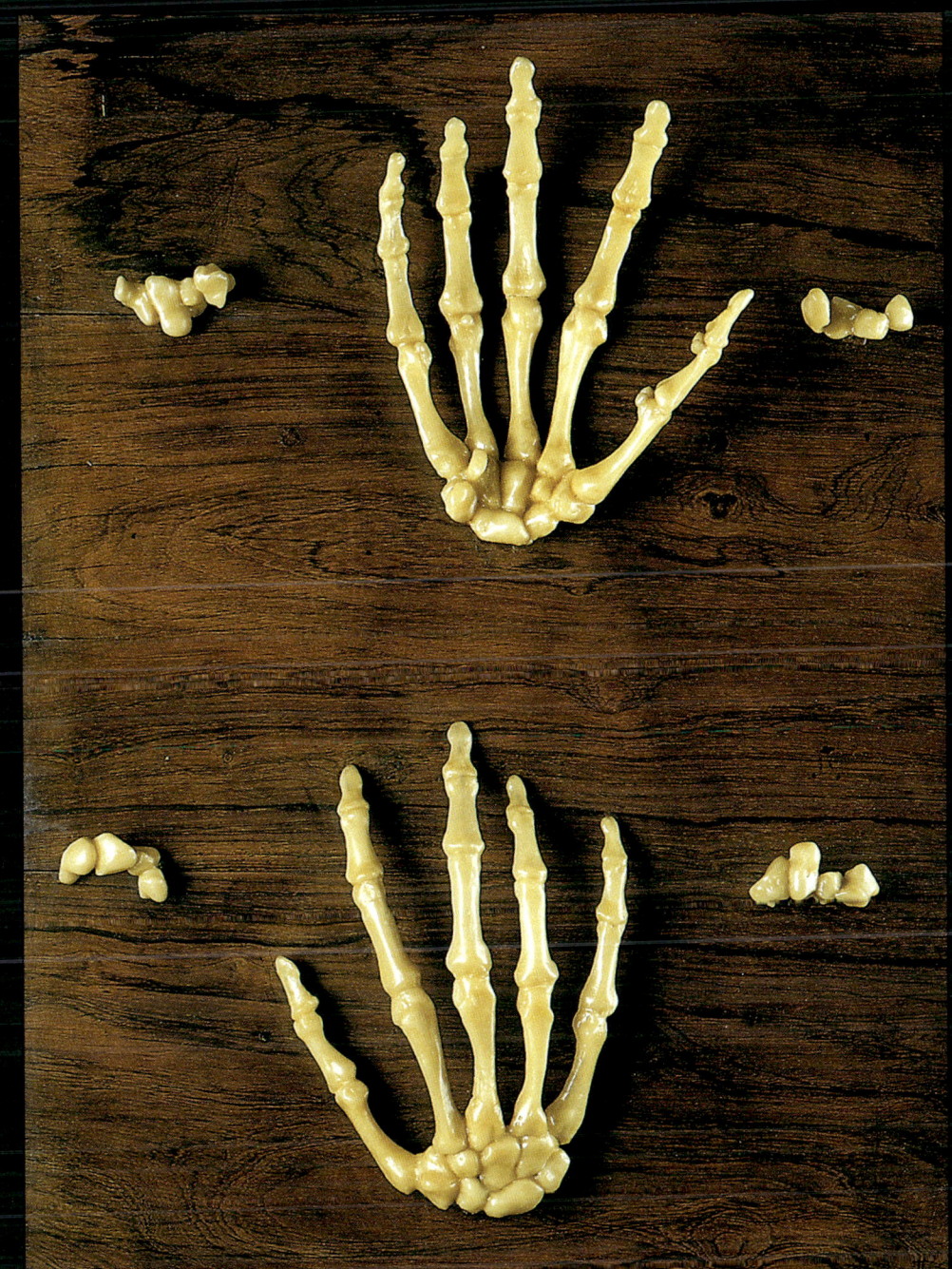

Ossa manus

Two views of the skeleton of the right hand

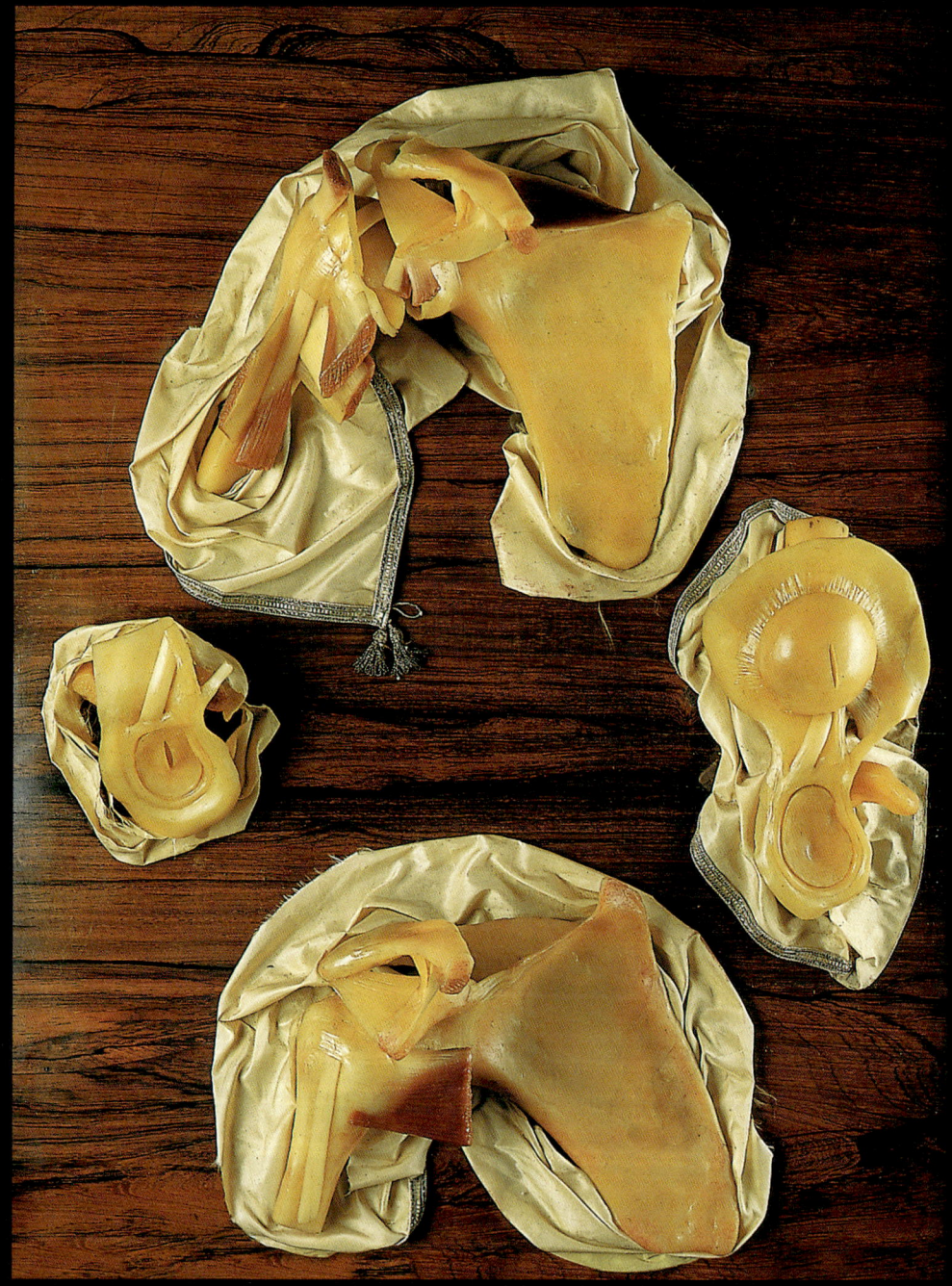

Articulatio humeri

The shoulder joint, different views

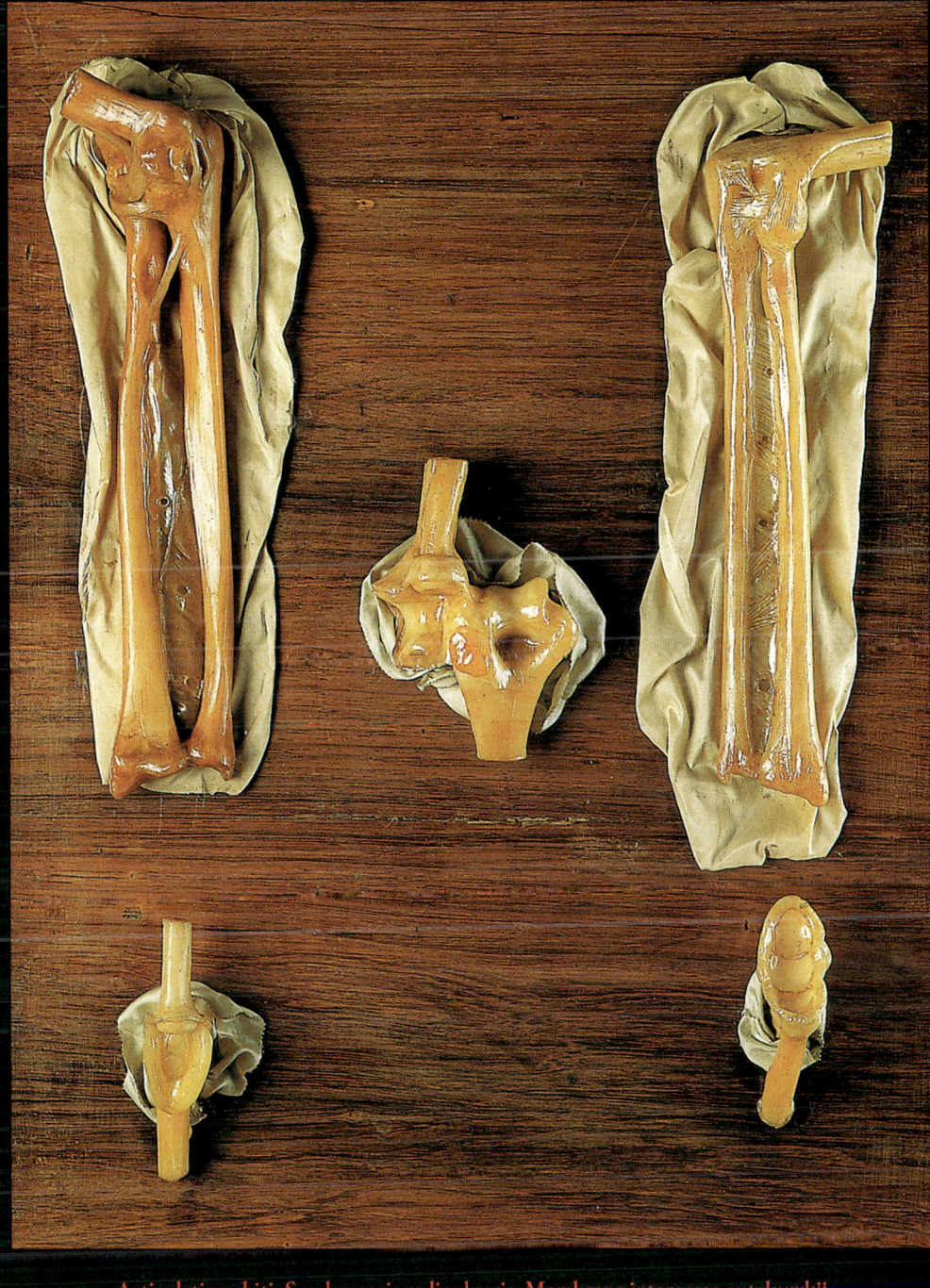

Articulatio cubiti, Syndesmosis radioulnaris, Membrana interossea antebrachii

Display showing various aspects of the elbow joint

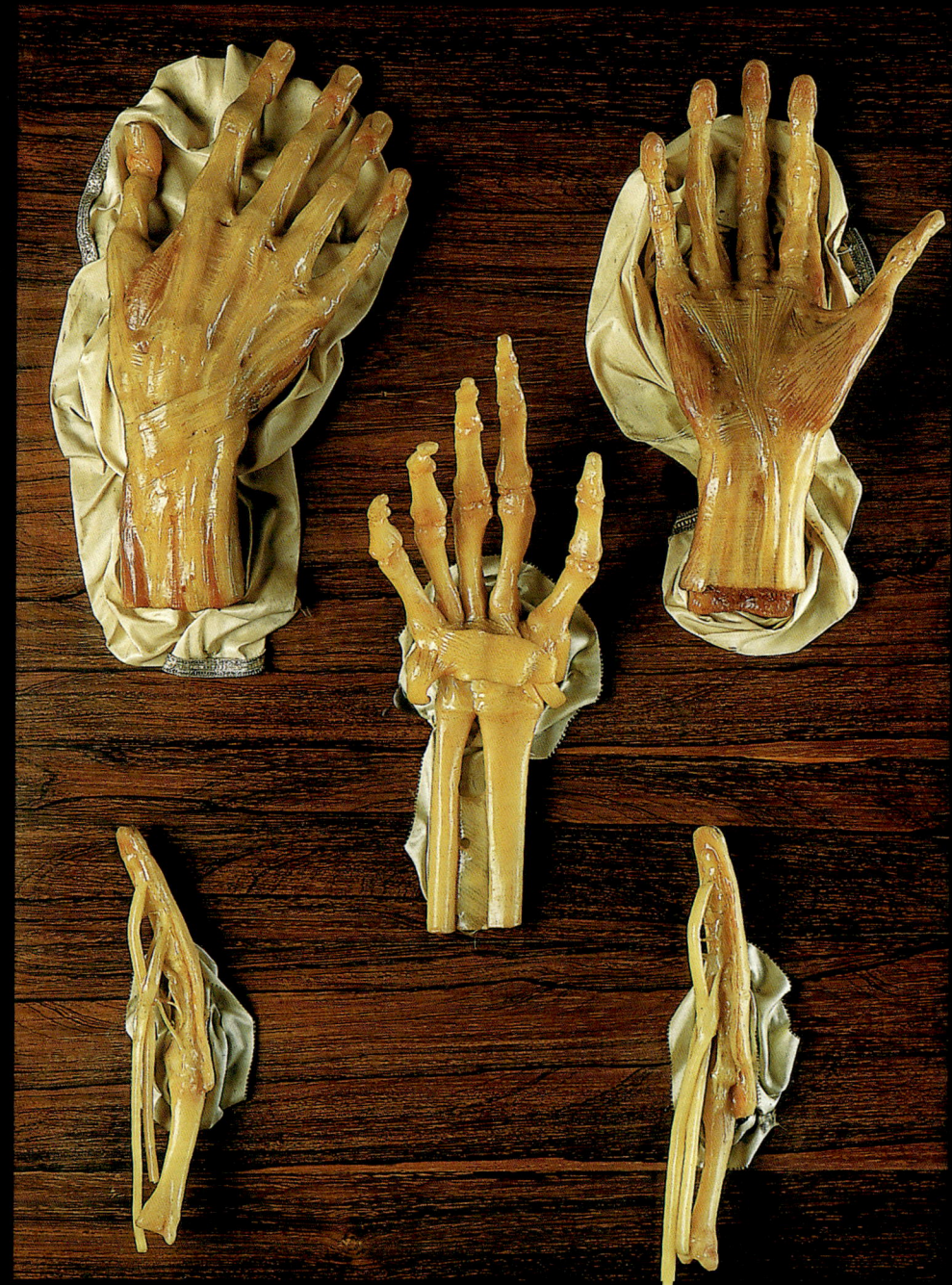

Tendines, Vaginae synoviales, Aponeurosis palmaris, Ligamenta et musculi manus

Tendons and tendon sheaths of the back of the hand. Tendinous sheet of palm, and the insertions of tendons of the flexor and extensor muscles of the fingers

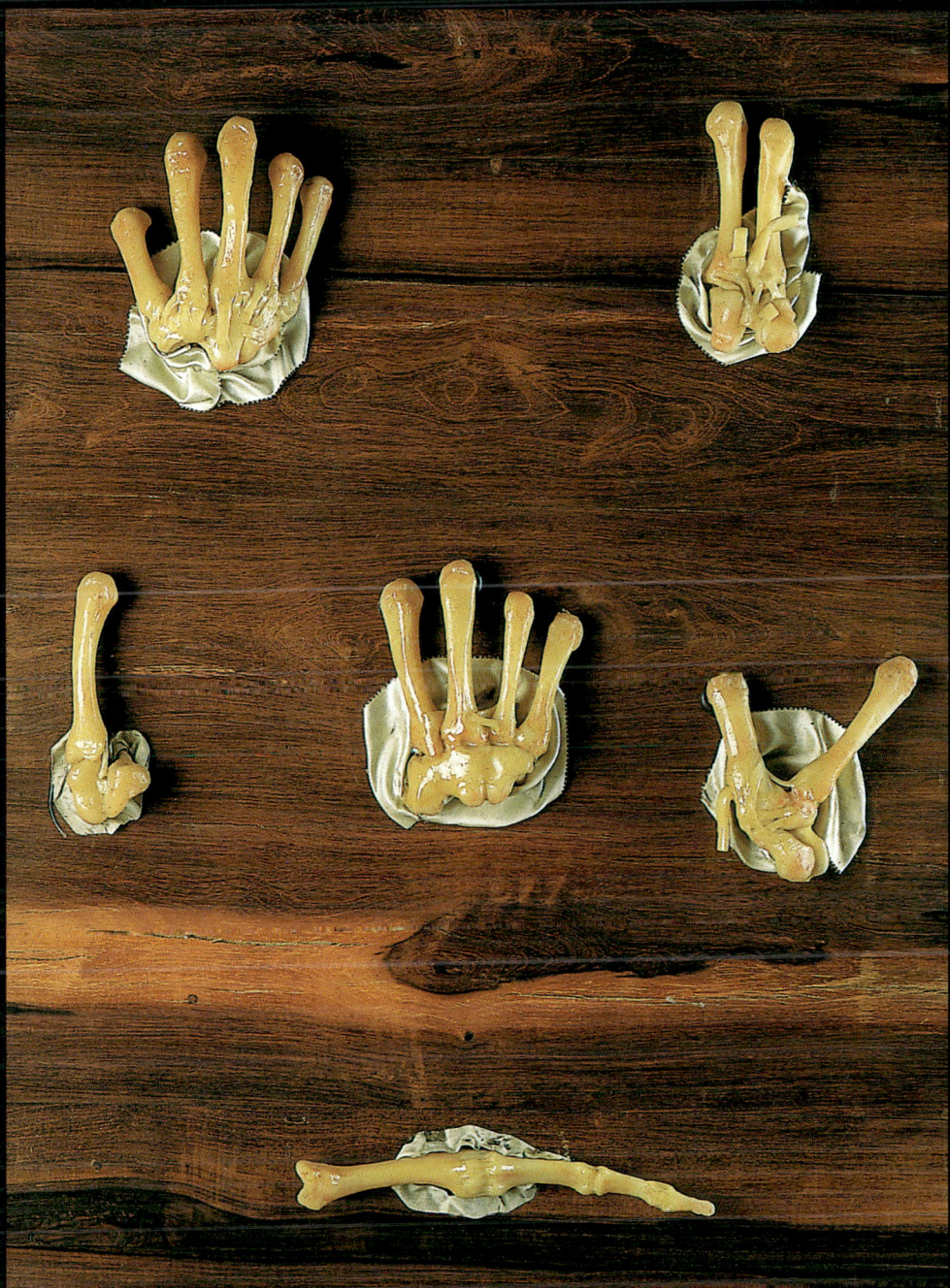

Articulatio carpometacarpea, Ligamenta carpometacarpea dorsalia et palmaria
Joints of the hand and middle finger

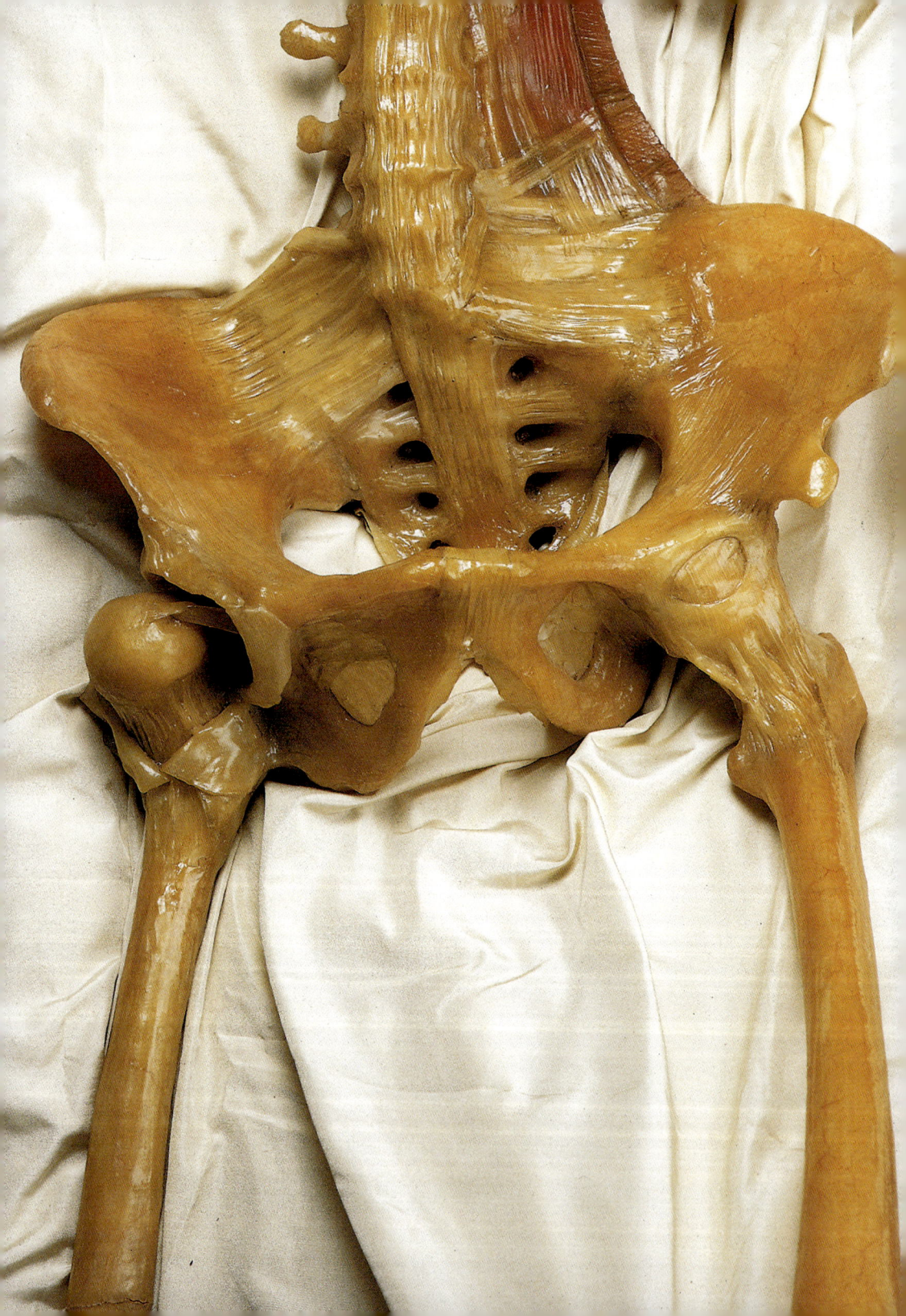

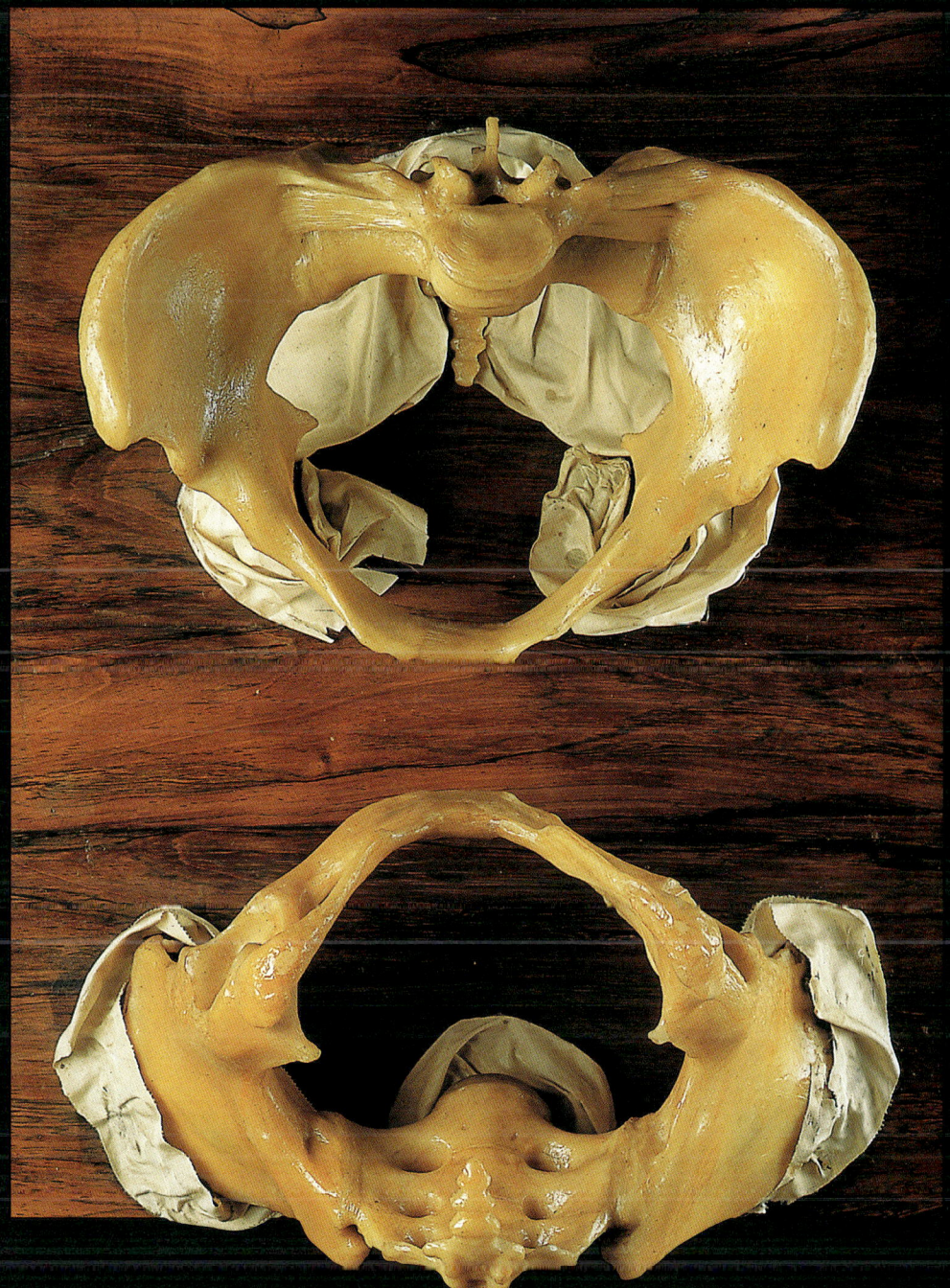

Cingulum membri inferioris, Os sacrum, Os coxae, Os coccygis

Pelvic girdle, consisting of the sacrum and two hip bones, seen from above
in the upper specimen and from below in the lower specimen

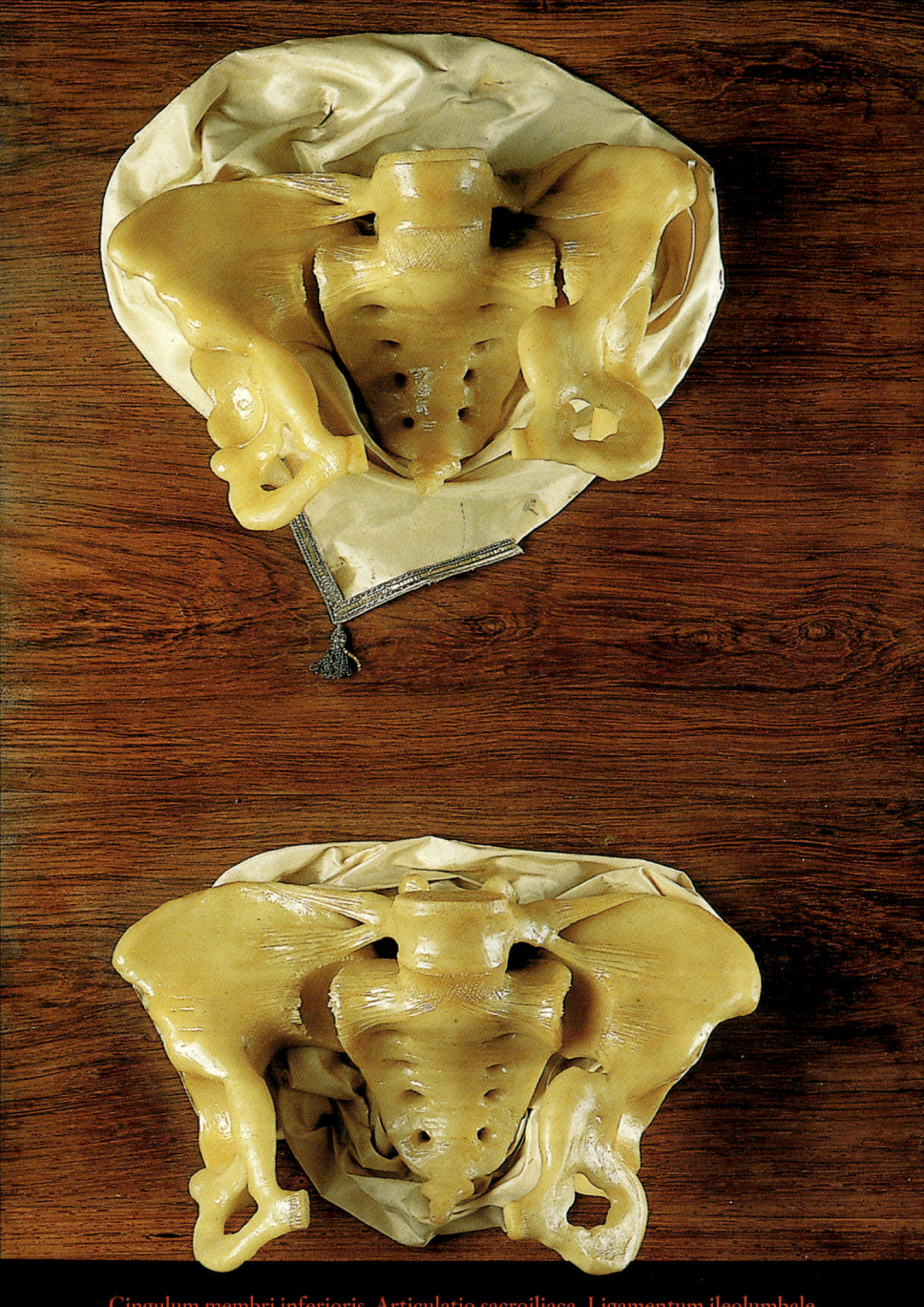

Cingulum membri inferioris, Articulatio sacroiliaca, Ligamentum ileolumbale

Pathologically deformed pelvis. Ligaments between sacrum,
iliac and pubic bones have been divided

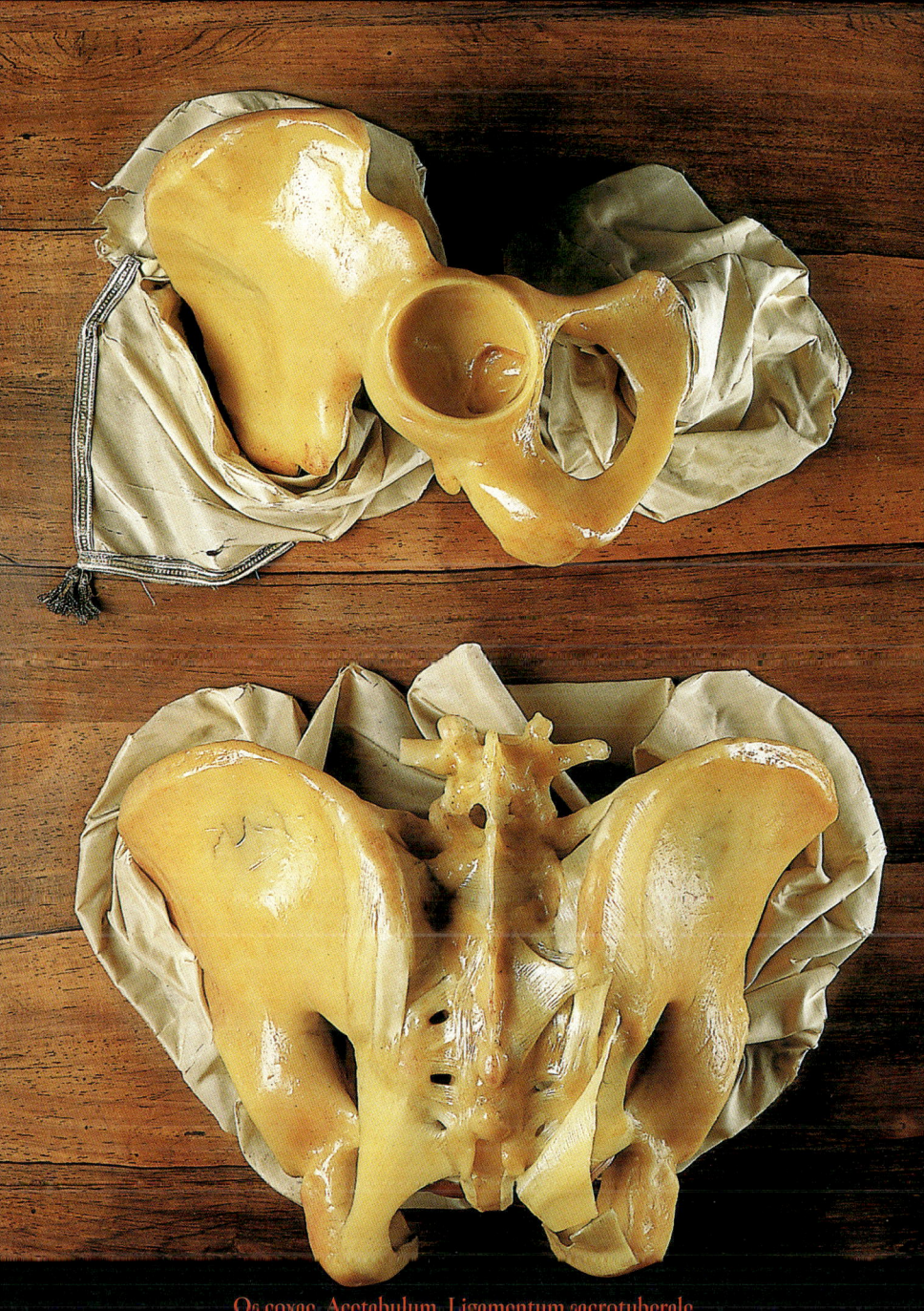

Os coxae, Acetabulum, Ligamentum sacrotuberale

Upper specimen: hip bone with socket for the head of the femur.
Lower specimen: pelvis with ligaments seen from behind

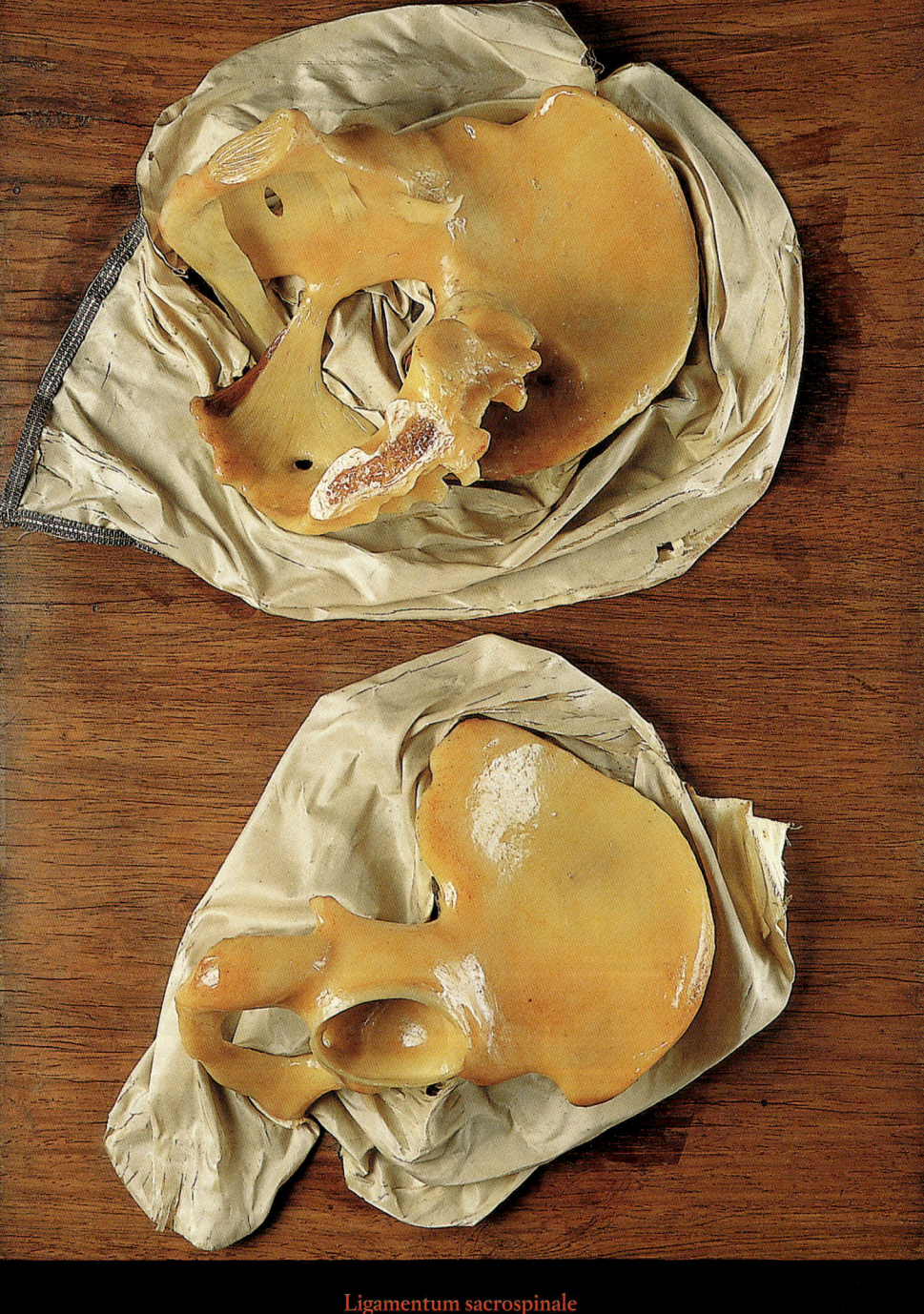

Ligamentum sacrospinale

Hip bone with its ligaments

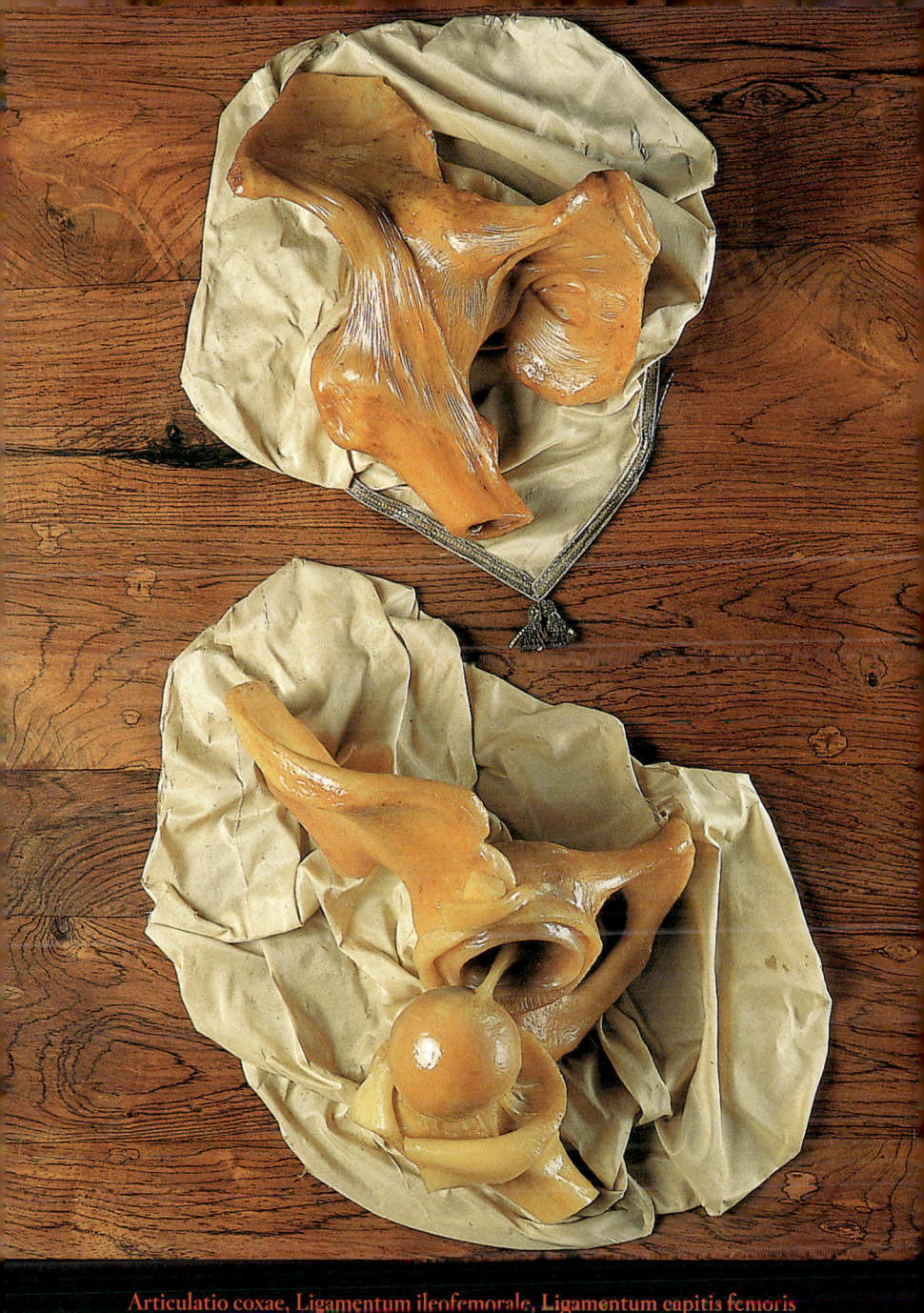

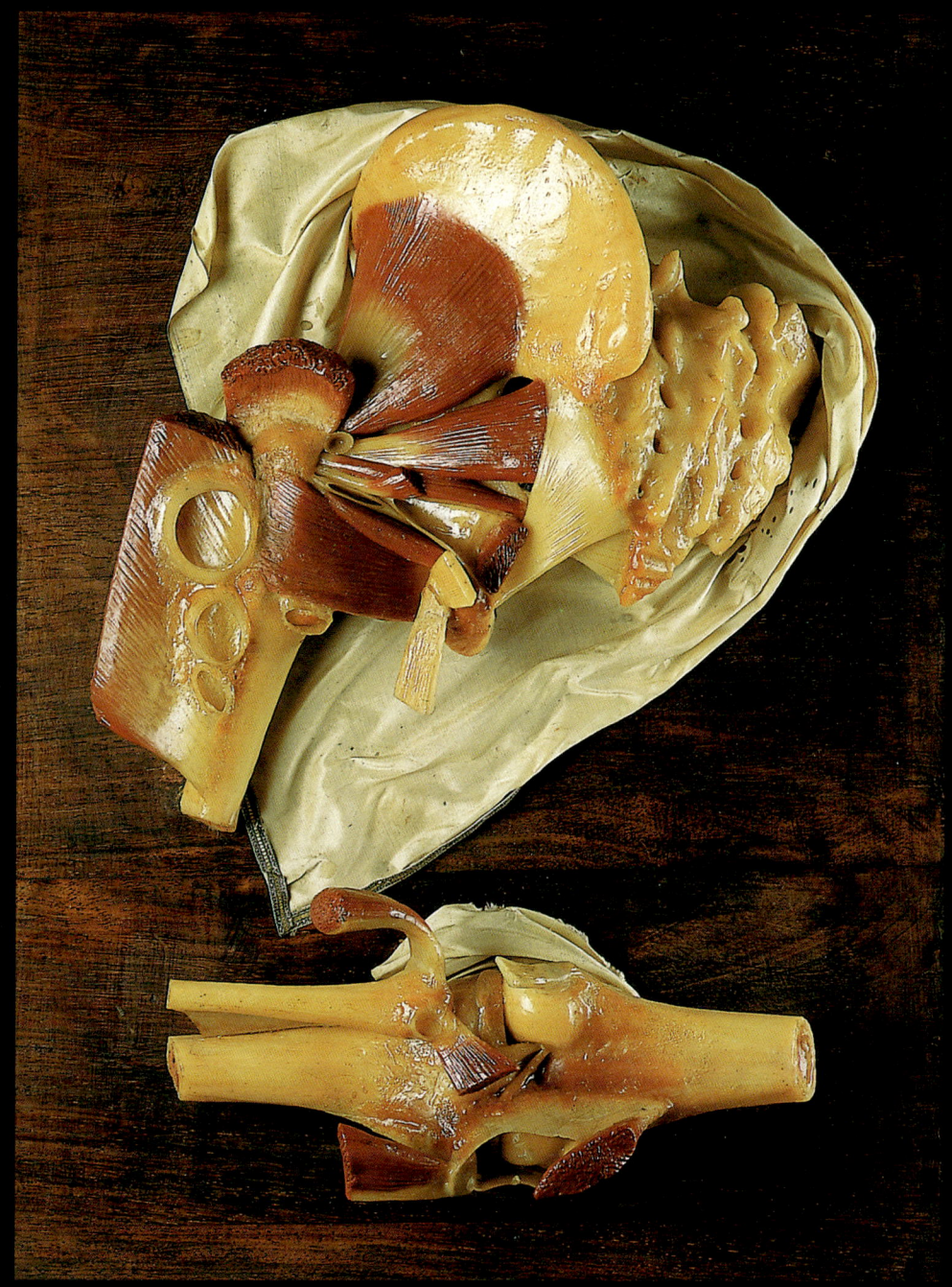

Articulatio coxae, Articulatio genus

Left hip and knee joints with muscle insertions, tendons, ligaments
and opened bursae seen from behind

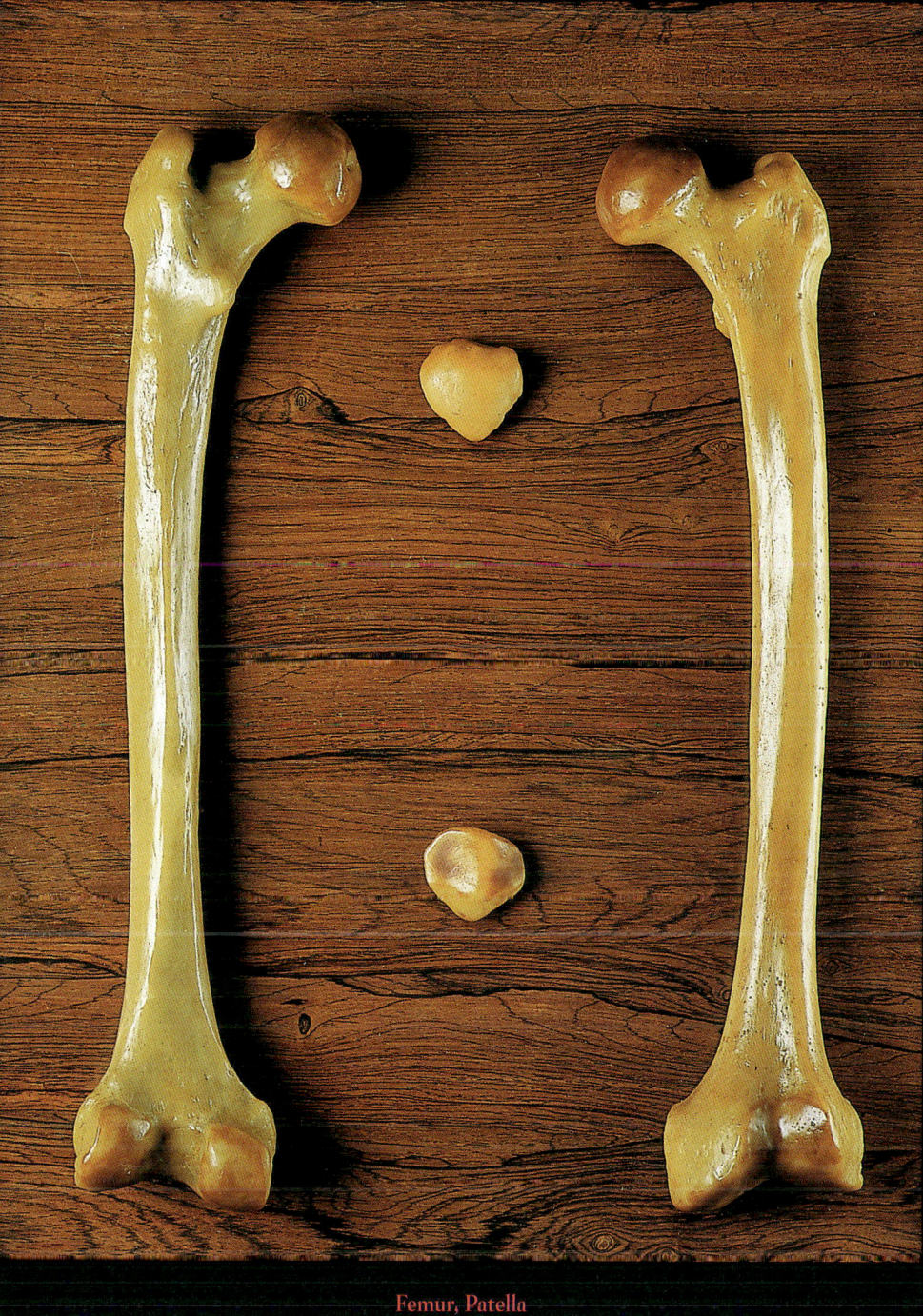

Femur, Patella

Left femur and kneecap seen from in front (left specimen) and from behind

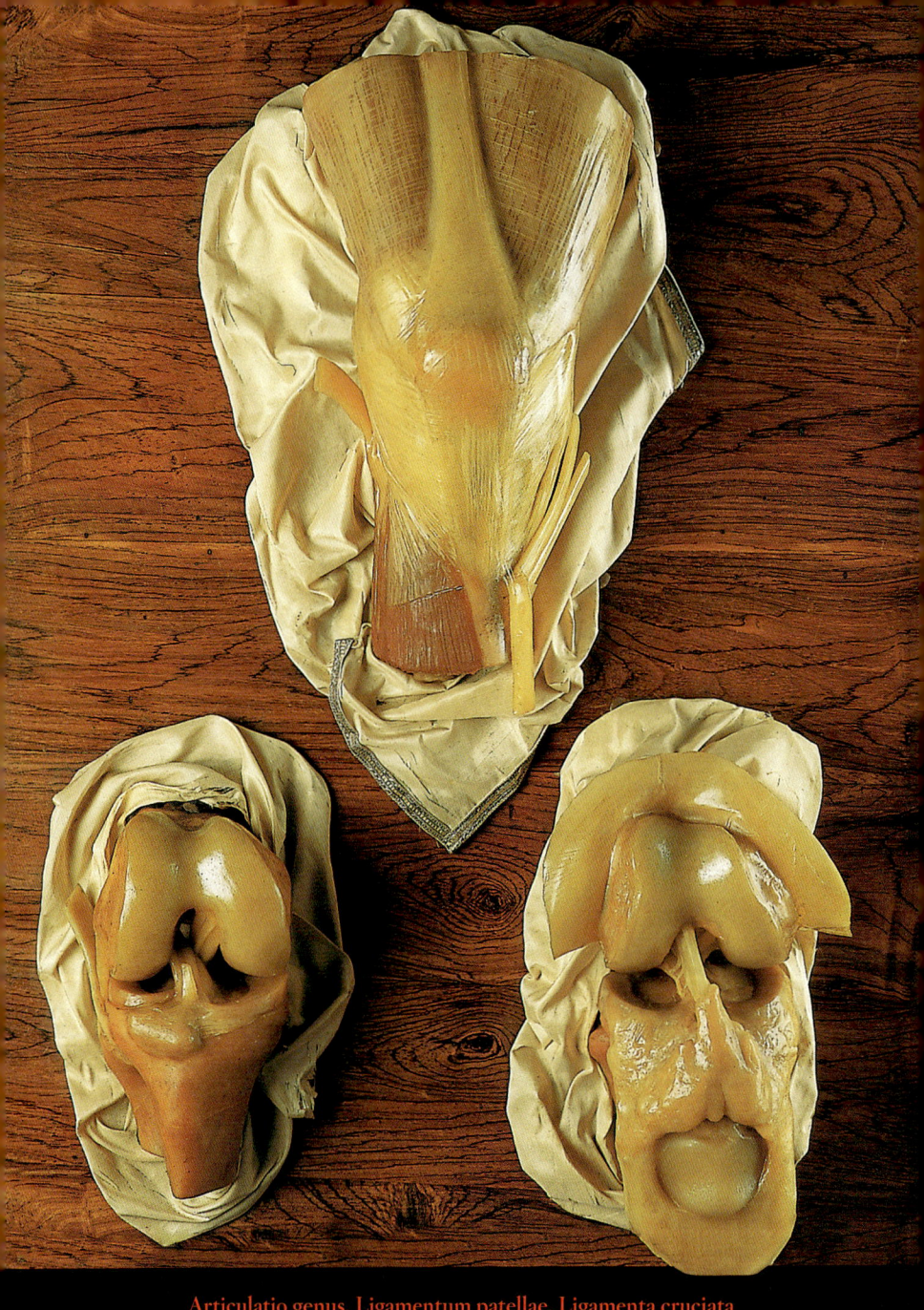

Articulatio genus, Ligamentum patellae, Ligamenta cruciata

Upper specimen: knee joint from in front. Lower specimen: the capsule of the knee joint
has been removed to show the cruciate ligaments and semilunar cartilages.

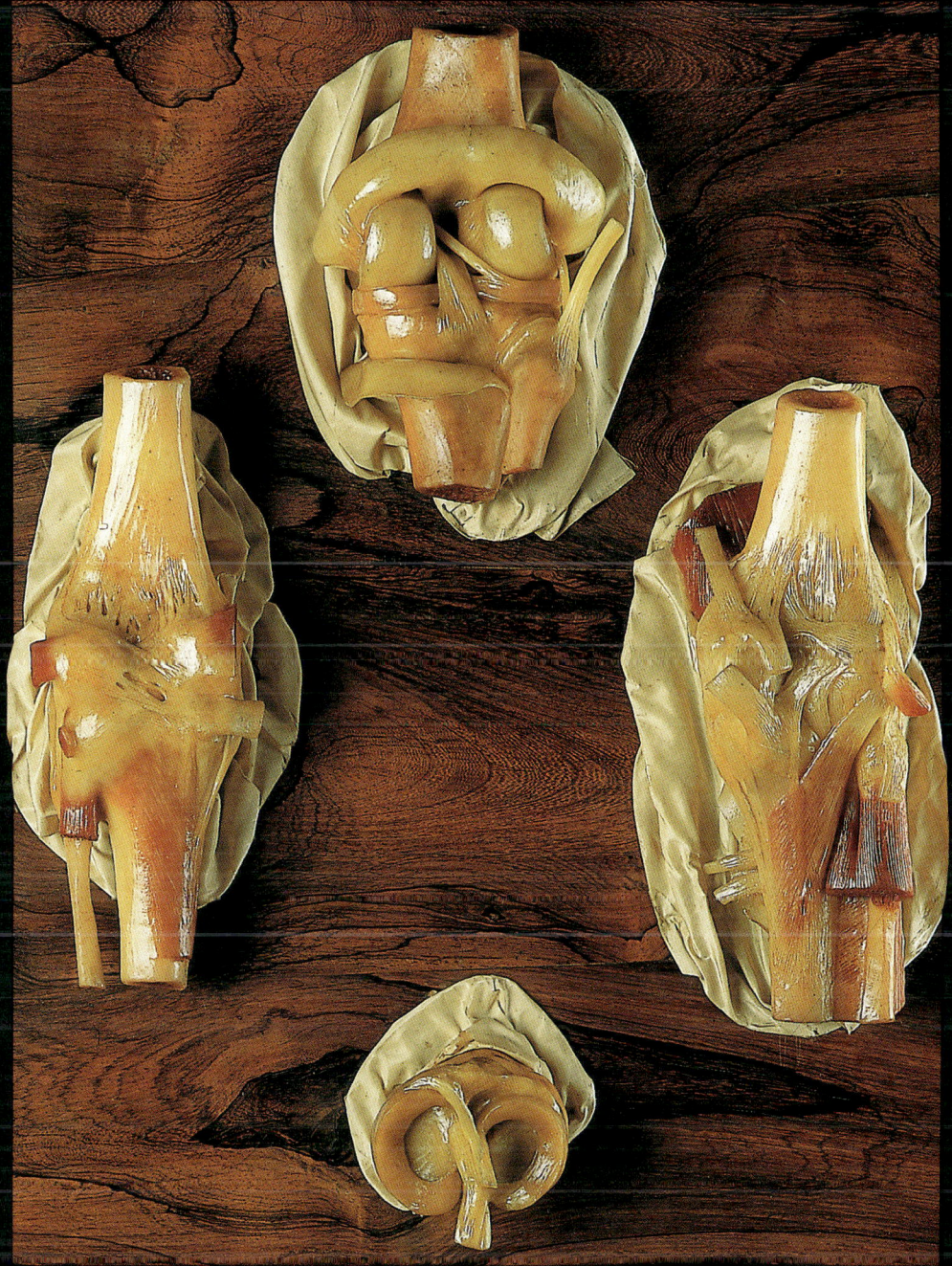

Articulatio genus, Meniscus medialis et lateralis

Upper specimen: left knee joint with ligaments, tendons and muscle insertions seen from behind.
Lower specimen: view of the semilunar cartilages and cruciate ligaments

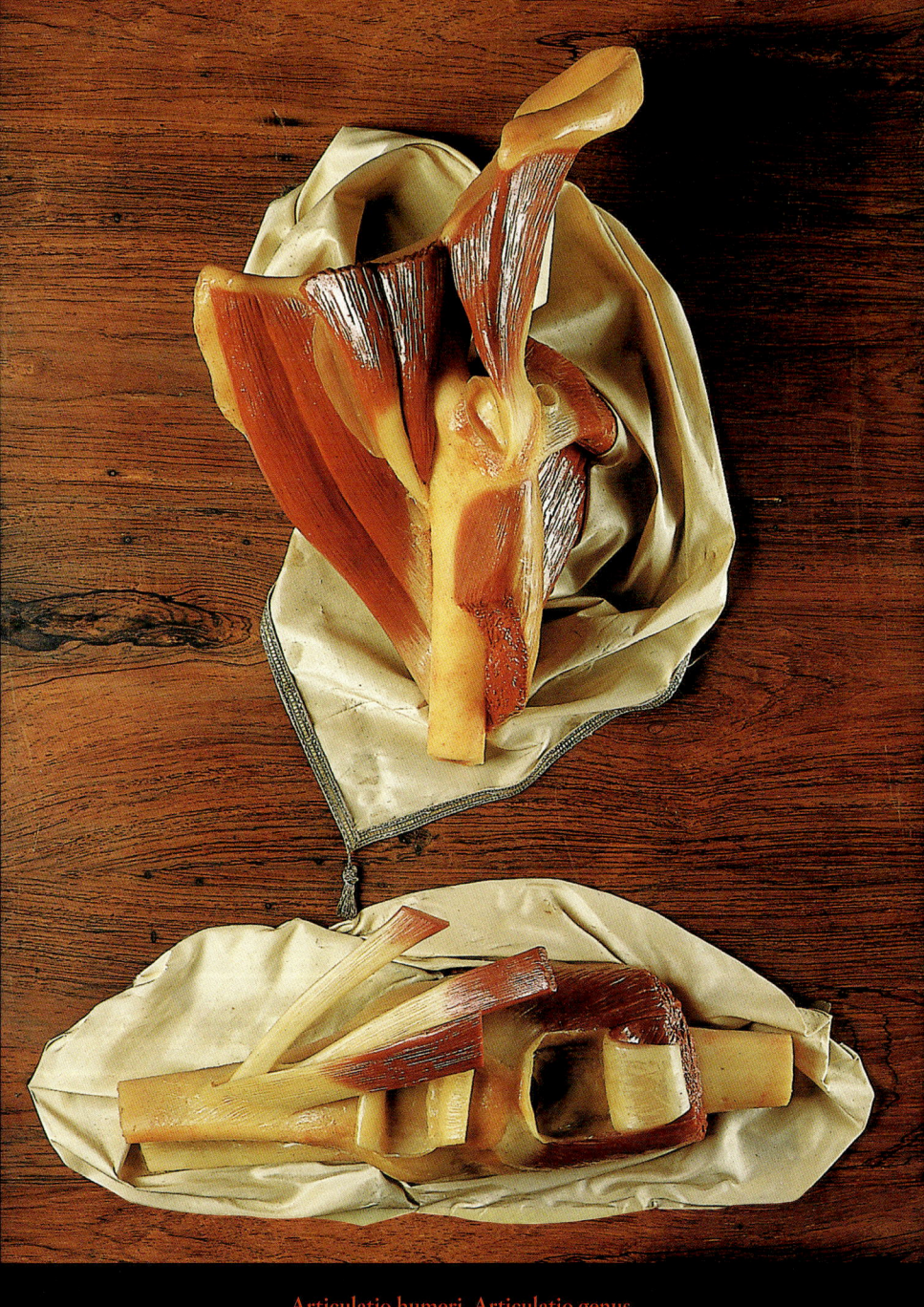

Articulatio humeri, Articulatio genus

Shoulder and knee joints with muscle insertions

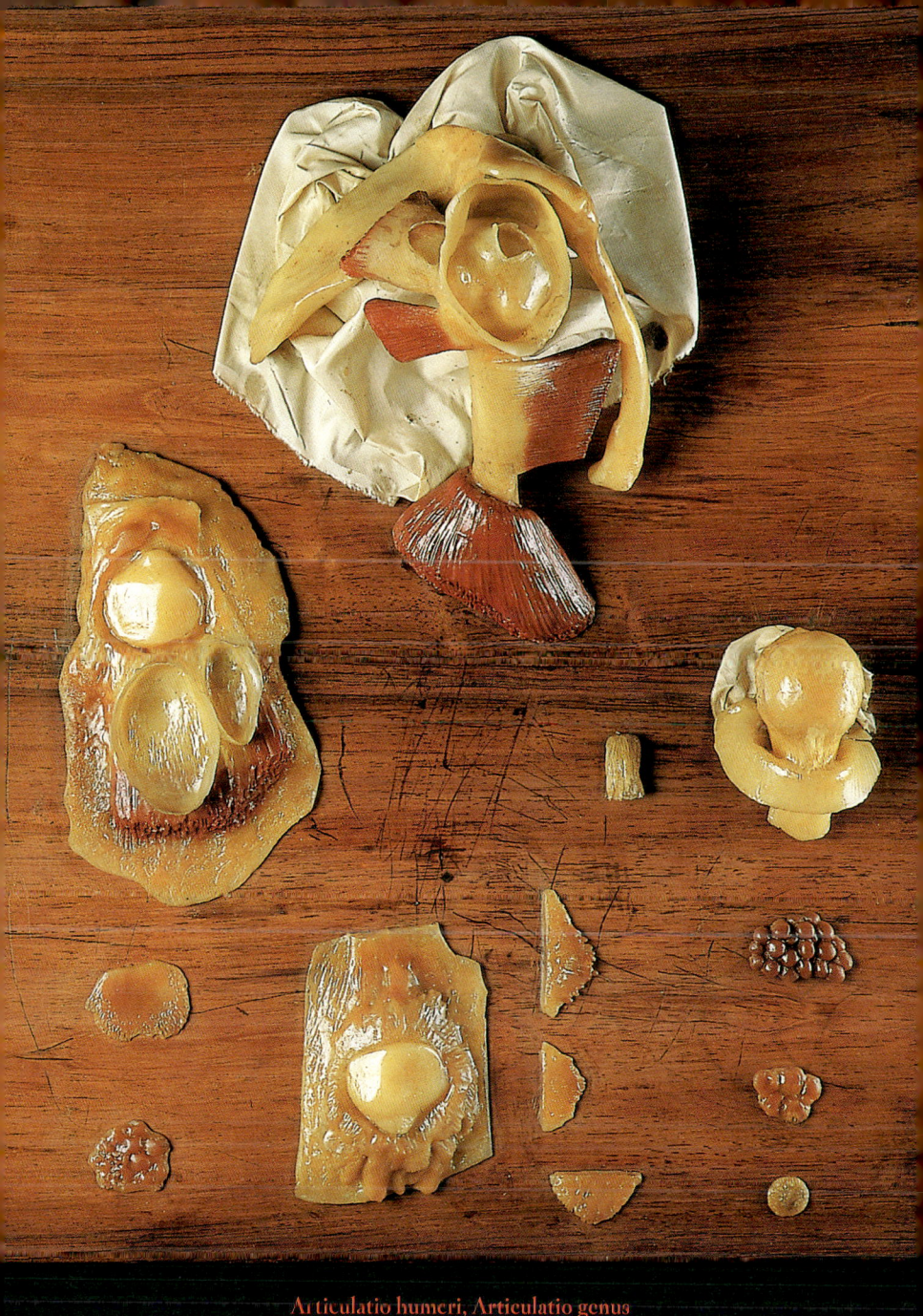

Articulatio humeri, Articulatio genus

Shoulder and knee joints: the joint components, joint capsule and adipose tissue

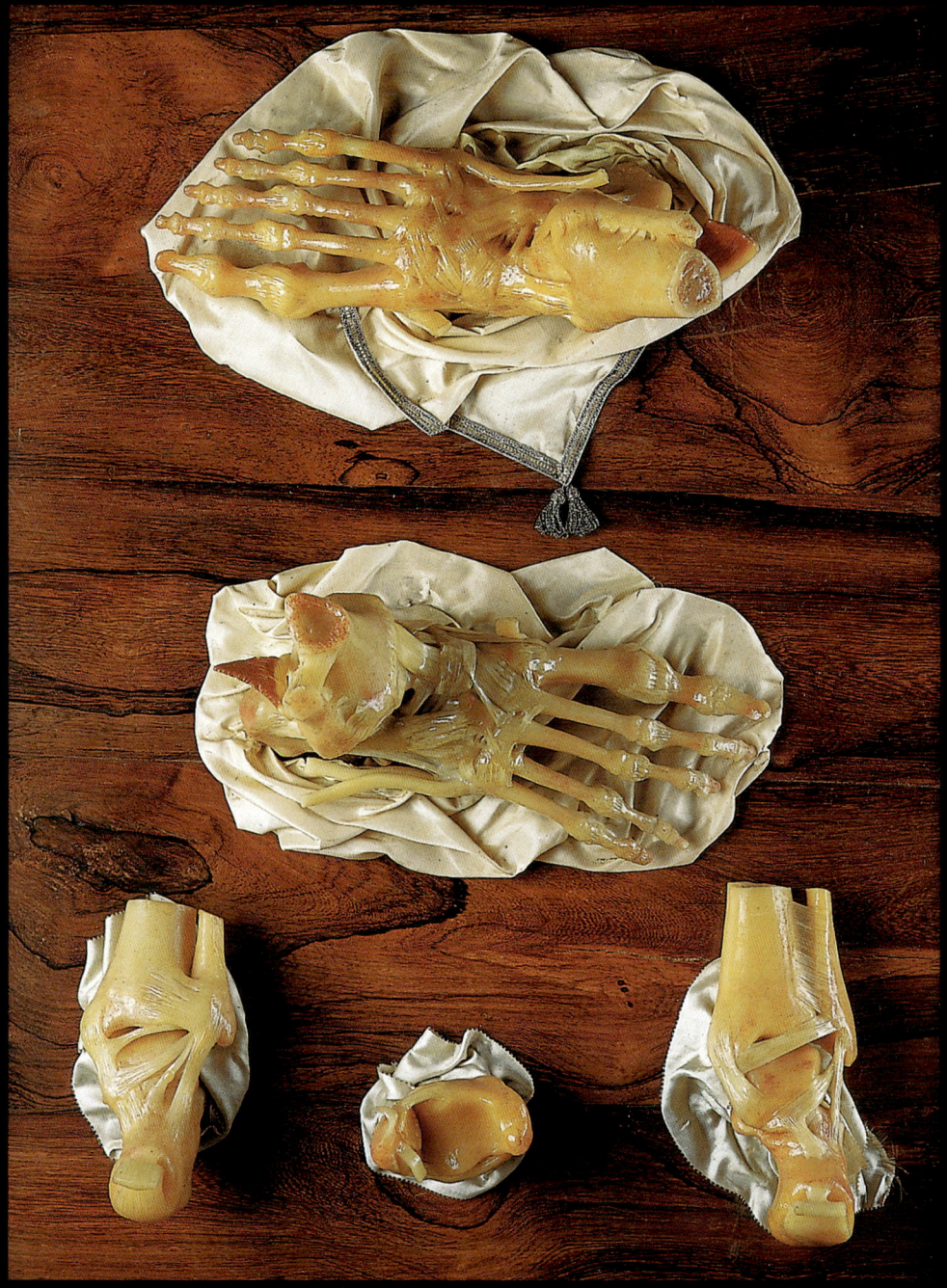

Ligamenta pedis

The ligaments of the ankle joint and foot

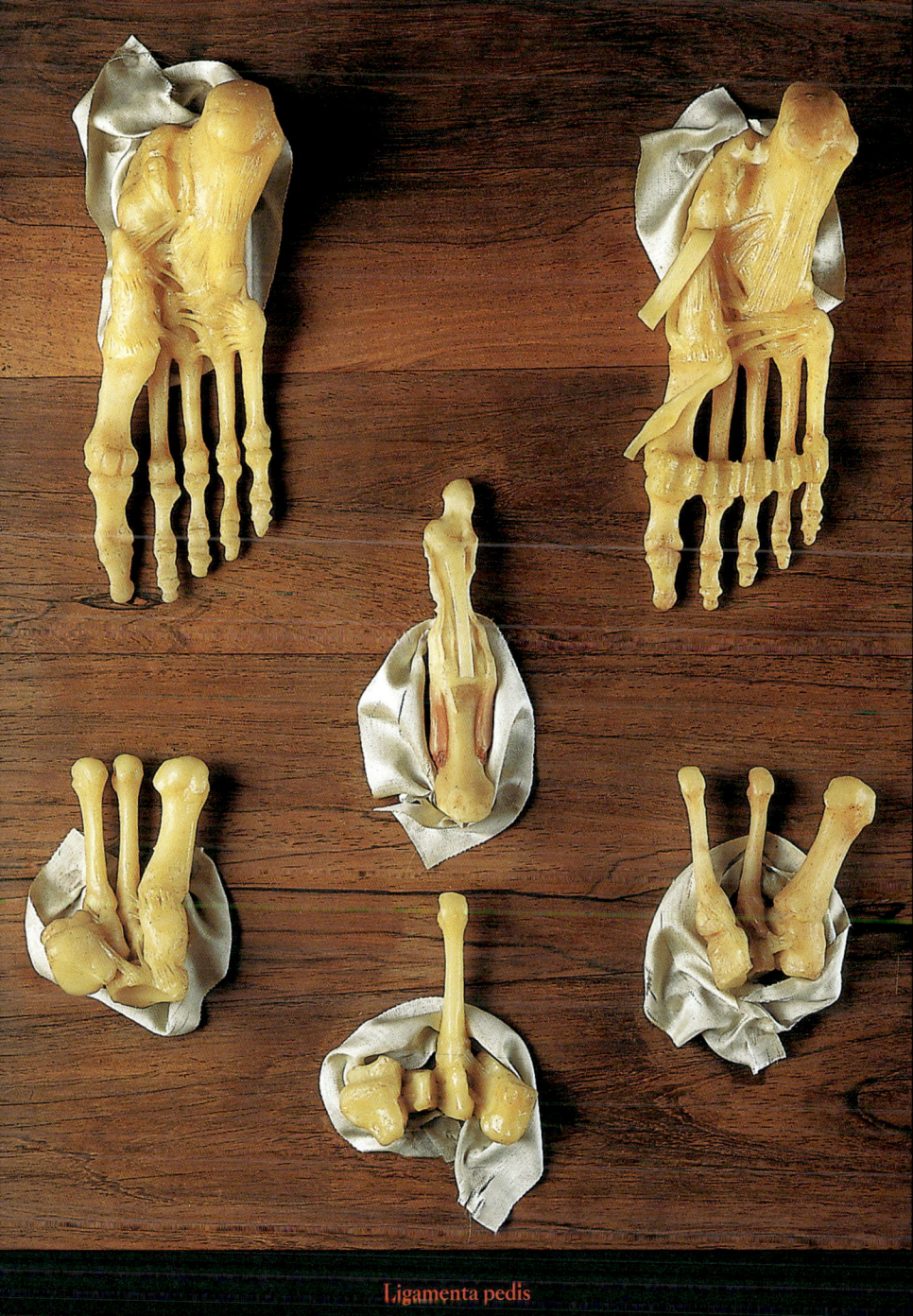

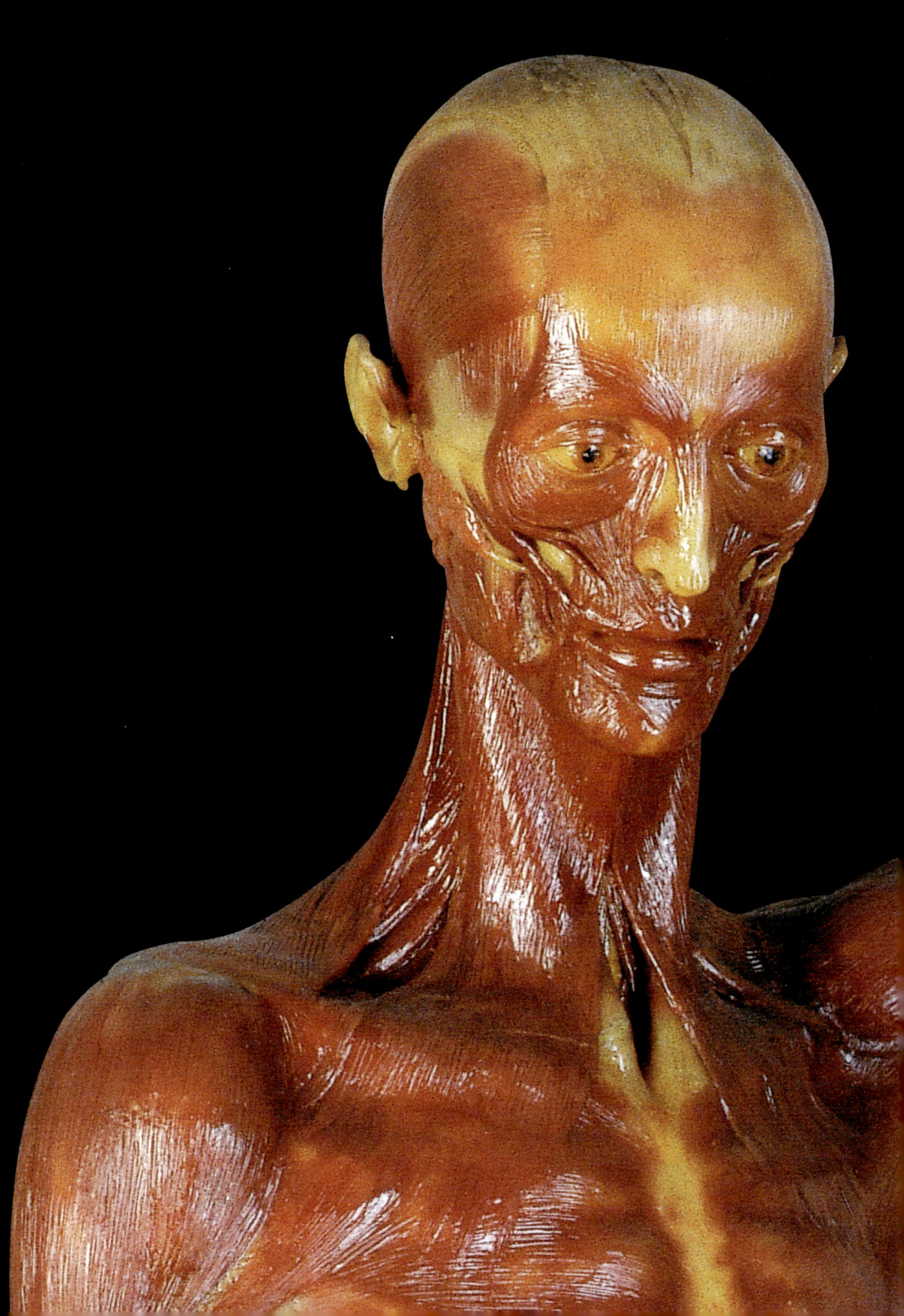

Myologia

Muscles | Muskeln | Muscles

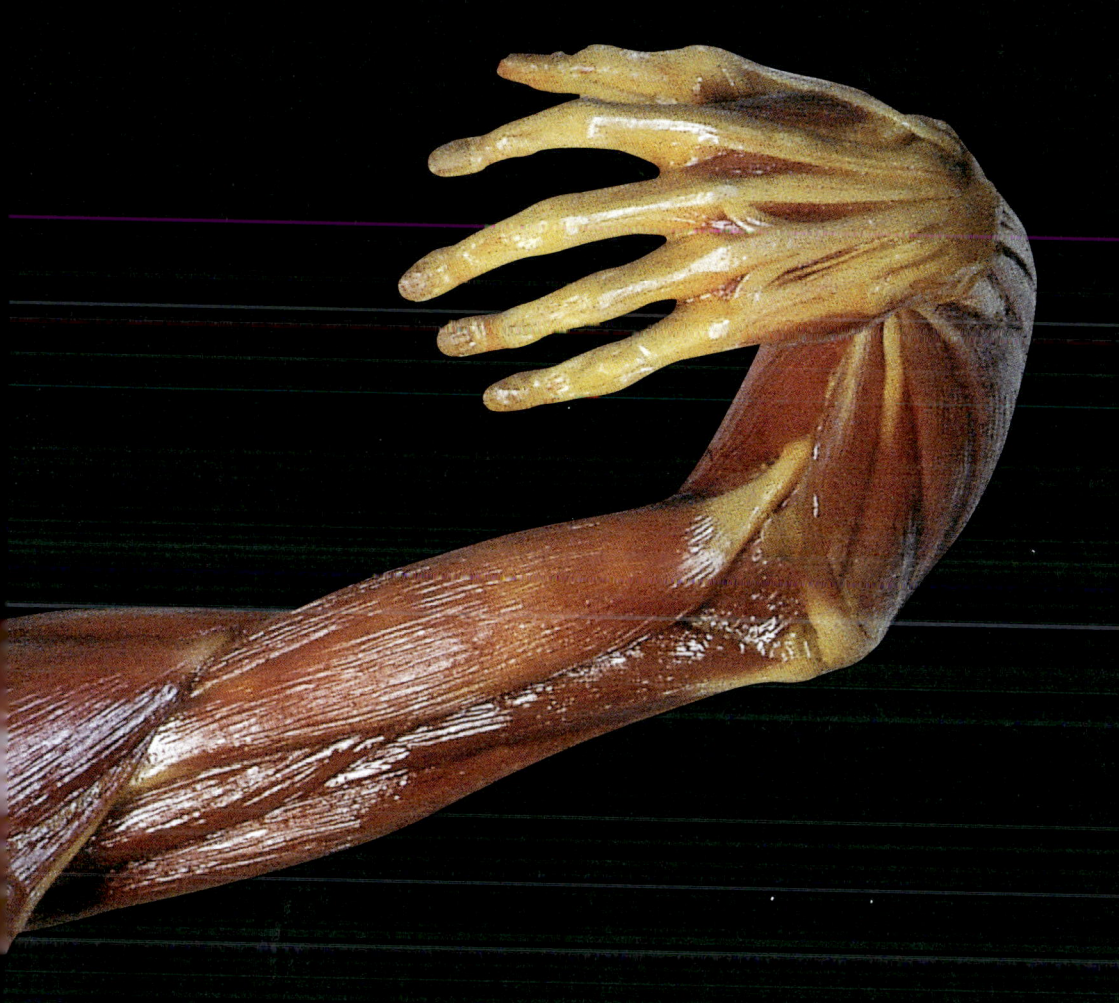

Muscles and Locomotor System

The ability of human beings to walk on two feet and the postures they take up when active or at rest necessitate highly sophisticated coordination between the different groups of muscles. Our central nervous system is responsible for the neuronal control of this complex conscious and unconscious activity, all of which is carried out by motor units standing in a particular functional relationship to one another and acting on the relevant joints as the individual muscles come into play together or in sequence. In order to produce controlled movement, those muscles which act together as agonists and antagonists must be finely tuned to work in accordance with each other, since only in this way does the overall achievement of harmonious movement become possible.

Posture, walking, grasping and eye movement are probably the kinds of movement most familiar to all people, and these can be extended to include the chewing of food, breathing and speech. The morphological basis of facial expression, which reaches the maximum variety only in our species, is provided by the numerous small muscles of the face. It is not, however, only these facial muscles but the entire musculature of the whole body, which contribute to the individual "body language" by which our conscious or unconscious nonverbal self-awareness and emotions are expressed.

This extensive and remarkable collection of exhibits certainly emphasises the extent and variety of our locomotor system, but it also illustrates the particular relationship between muscles and bones, and thus provides a key to the understanding of both simple and complex activity. The models depicting the musculature of the hand, face, eyes and larynx are extraordinarily impressive, and their representation in this medium offers a clear insight into the finely graded and highly differentiated movement of which these organs are capable.

Pages 112–113:
Detail from page 118

Muskeln und Bewegungsapparat

Der aufrechte Gang des Menschen, seine Körperhaltung in Ruhe sowie seine Körperbe-
wegungen erfordern ein überaus differenziertes Zusammenspiel verschiedener Muskelgruppen.
Unser zentrales Nervensystem übernimmt dabei die neuronale Steuerung der komplexen be-
wussten und unbewussten Bewegungsabläufe. Alle Bewegungsabläufe werden von funktionell
in Beziehung stehenden Muskelgruppen ausgeführt, die über mehrere Gelenke hinweg ziehen
können und deren individuelle Muskeln im direkten Zusammenspiel bzw. nacheinander arbeiten. Für den harmonischen Bewegungsablauf sind die agonistisch und antagonistisch arbeiten-
den Muskeln in ihrer Aktivität immer aufeinander abgestimmt. Nur so entsteht der Eindruck
eines harmonischen Bewegungsablaufs.

Haltungs-, Stütz-, Bewegungs-, Greif- und Augenmotorik sind für den Laien die wohl
bekanntesten Bewegungsformen. Einzubeziehen sind ferner die Kaumotorik bei der Nah-
rungsaufnahme, die Atmungsmotorik sowie die Sprachmotorik. Morphologische Grundlage
der Mimik sind die zahlreichen kleinen Gesichtsmuskeln, die in dieser Vielfalt ausschließlich
im menschlichen Gesicht vorkommen. Aber nicht nur diese mimischen Muskeln, sondern alle
Muskeln unseres Körpers haben ihren funktionellen Anteil an der individuellen Körpersprache,
mit der wir bewusst oder unbewusst nonverbal unser Selbstwertgefühl und unsere Emotionen
ausdrücken.

Die vielen außergewöhnlichen Exponate vermitteln nicht nur ein umfassendes Bild von
der Formenvielfalt der Muskeln unseres Körpers, sondern geben auch Aufschluss über ihre spezielle Anordnung am Skelett und tragen so zu einem funktionellen Verständnis für einfache und
komplexe Bewegungsabläufe bei. Besonders beeindruckend sind die Modelle der Muskeln von
Hand, Gesicht, Augen und Kehlkopf, deren plastische Darstellung ein umfassendes Verständnis
der fein abgestuften und differenzierten Bewegungen ermöglicht.

Les muscles et l'appareil locomoteur

Tous les mouvements du corps humain, et en particulier la marche bipède et l'attitude érigée au repos, nécessitent la mise en jeu coordonnée de plusieurs groupes musculaires. Le système nerveux central prend en charge la régulation des mouvements complexes volontaires et involontaires. Des groupes musculaires fonctionnellement interdépendants, pouvant croiser plusieurs articulations, sont mis en jeu lors de l'exécution des mouvements. Chaque muscle participe à l'ensemble du mouvement ou prend le relais d'un autre. Le déroulement de mouvements harmonieux nécessite une coordination permanente des muscles agonistes et antagonistes.

Les aspects les plus connus de la motricité sont pour le non-spécialiste le maintien d'une position, le déplacement dans l'espace, la préhension et les mouvements de l'œil ; la mastication, les mouvements respiratoires, ou la parole articulée sont plus rarement pris en considération. Les nombreux petits muscles de la face, particulièrement développés dans l'espèce humaine, sont les supports morphologiques de la mimique. Ces muscles de la mimique ne sont pas les seuls qui interviennent dans le langage corporel, mais tous les muscles de l'organisme participent à la manifestation consciente ou inconsciente de notre personnalité et de nos émotions.

Le nombre particulièrement important des pièces de cette série ne donne pas seulement un aperçu de la variété des formes musculaires, mais renseigne aussi sur l'agencement particulier des muscles par rapport au squelette ; ces rapports permettent de comprendre la fonction de chaque muscle dans les mouvements élémentaires ou complexes. Particulièrement remarquables sont les modèles concernant la main, la face, les yeux ou le larynx, dont la représentation plastique permet la compréhension globale de mouvements différenciés fins.

Detail from page 274–275

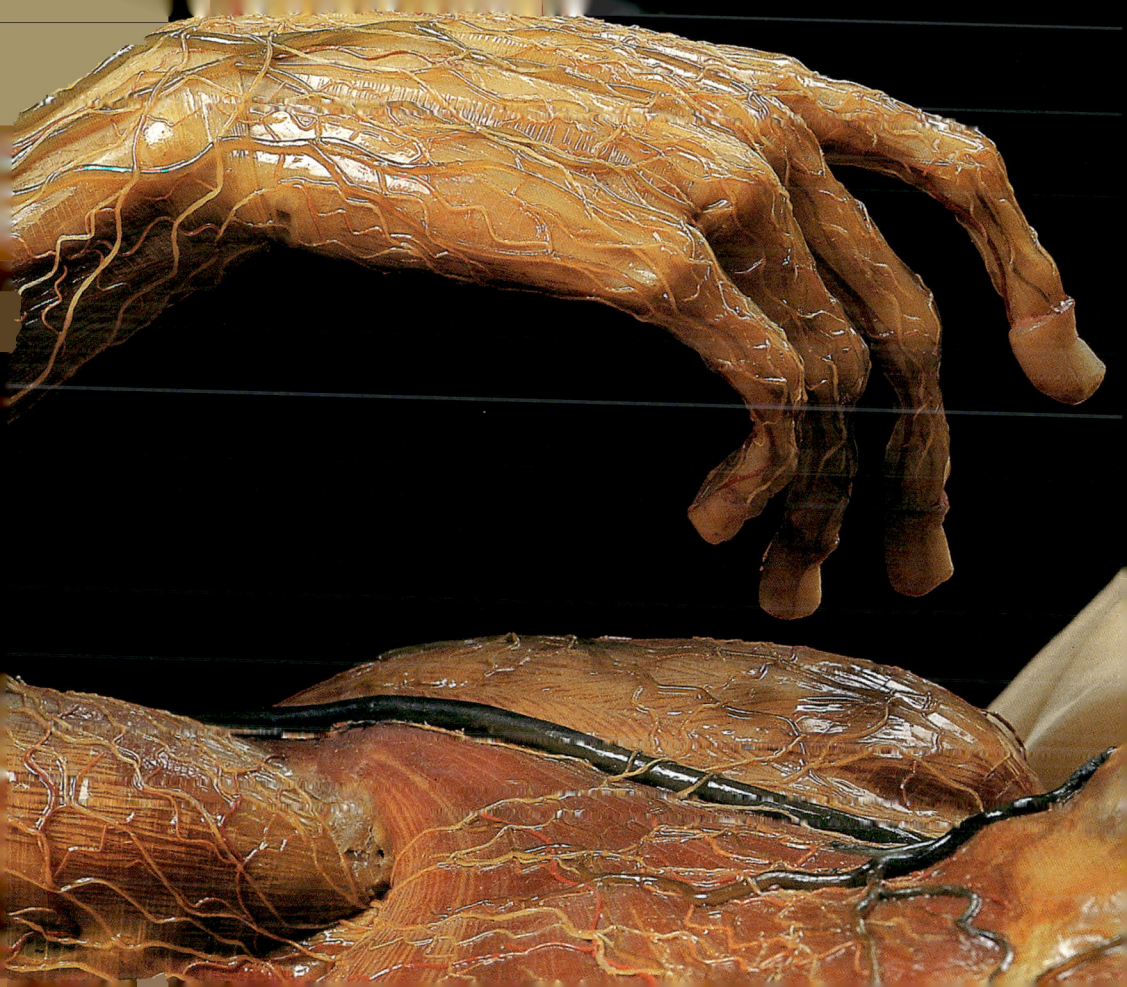

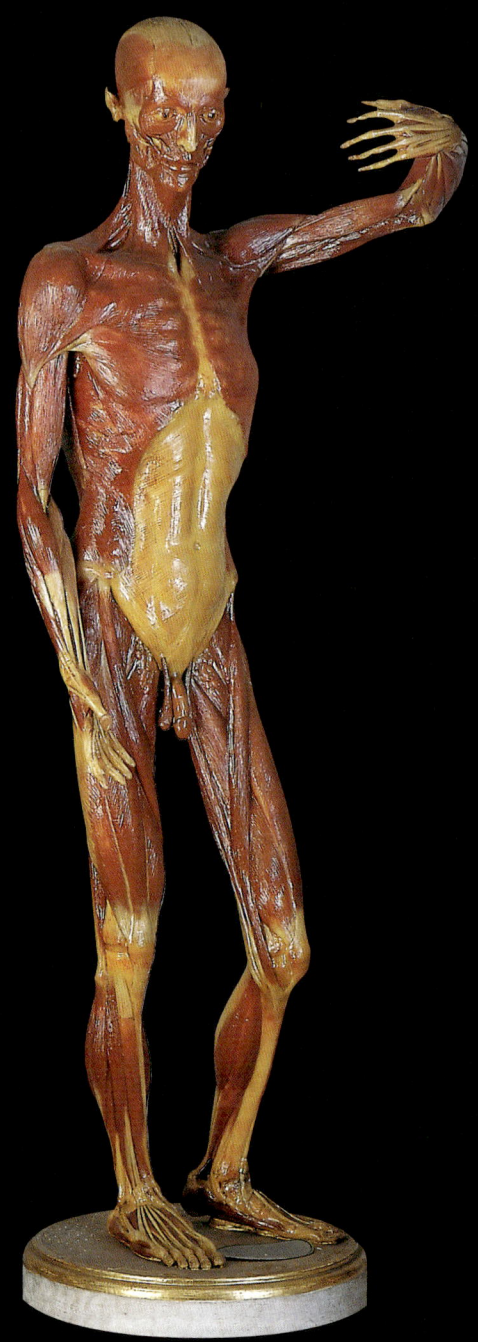

Musculi faciales, Musculi colli, Musculi thoracis, Musculi abdominis,
Musculi membri superioris, Musculi membri inferioris
Whole body specimen displaying the superficial muscles

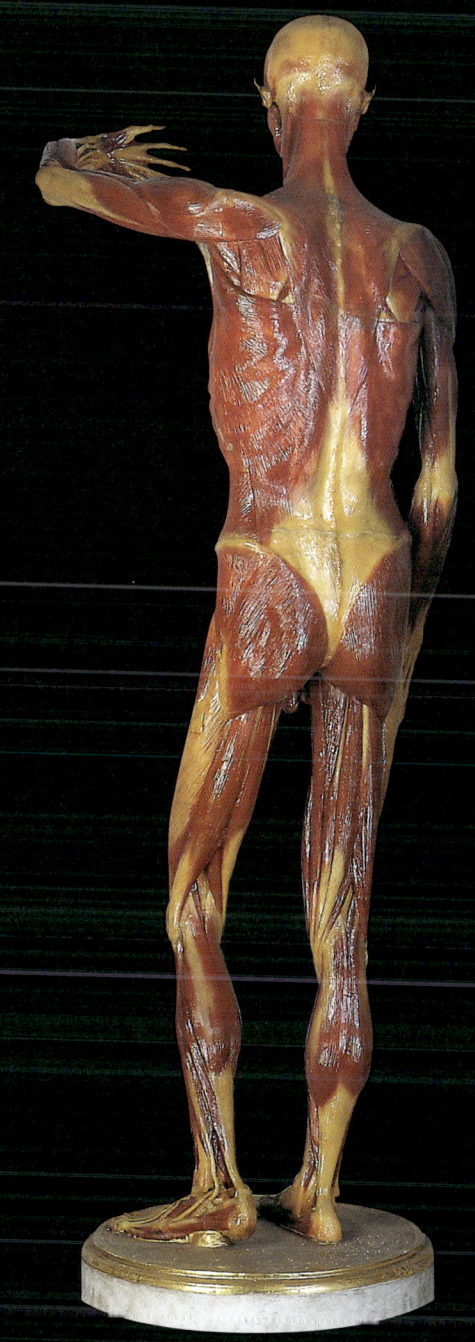

Musculi faciales, Musculi colli, Musculi thoracis, Musculi abdominis, Musculi membri superioris, Musculi membri inferioris

Whole body specimen displaying the superficial muscles

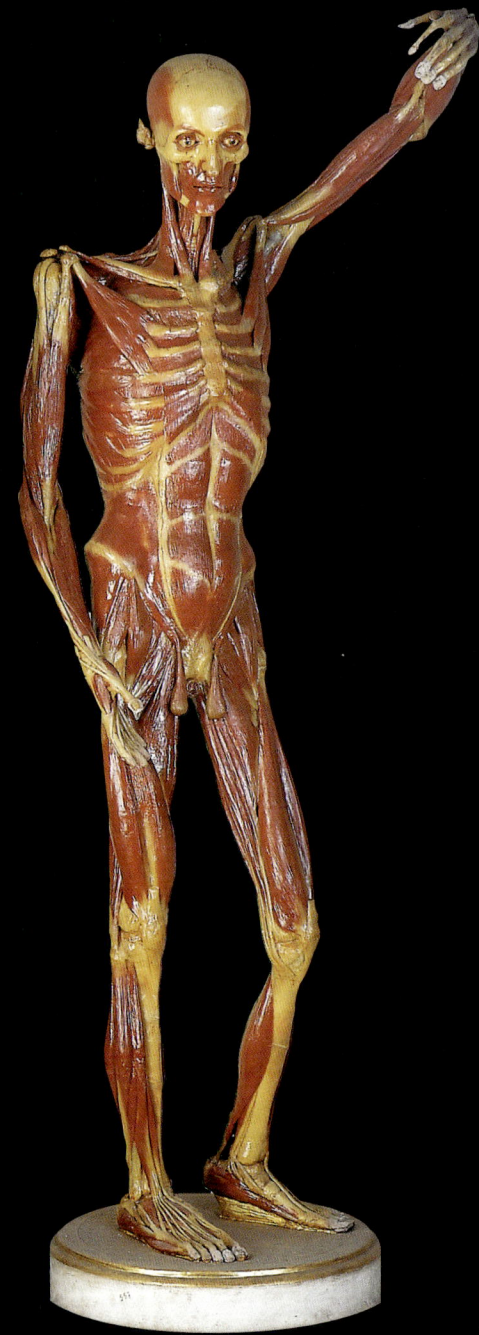

Musculi faciales, Musculi colli, Musculi thoracis, Musculi abdominis,
Musculi membri superioris, Musculi membri inferioris, Musculi dorsi

Whole body specimen displaying the superficial and deep muscles

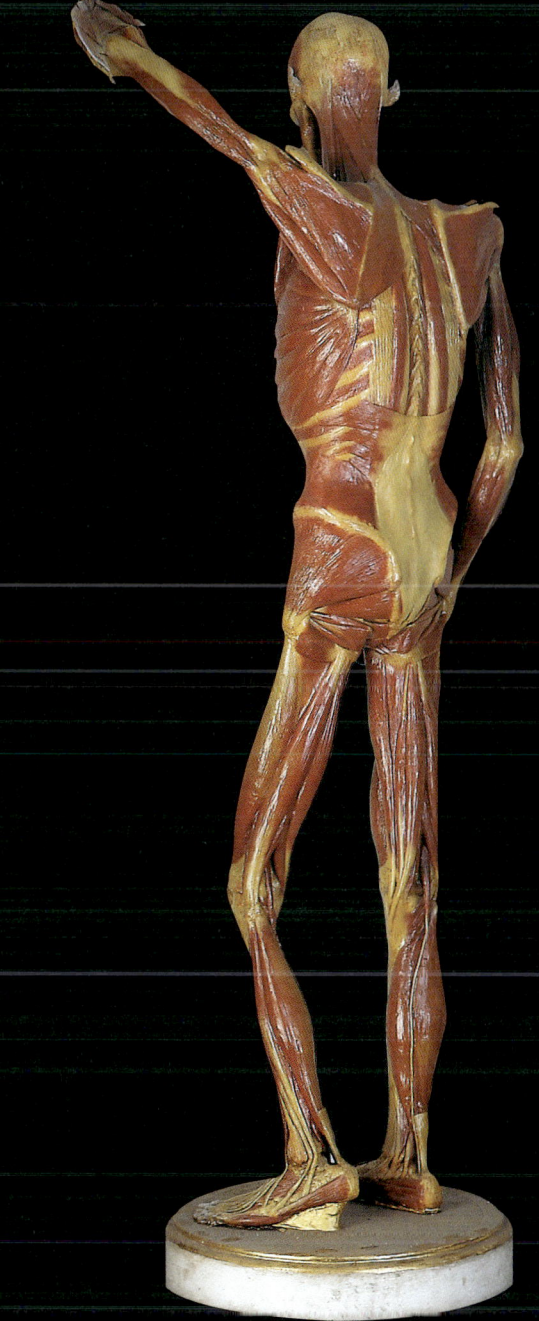

Musculi faciales, Musculi colli, Musculi thoracis, Musculi abdominis,
Musculi membri superioris, Musculi membri inferioris, Musculi dorsi

Whole body specimen displaying both the superficial and deep muscles

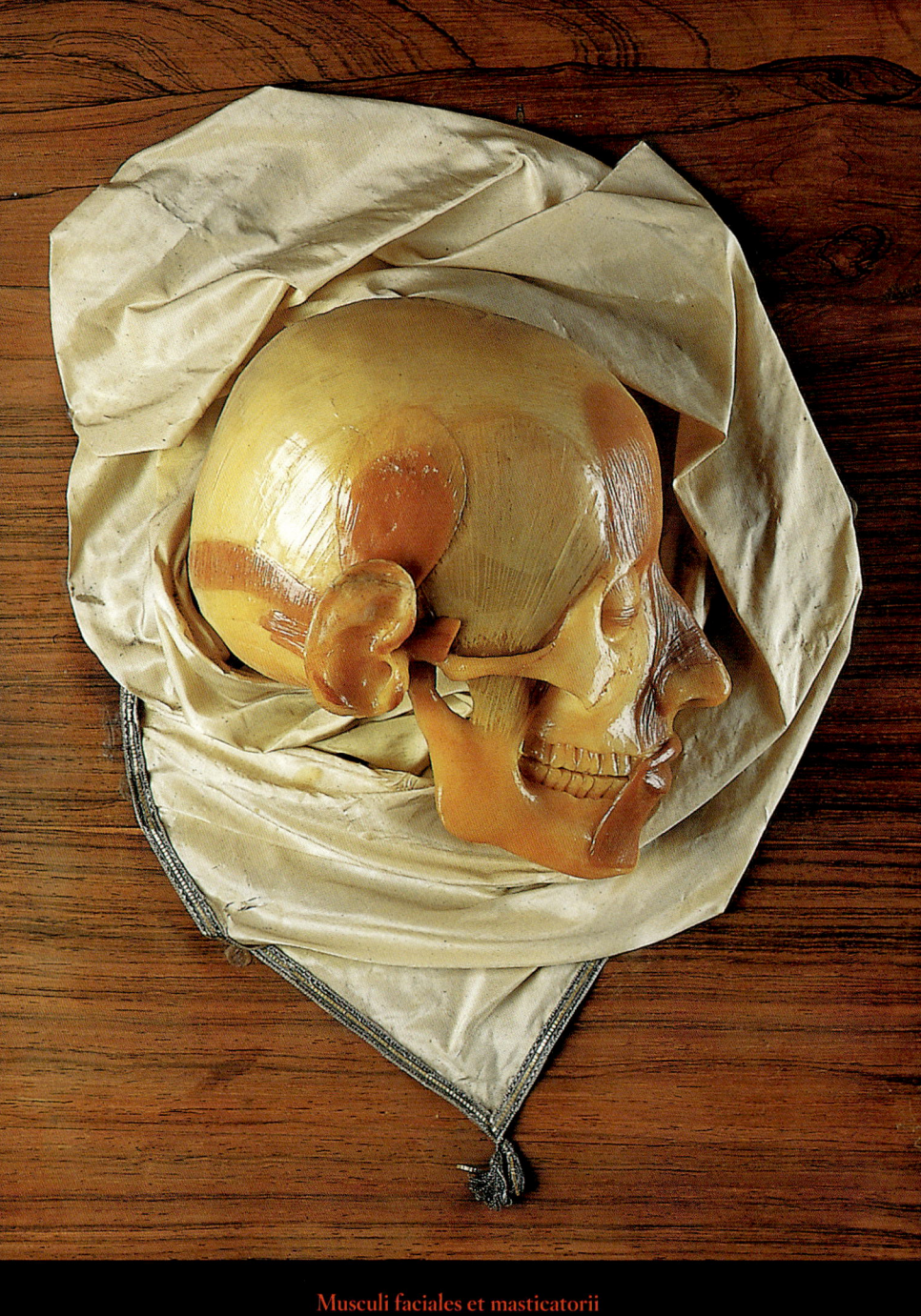

Muscles of facial expression, muscles of mastication

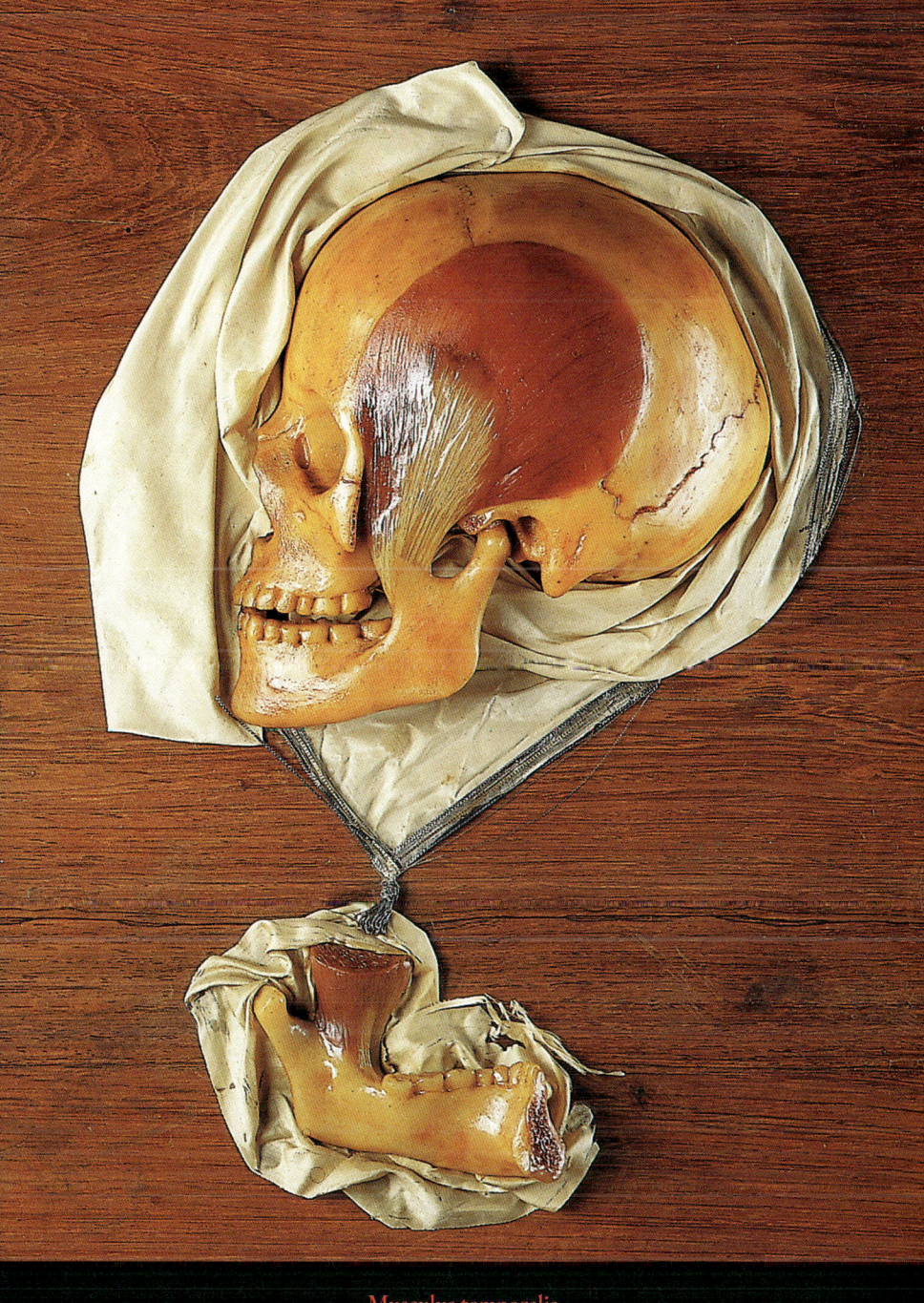

Musculus temporalis

Views of the temporalis muscle. The lower specimen shows the attachment
of this muscle of mastication to the lower jaw

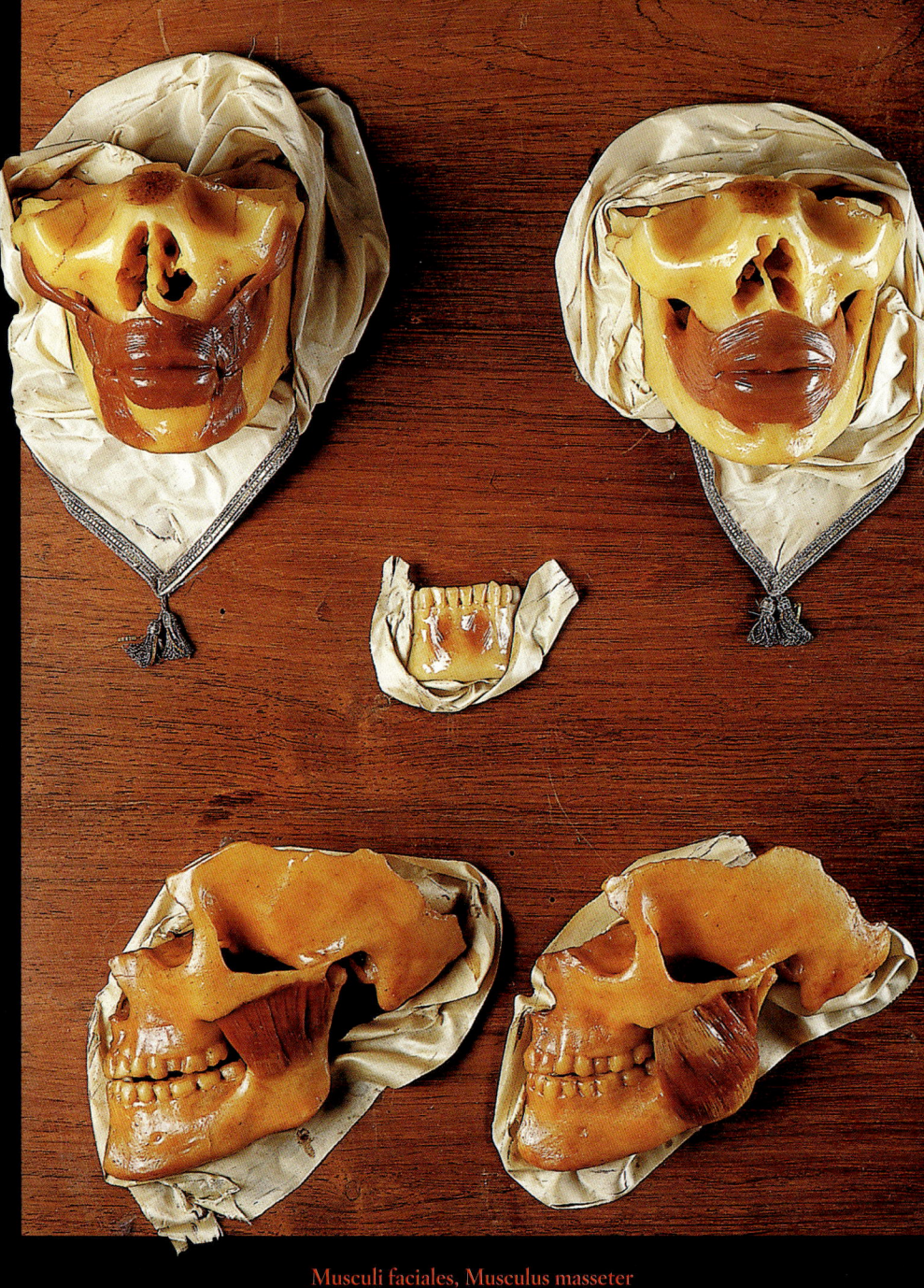

Musculi faciales, Musculus masseter

Muscles of facial expression (upper specimens) and muscle of mastication (lower specimens)

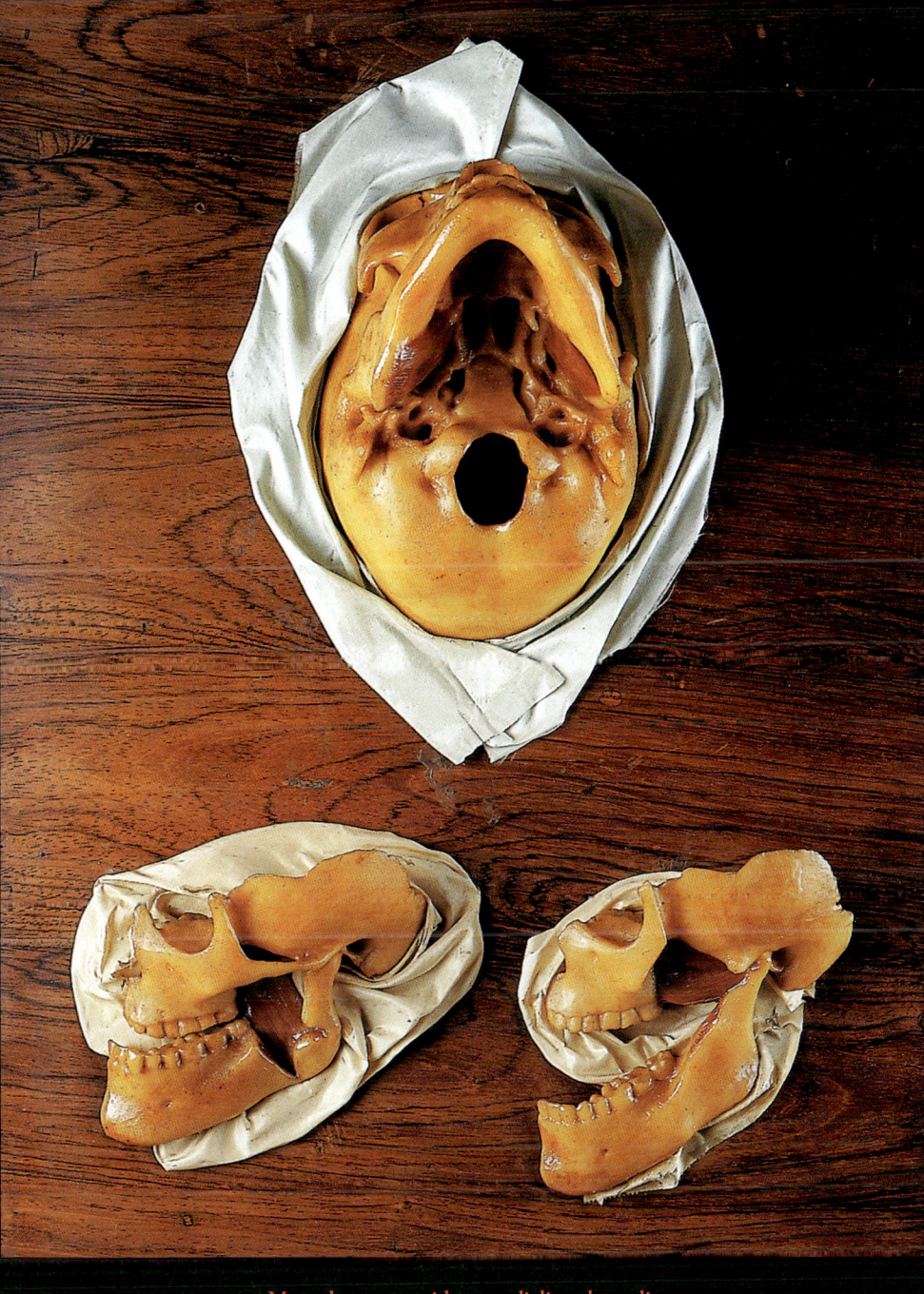

Musculus pterygoideus medialis et lateralis
The more deeply placed muscles of mastication

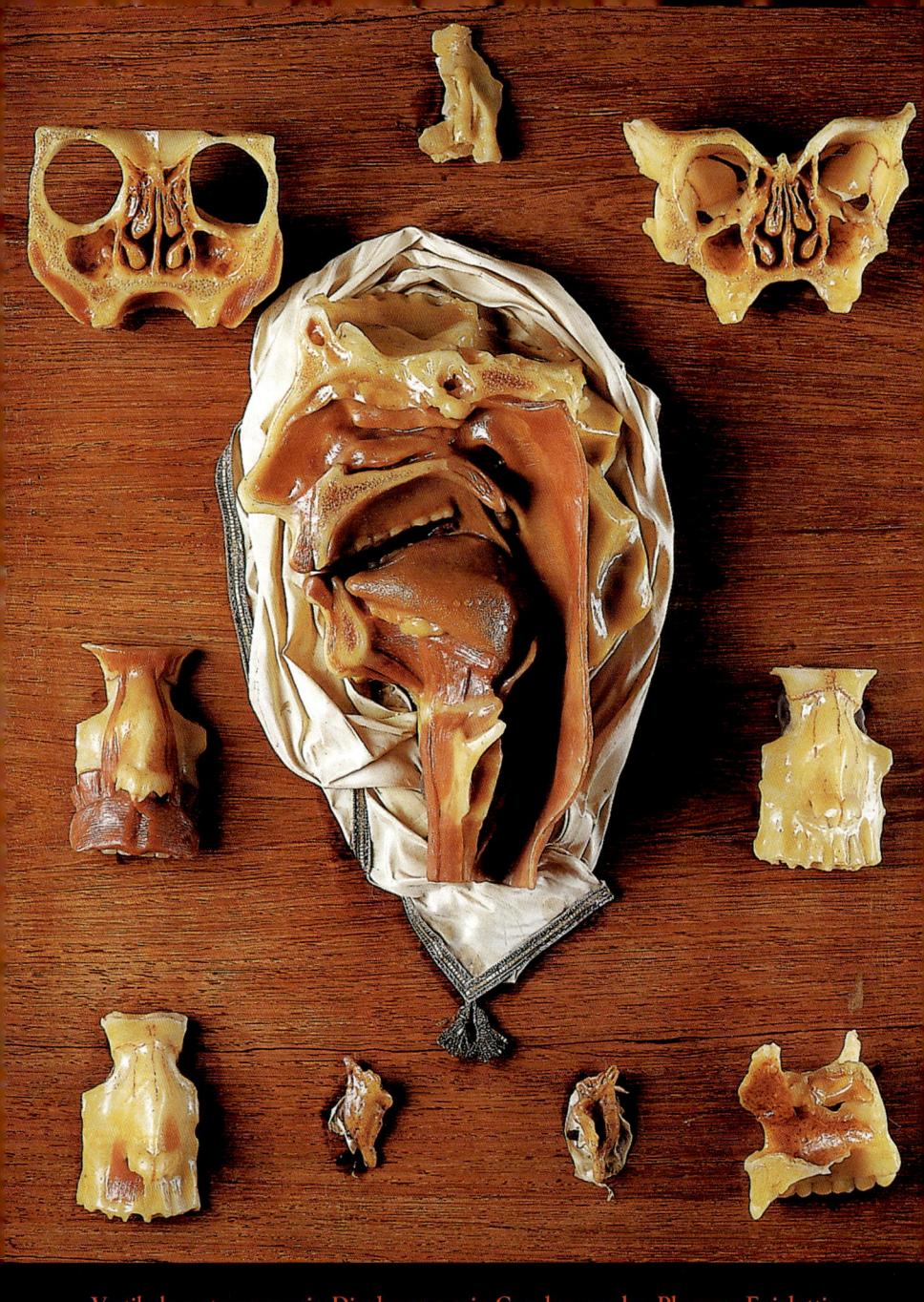

Vestibulum et cavum oris, Diaphragma oris, Conchae nasales, Pharynx, Epiglottis

Middle specimen: longitudinal section through the facial skeleton showing the oral cavity and throat.
Above and to the side one can see the nasal conchae, orbital cavities and maxillary sinuses.

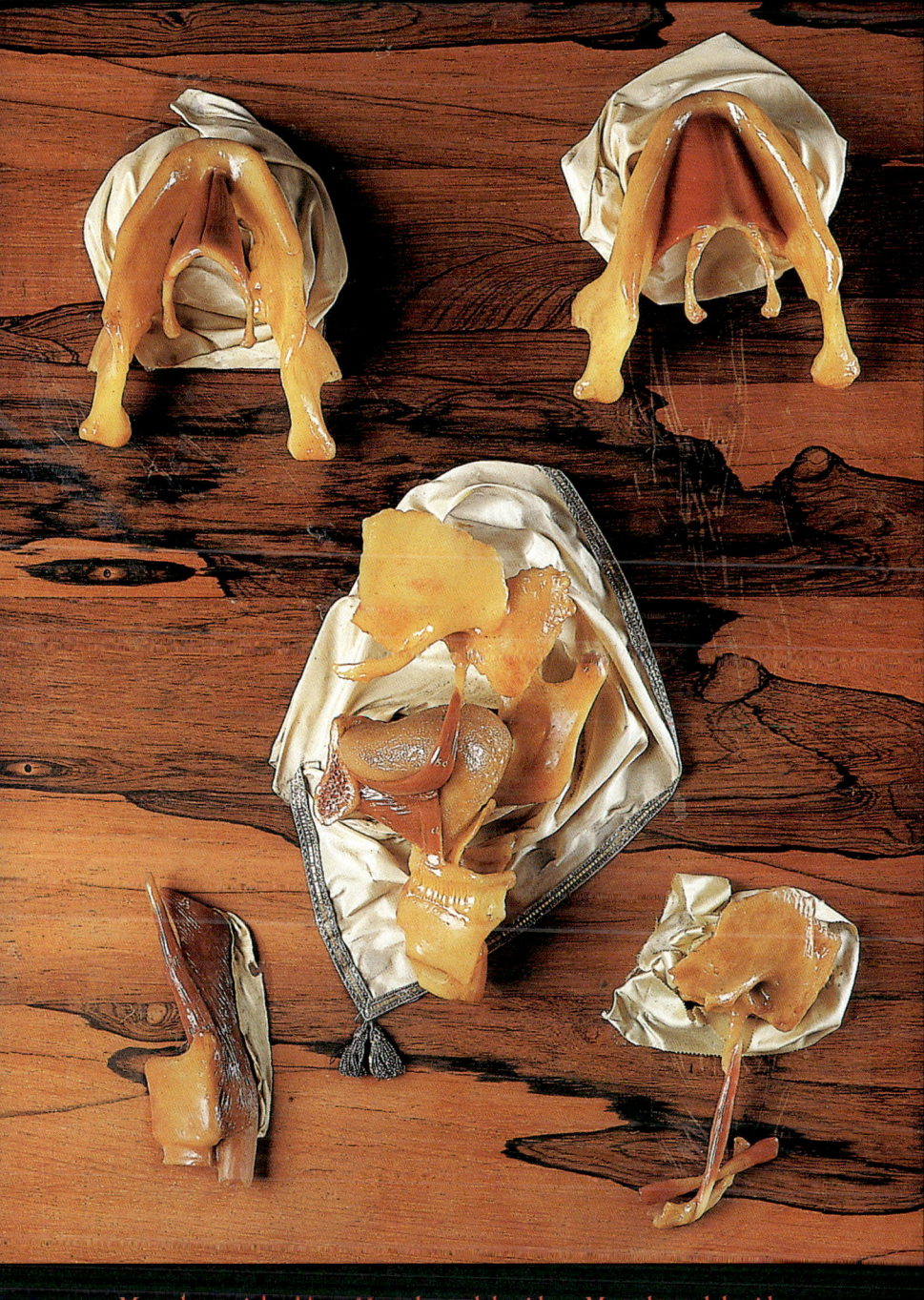

Musculus geniohyoideus, Musculus mylohyoideus, Musculus stylohyoideus
Muscles attached to the hyoid bone

127

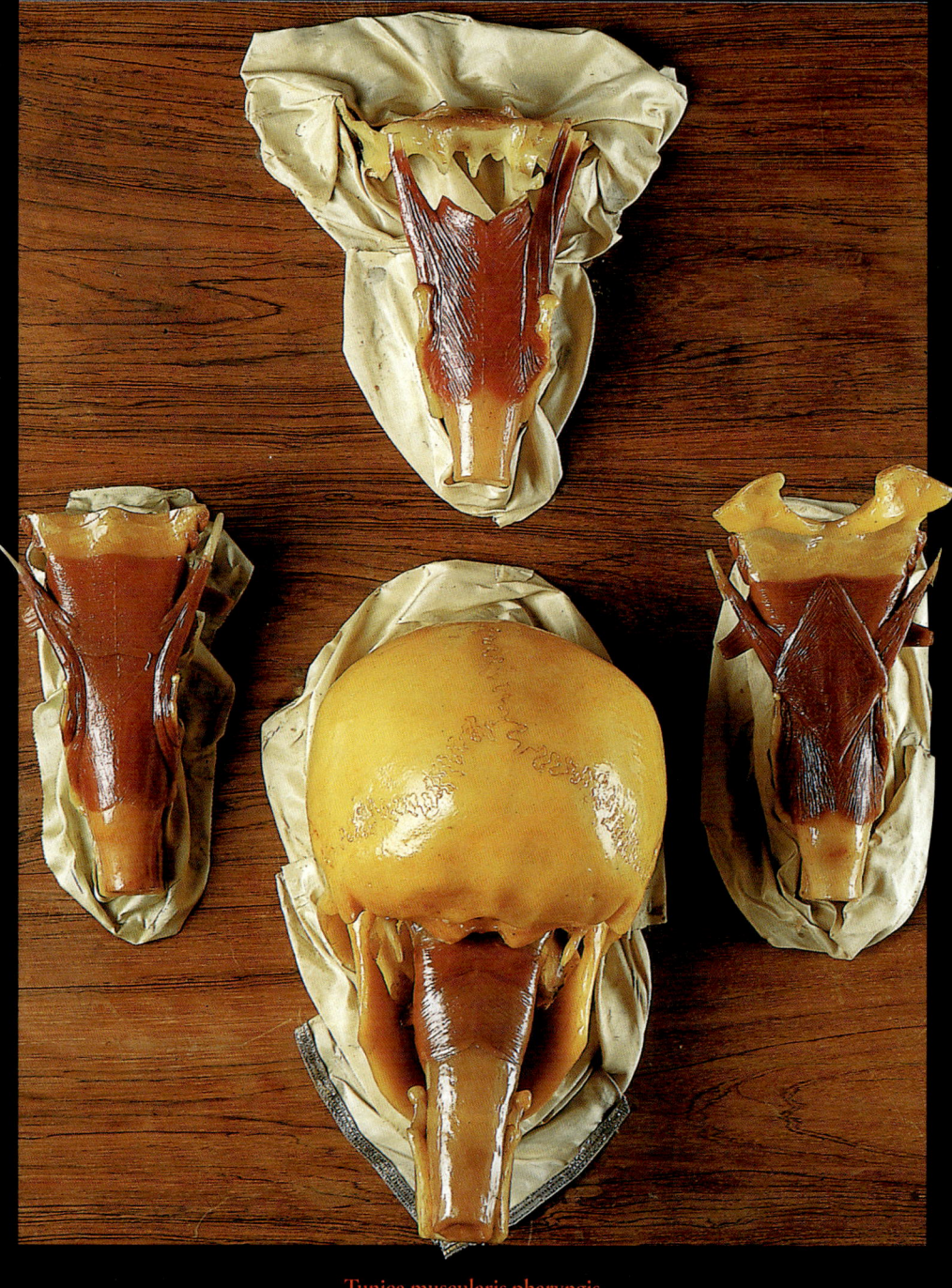

Tunica muscularis pharyngis

Muscles of the throat seen from behind to show the arrangement of their various layers

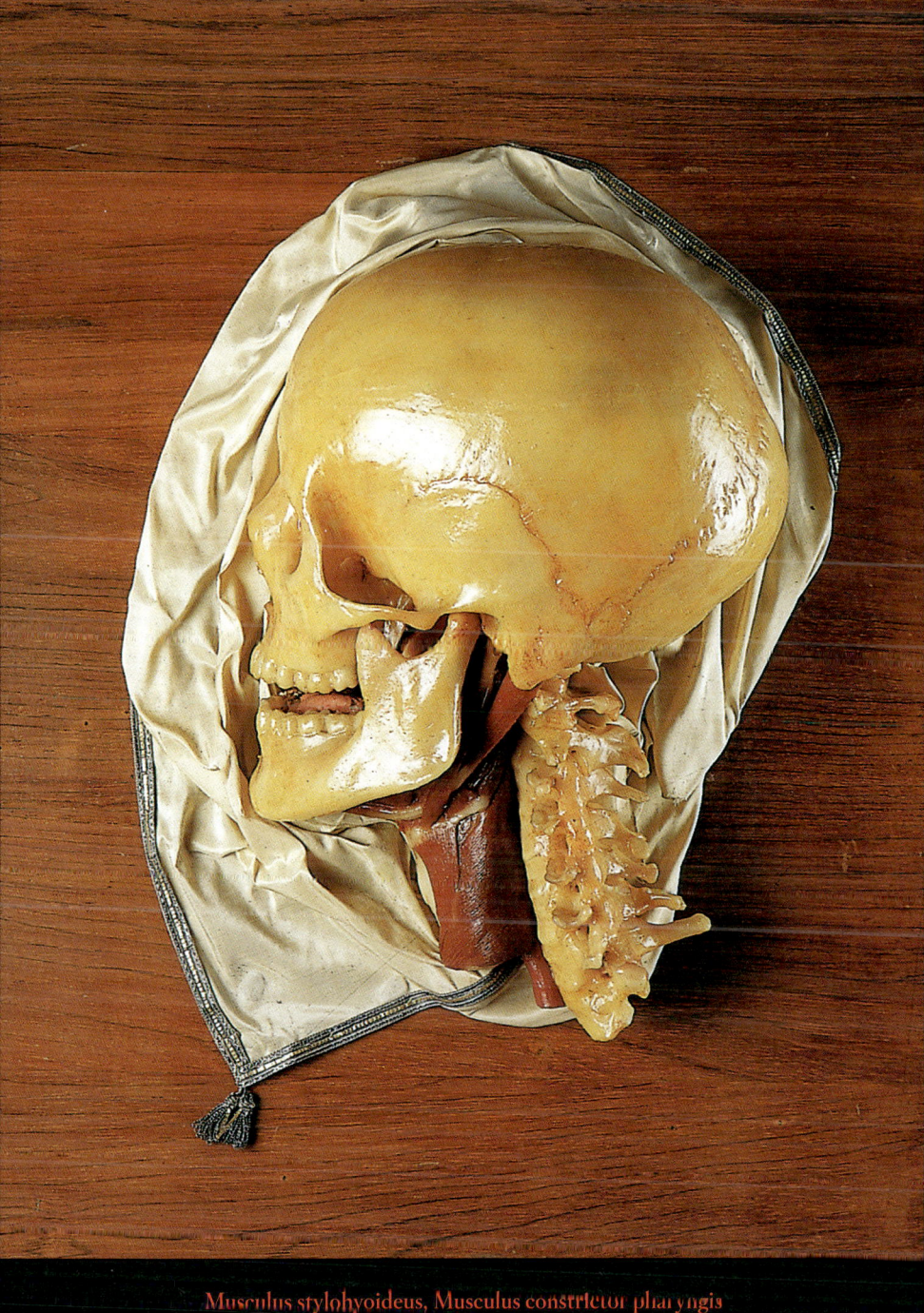

Musculus stylohyoideus, Musculus constrictor pharyngis
Muscles of the throat seen from the side

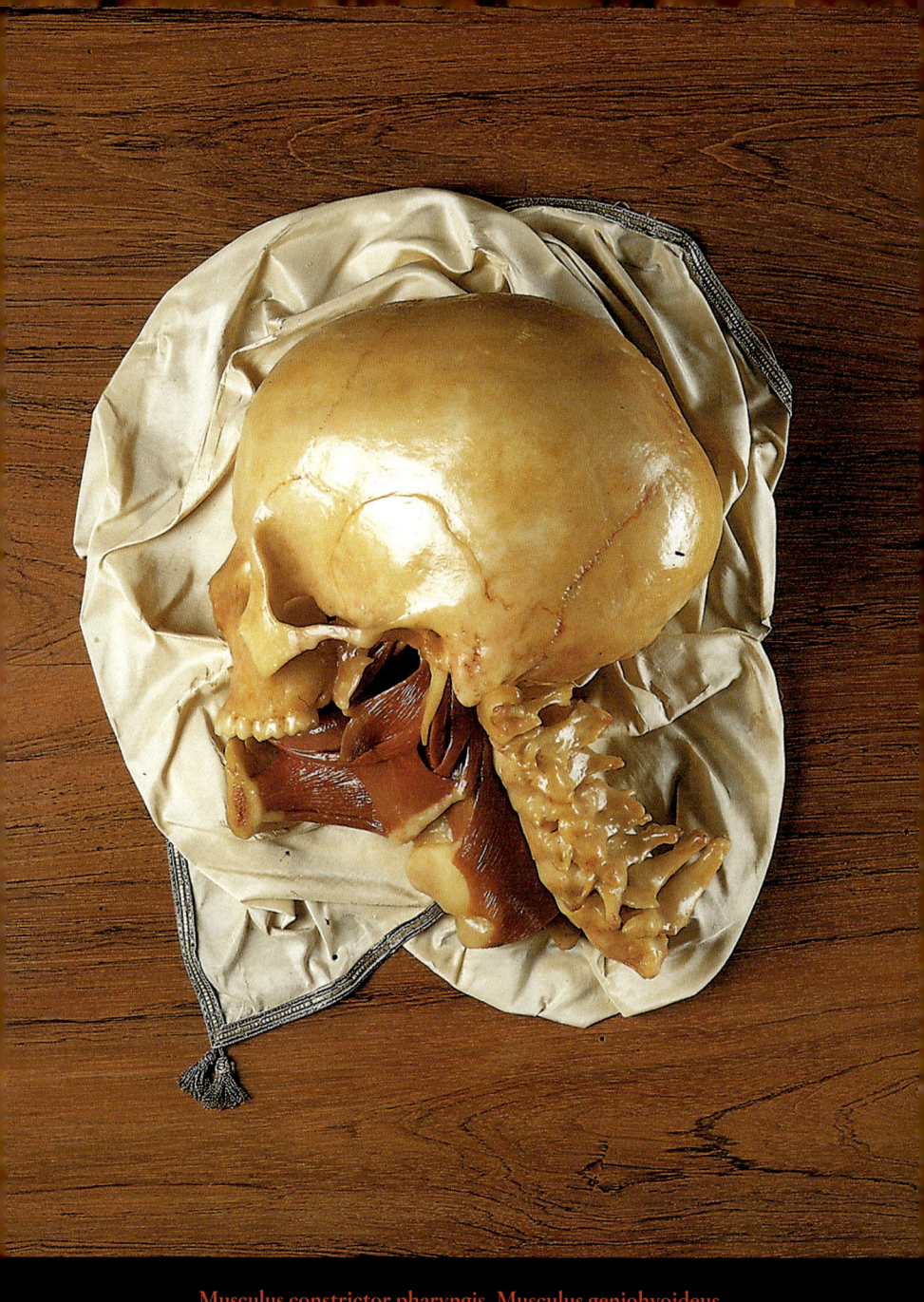

**Musculus constrictor pharyngis, Musculus geniohyoideus,
Musculus genioglossus, Musculus hyoglossus**

Deeper layer of the muscles of the floor of the mouth, tongue and throat seen from the side

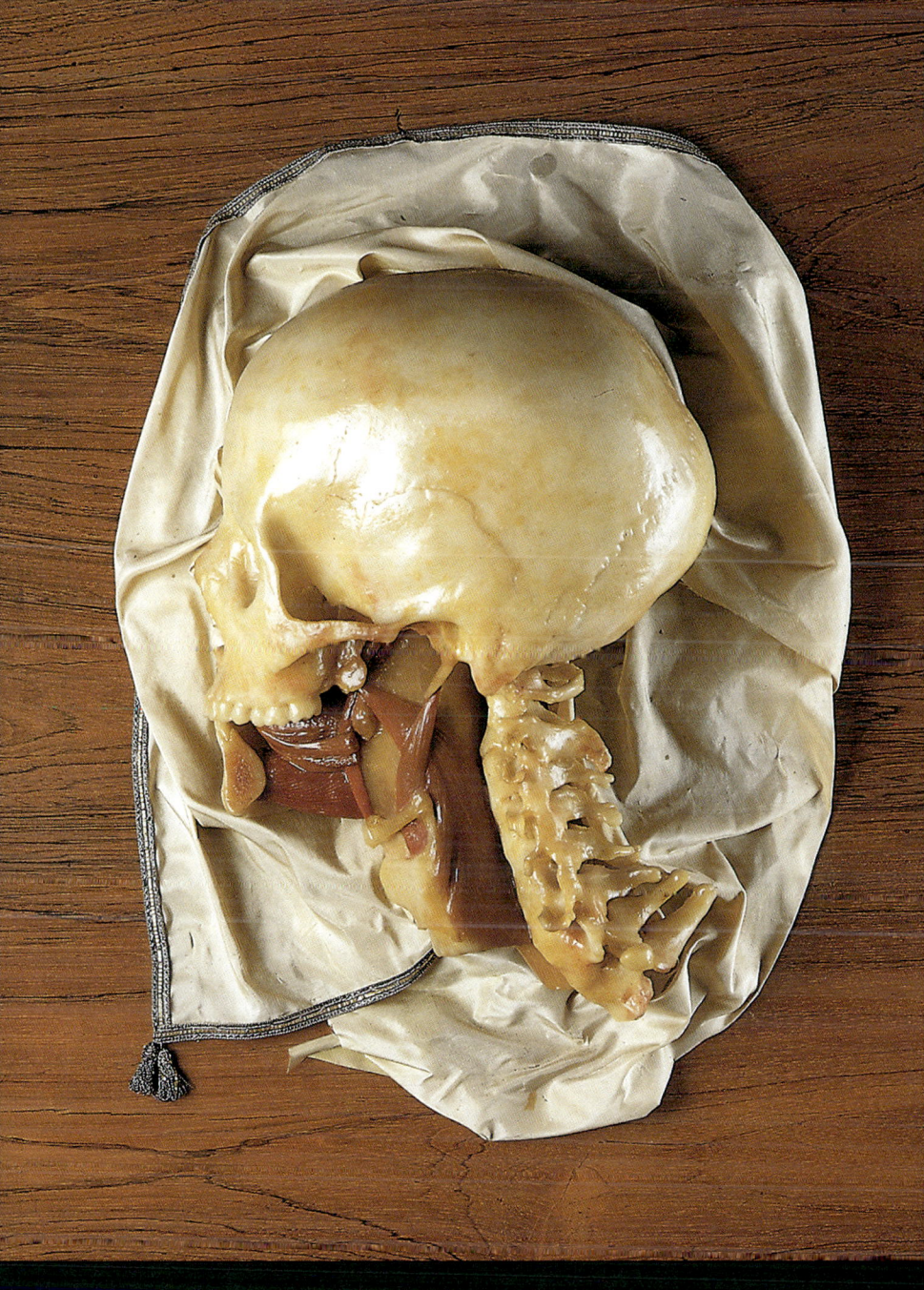

Musculus genioglossus, Musculus stylohyoideus, Musculus constrictor pharyngis

The deeper layer of muscles of the tongue, floor of the mouth and pharynx seen from the side
after removal of part of the bone of the lower jaw

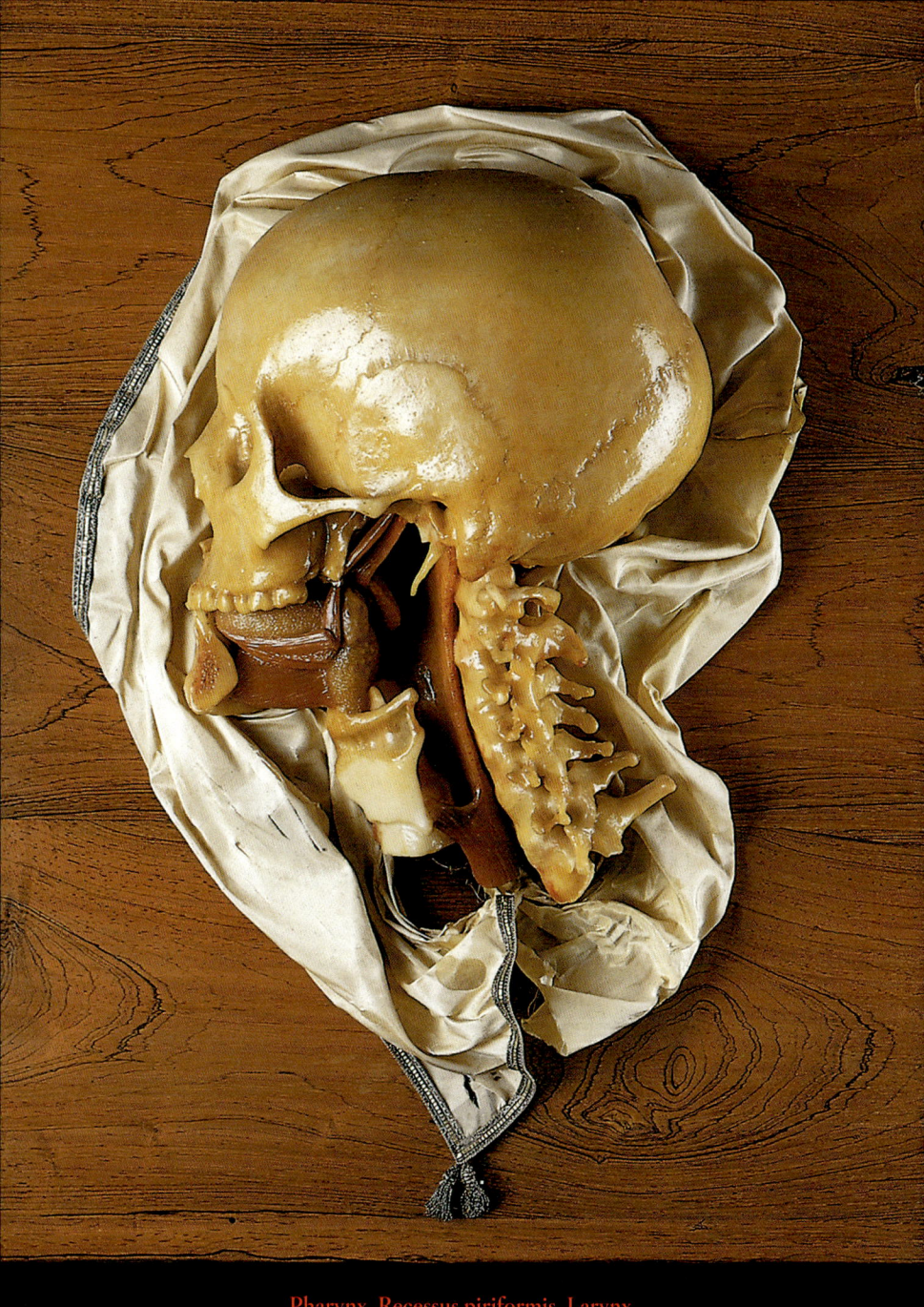

Pharynx, Recessus piriformis, Larynx

View from the side into the opened pharynx. Parts of the cartilaginous skeleton
of the larynx can be seen

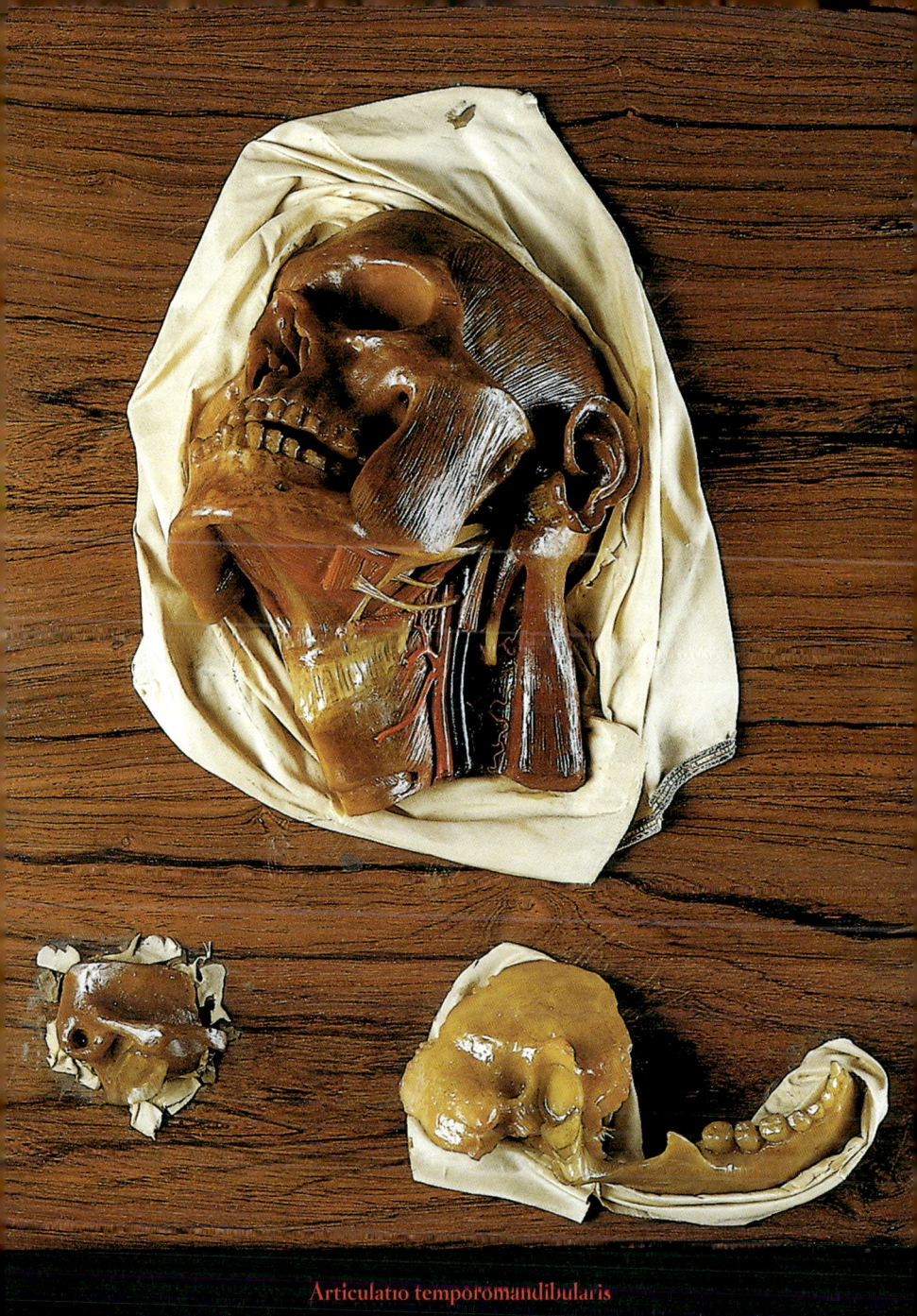

Articulatio temporomandibularis

Display showing the floor of the mouth, muscles of mastication and jaw joint

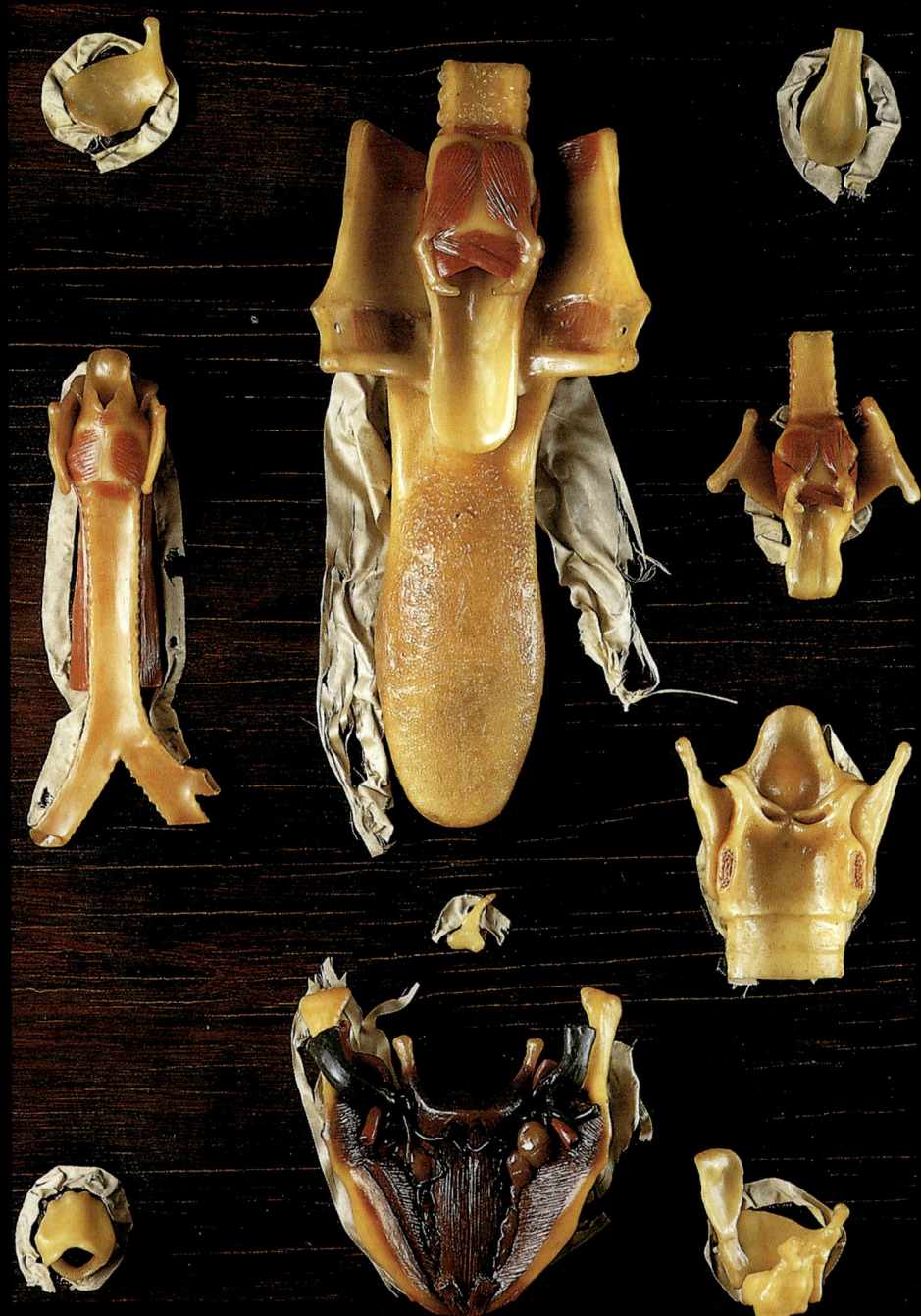

Larynx, Trachea, Lingua, Musculus cricoarytenoideus posterior, Epiglottis, Plica vocalis

Display showing the larynx with ligaments and muscles, the trachea (left specimen)
and the tongue (middle specimen)

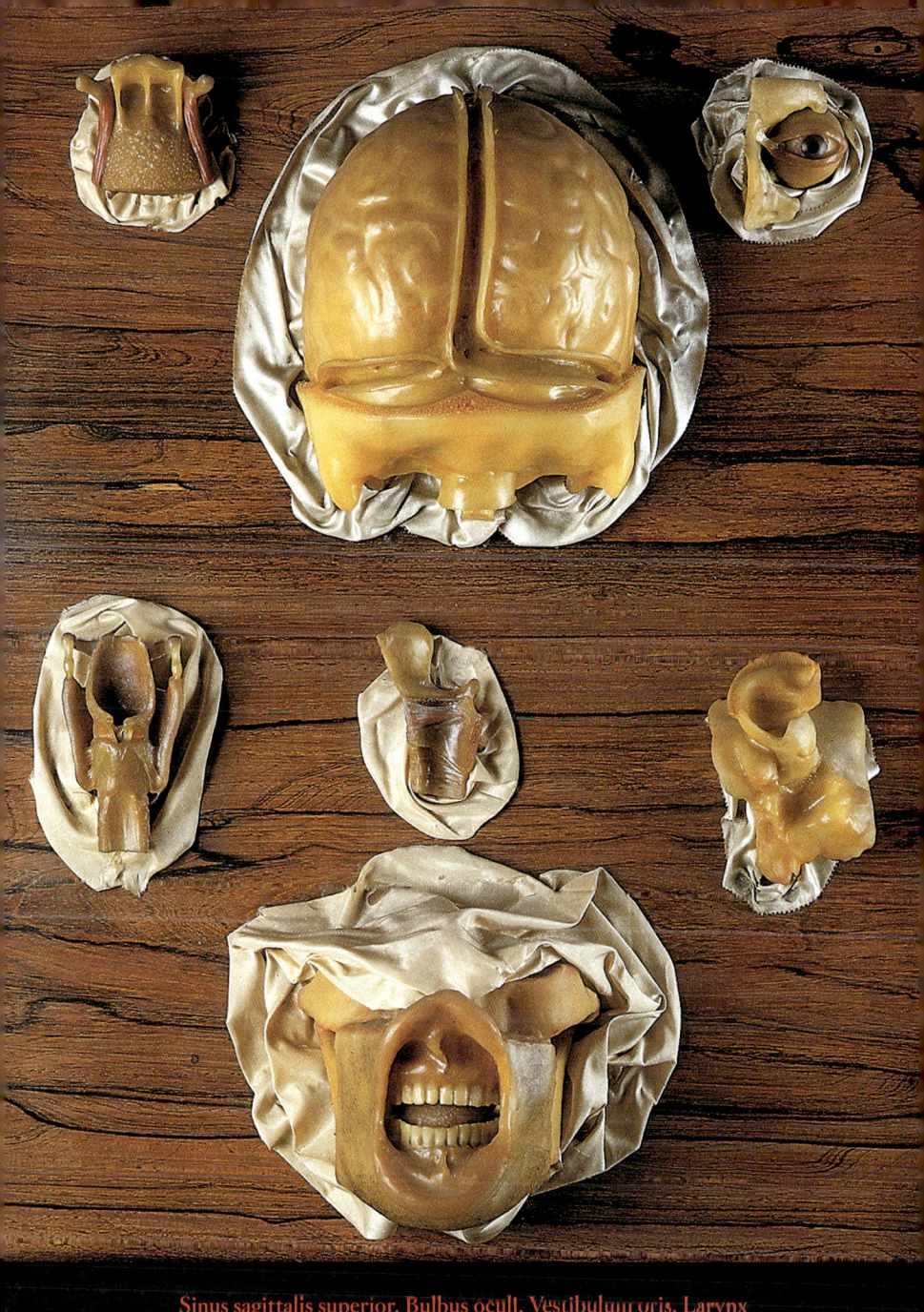

Sinus sagittalis superior, Bulbus oculi, Vestibulum oris, Larynx

Selected specimens from the head and neck region

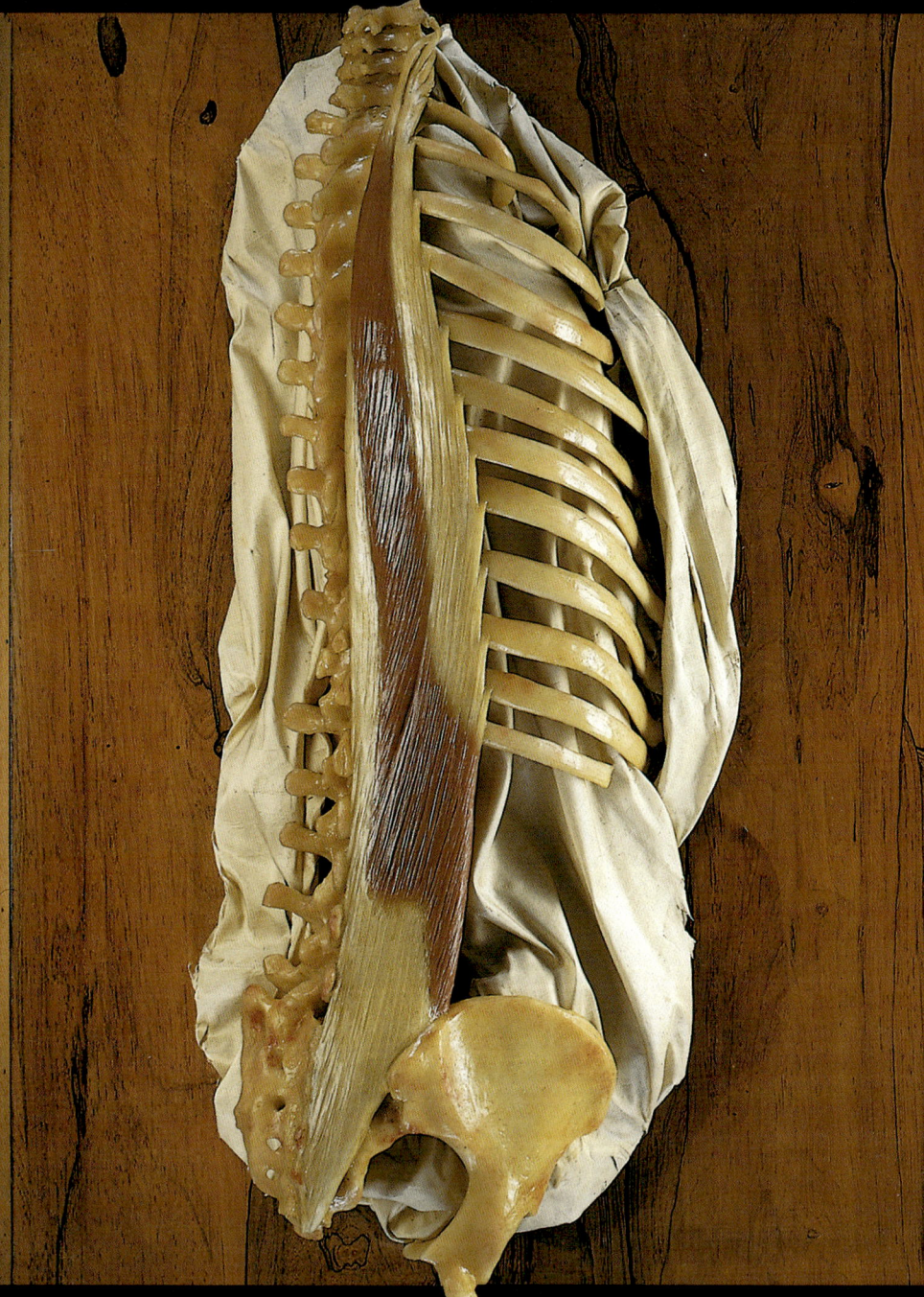

Musculus longissimus

Display showing some muscles of the back

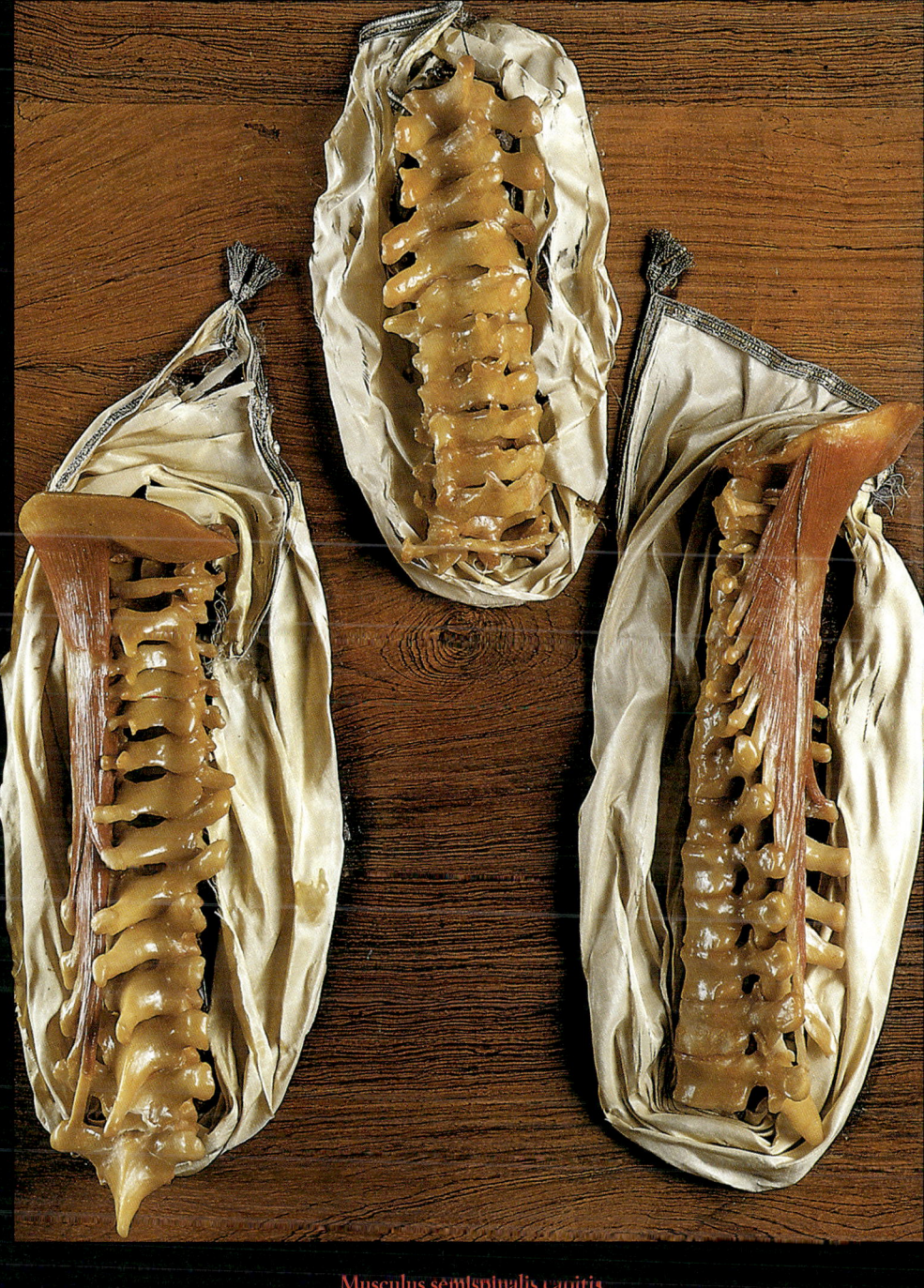

Musculus semispinalis capitis

Display showing a deep extensor muscle of the back

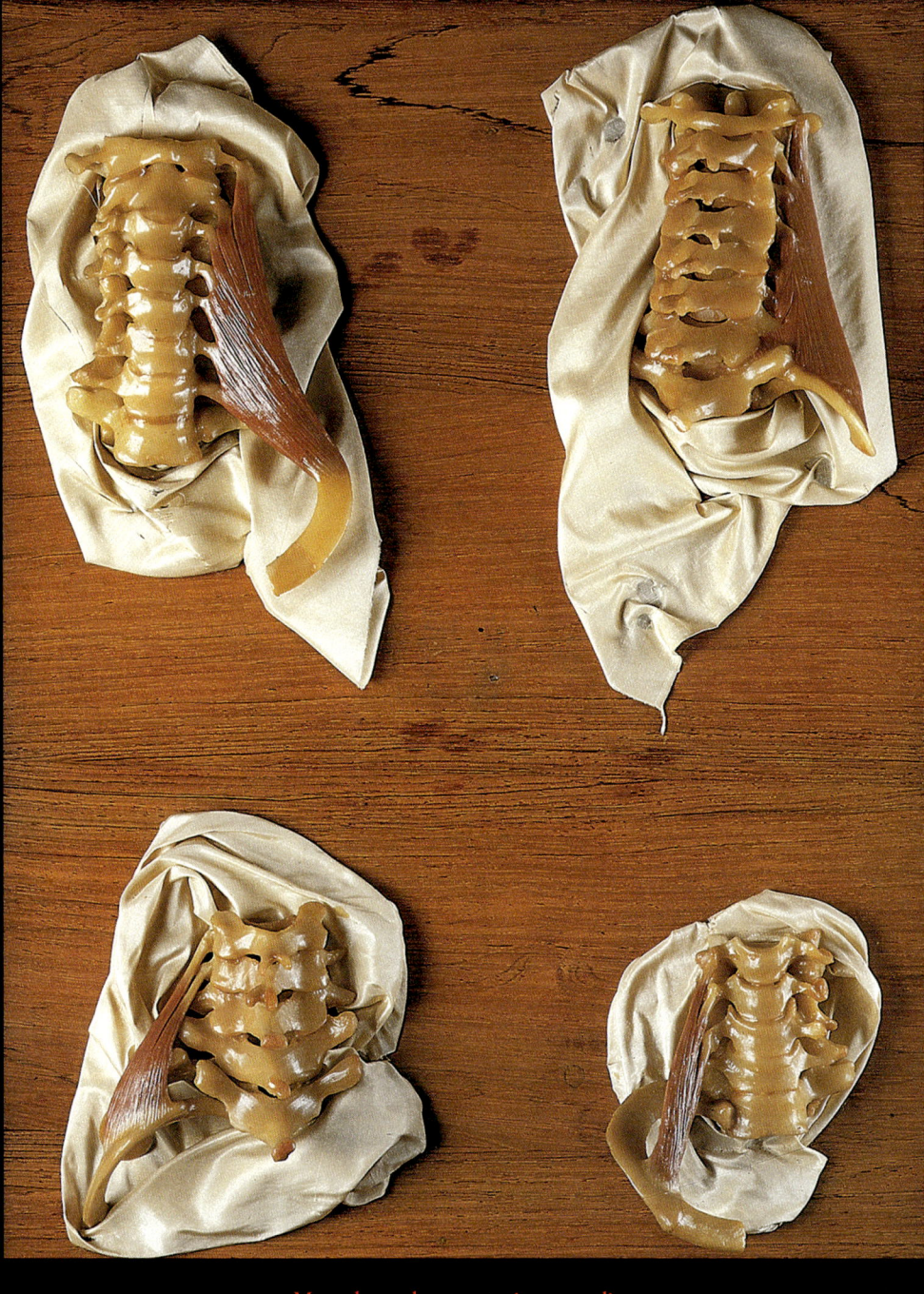

Musculus scalenus anterior et medius

Display showing deep muscles at the side of the neck

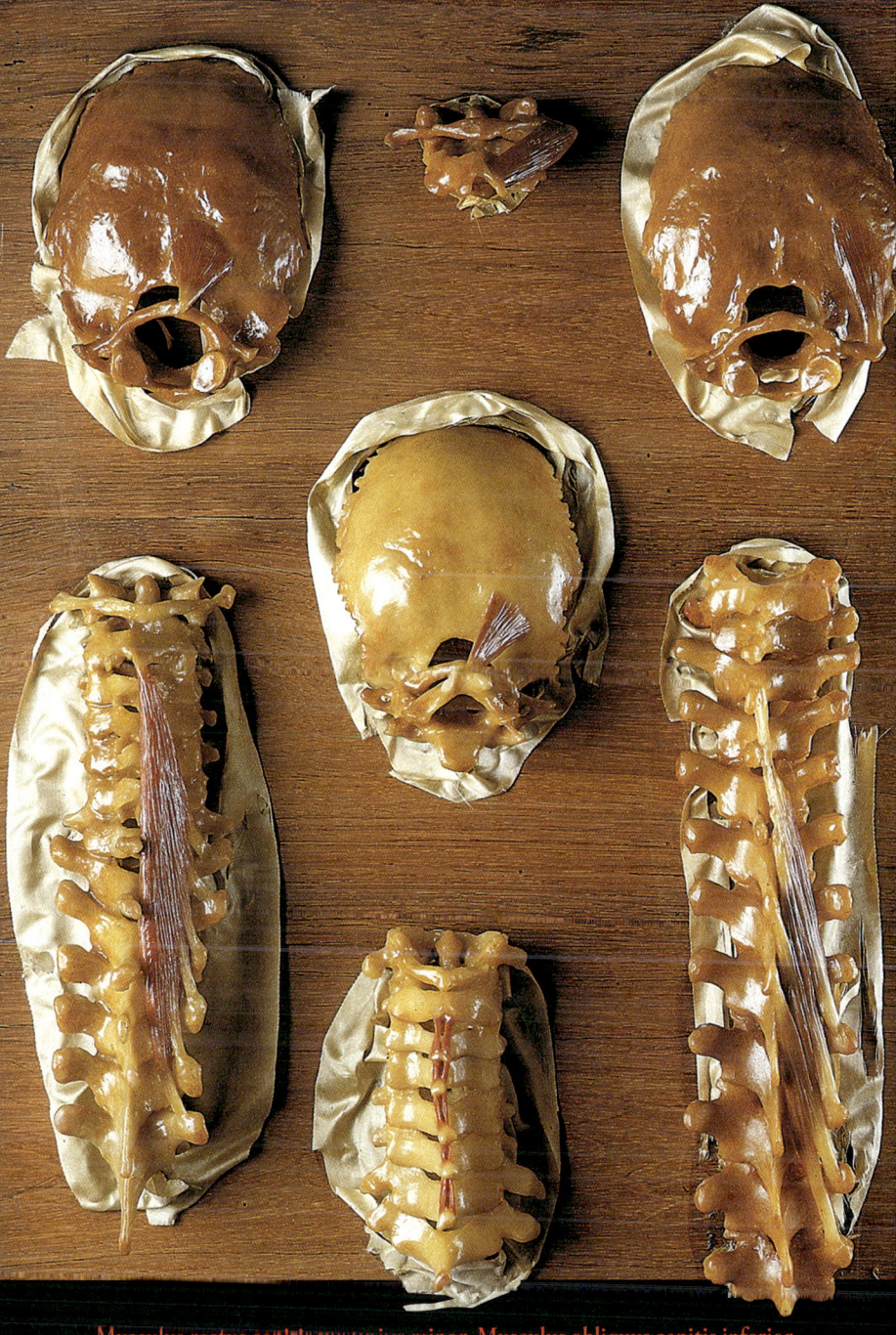

Musculus rectus capitis posterior minor, Musculus obliquus capitis inferior, Musculi interspinales, Musculi rotatores

Display showing part of the deep musculature of the neck and back

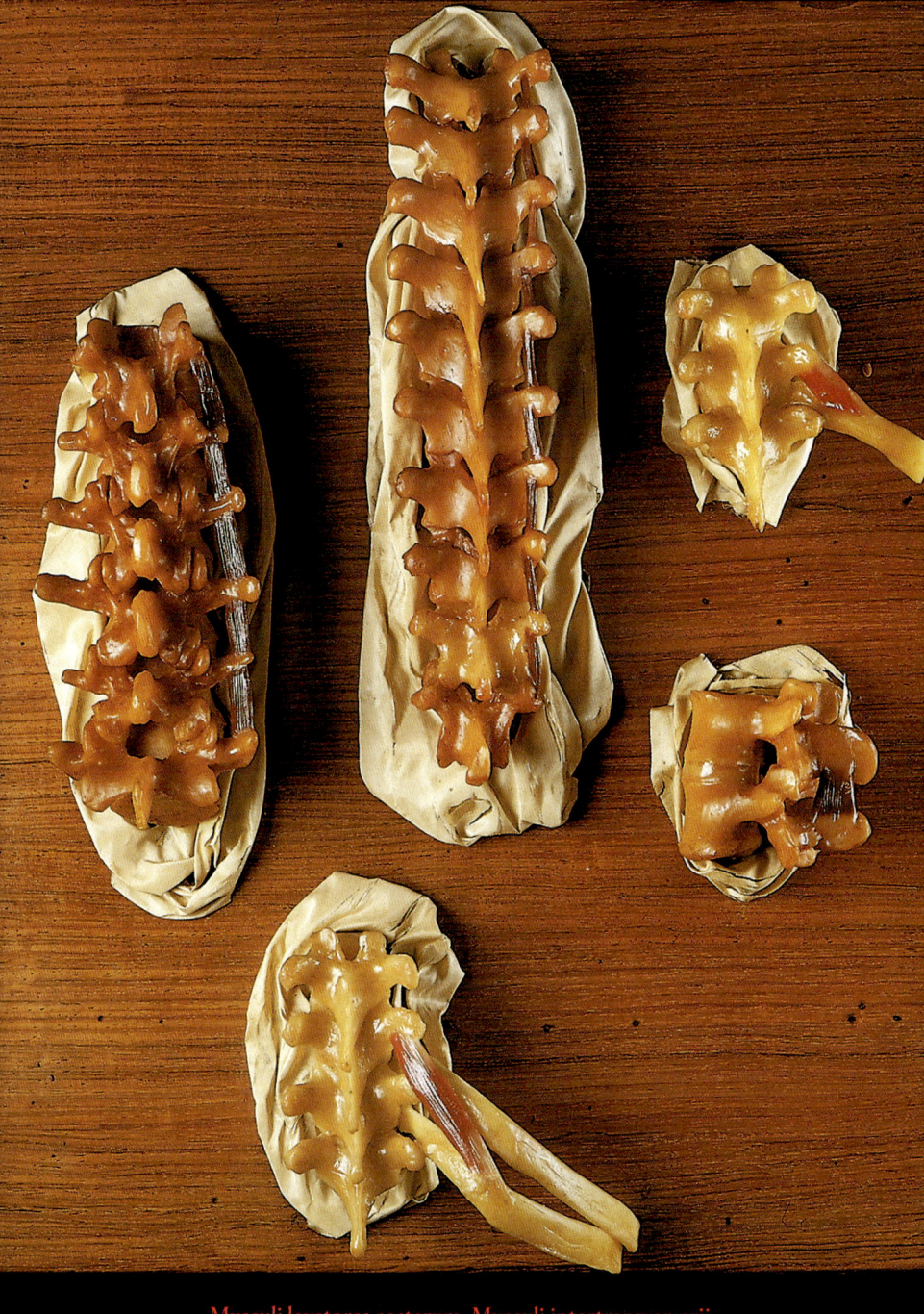

Musculi levatores costarum, Musculi intertransversarii

Display showing part of the deep musculature of the neck and back

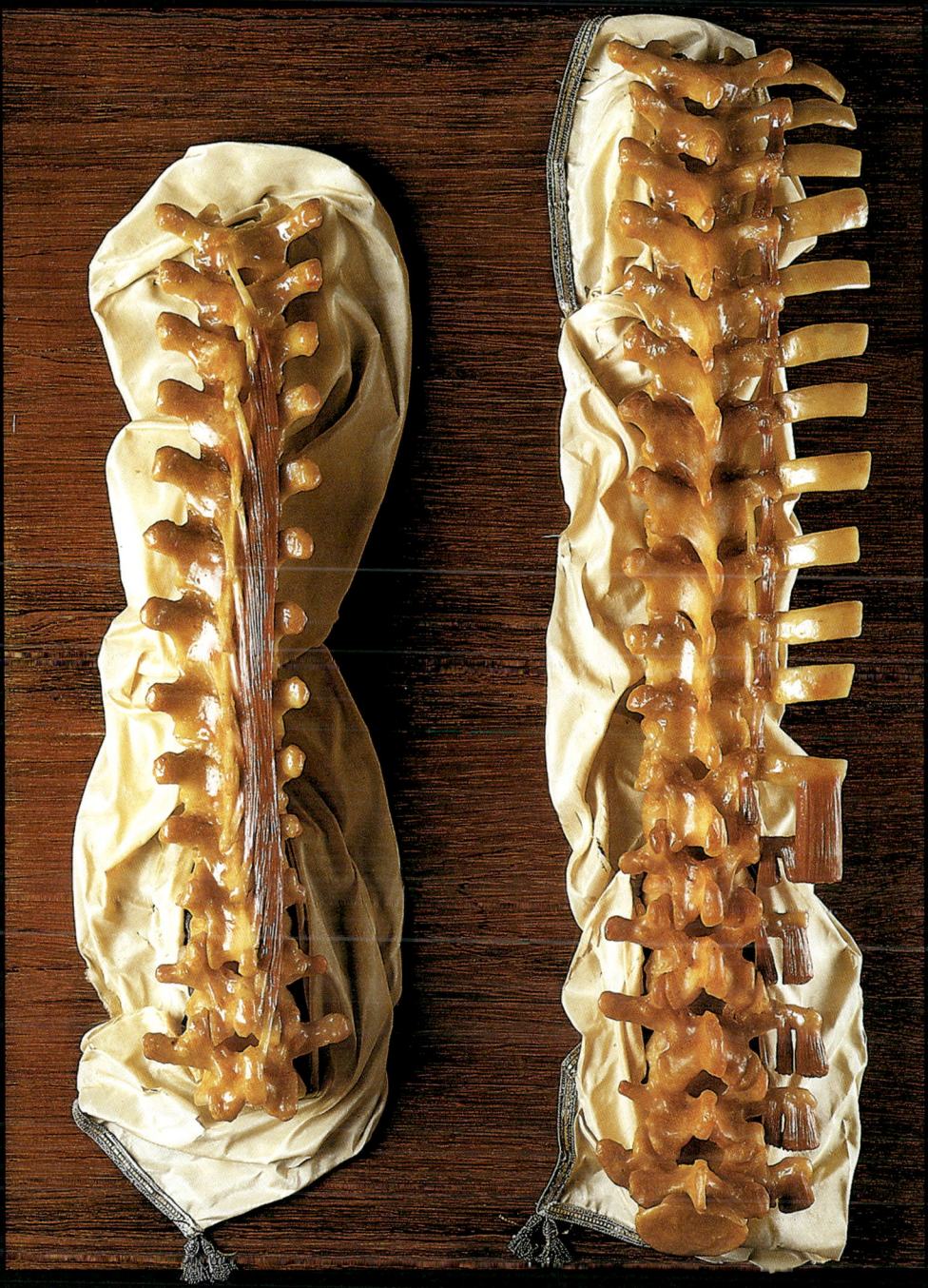

Musculus spinalis, Musculi intertransversarii

Display showing deep muscles of the back in the thoracic
and lumbar regions of the vertebral column

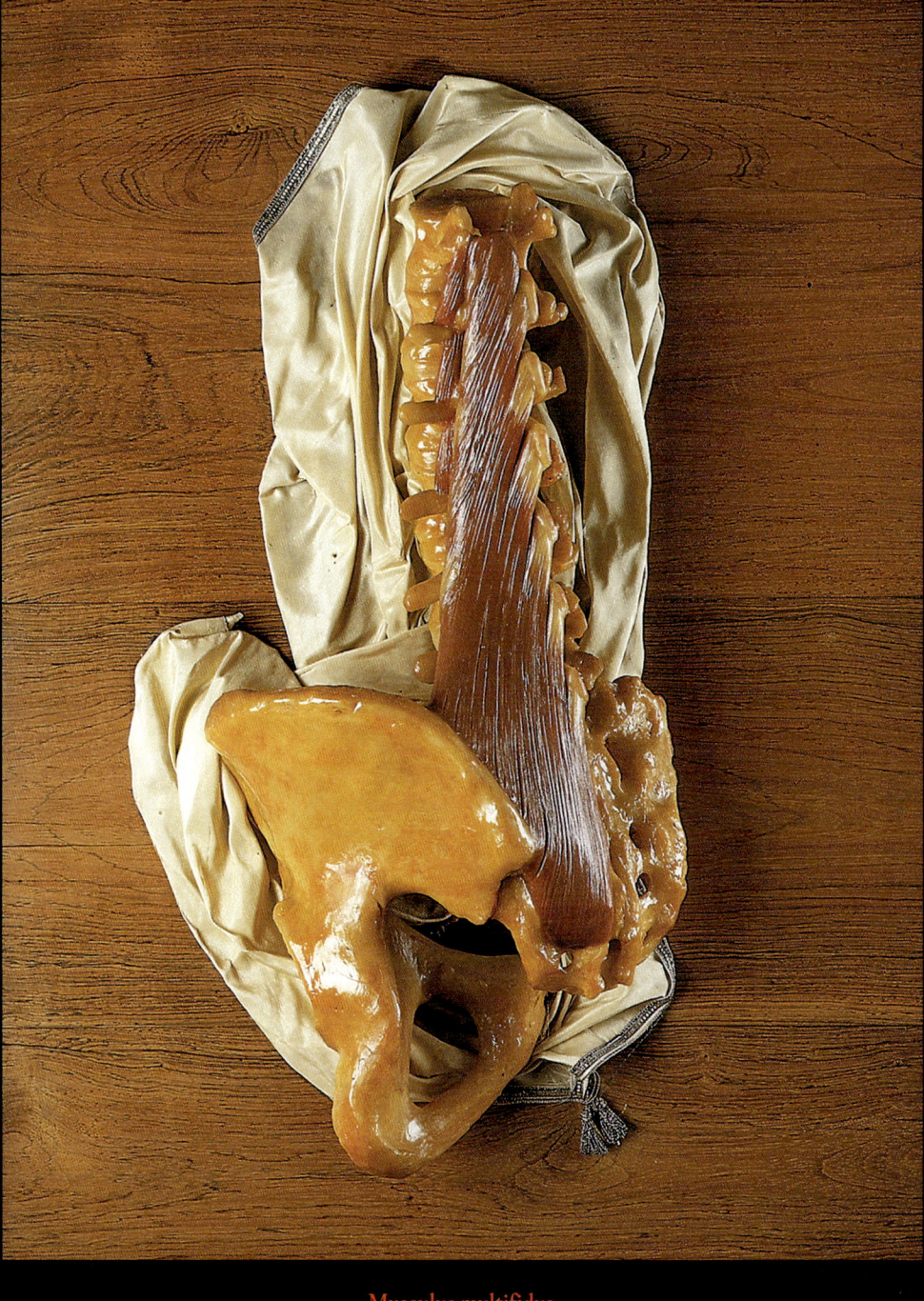

Musculus multifidus

Deep muscles of the back in the lumbar region of the vertebral column

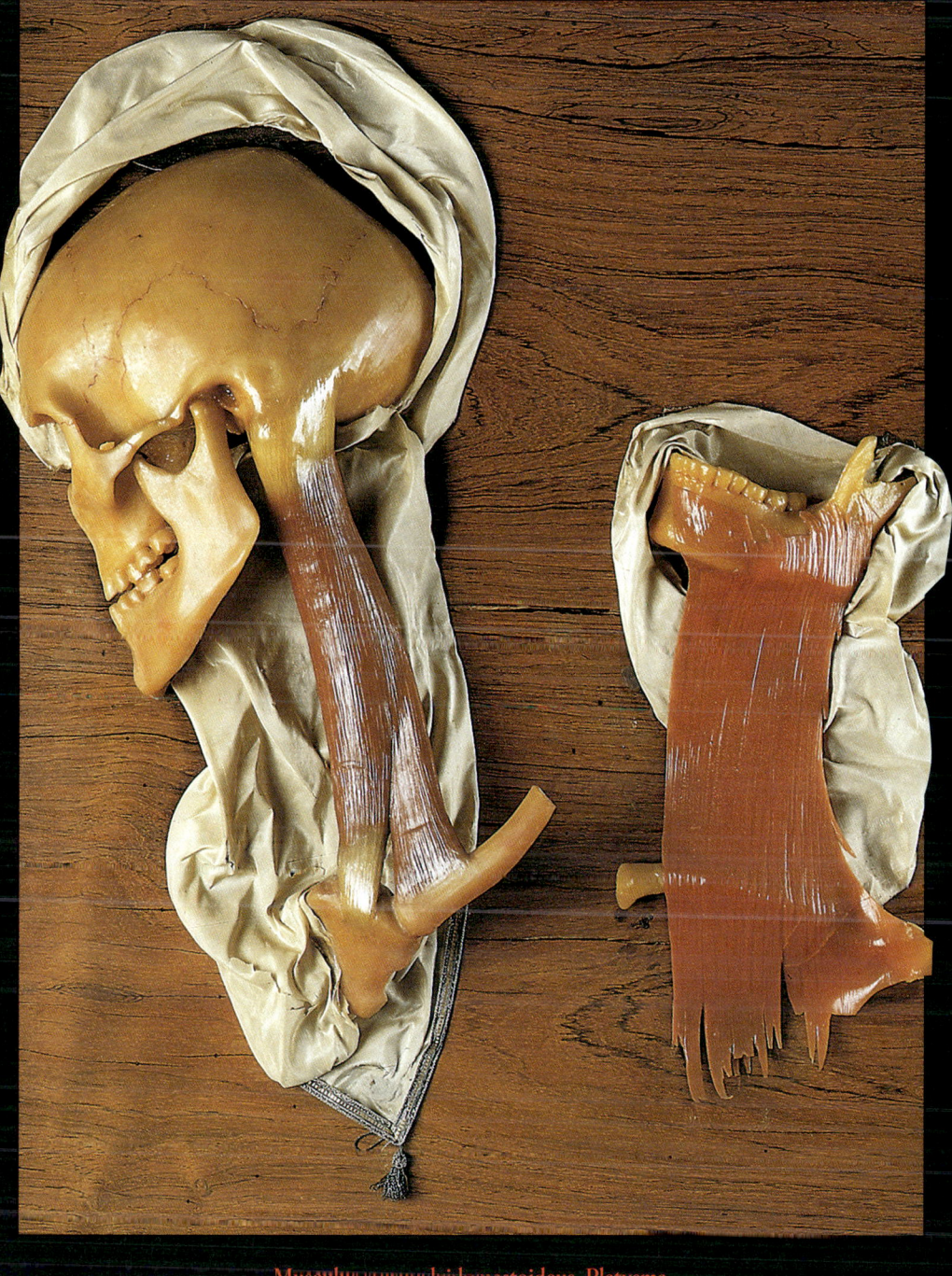

Musculus sternocleidomastoideus, Platysma

A muscle which rotates the head (left specimen) and one which lies in the skin
of the neck (right specimen)

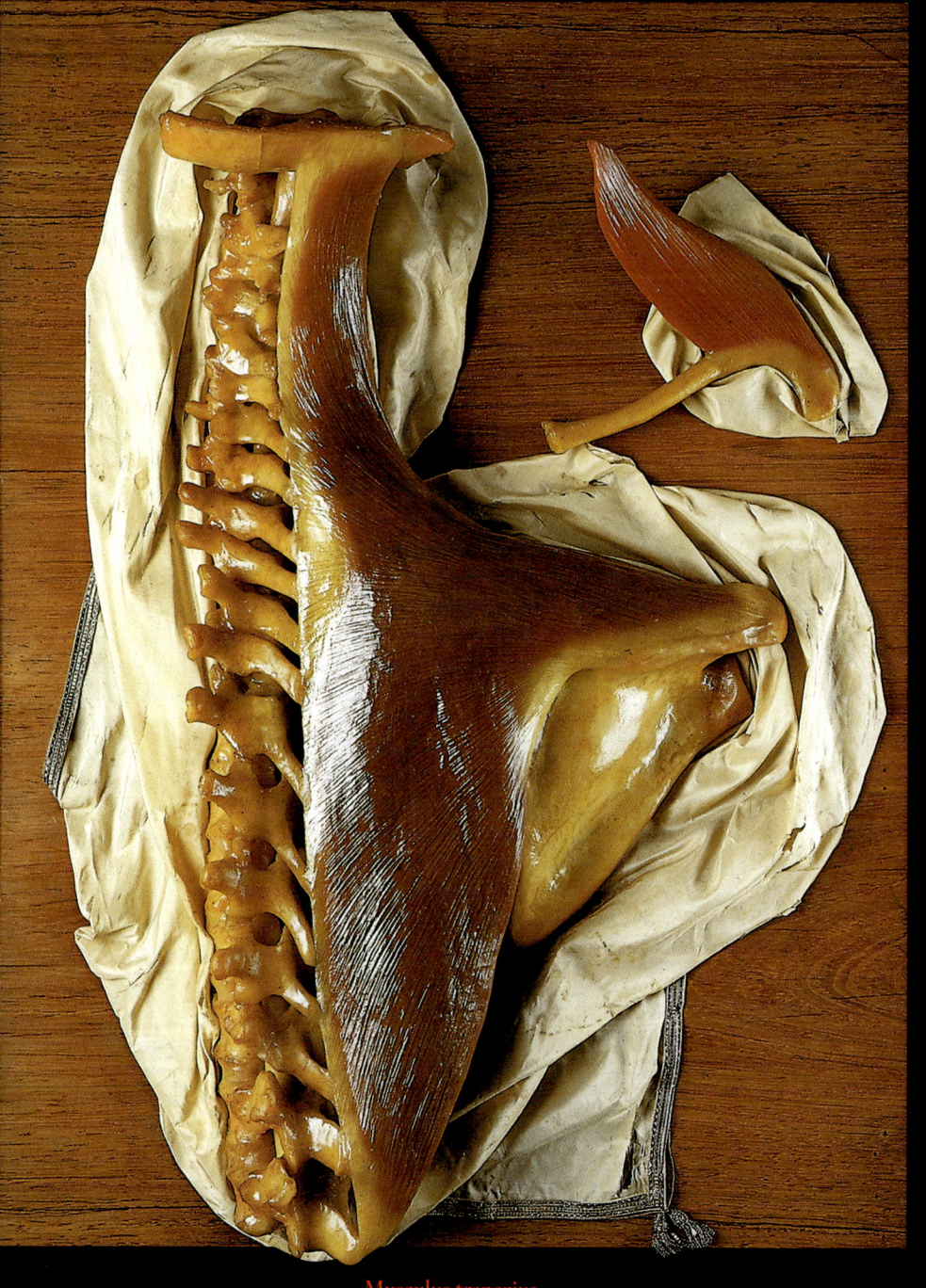

Musculus trapezius

View of the trapezius muscle

Detail from page 212 →

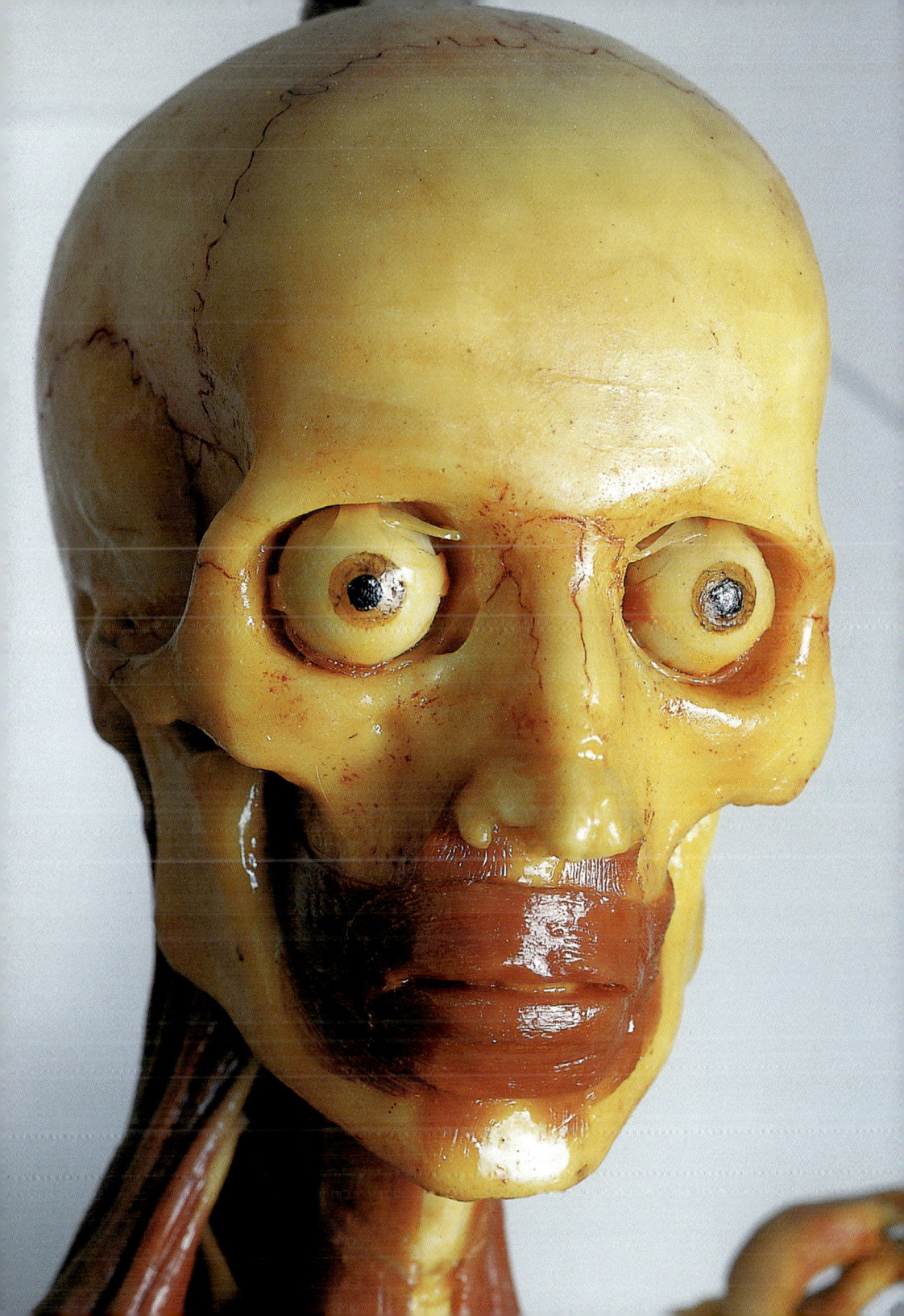

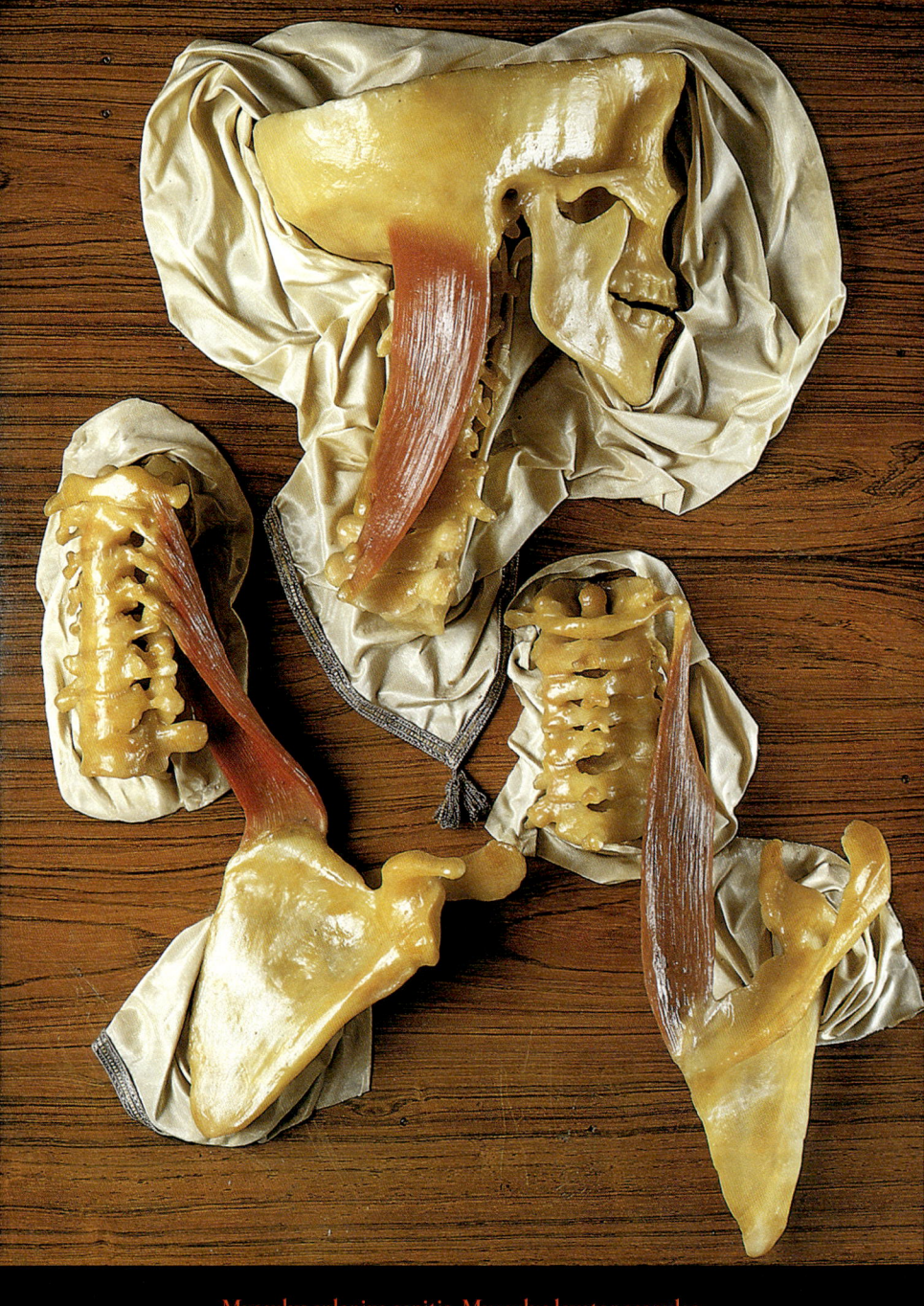

Musculus splenius capitis, Musculus levator scapulae

View of the flat muscles of the back

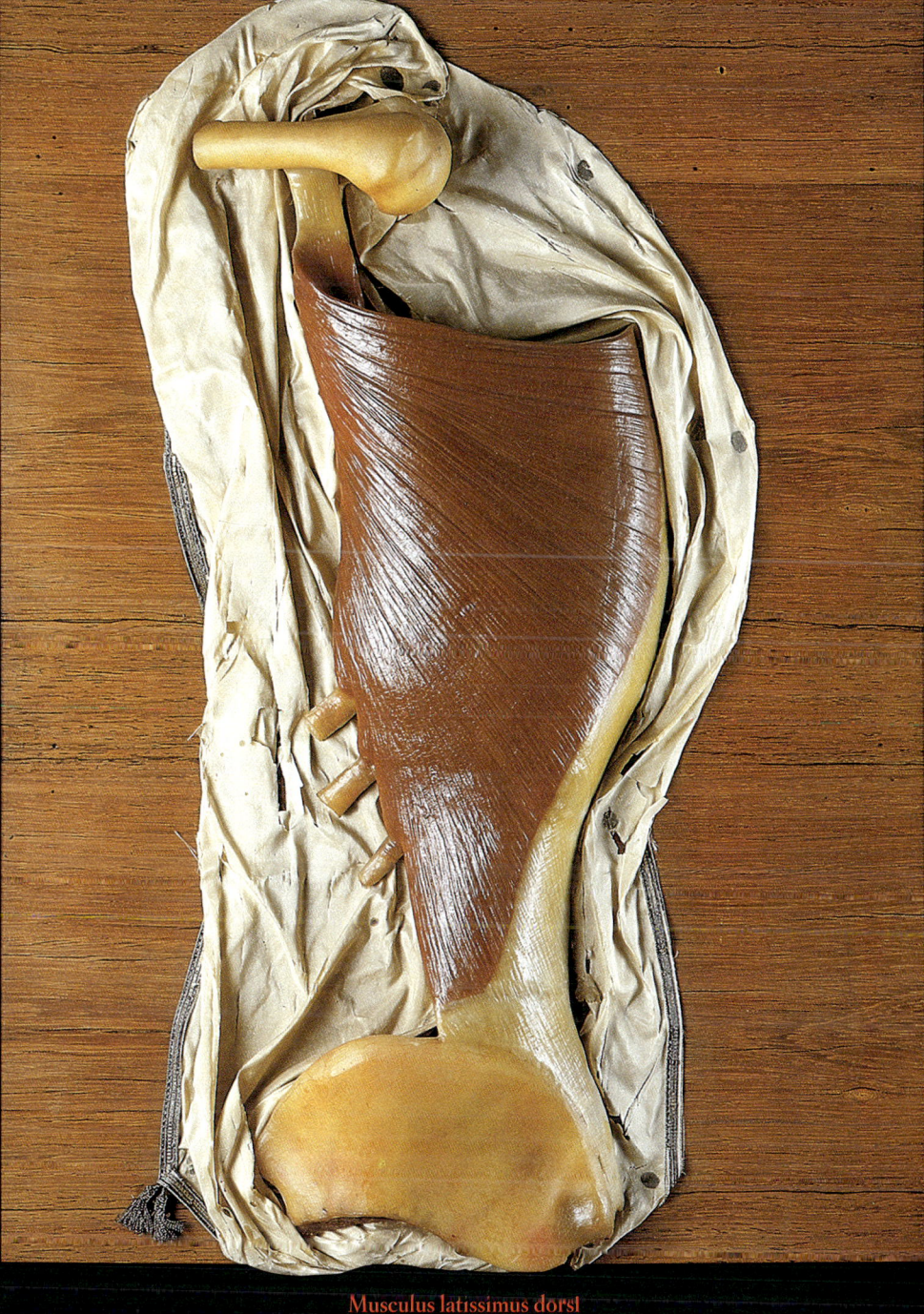

Musculus latissimus dorsi

View of a superficial muscle of the back

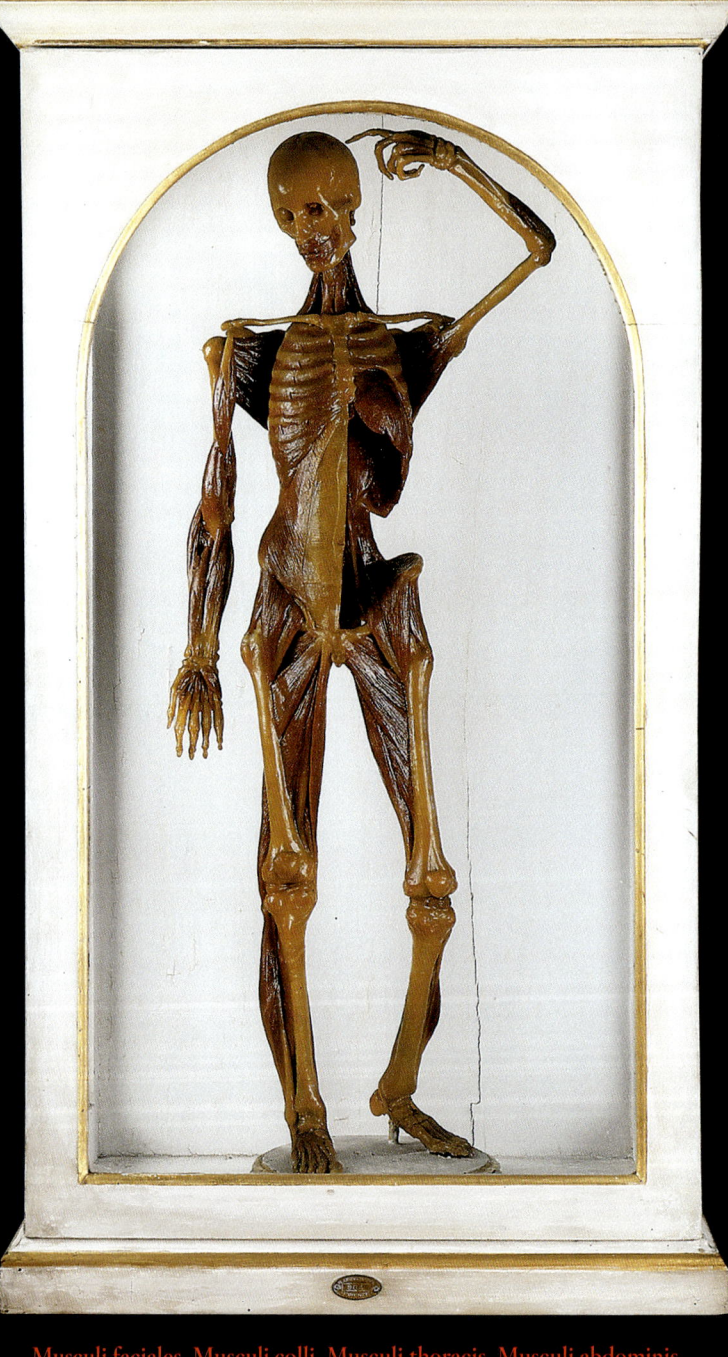

Musculi faciales, Musculi colli, Musculi thoracis, Musculi abdominis, Musculi membri superioris, Musculi membri inferioris
Whole body specimen with the deep layer of muscles displayed

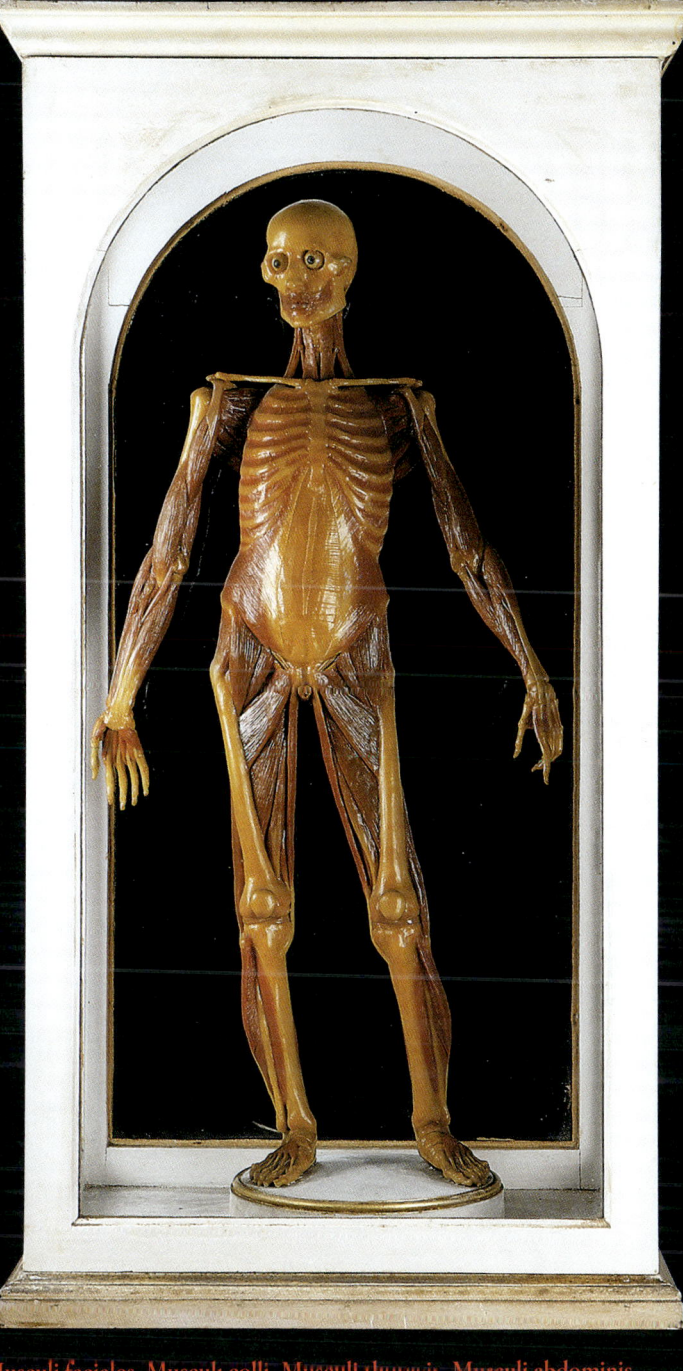

Musculi faciales, Musculi colli, Musculi thoracis, Musculi abdominis, Musculi membri superioris, Musculi membri inferioris

Whole body specimen with the deep layer of muscles displayed

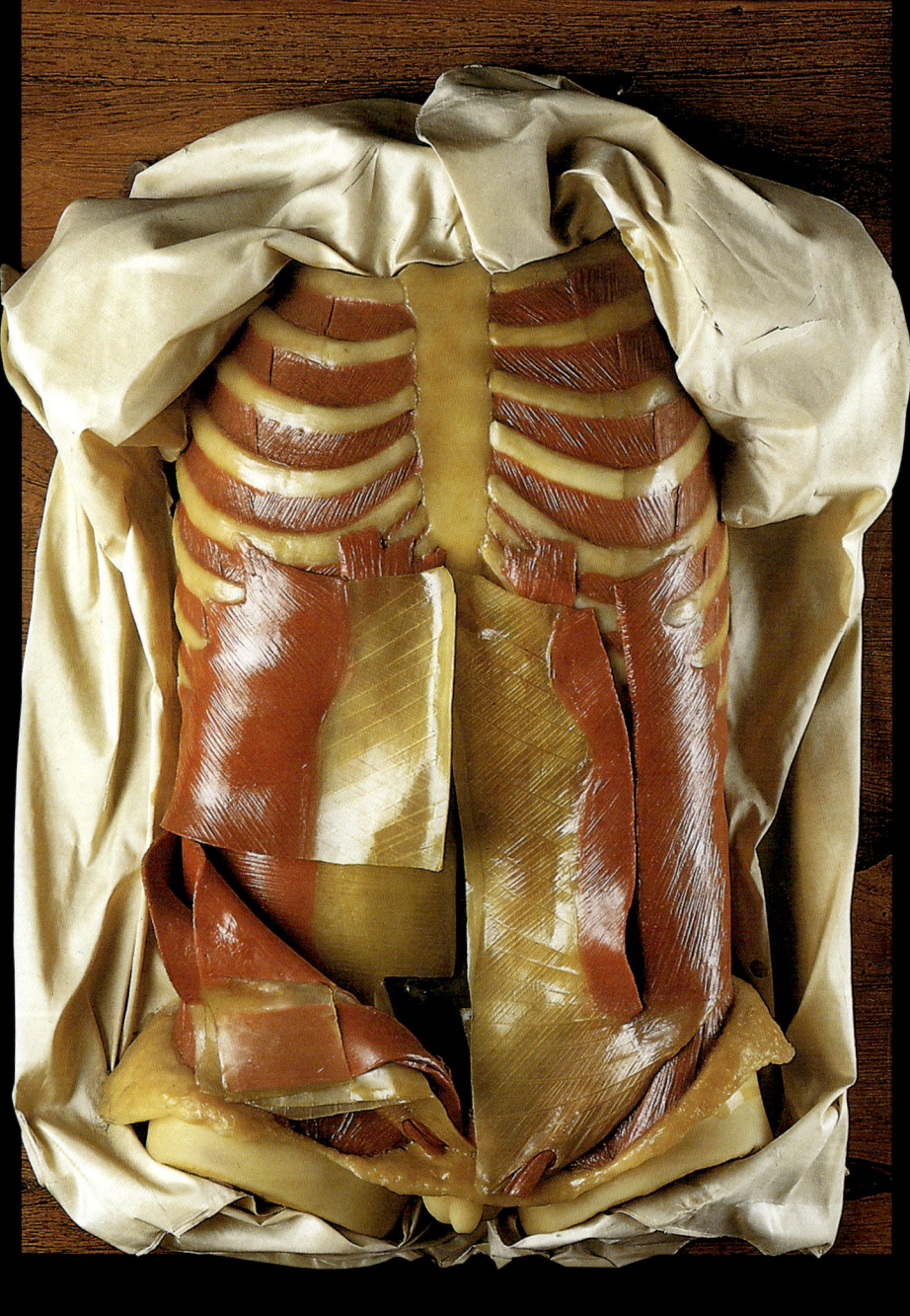

Musculi abdominis

Superficial and deep muscles of the anterior abdominal wall

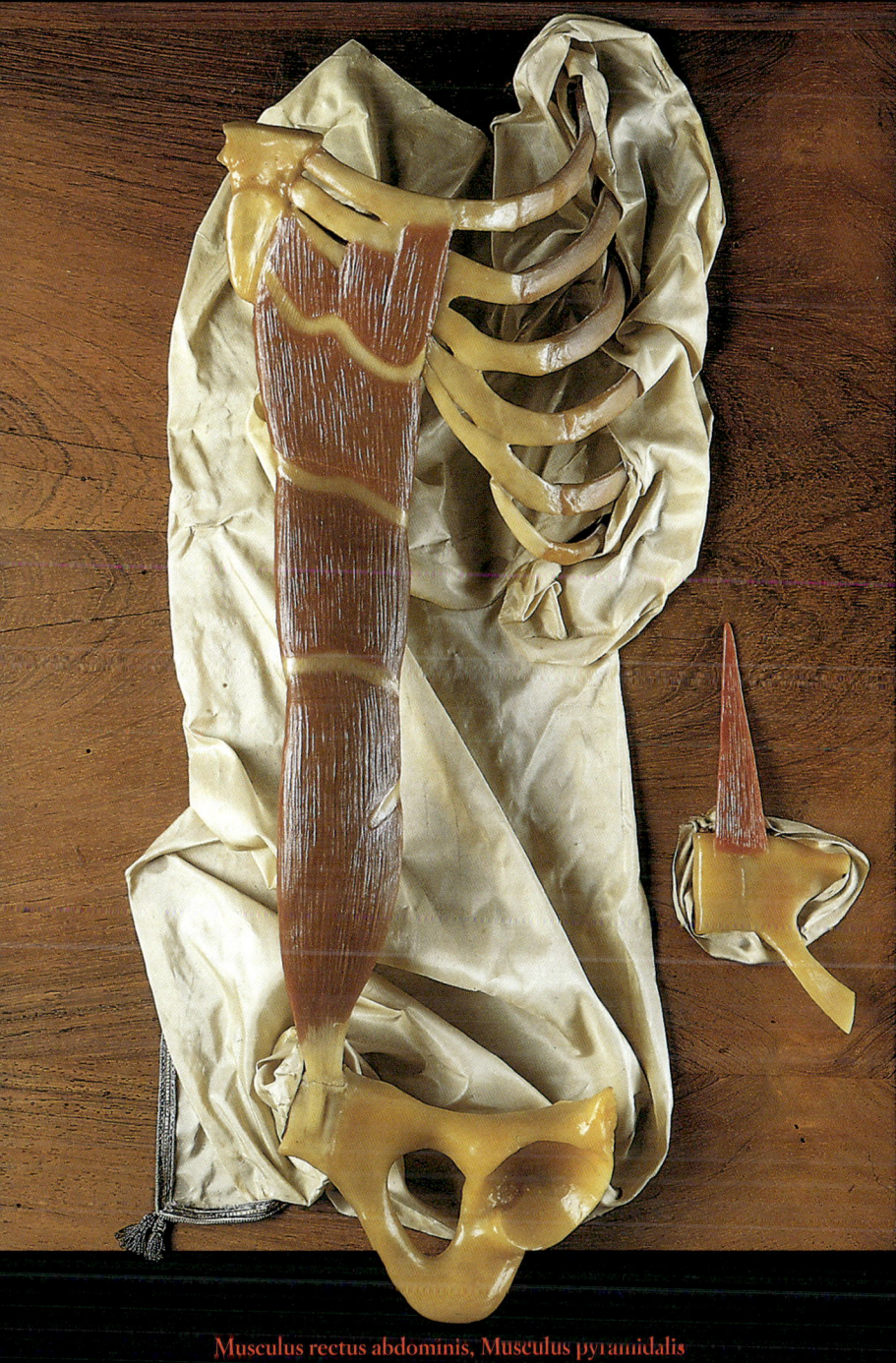

Musculus rectus abdominis, Musculus pyramidalis
View of the straight abdominal muscle with its intermediate tendons

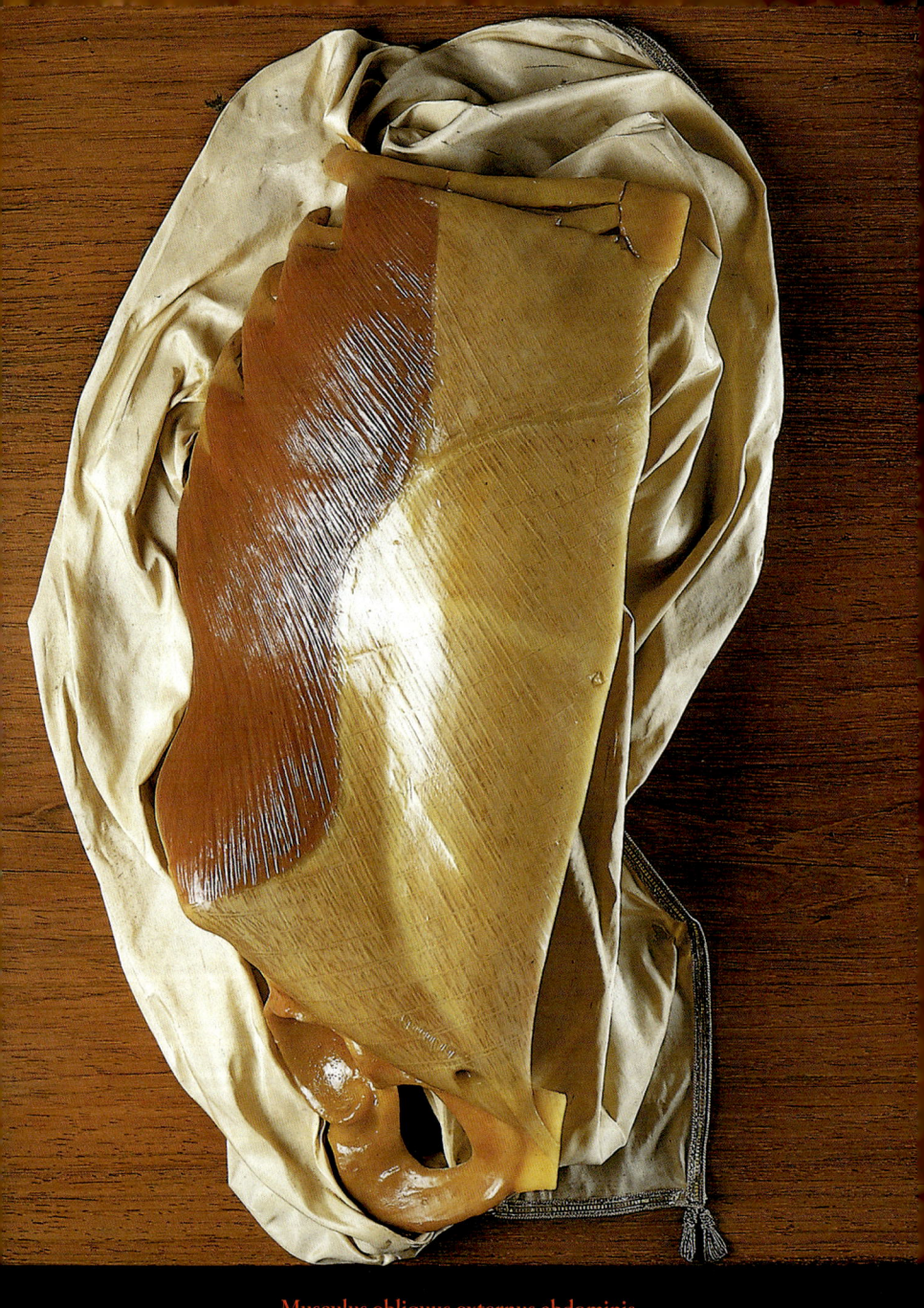

Musculus obliquus externus abdominis

The external oblique abdominal muscle together with its tendinous sheet

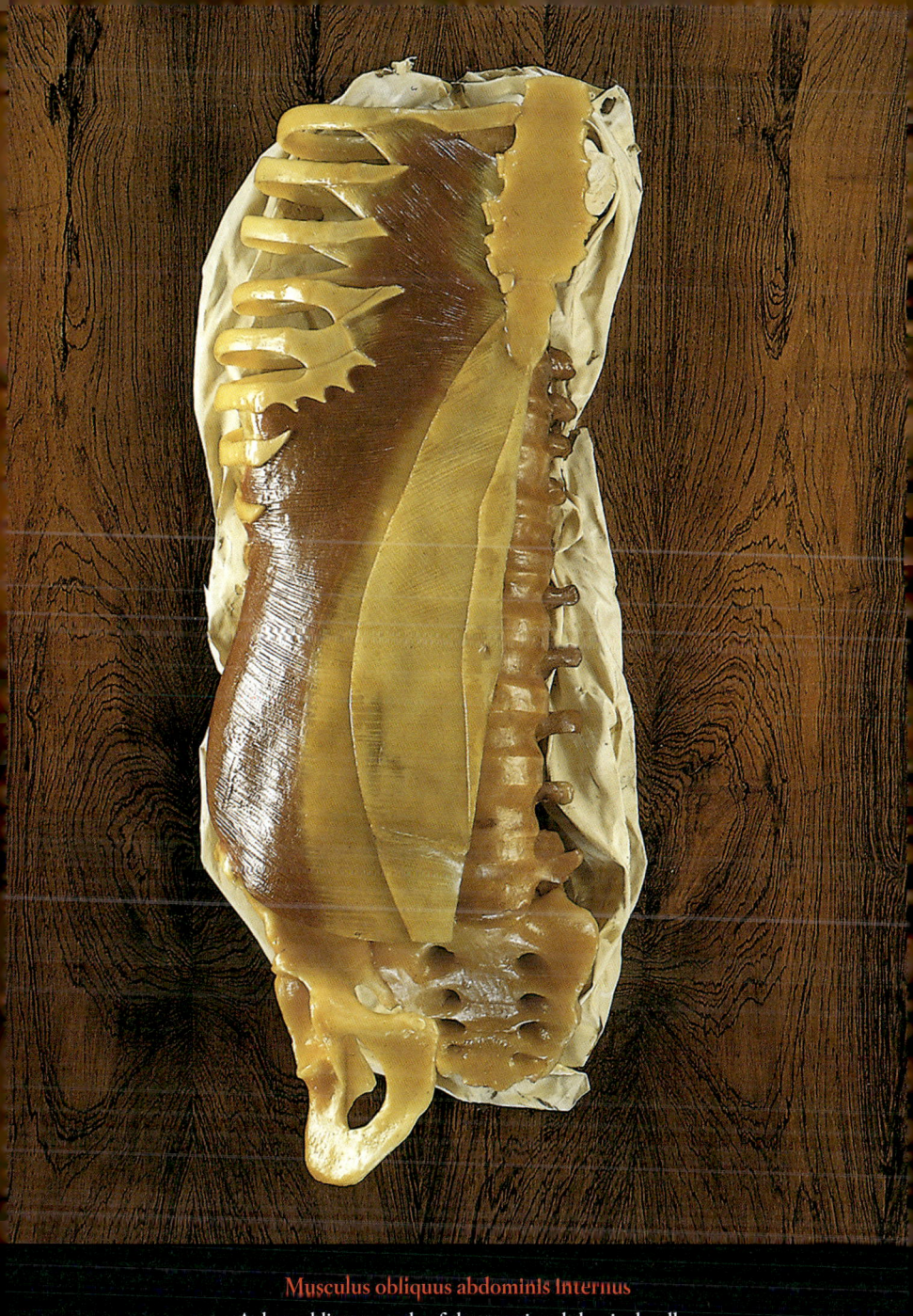

Musculus obliquus abdominis internus

A deep oblique muscle of the anterior abdominal wall

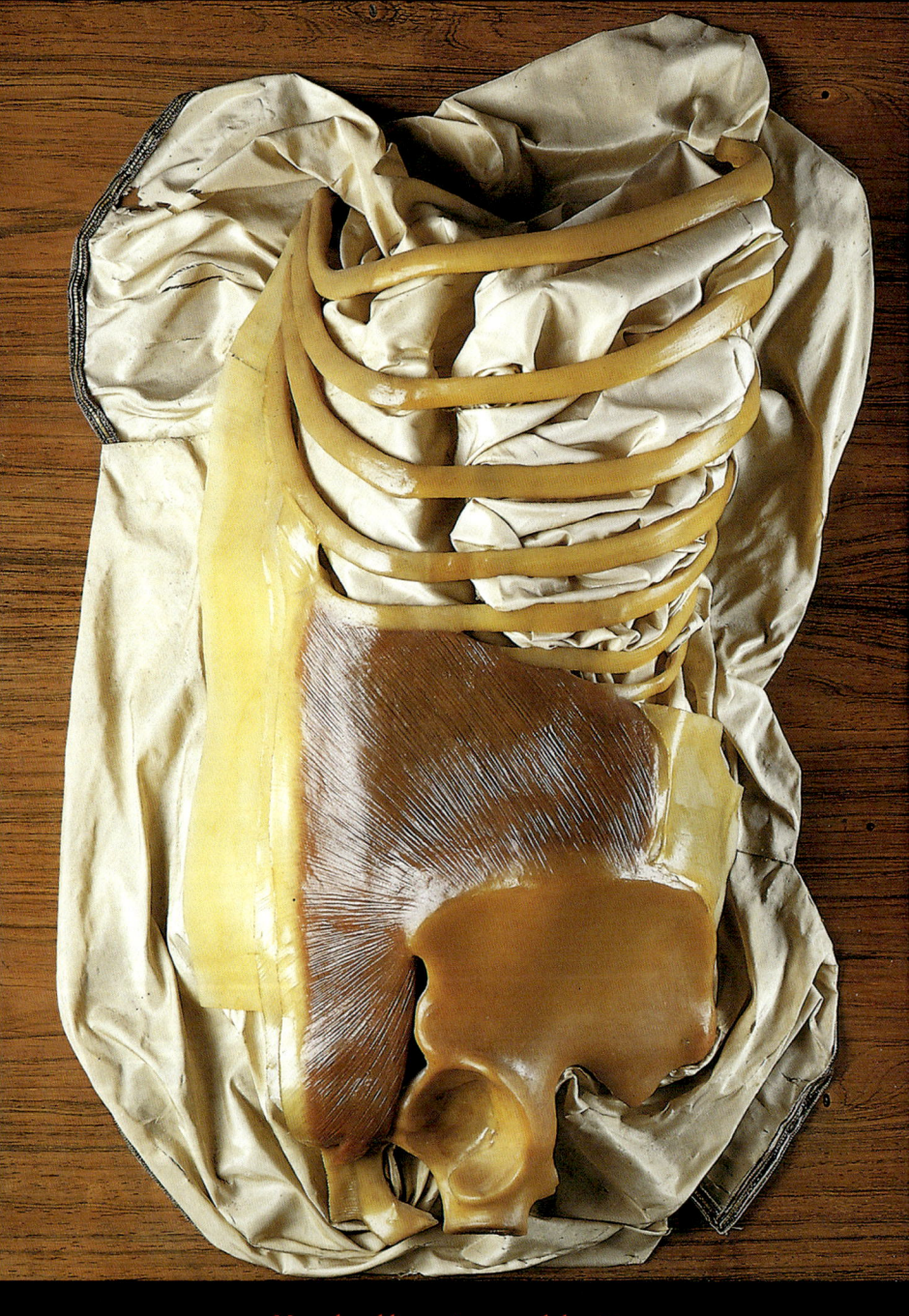

Musculus obliquus internus abdominis

The internal oblique abdominal muscle seen from the side.
Below: the hip bone with its joint socket

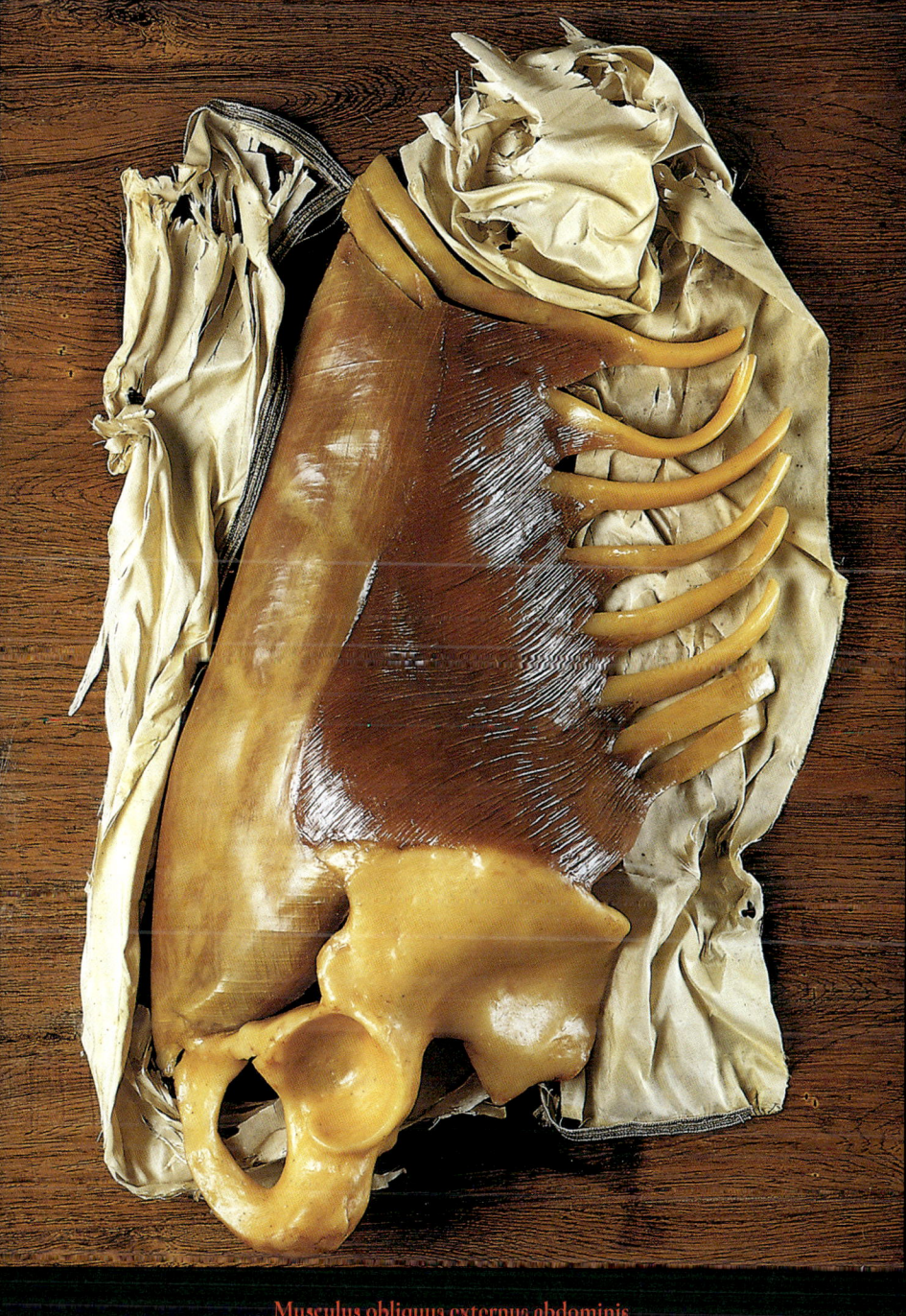

Musculus obliquus externus abdominis

The external oblique abdominal muscle seen from the side

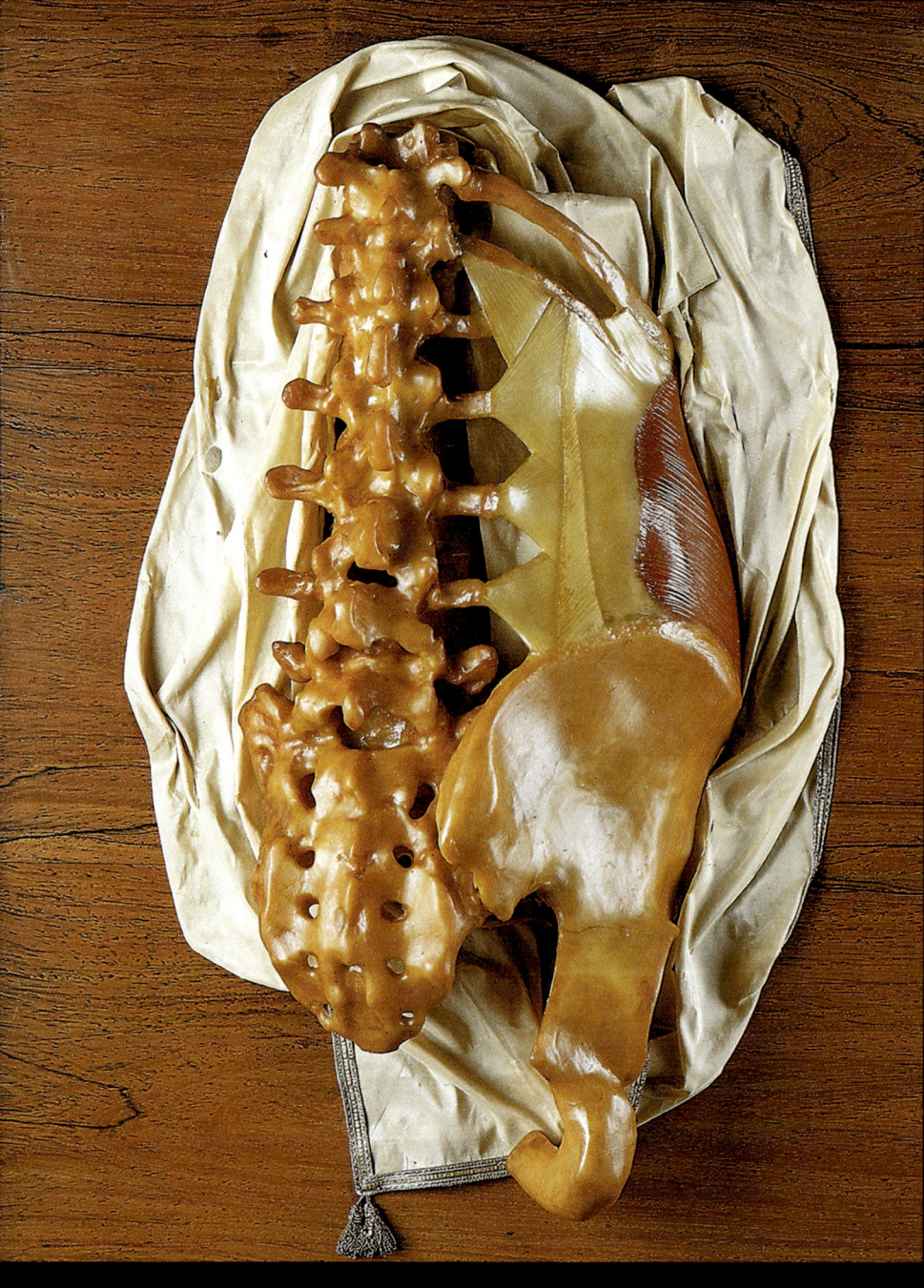

The transverse abdominal muscle seen from behind

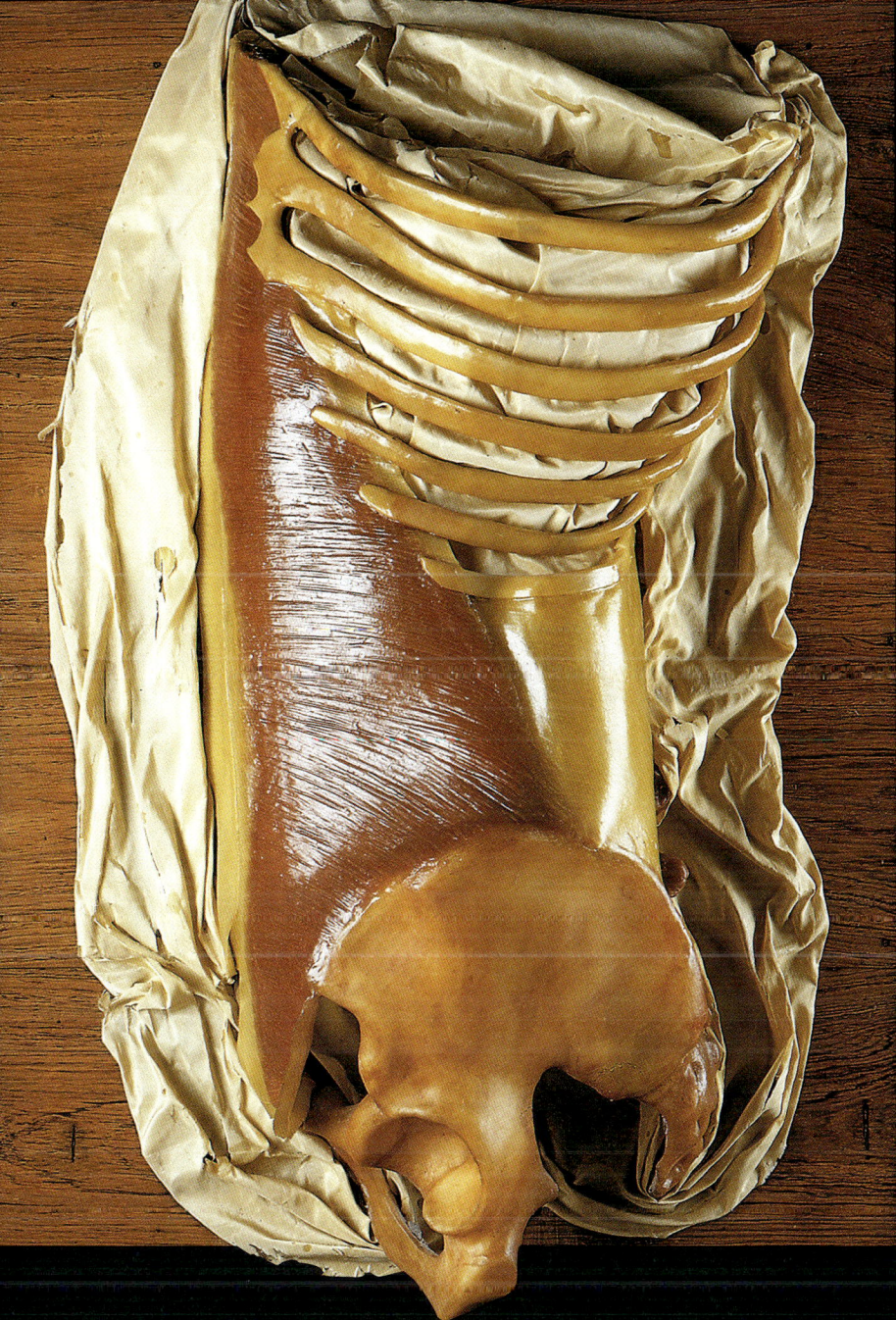

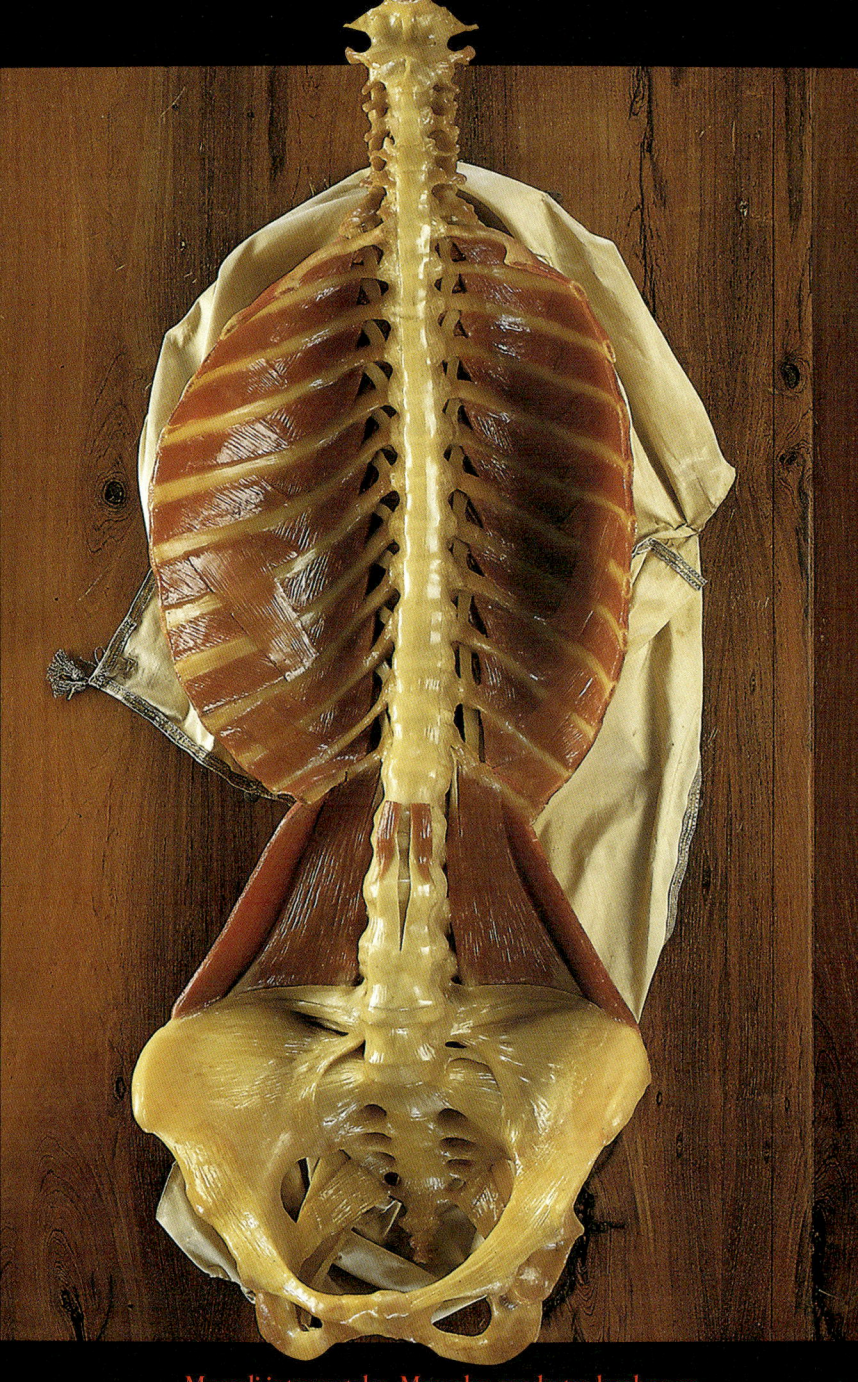

Musculi intercostales, Musculus quadratus lumborum

The posterior thoracic and abdominal walls seen from in front

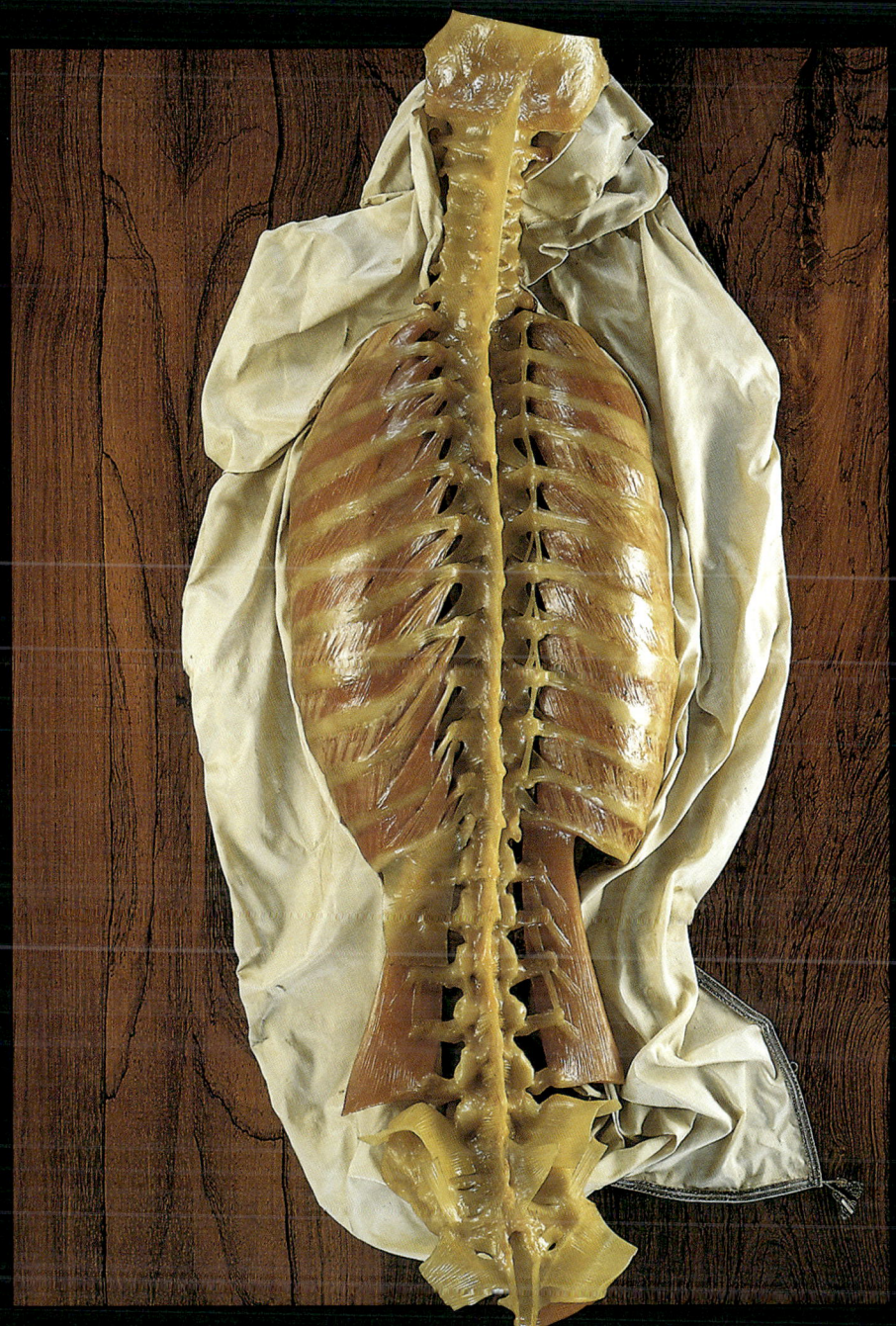

Musculi intercostales

The intercostal muscles seen from behind

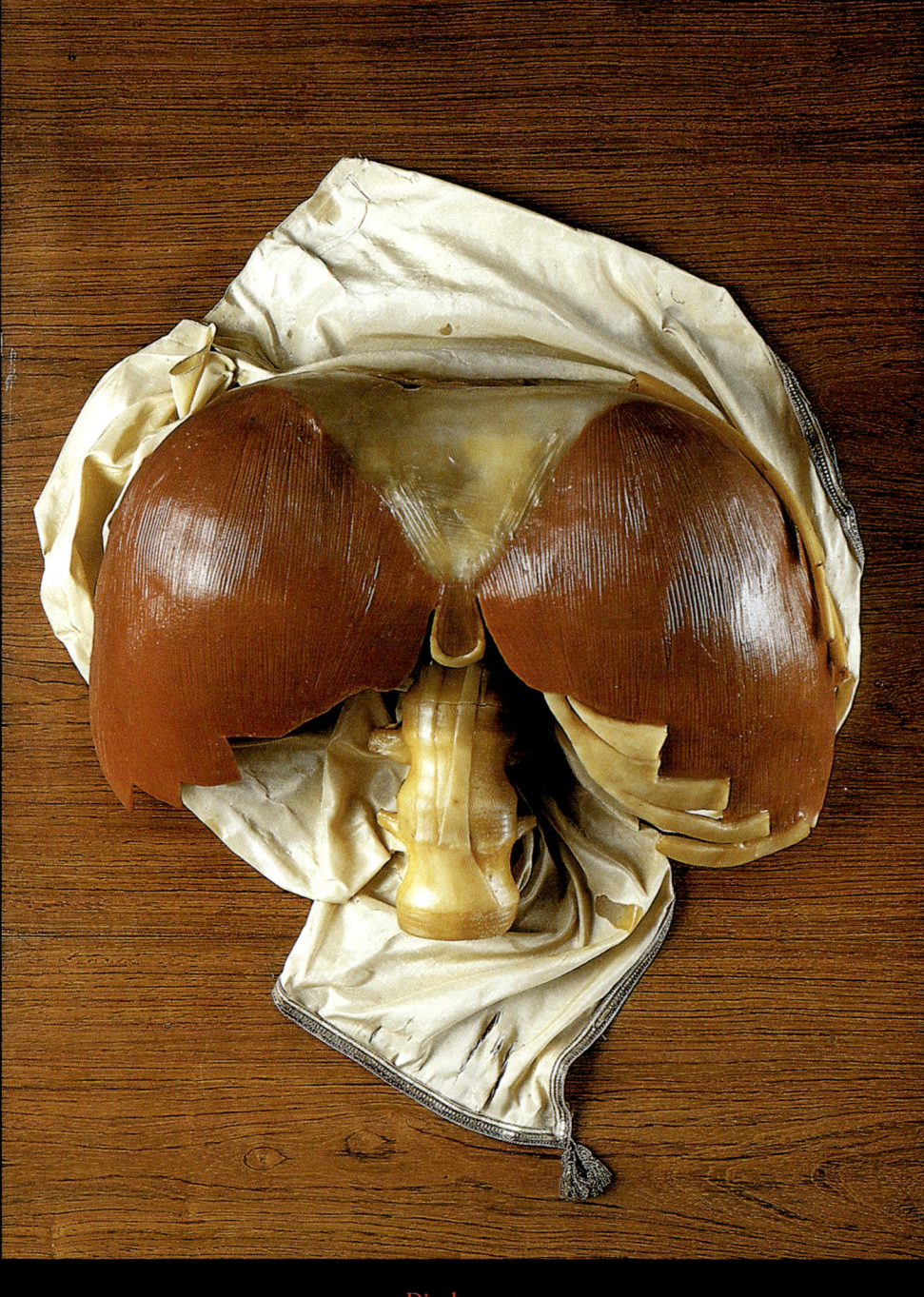

Diaphragma

The diaphragm seen from in front

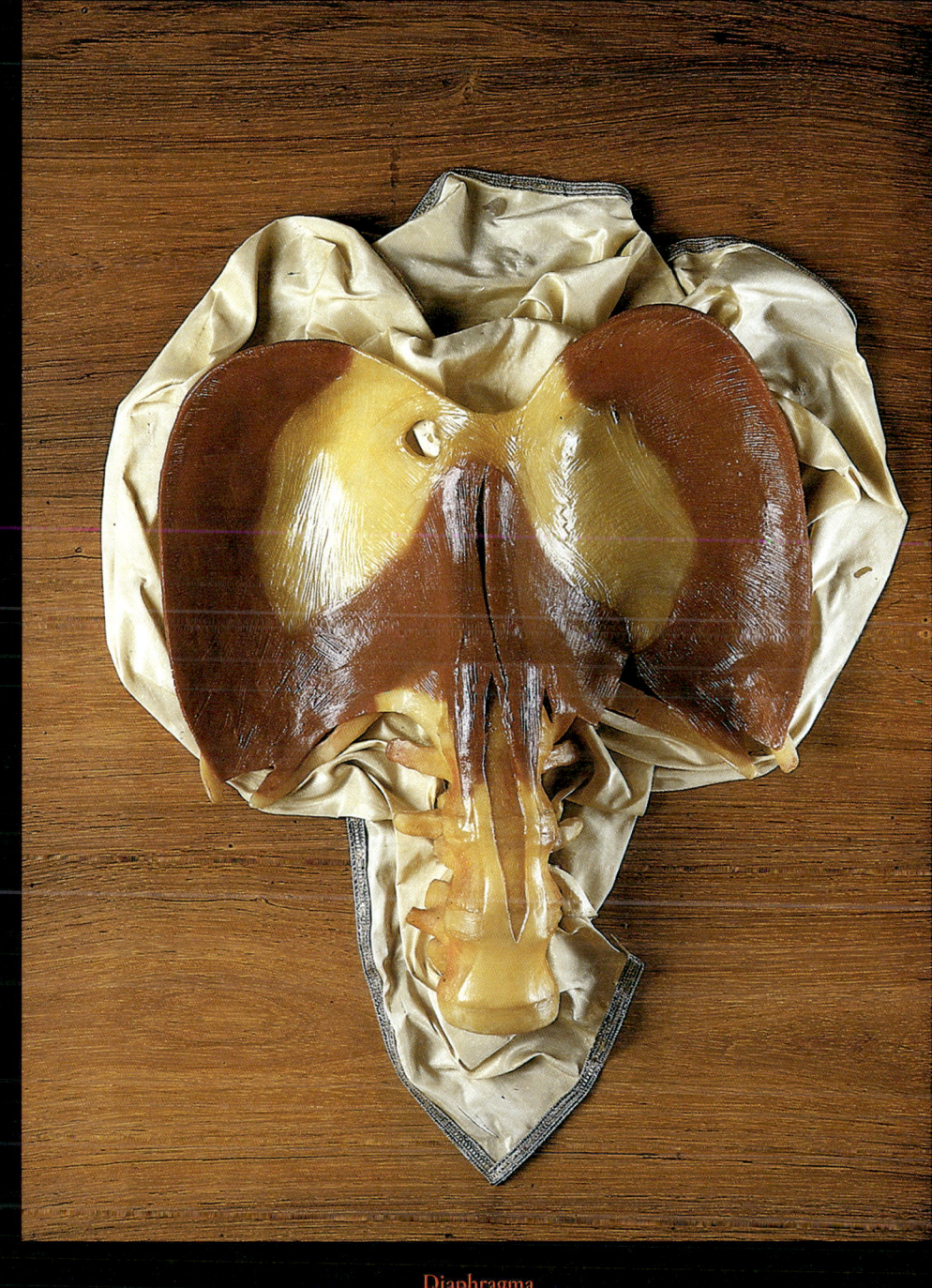

Diaphragma

View of the abdominal surface of the diaphragm showing the openings for esophagus,
aorta and inferior vena cava

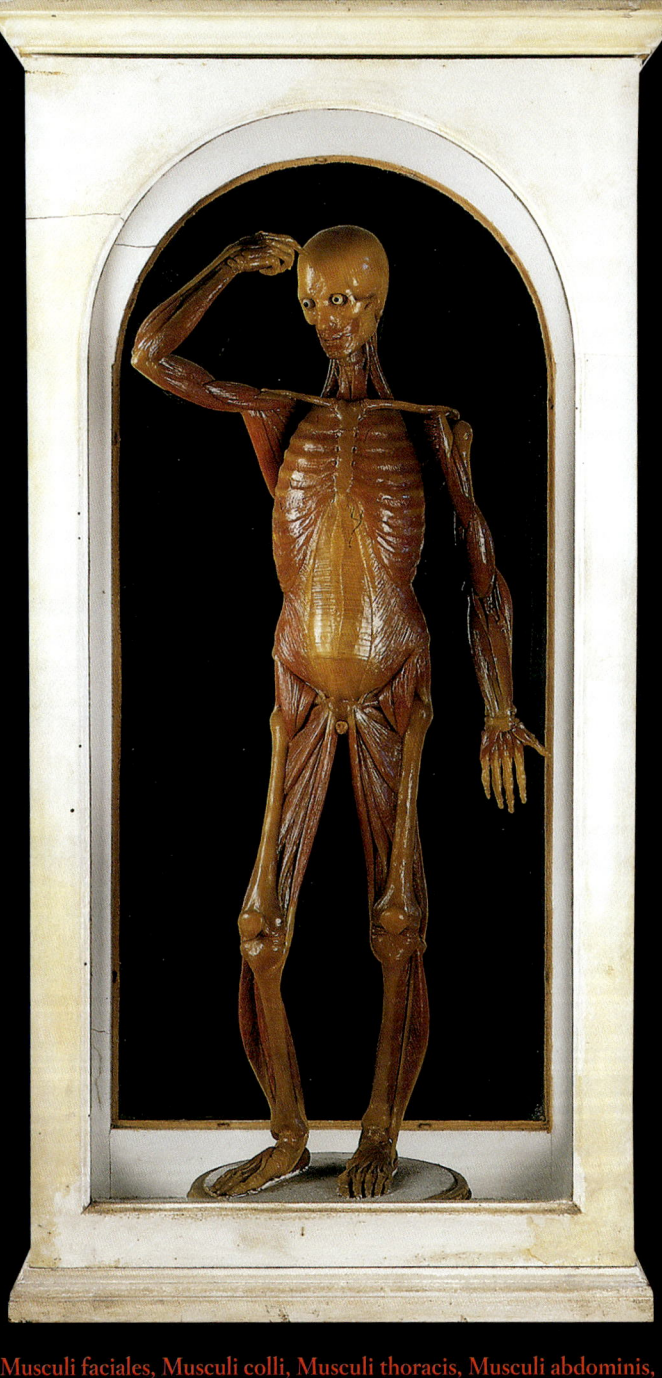

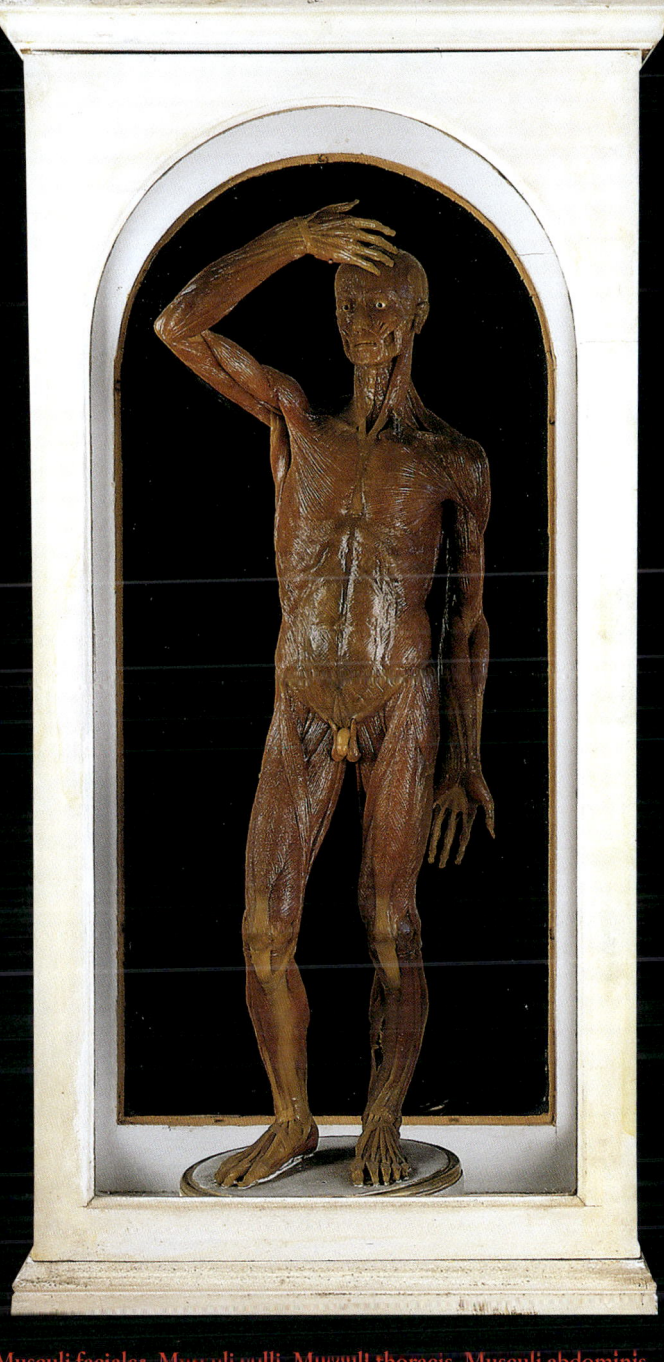

Musculi faciales, Musculi colli, Musculi thoracis, Musculi abdominis, Musculi membri superioris, Musculi membri inferioris

Whole body specimen displaying the superficial muscles

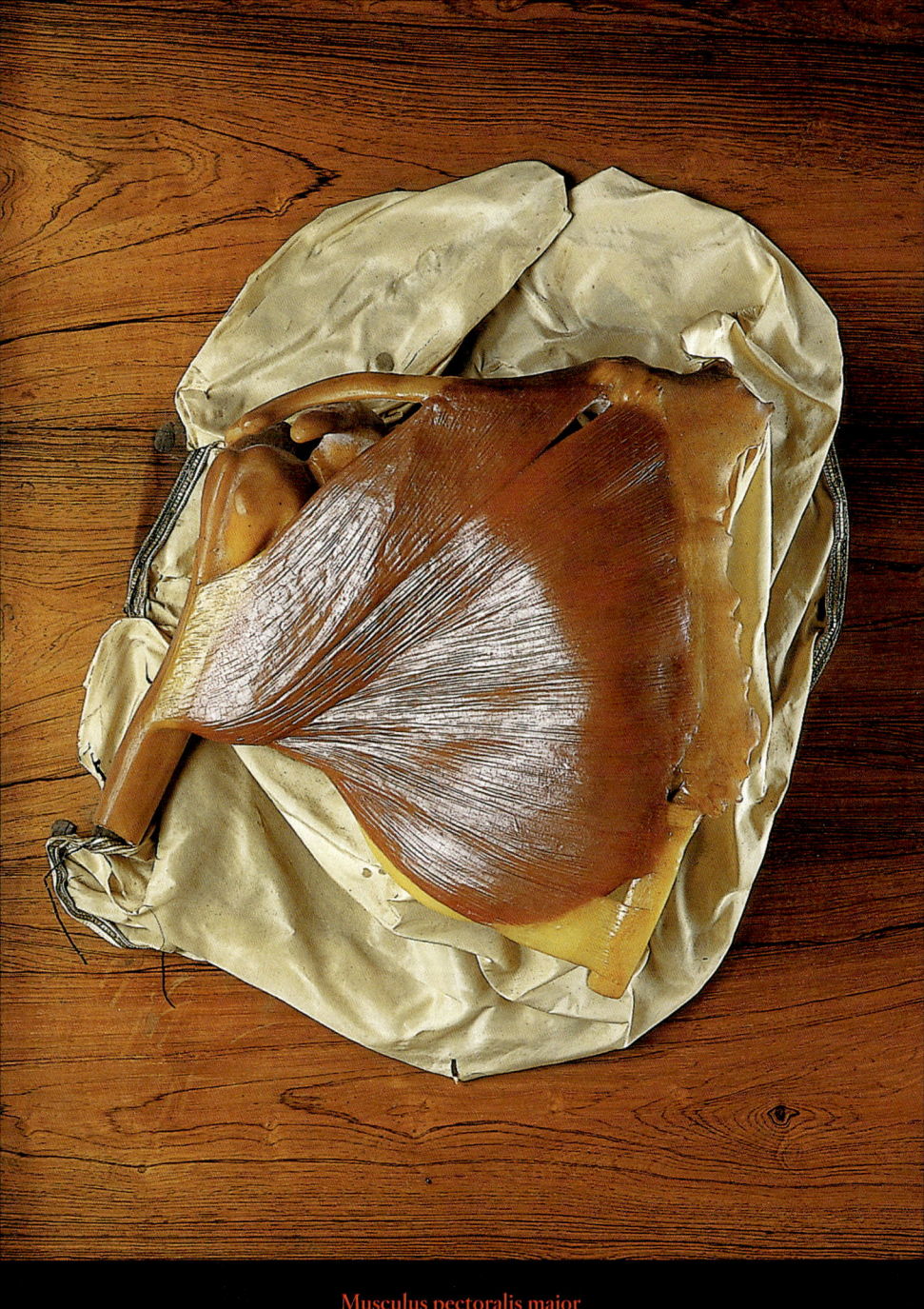

Musculus pectoralis major

The large muscle of the chest seen from in front

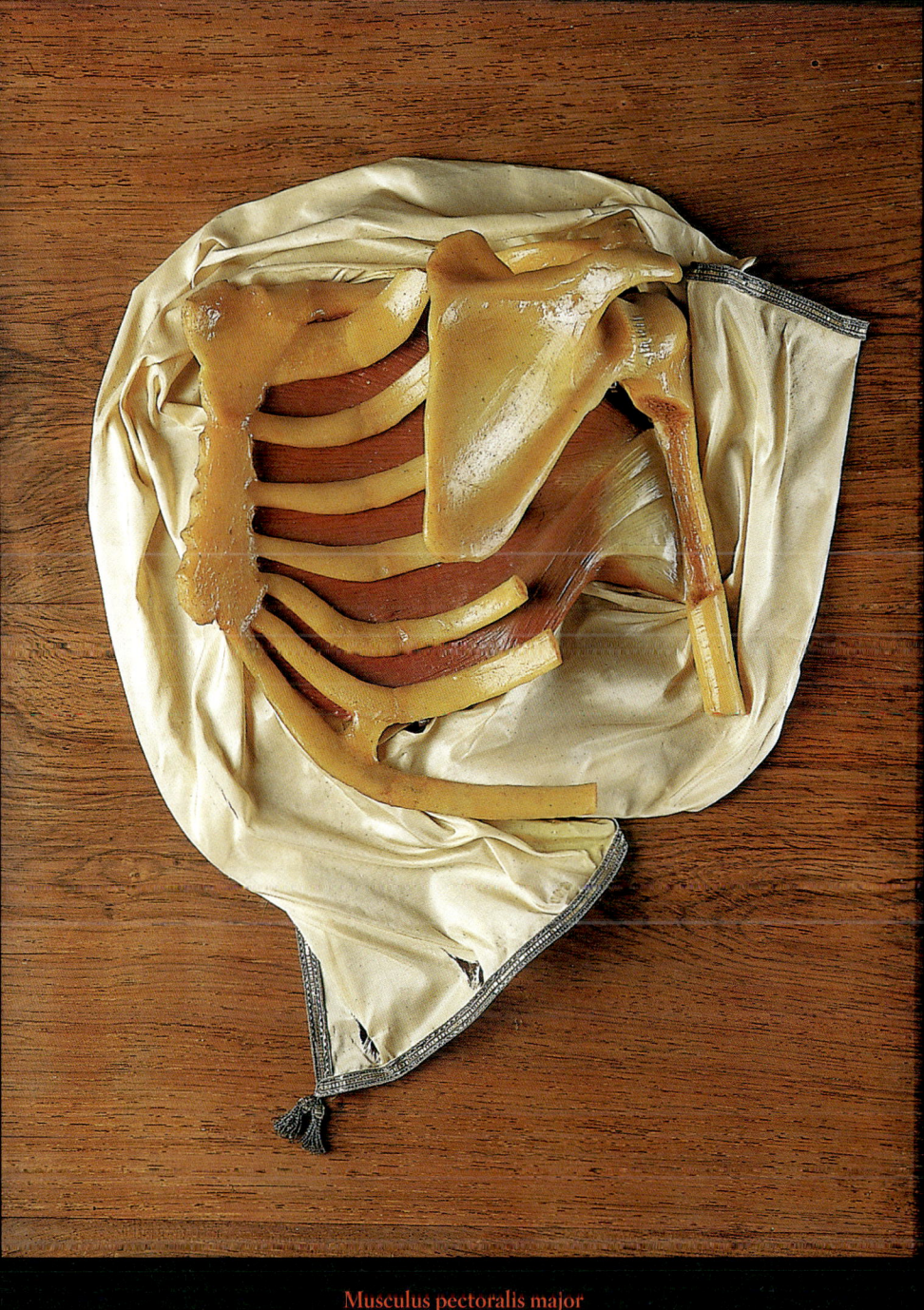

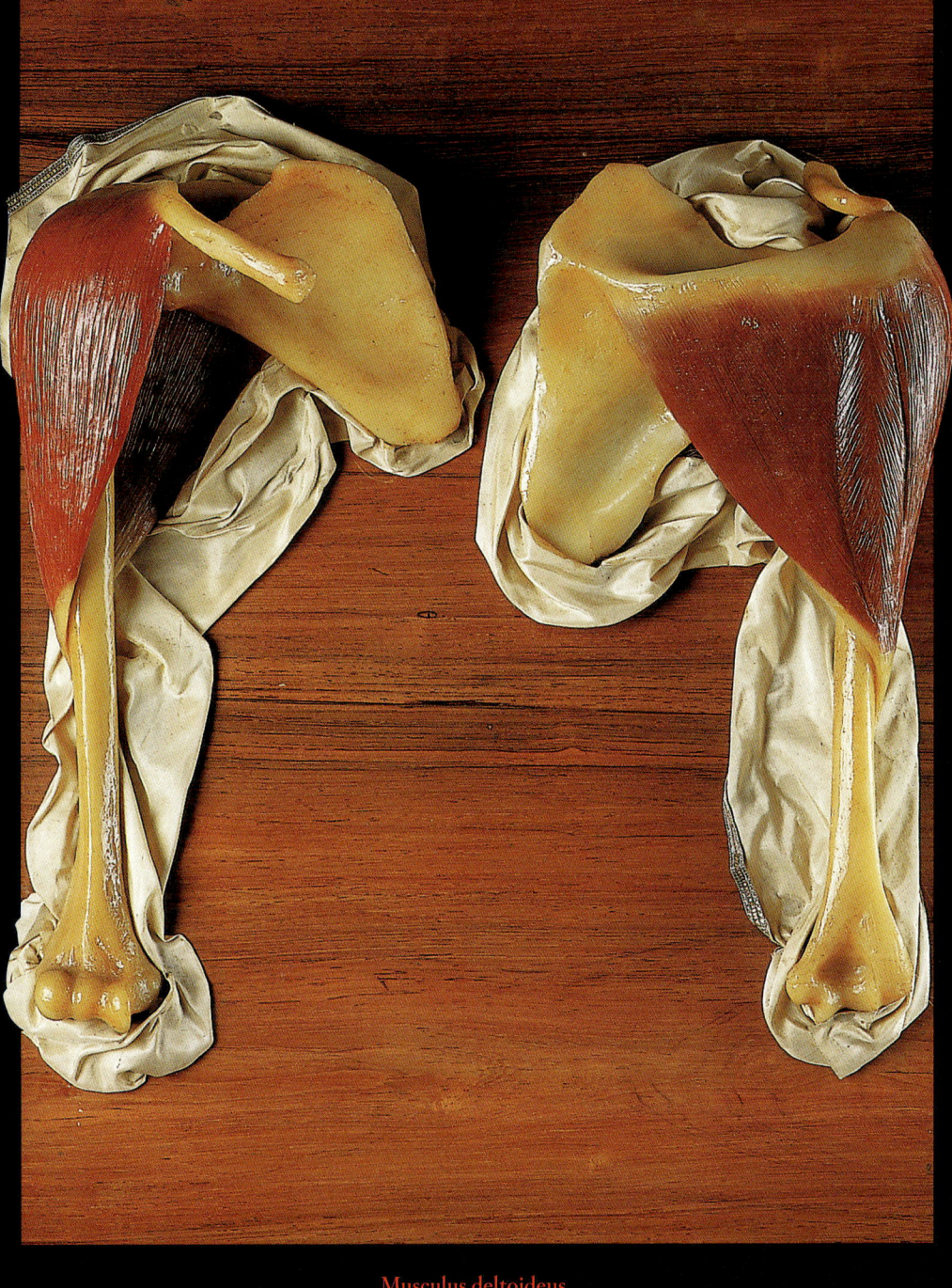

Musculus deltoideus

The deltoid muscle

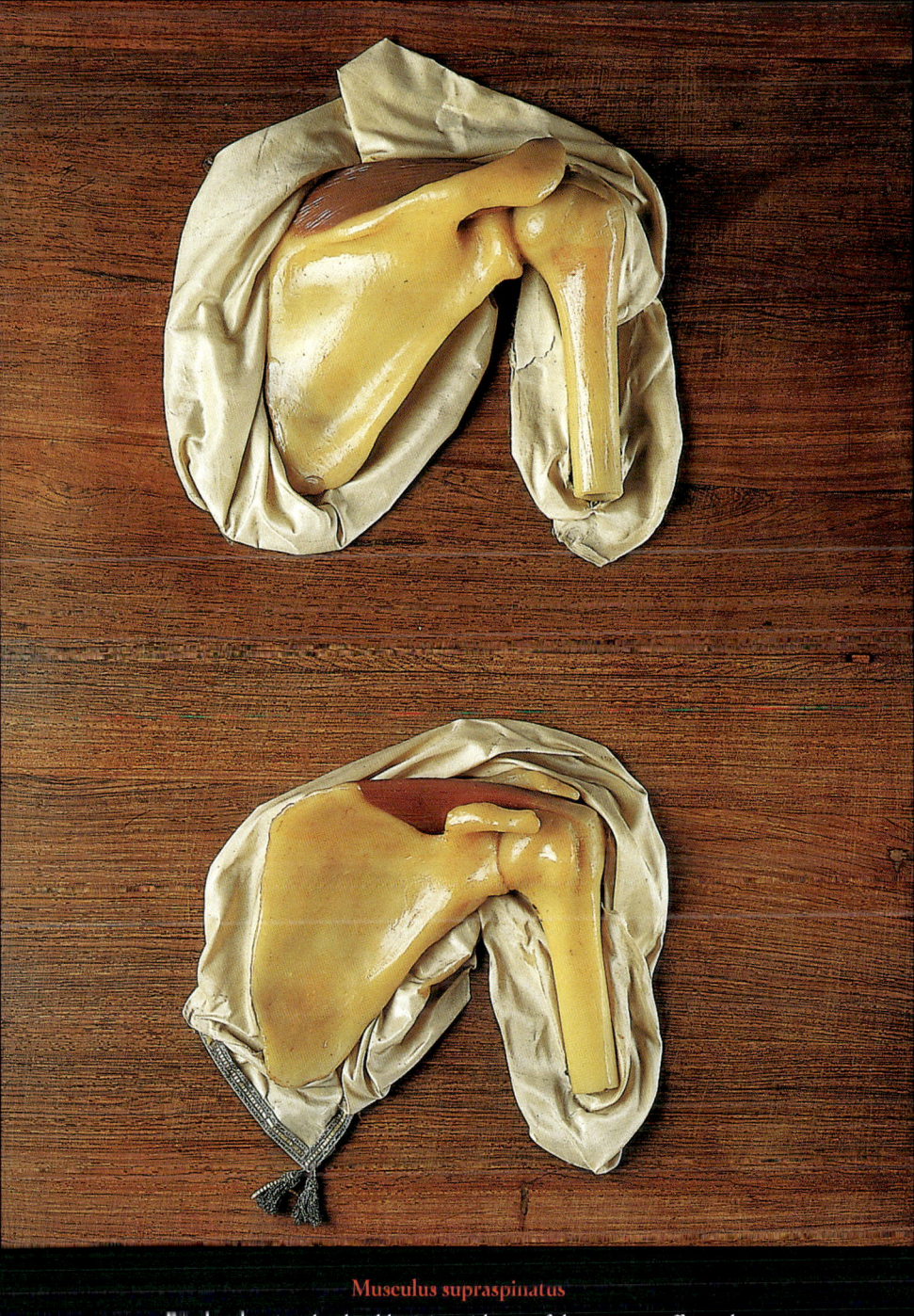

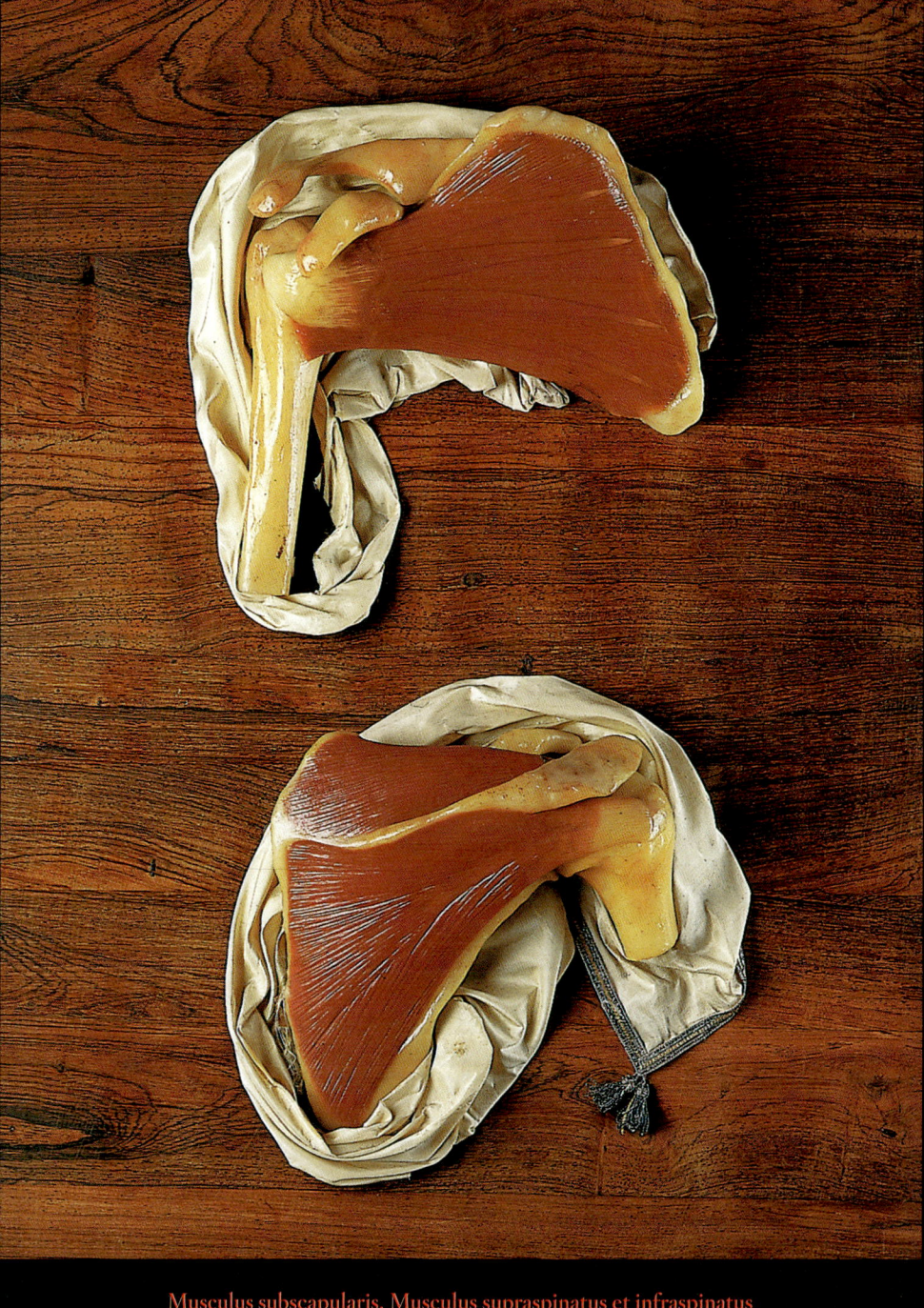

Musculus subscapularis, Musculus supraspinatus et infraspinatus
Display showing shoulder muscles

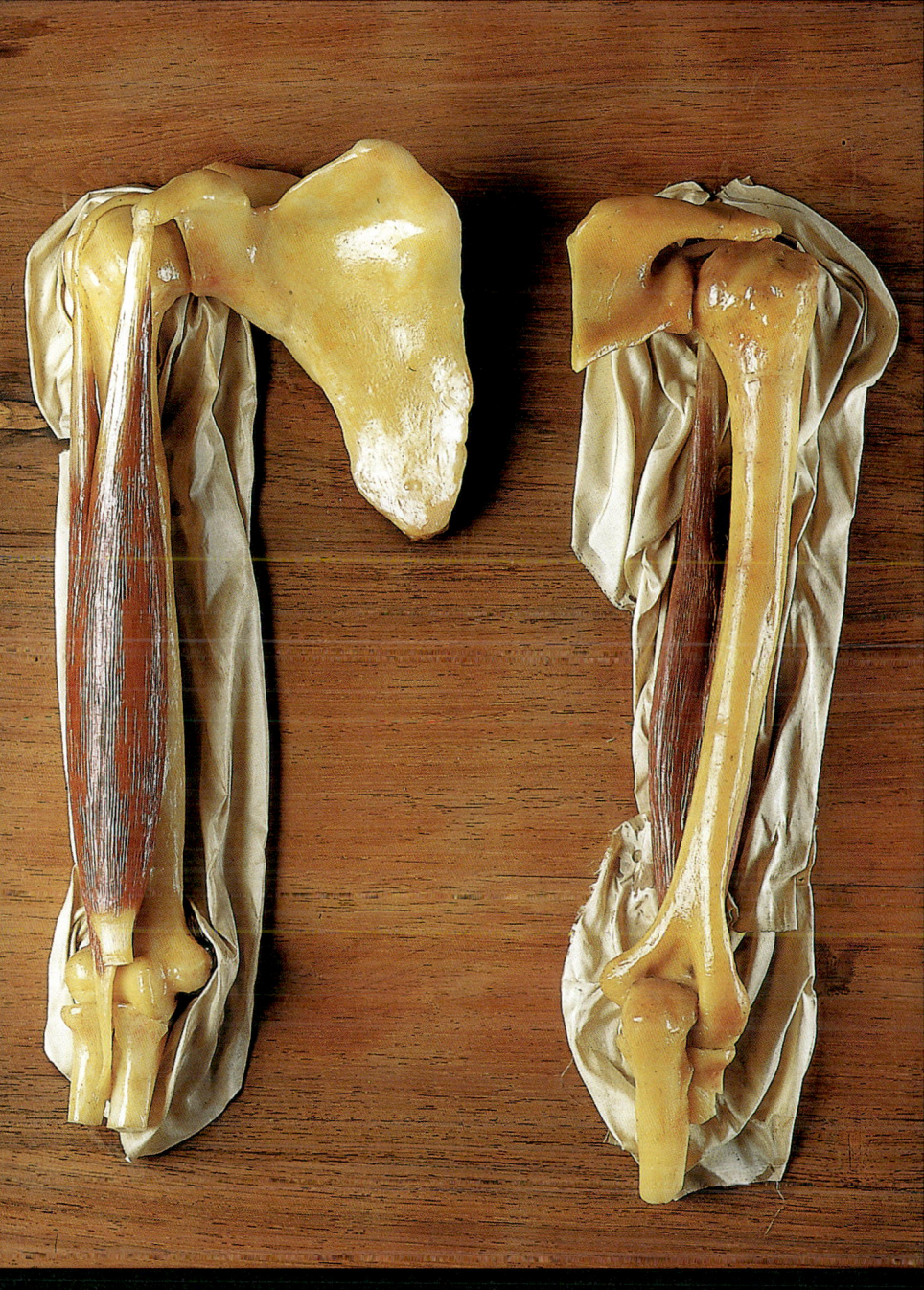

Musculus biceps brachii, Musculus coracobrachialis

View of the biceps muscle from in front (left specimen) and from behind (right specimen)

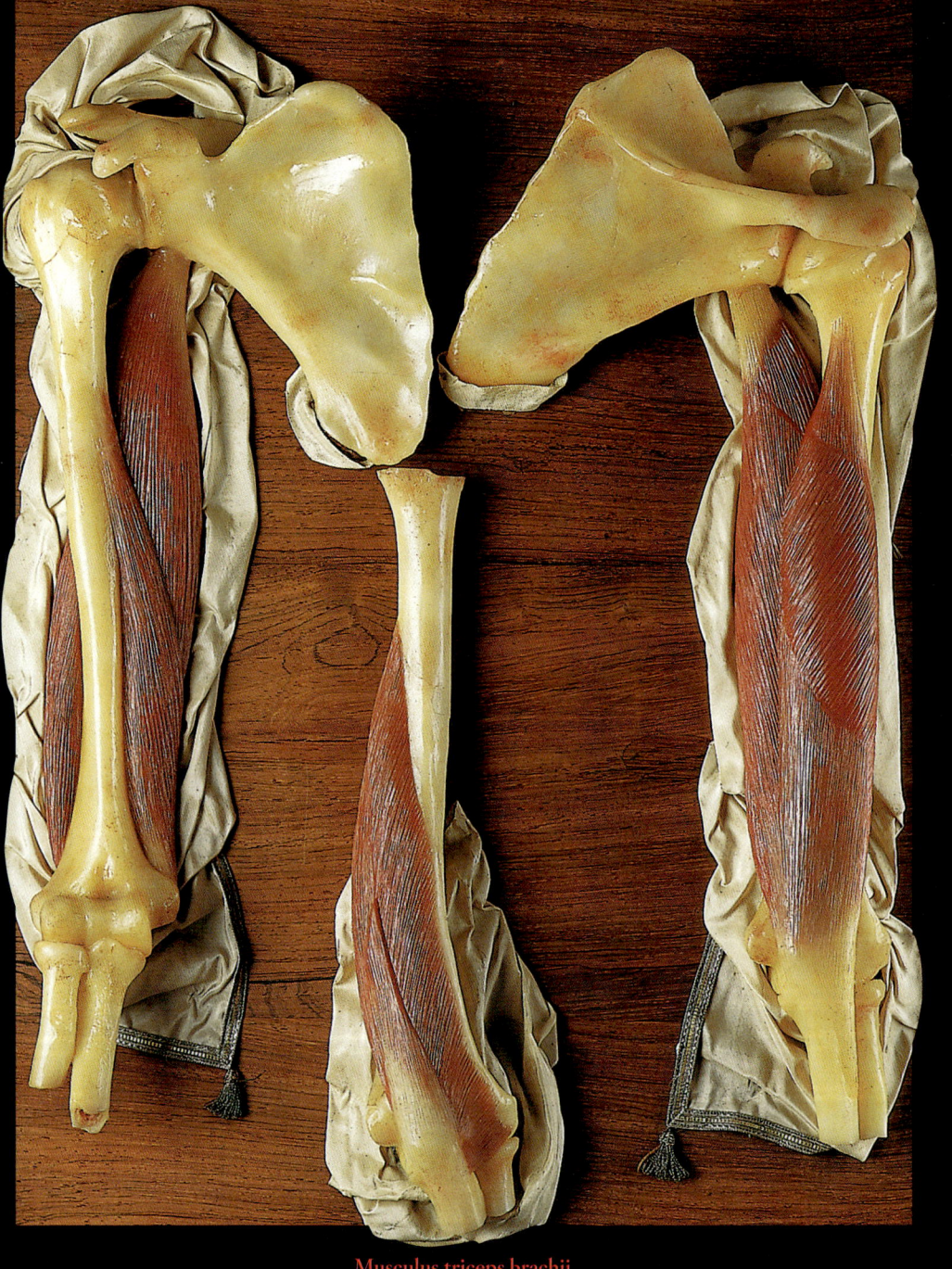

Musculus triceps brachii

Display showing the extensor muscle of the upper arm

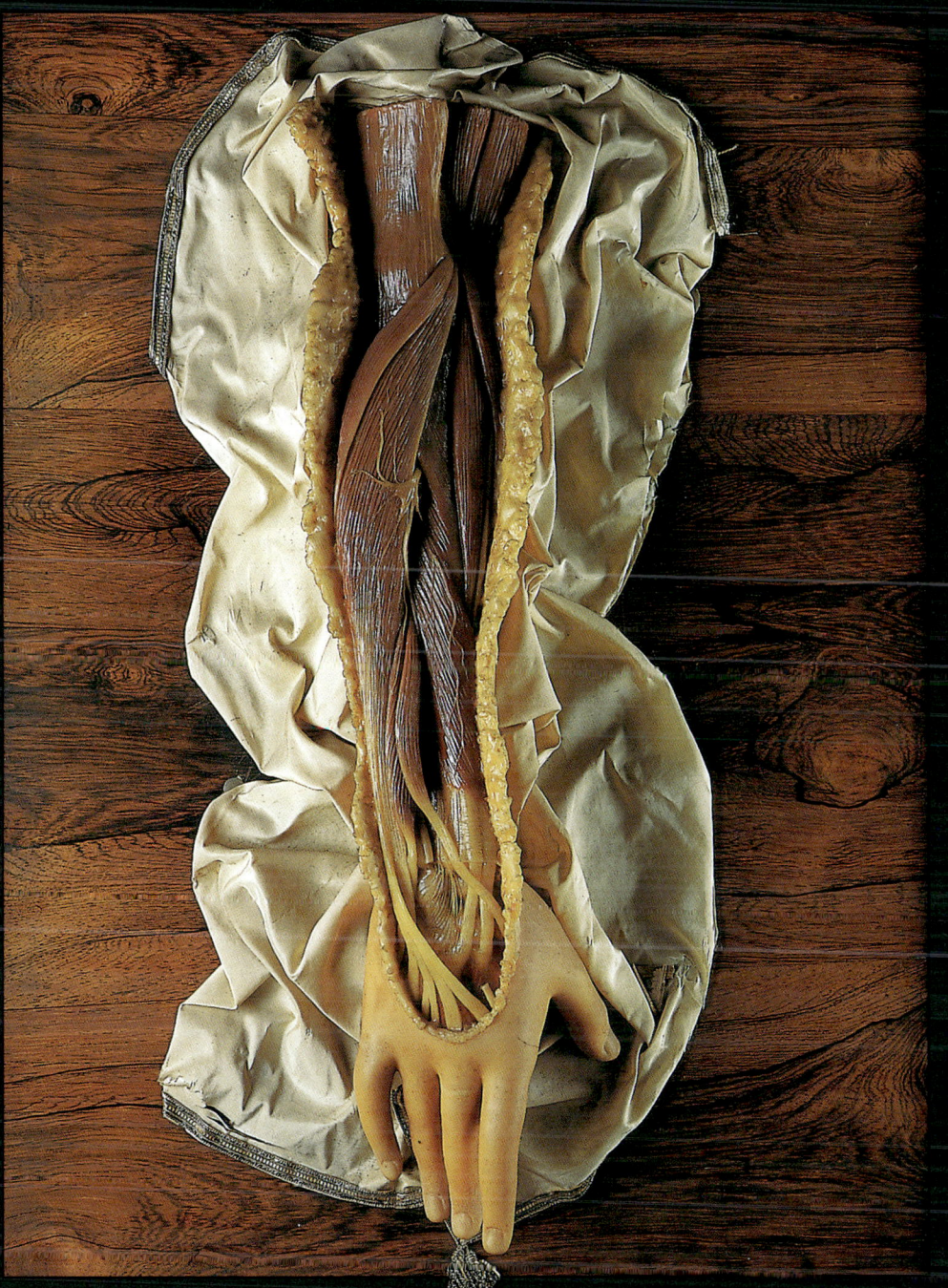

Musculus extensor digitorum, Musculus extensor pollicis longus
The extensor muscles of the hand and thumb and their tendons

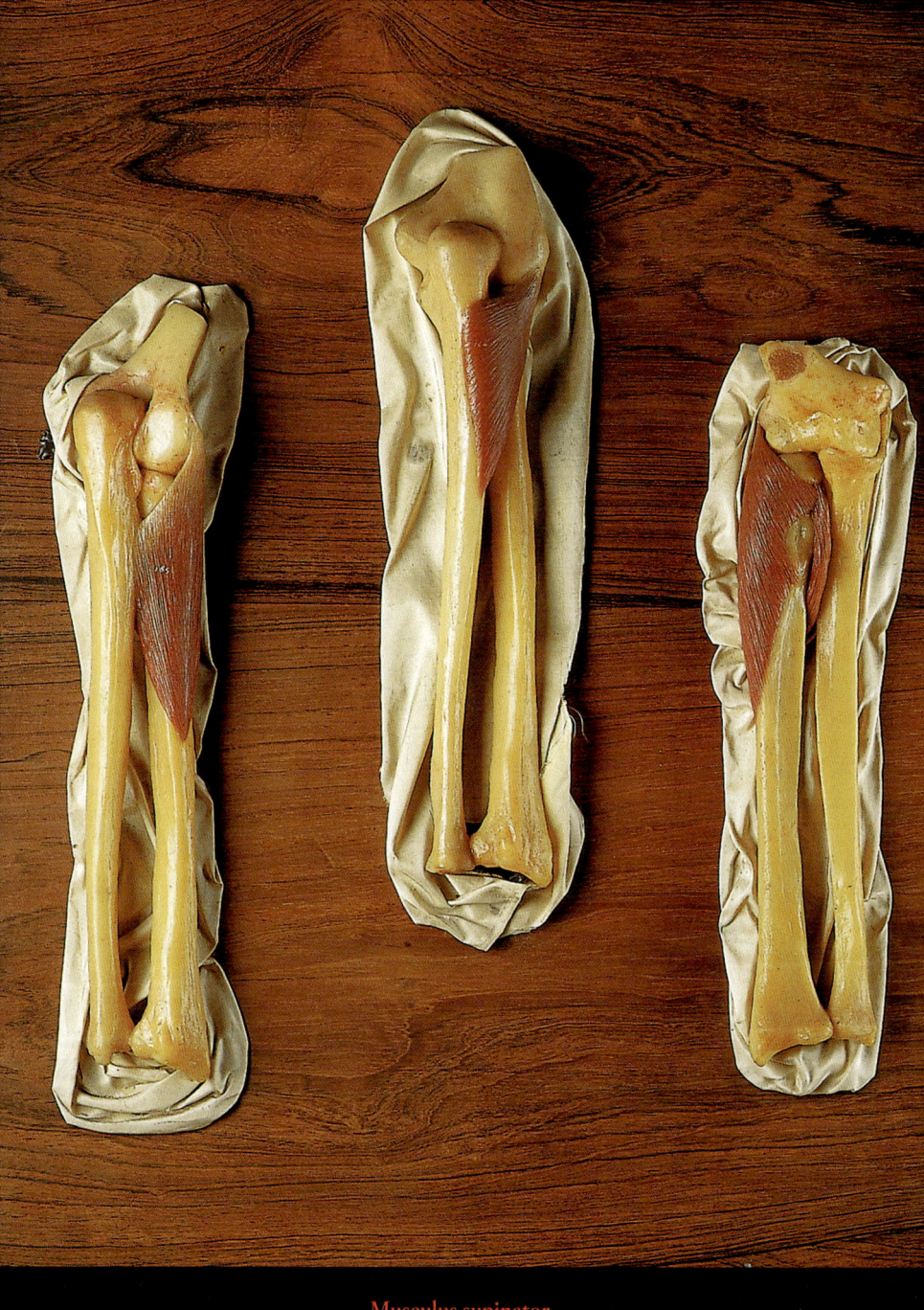

Musculus supinator

A muscle which contributes to the rotation of the forearm

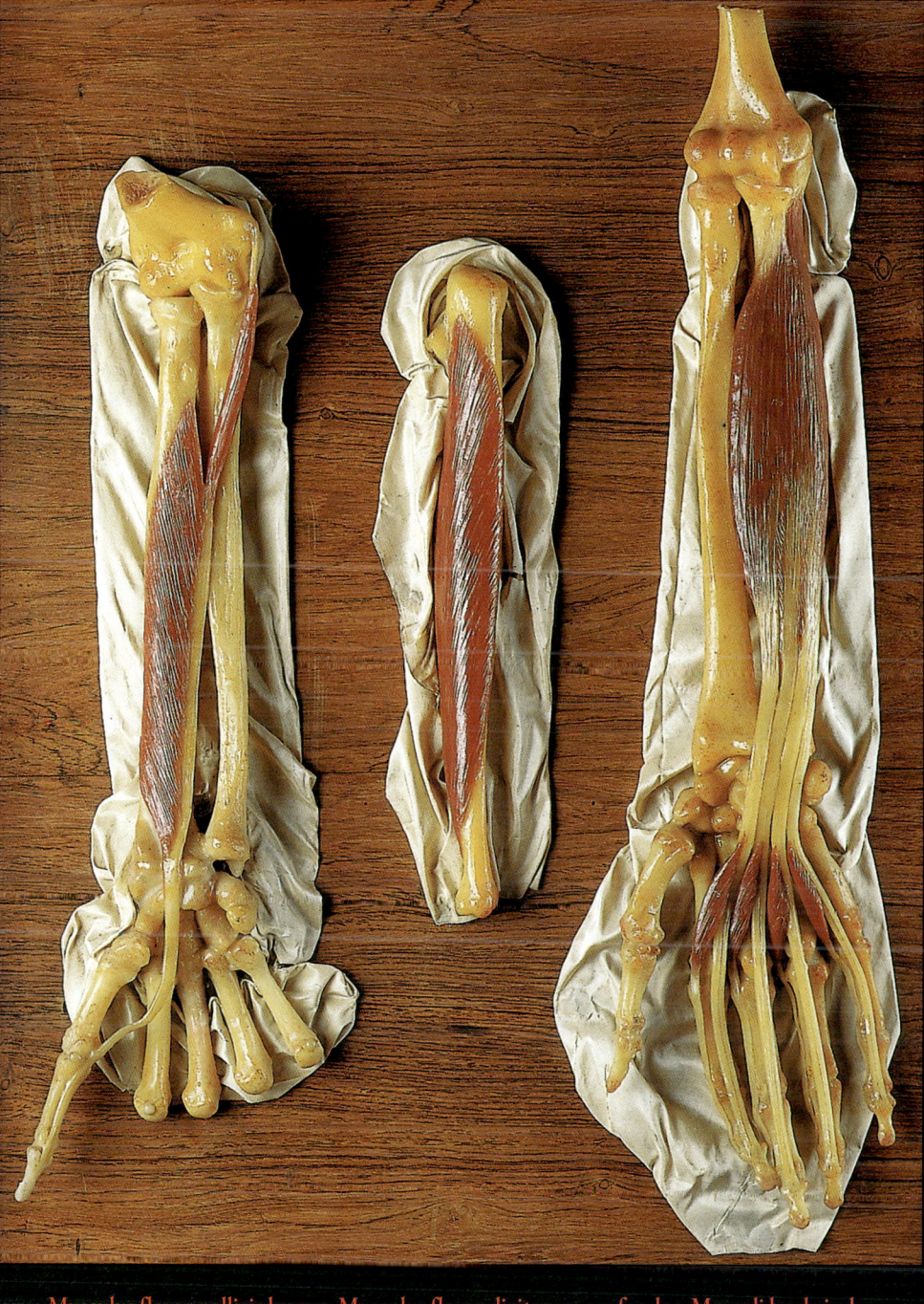

Musculus flexor pollicis longus, Musculus flexor digitorum profundus, Musculi lumbricales
Display showing the long flexor muscles of the fingers with their tendons

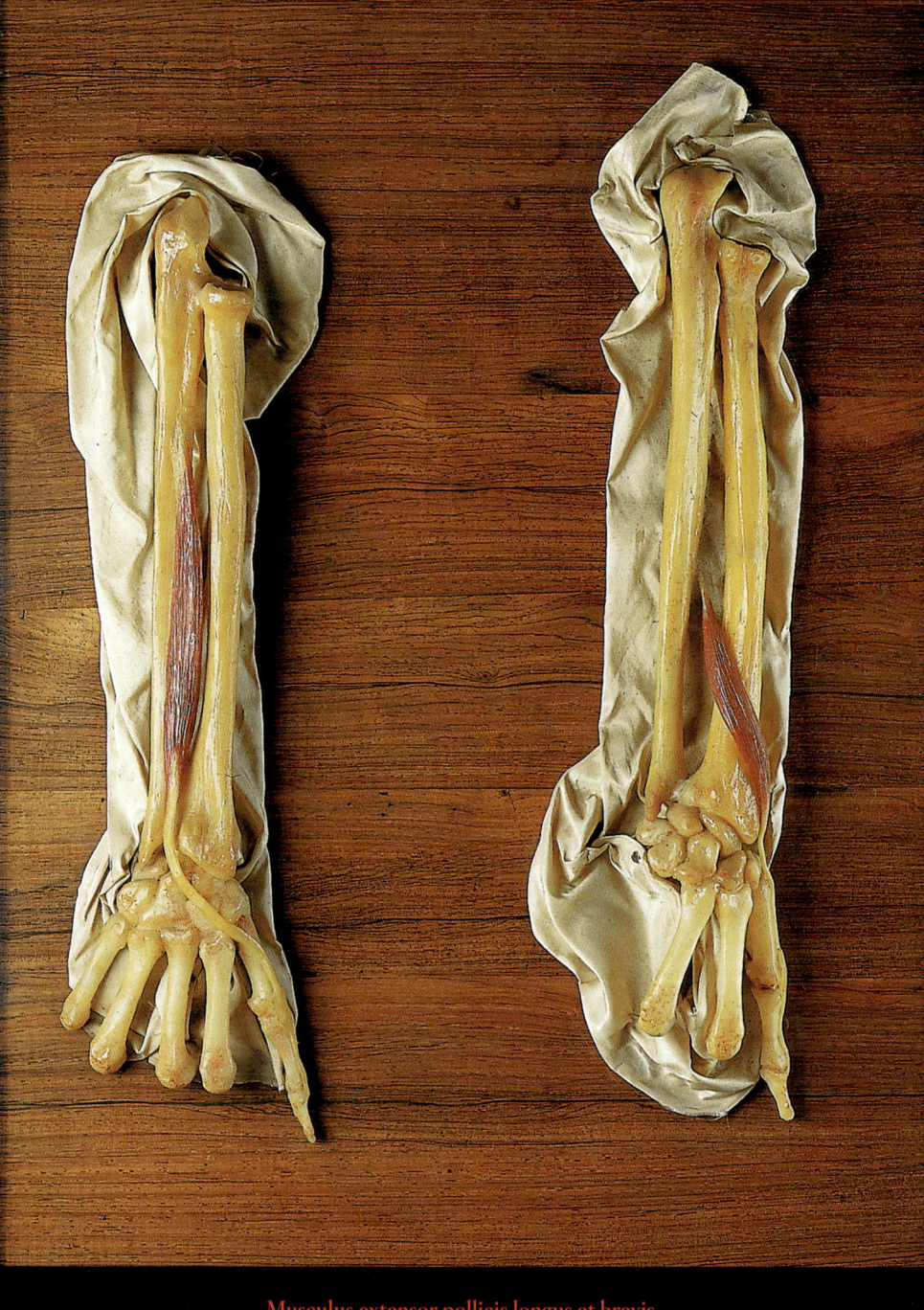

Musculus extensor pollicis longus et brevis

The long extensor muscle of the thumb

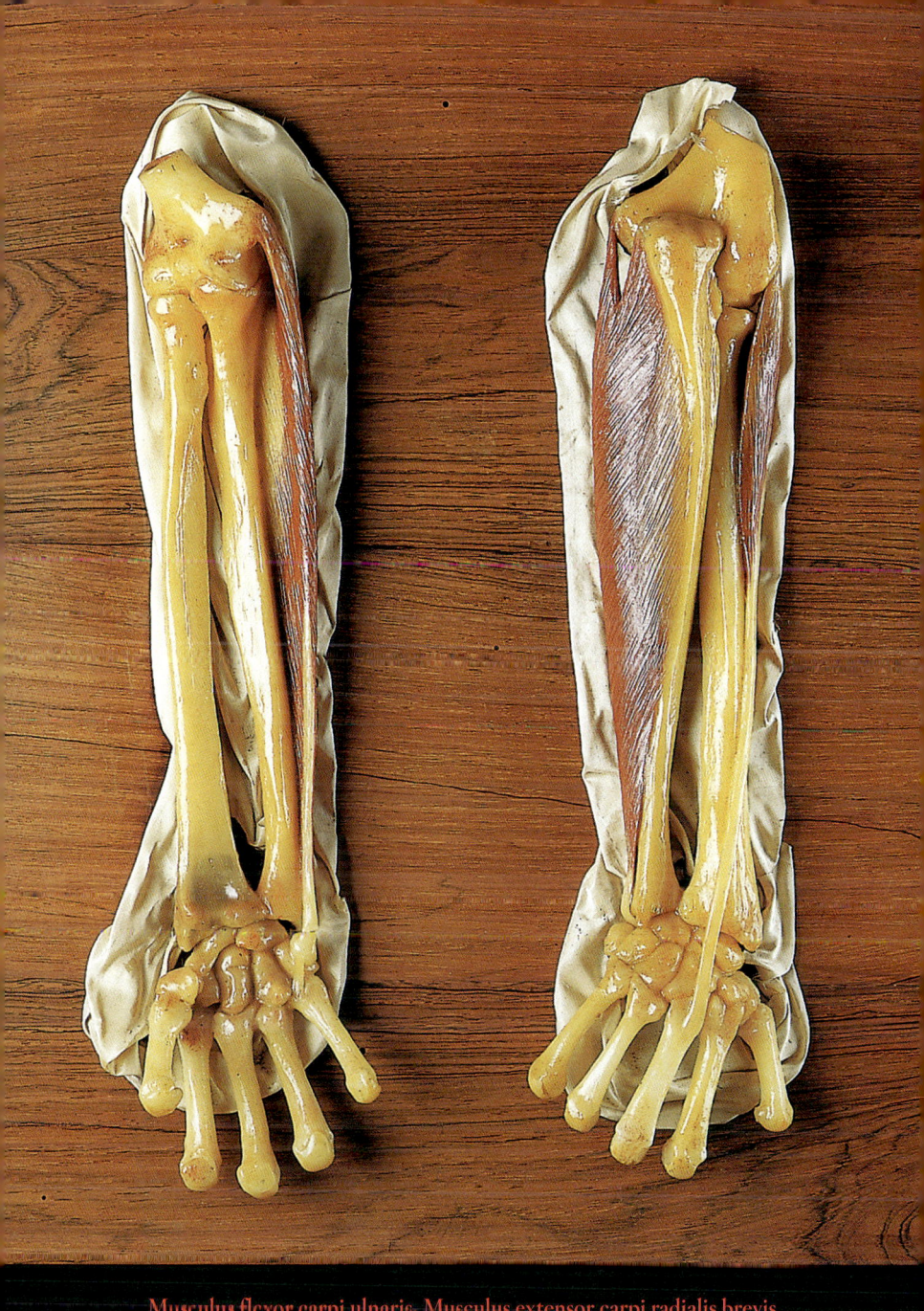

Musculus flexor carpi ulnaris, Musculus extensor carpi radialis brevis.

Display showing forearm muscles which contribute to flexion and extension at the wrist joint.

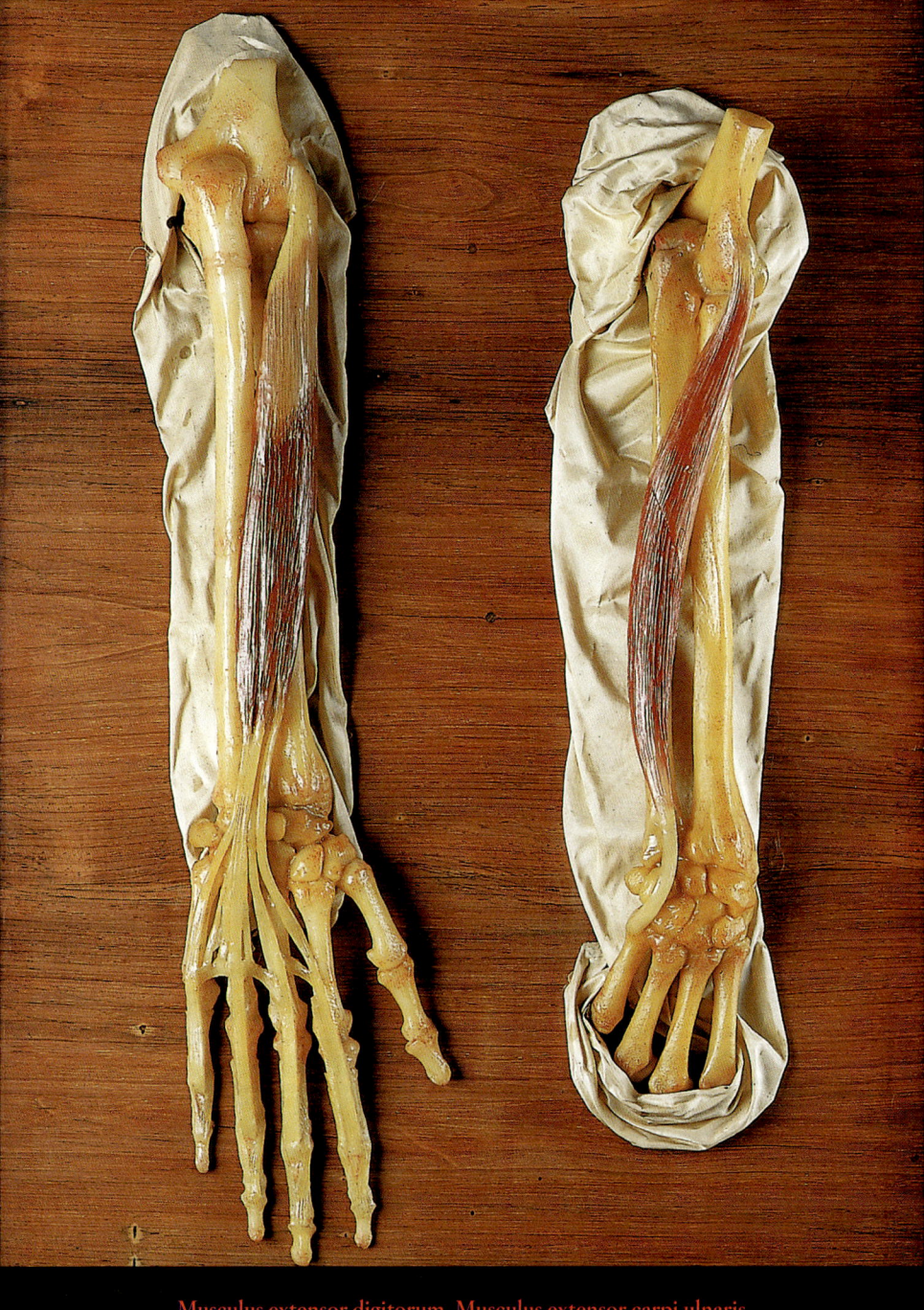

Musculus extensor digitorum, Musculus extensor carpi ulnaris

Display showing muscles bringing about extension of the fingers and
sideways movement of the hand

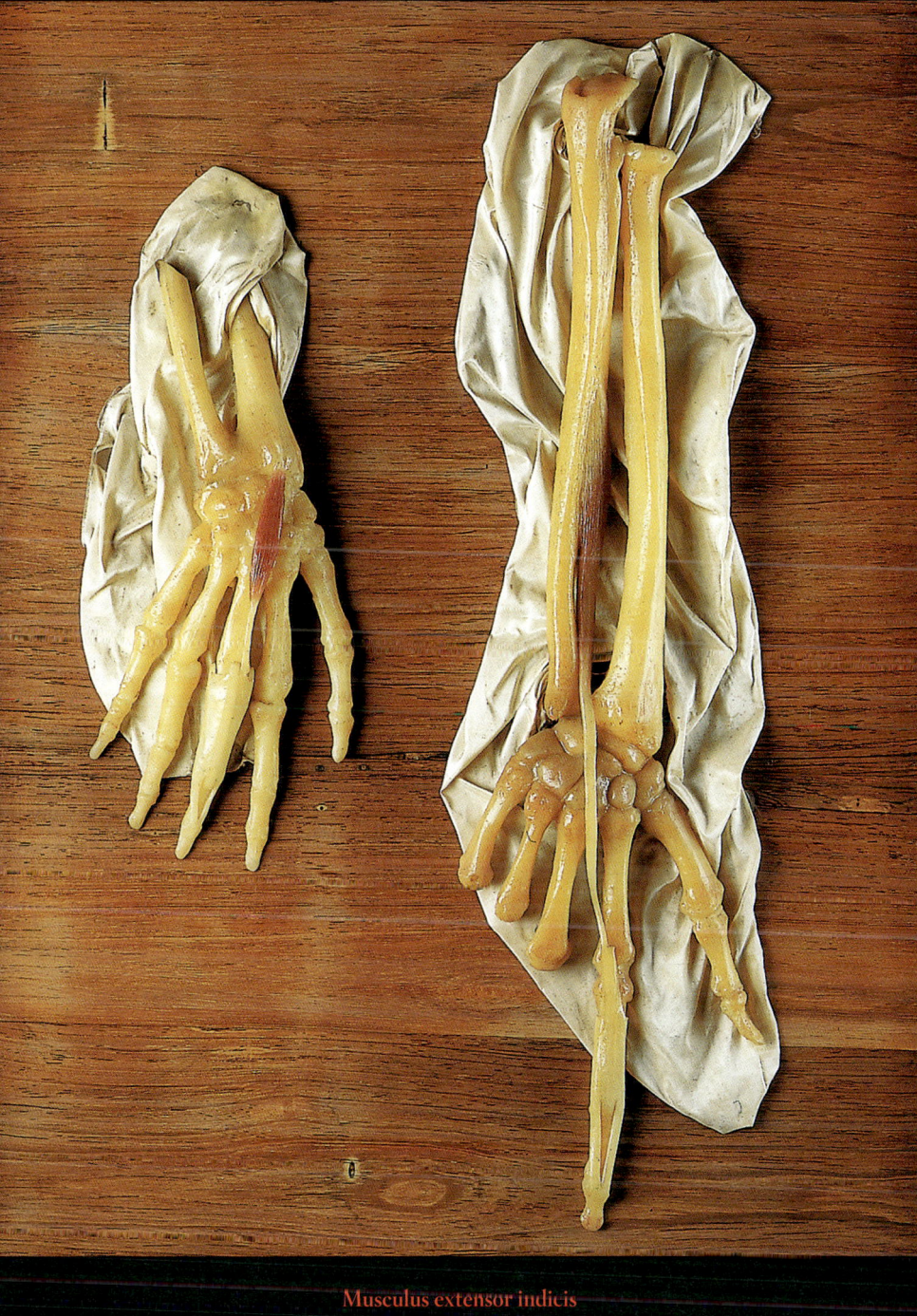

Musculus extensor indicis

The extensor muscle of the index finger

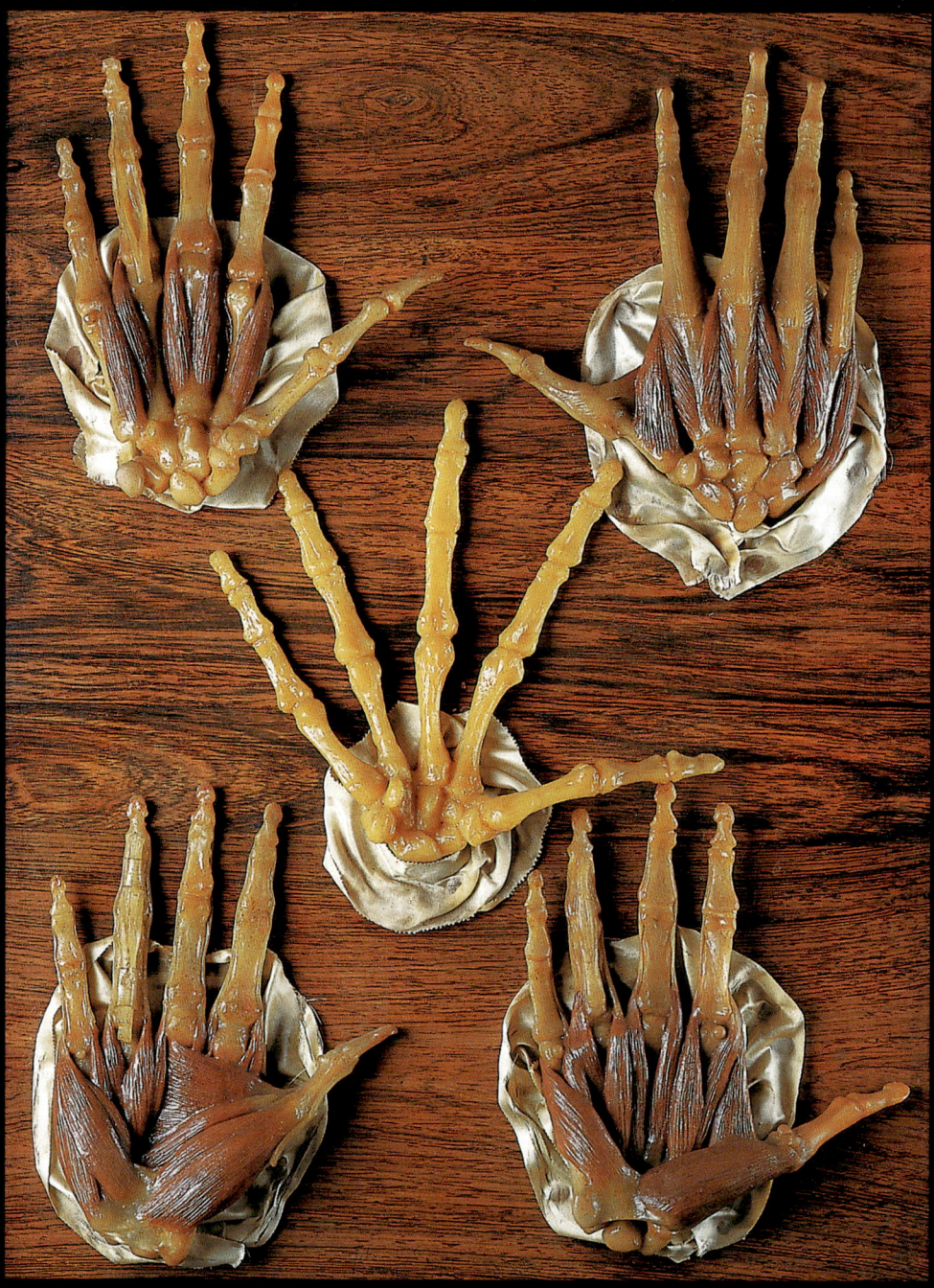

Musculi interossei manus, Musculus abductor pollicis brevis, Musculus flexor pollicis brevis,
Musculus adductor pollicis, Musculus abductor digiti minimi

Display showing the deep muscles in the hollow of the hand which bring
about fine movements of the fingers

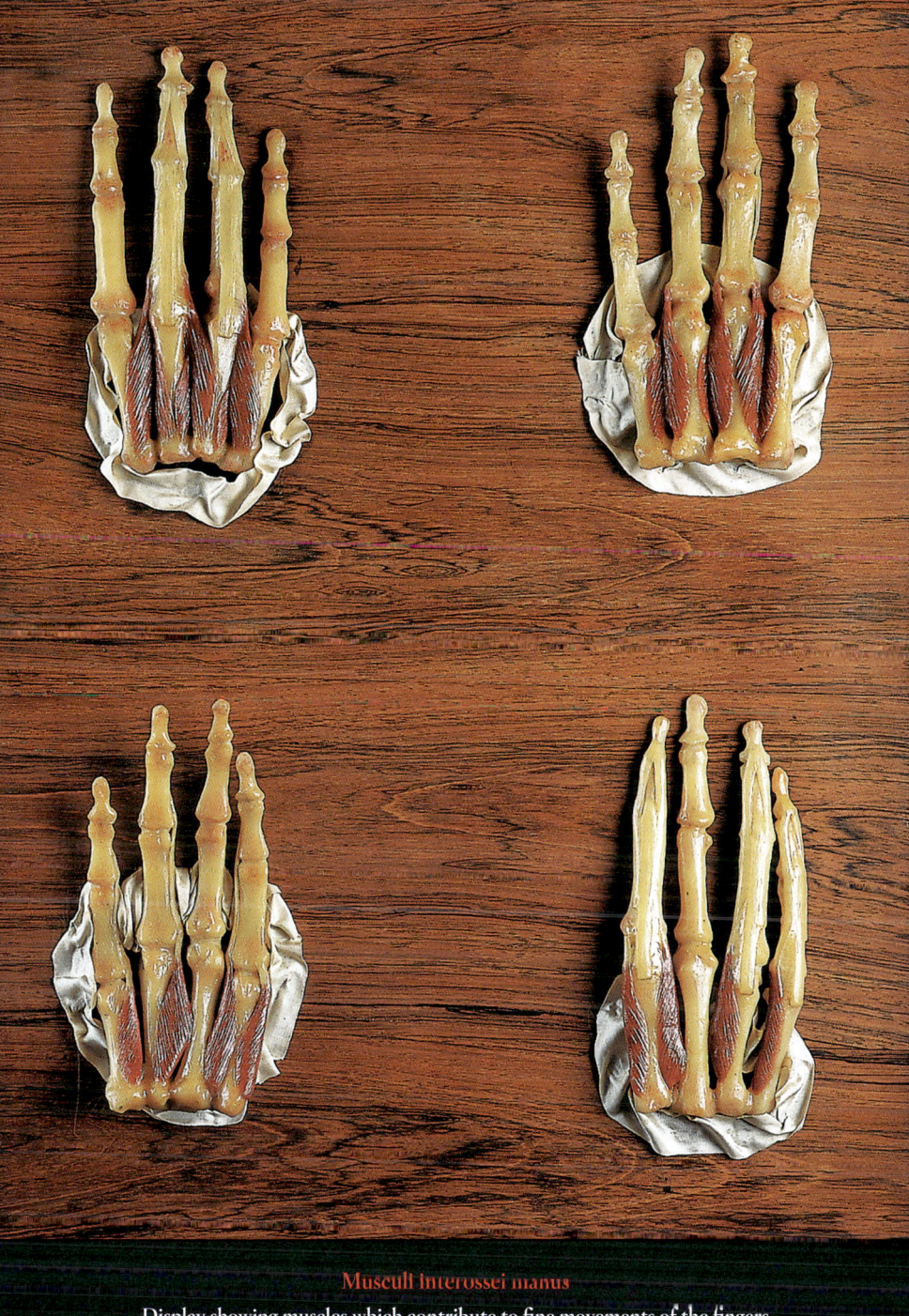

Musculi interossei manus

Display showing muscles which contribute to fine movements of the fingers

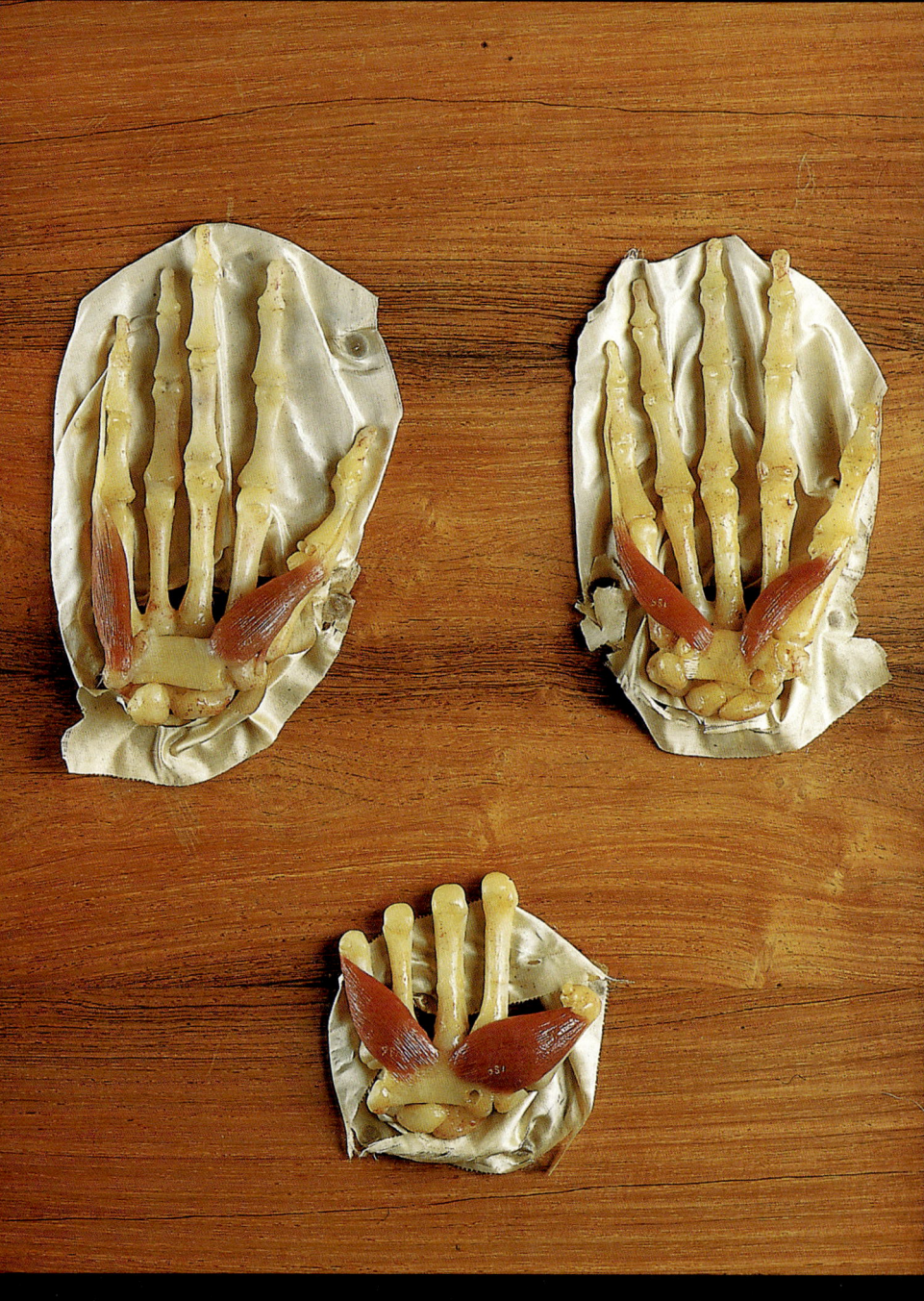

Display showing the muscles of the ball of the thumb and little finger

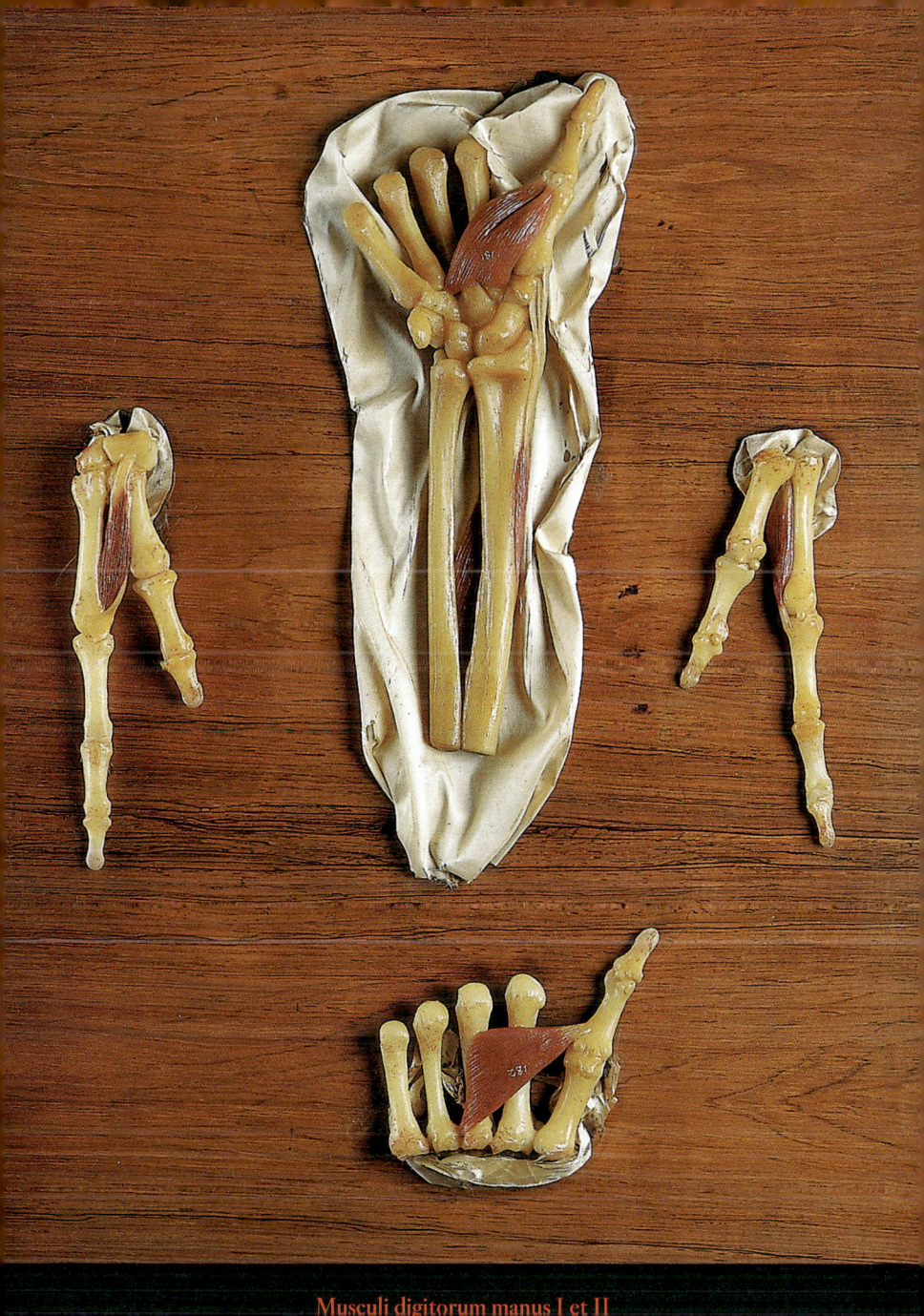

Musculi digitorum manus I et II
Musculus adductor pollicis, Musculus flexor pollicis brevis, Musculi interossei

Display showing the muscles responsible for fine movements of the index finger and thumb

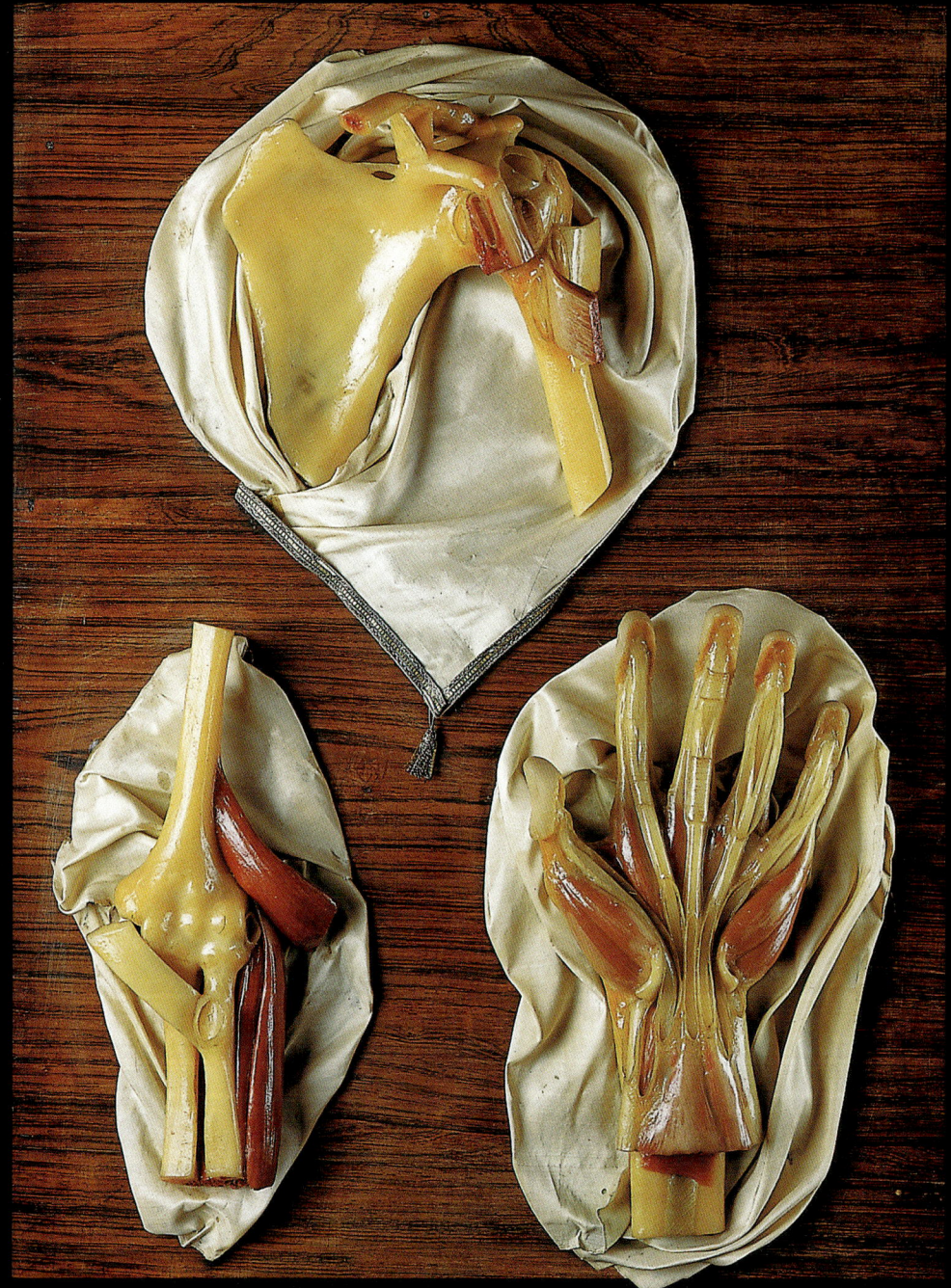

Articulatio humeri, Articulatio cubiti, Vaginae synoviales digitorum manus
Display showing the joints of the shoulder, elbow and wrist with muscle insertions and tendons

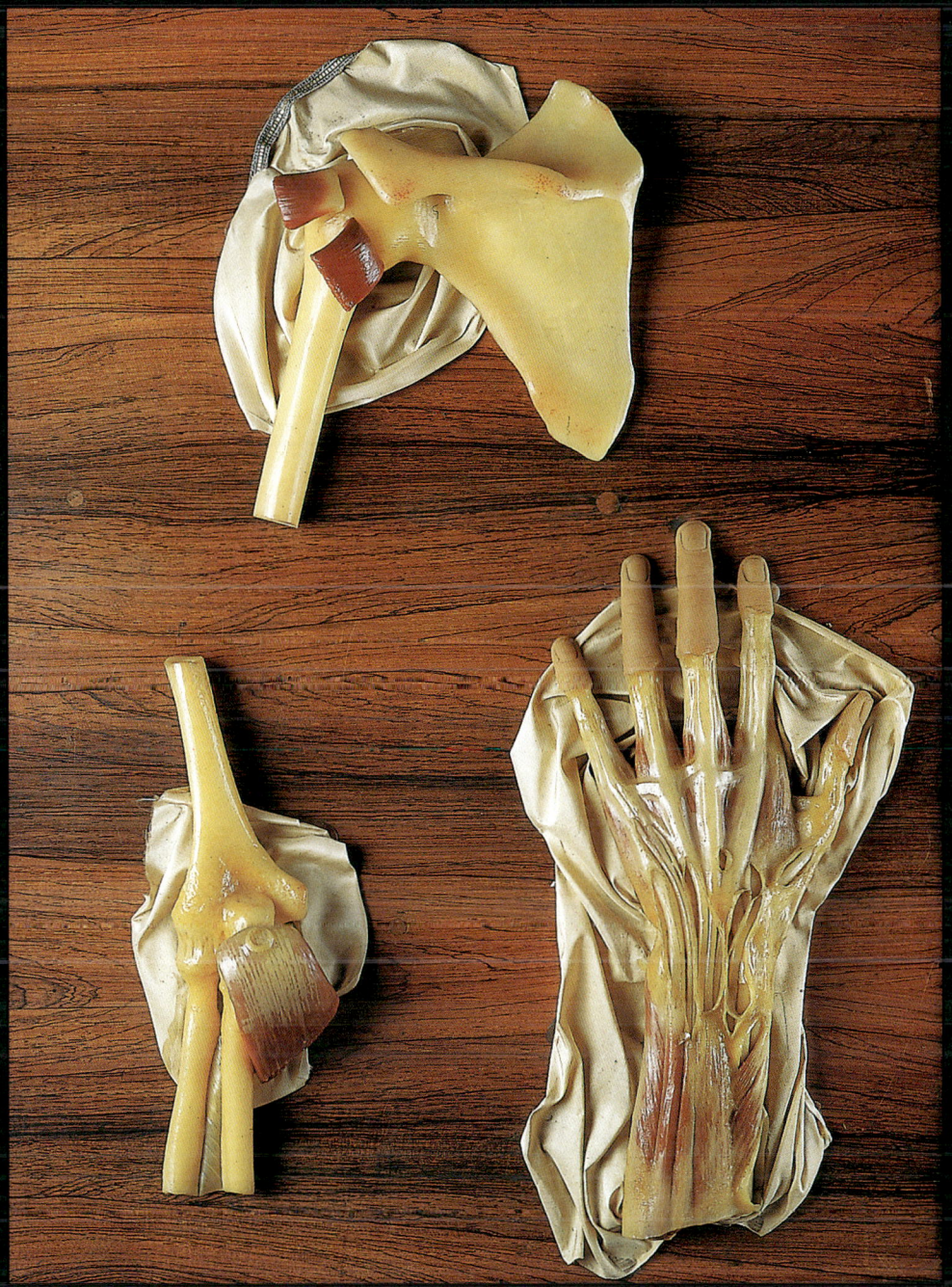

Articulatio humeri, Articulatio cubiti, Vaginae synoviales digitorum manus
Display showing the joints of the shoulder, elbow and wrist with muscle insertions and tendons

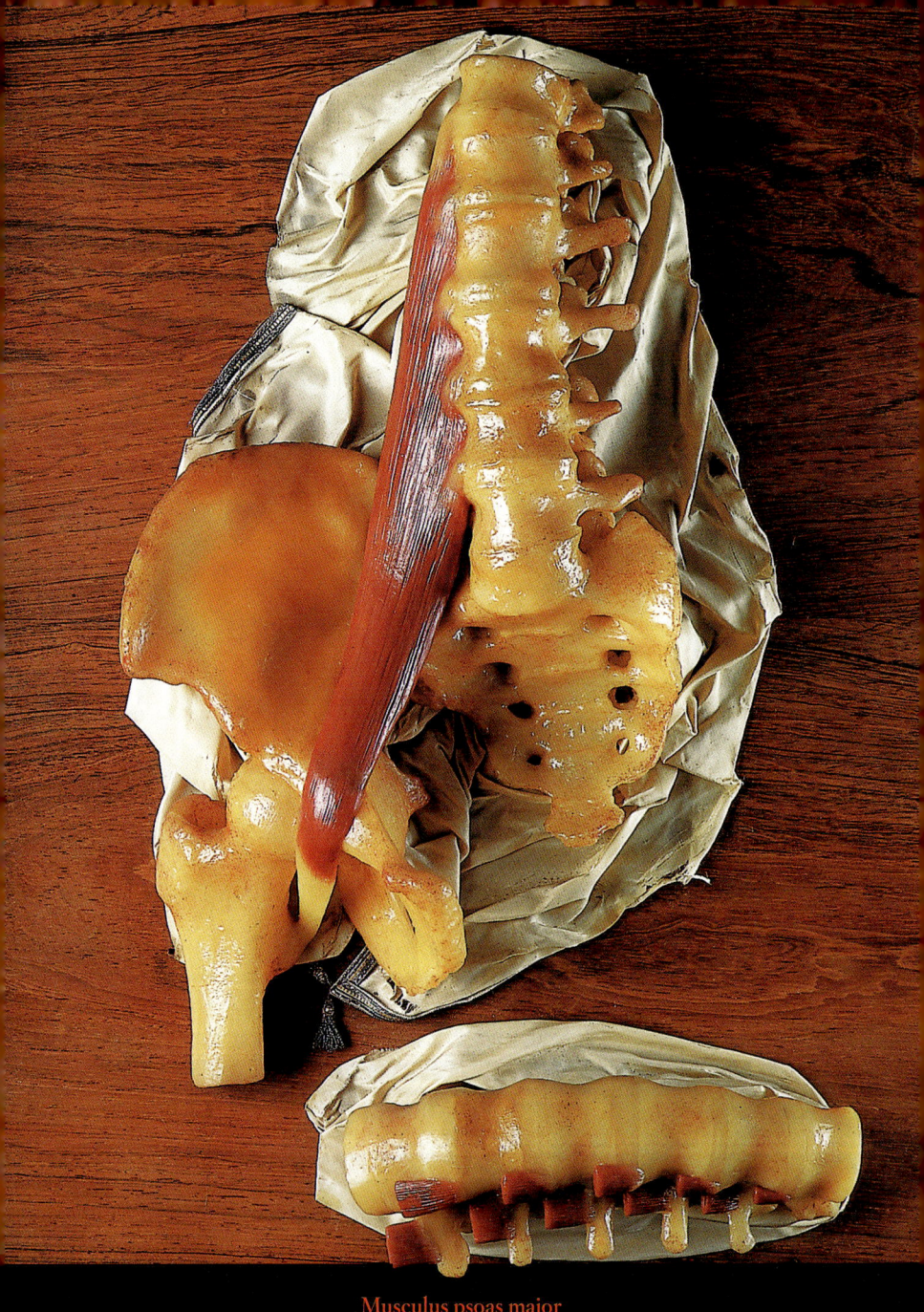

Musculus psoas major

A large muscle running downwards from the lumbar vertebral column viewed from in front

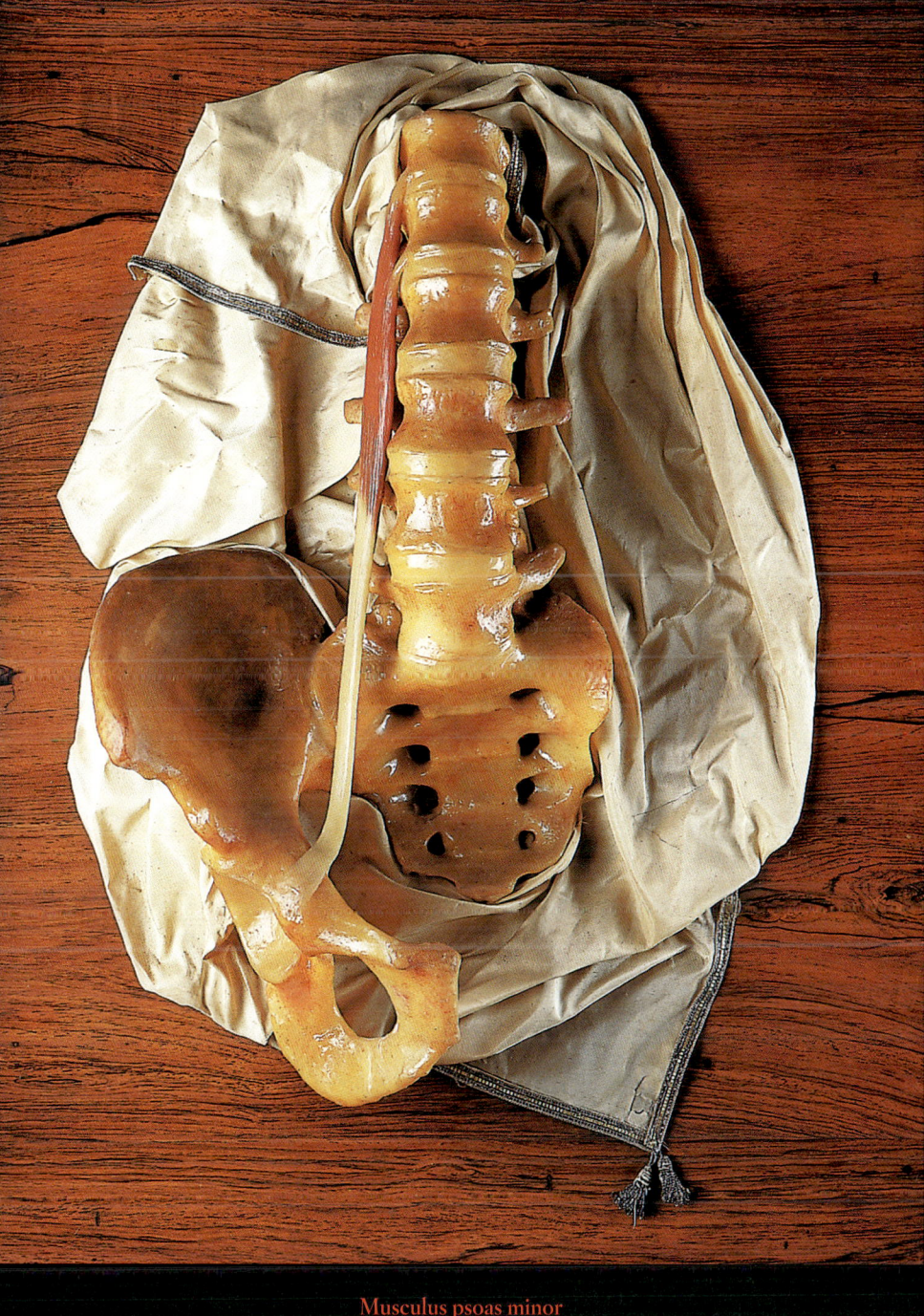

Musculus psoas minor

A small muscle running downwards from the lumbar vertebral column seen from in front

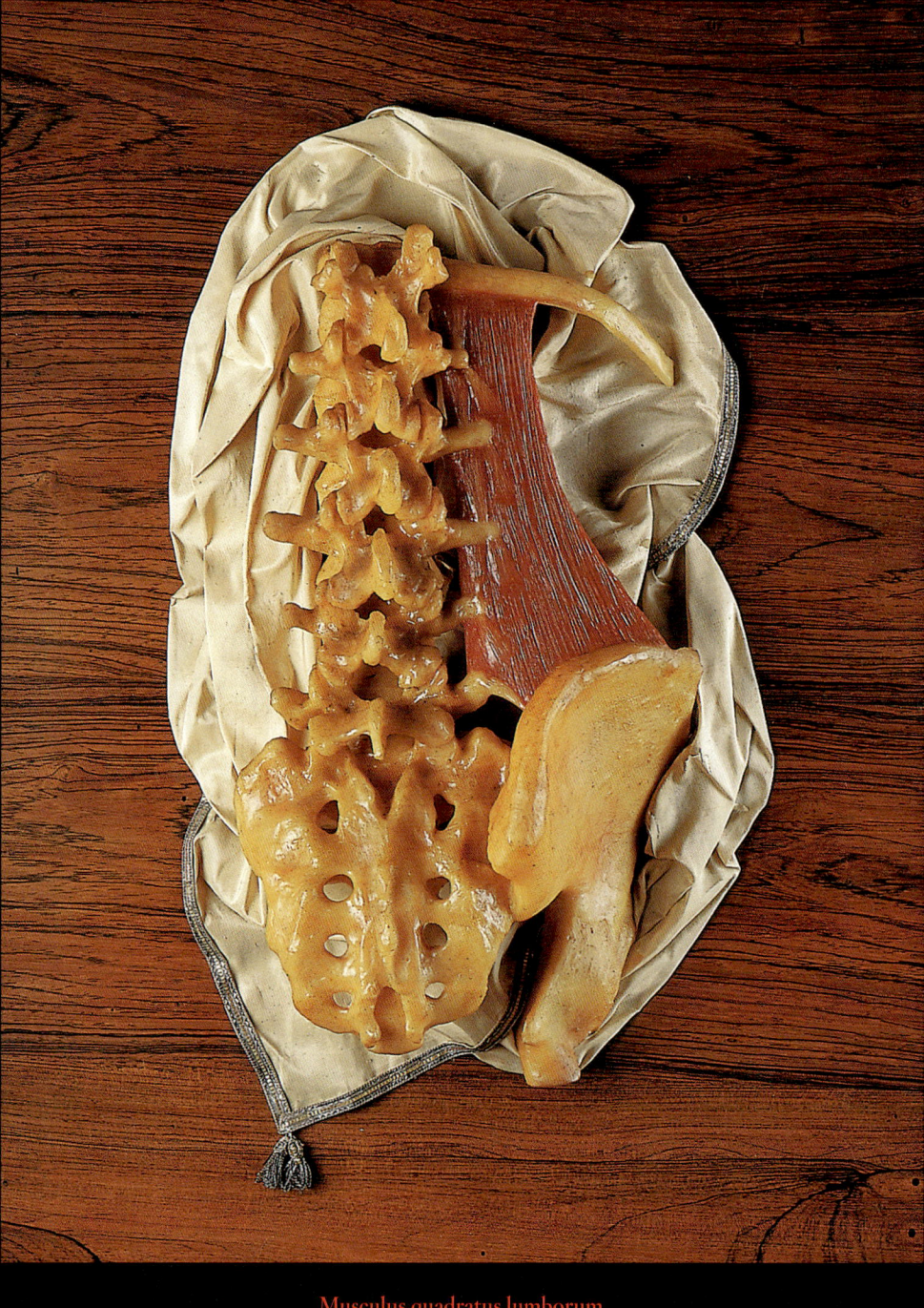

Musculus quadratus lumborum

A muscle of the posterior abdominal wall seen from behind

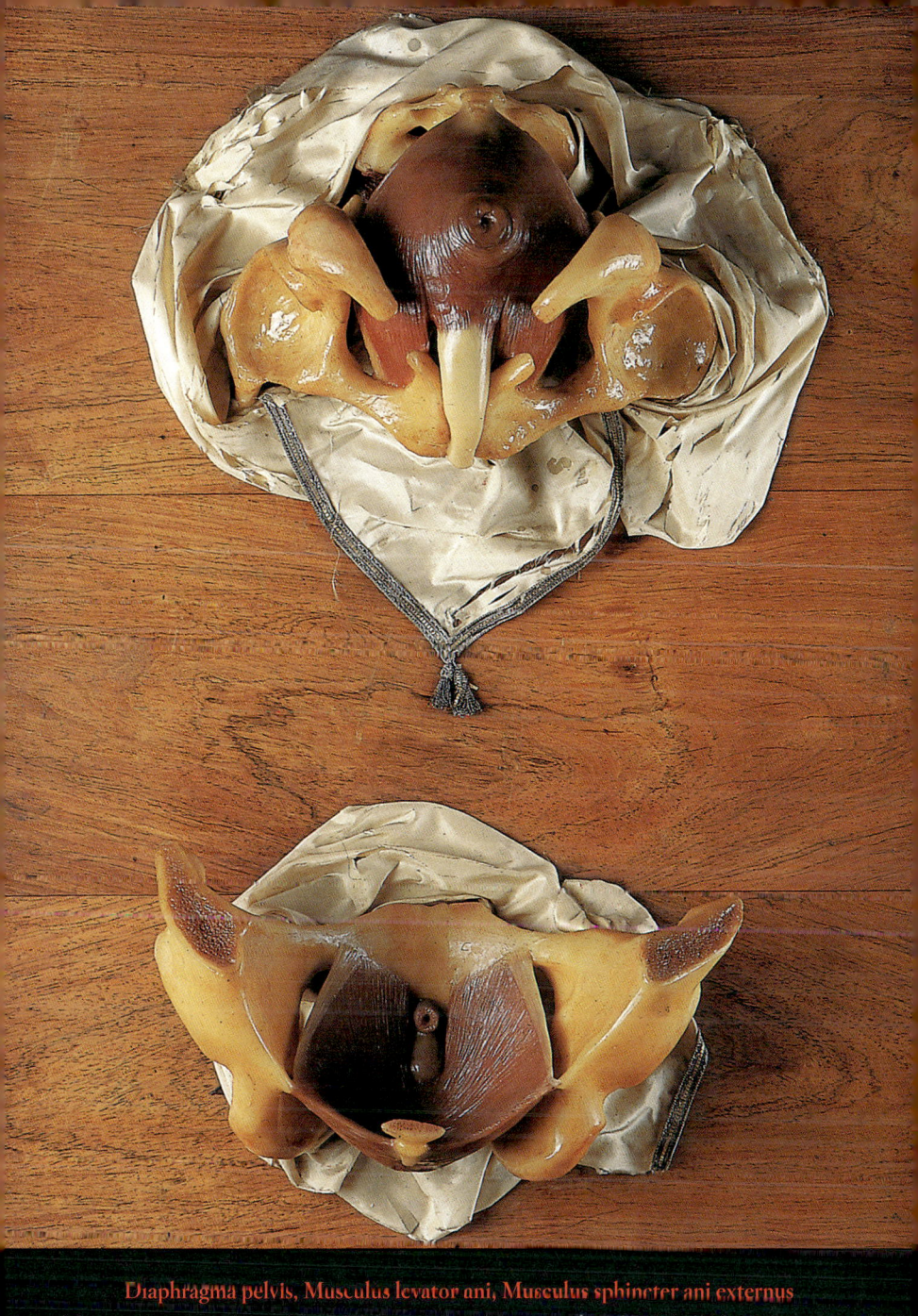

Diaphragma pelvis, Musculus levator ani, Musculus sphincter ani externus

View of the muscles of the floor of the pelvis from below (upper specimen) and from above (lower specimen) also showing the circular course of the muscles of the external sphincter

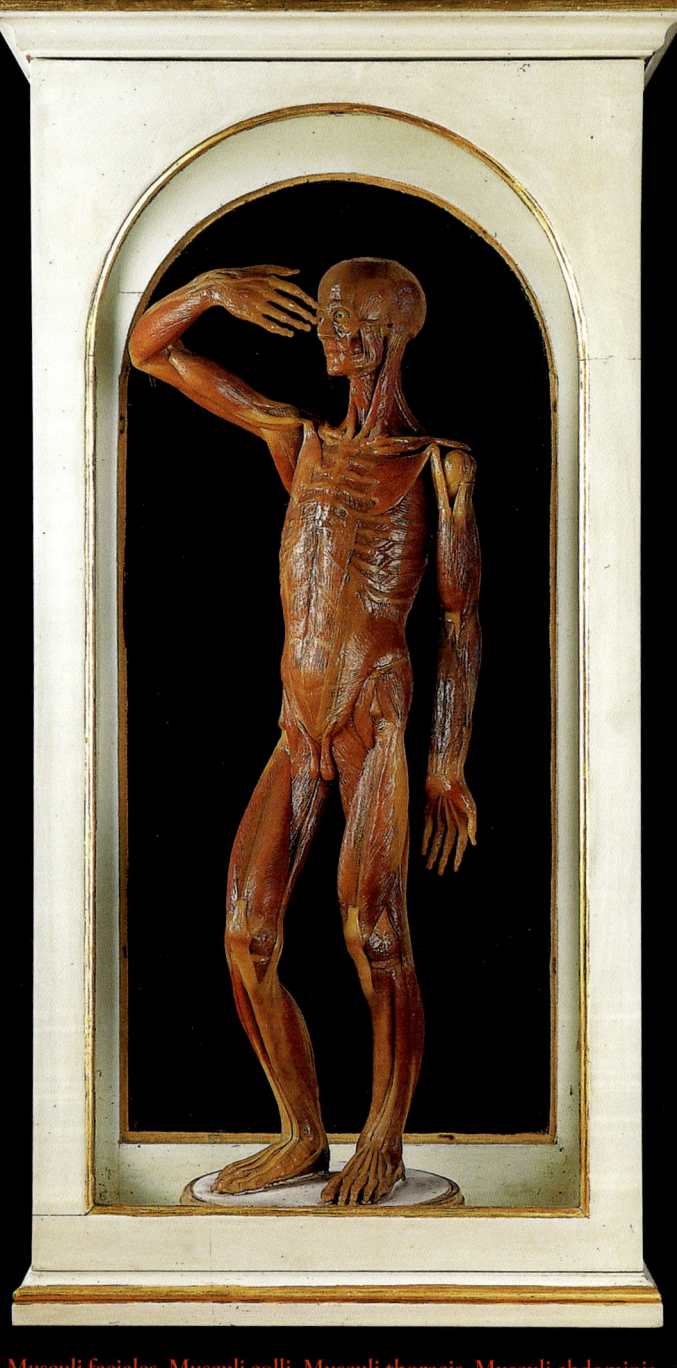

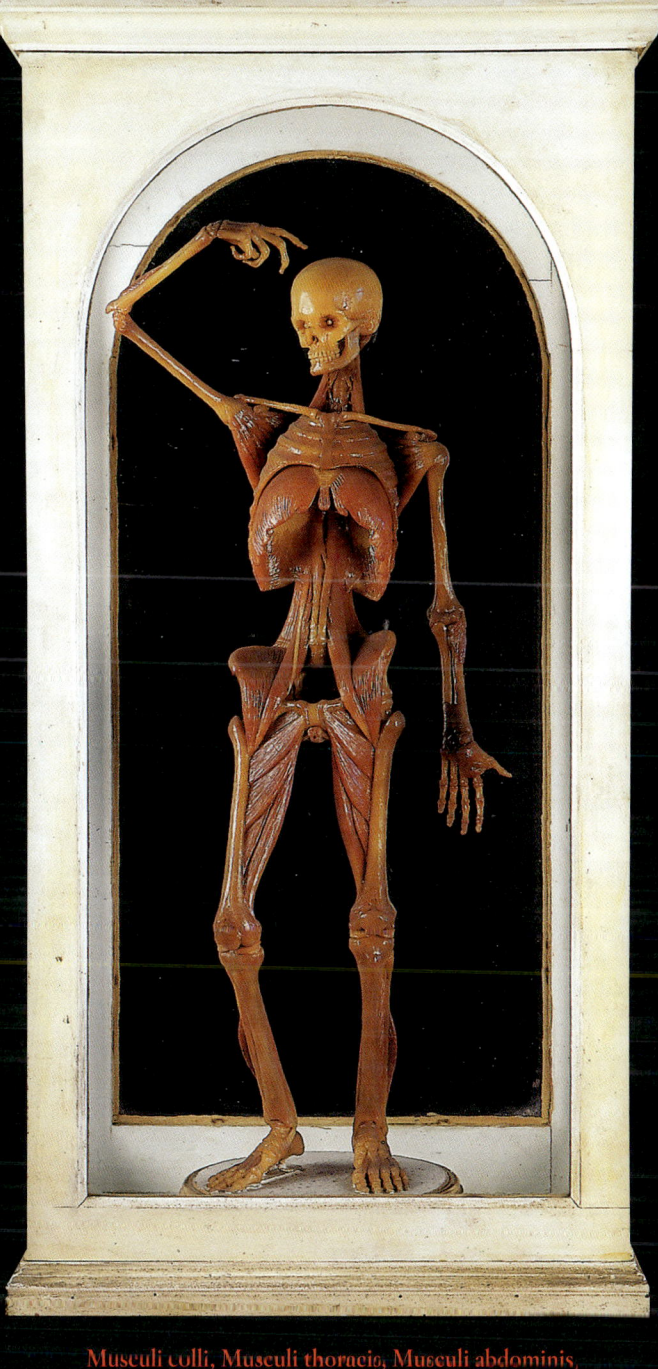

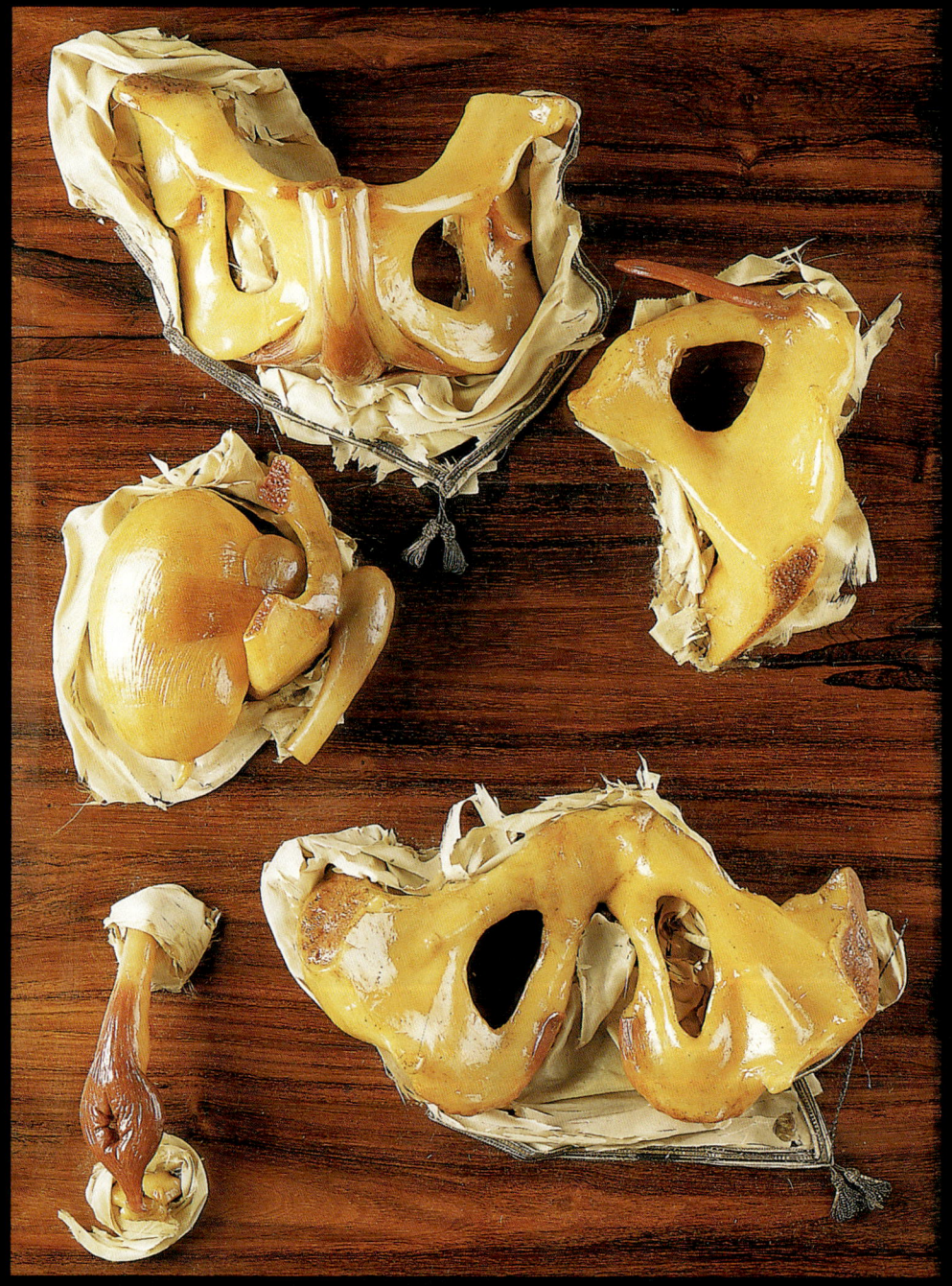

Cingulum membri inferioris

Several views of the bony pelvis and pelvic floor

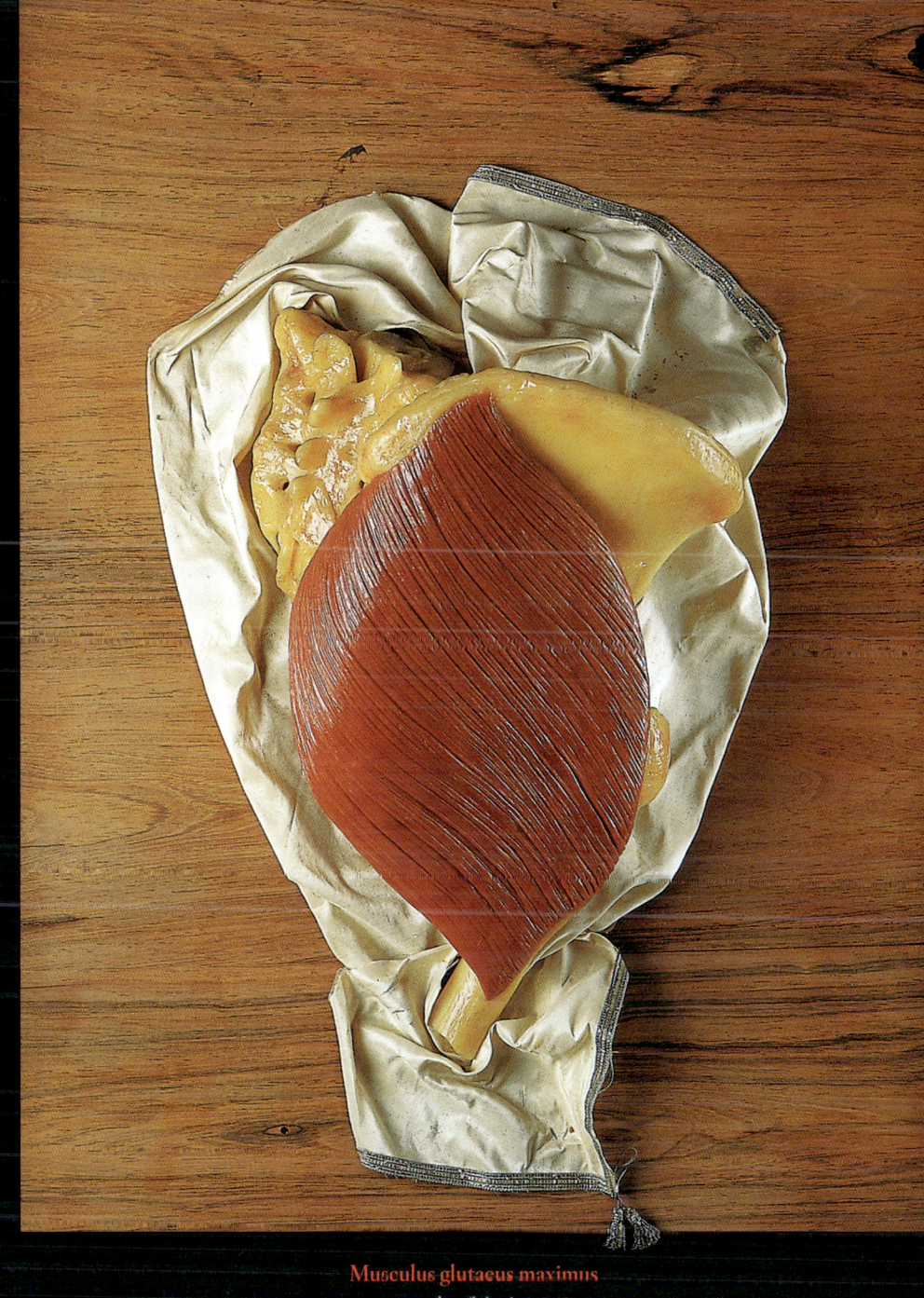

Musculus glutaeus maximus

The large outermost muscle of the buttock seen from the side

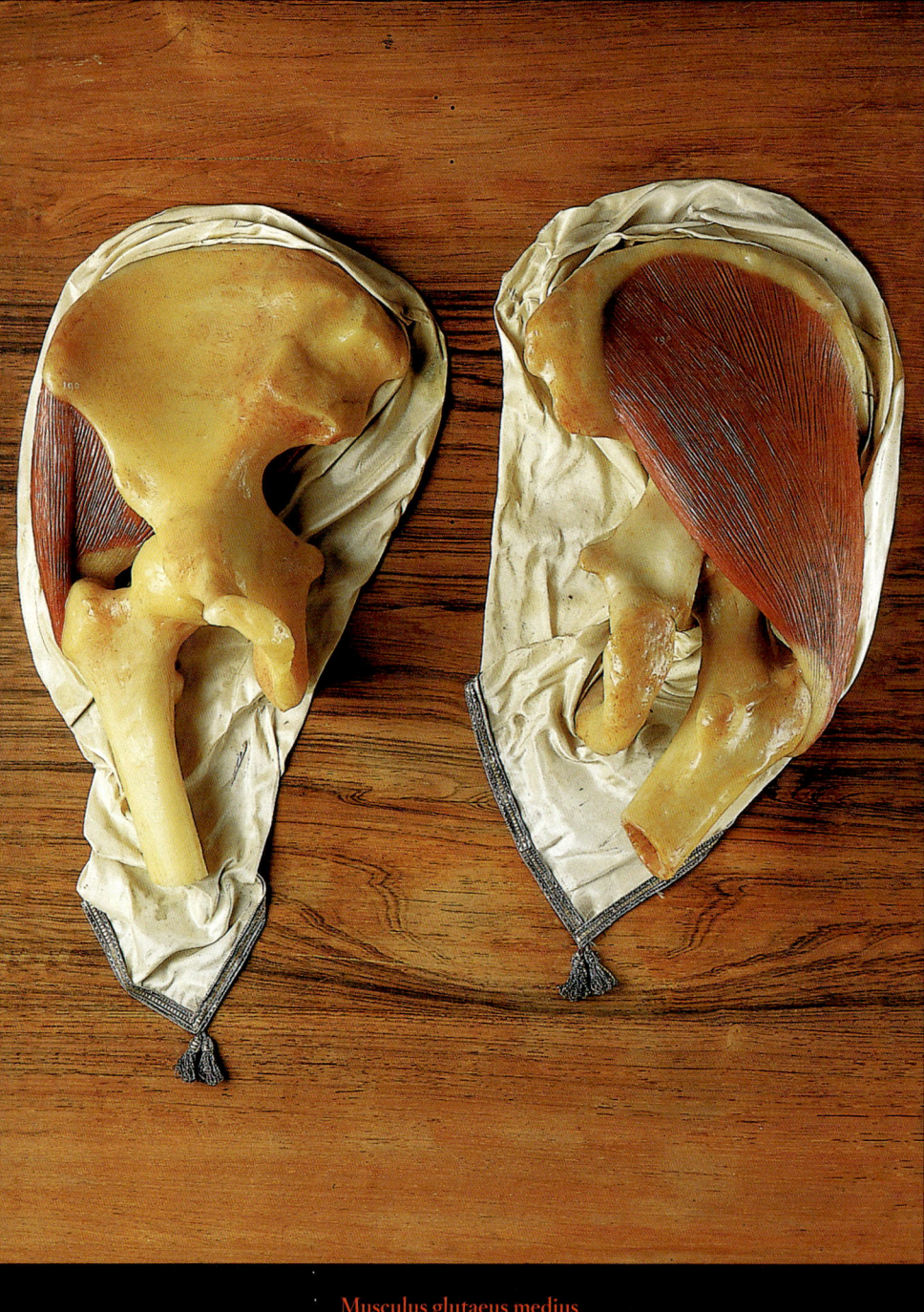

Musculus glutaeus medius

Display showing part of the muscles of the middle layer of the buttock

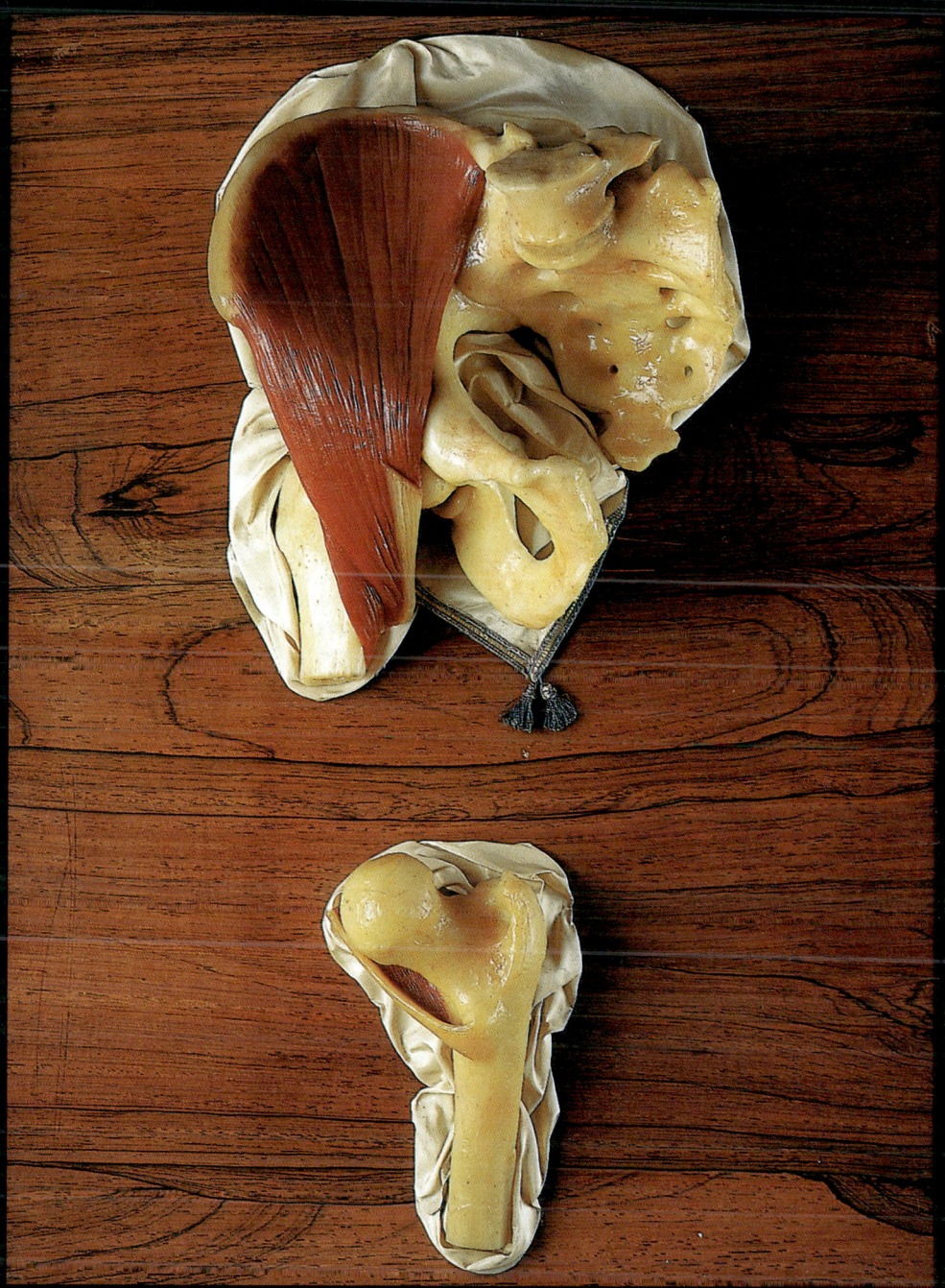

Musculus iliacus

A muscle attached to the inside of the wing of the iliac bone

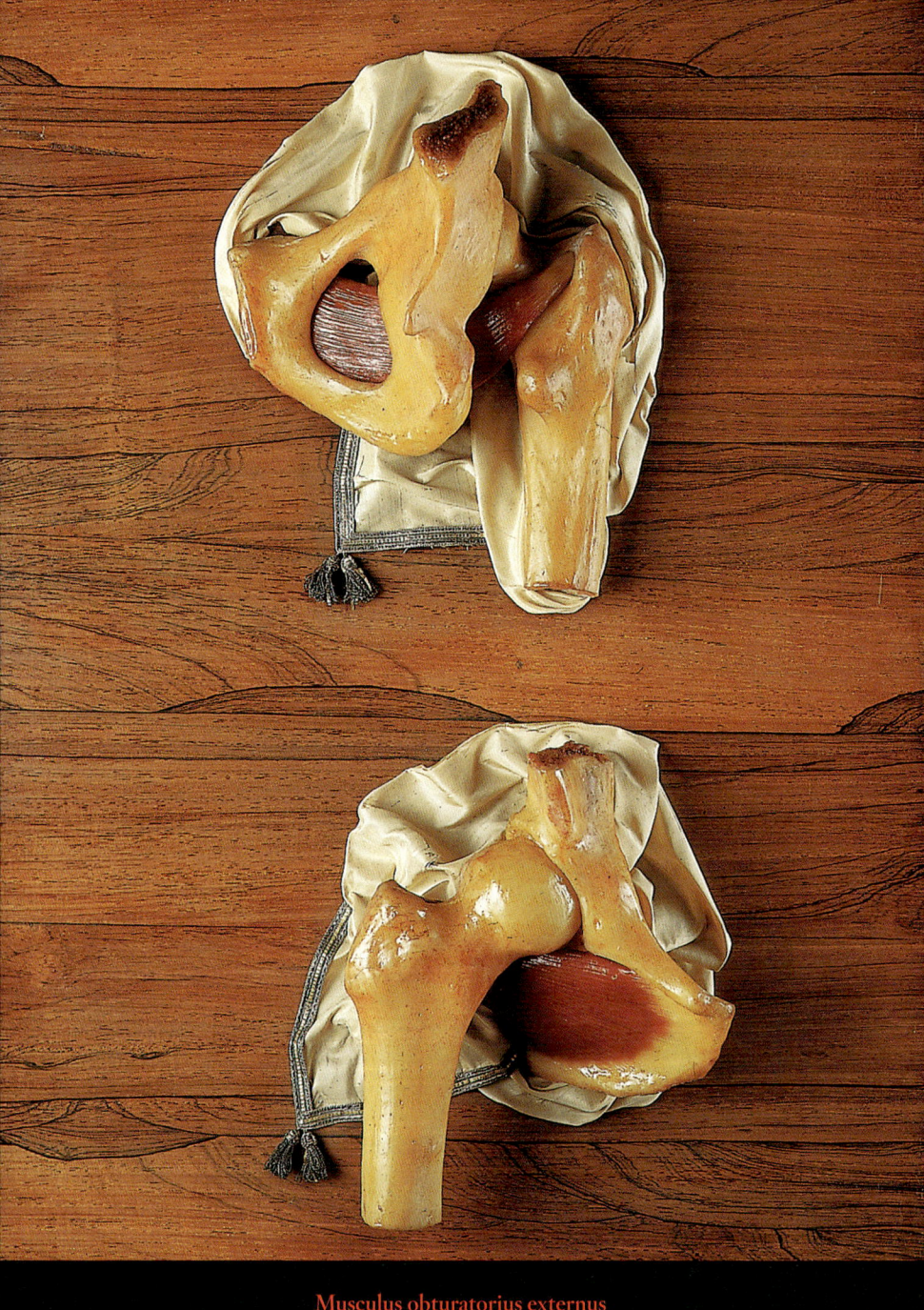

Musculus obturatorius externus

A muscle which contributes to rotation at the hip joint

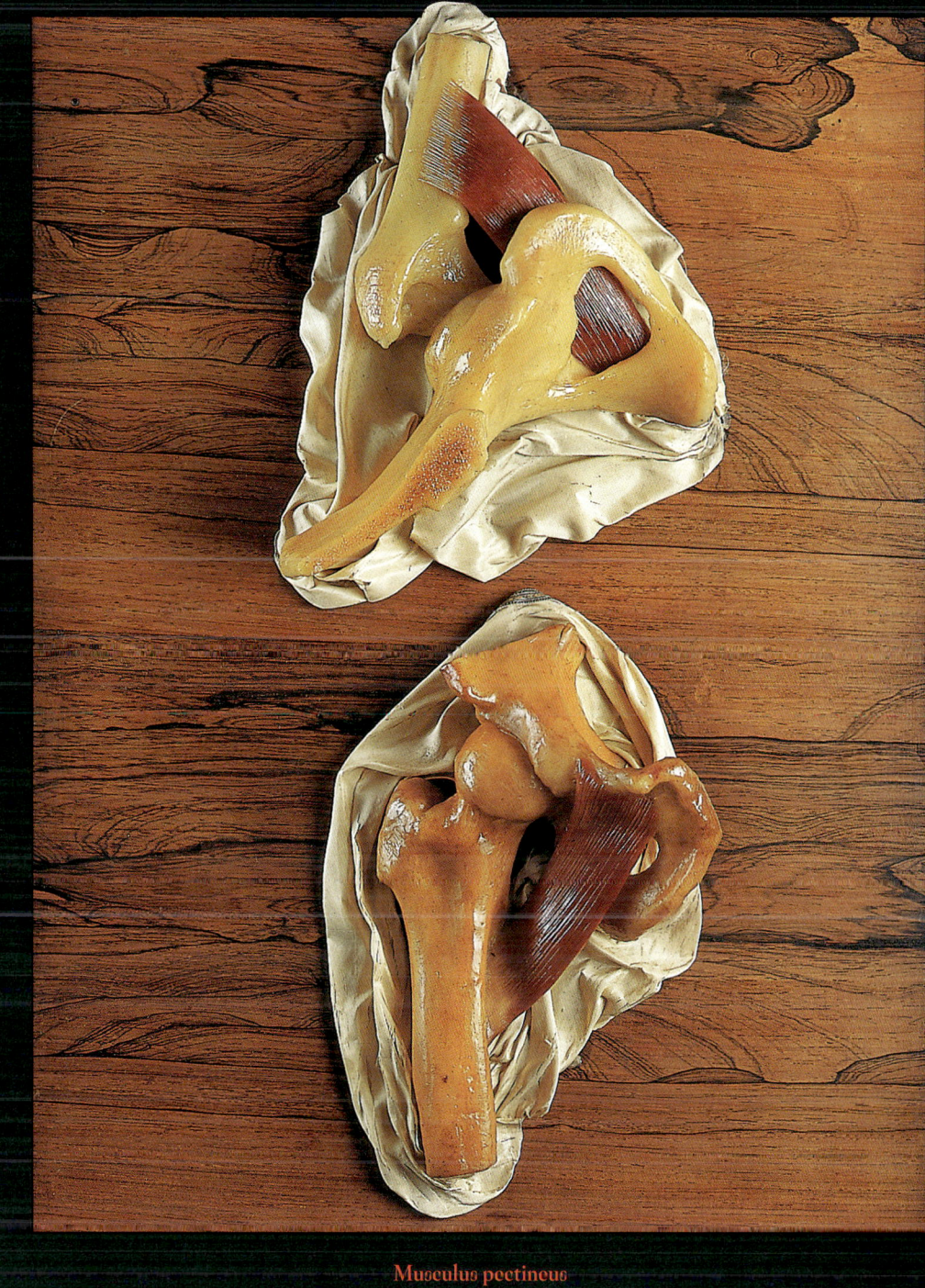

Musculus pectineus

A deeply placed muscle of the thigh

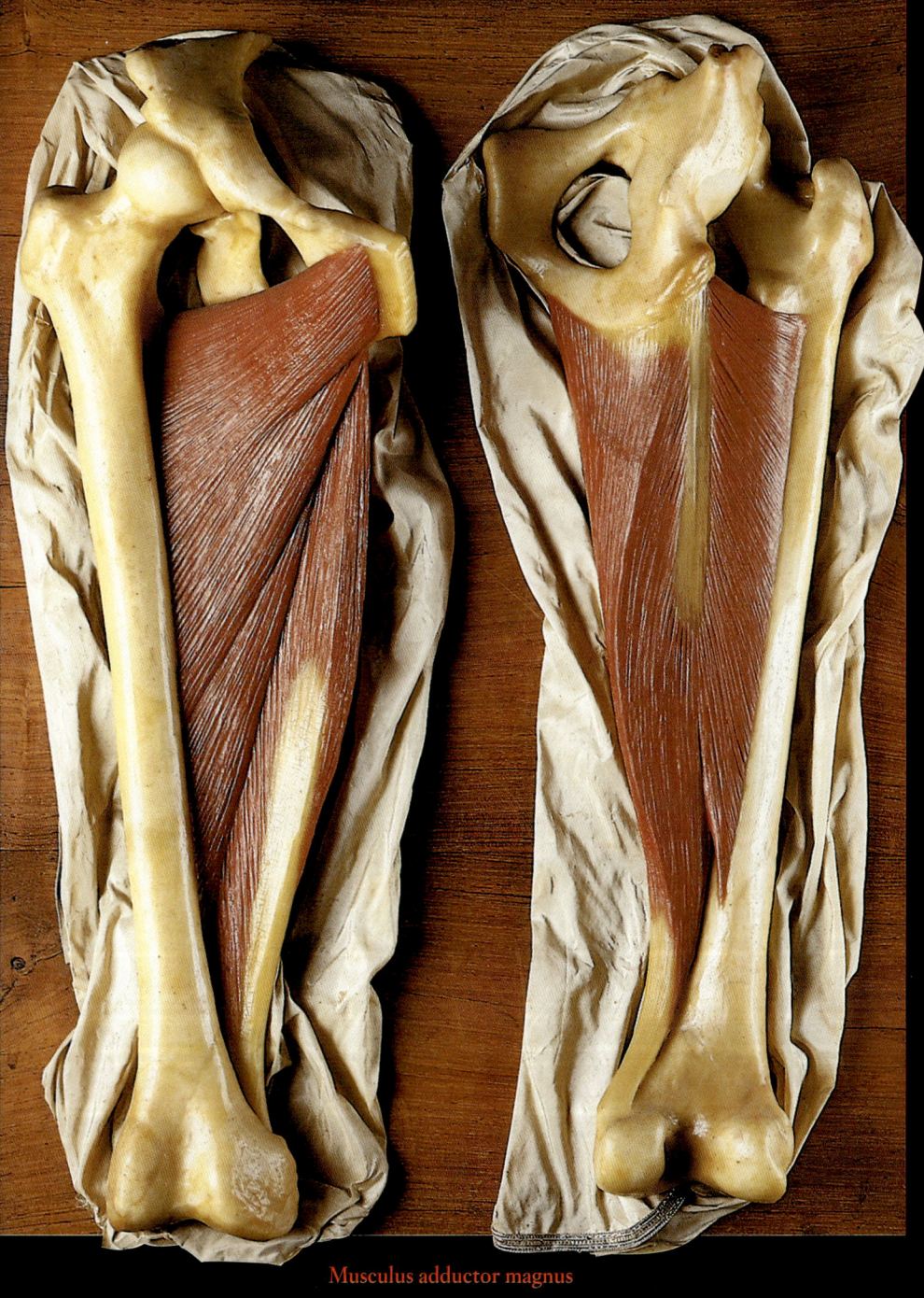

Musculus adductor magnus

A muscle on the inside of the thigh

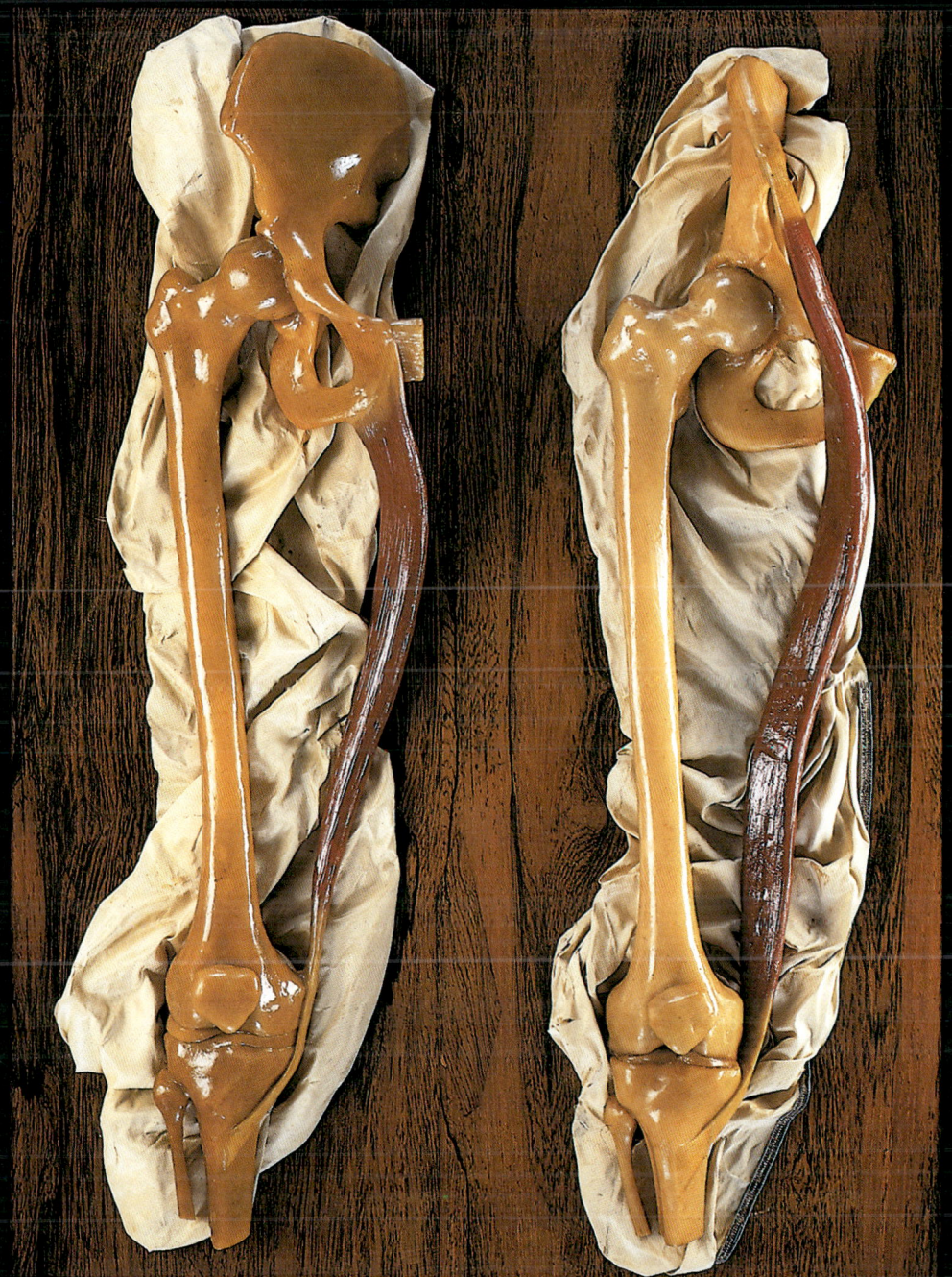

Musculus gracilis, Musculus sartorius
Display showing muscles of the thigh

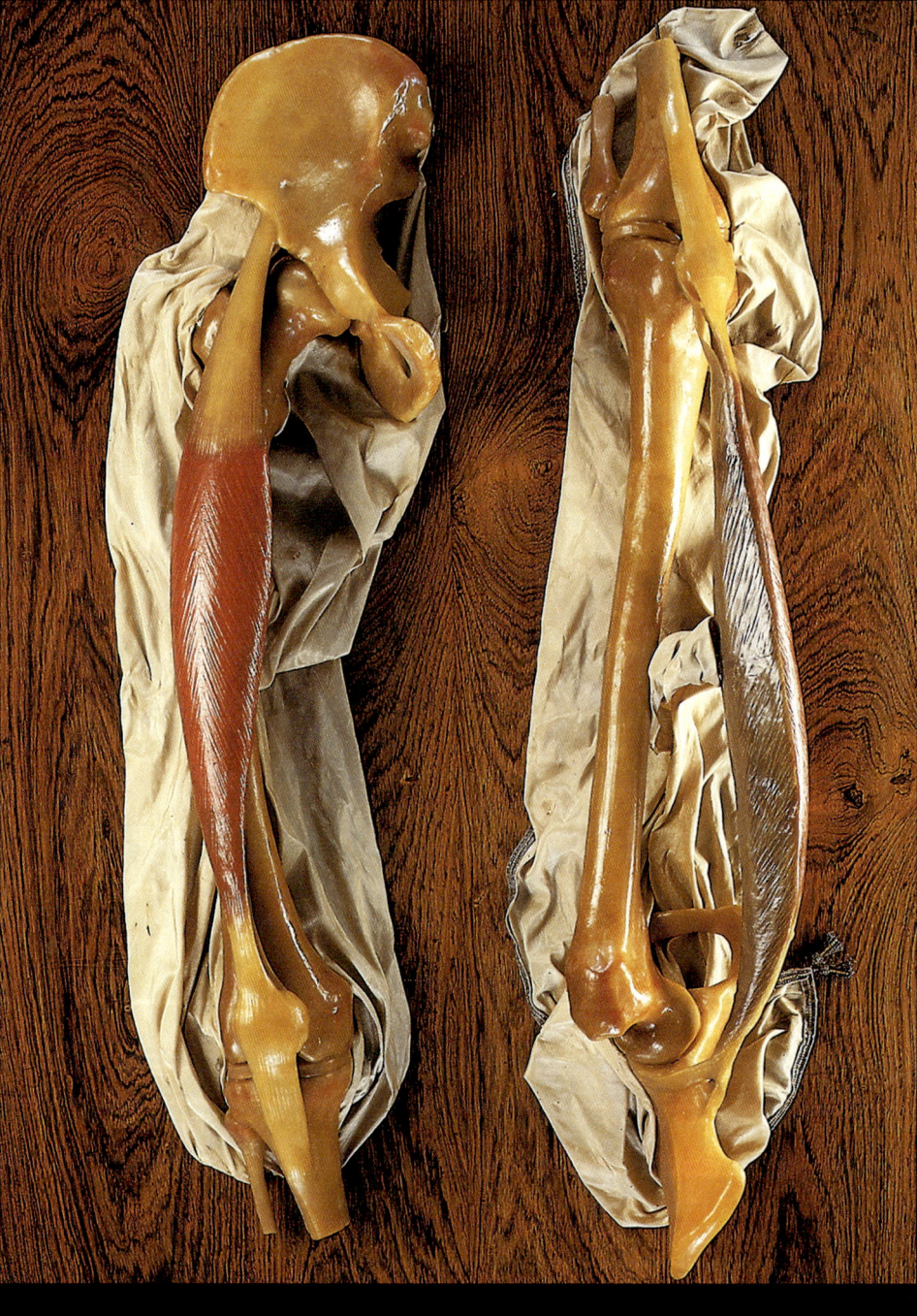

Musculus rectus femoris

Two views of a muscle at the front of the thigh

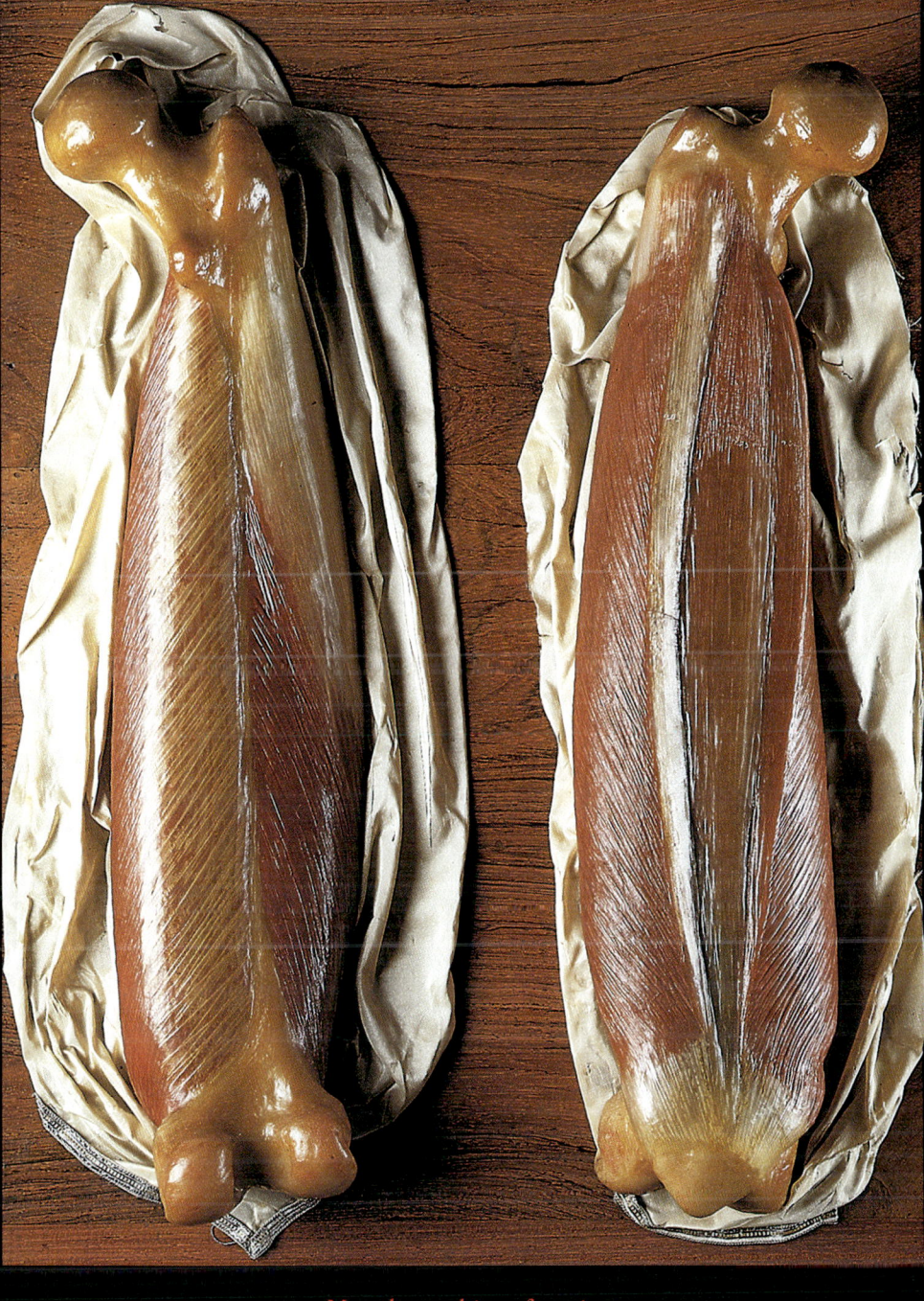

Musculus quadriceps femoris

Display showing the deeper parts of the four-headed muscle of the thigh

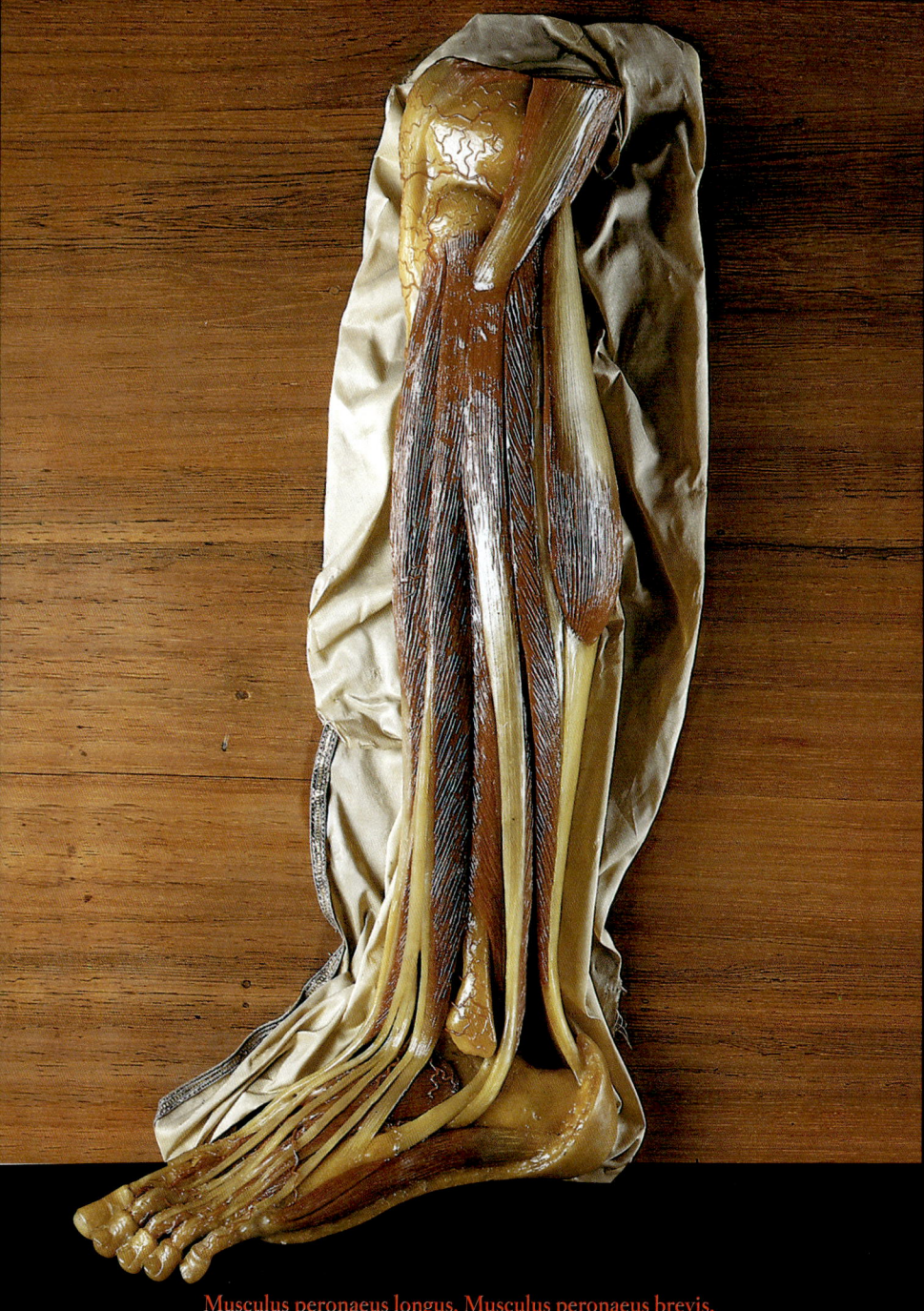

Musculus peronaeus longus, Musculus peronaeus brevis,
Musculus extensor digitorum longus, Musculus triceps surae

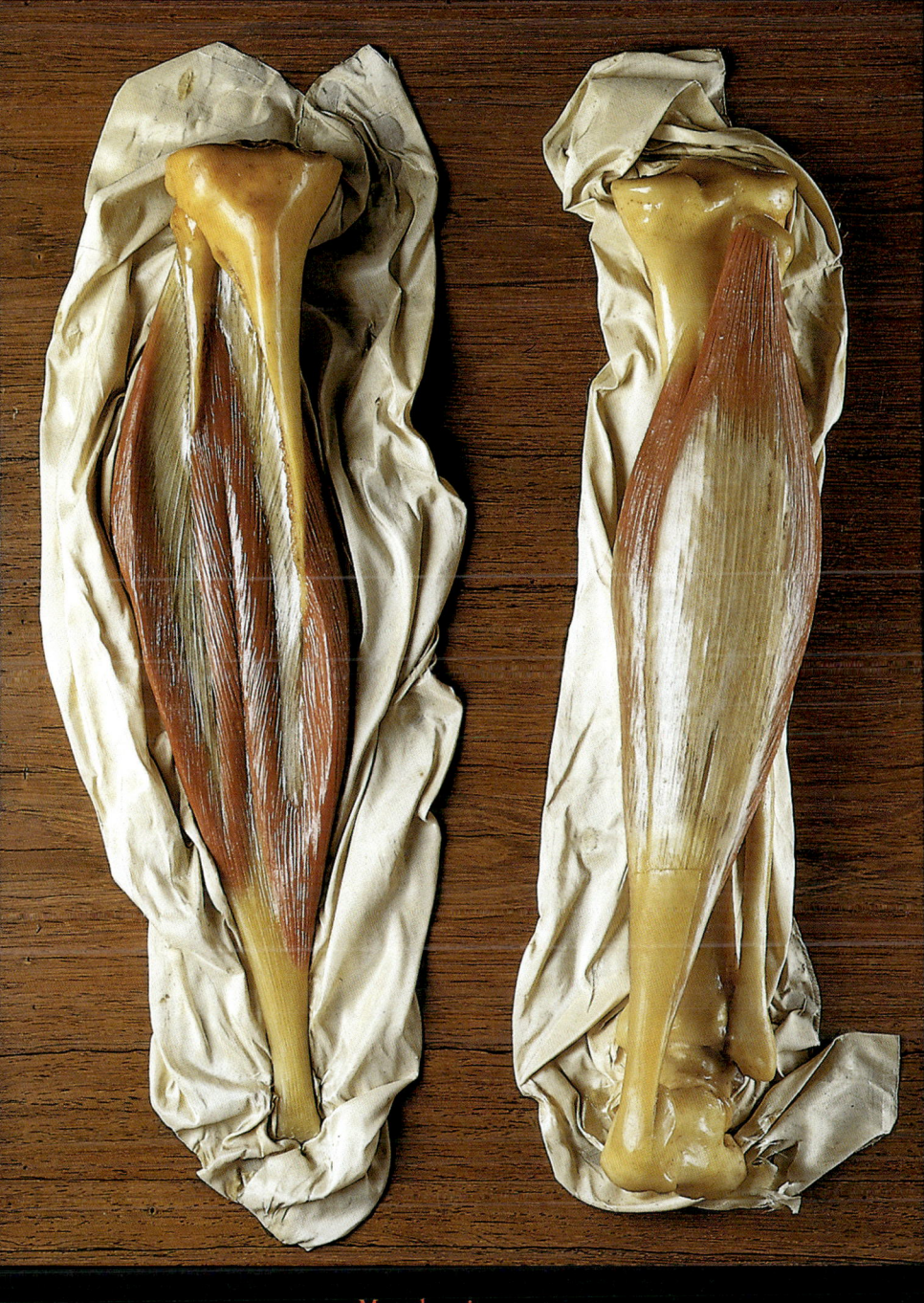

Musculus triceps surae

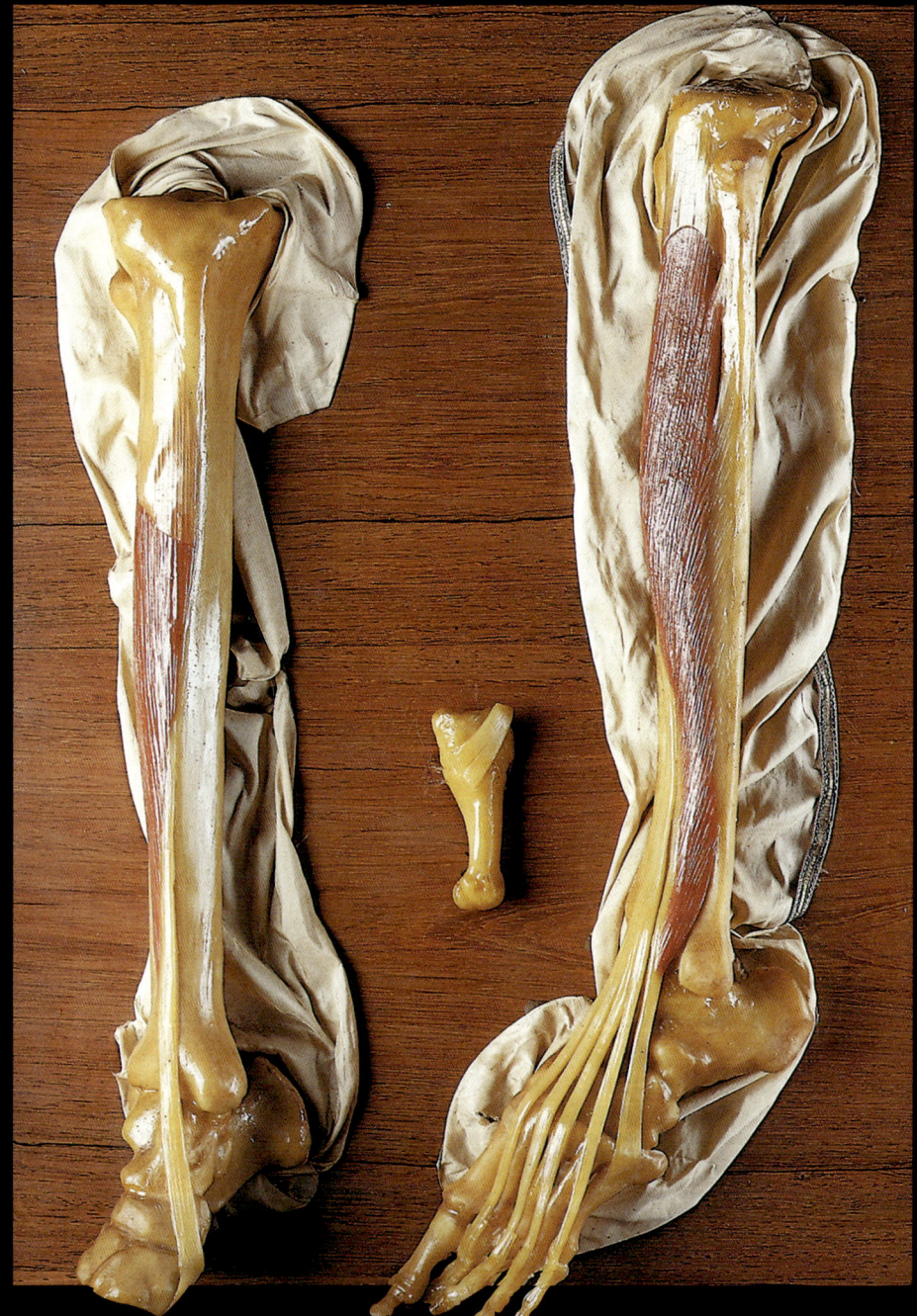

Musculus tibialis anterior, Musculus extensor digitorum longus

The long extensor muscle of the toes

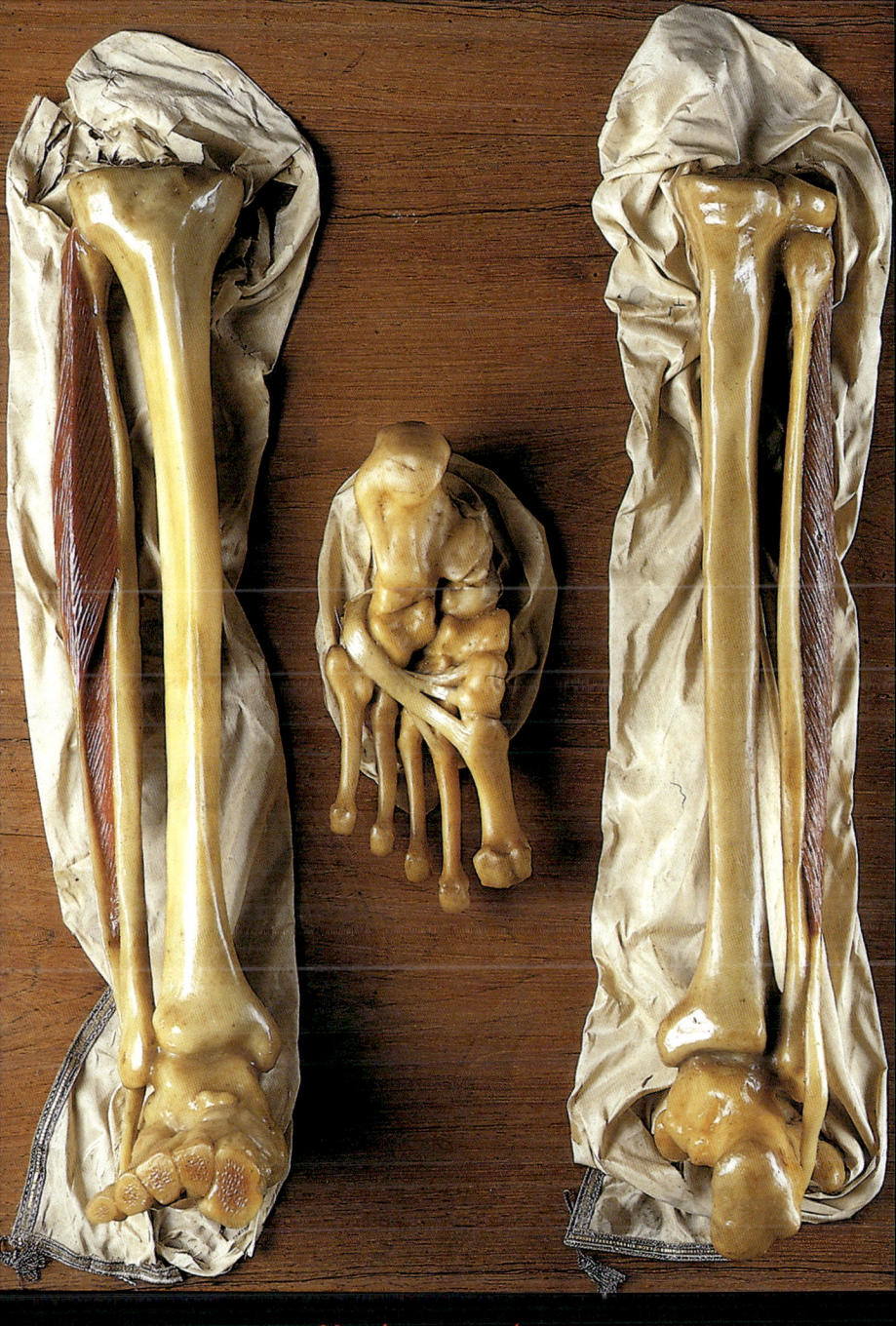

Musculus peronaeus longus

A long muscle on the outer side of the lower leg which helps to maintain the tension
of the arch of the foot

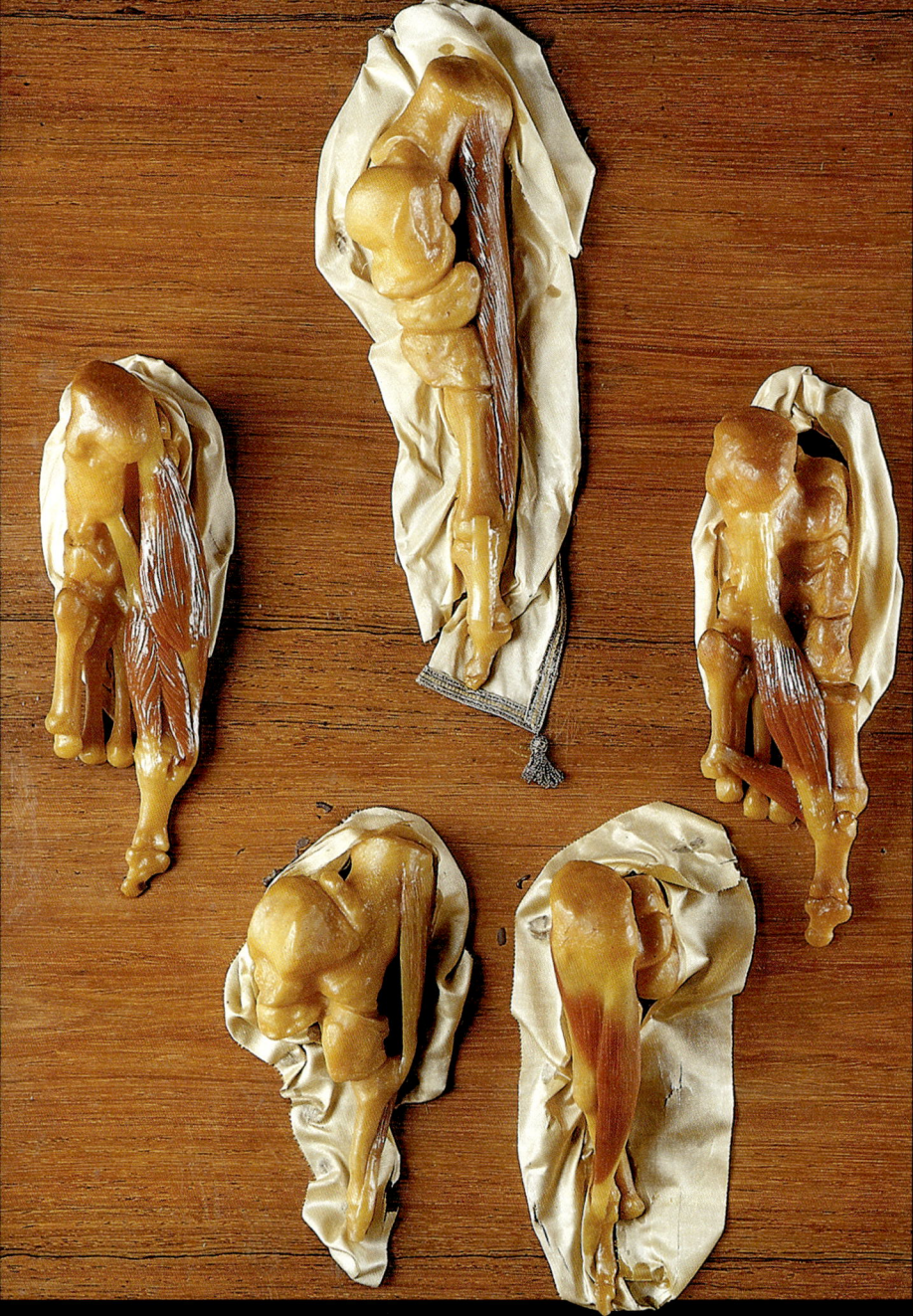

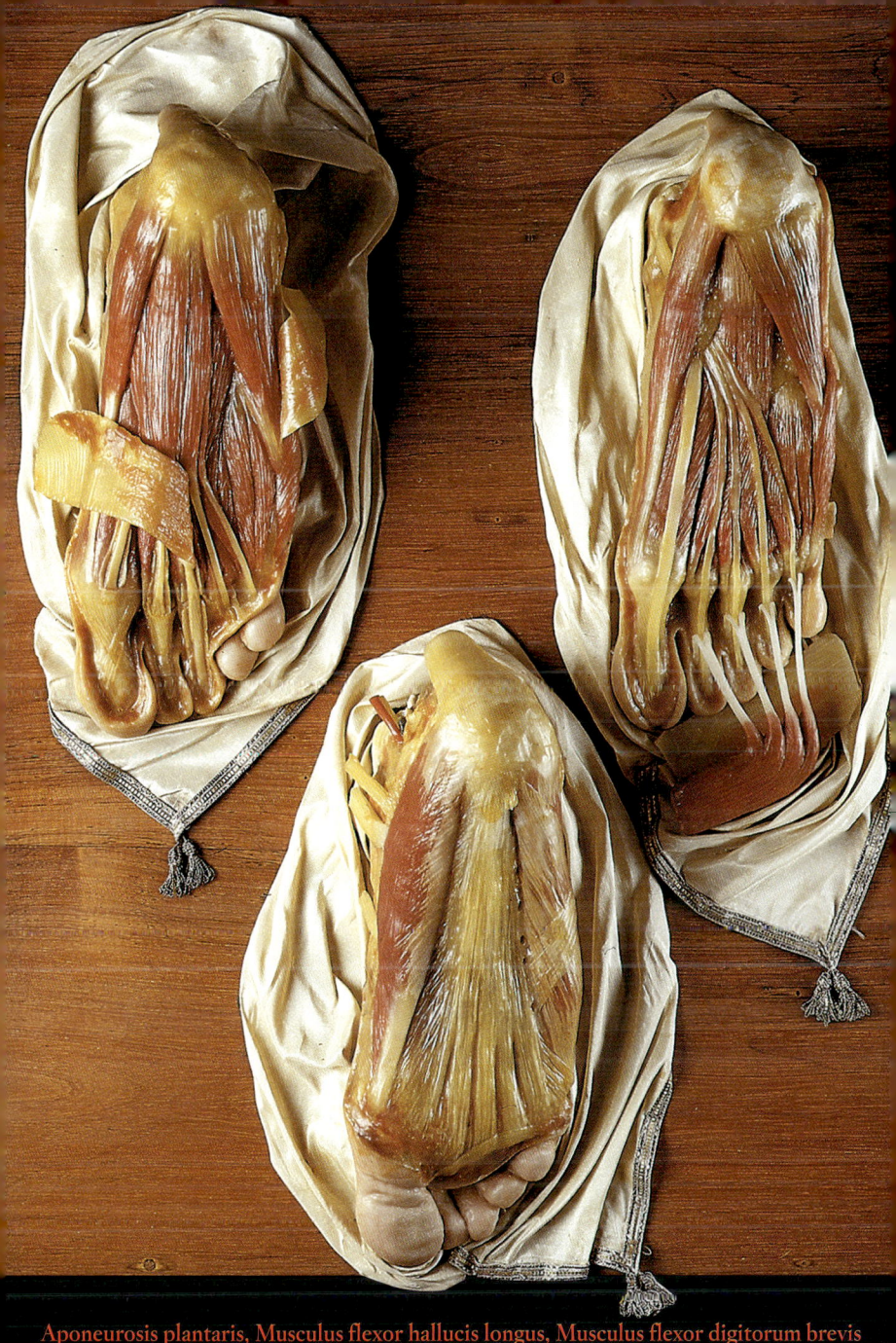

Aponeurosis plantaris, Musculus flexor hallucis longus, Musculus flexor digitorum brevis

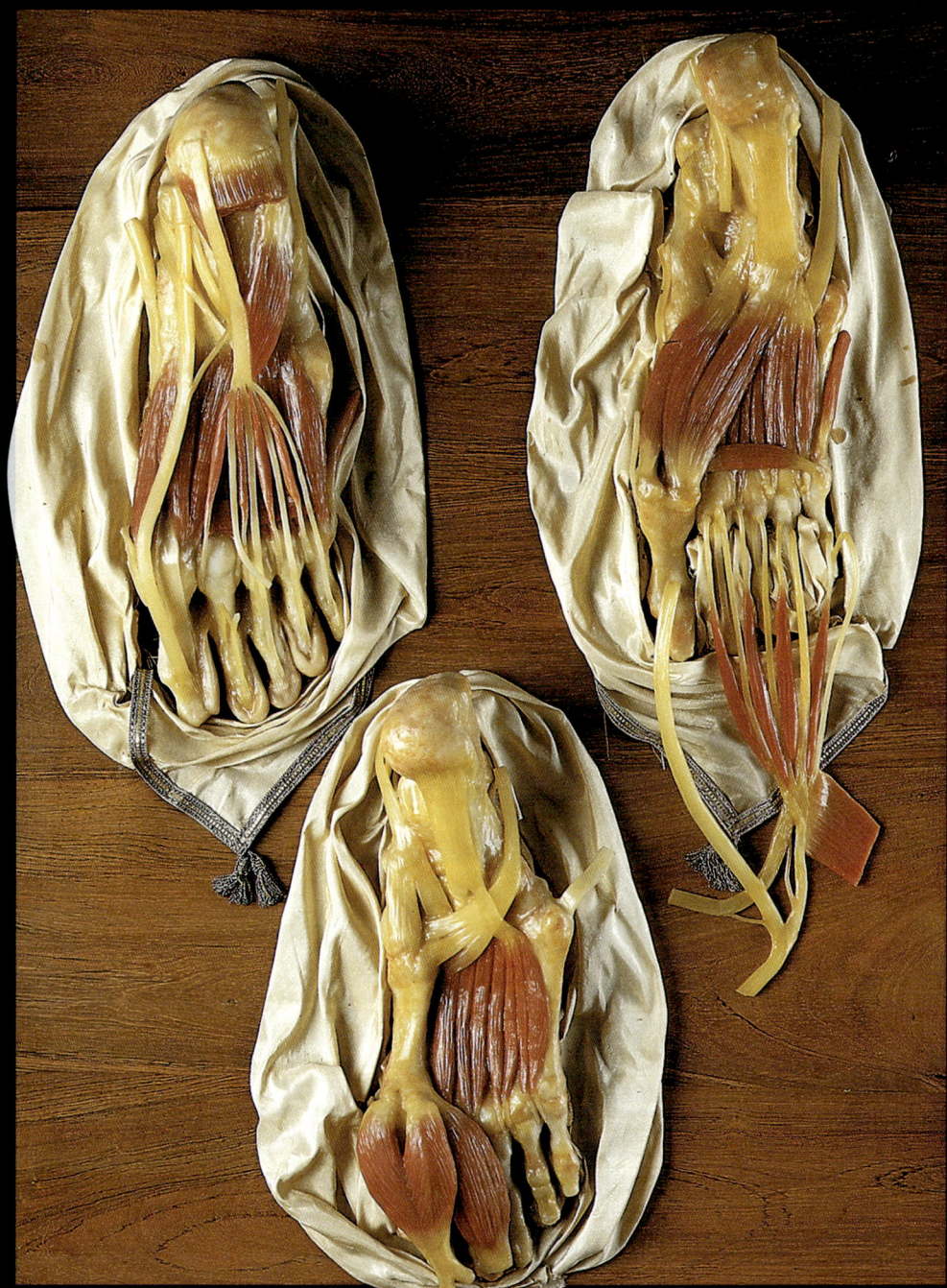

Musculi lumbricales, Musculi interossei plantares
Display showing muscles and tendons in the different layers of the sole of the foot

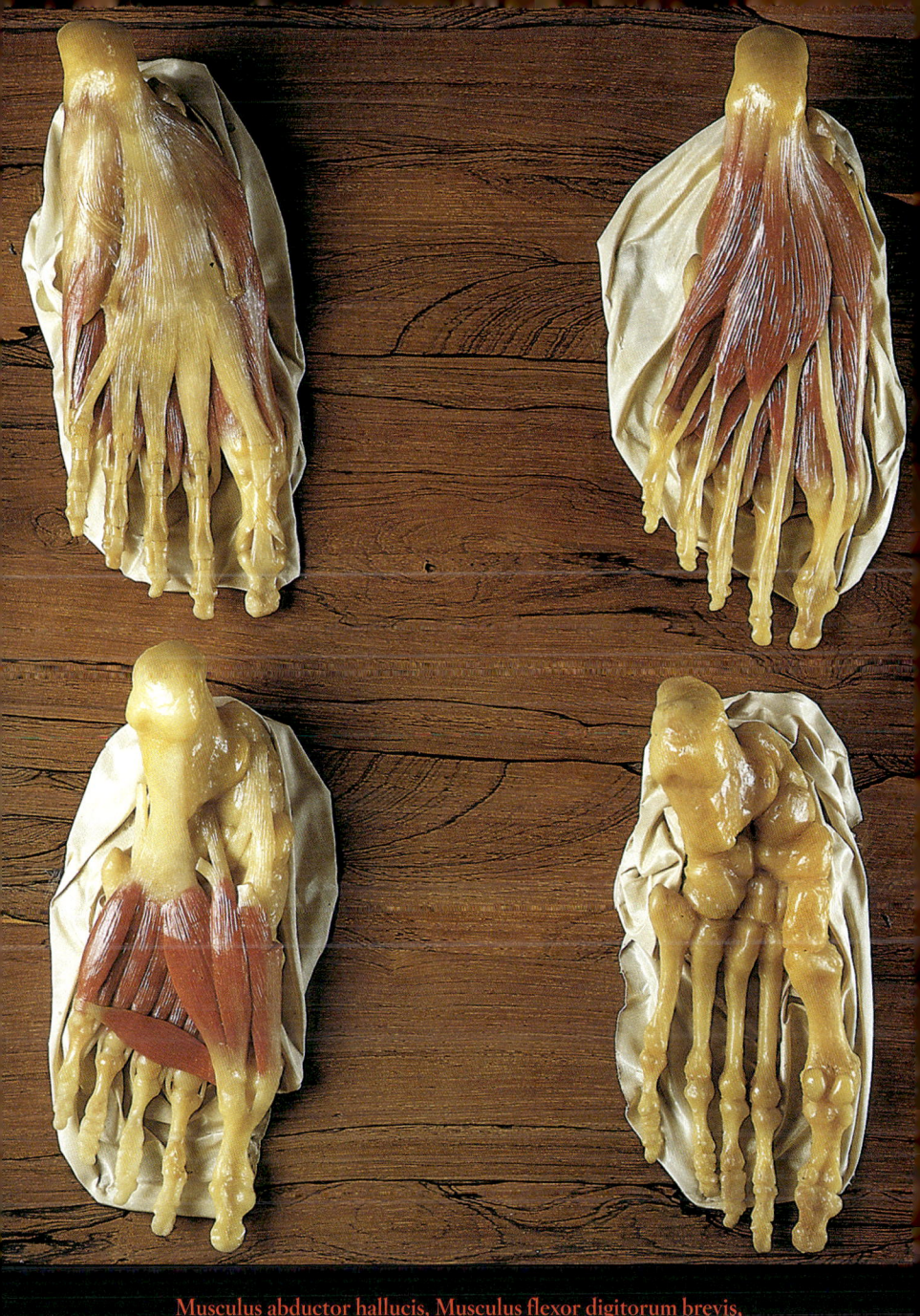

Musculus abductor hallucis, Musculus flexor digitorum brevis,
Musculus flexor hallucis brevis

Display showing muscles and tendons in the different layers of the sole of the foot

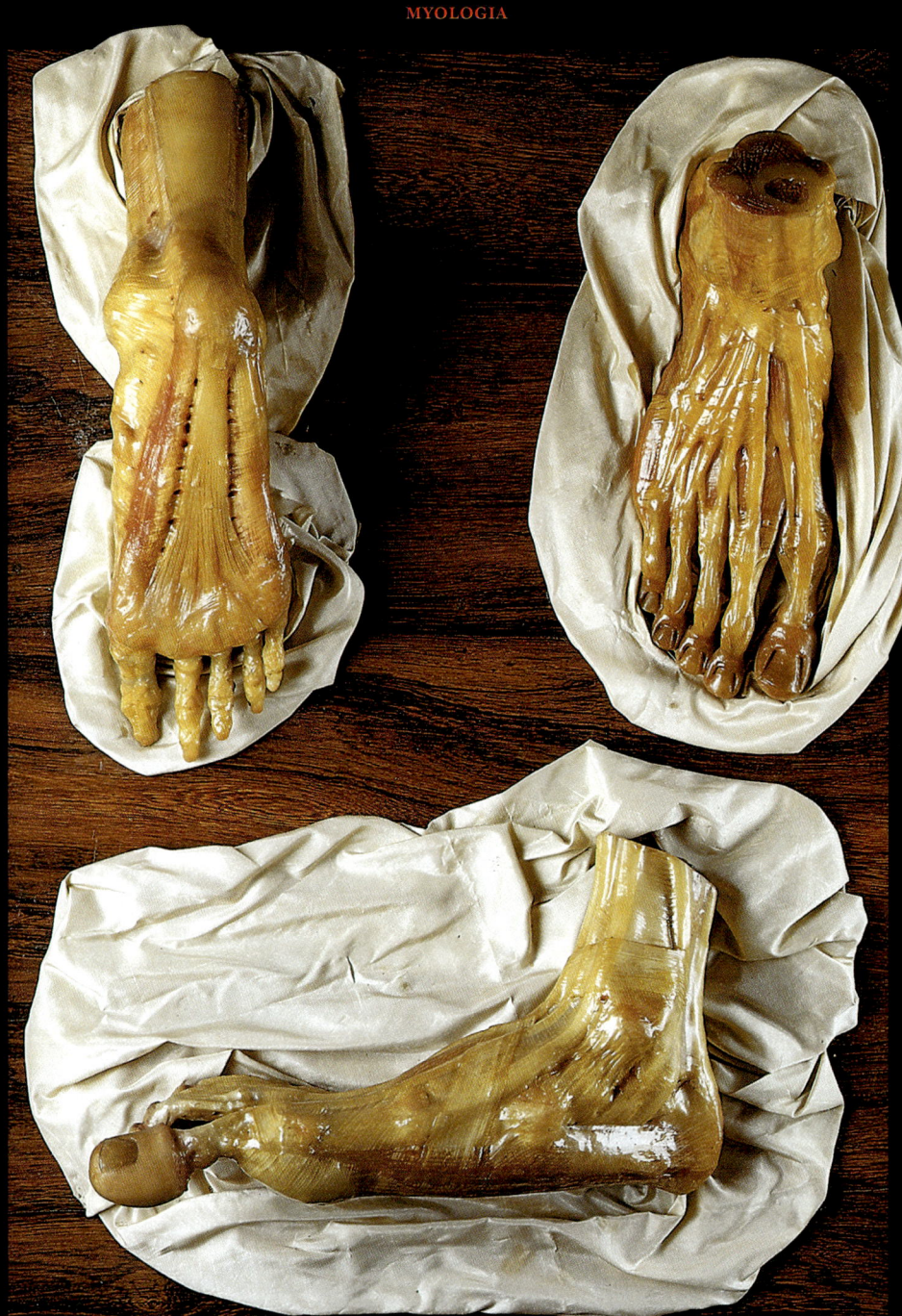

Retinacula musculi extensorum et flexorum

The superficial tendons and ligaments of the foot

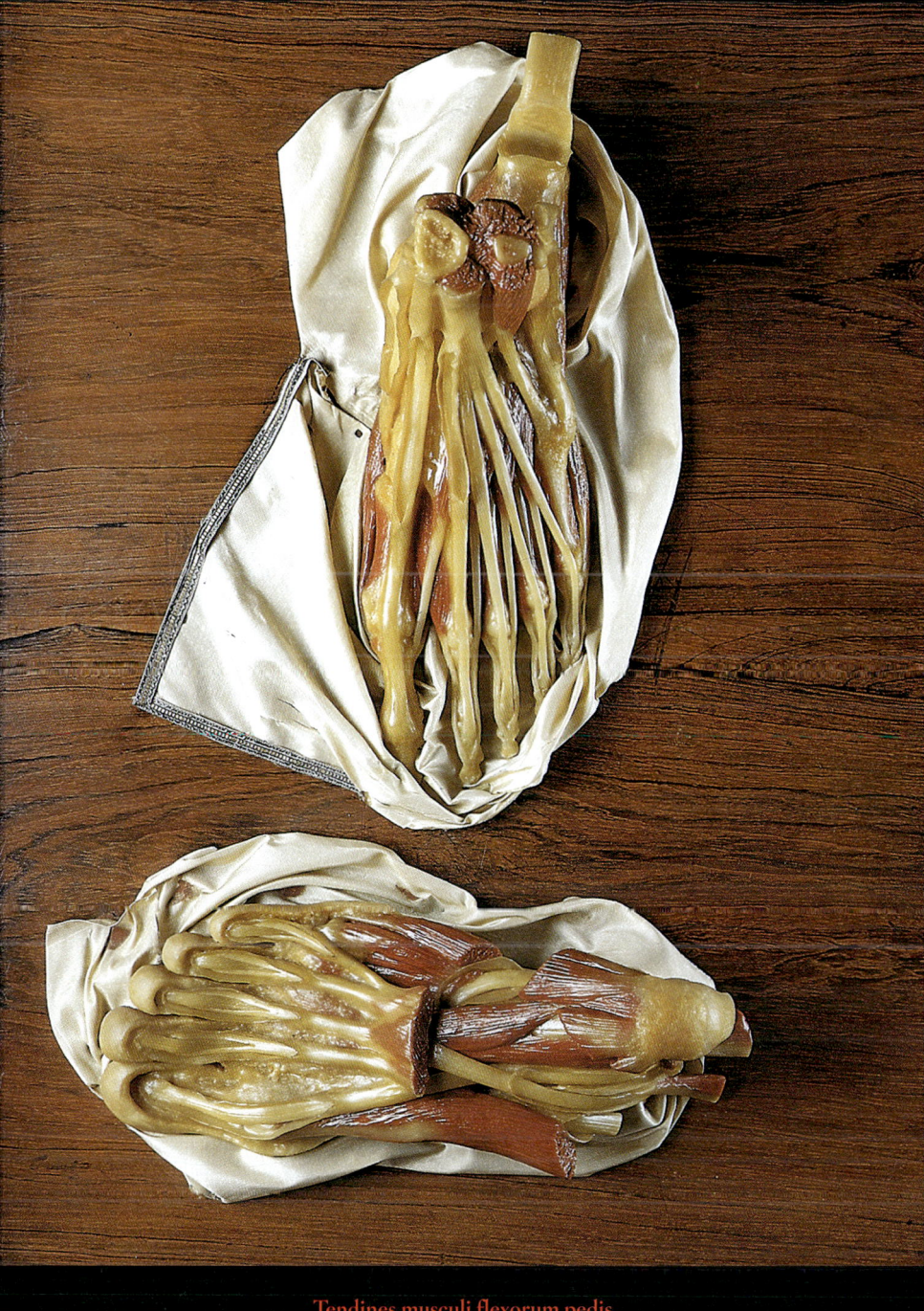

Tendines musculi flexorum pedis

Display showing the tendons in the back of the foot (upper specimen)
and sole of the foot (lower specimen)

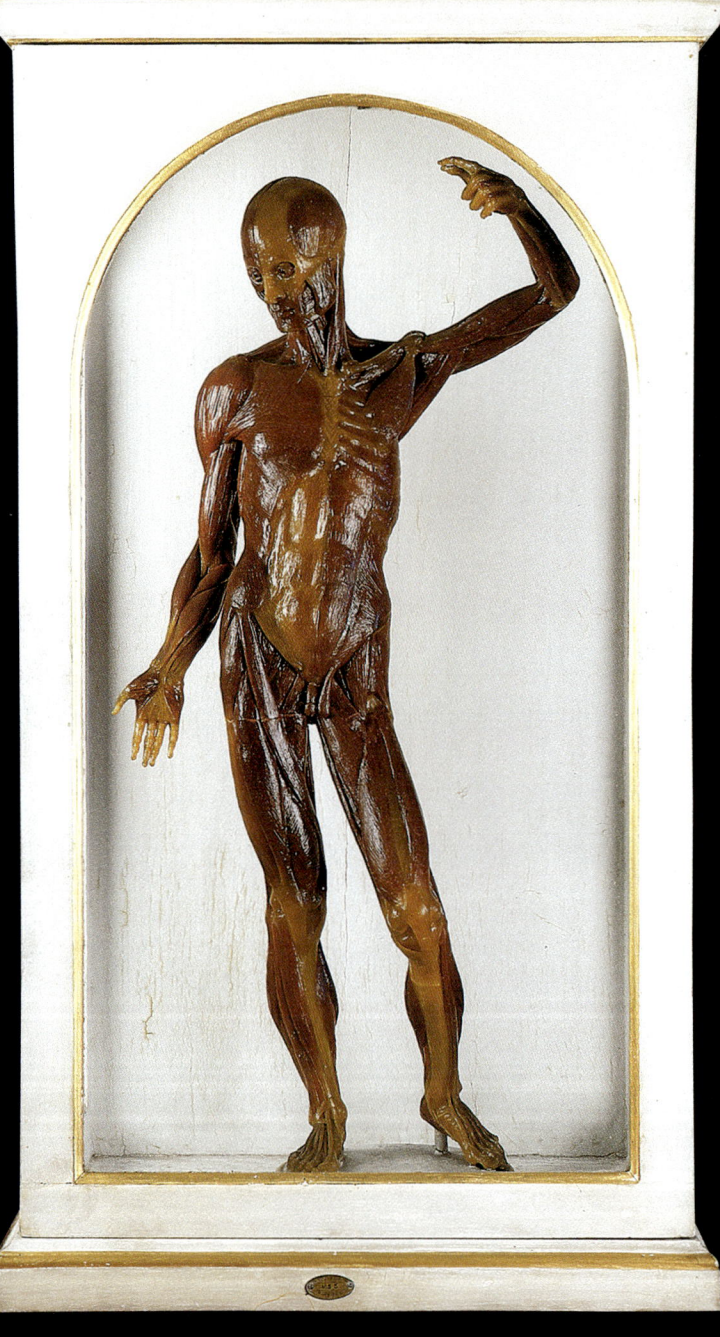

Musculi faciales, Musculi colli, Musculi thoracis, Musculi abdominis, Musculi membri superioris, Musculi membri inferioris

Whole body specimen displaying the superficial muscles

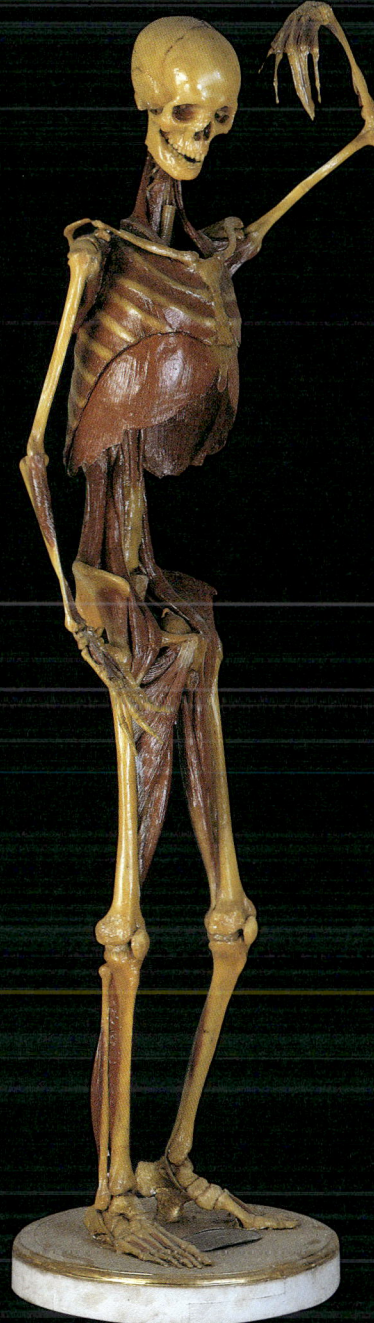

Whole body specimen with the deep layers of muscles

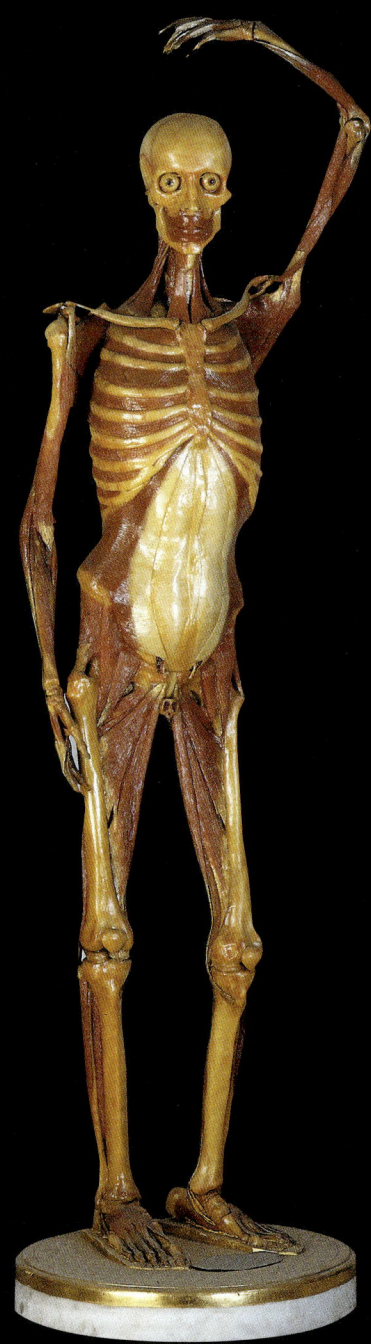

Musculi faciales, Musculi colli, Musculi thoracis, Musculi abdominis,
Musculi membri superioris, Musculi membri inferioris

Whole body specimen with the deep layers of muscles

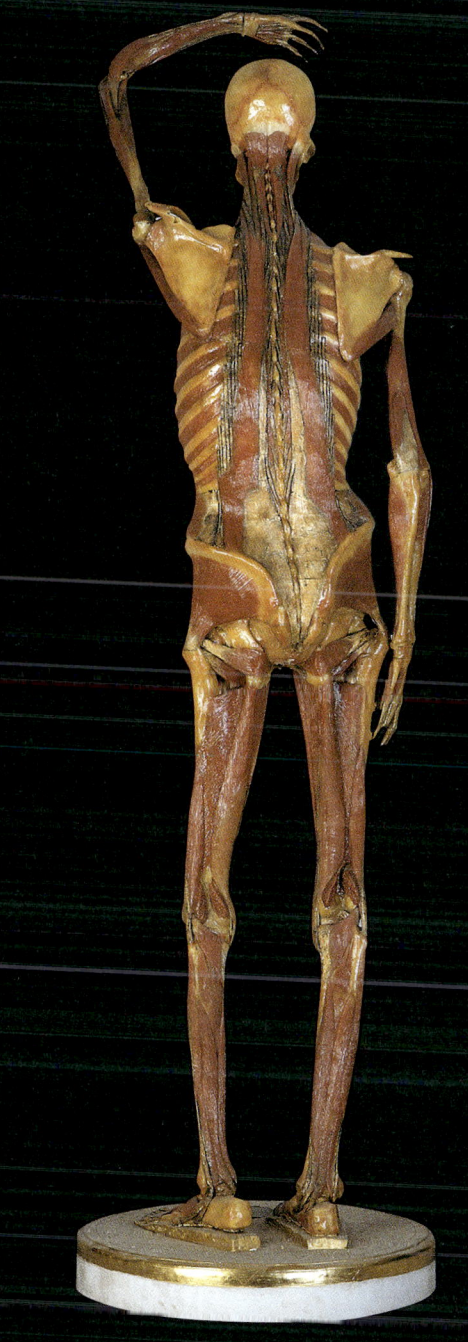

Musculi colli, Musculi thoracis, Musculi abdominis,
Musculi membri superioris, Musculi membri inferioris, Musculi dorsi

Whole body specimen with the deep layers of muscles

Systema cardiovasculare

Heart and Blood Circulation
Herz und Kreislauf
Cœur, artères et veines

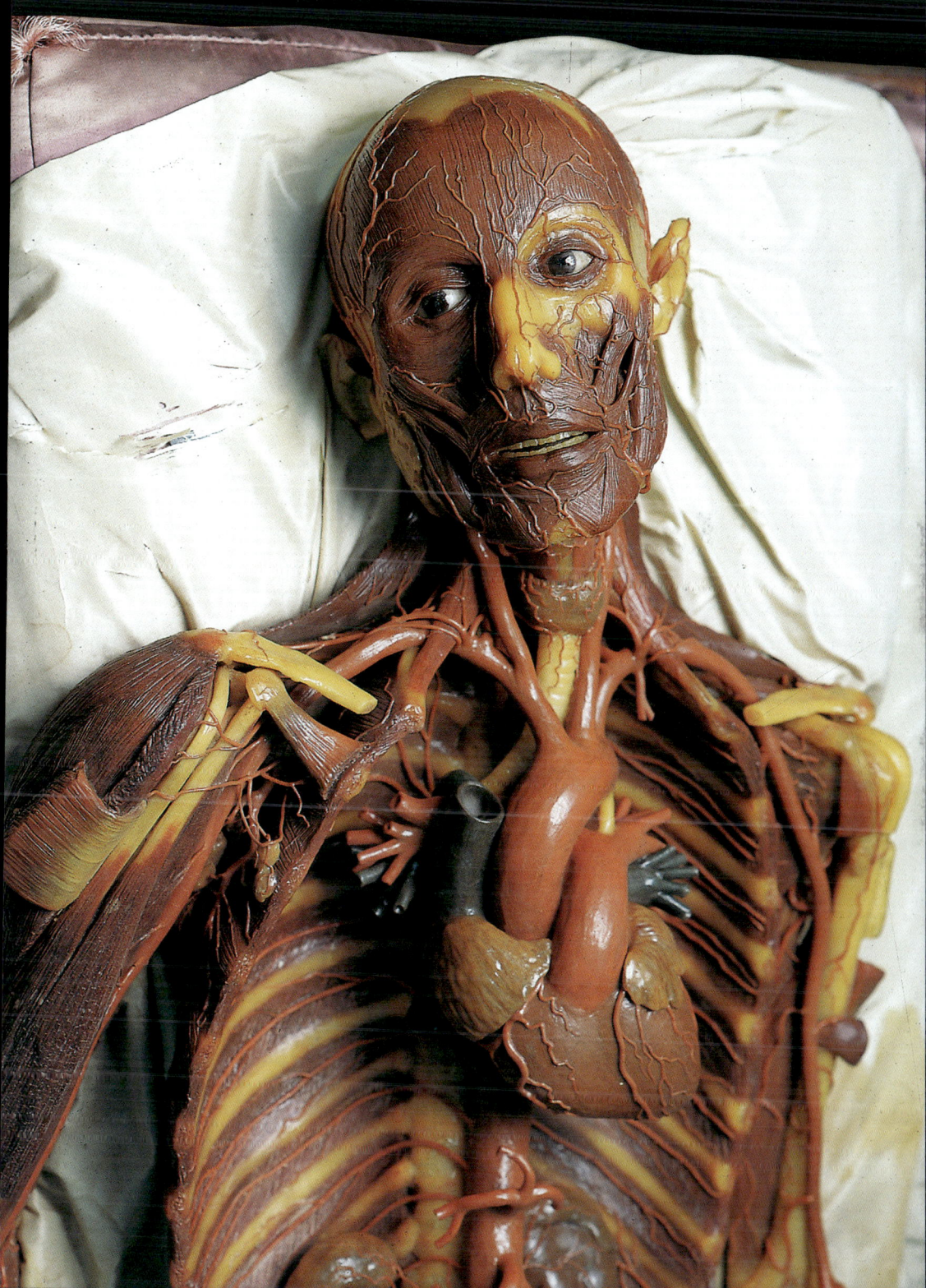

Heart and Blood Circulation

Of all those organs which are essential to life, the heart has long occupied a unique position in our cultural history, having been accepted as the seat of the soul, and of love, feeling and the spirit. In radical contrast to this view, these wax models bear impressive witness to the state of knowledge in topographical anatomy during the 17th century. As early as 1628, the English physician William Harvey (1578–1657) discovered the relationship between the heart and circulation. Harvey's revolutionary theory of the blood circulation upset medical ideas which had been taken for granted for centuries, but it was generally accepted by the end of the 18th century.

The selection of specimens was related to the corresponding woodcuts and copperplate illustrations of the 16th and 17th centuries. These models represent many aspects of the cardiovascular system, such as the position of the heart between the lungs within the thorax, its relationship to the diaphragm, the pericardium with its nerves and vessels and the internal structure of both atria and ventricles. In some of the models one can look through openings in the anterior wall of the heart and see the position of the atrioventricular and semilunar valves on both sides of the organ.

The great vessels entering and leaving the heart and the vessels supplying the head and neck are shown in their relationship to the muscles, and the arteries and veins are clearly distinguished by their colouring. The whole distribution of the arterial system, including the large branches on the anterior abdominal wall, in the limbs and between the dissected muscles of the pelvis, can be easily followed.

Page 215:
Detail from pages 220–221

Herz- und Blutkreislauf

Das Herz nimmt kulturgeschichtlich unter allen lebenswichtigen Organen eine Sonderstellung ein, da es lange Zeit als Sitz der Seele, der Liebe, der Gefühle und des Geistes angesehen wurde. Die Wachspräparate zeugen im Gegensatz zu dieser Betrachtungsweise eindrucksvoll vom Kenntnisstand der makroskopischen Anatomie des 17. Jahrhunderts. Bereits 1628 hatte der englische Anatom William Harvey (1578–1657) das Zusammenspiel zwischen Herz und Kreislauf entdeckt. Harveys revolutionäre Kreislauftheorie, die ein über Jahrhunderte gültiges Medizinkonzept ins Wanken gebracht hatte, war im ausgehenden 18. Jahrhundert allgemein anerkannt.

Die Auswahl der Präparate des Herzens orientiert sich an entsprechenden Holzschnitten und Kupferstichen des 16. und 17. Jahrhunderts. An diesen Modellen werden die vielfältige Aspekte des Herzens und des Kreislaufsystems herausgearbeitet, wie etwa die topographische Lage des Herzens im Brustkorb zwischen den Lungenflügeln, die Lagebeziehung zum Zwerchfell, der Herzbeutel mit seinen Gefäßen und Nerven, das Innenrelief der beiden Herzvorhöfe und -kammern. Die Gefäßversorgung des Herzens mit Arterien, Venen und Lymphgefäßen sowie die Nervenversorgung werden in unterschiedlichen Perspektiven dargestellt. In einzelnen Modellen kann der Betrachter durch die aufgeschnittene Herzvorderwand in das Innere des rechten und linken Herzens blicken und die Position der Segel- und Taschenklappen einsehen.

Die großen herznahen Blutgefäße sowie die versorgenden Gefäße der Hals- und Kopfregion sind in ihrer natürlichen Lage in den Muskellogen wiedergegeben, wobei die arteriellen von den venösen Gefäßen durch unterschiedliche Farbgebung klar unterschieden sind. Der gesamte Verlauf der arteriellen Gefäße, einschließlich größerer Verzweigungen an der vorderen Bauchwand, den Extremitäten sowie dem Becken zwischen den aufpräparierten Muskelgruppen, lässt sich klar verfolgen.

Le cœur et l'appareil circulatoire

Le cœur occupe une place toute particulière, car il a longtemps été considéré dans l'histoire des civilisations comme le siège de l'âme et de l'esprit, de l'amour et des émotions. En opposition à ces conceptions, les modèles en cire témoignent de façon remarquable de l'état des connaissances exactes de l'anatomie macroscopique du XVIIe siècle. En 1628, l'anatomiste anglais William Harvey (1578–1657) découvre l'interdépendance du cœur et de l'appareil circulatoire ; sa théorie révolutionnaire de la circulation du sang, ébranlant un concept médical multiséculaire, est globalement admise à la fin du XVIIIe siècle.

Le choix des préparations du cœur se réfère à des gravures correspondantes, sur bois ou sur cuivre, des XVIe et XVIIe siècles. Sur les modèles en cire ressortent de nombreux aspects de la morphologie du cœur et des vaisseaux : topographie du cœur à l'intérieur de la cavité thoracique entre les deux poumons, position par rapport au diaphragme, disposition du péricarde accompagné de vaisseaux et nerfs, morphologie des cavités des oreillettes et des ventricules. La vascularisation artérielle, veineuse et lymphatique du cœur, ainsi que l'innervation, sont représentées sous différents angles. Sur des modèles du cœur isolé, une ouverture de la paroi cardiaque antérieure permet de découvrir les reliefs des cavités droites et gauches et la position des valvules cardiaques.

Les gros vaisseaux de la base du cœur, ainsi que les troncs vasculaires de la tête et du cou, sont représentés dans leur position anatomique en relation avec les loges musculaires. La distinction entre artères et veines est clairement donnée par une coloration spécifique, rouge ou bleue, de chaque type de vaisseau. Le trajet des artères entre les groupes musculaires disséqués, ainsi que le trajet des grosses branches collatérales, peut être suivi sur toute leur longueur au niveau de la paroi abdominale, du bassin et des membres.

Detail from pages 412–413

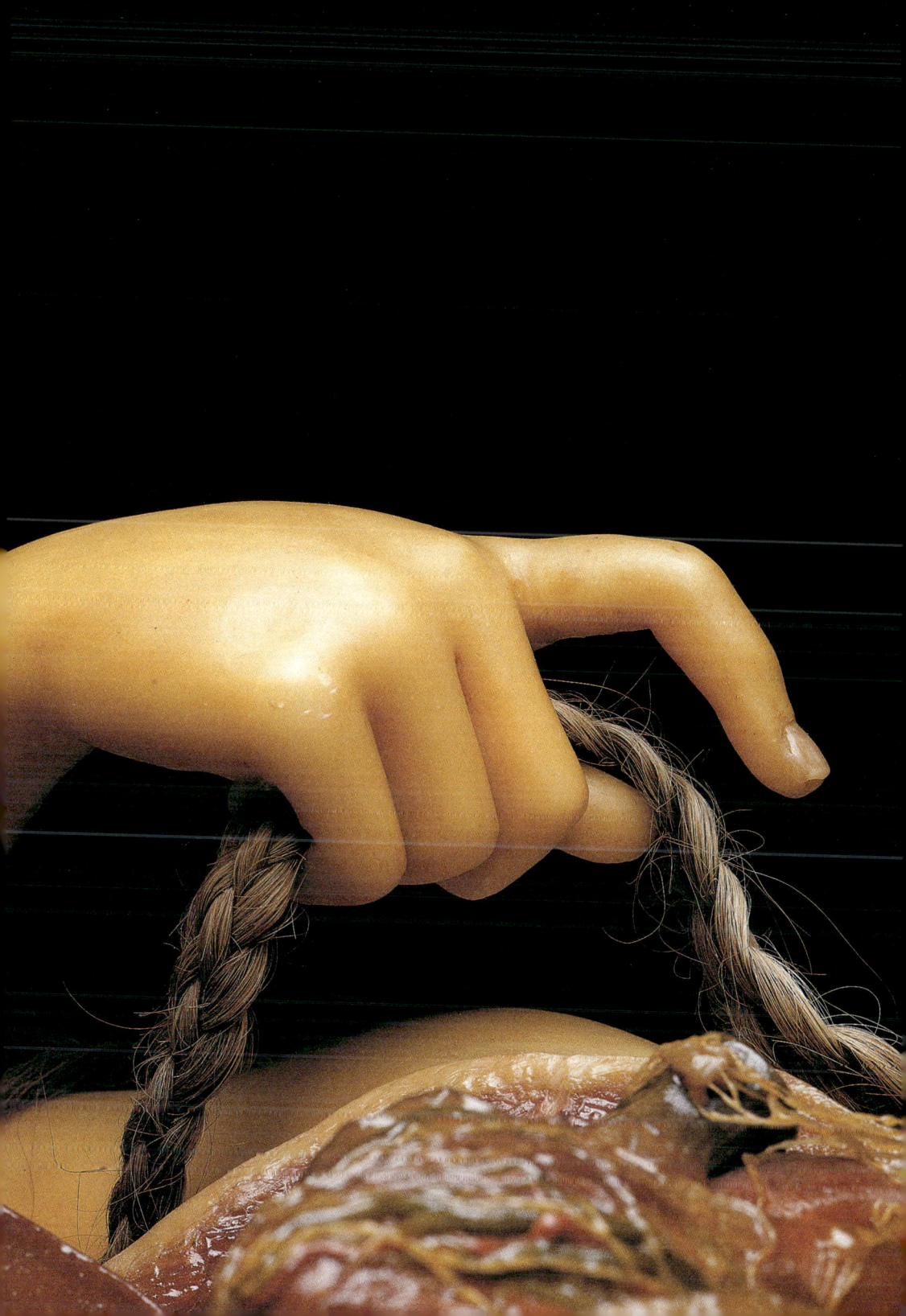

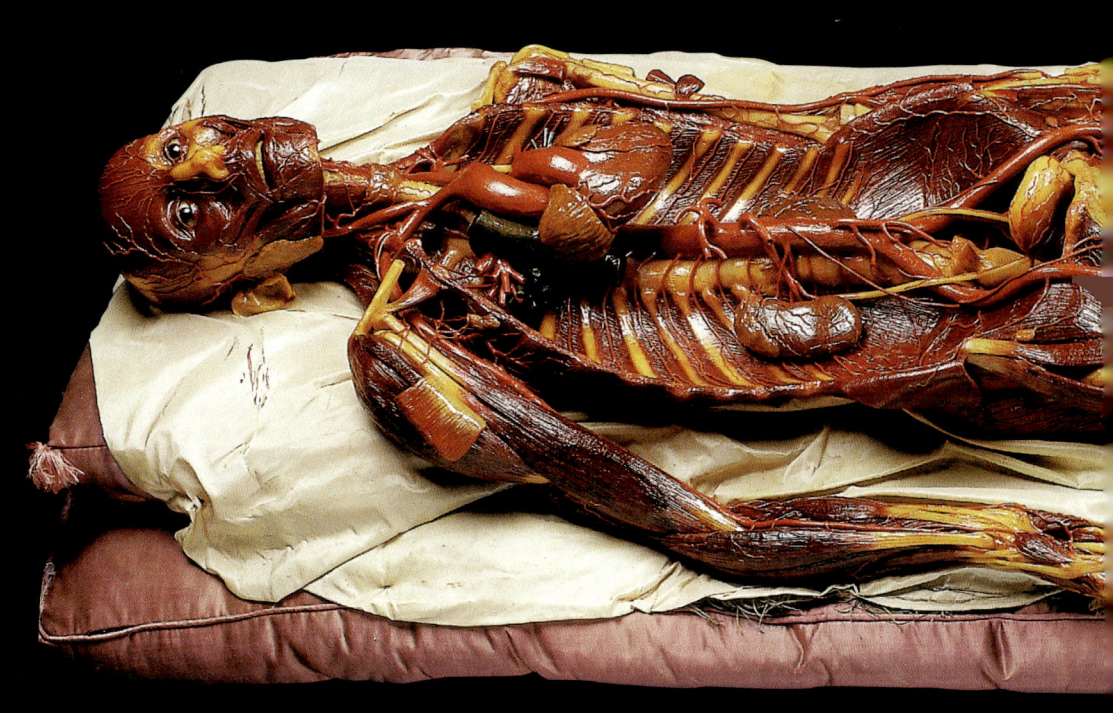

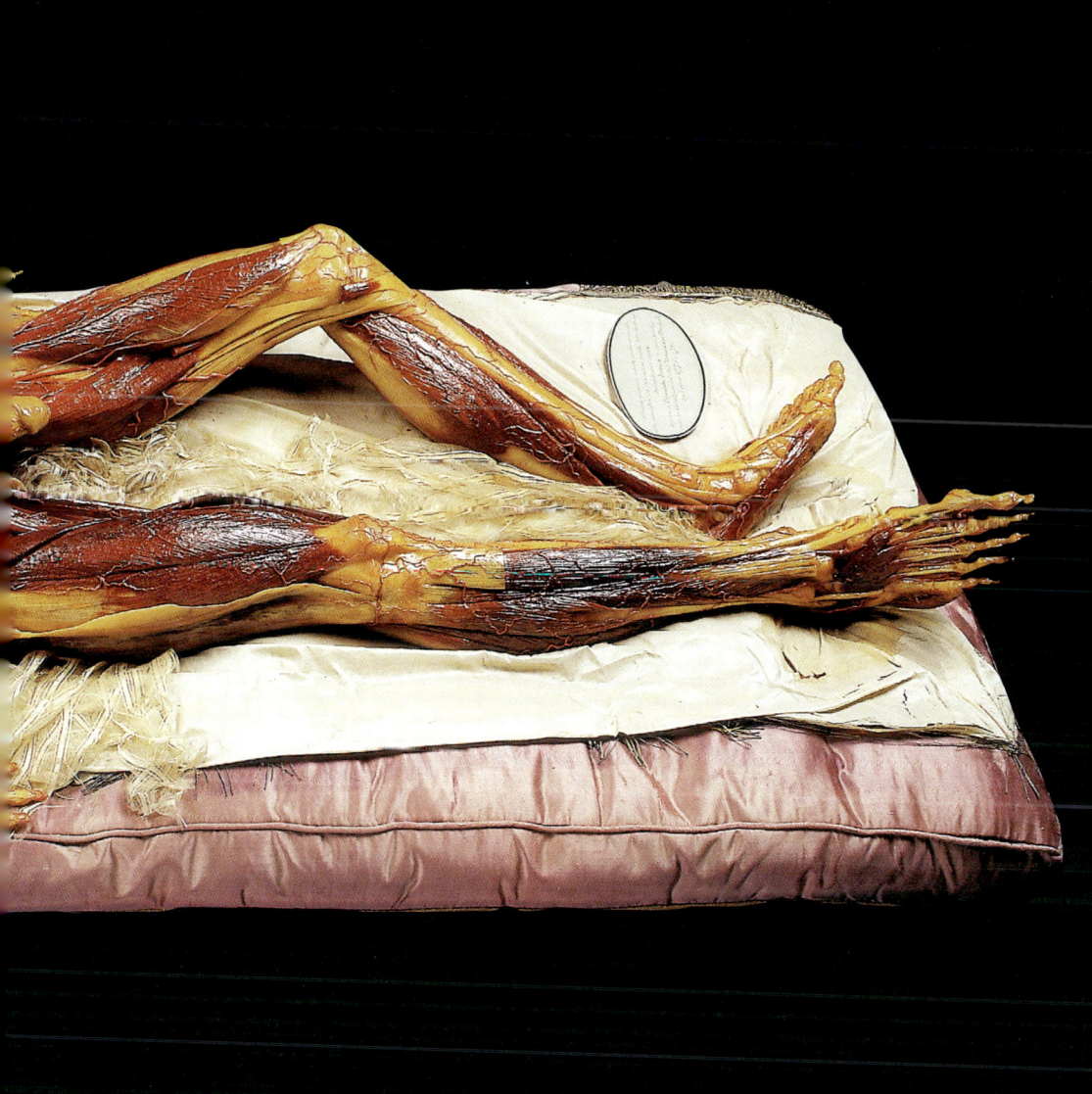

Arteriae

Whole body specimen with the arteries displayed

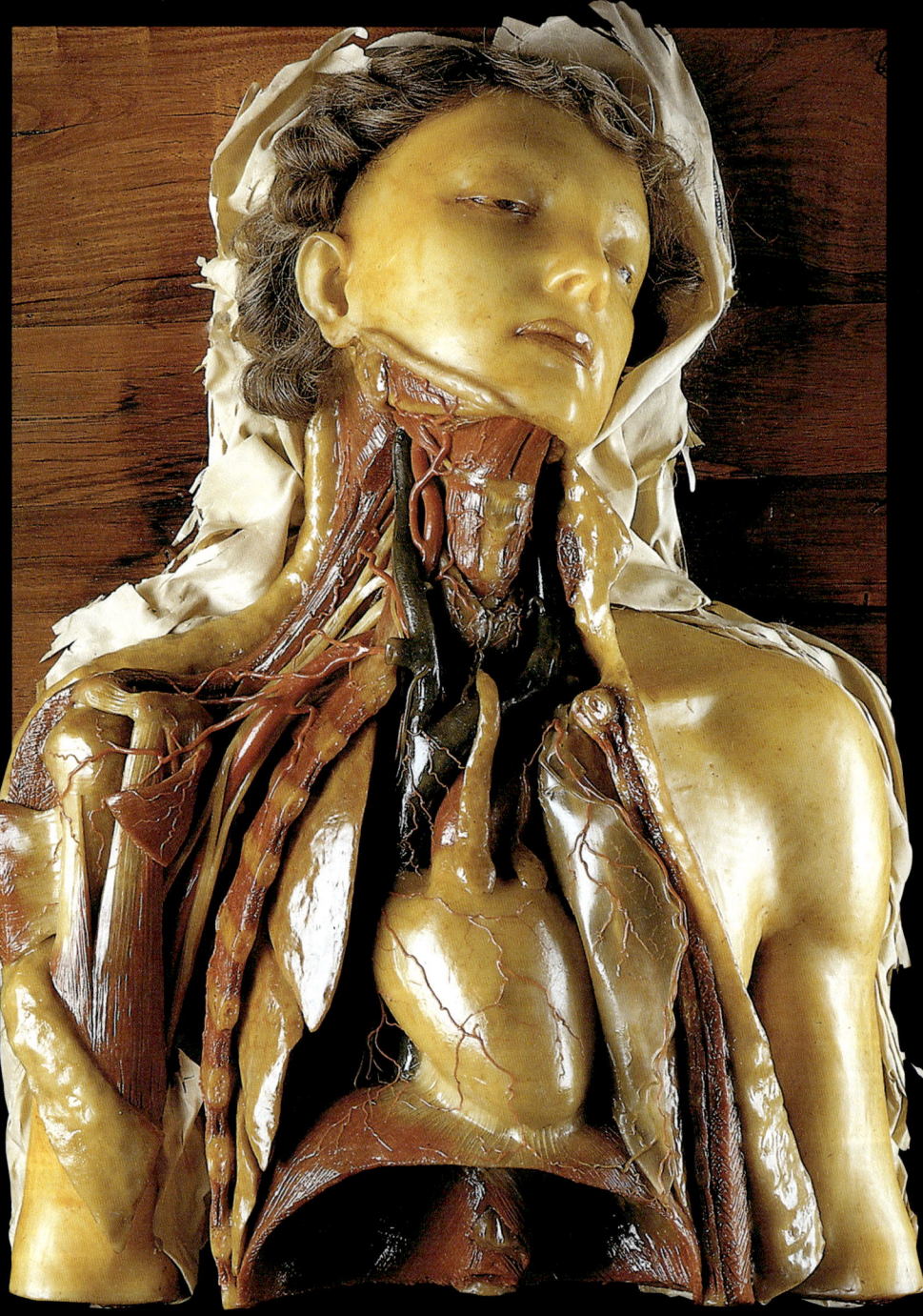

Larynx, Glandula thyroidea, Thorax, Mediastinum, Cor, Thymus

Thoracic cavity laid open to show the heart in its pericardium, the thymus gland and the great vessels.
The lungs have been displaced sideways

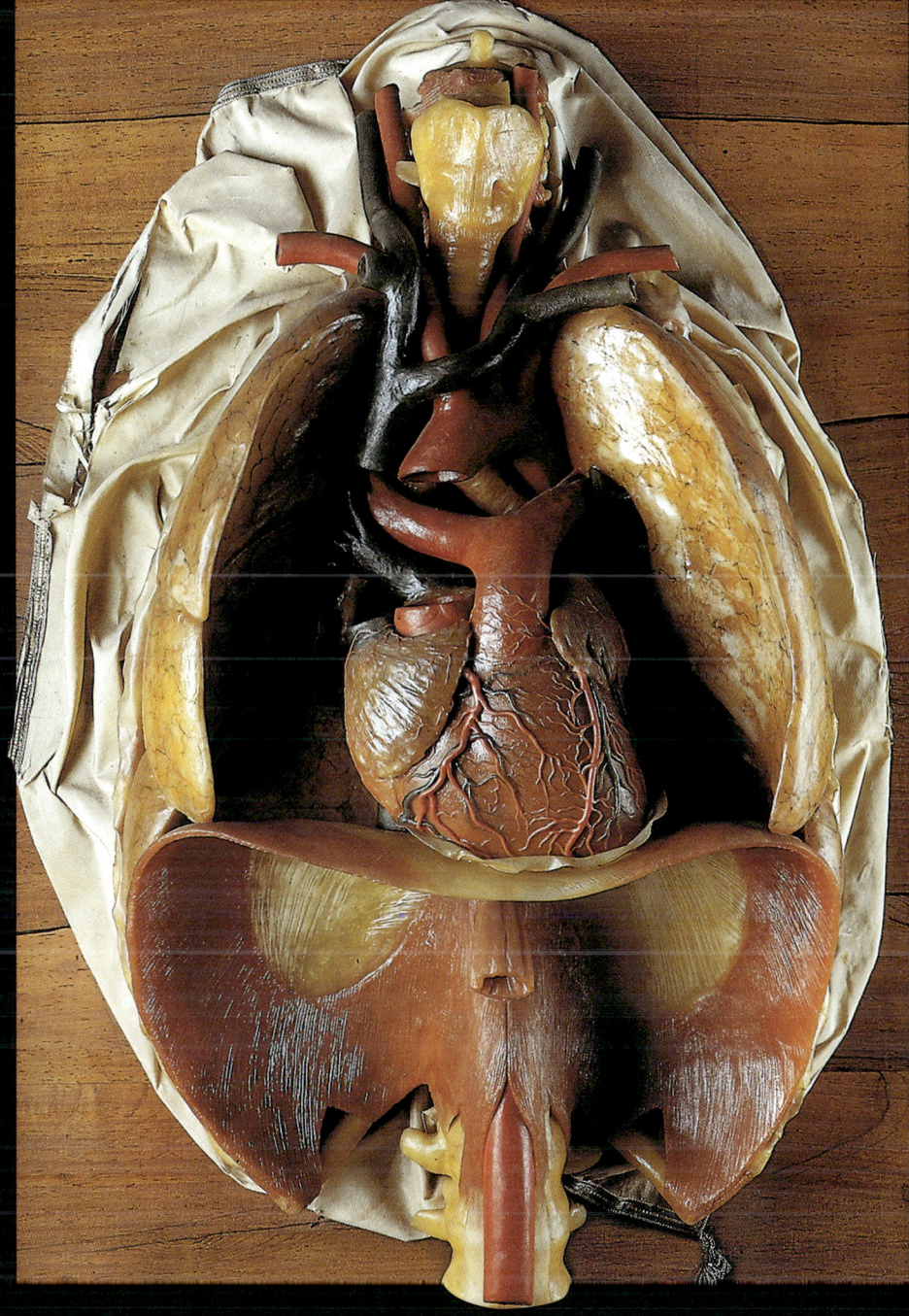

Thorax, Mediastinum, Diaphragma, Cor, Pulmones

The heart displayed in its natural position in the thoracic cavity.
The pericardium has been removed and the lungs displaced sideways

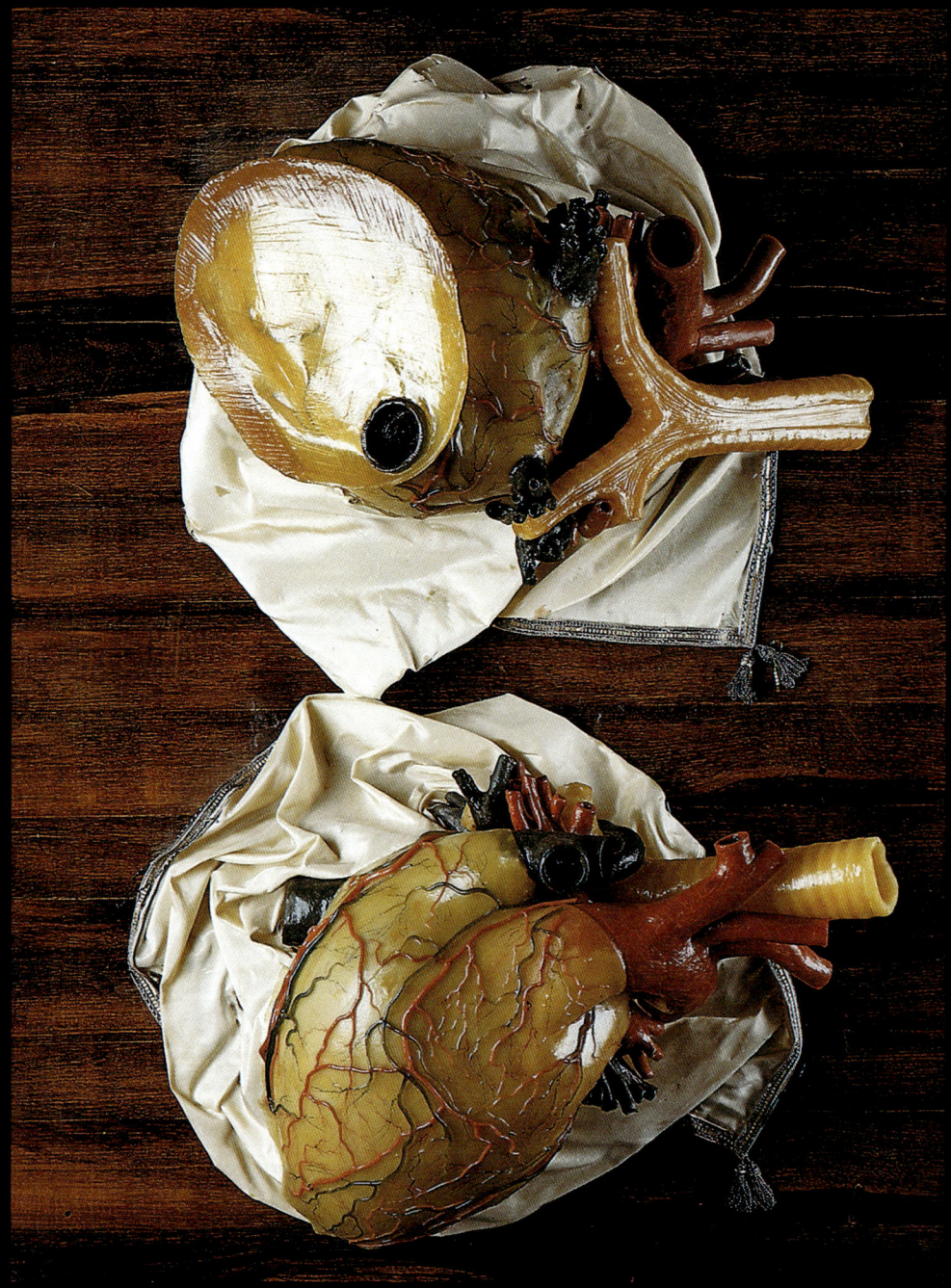

Cor, Aorta, Trachea, Bronchi principales, Diaphragma, Vena cava inferior

Specimen showing the heart with the aorta, the trachea and the main bronchi. In the upper specimen a part
of the diaphragm has been left behind. The specimen is displayed as if viewed from the inside
of the abdominal cavity. The lower specimen shows the anterior surface of the heart

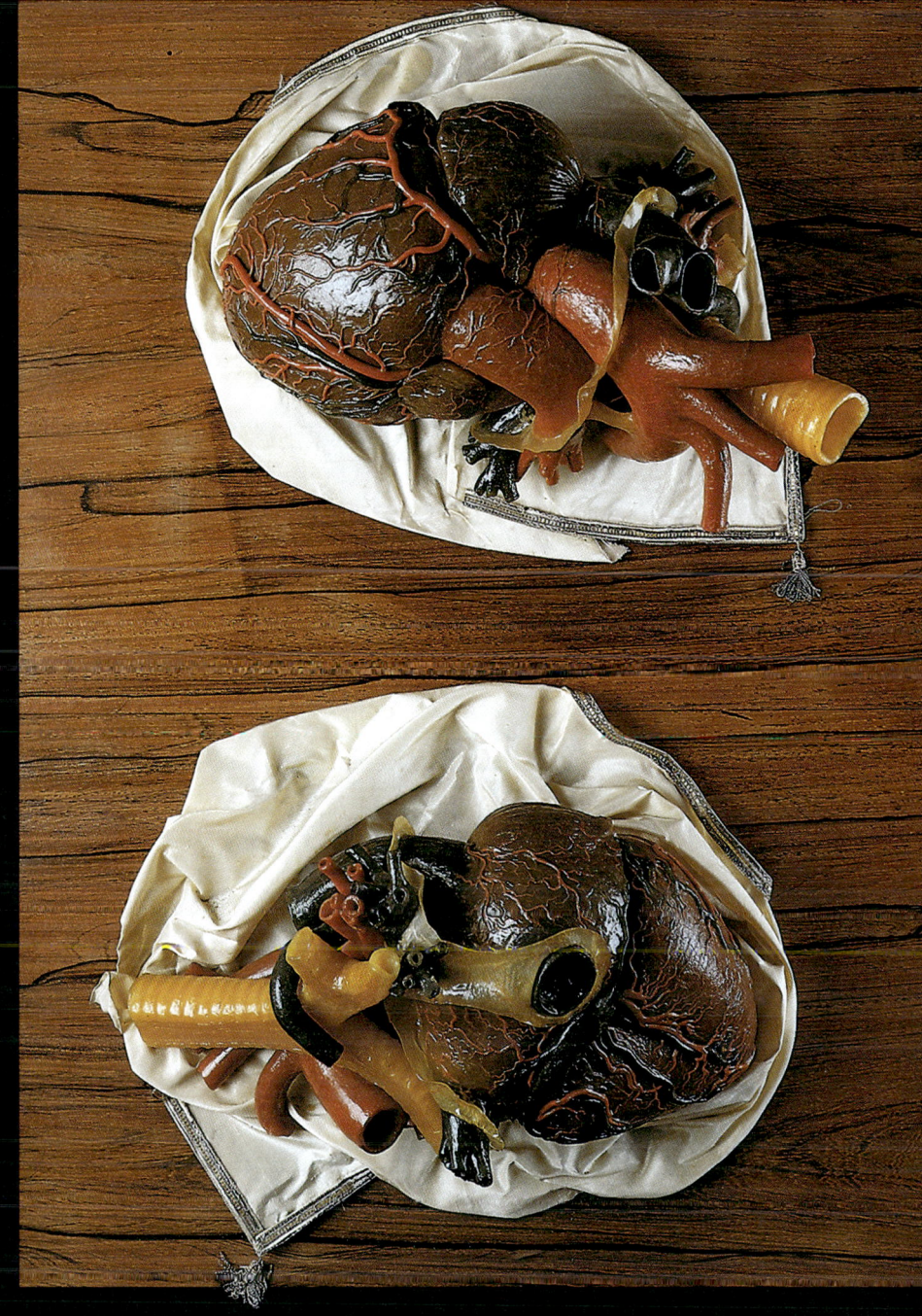

Cor, Pericardium, Arteria coronaria dextra, Ramus interventricularis anterior, Ligamentum arteriosum Botalli

Specimens representing various aspects of the heart. The upper specimen shows the anterior surface of the heart with the coronary arteries

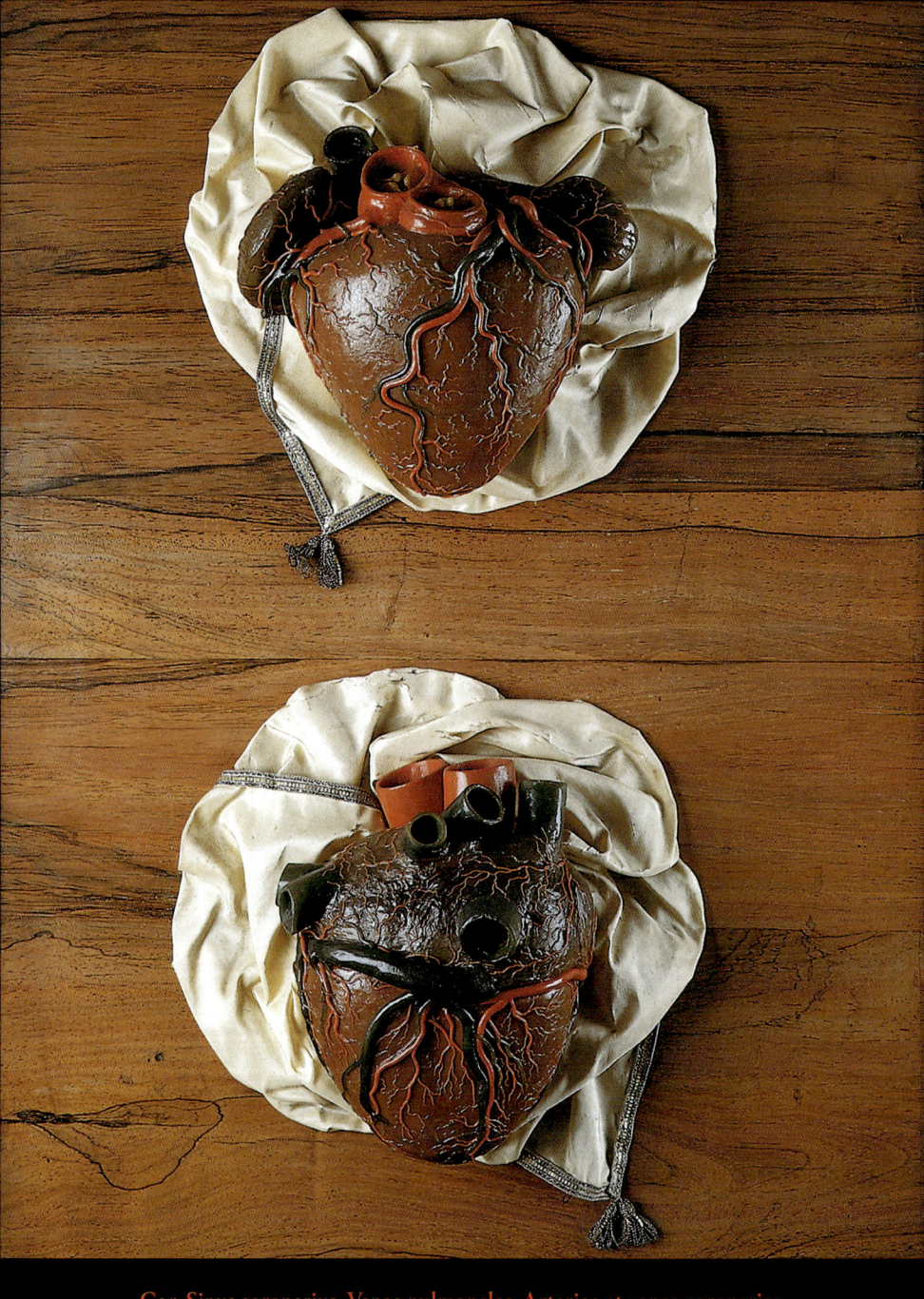

Cor, Sinus coronarius, Venae pulmonales, Arteriae et venae coronariae

Upper specimen: anterior aspect of the heart seen from the left. A branch of the left coronary artery.
Lower specimen: base of the heart lying on the diaphragm

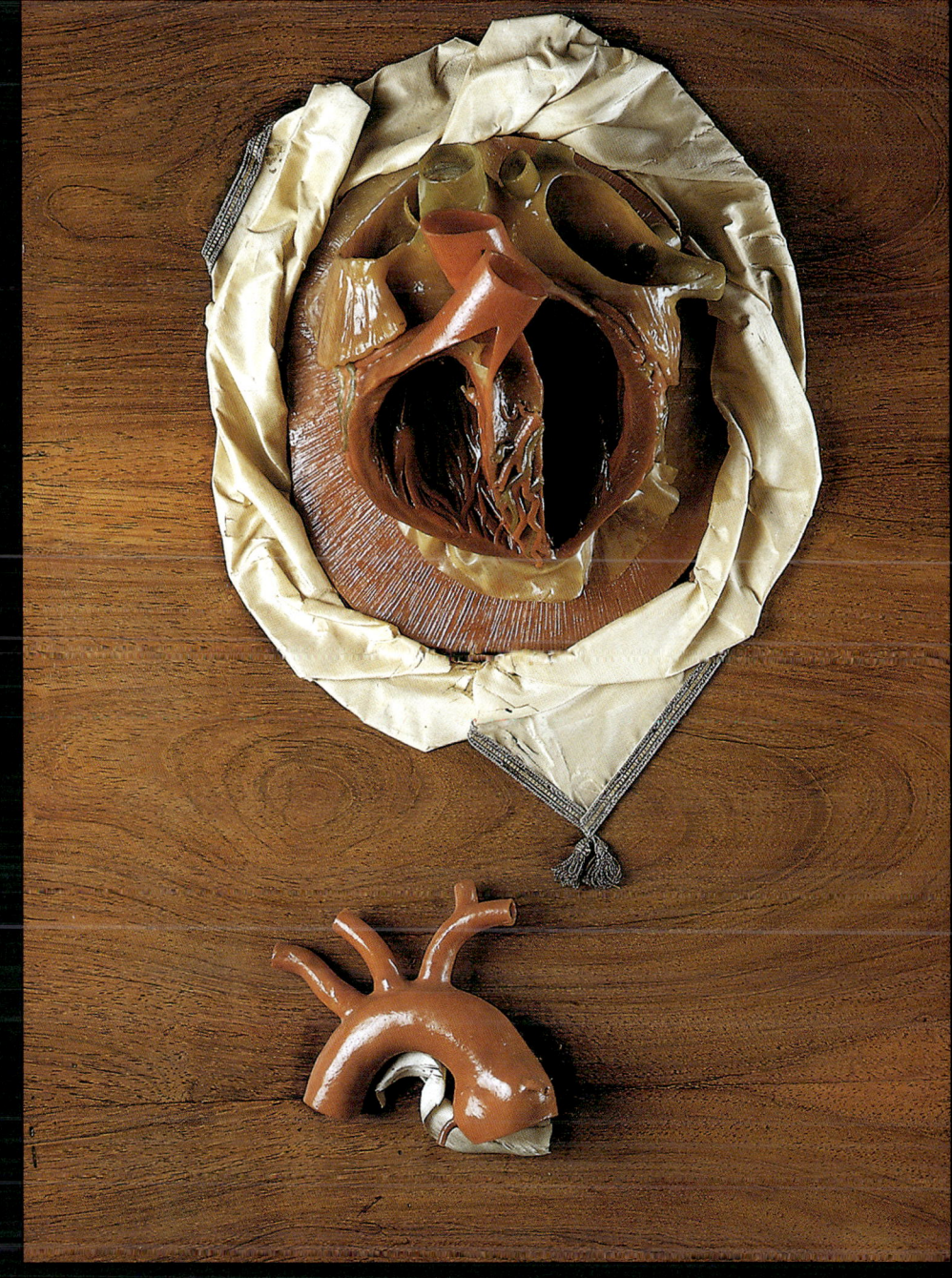

Cor, Septum interventriculare, Arcus aortae

View from in front into the opened right and left ventricles, below the aortic arch

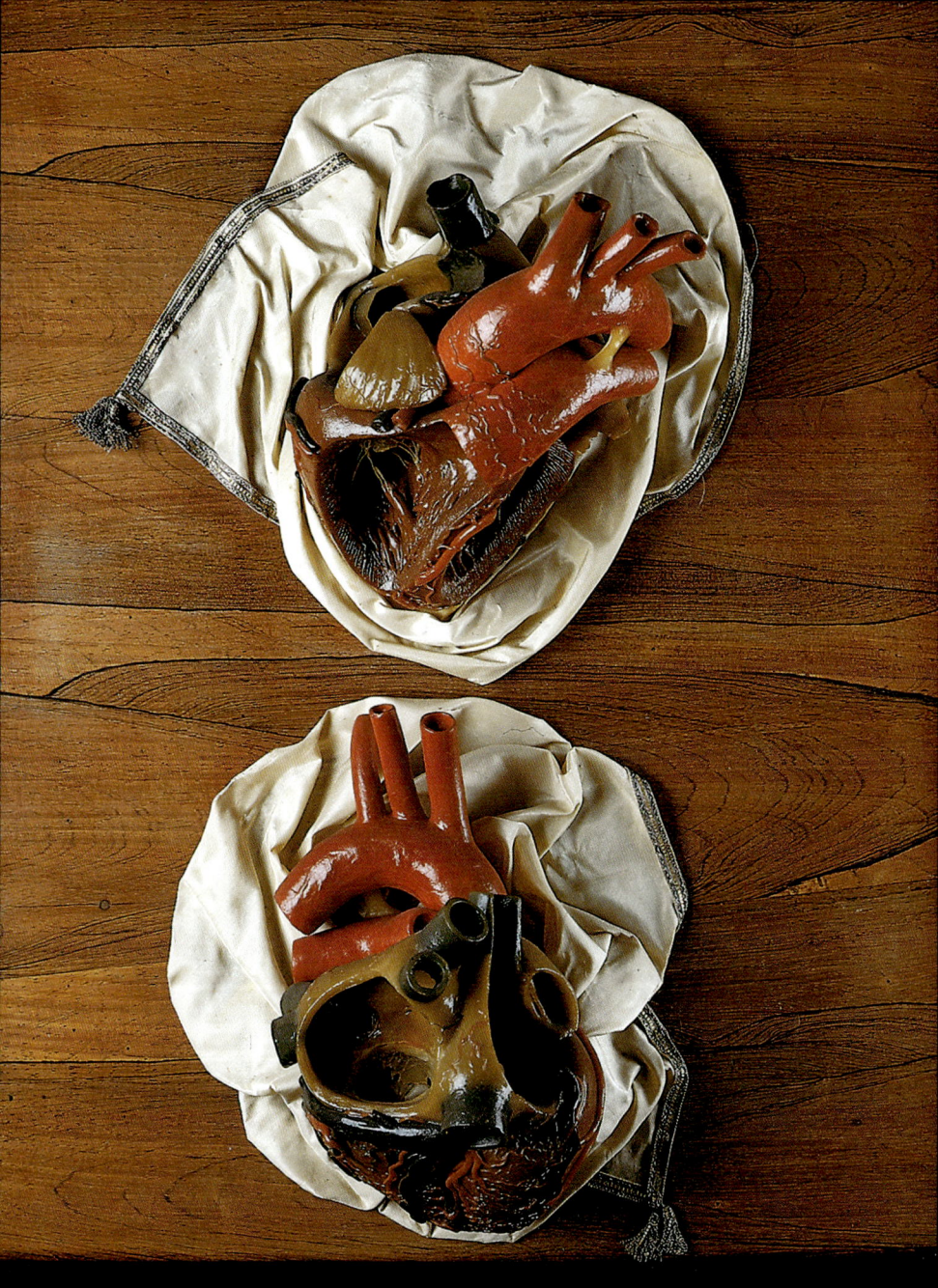

Cor, Arteria pulmonalis, Arcus aortae, Ligamentum arteriosum Botalli

Various specimens of the heart with the ventricles laid open (above) and the atria (below).
The ligamentous structure (depicted in yellow) connecting the pulmonary artery
with the arch of the aorta is a remnant of the foetal circulation

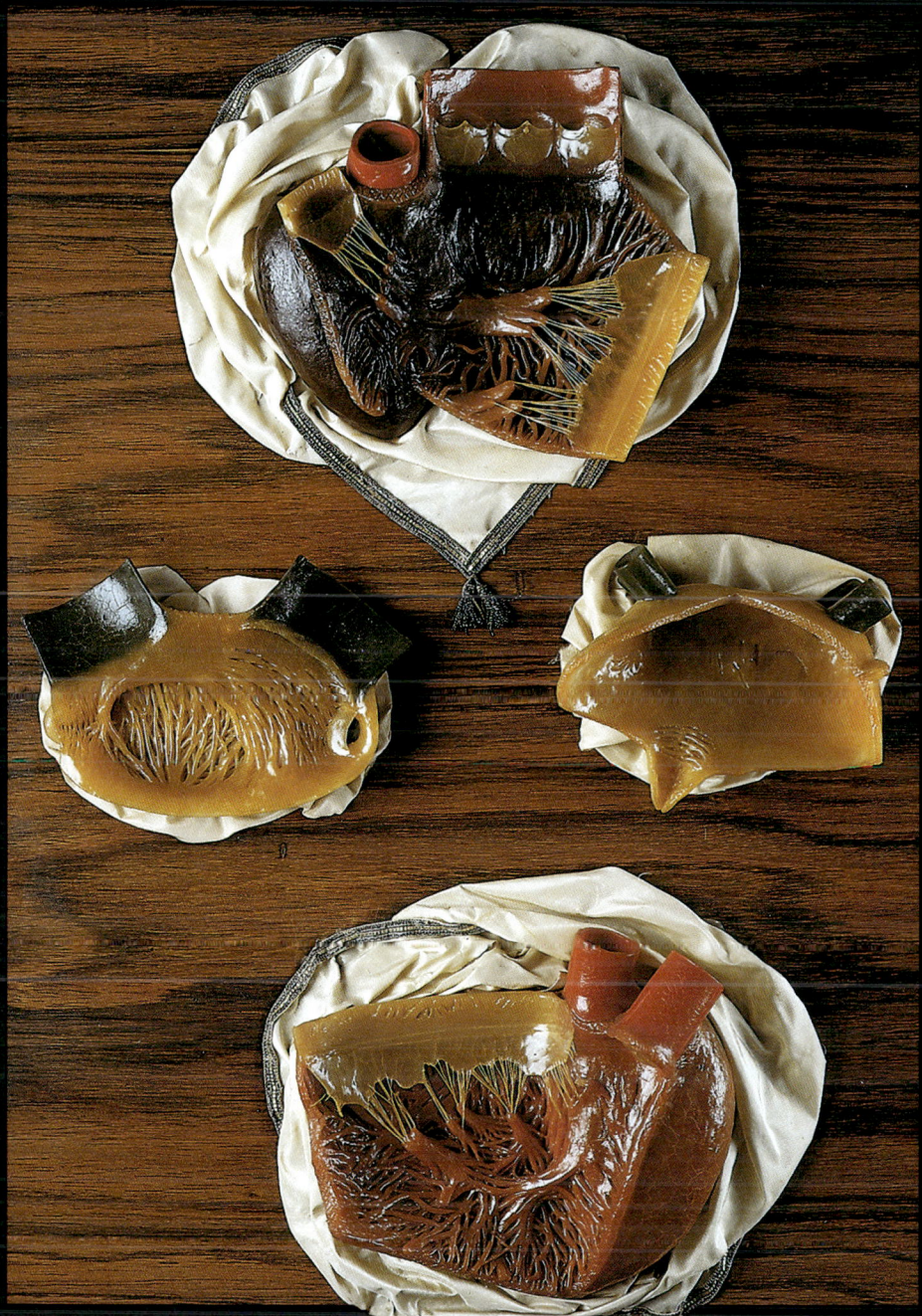

Cor, Valvula bicuspidalis, Valvula tricuspidalis, Musculi papillares, Musculi pectinati

The upper and lower specimens depict the heart valves and the internal form of the ventricular muscle.
The two specimens in the middle show the atrial musculature

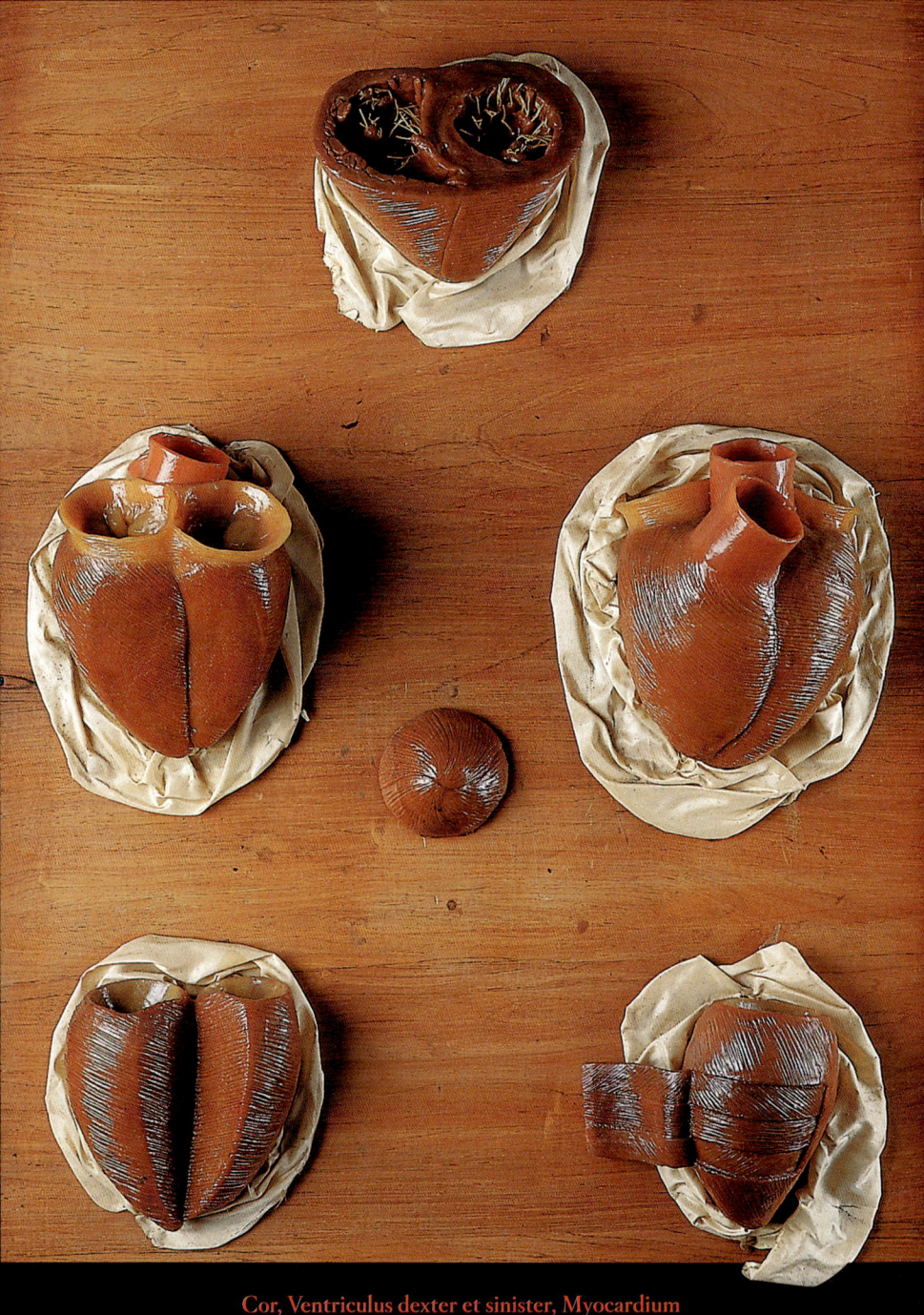

Cor, Ventriculus dexter et sinister, Myocardium

Various views showing the arrangement of the fibres of the ventricular muscle

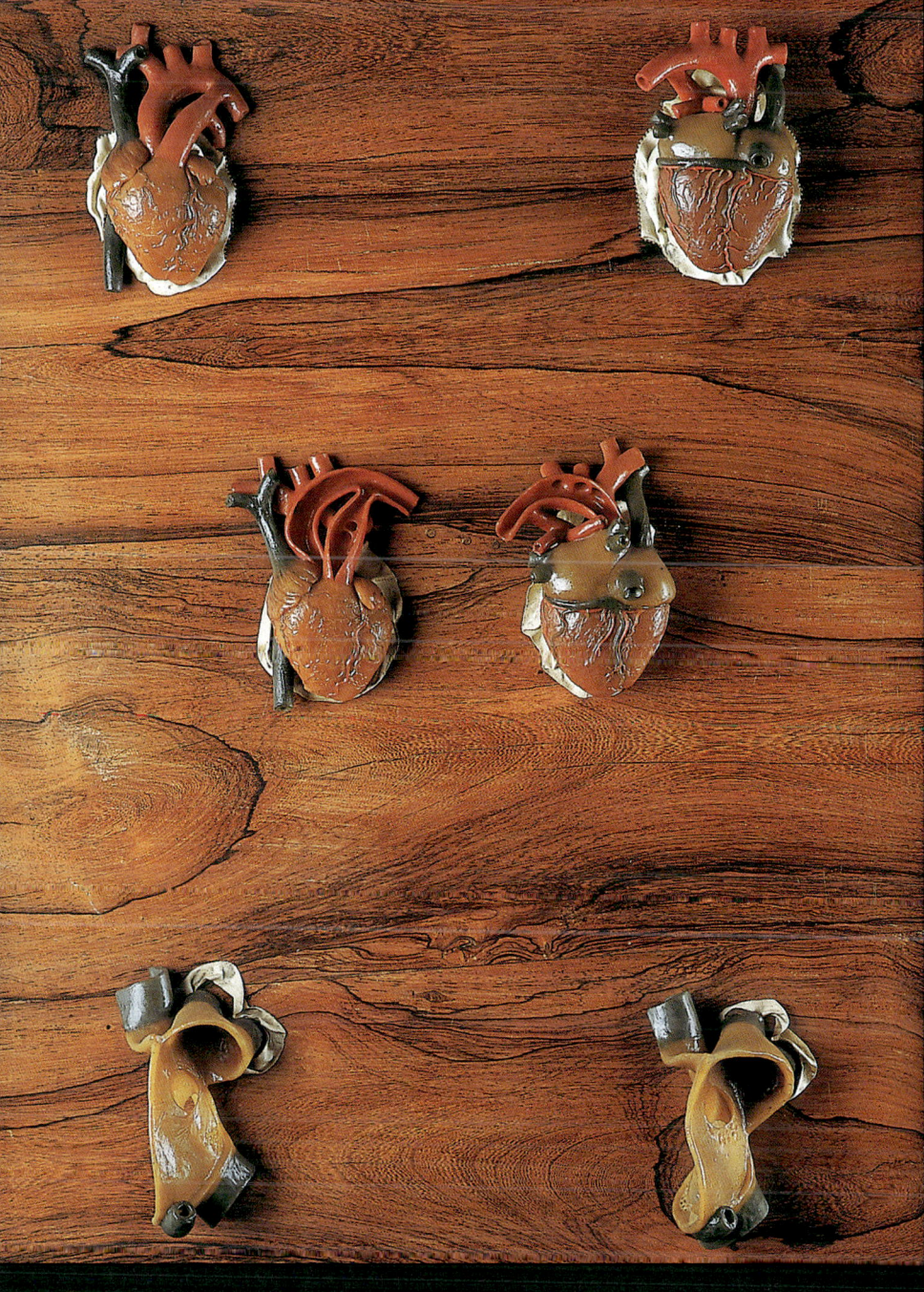

Demonstration of the foetal heart showing the bypass between the pulmonary artery
and the aortic arch, and also between the two atria (lower specimen)

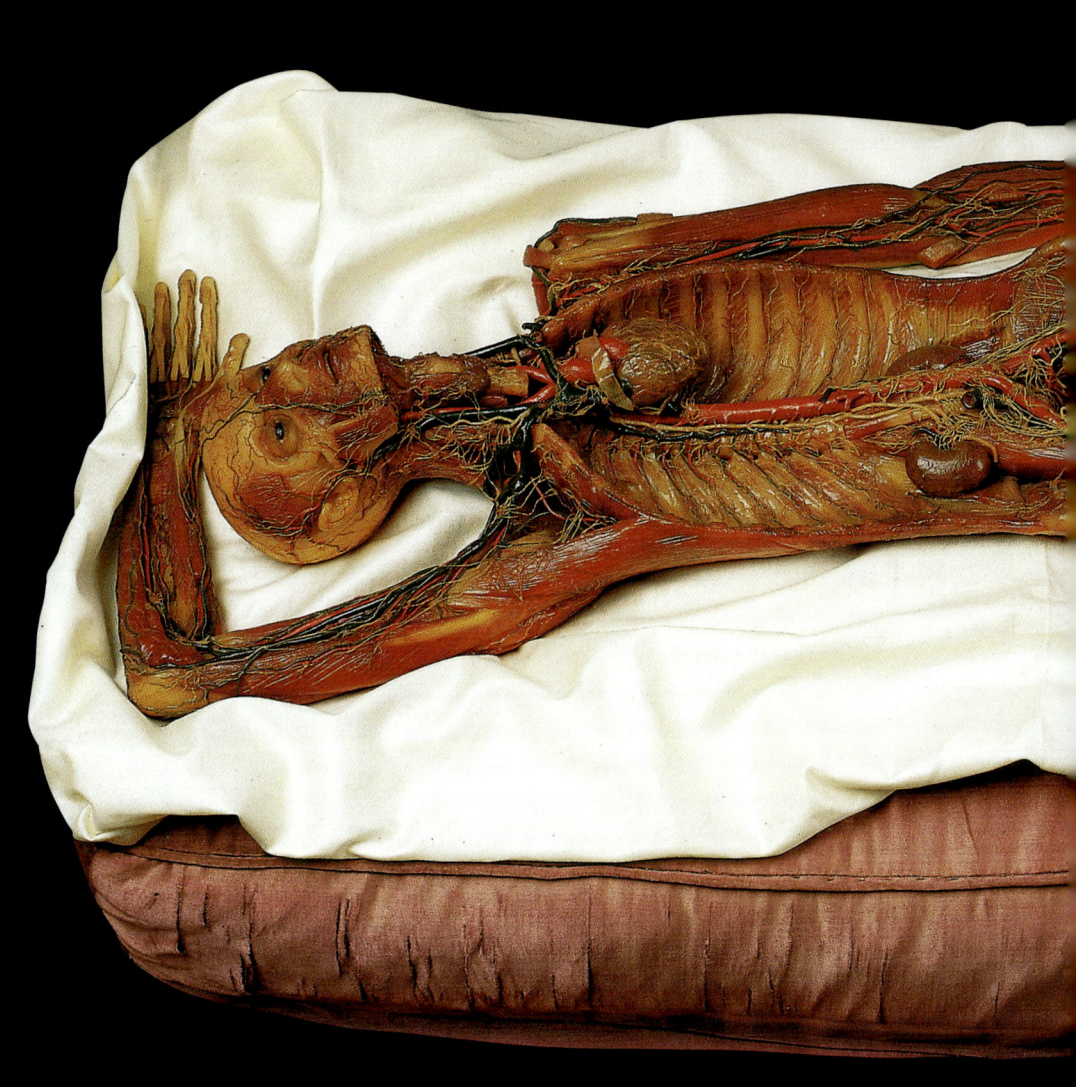

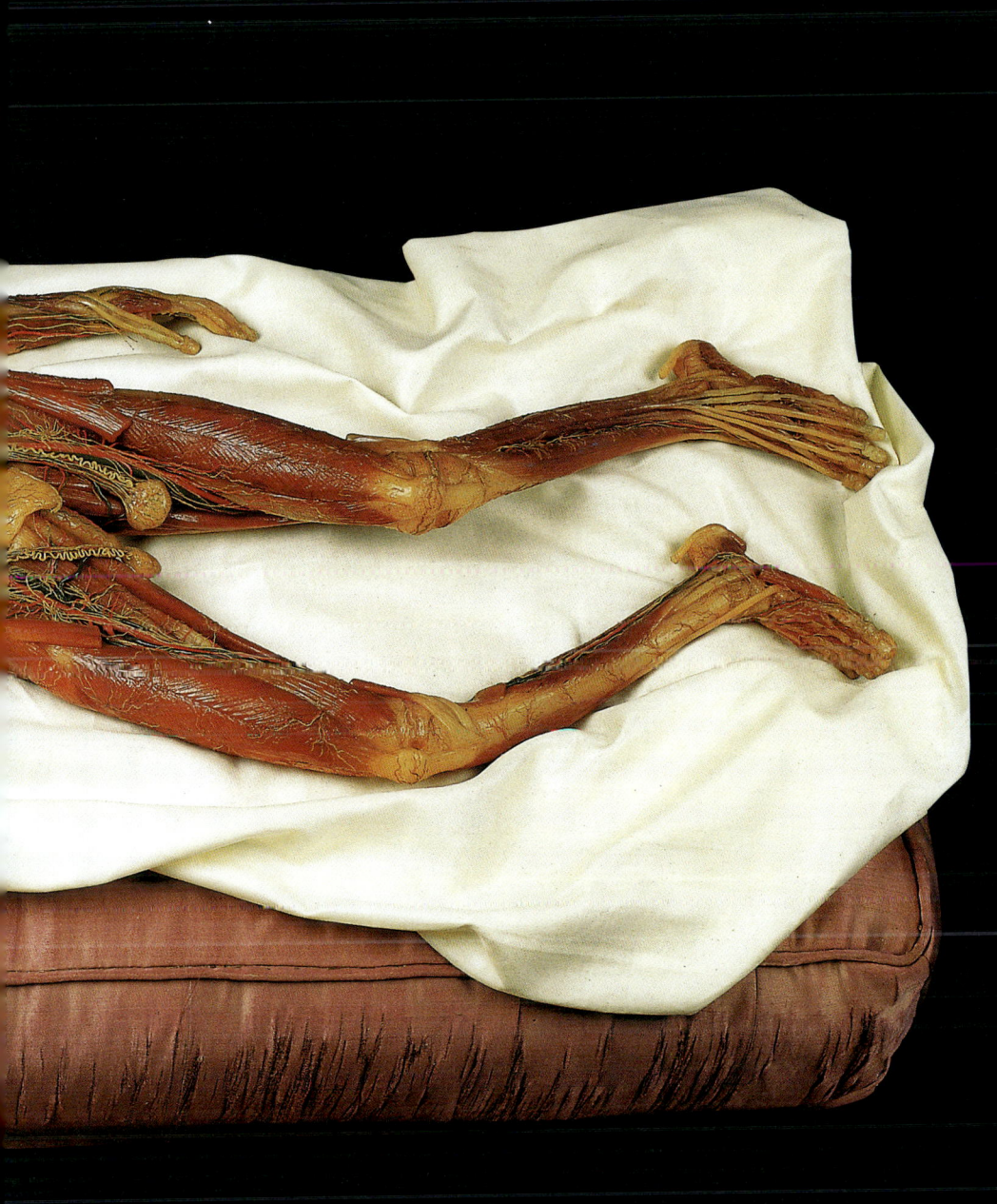

Vasa sanguinea, Vasa lymphatica
Whole body specimen showing sections of the vascular system

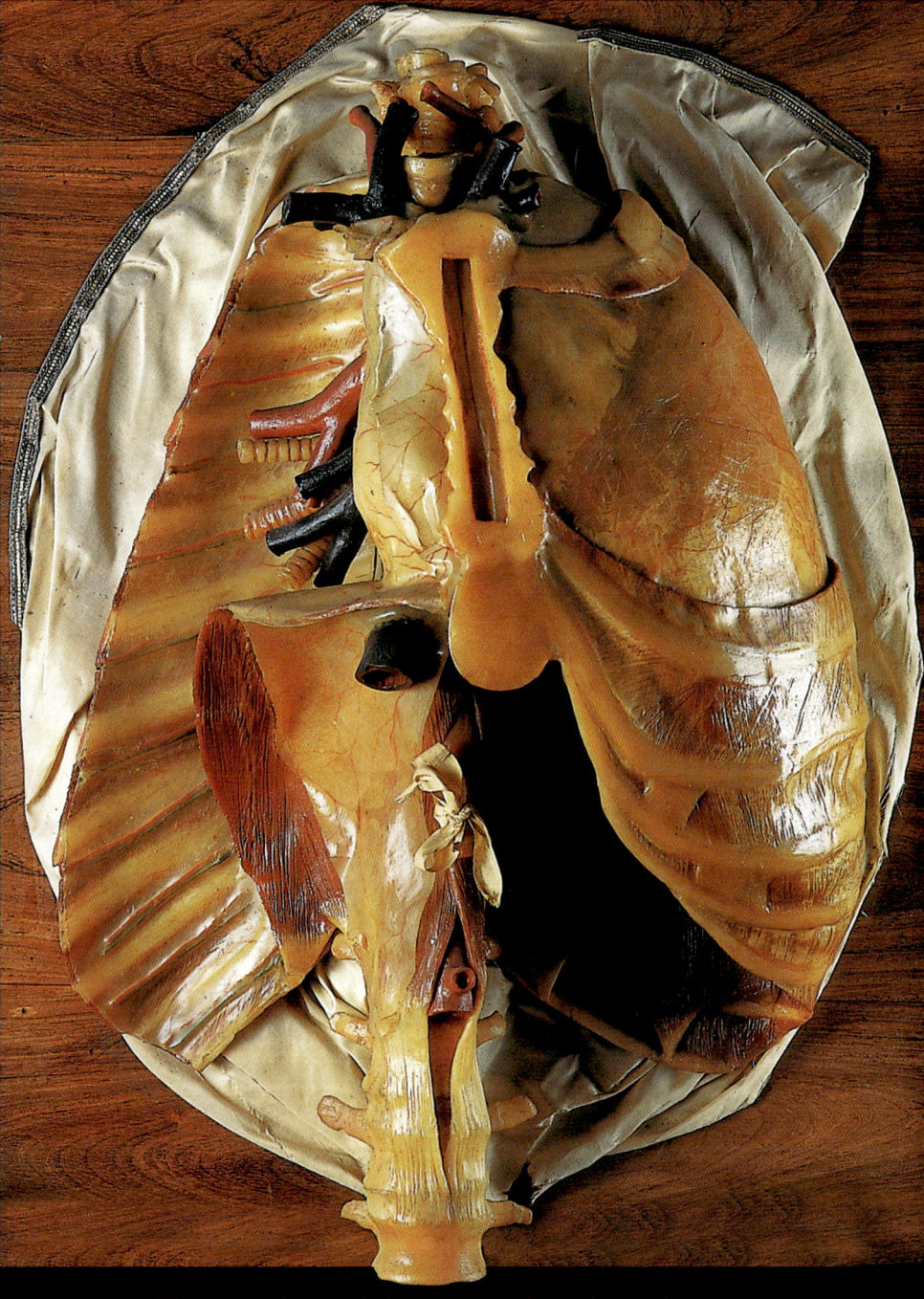

Thorax, Diaphragma, Pulmo sinister, Hilum pulmonis dextris, Cor

Thoracic cavity laid open and seen from in front. On the left the heart can be seen in the pericardium.
Behind are lying the branches of the pulmonary arteries and veins and the bronchi

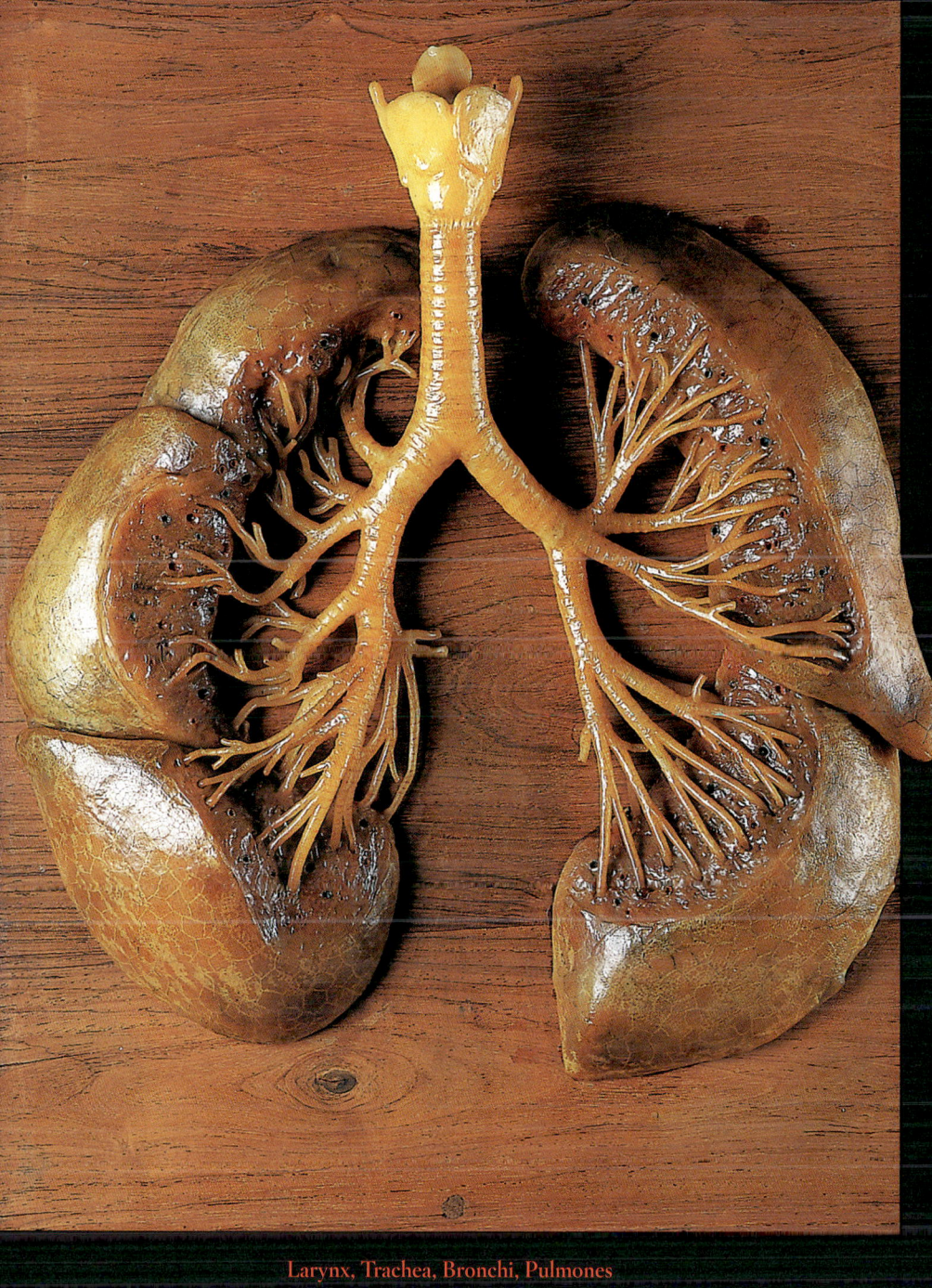

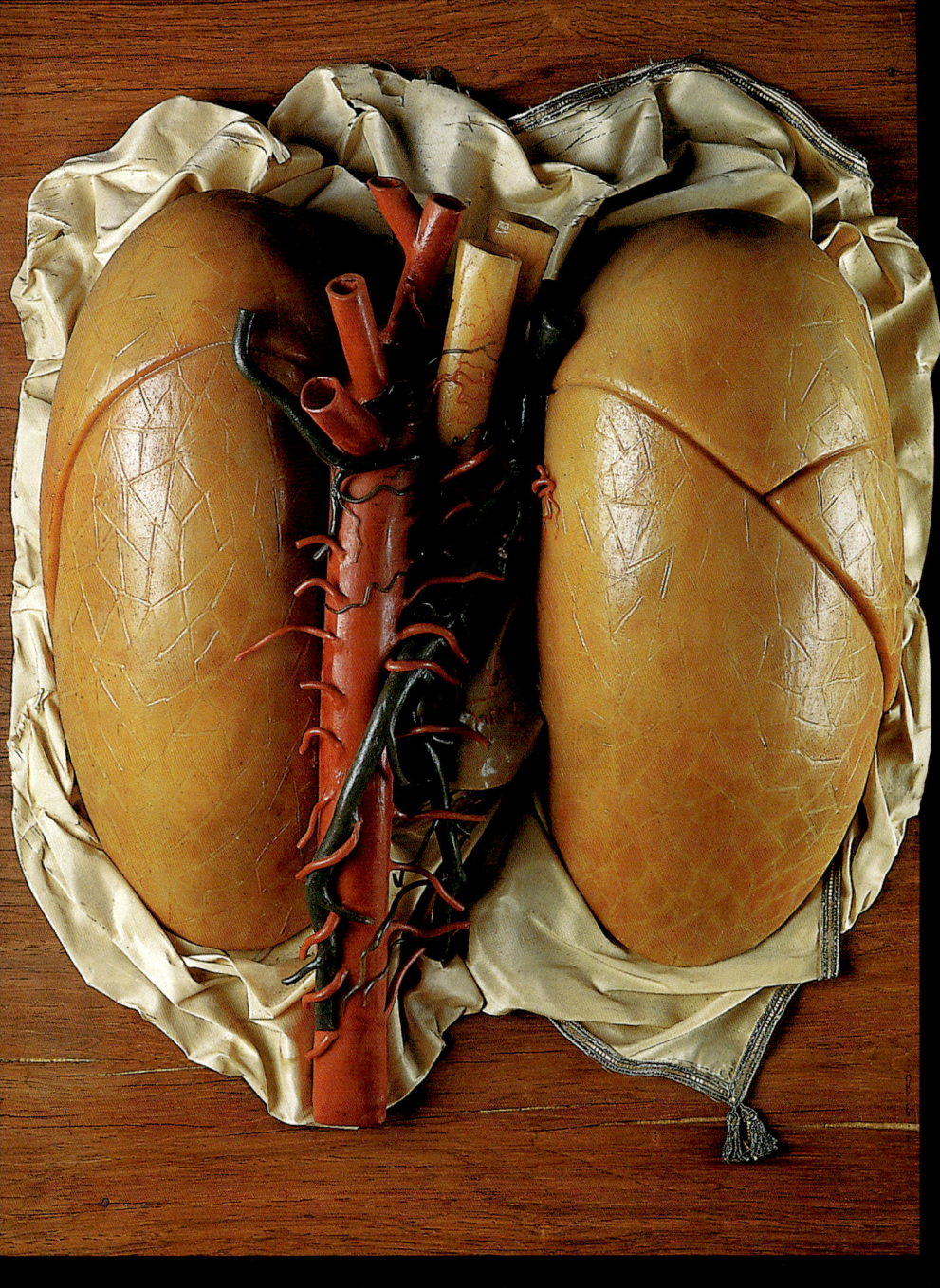

Pulmones, Aorta thoracica

The left and right lungs with their respective two (left) or three lobes (right) seen from behind.
The aorta and its intercostal branches can also be seen

236

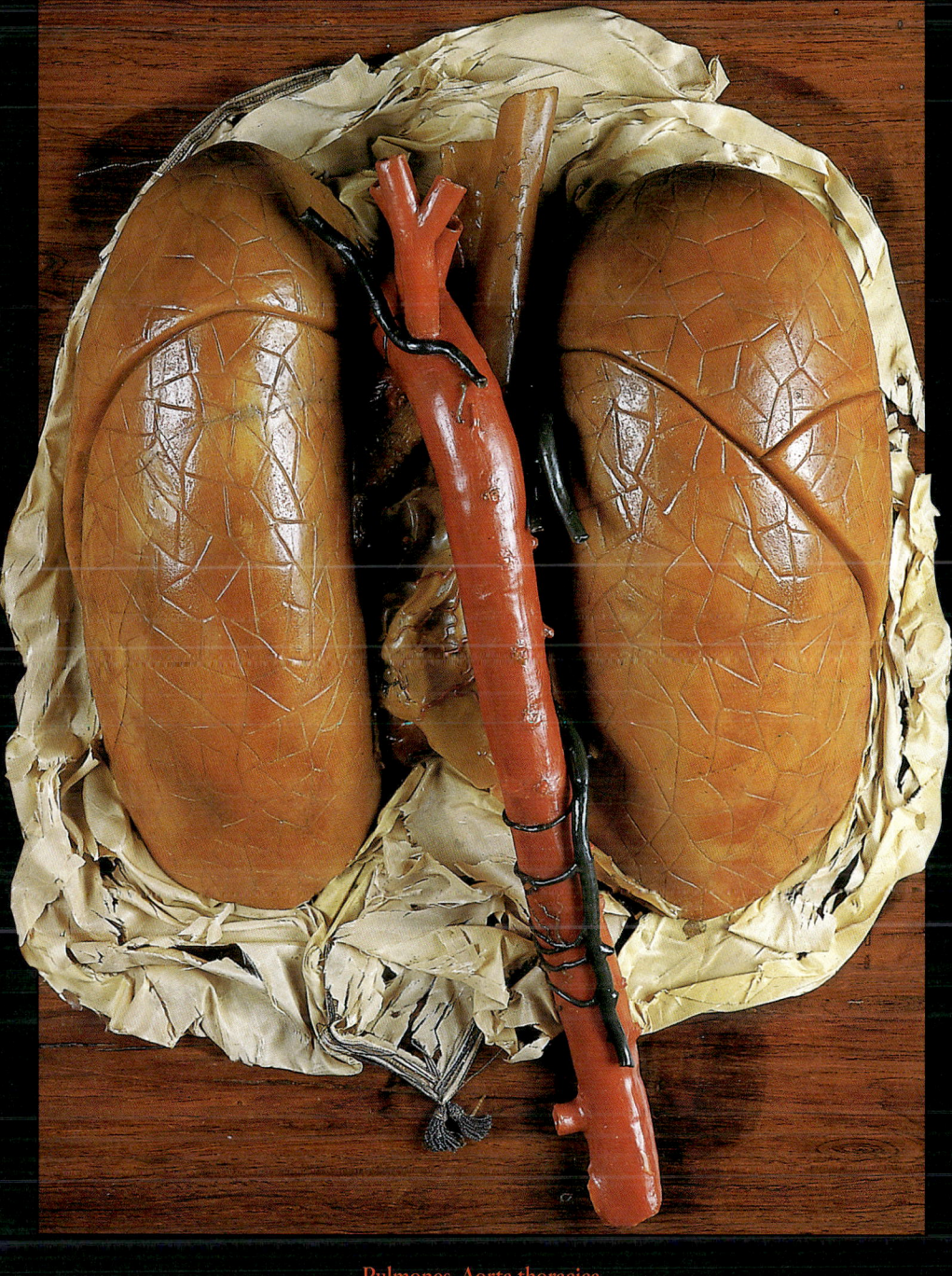

Pulmones, Aorta thoracica

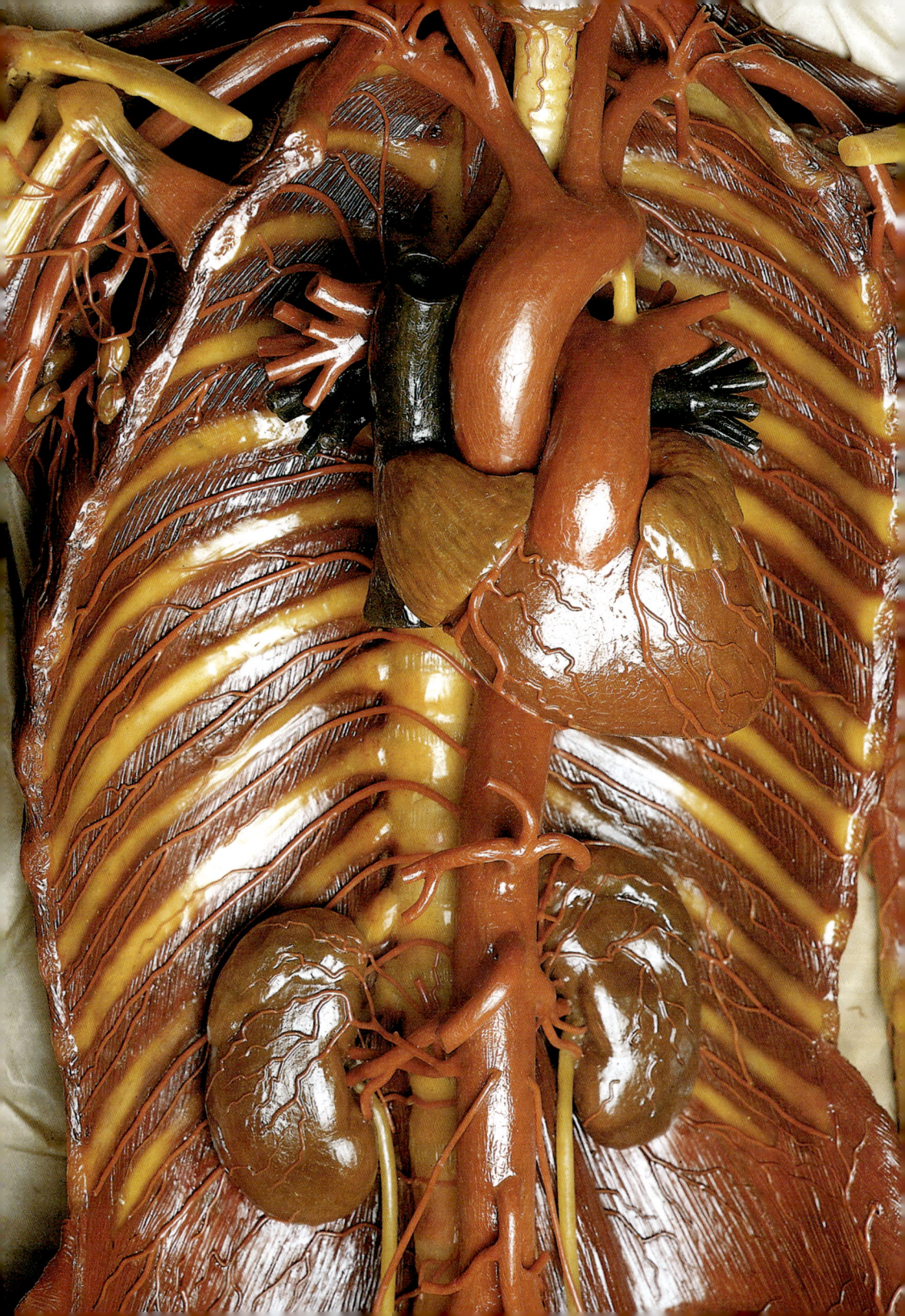

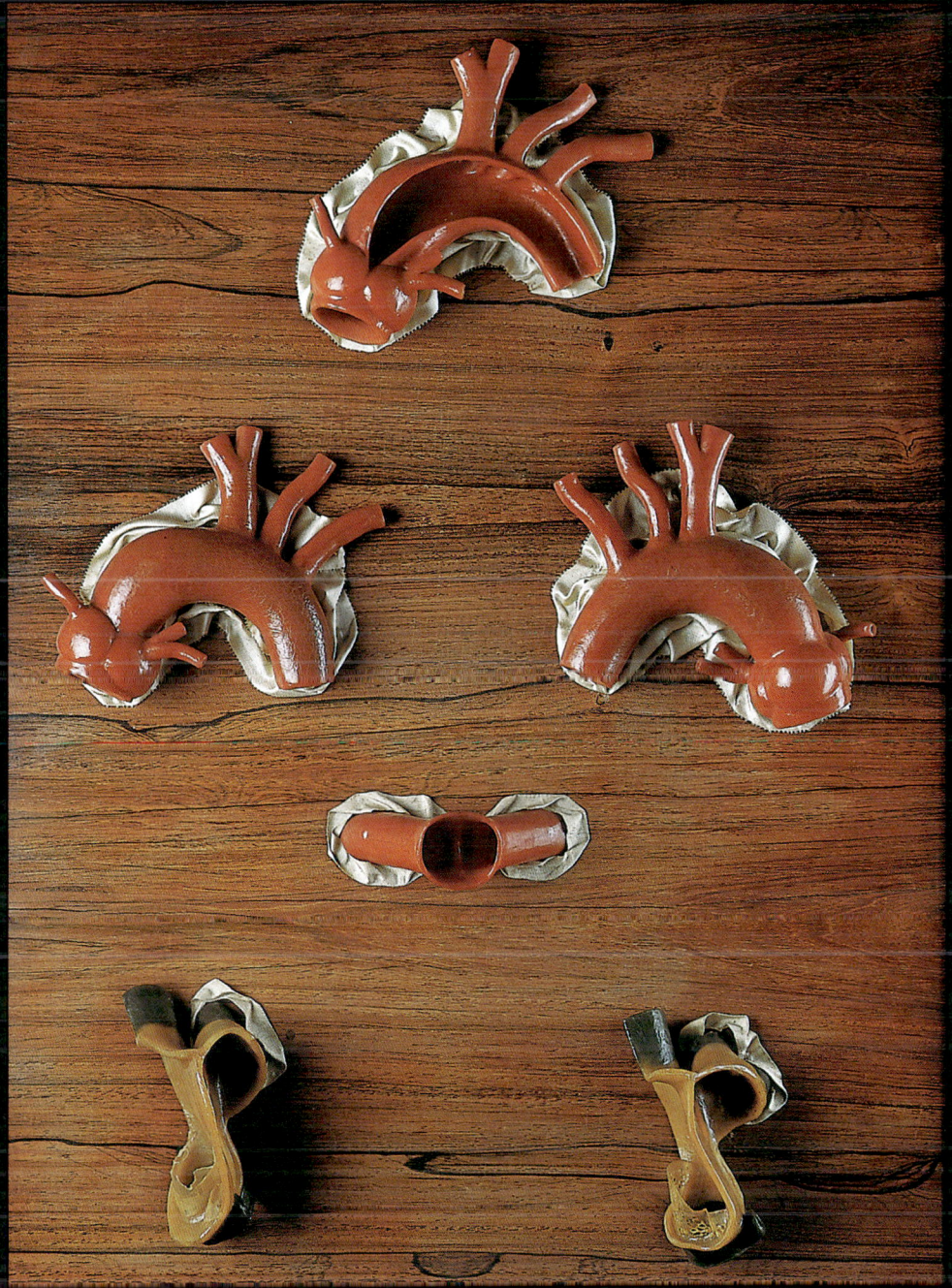

Arcus aortae, Bulbus aortae

Various views of the aortic arch. The coronary arteries arise from the aortic bulb

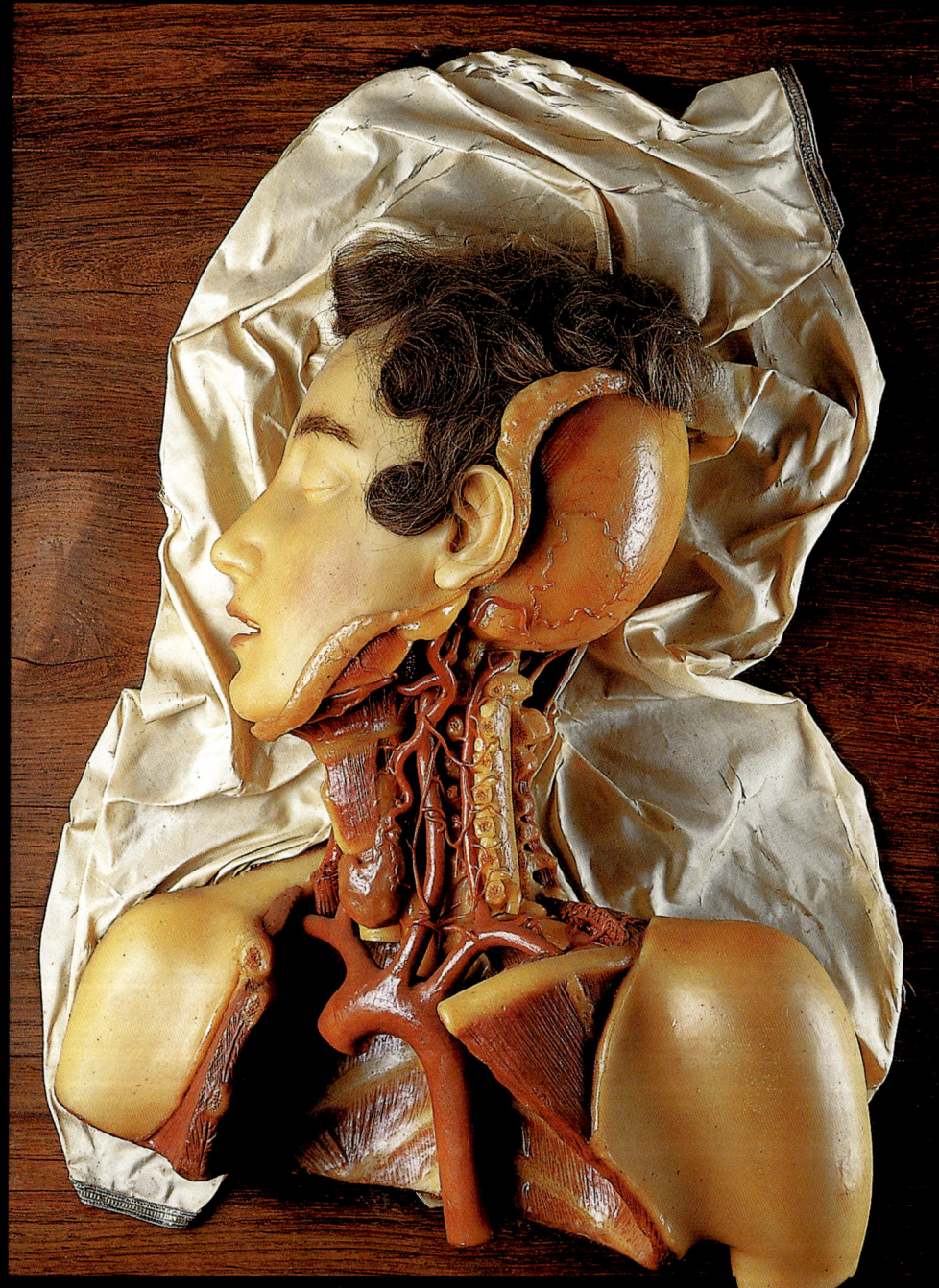

Arteria carotis communis, Arteria carotis externa, Arteria subclavia, Arteria vertebralis

The aorta and the large arteries in the neck region

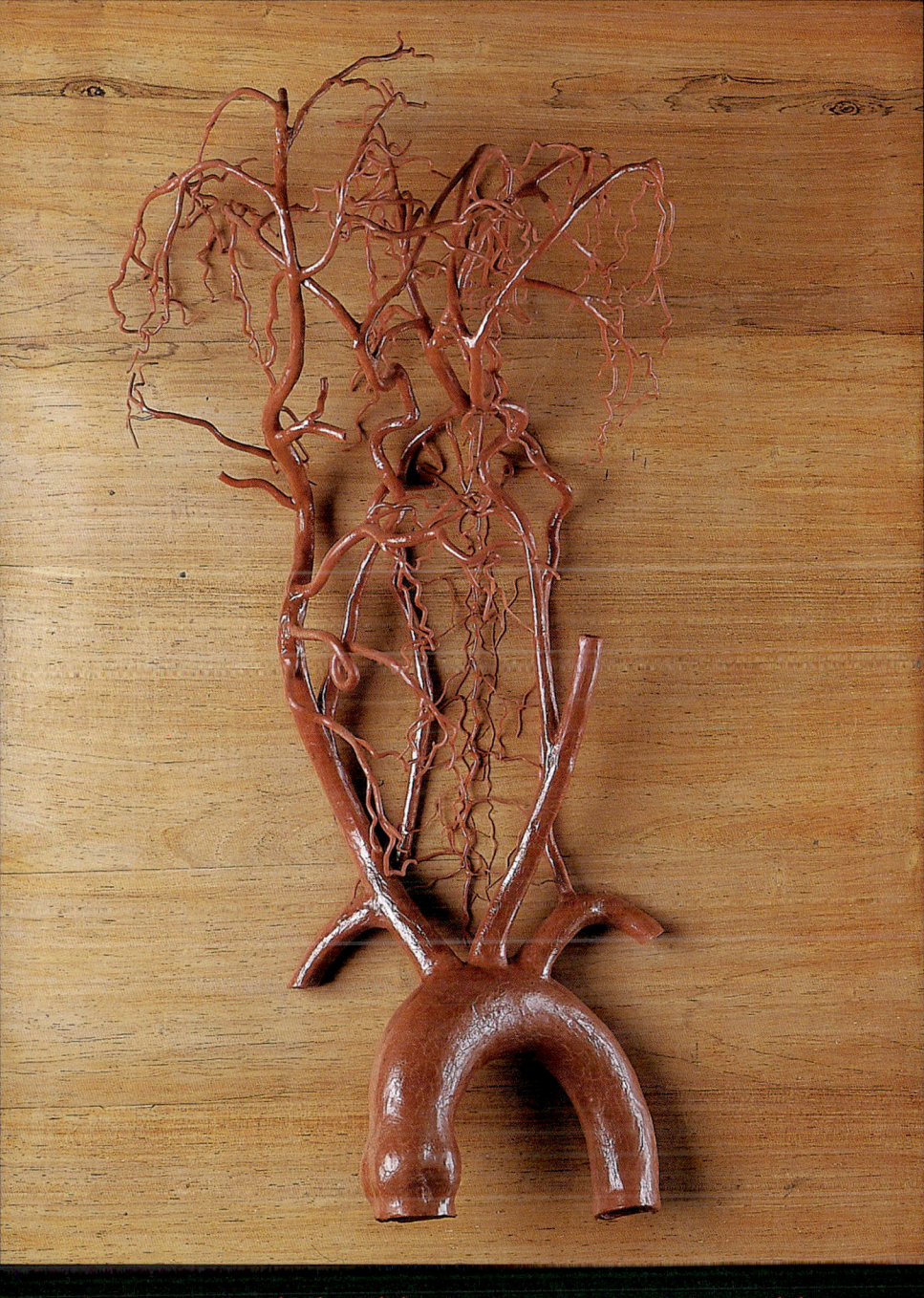

Specimen showing the arteries of the head and neck. On the left side of the model
the carotid artery with its branches can be seen taking origin from the aortic arch

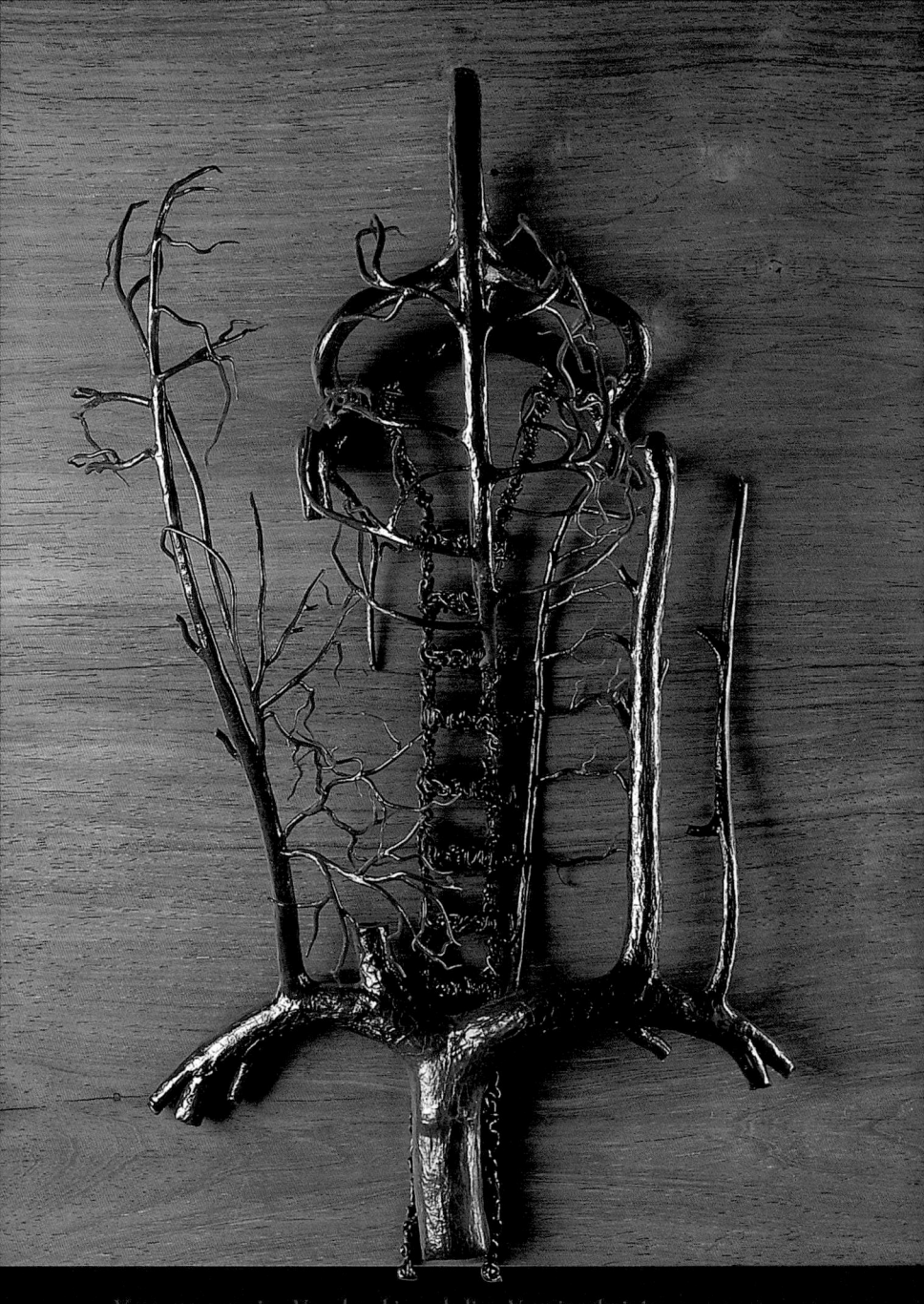

Vena cava superior, Vena brachiocephalica, Vena jugularis interna et externa,
Sinus sagittalis superior, Sinus transversus, Sinus sigmoideus, Plexus venosus vertebralis internus

The veins of the head and neck

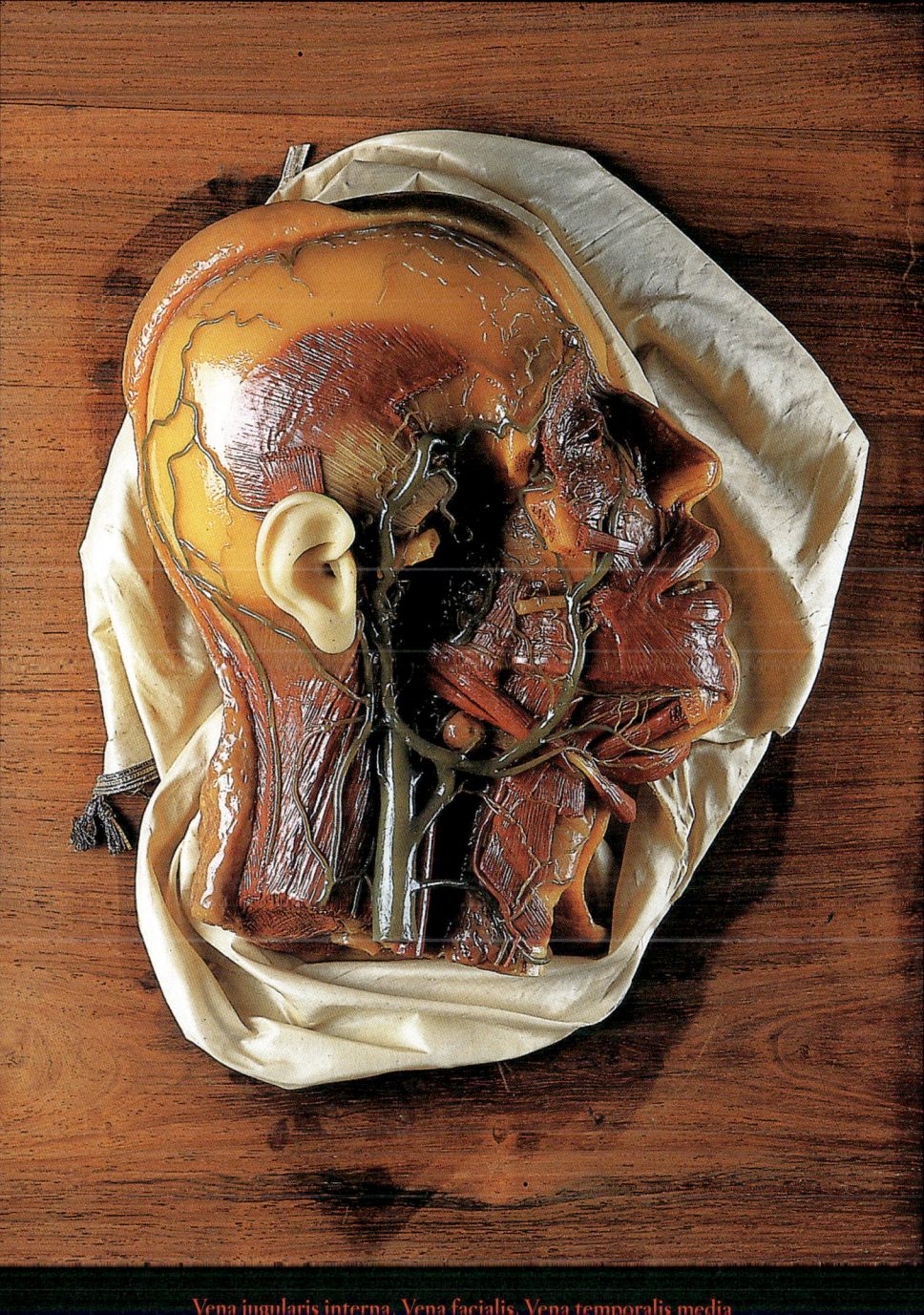

The veins in the region of the neck and face

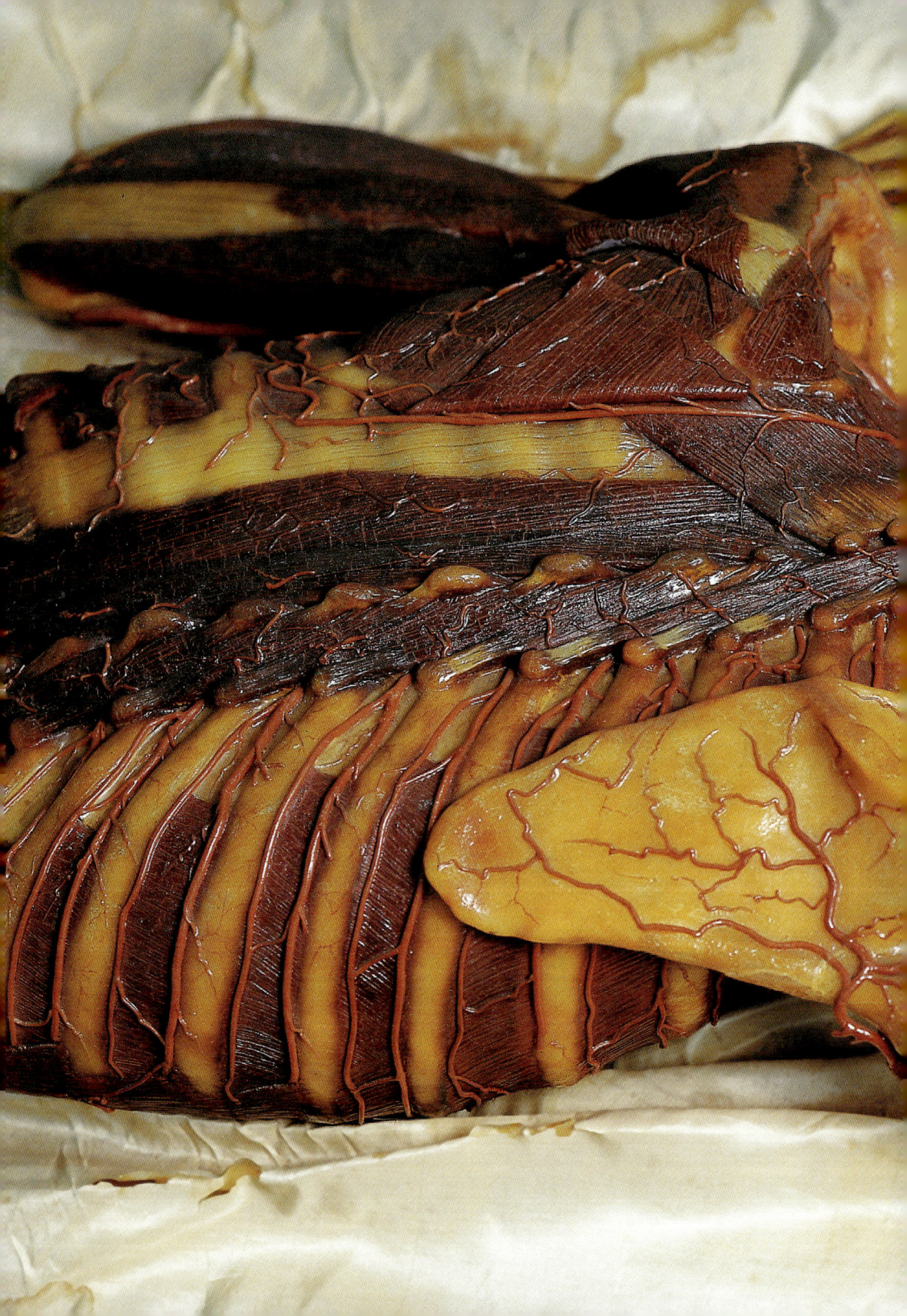

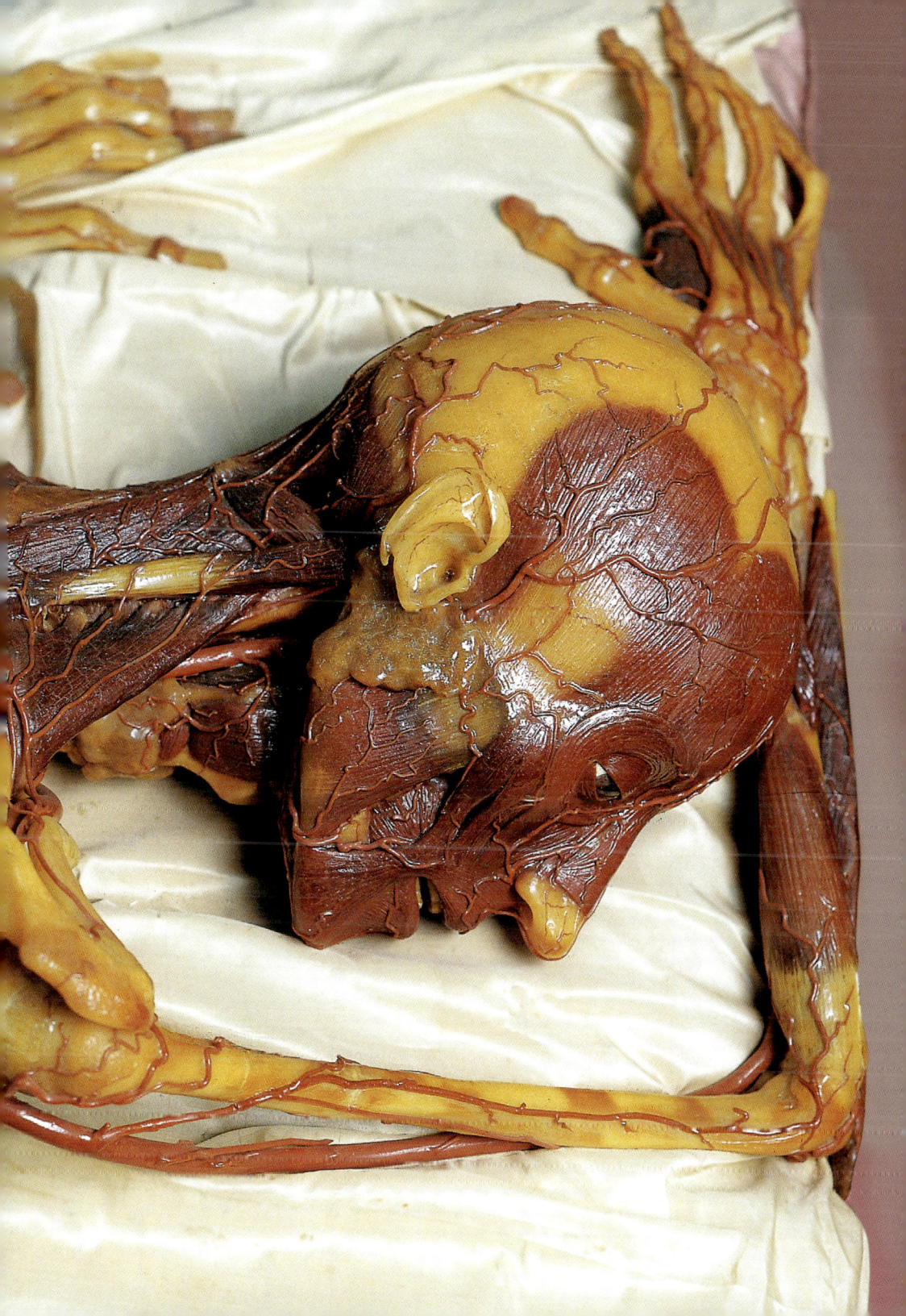

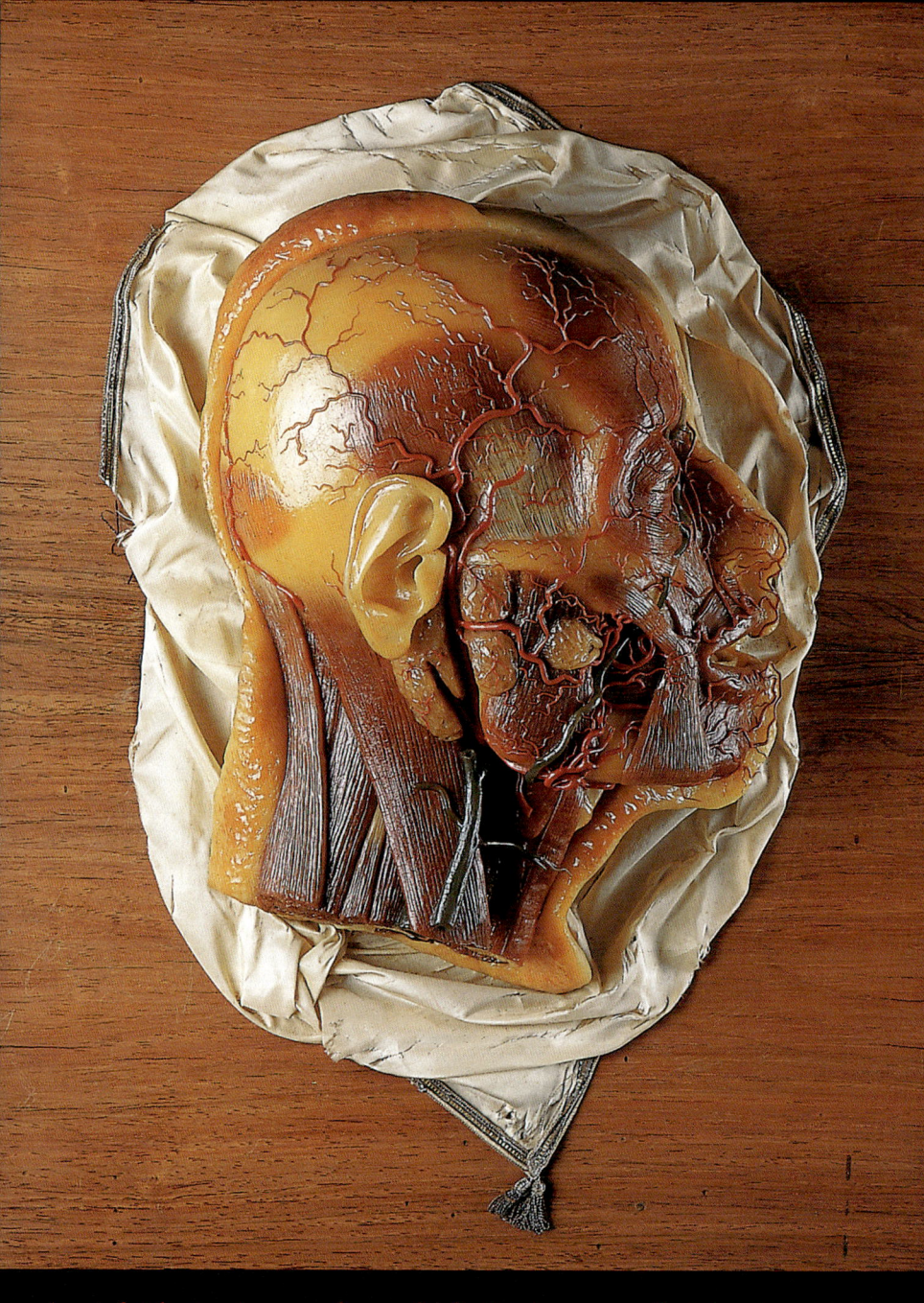

Arteria facialis, Arteria temporalis superficialis, Arteria transversa facei, Arteria occipitalis,
Vena retromandibularis, Vena facialis, Vena jugularis externa

The superficial arteries and veins of the head and neck

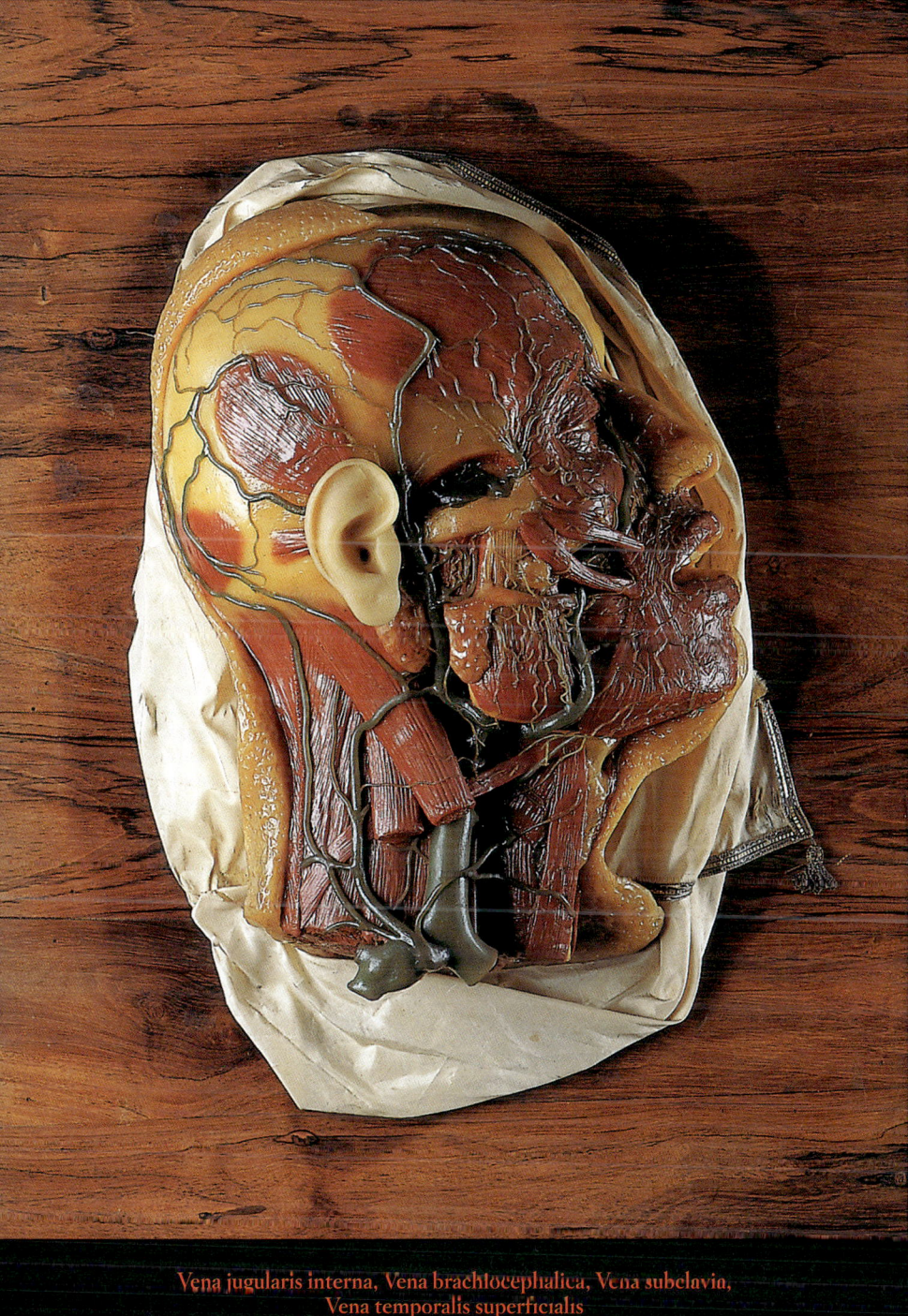

Vena jugularis interna, Vena brachiocephalica, Vena subclavia,
Vena temporalis superficialis

The veins in the region of head and neck

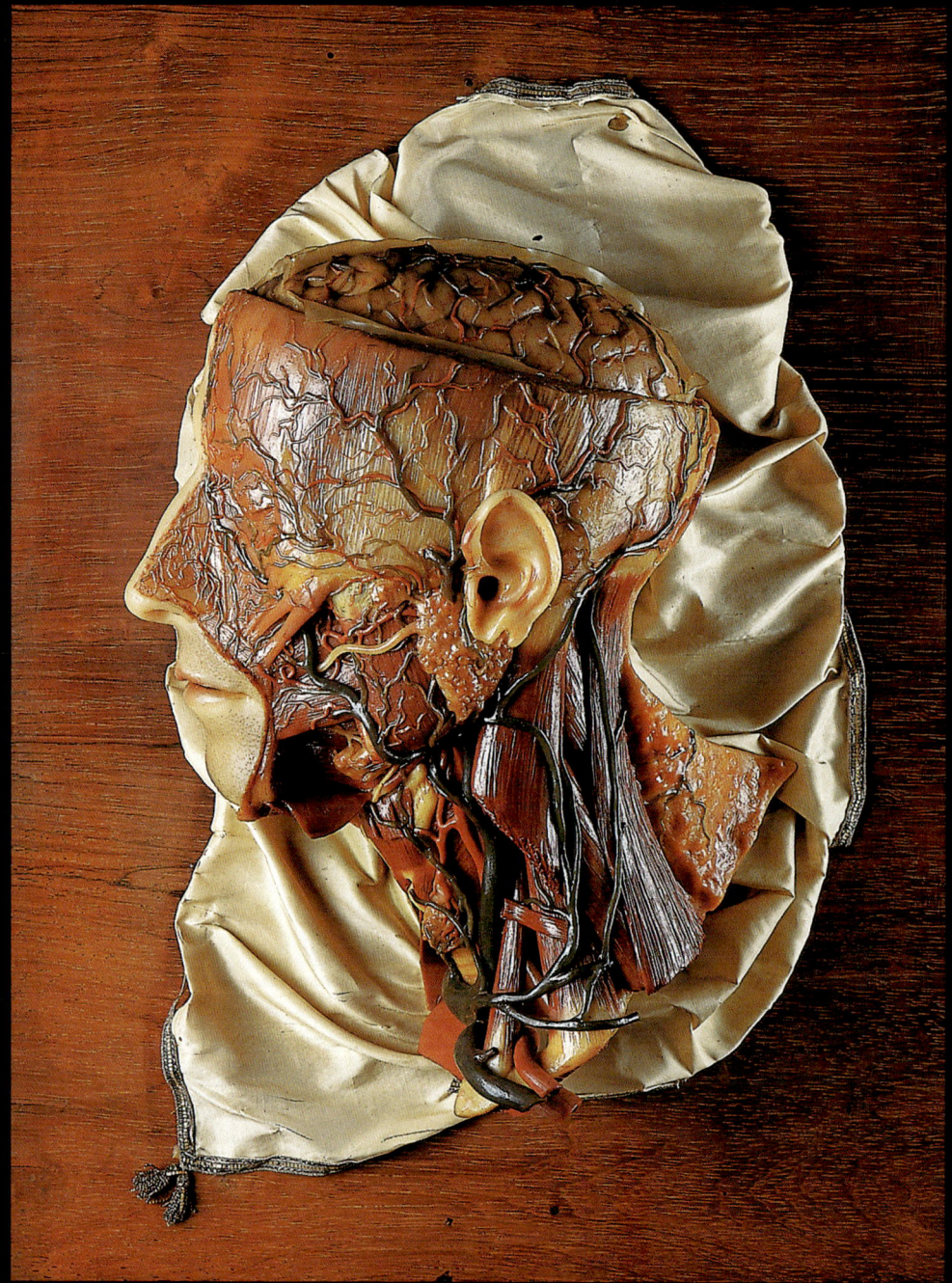

Vena jugularis interna, Vena facialis, Vena occipitalis

The superficial veins of head and neck. The parotid gland and its duct can be seen lying in front of the ear

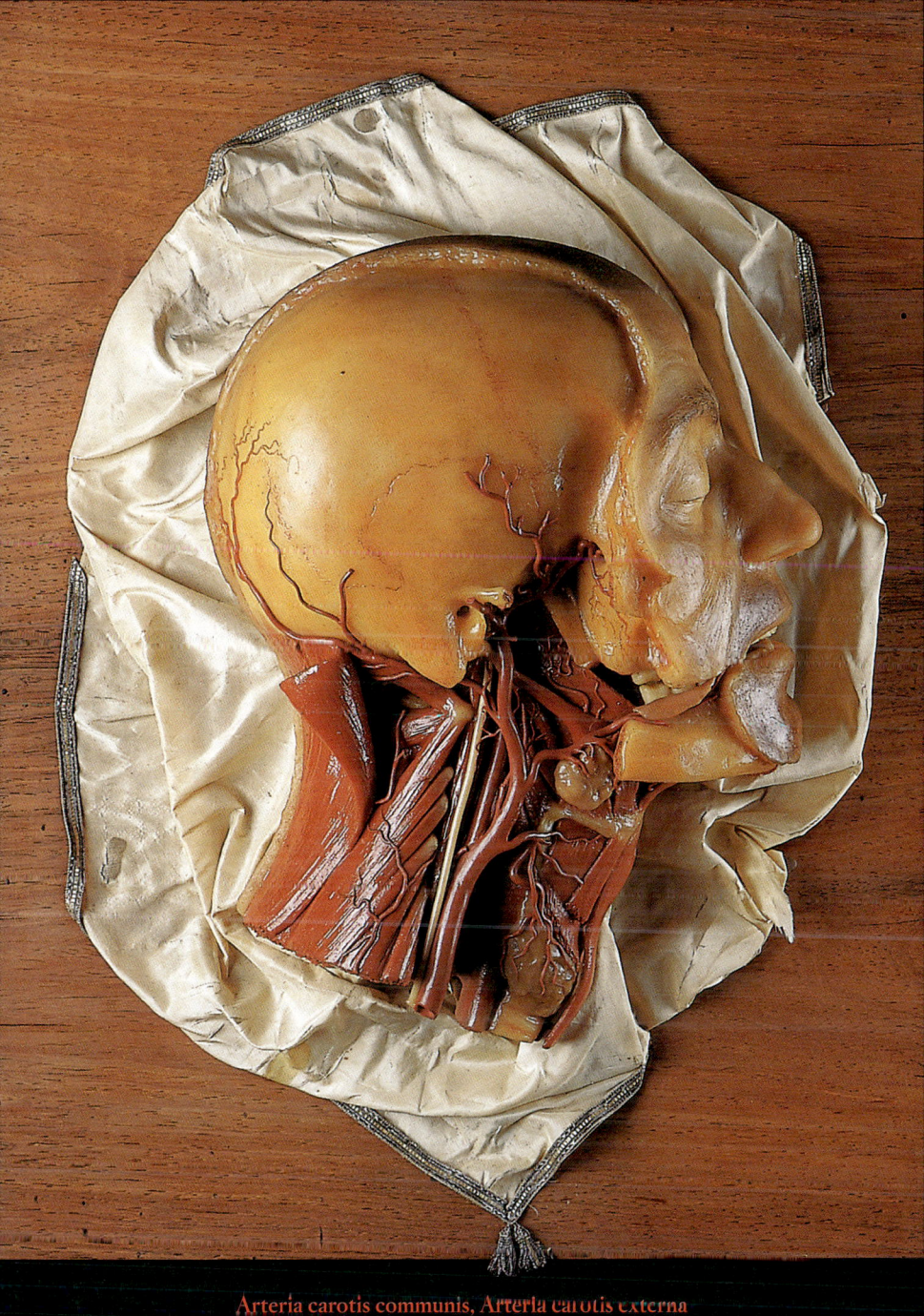

Arteria carotis communis, Arteria carotis externa

The carotid artery and the vagus nerve

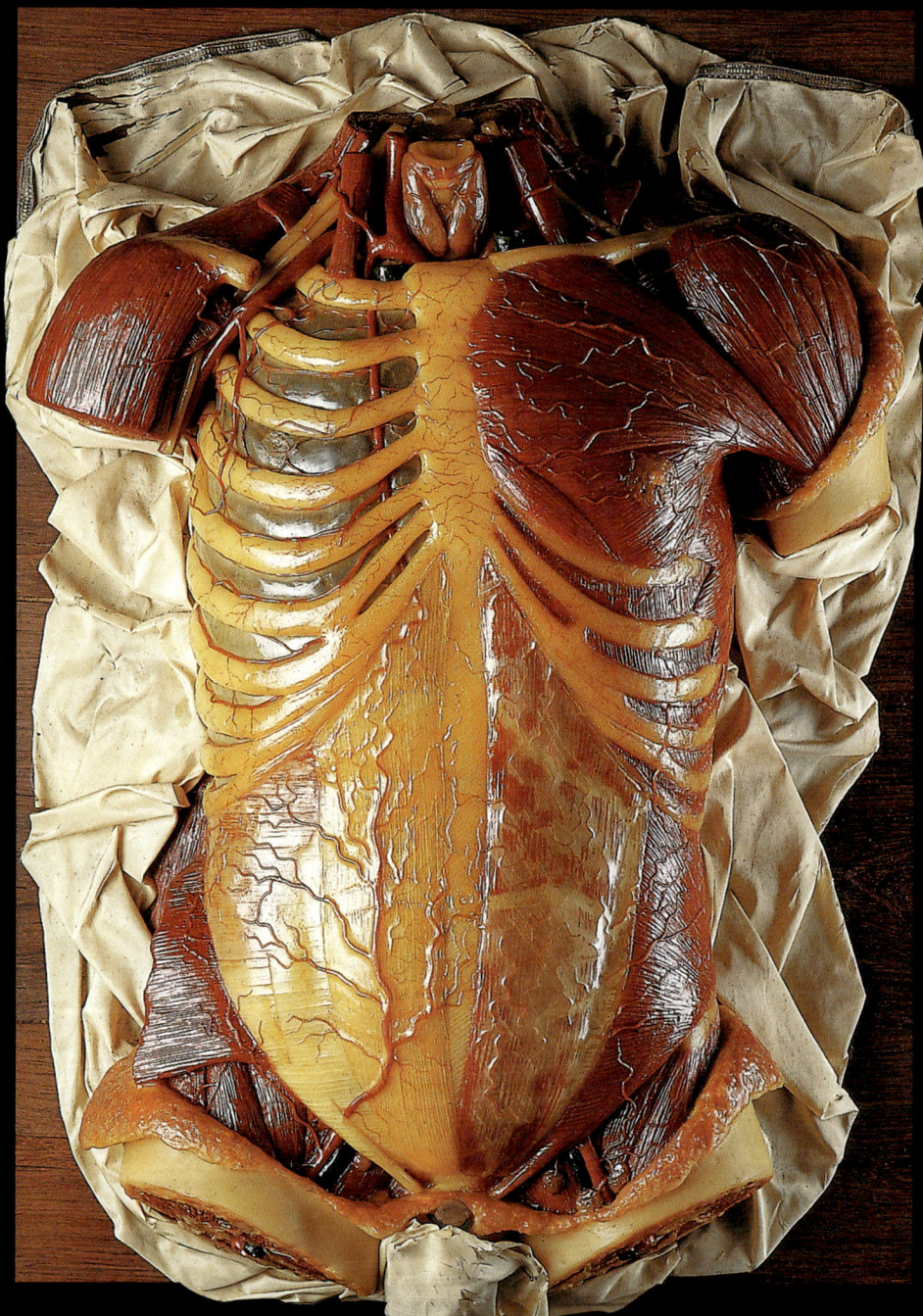

Arteria thoracica interna, Arteria epigastrica inferior, Arteria iliaca externa

The arteries of the anterior thoracic and abdominal walls

Detail from pages 274–275 →

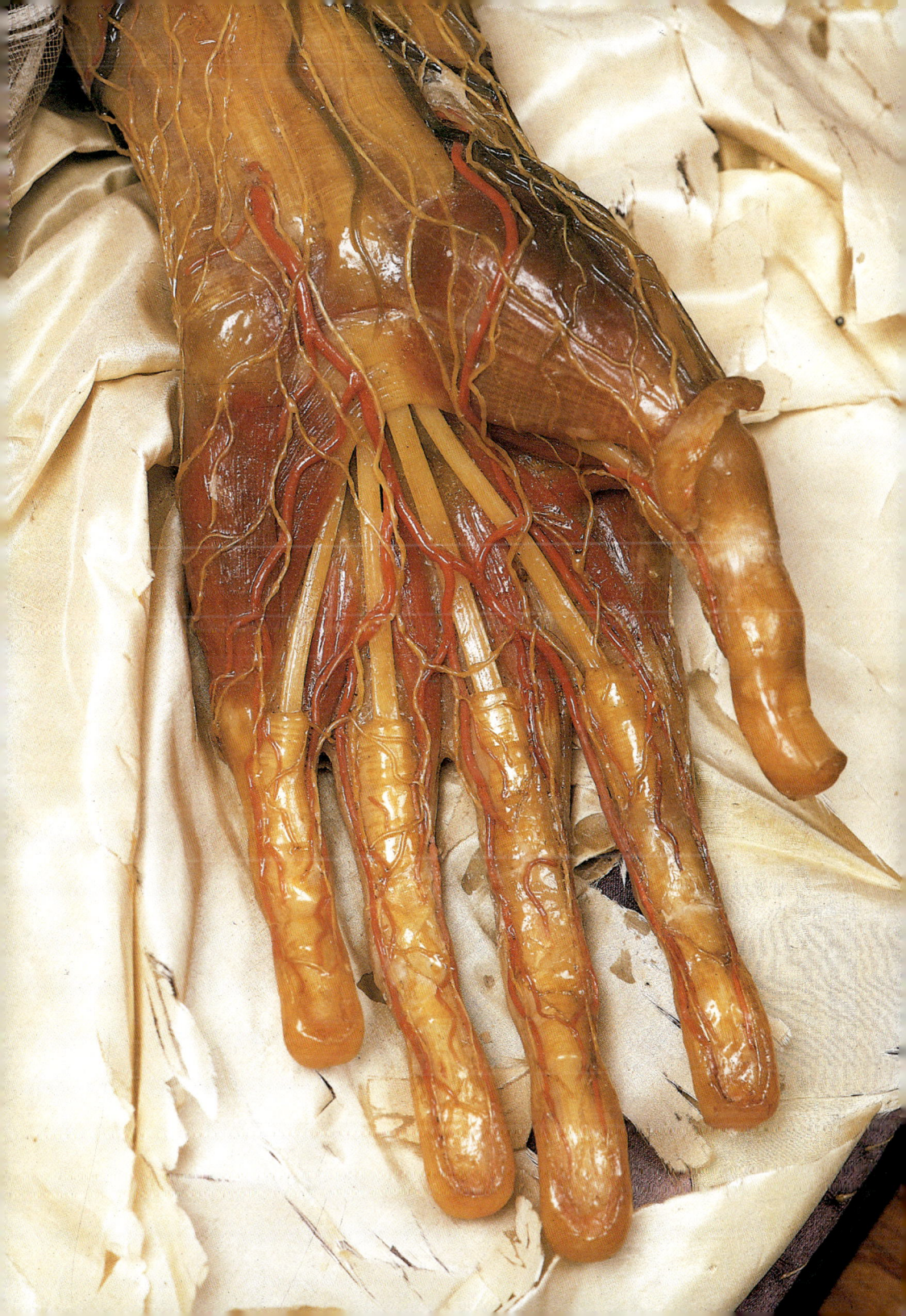

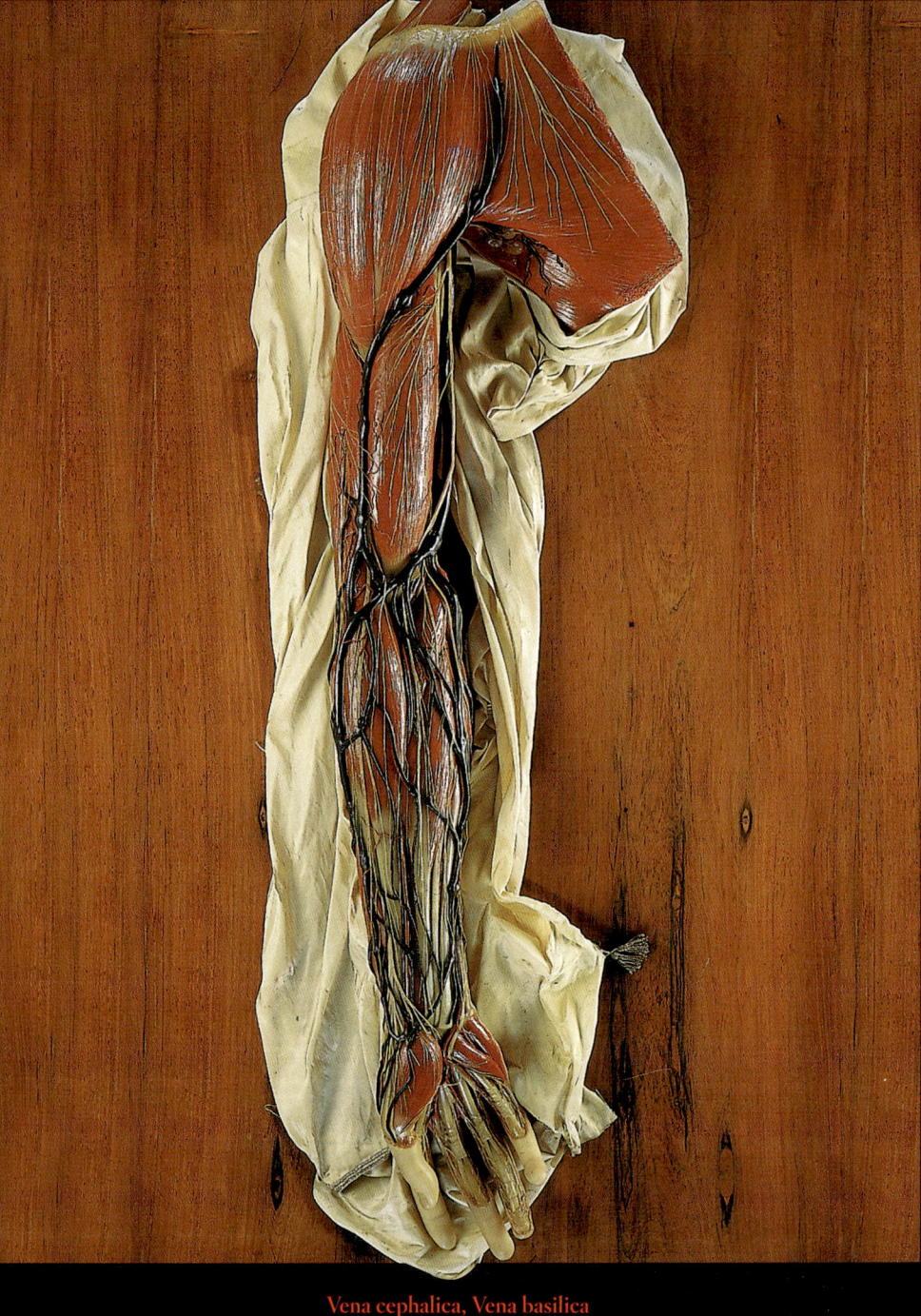

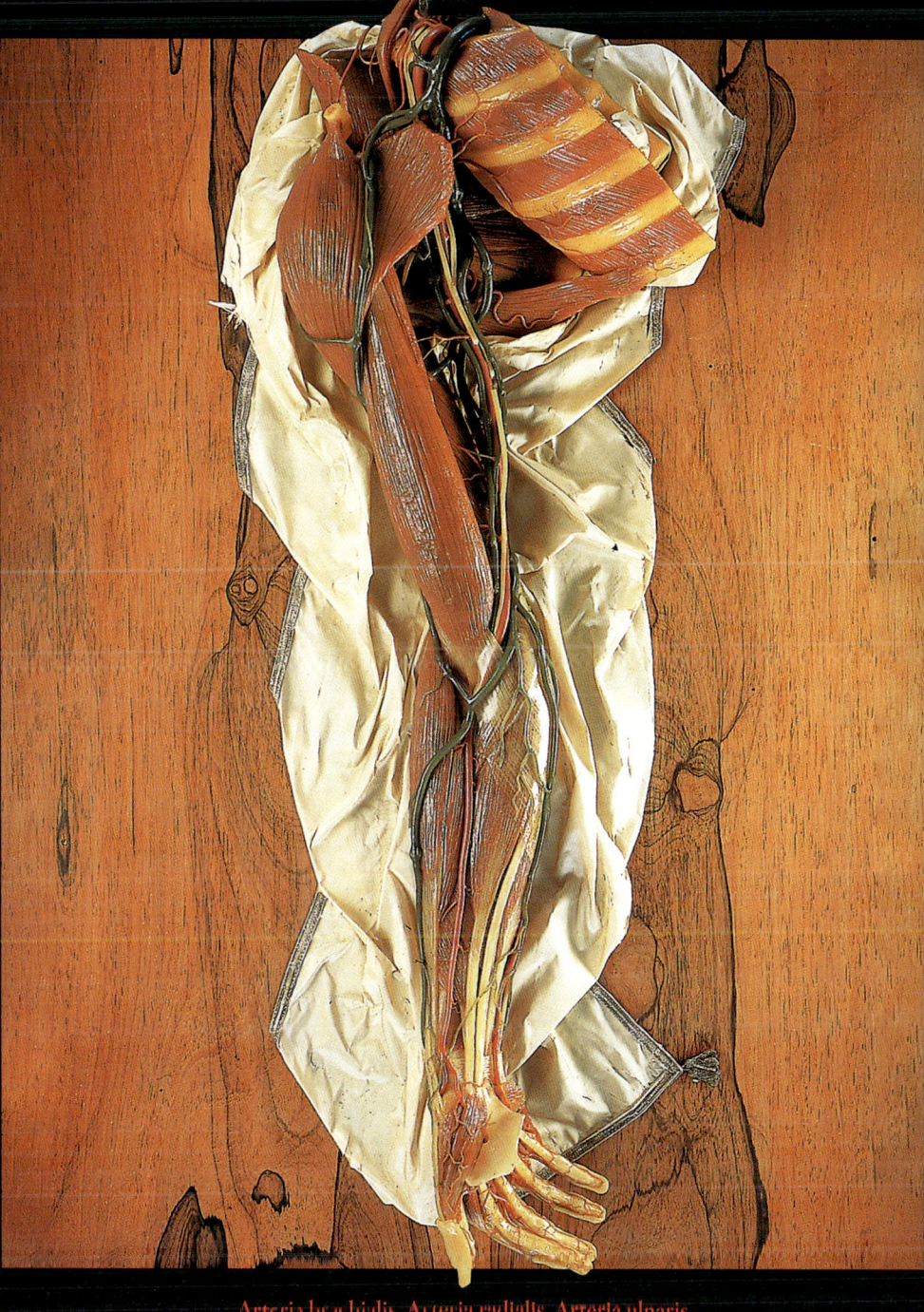

Arteria brachialis, Arteria radialis, Arteria ulnaris

The muscles, vessels and nerves of the arm

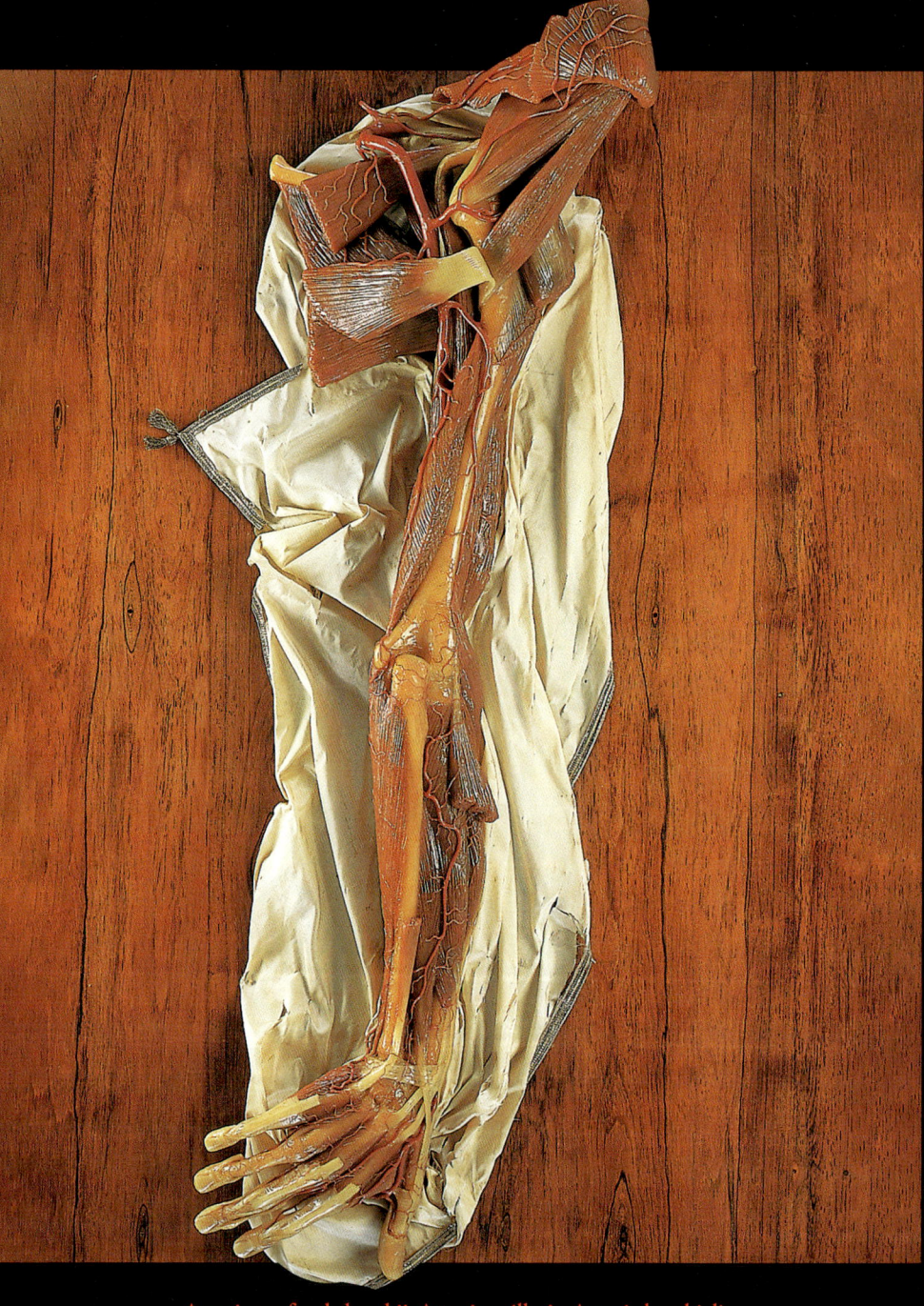

Arteria profunda brachii, Arteria axillaris, Arteria brachialis

The deep arteries of the upper arm and forearm

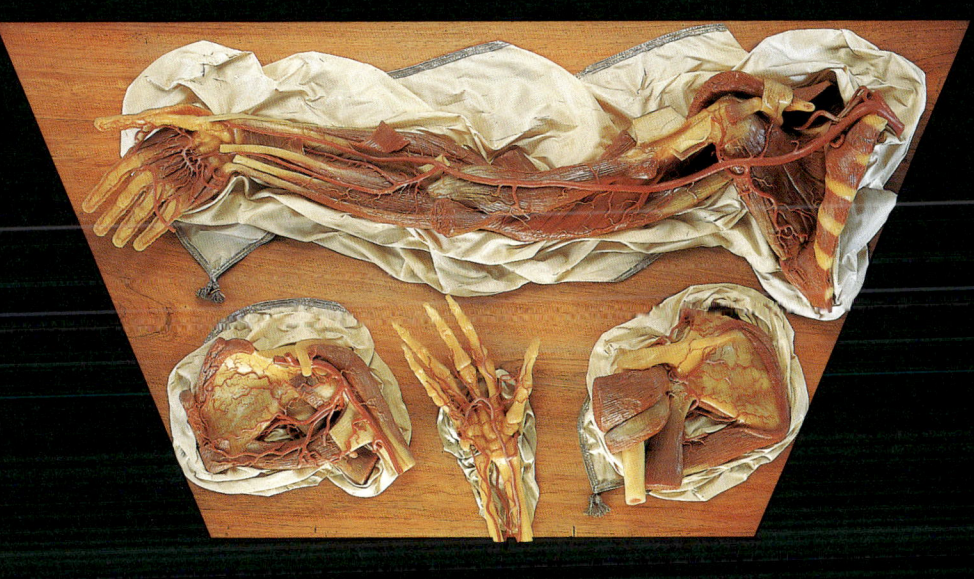

Arteria subclavia, Arteria axillaris, Arteria brachialis, Arteria radialis,
Arteria ulnaris, Arcus palmaris

The arteries supplying the shoulder girdle, arm and hand

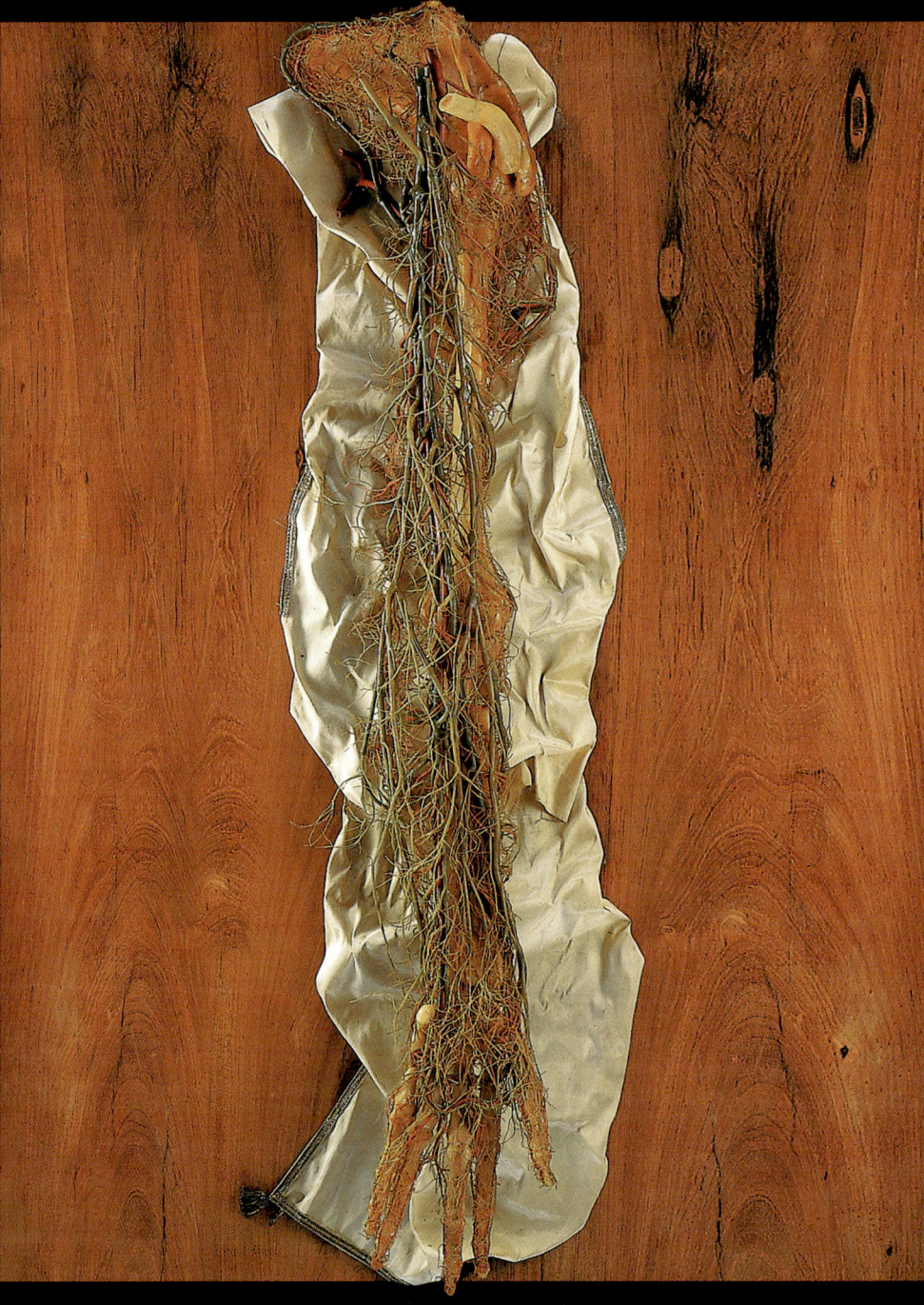

Arteriae, venae et nervi membri superioris

The arteries, veins and nerves of the arm

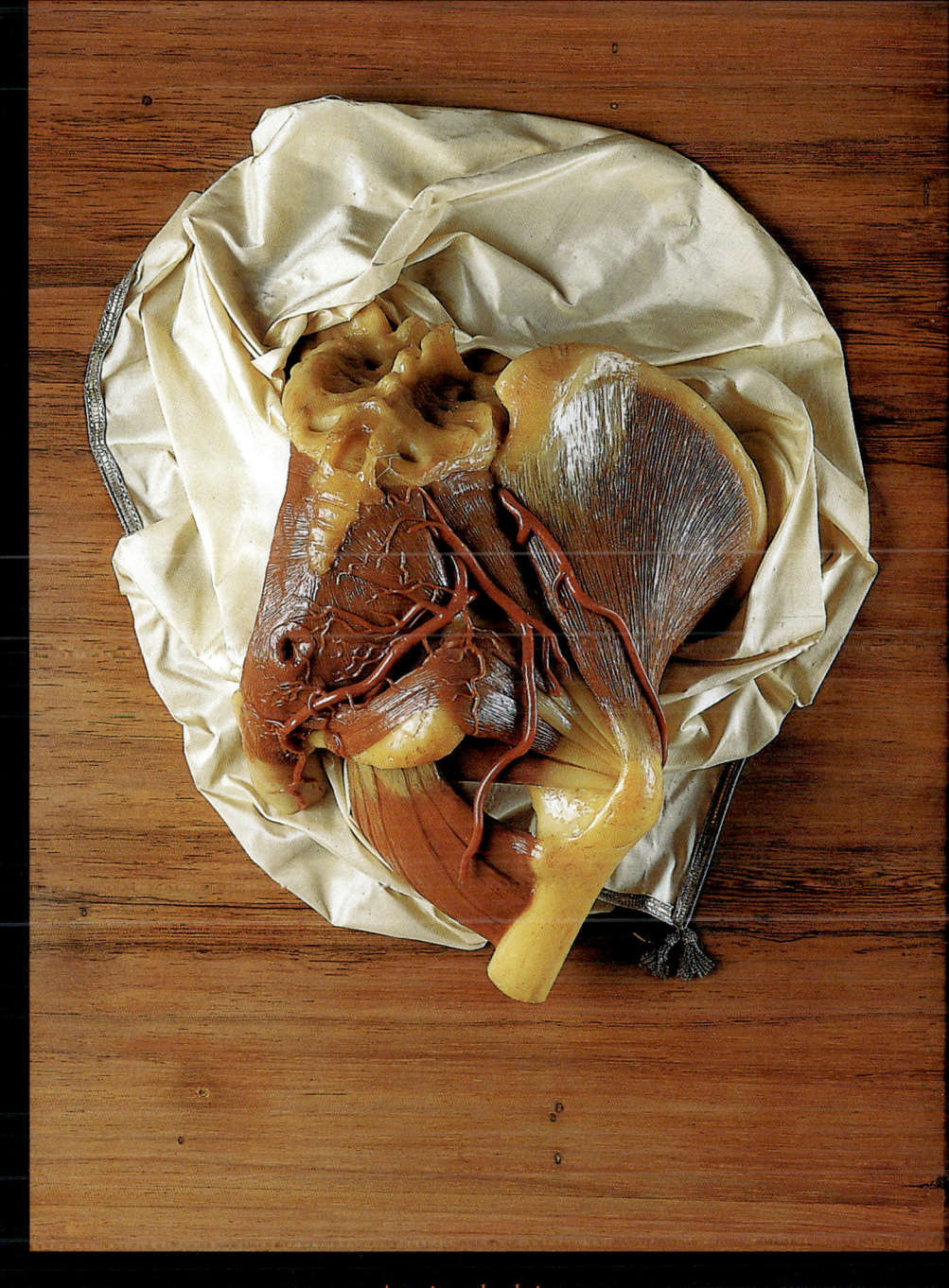

Arteria pudenda interna

The arteries associated with the pelvic floor

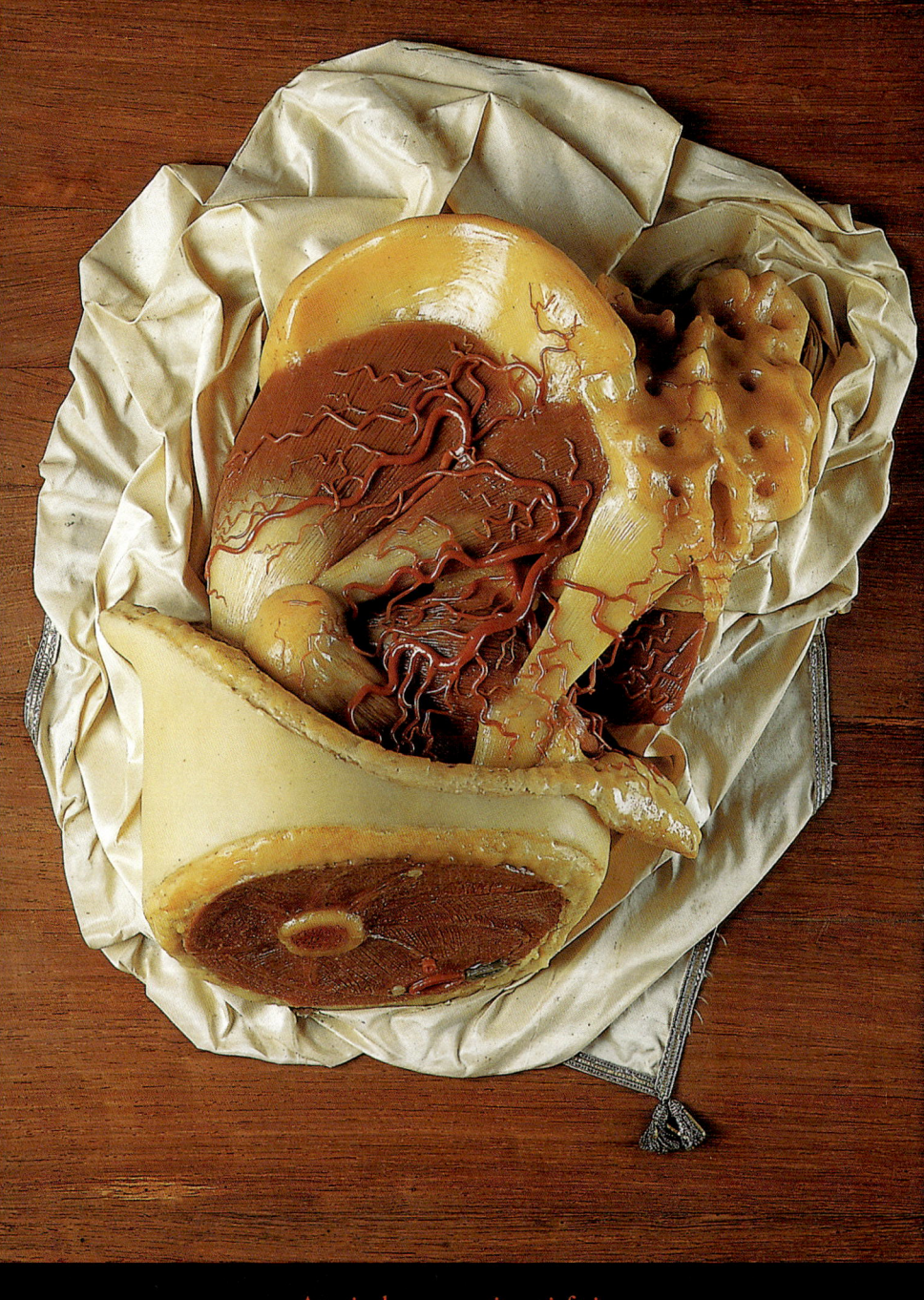

Arteria glutaea superior et inferior

The arteries in the region of the buttock after removal of the large gluteal muscle

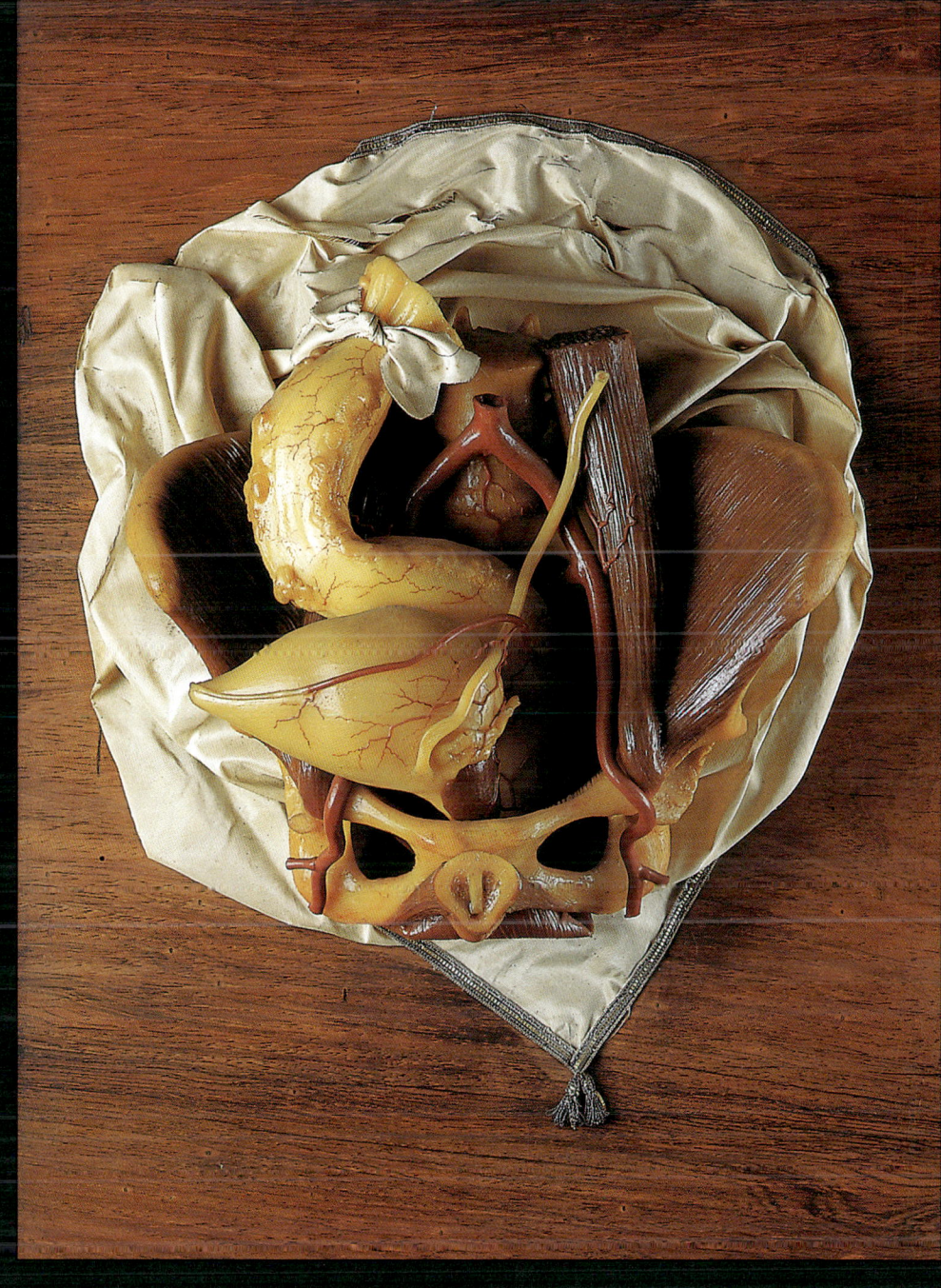

Arteria iliaca communis, Arteria iliaca externa, Arteria femoralis

View from in front of the organs and arteries of a male pelvis. The rectum has been ligated above, and the urinary bladder displaced to one side to show the entry of the ureter

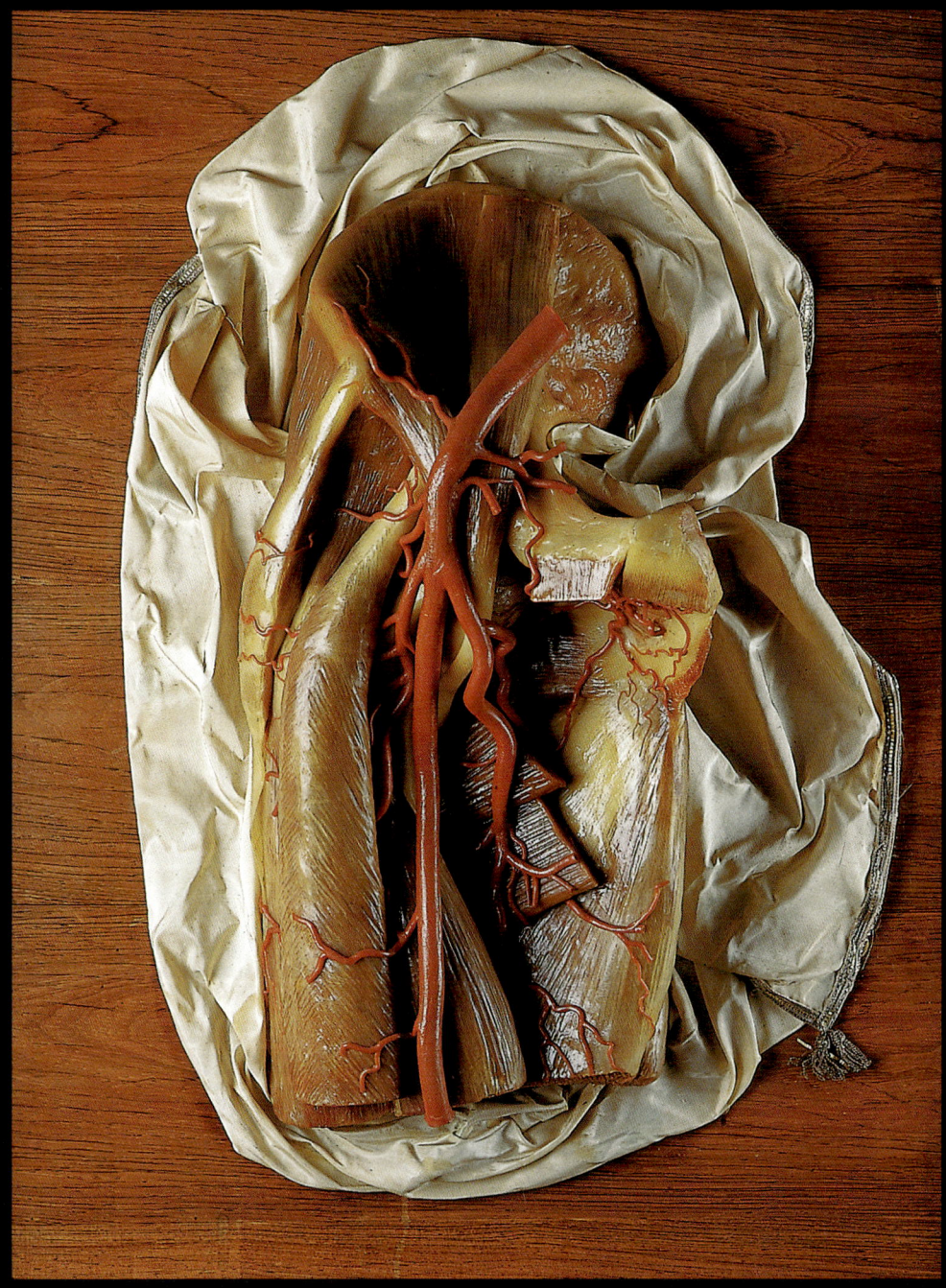

Arteria iliaca externa, Arteria femoralis, Arteria femoralis profunda

The deep course of the femoral artery

Detail from pages 220–221 →

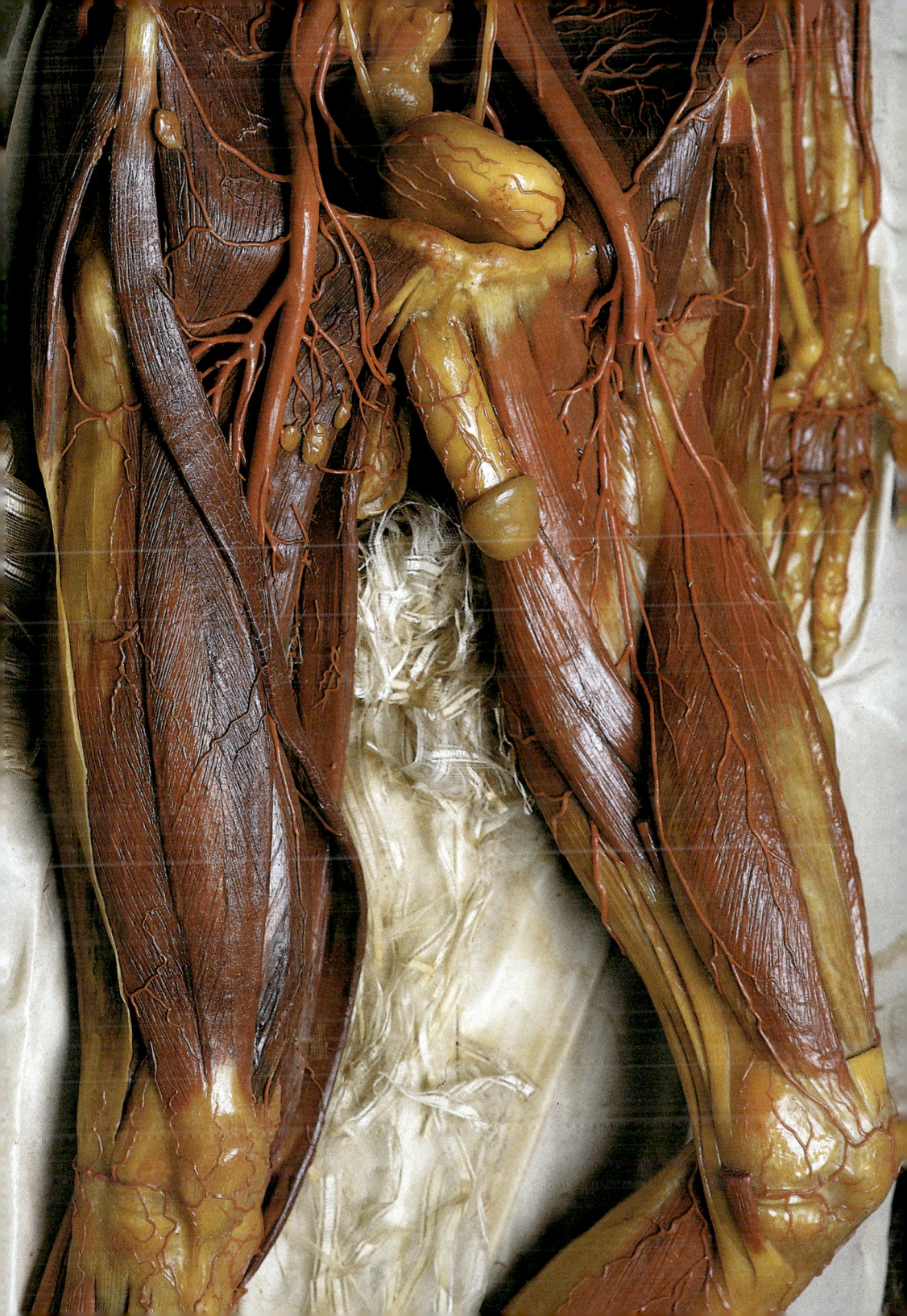

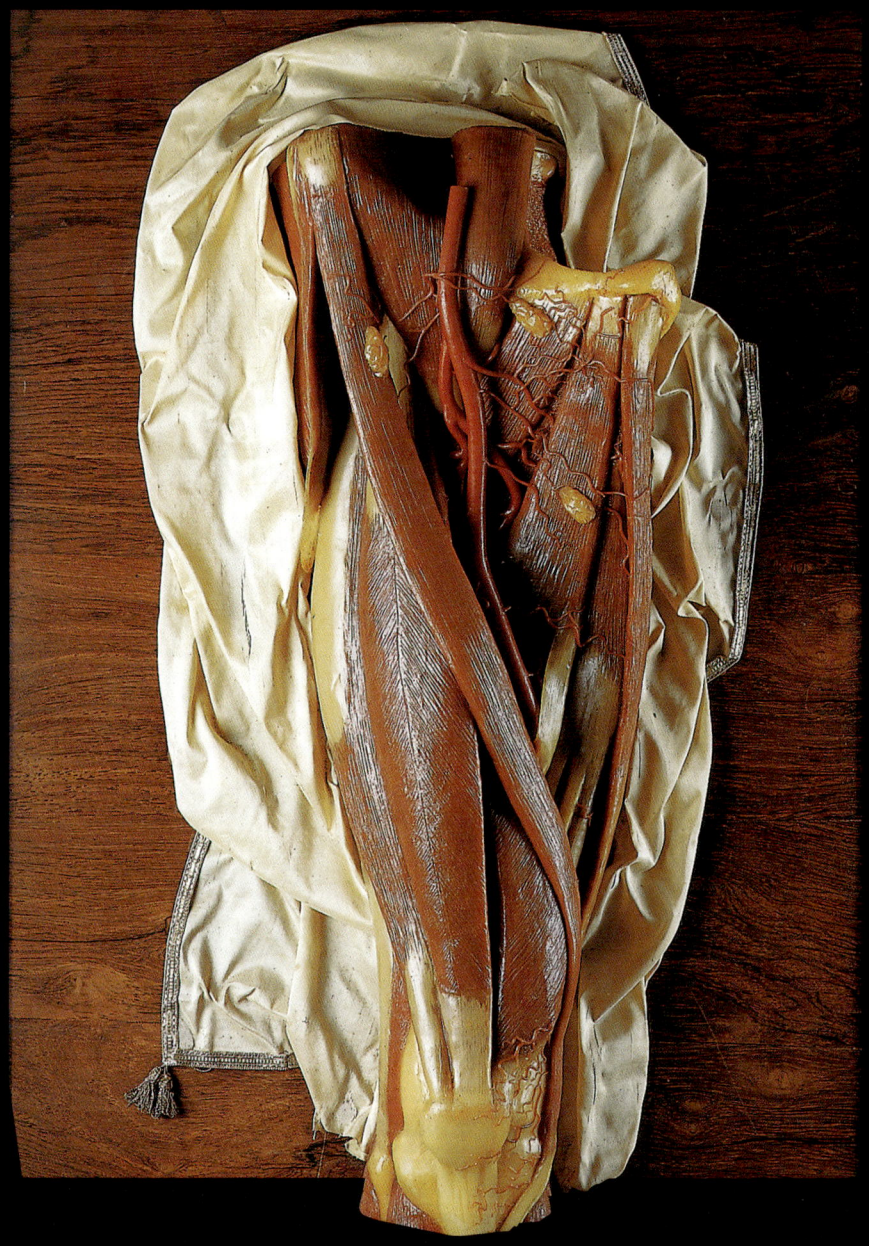

Arteria femoralis

The femoral artery

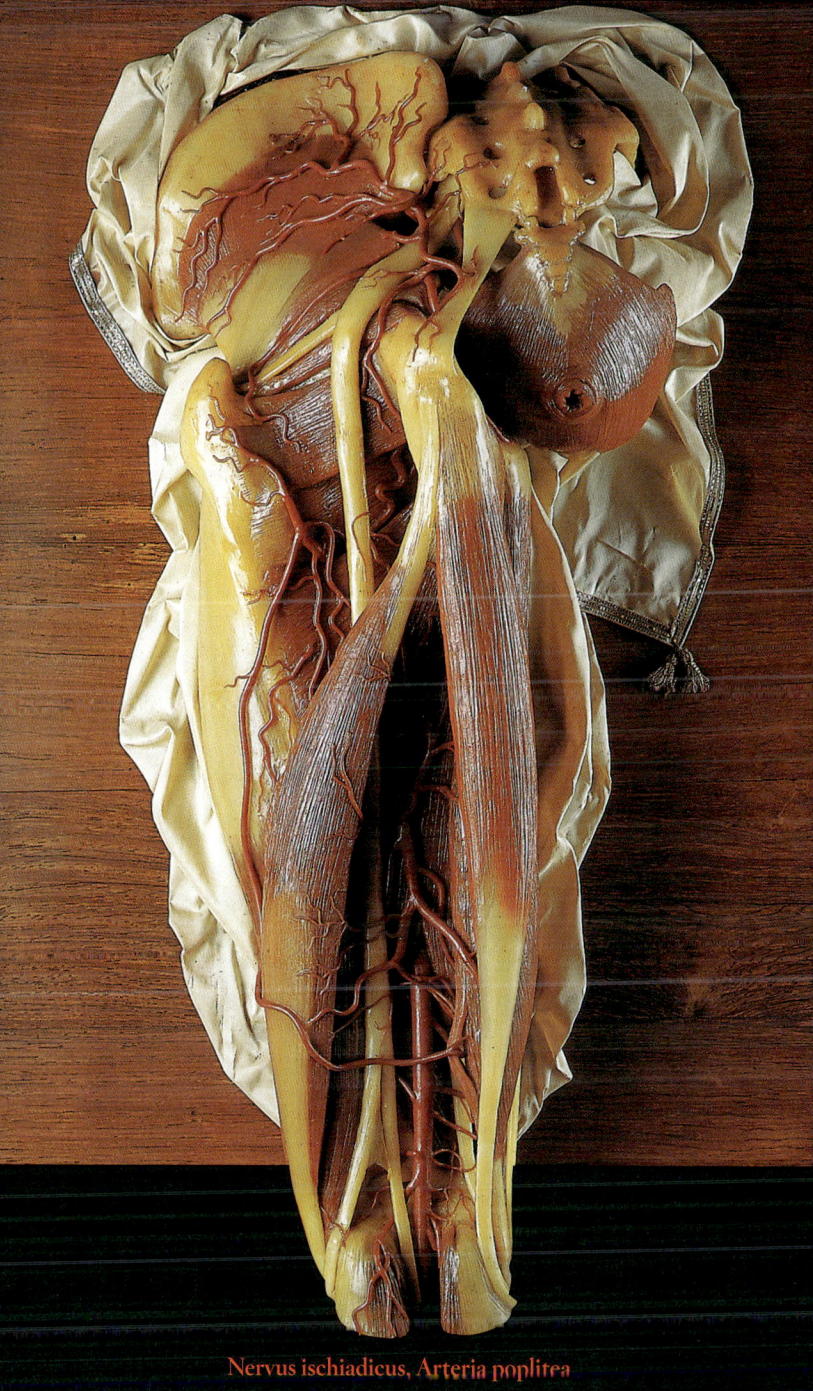

Nervus ischiadicus, Arteria poplitea

The thigh seen from behind, showing the sciatic nerve and popliteal artery

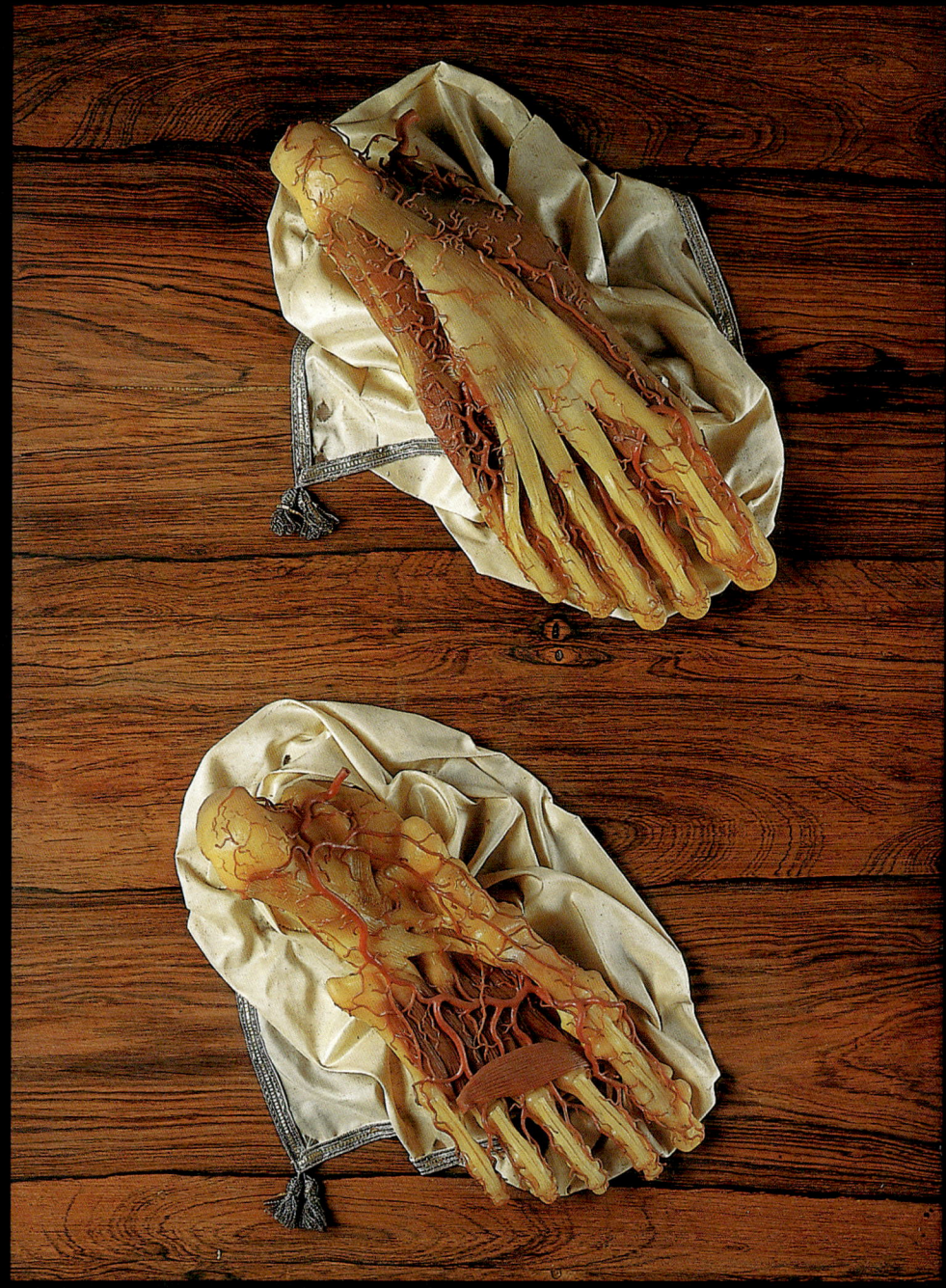

Arteria plantaris medialis et lateralis

The arteries of the sole of the foot

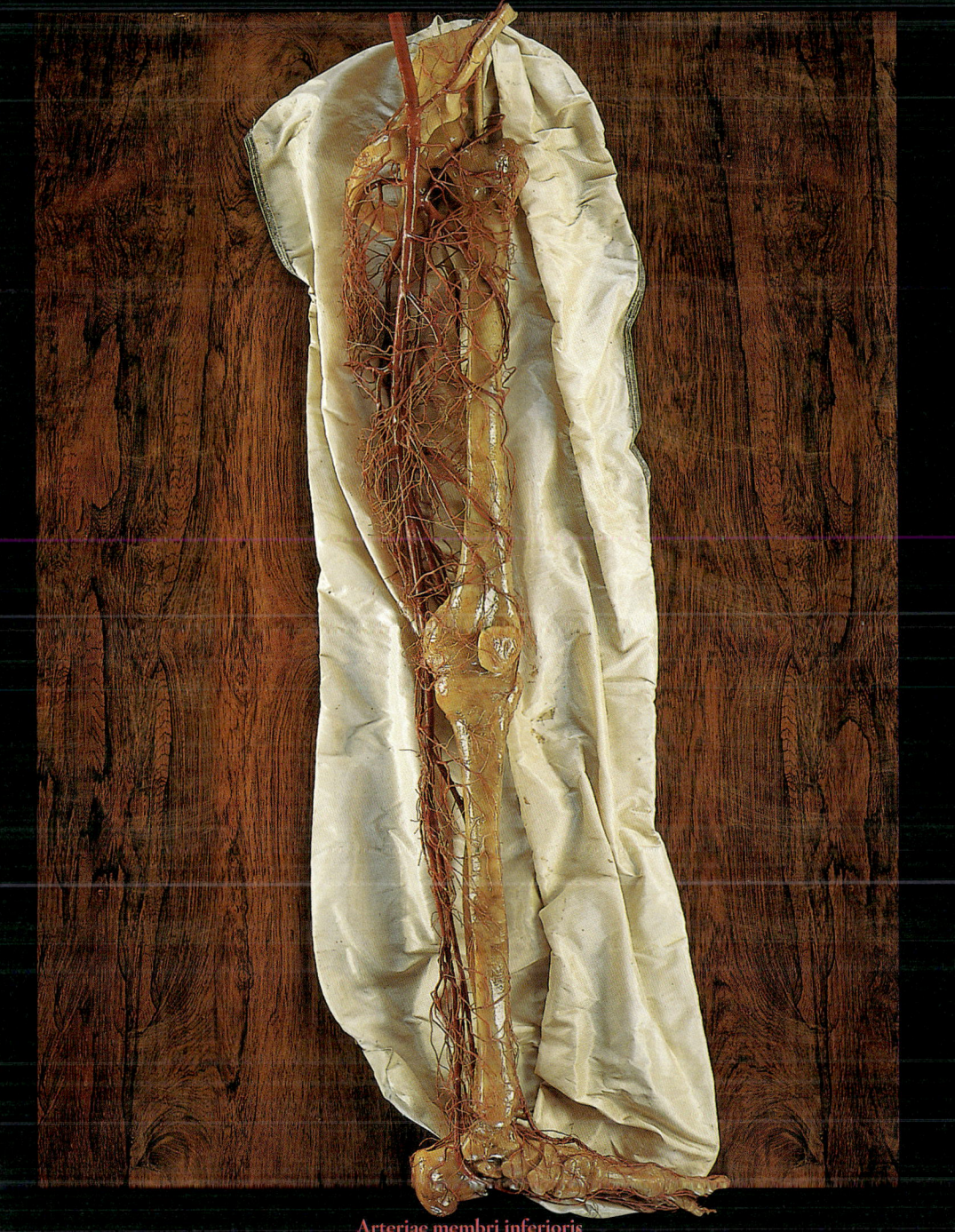

Arteriae membri inferioris
Specimen showing the arterial supply of the lower limb

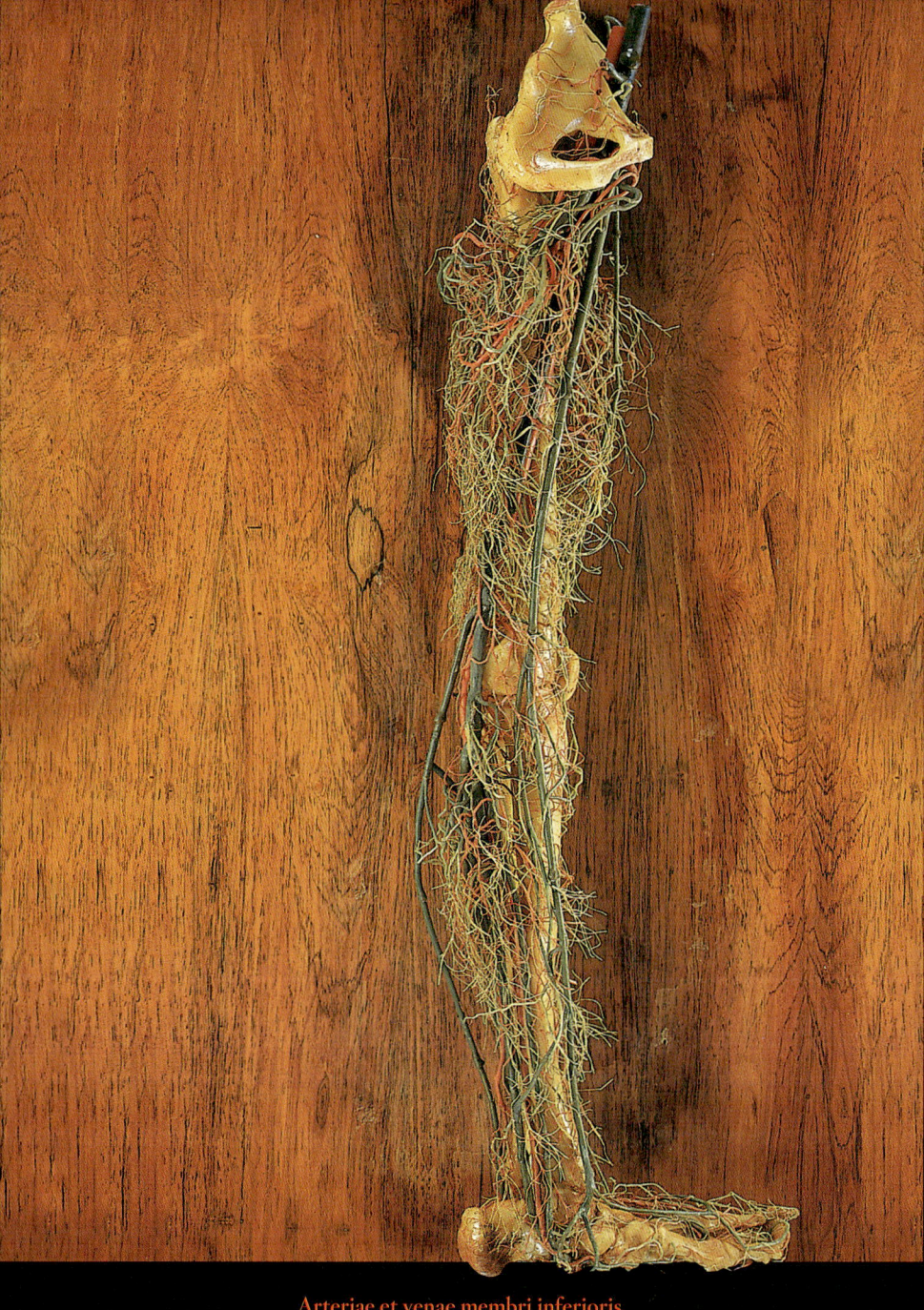

Arteriae et venae membri inferioris

The blood vessels of the lower limb

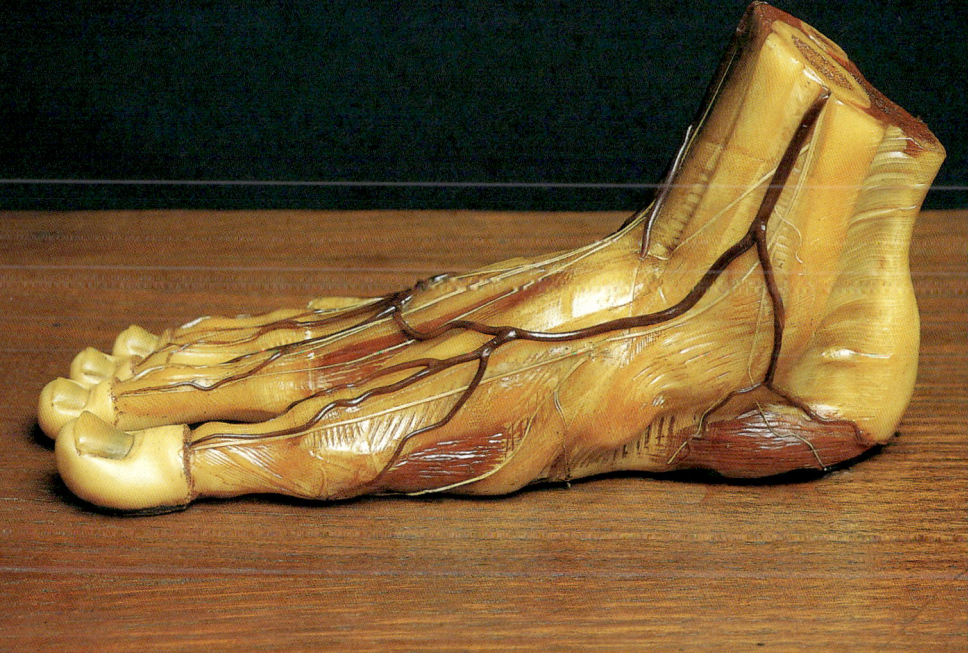

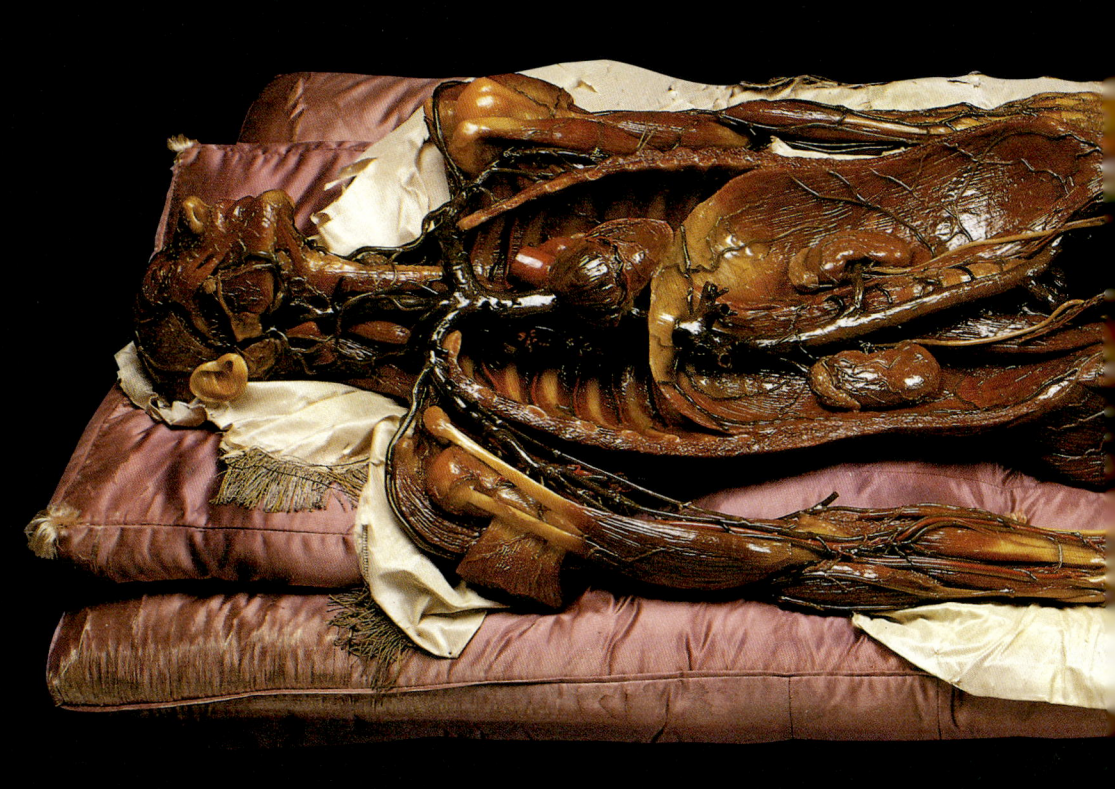

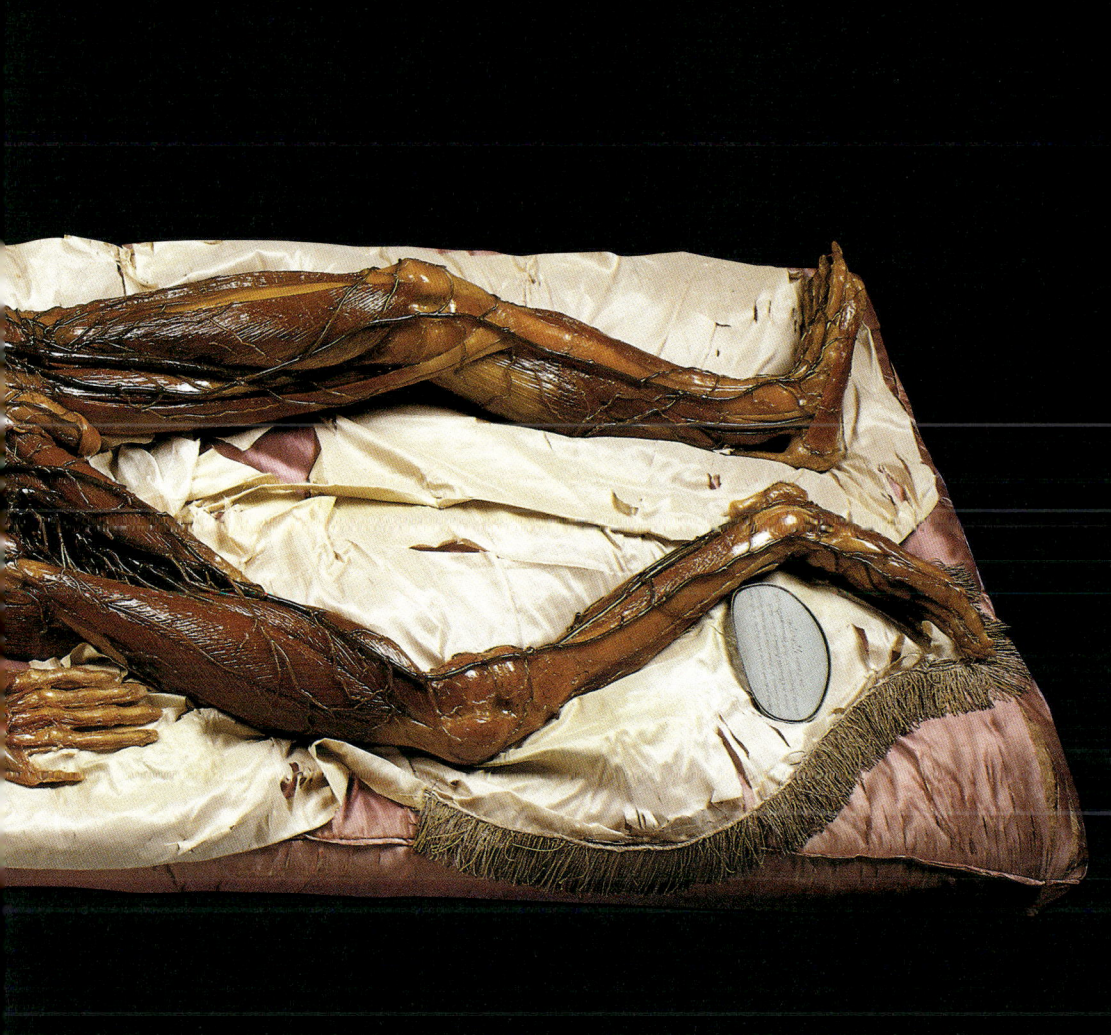

Venae

Whole body specimen with the veins displayed

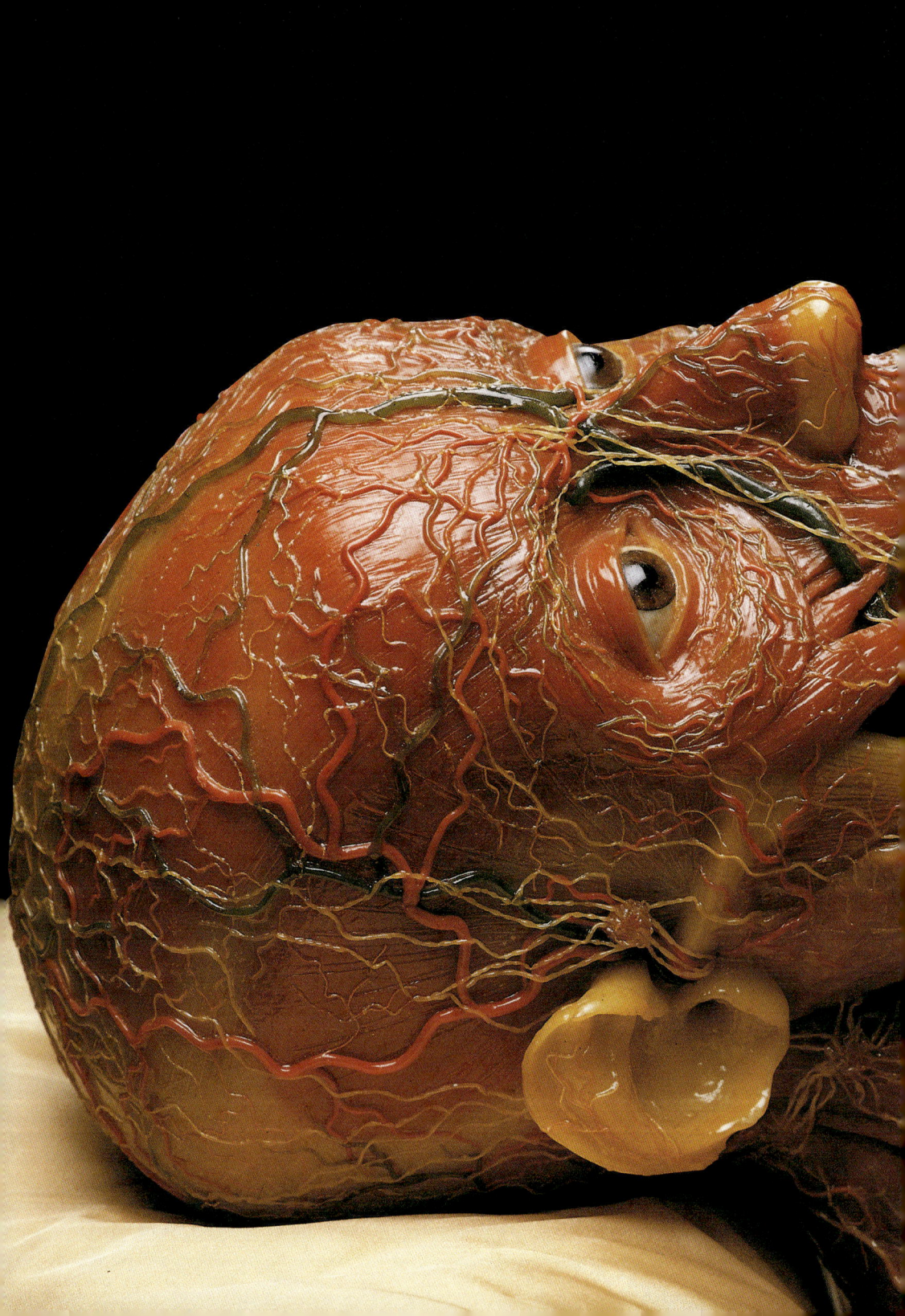

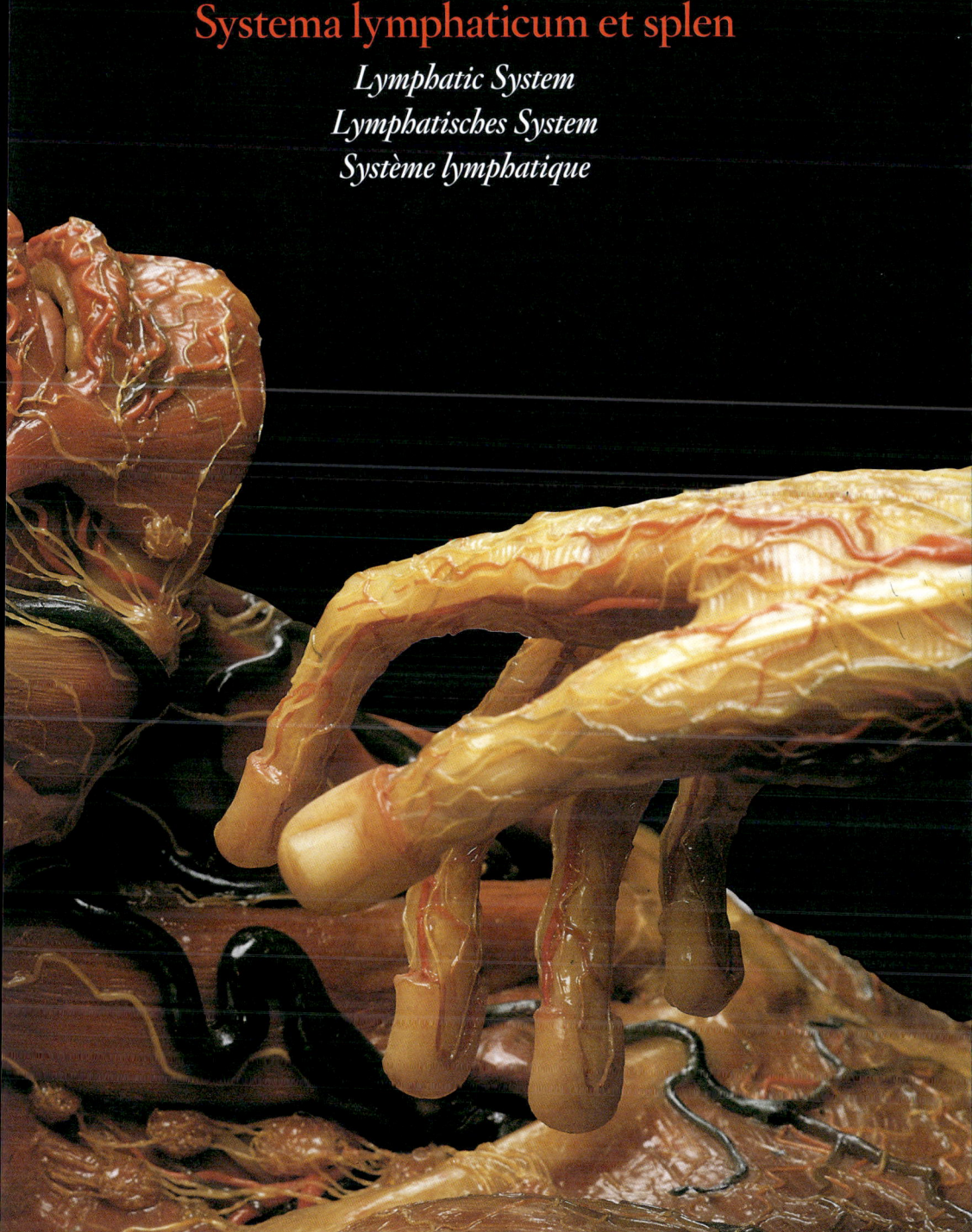

Systema lymphaticum et splen

Lymphatic System
Lymphatisches System
Système lymphatique

Lymphatic System

The lymphatic vessels, and especially the lymphatic organs such as the thymus, spleen and lymph nodes, play a central role in the immune system of our entire organism. In the 17th century Gaspare Aselli (1581–1626) had already observed the lymphatic vessels in the mesentery of a dog dissected immediately after feeding. Following the ingestion of food fats in the form of very fine droplets in the delicate lymph vessels of the intestine, the latter turned white and thus became visible. At the end of the 18th century the Italian anatomist Paolo Mascagni (1755–1815) used a sophisticated mercury injection technique to demonstrate that all organs possess their own widely branching lymph vessels with their associated regional lymph nodes. He described and named most of the vessels and nodes which are familiar to us today.

The models shown here summarise most impressively what was then the most up-to-date anatomical knowledge of that time. Without any doubt, these specimens must count as outstandingly graphic achievements.

Lymphatisches System

Das Lymphgefäßsystem und insbesondere die lymphatischen Organe, wie beispielsweise Thymus, Milz und Lymphknoten, nehmen eine zentrale Stellung in dem komplexen Abwehrsystem unseres Gesamtorganismus ein. Bereits Gaspare Aselli (1581–1626) entdeckte im 17. Jahrhundert die Lymphgefäße im Darmgekröse eines unmittelbar nach der Mahlzeit sezierten Hundes. Durch die Aufnahme der Nahrungsfette in Form feinster Tröpfchen in die sehr feinen Darmlymphgefäße waren diese weißlich gefärbt und dadurch sichtbar. Ende des 18. Jahrhunderts gelang es dem italienischen Anatom Paolo Mascagni (1755–1815) mittels einer raffiniert eingesetzten Quecksilber-Injektionstechnik zu zeigen, dass alle Organe über ein eigenständiges und weit verzweigtes Lymphgefäßsystem mit regionalen Lymphknoten verfügen. Er beschrieb und benannte die meisten der uns heute bekannten Lymphgefäße und Lymphknoten.

Die hier gezeigten Modelle fassen diese damals hochaktuellen anatomischen Erkenntnisse zusammen. Zweifelsohne gehören diese Präparate zu den herausragenden darstellerischen Leistungen.

Système lymphatique

Le système lymphatique, et plus particulièrement les organes lymphatiques tels que la rate, les nœuds ou ganglions lymphatiques, et chez l'enfant le thymus, occupe une place primordiale dans la défense de l'organisme. Au XVIIᵉ siècle, Gaspare Aselli (1581–1626) avait découvert les vaisseaux lymphatiques dans le mésentère d'un chien disséqué pendant la digestion, le transport de fines gouttelettes de graisse alimentaire par les vaisseaux lymphatiques de l'intestin les rendant lactescents et donc visibles. À la fin du XVIIIᵉ siècle, l'anatomiste italien Paolo Mascagni (1755–1815) réussit à démontrer, grâce à une technique d'injection élaborée de mercure, que tous les organes sont tributaires d'une même circulation lymphatique, largement ramifiée, disposant de nœuds ou ganglions lymphatiques dans chaque région ; il a décrit et dénommé la plupart des vaisseaux et nœuds lymphatiques connus de nos jours.

Les connaissances anatomiques de cette époque, toujours valables actuellement, sont résumées de manière impressionnante par cette série de modèles en cire ; ces préparations font sans aucun doute partie des œuvres les plus représentatives de la collection.

Pages 270–271:
Detail from pages 274–275

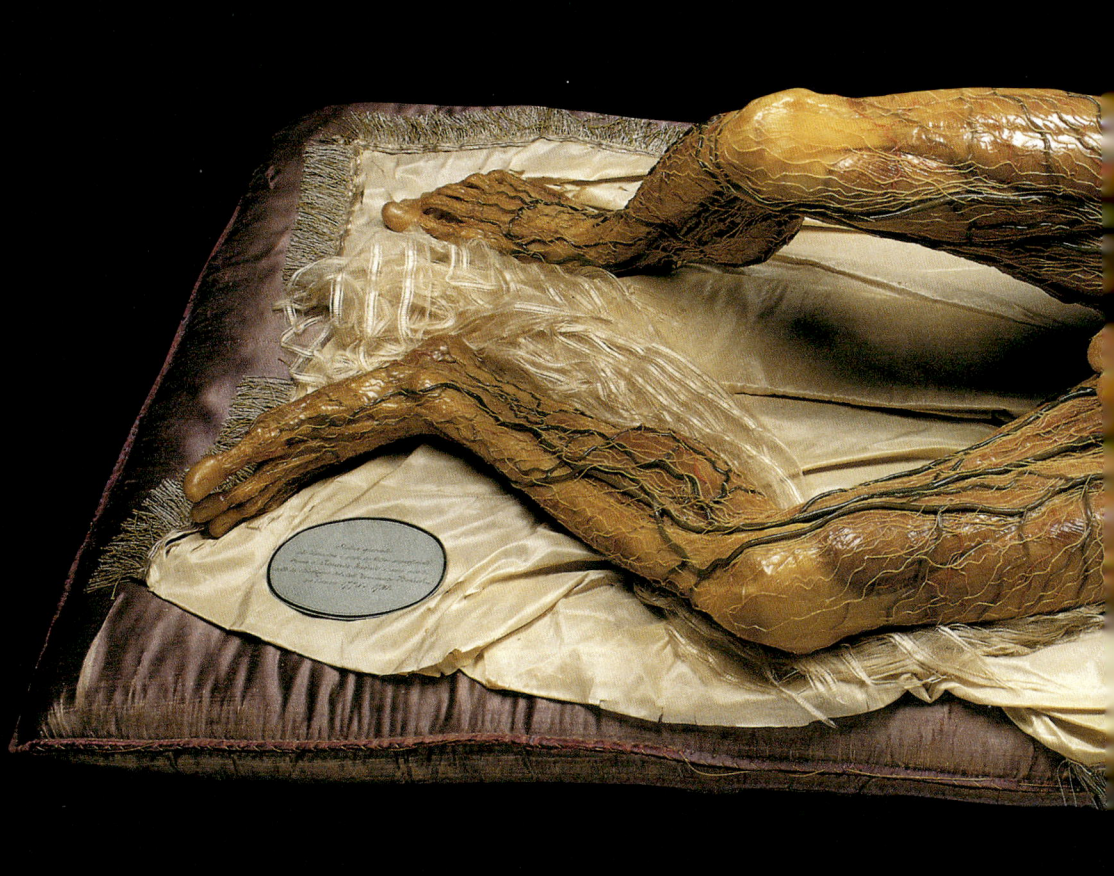

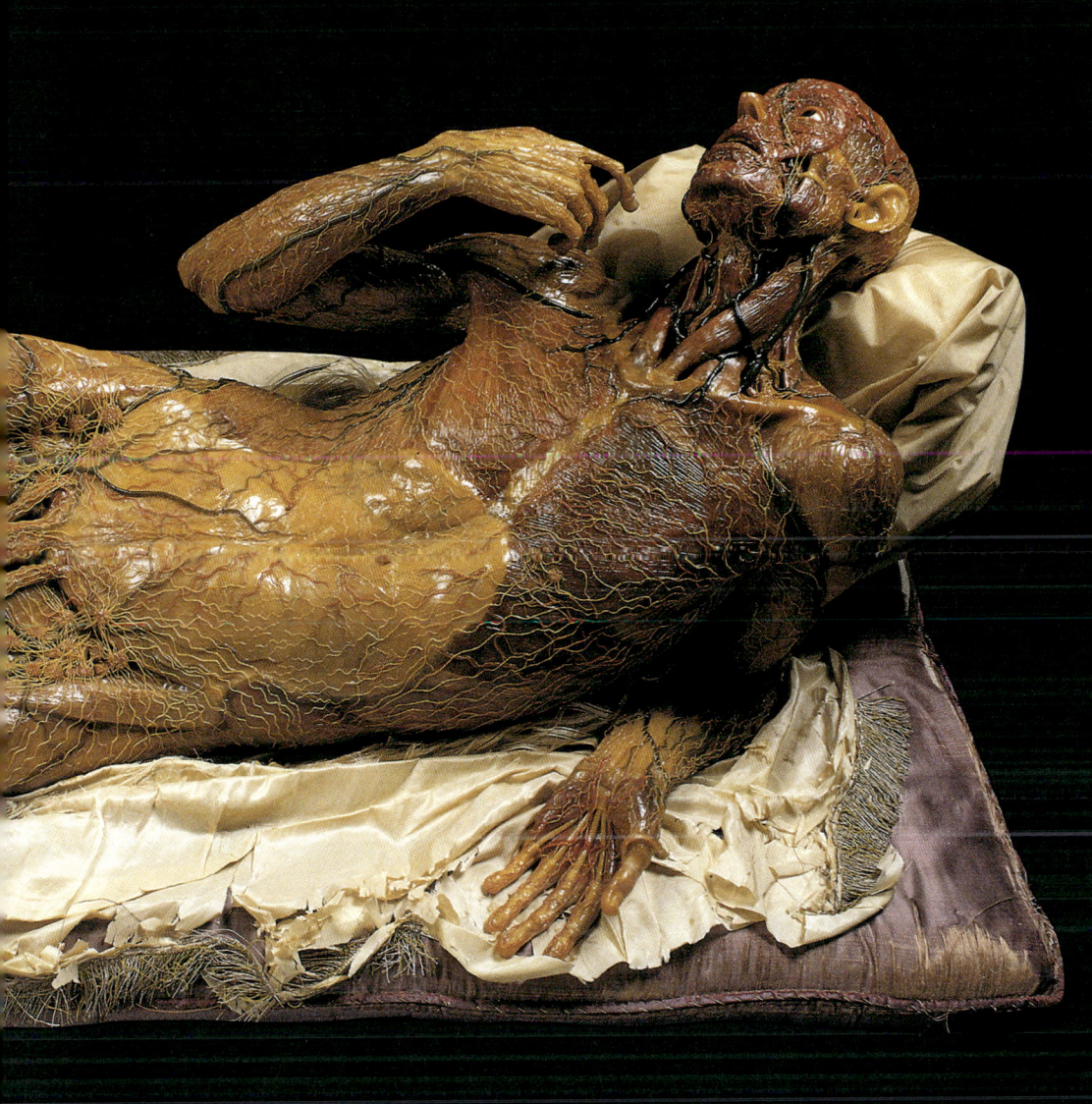

Whole body specimen with the superficial veins and lymphatic vessels displayed

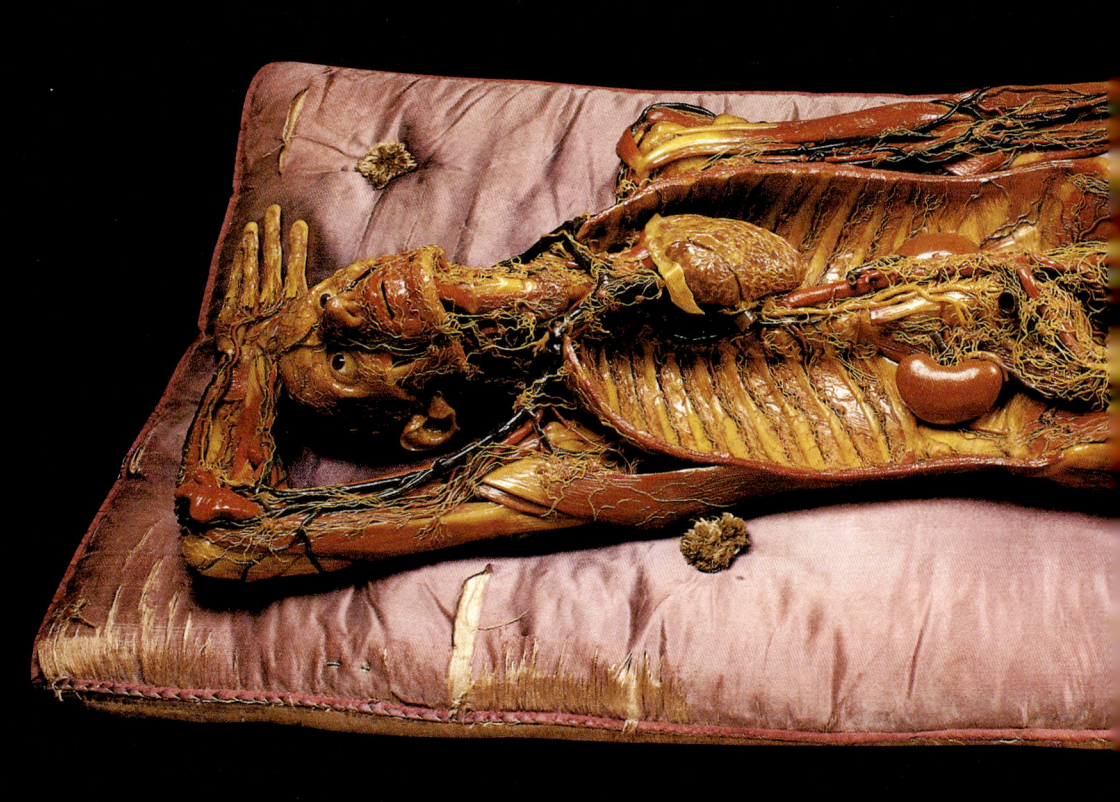

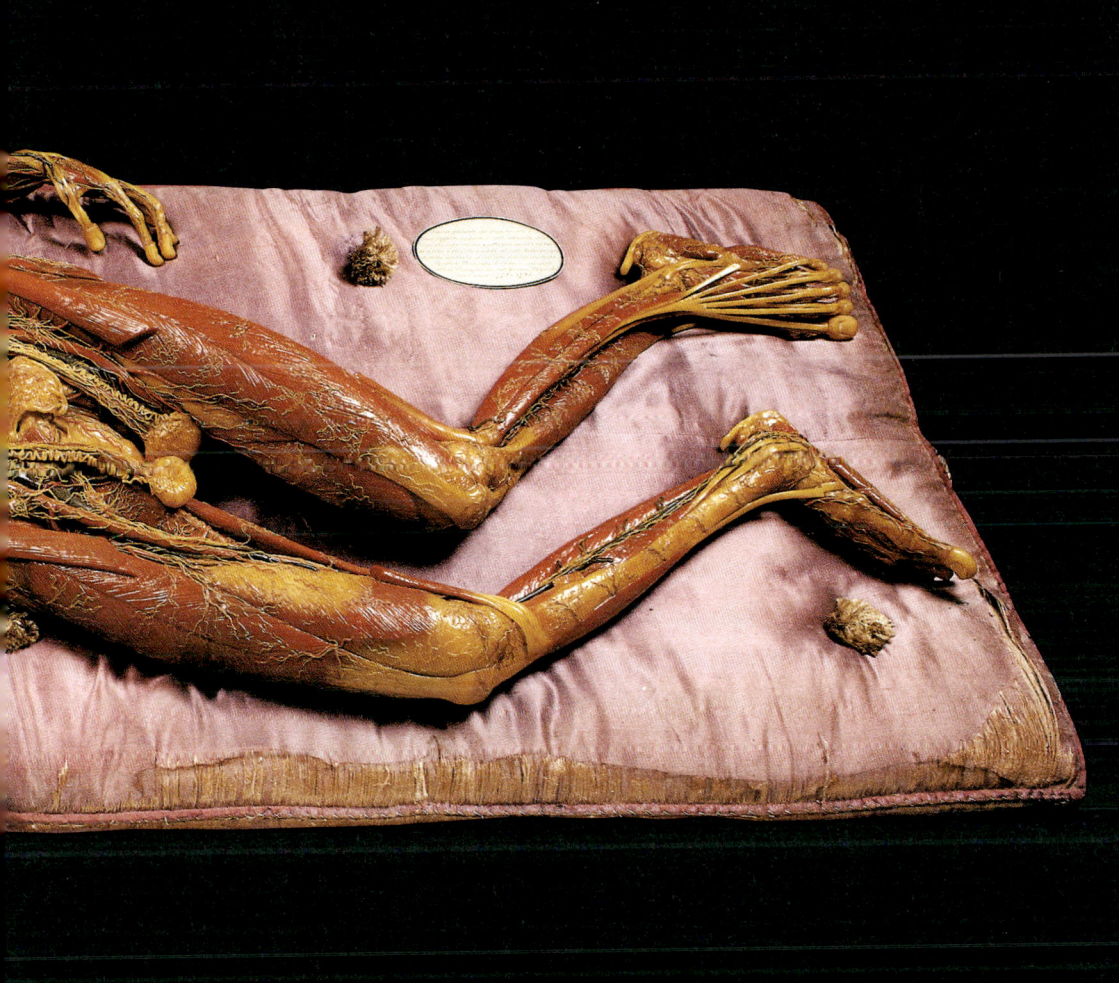

Whole body specimen showing the lymphatic vessels

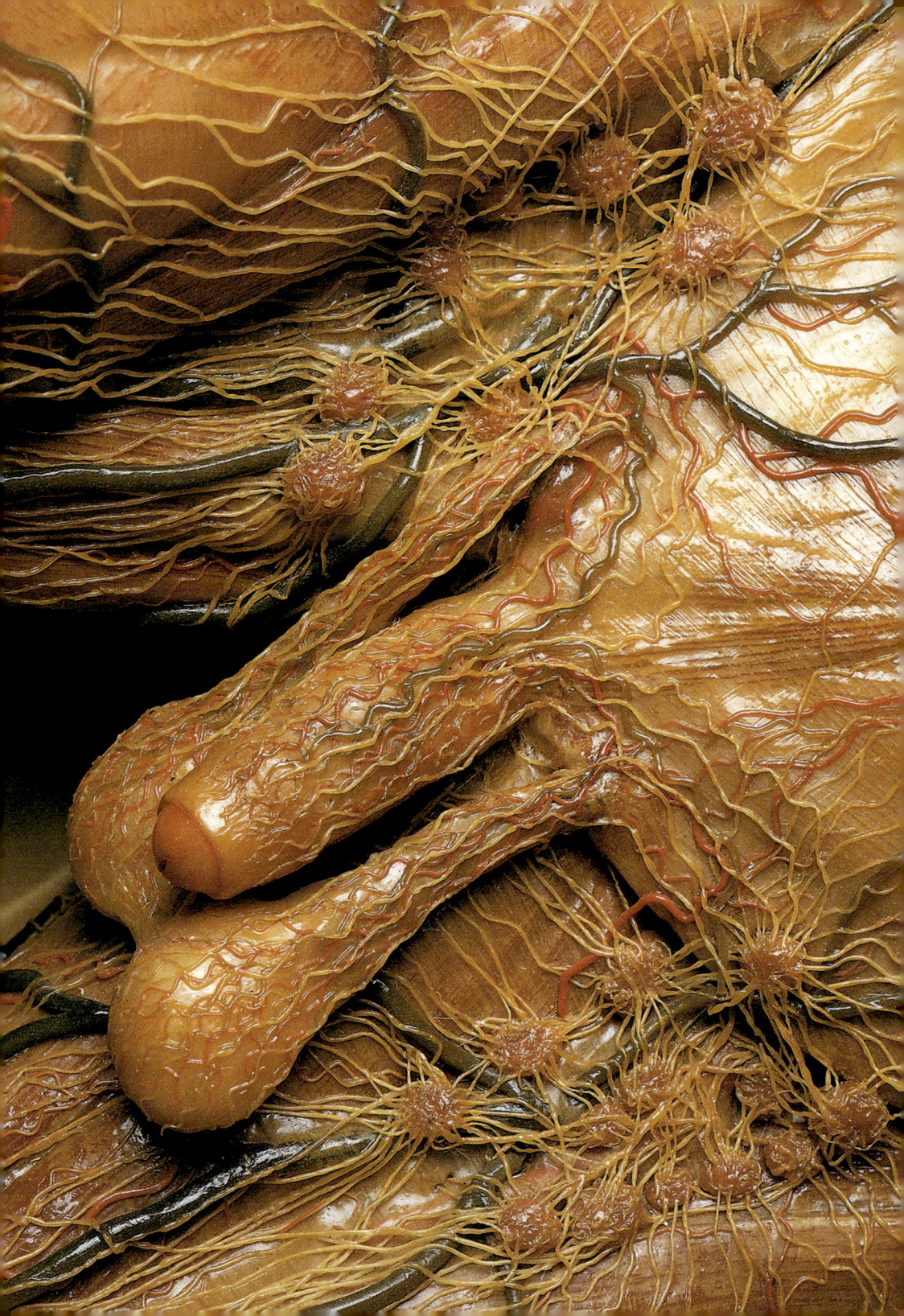

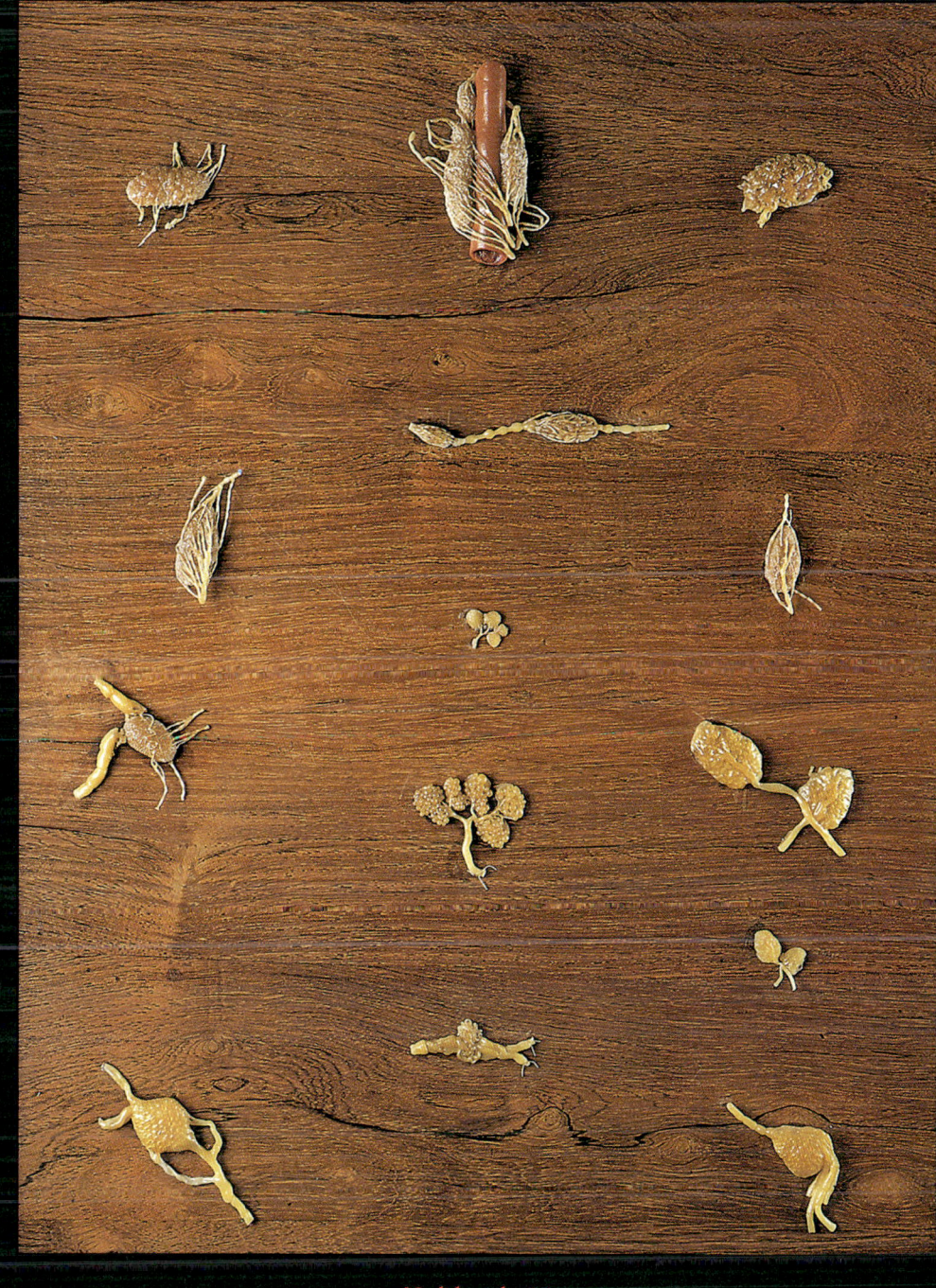

← Detail from pages 274–275

Nodi lymphatici

Different views of lymph nodes with efferent and afferent lymphatic vessels

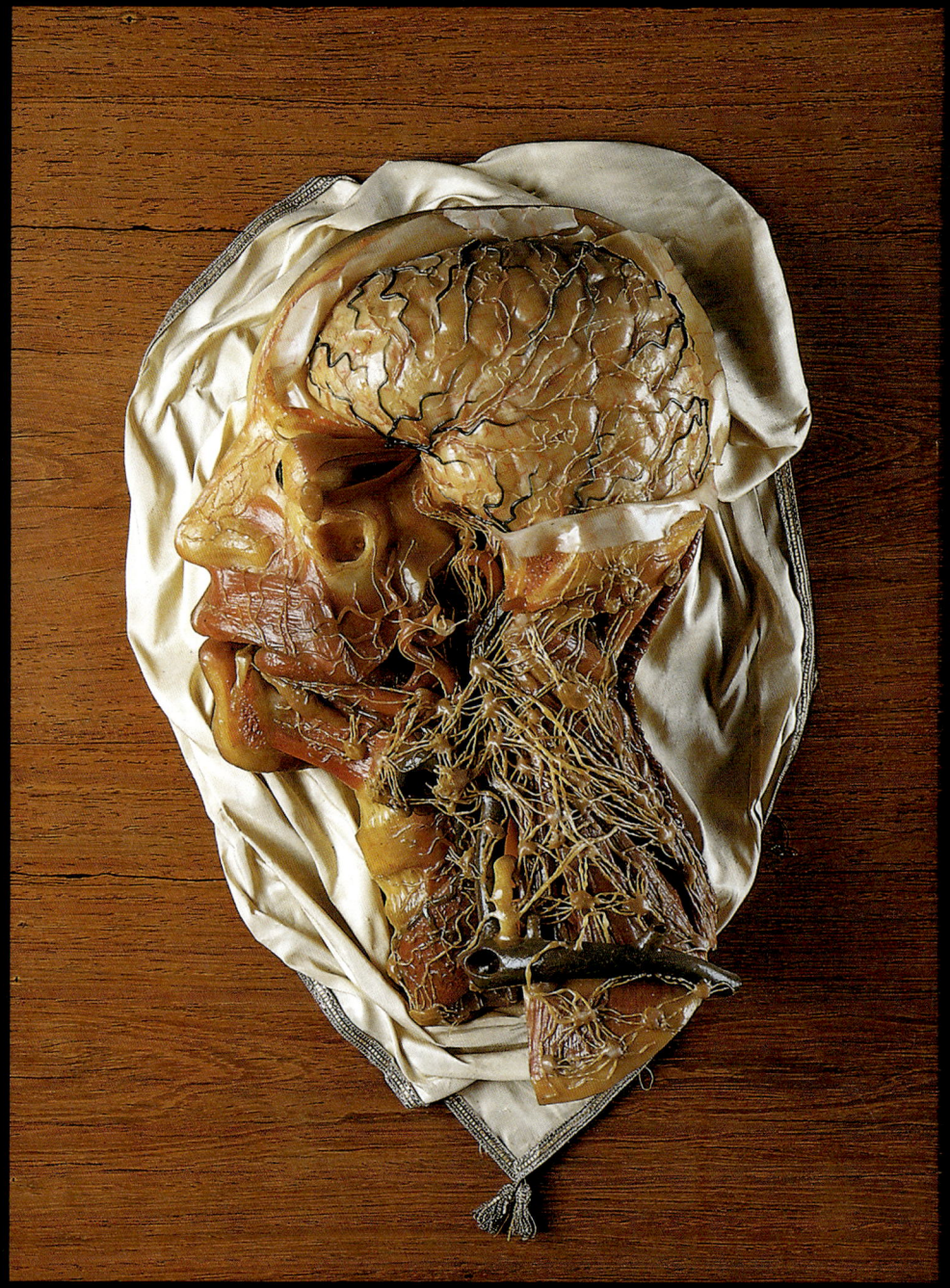

Vasa lymphatica, Nodi lymphatici submandibulares, Nodi lymphatici submentales,
Nodi lymphatici retroauriculares, Nodi lymphatici cervicales superficiales

Lymphatic vessels and nodes of the head and neck. The skull has been laid open to reveal the pia-arachnoid
and the brain. The lymphatic vessels of the pia-arachnoid as displayed in the specimen do not exist

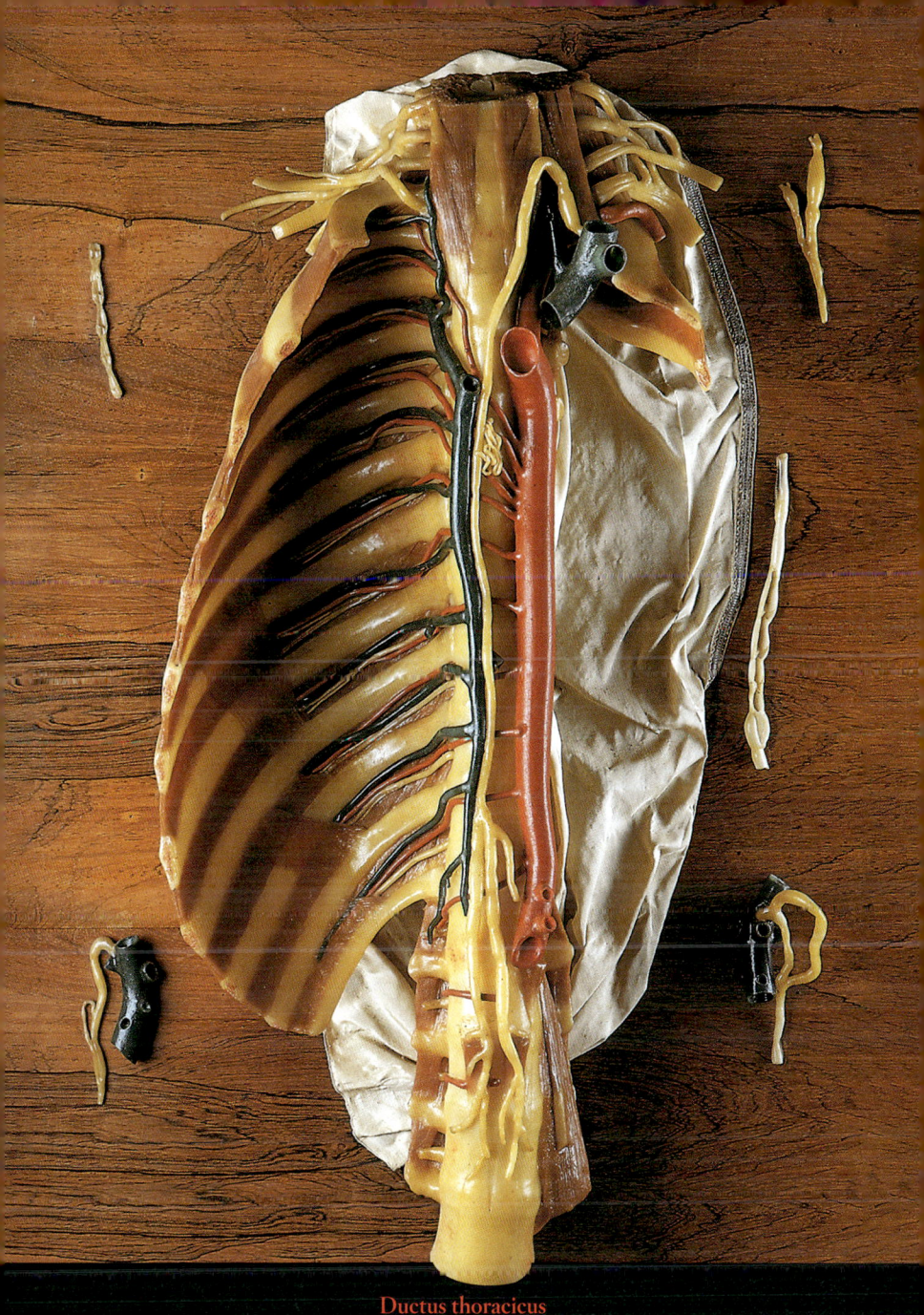

Ductus thoracicus

View into the thoracic cage to show the thoracic duct (the main collecting lymphatic vessel), its tributaries and its opening into the great vein on the left side of the neck

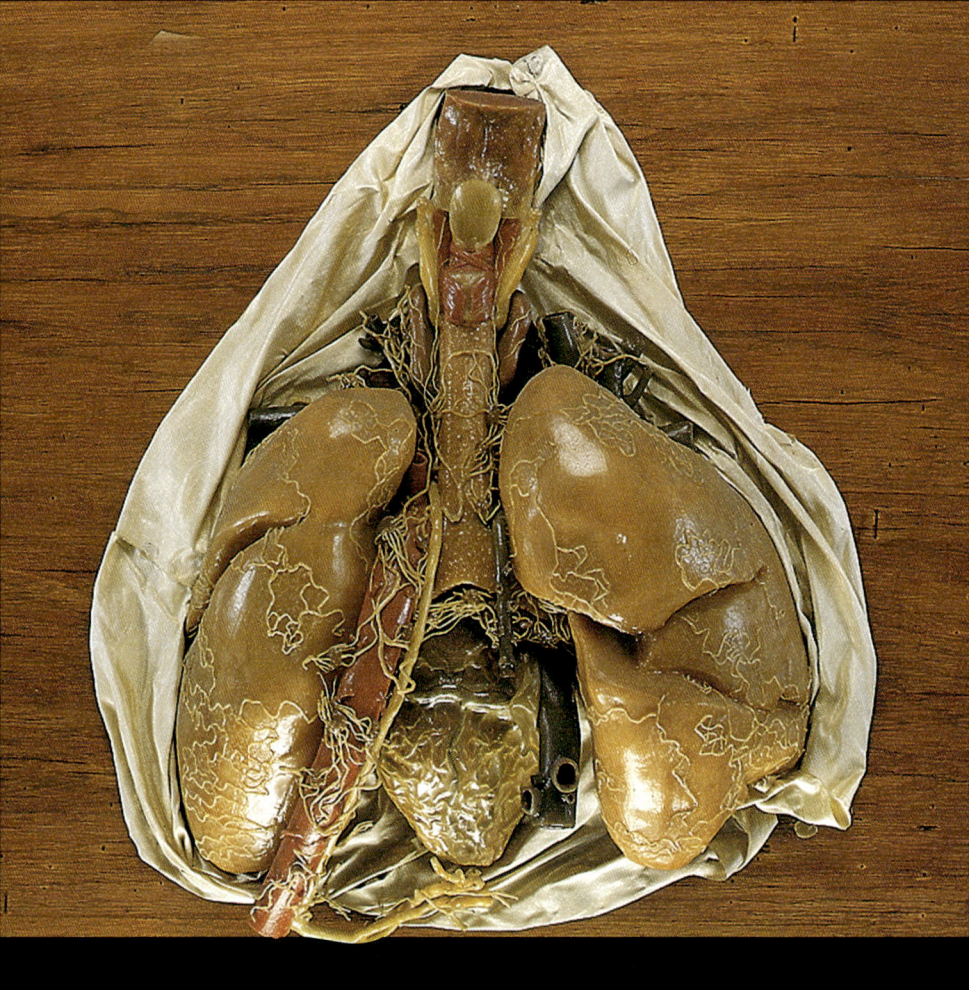

Ductus thoracicus, Vasa lymphatica pulmones superficiales

View from behind into the thoracic cavity to show the course of the lymphatic vessels and the thoracic duct

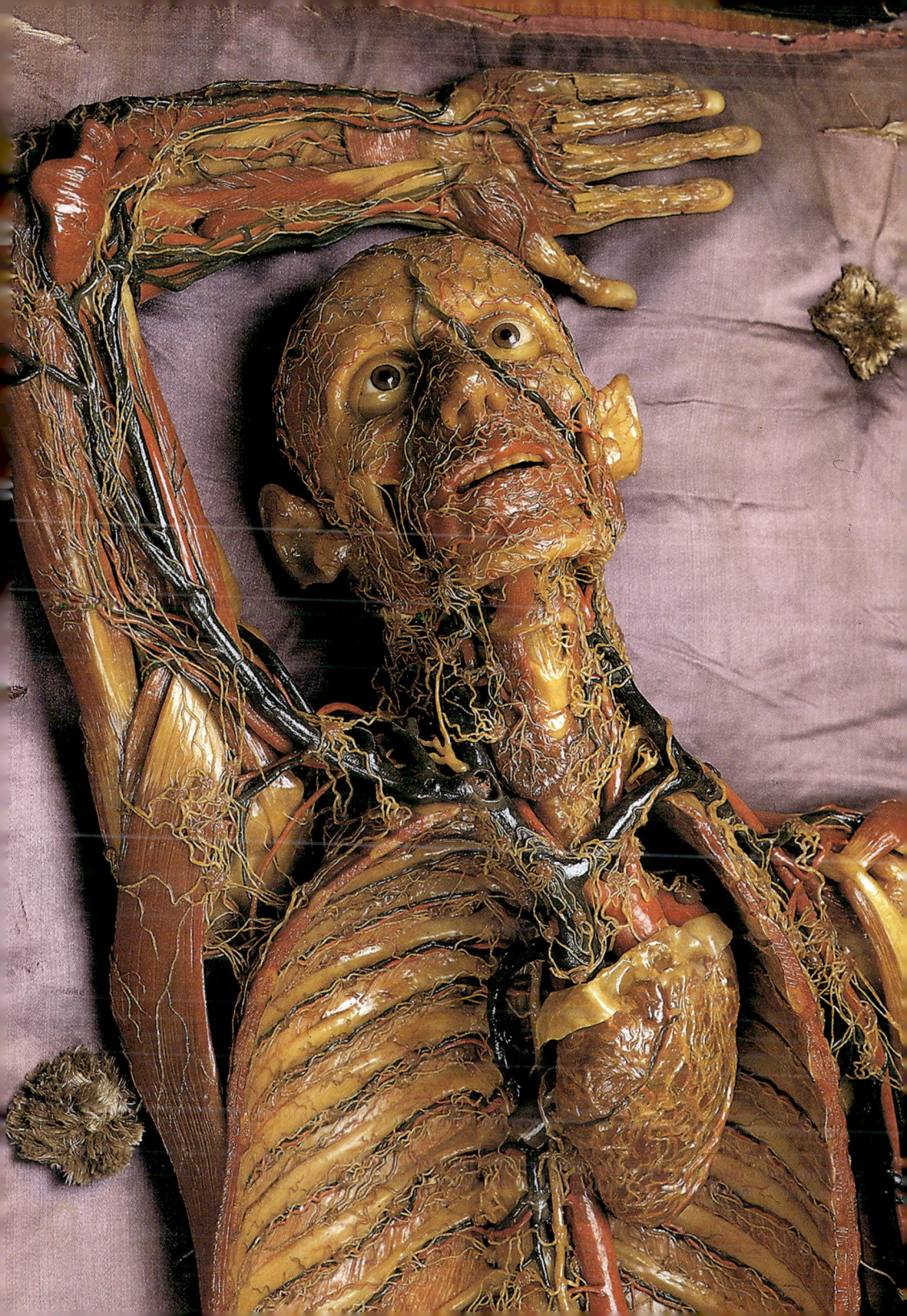

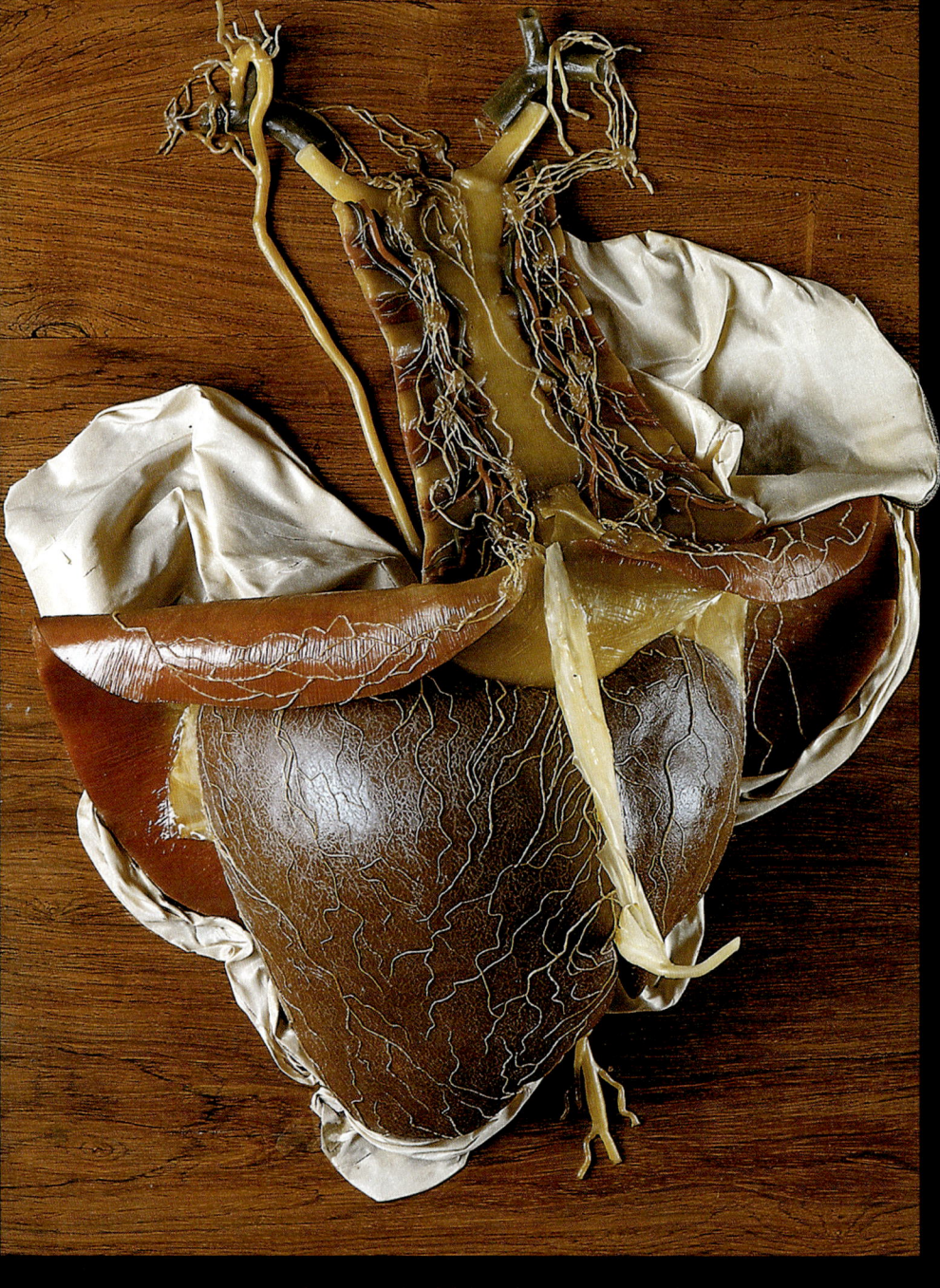

Ductus thoracicus, Noduli lymphatici retrosternalis

Lymphatic vessels of the liver and on the internal surface of the breastbone.
The thoracic duct can be discerned as a pale cord

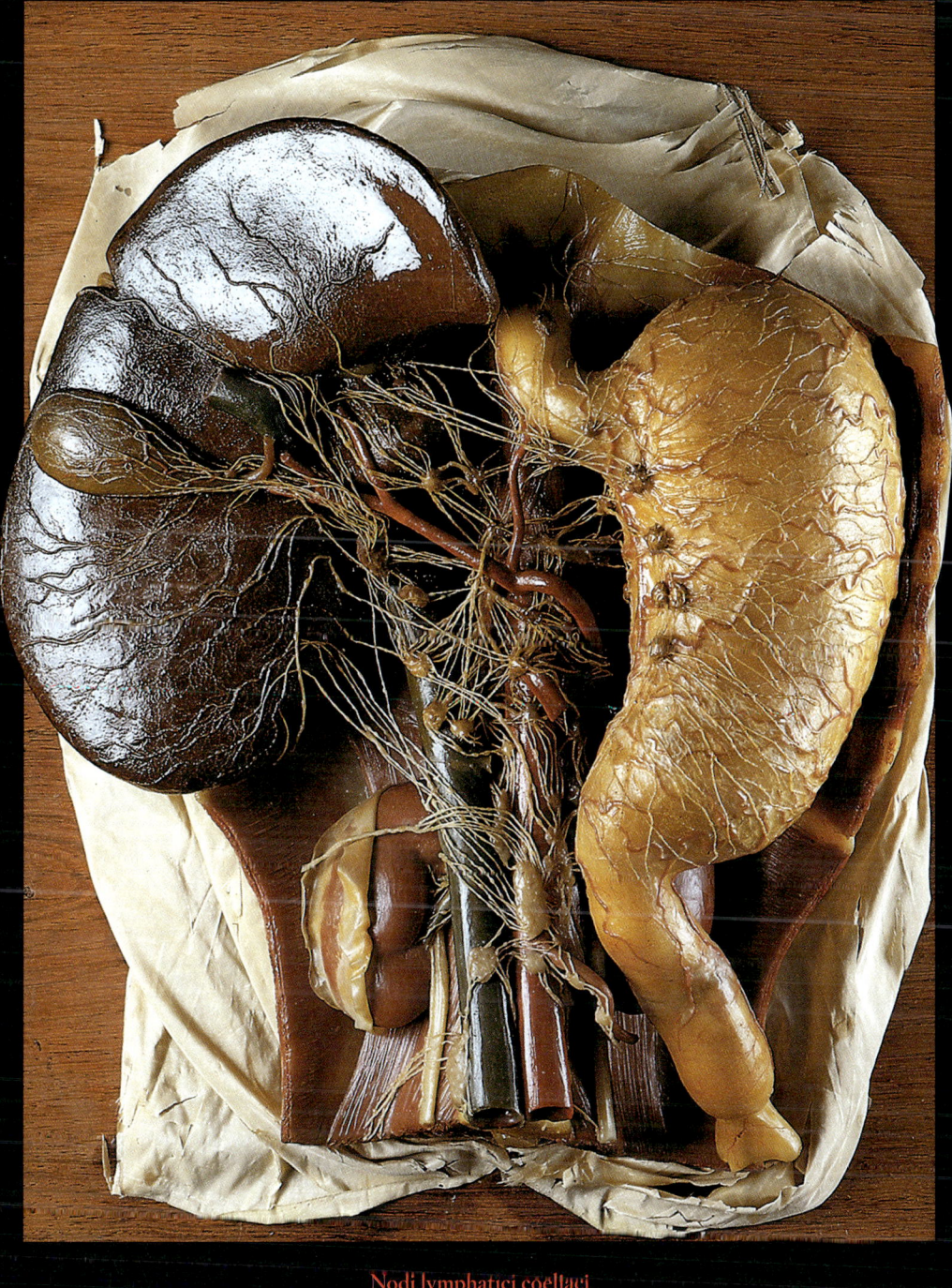

Nodi lymphatici coeliaci

Lymphatic vessels and nodes around the organs in the upper part of the abdominal cavity

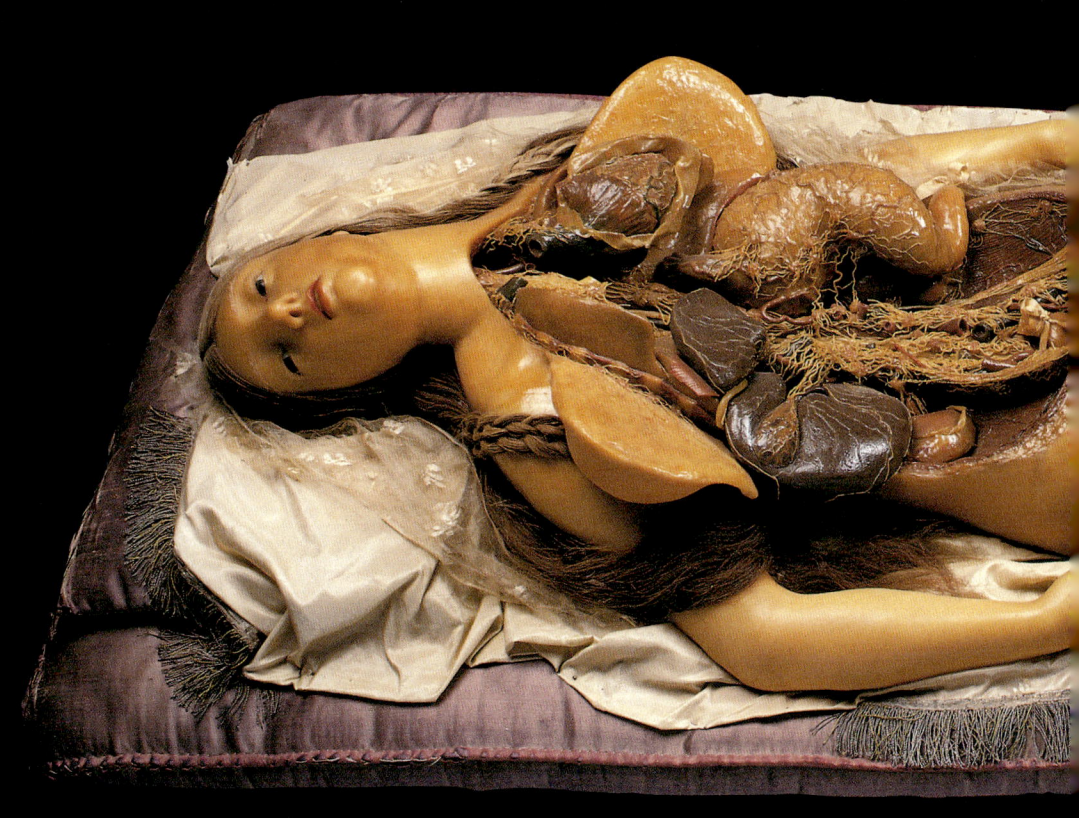

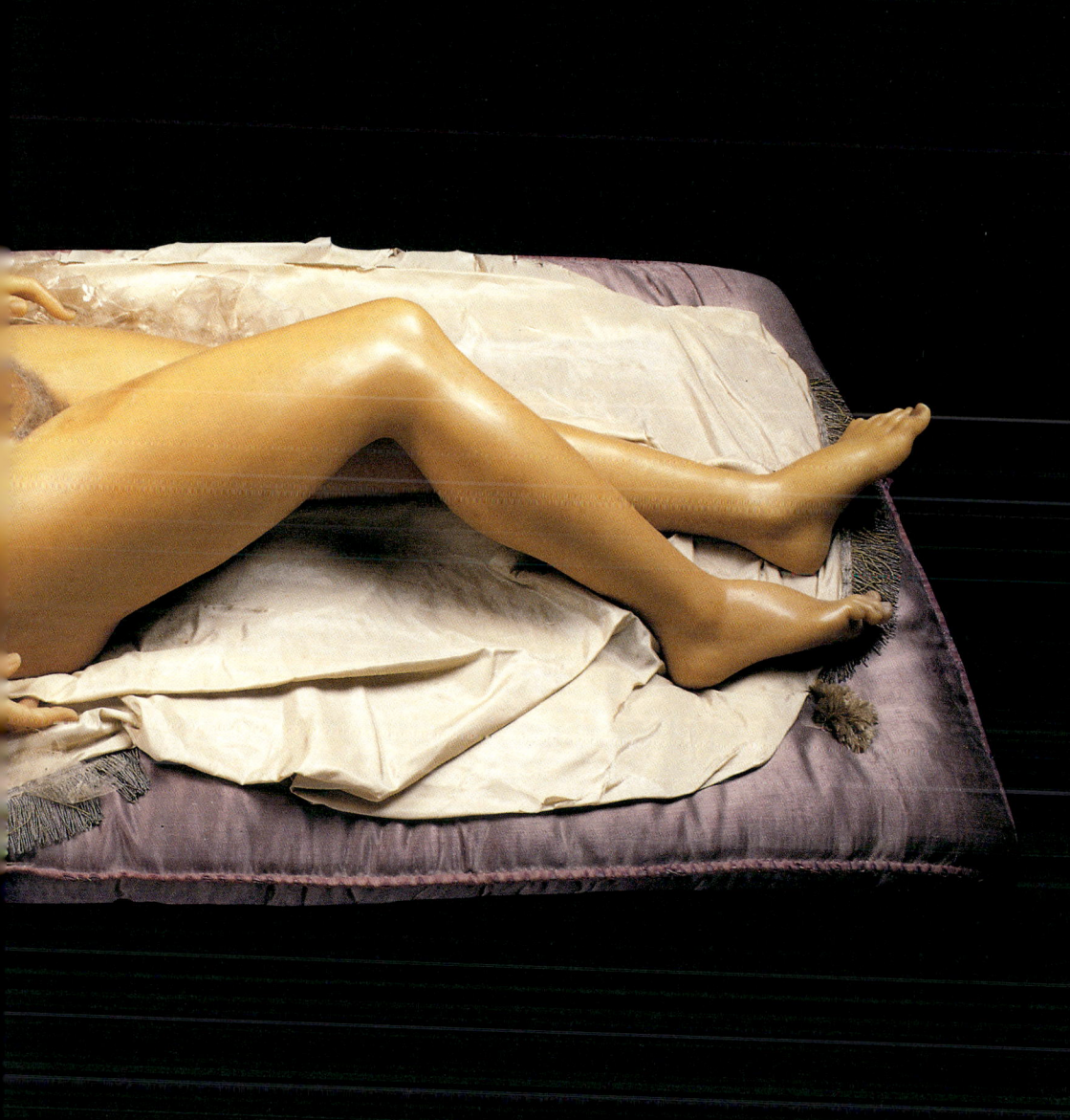

Whole body specimen showing the lymphatic vessels of the thoracic and abdominal cavities

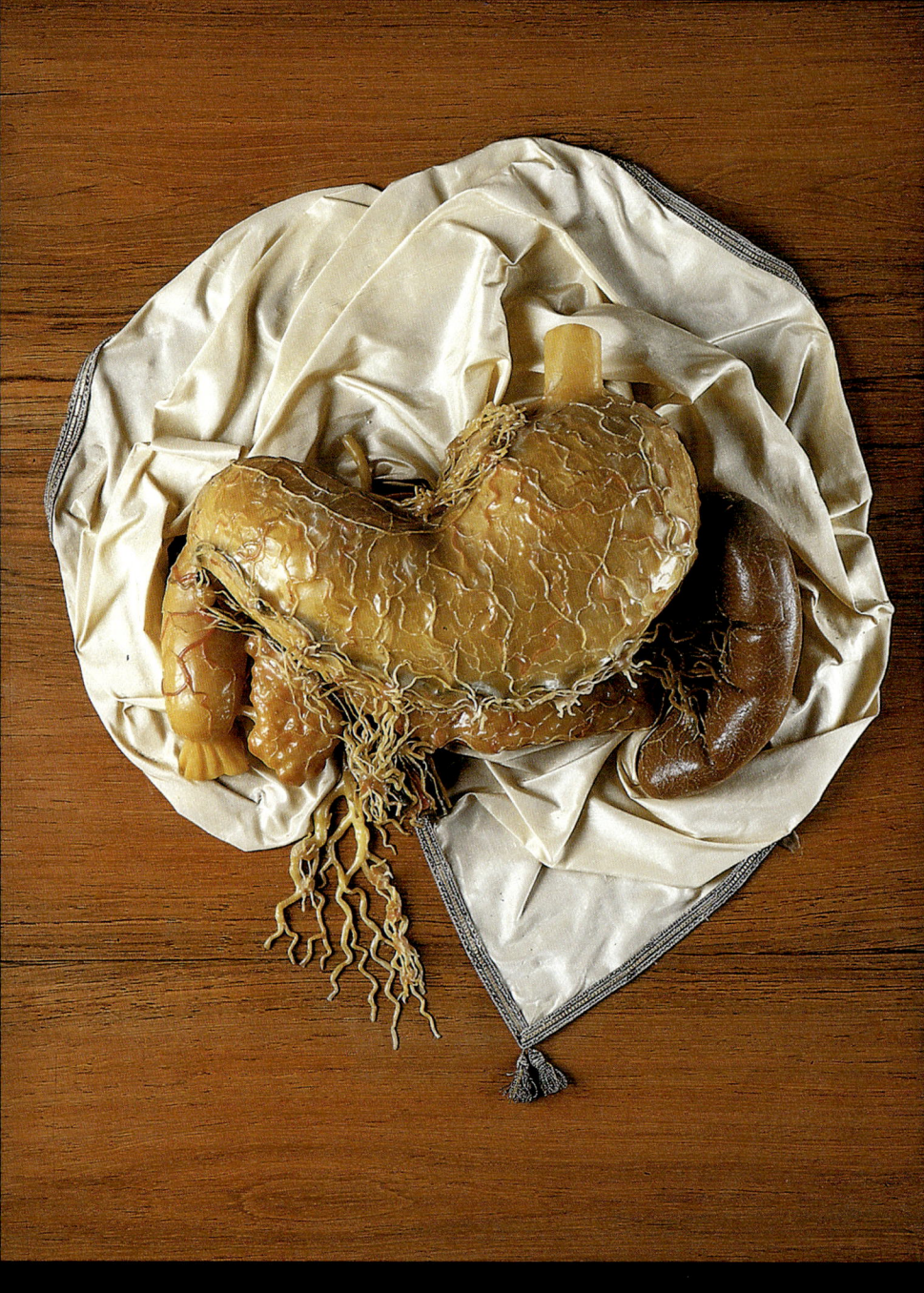

Nodi lymphatici gastroomentalis

Lymphatic vessels on the anterior surface of the stomach

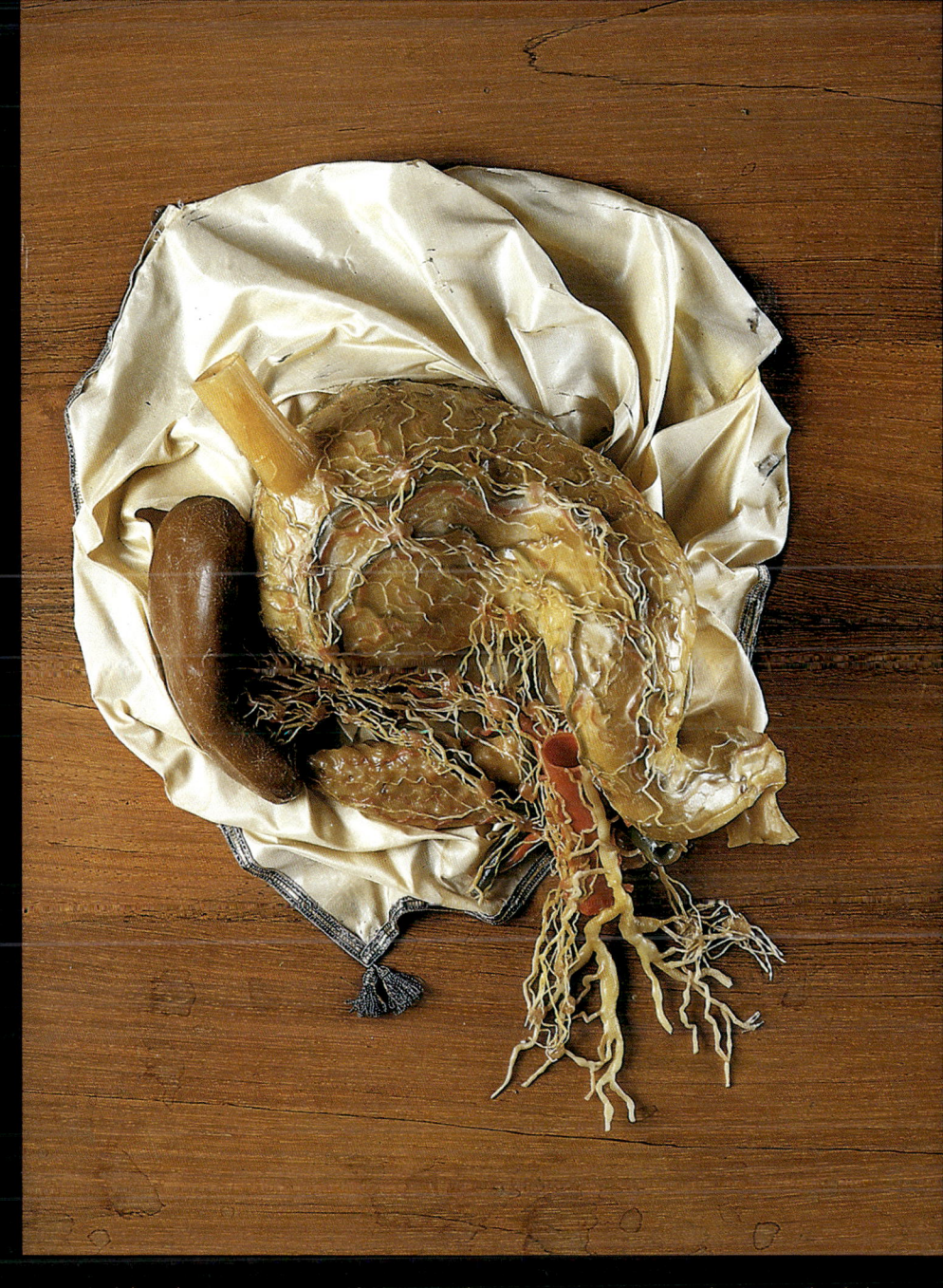

Nodi lymphatici gastroomentalis, Ductus thoracicus, Trunci intestinales, Trunci lumbales

Lymph nodes and lymphatic vessels on the posterior surface of the stomach,
and the thoracic duct receiving the main lymphatic tributaries

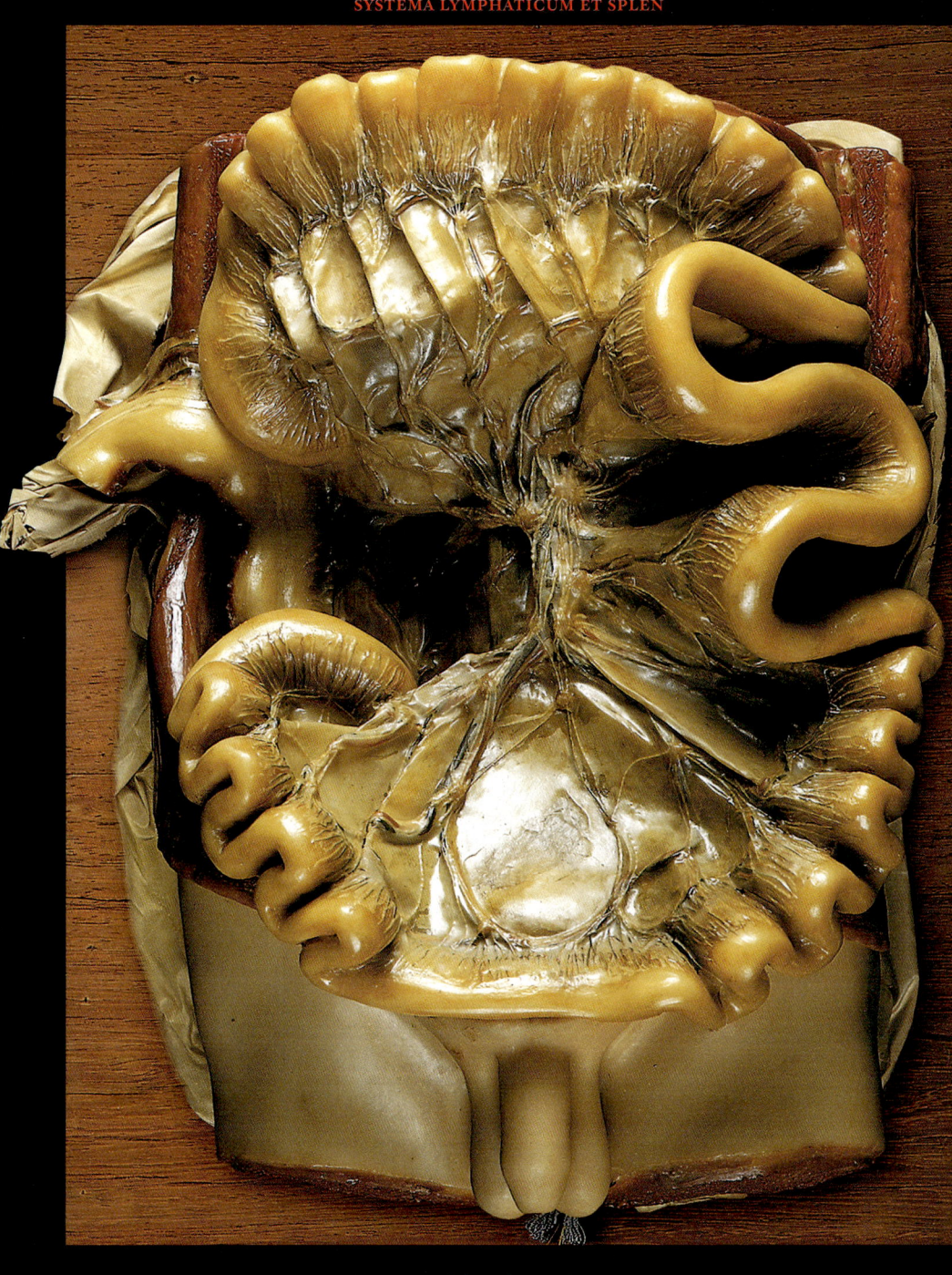

Vasa et nodi lymphatici ilei et jejuni

Lymph nodes and lymphatic vessels of the small intestine

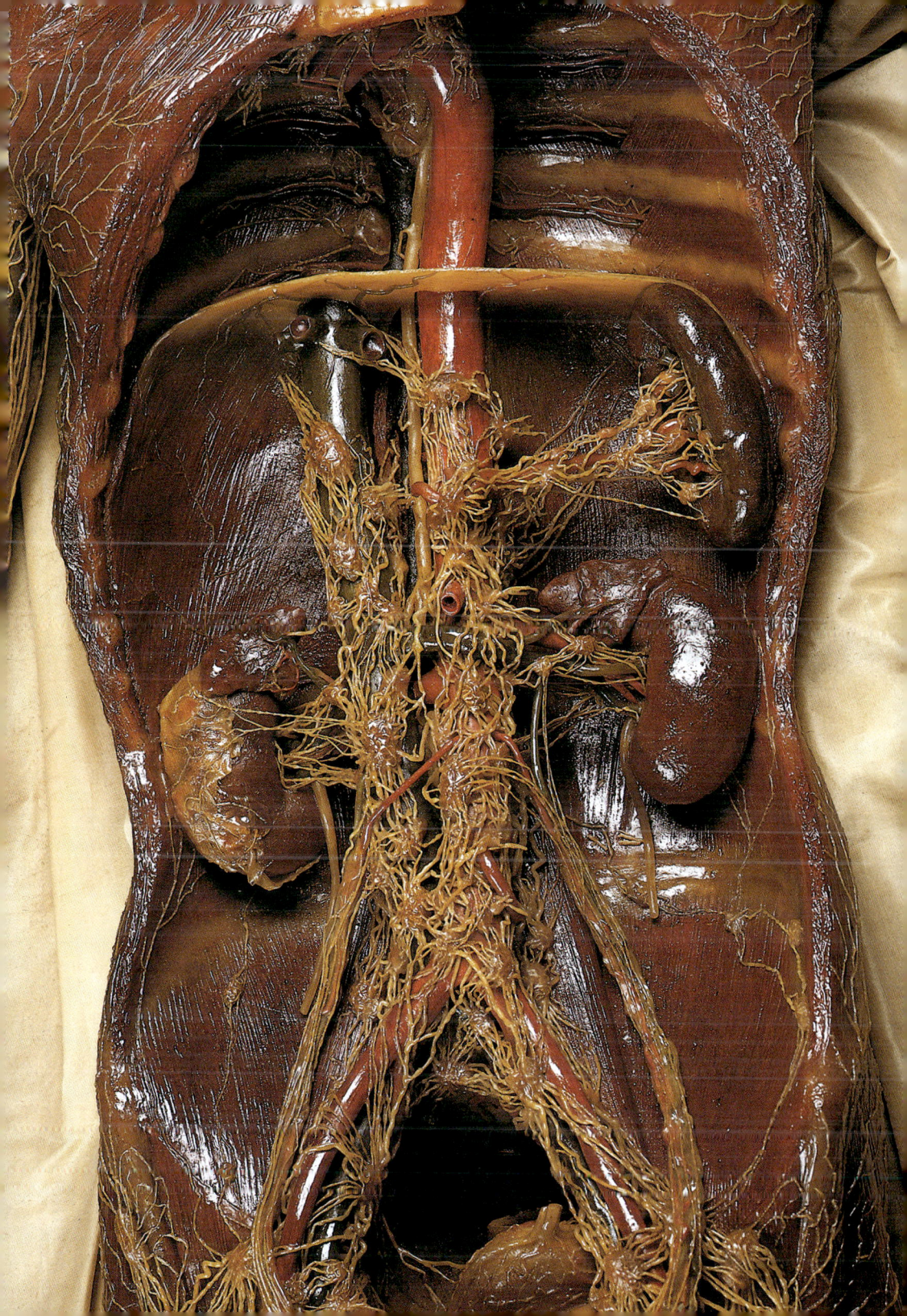

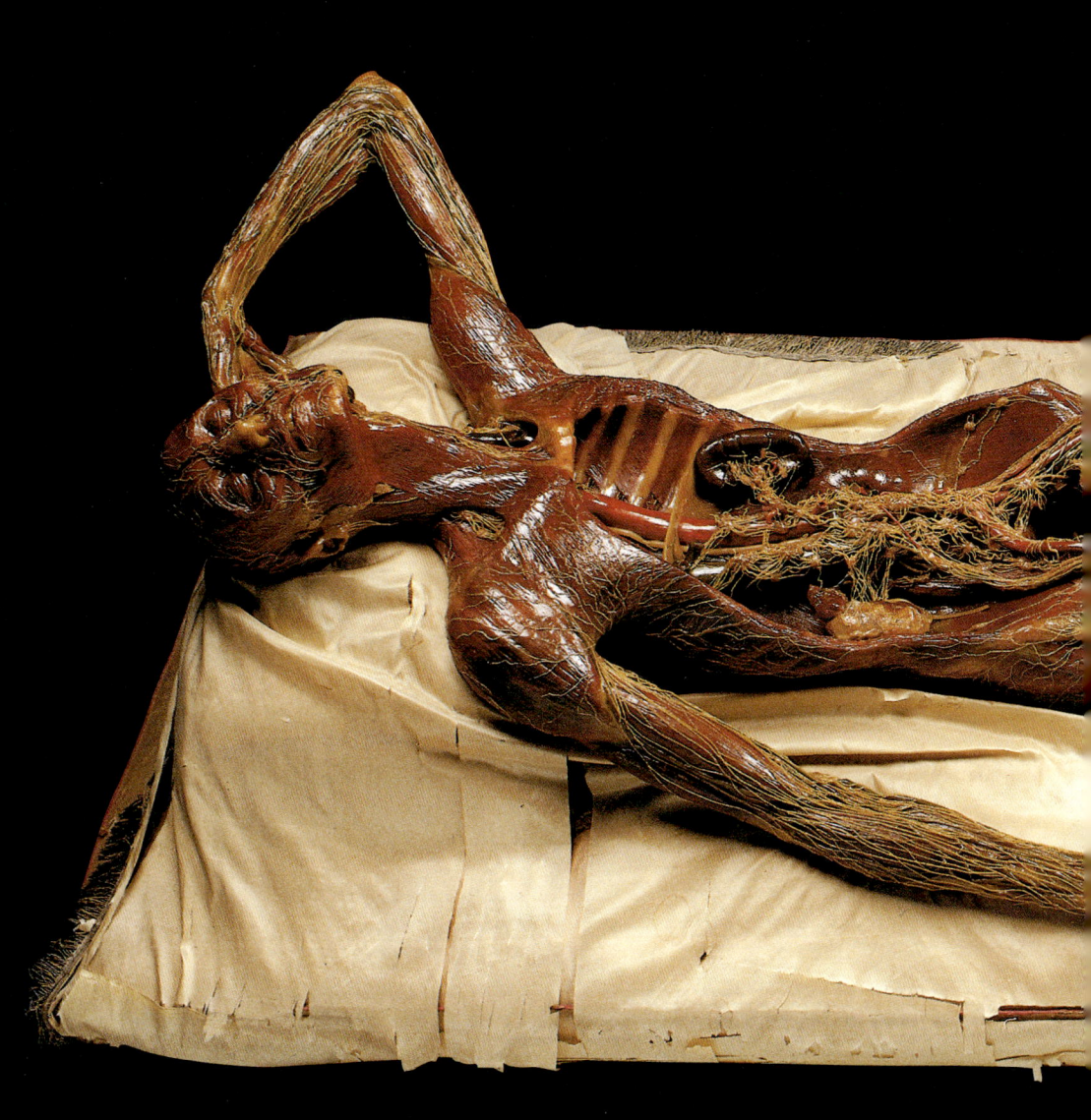

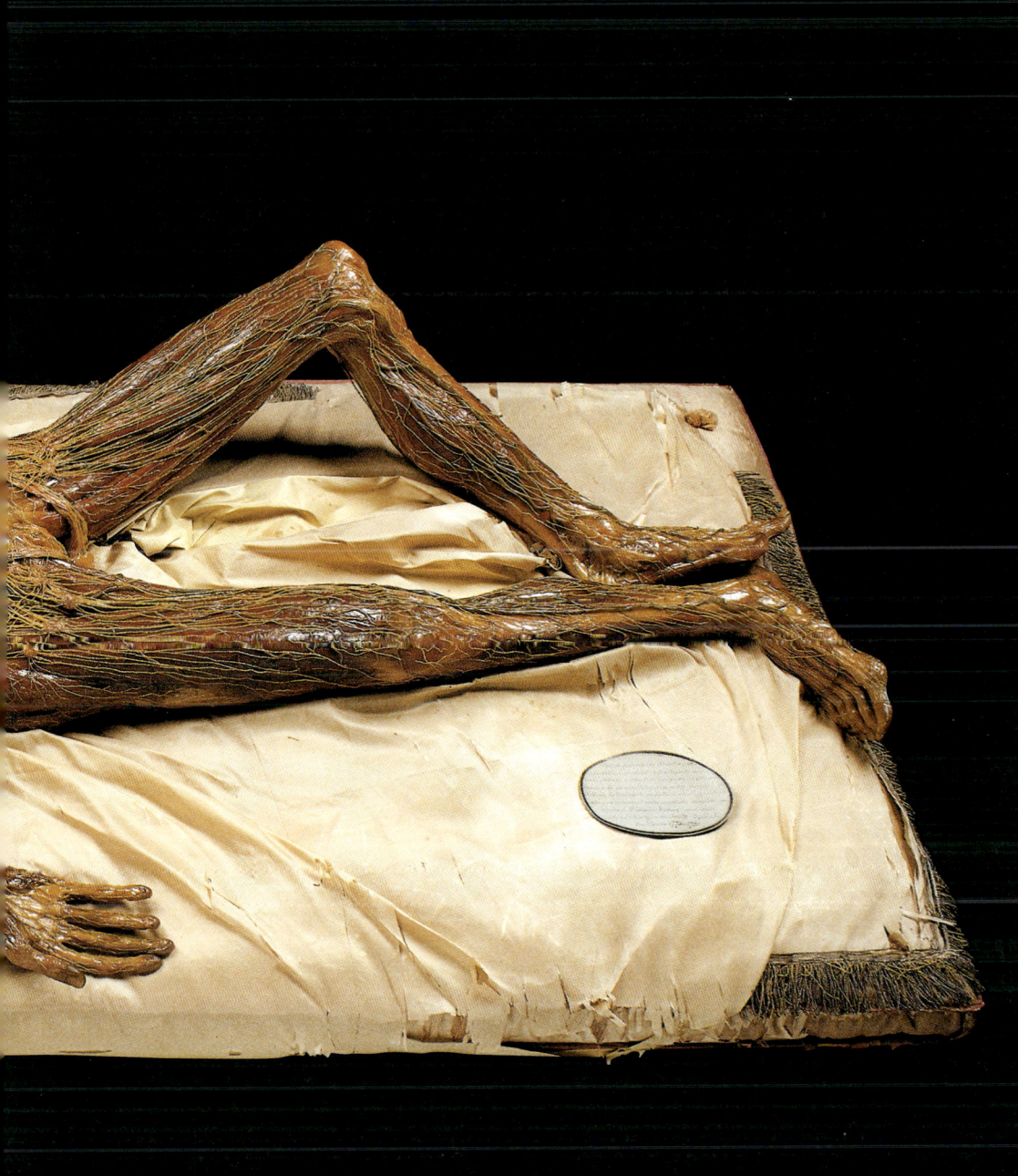

Systema lymphaticum

Whole body specimen showing the lymphatic vessels

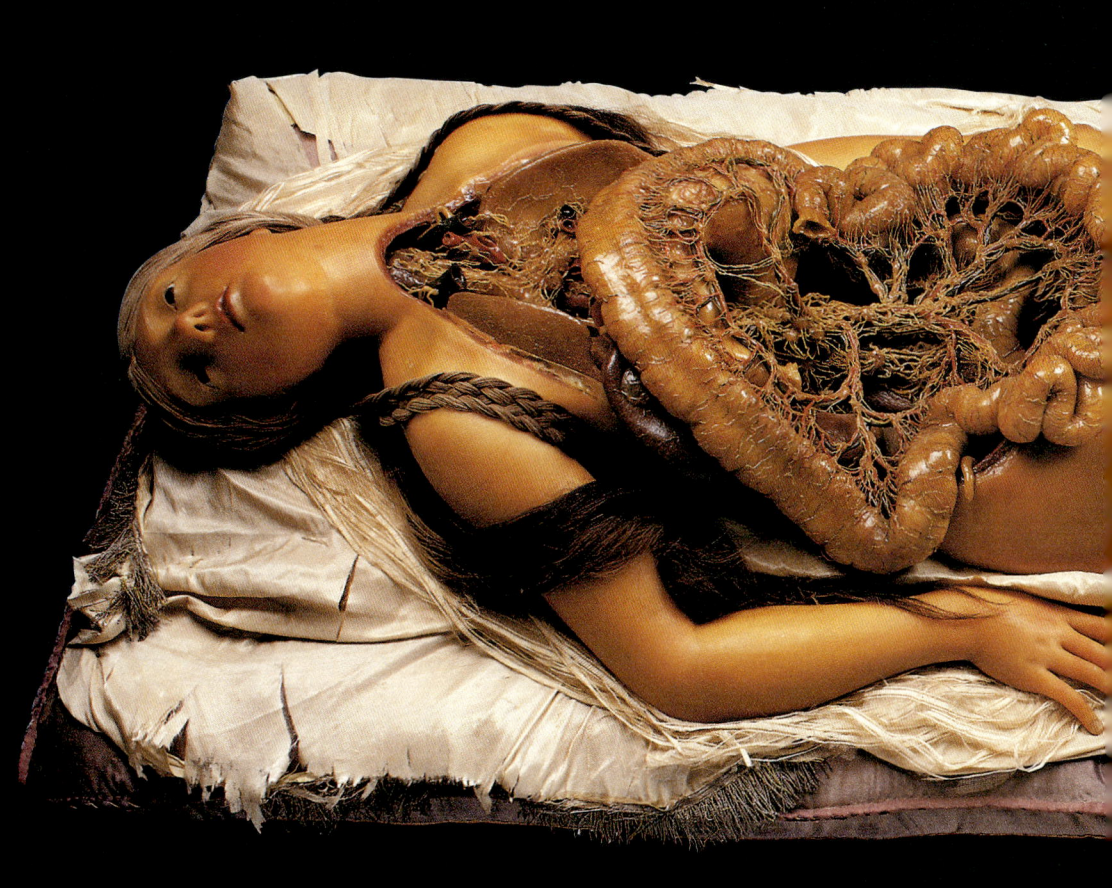

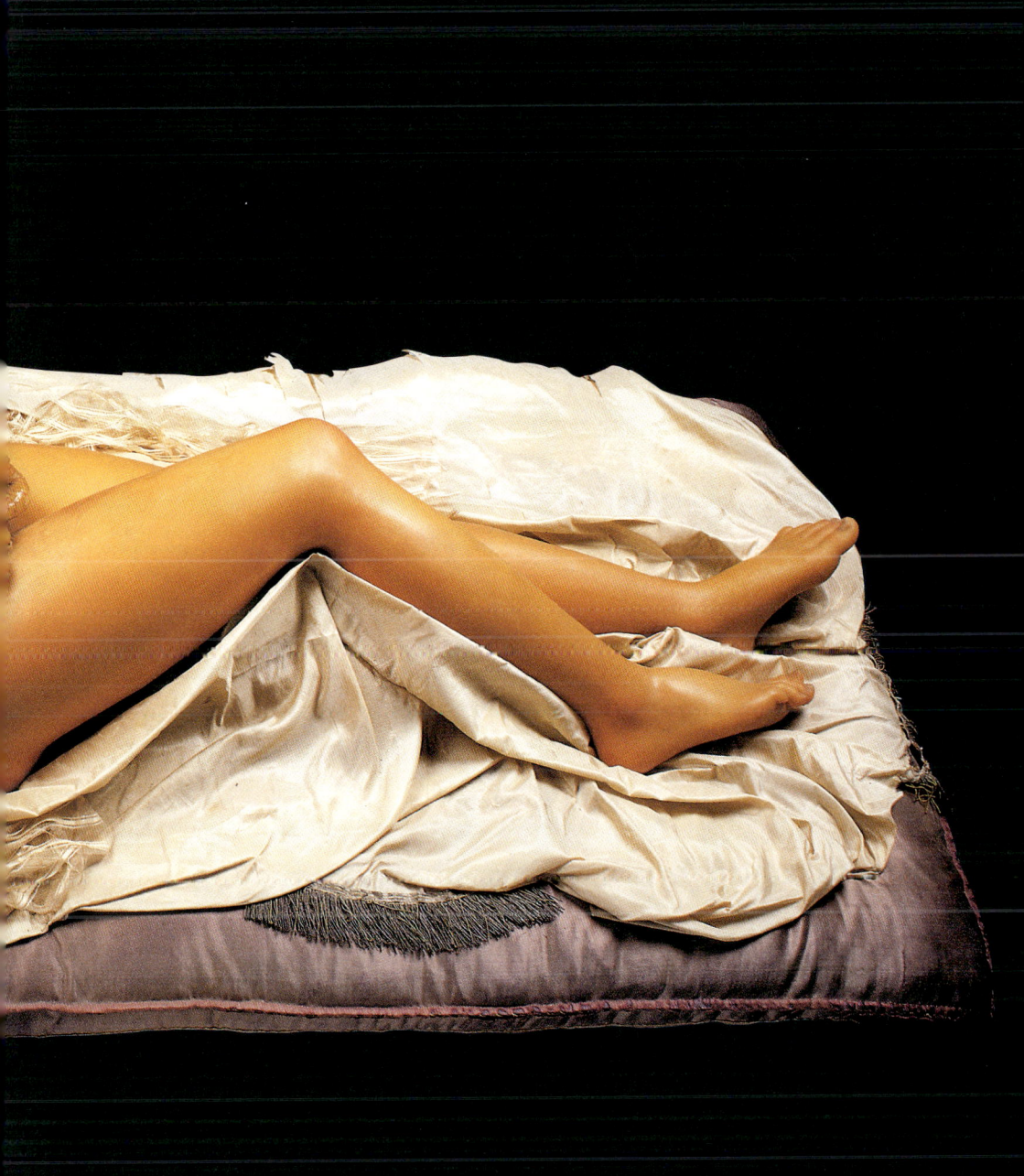

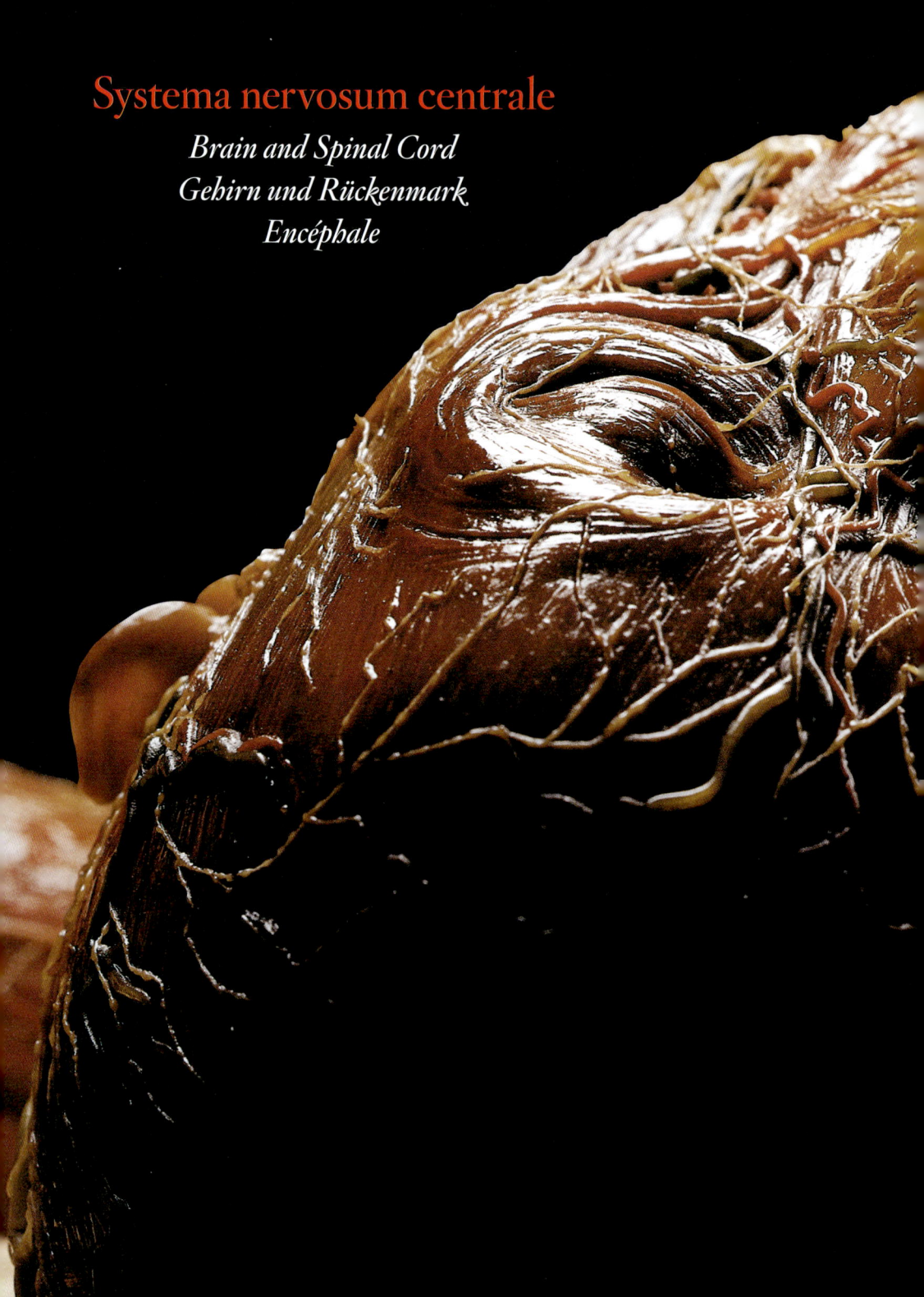

Systema nervosum centrale

Brain and Spinal Cord
Gehirn und Rückenmark
Encéphale

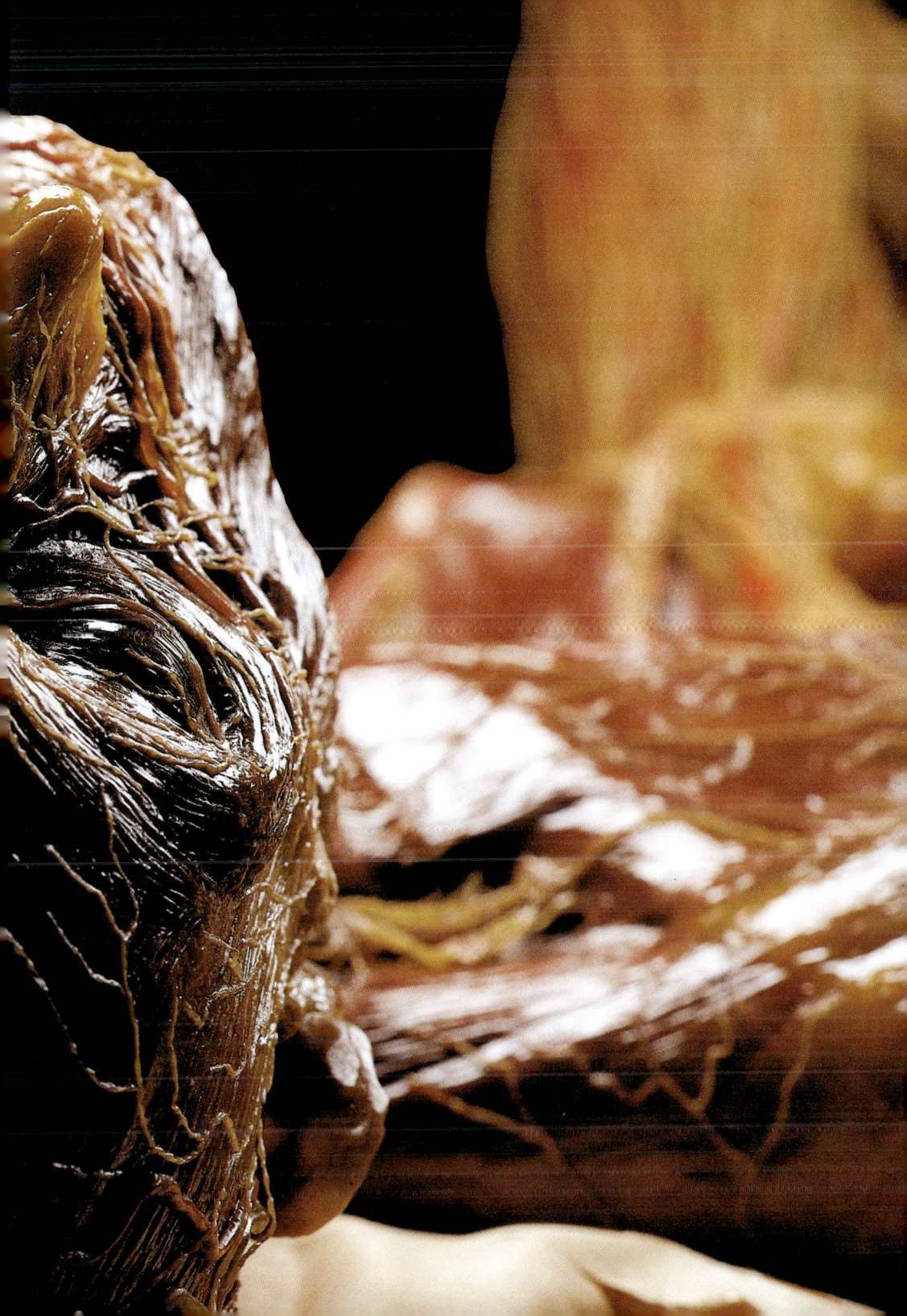

The Brain

The uniqueness of every human being is founded in his soul, i.e., his mental, religious, cultural, social and motor capabilities, which are carried out and controlled by the neuronal networks in our brains. The nerve cells are the morphological basis of the complex activities of the brain, their processes (dendrites, axons) and specialist sites of contact (synapses) serve to conduct and transmit information. Whereas two centuries ago only the vaguest concepts of their significance existed, we, by applying our modern imaging techniques, are able today to obtain detailed morphological and functional information about the individual brain without opening the skull. According to contemporary medical knowledge, the complete and irreversible cessation of the function of the cerebrum and brainstem as the site of complex and vital cerebral processes is the criterion of the individual's death. The following illustrations display the various aspects of the anatomy of the brain: an intact infant's skull after removal of the skin and muscles of the head showing the still patent anterior fontanelle, a skull laid open to reveal the connective tissue coverings of the brain (the hard and soft meninges) with their arteries and veins, the vessels on the upper and lower surfaces of the brain, including the emerging nerves and their course through the openings in the base of the skull. In contrast to these still relevant insights, the cerebral gyri bear little relationship to the original specimen, since in most models they are merely stylised. In most of the sections the visible border between the dark outer cortex and the internal white matter has been artificially emphasised.

Of great importance is the display of the points of attachment of the cranial nerves to the brainstem, as well as the various fluid-filled internal chambers (ventricles) of the brain, the position and extent of which can be very well studied in the specimens. These cerebral ventricles used to be of great interest, because they were thought during the 17th century to be the site of the spirit and the soul.

Many of the models are anatomically correct and impress one also by their great artistic merit. However, there are also certain exhibits seen from unusual points of view, the significance and meaning of which may not always be easy for the non-professional to grasp.

Pages 296–297:
Detail from pages 292–293

G. G. Zumbo:
Anatomist's model of a head (cf. pp. 13, 41, 302)

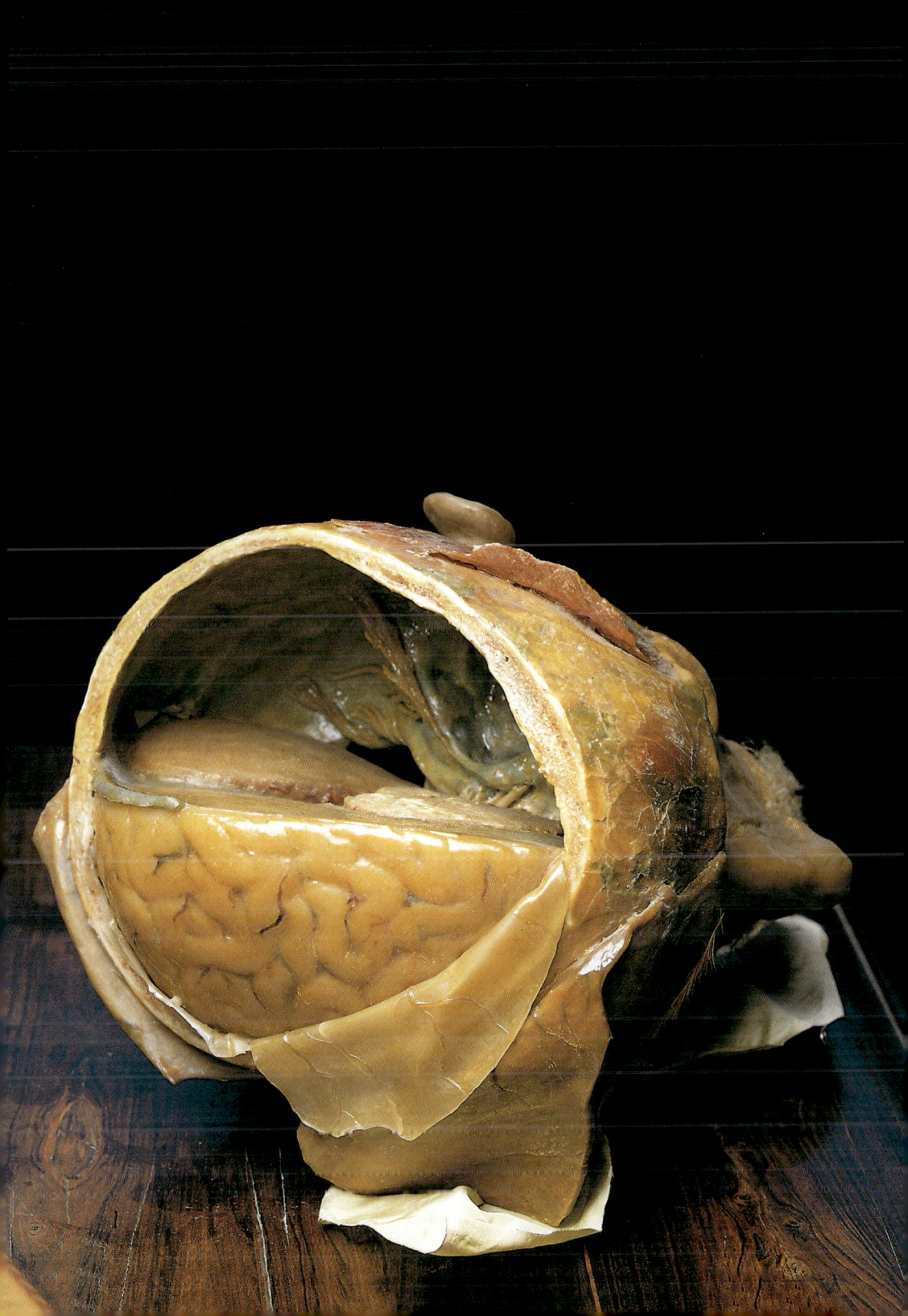

Das Gehirn

Die Einzigartigkeit eines jeden Menschen ist begründet in seiner Seele, d. h. in seinen geistigen, religiösen, kulturellen, sozialen und motorischen Fähigkeiten, die von den neuronalen Netzwerken in unserem Gehirn getragen und gesteuert werden. Die morphologische Grundlage dieser komplexen Hirnleistungen bilden die Nervenzellen. Mit ihren Ausläufern und über spezielle Kontake, den Synapsen, dienen sie der Informationsleitung und -übertragung. Während vor 200 Jahren nur vage Vorstellungen über ihre Funktionsweise bestanden, können wir heute mit modernen bildgebenden Verfahren am ungeöffneten Gehirn detaillierte Aussagen über die komplexe Gestalt des individuellen Gehirns machen und funktionelle Teilleistungen erfassen. Nach den heutigen medizinischen Erkenntnissen gilt der endgültige Ausfall der Funktion von Großhirn und Stammhirn als den Regionen komplexer und lebensnotwendiger Hirnleistungen als Todeszeichen des Menschen.

Auf den nachfolgenden Tafeln sind die verschiedensten anatomischen Ansichten zur Hirnanatomie zu sehen: ein intakter Säuglingsschädel nach Wegnahme der Kopfhaut und Kopfmuskeln mit der großen vorderen, noch offenen Fontanelle, der eröffnete Schädel mit der Aufsicht auf die Bindegewebshüllen des Gehirns (harte und weiche Hirnhaut) mit versorgenden arteriellen und venösen Blutgefäßen, die Gefäße auf der Hirnoberfläche und Hirnunterfläche, Hirnhälften mit austretenden Hirnnerven einschließlich ihrer Verläufe durch die Knochenkanäle des Schädels. Im Gegensatz zu diesen bis heute gültigen Erkenntnissen haben die in den Modellen gezeigten Hirnwindungen nur wenig oder keinen Bezug zum Originalpräparat. In den meisten Schnittpräparaten ist die sichtbare Grenze zwischen der dunkleren Hirnrinde und dem weißlichen Hirnmark herausgearbeitet.

Thematische Schwerpunkte bilden die Darstellung der Austrittsstellen der Hirnnerven am Hirnstamm sowie die verschiedenen flüssigkeitsgefüllten Innenräume (Ventrikel) des Gehirns, deren Lage und Ausdehnung an den Präparaten sehr gut studiert werden können. Diese Hirnventrikel waren von großem allgemeinen Interesse, weil man im 17. Jahrhundert in ihnen den Sitz des Geistes und der Seele vermutete.

Zahlreiche Modelle sind anatomisch korrekt und beeindrucken gleichzeitig durch ihre hohe künstlerische Qualität. Daneben finden sich aber auch Exponate in ungewöhnlicher perspektivischer Ansicht, deren Bedeutung und Aussage dem Laien nur schwer zugänglich ist.

L'encéphale

C'est son esprit qui donne à chaque être humain son caractère unique ; ces capacités spirituelles, religieuses, culturelles, sociales ou motrices. Les cellules nerveuses sont la base morphologique de ces fonctions cérébrales complexes ; par leurs prolongements et à travers des zones de contact spécifiques, les synapses, elles véhiculent et transmettent les informations. Il est possible aujourd'hui, par des techniques d'imagerie, de mettre en évidence non seulement des détails de l'architecture complexe du cerveau d'un individu mais encore des localisations fonctionnelles particulières. Dans l'état actuel des connaissances médicales, l'arrêt total et définitif des fonctions du cerveau et du tronc cérébral signe la mort de l'individu.

Les vues les plus variées de l'anatomie du cerveau sont représentées dans les planches suivantes. Sur un crâne de nouveau-né, après ablation de la peau et des muscles de la tête, la grande fontanelle apparaît encore ouverte, et les méninges cérébrales (dure-mère et pie-mère) sont pourvues de leur vascularisation artérielle et veineuse. Plus loin, les vaisseaux de la convexité et de la base du cerveau sont au contact des structures nerveuses. D'un hémi-cerveau partent les nerfs crâniens, dont le trajet peut être suivi jusque dans les canaux osseux de la base du crâne.

La représentation des circonvolutions cérébrales n'a que peu ou pas de relation avec la réalité anatomique, contrairement à de nombreux autres détails encore valables aujourd'hui ; les circonvolutions ont souvent été stylisées. Dans la plupart des coupes, la limite entre la substance blanche et la substance grise du cerveau a été mise en évidence.

Une attention particulière a été portée à la représentation des points d'émergence des nerfs crâniens au niveau du tronc cérébral, et à la représentation des cavités cérébrales, ou ventricules, contenant le liquide cérébro-spinal, dont la disposition et le développement peuvent être bien étudiés sur ces modèles. Les ventricules cérébraux ont suscité un grand intérêt car ils étaient considérés, au XVIIe siècle, comme le siège de l'esprit et de l'âme.

De nombreux modèles sont les témoins fidèles de la réalité anatomique et sont remarquables par leur grande qualité artistique. À côté de ces pièces de qualité, il en existe quelques-unes dont les vues perspectives inhabituelles en rendent l'interprétation difficilement accessible au non-initié.

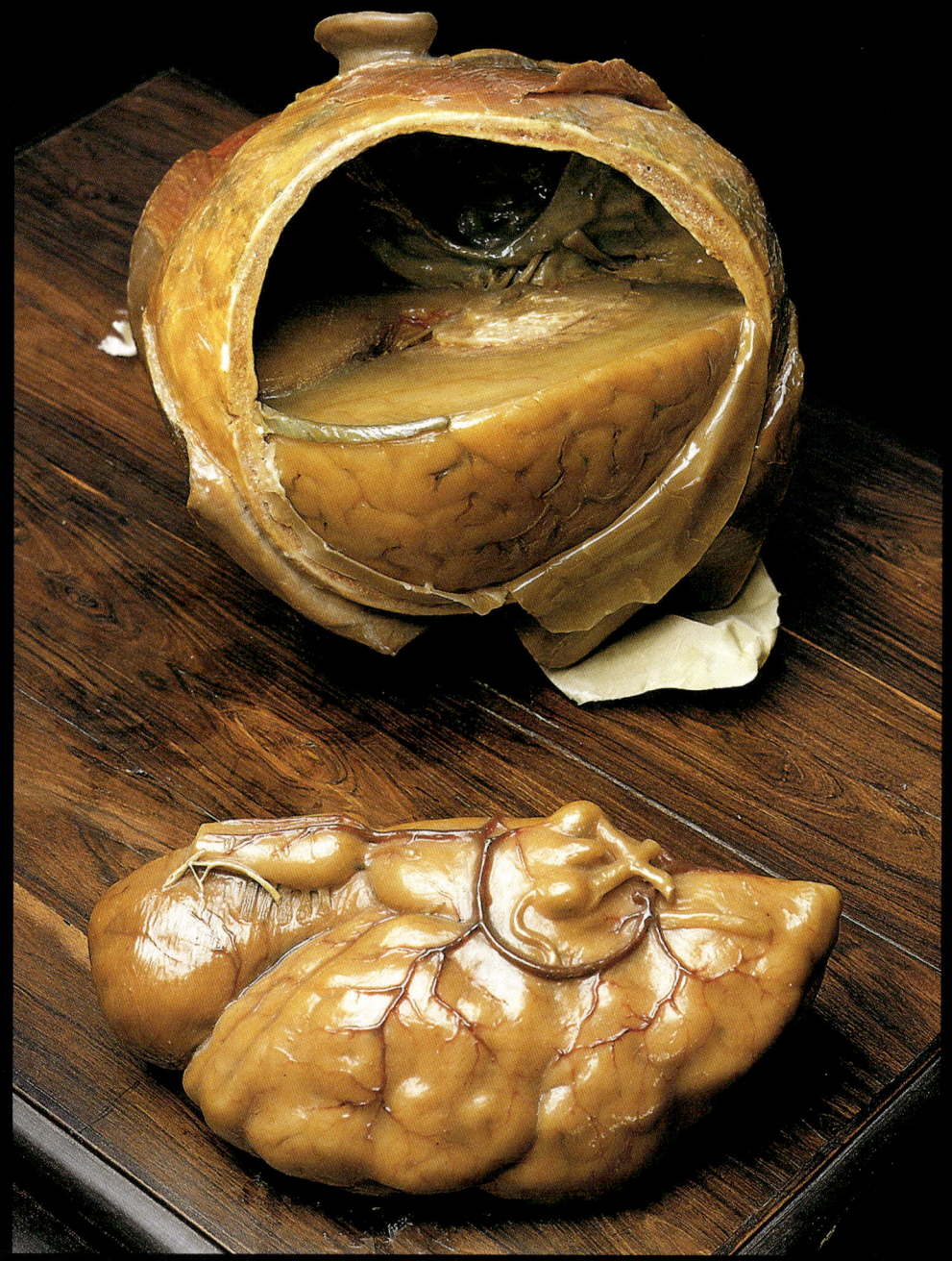

Cranium, Cerebrum

Specimen of a head. The calvaria has been removed and half of the brain can be taken out

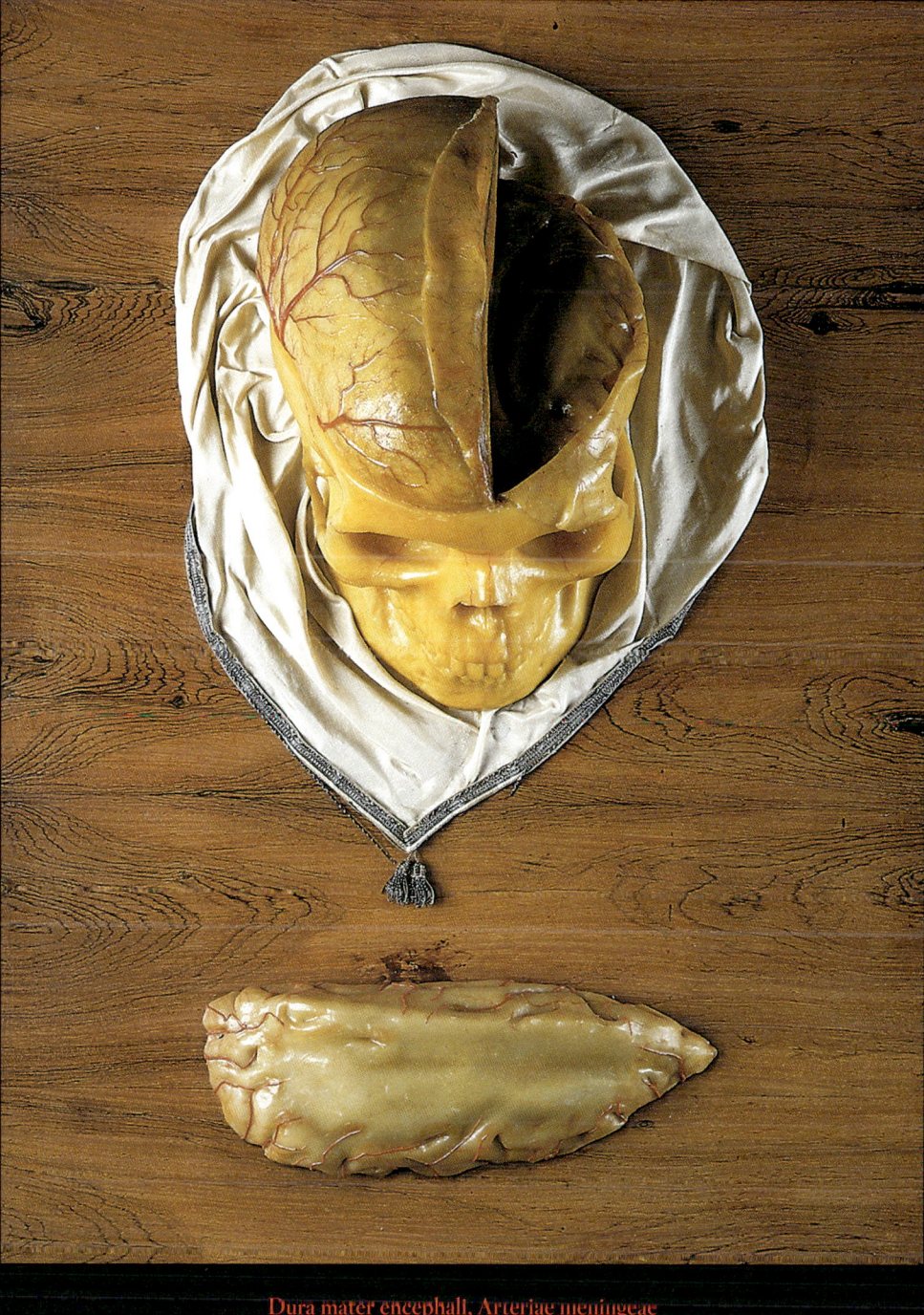

Specimen showing the dura mater with its arteries. The left half of the brain has been removed

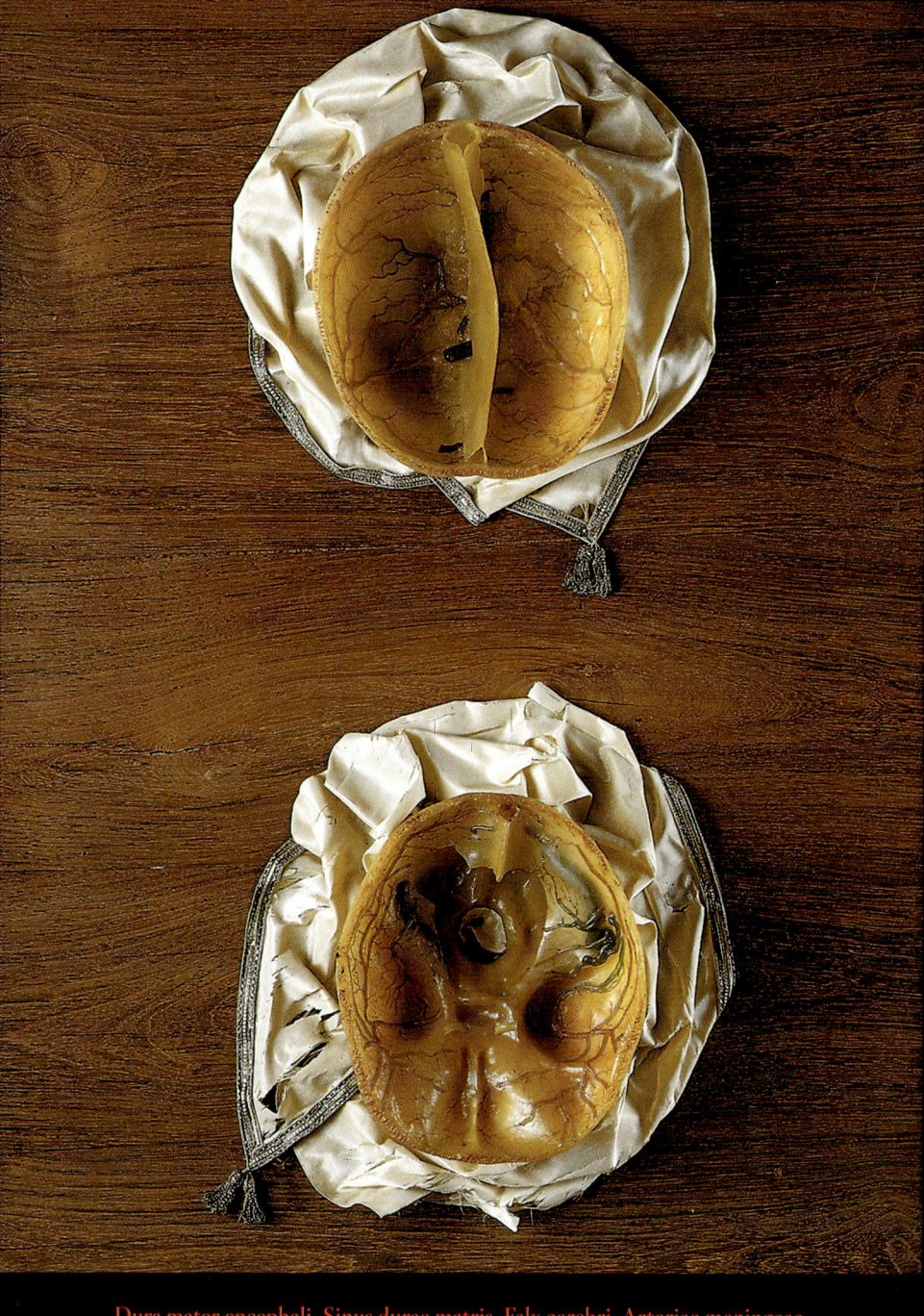

Dura mater encephali, Sinus durae matris, Falx cerebri, Arteriae meningeae

Internal aspects of the calvaria (upper specimen) and base of skull after removal of the brain.
View of the dura mater and its blood vessels. In the midline of the upper specimen one
can see the sickle-shaped falx of the cerebrum

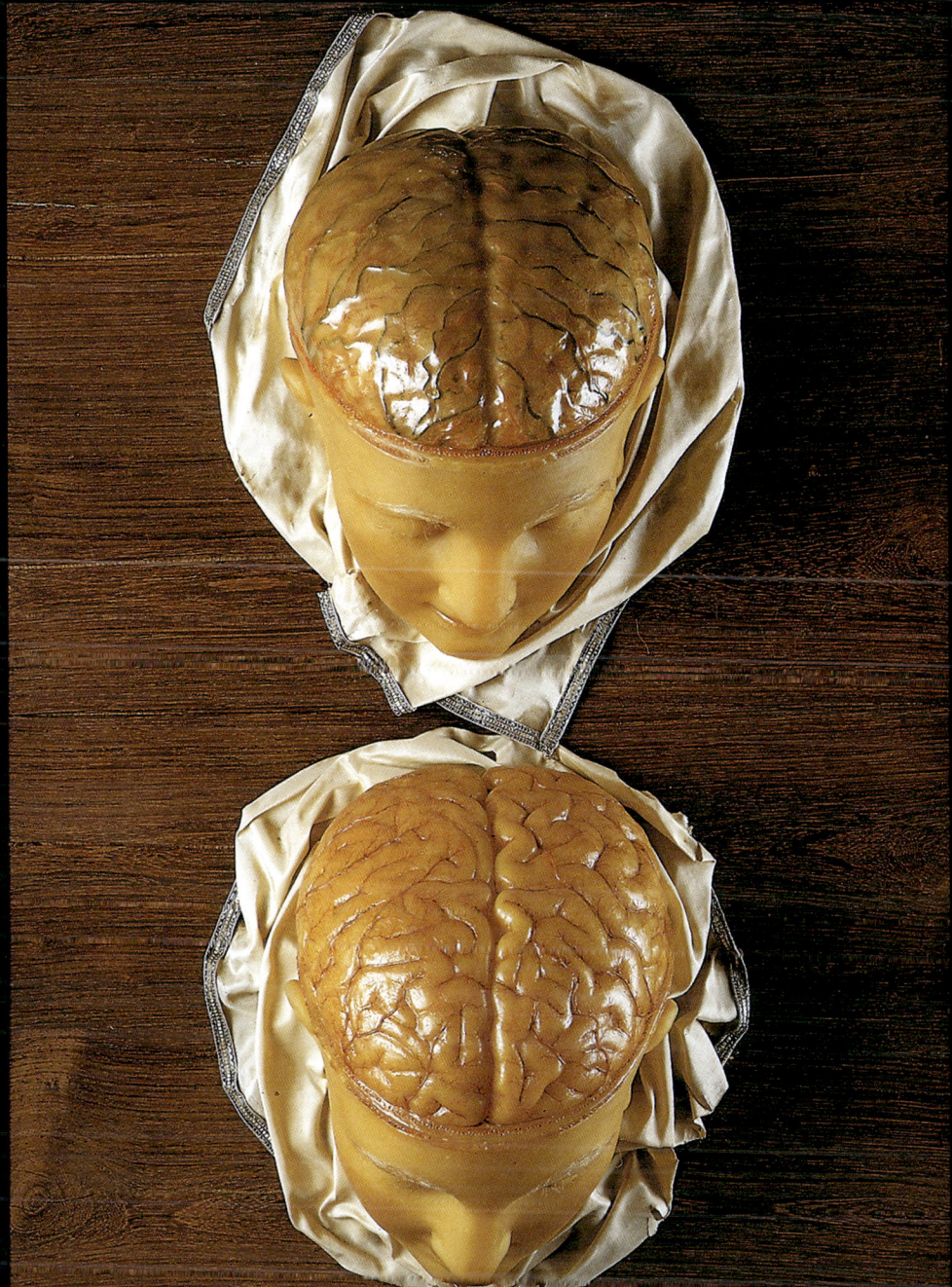

Leptomeninx, Arachnoidea, Gyri, Sulci, Fissura longitudinalis

View of the brain after removal of the calvaria and dura mater. In the upper specimen the soft meninges with their vessels can be seen. The convolutions of the brain surface can be clearly seen in the lower specimen after removal of the meninges

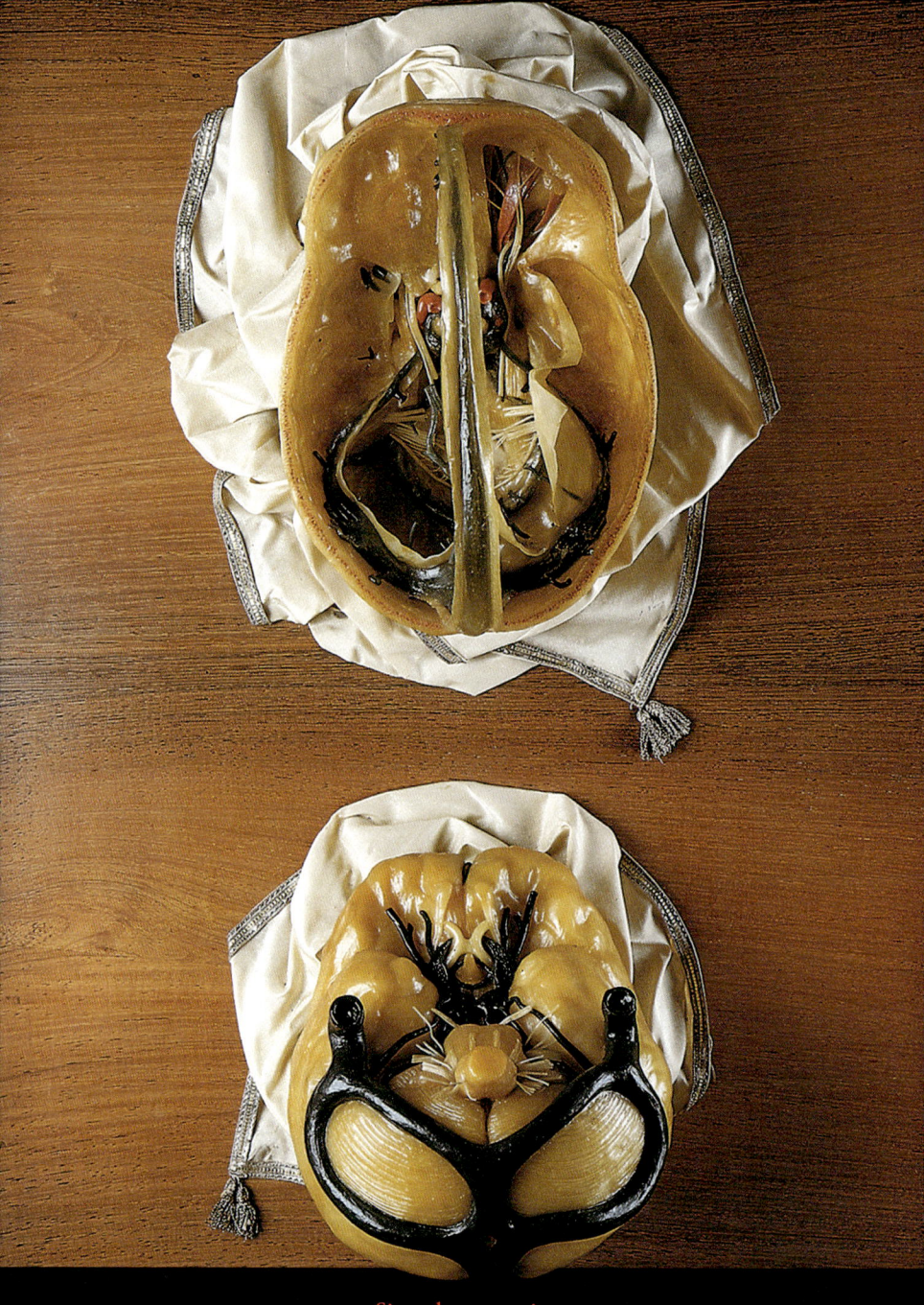

Sinus durae matris

The venous sinuses of the dura mater

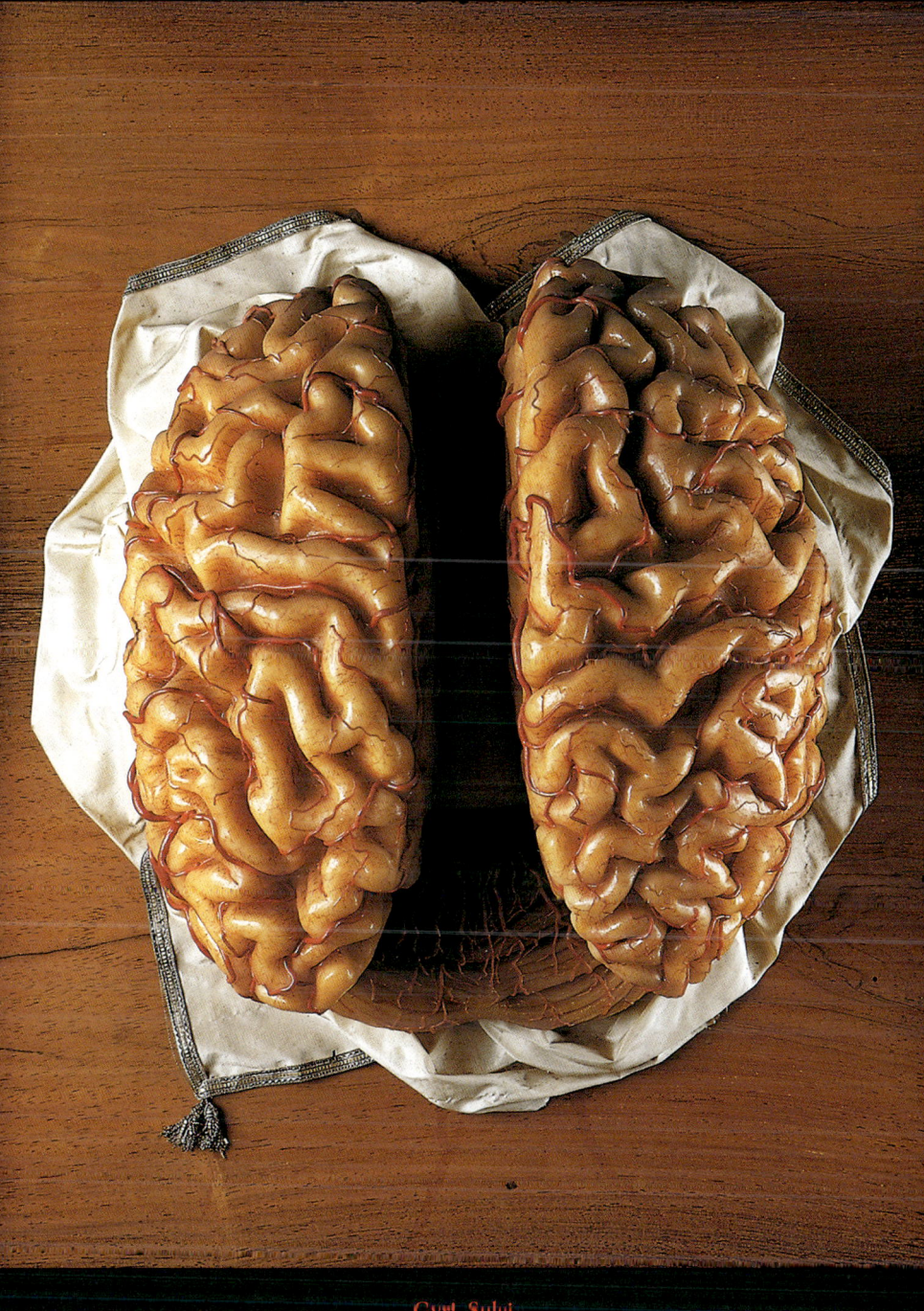

Gyri, Sulci

Specimen of the brain with its superficial arteries. The cerebrum has been divided in the midline
and the two halves pushed aside to reveal the cerebellum (below)

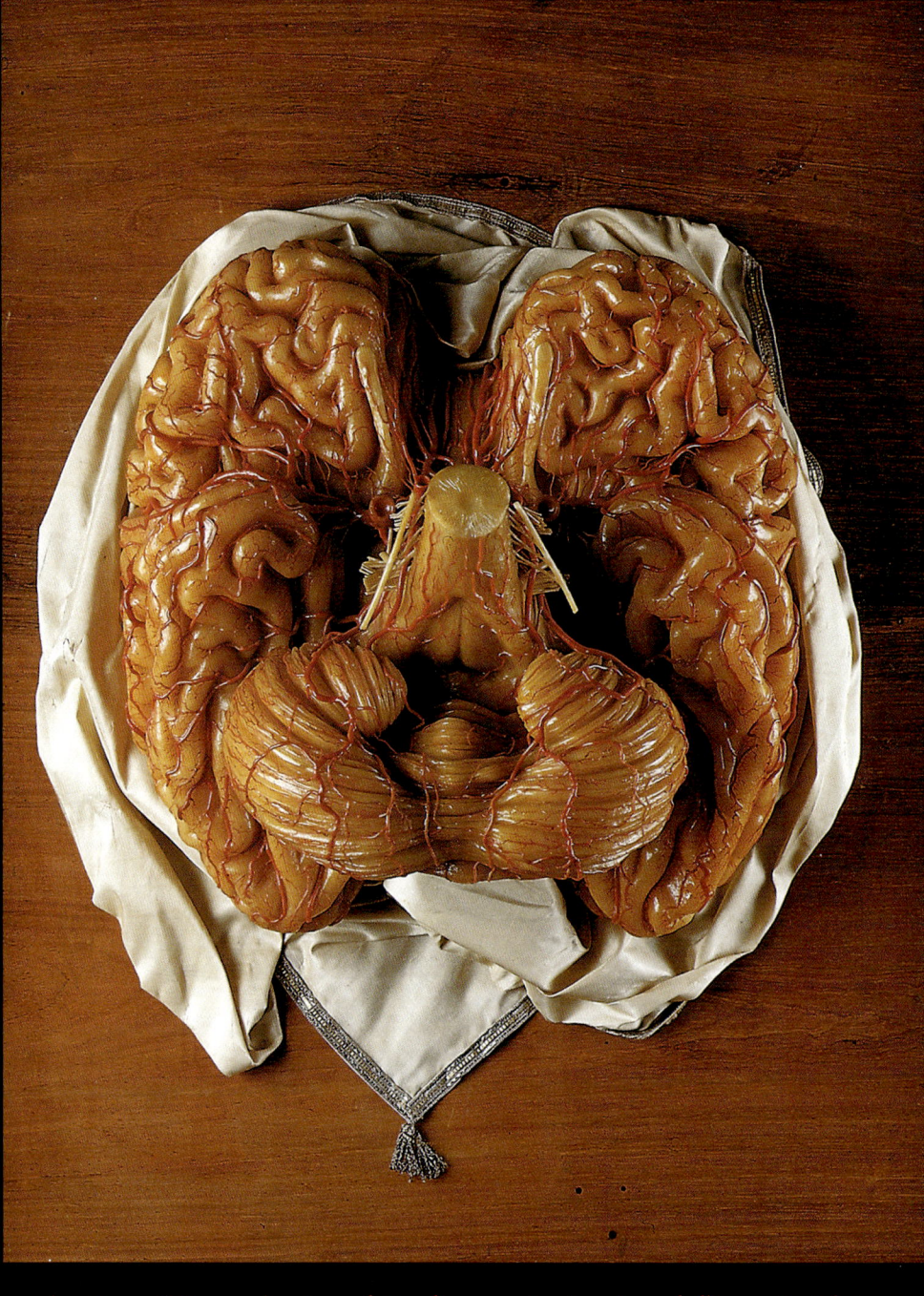

Arteria carotis interna, Arteria cerebri media et anterior, Arteria cerebelli posterior inferior

Specimen showing the various arteries at the base of the brain

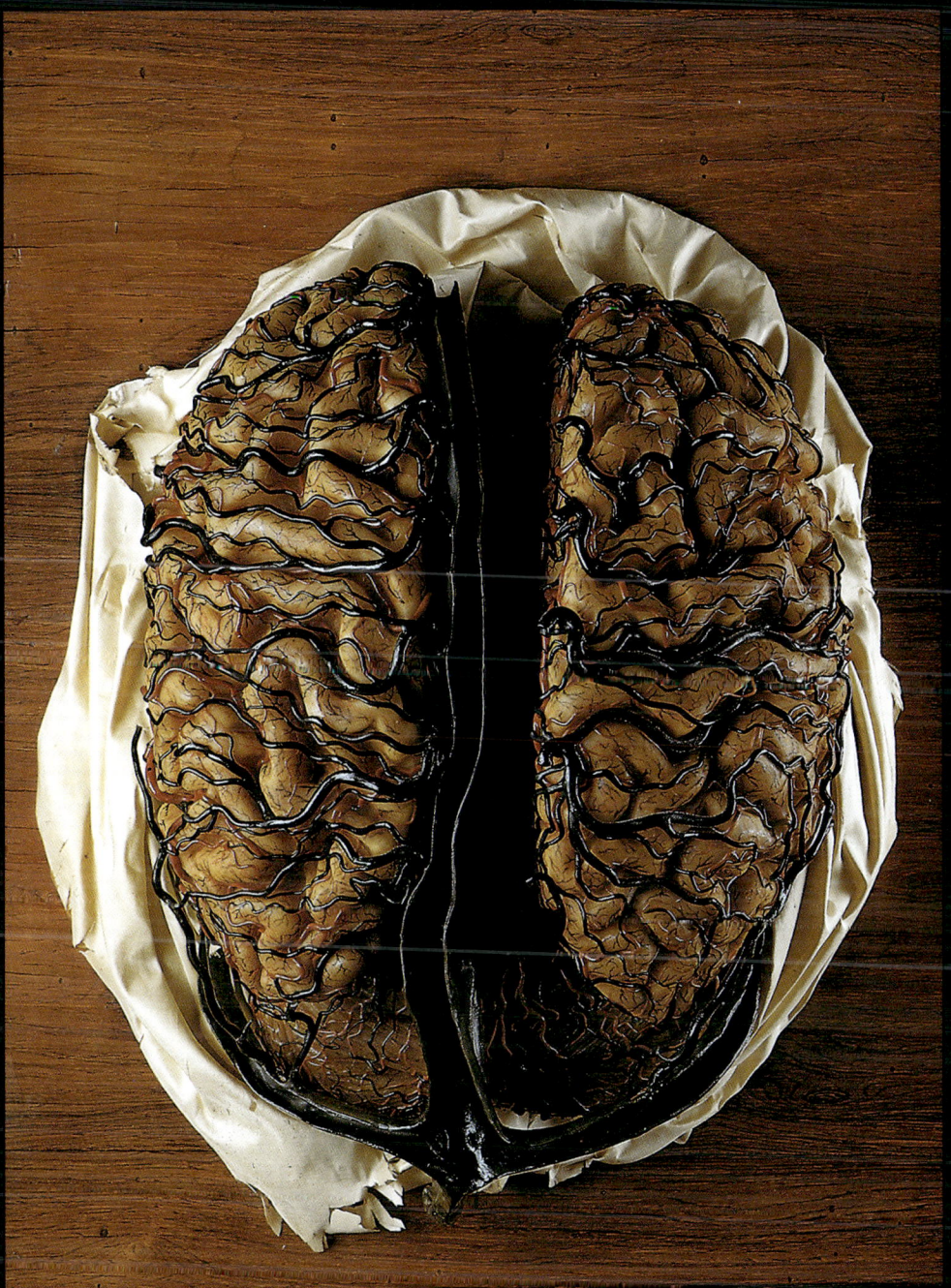

Venae cerebri superficiales, Sinus sagittalis superior, Confluens sinuum, Sinus transversus

The superficial veins of the brain and their openings into the superior sagittal sinus
within the falx of the cerebrum

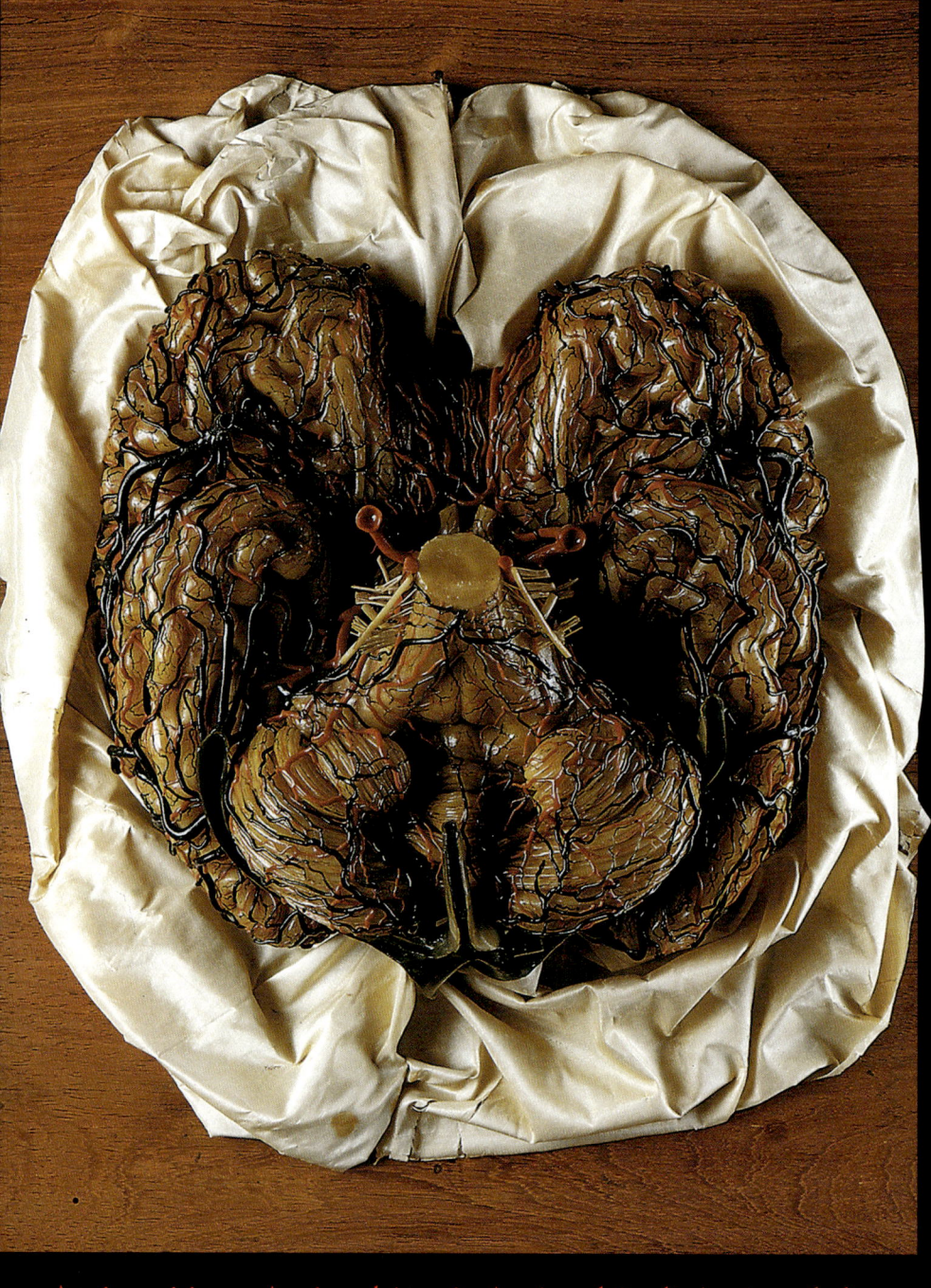

Arteria carotis interna, Arteria cerebri anterior, Arteria cerebri media, Arteria vertebralis

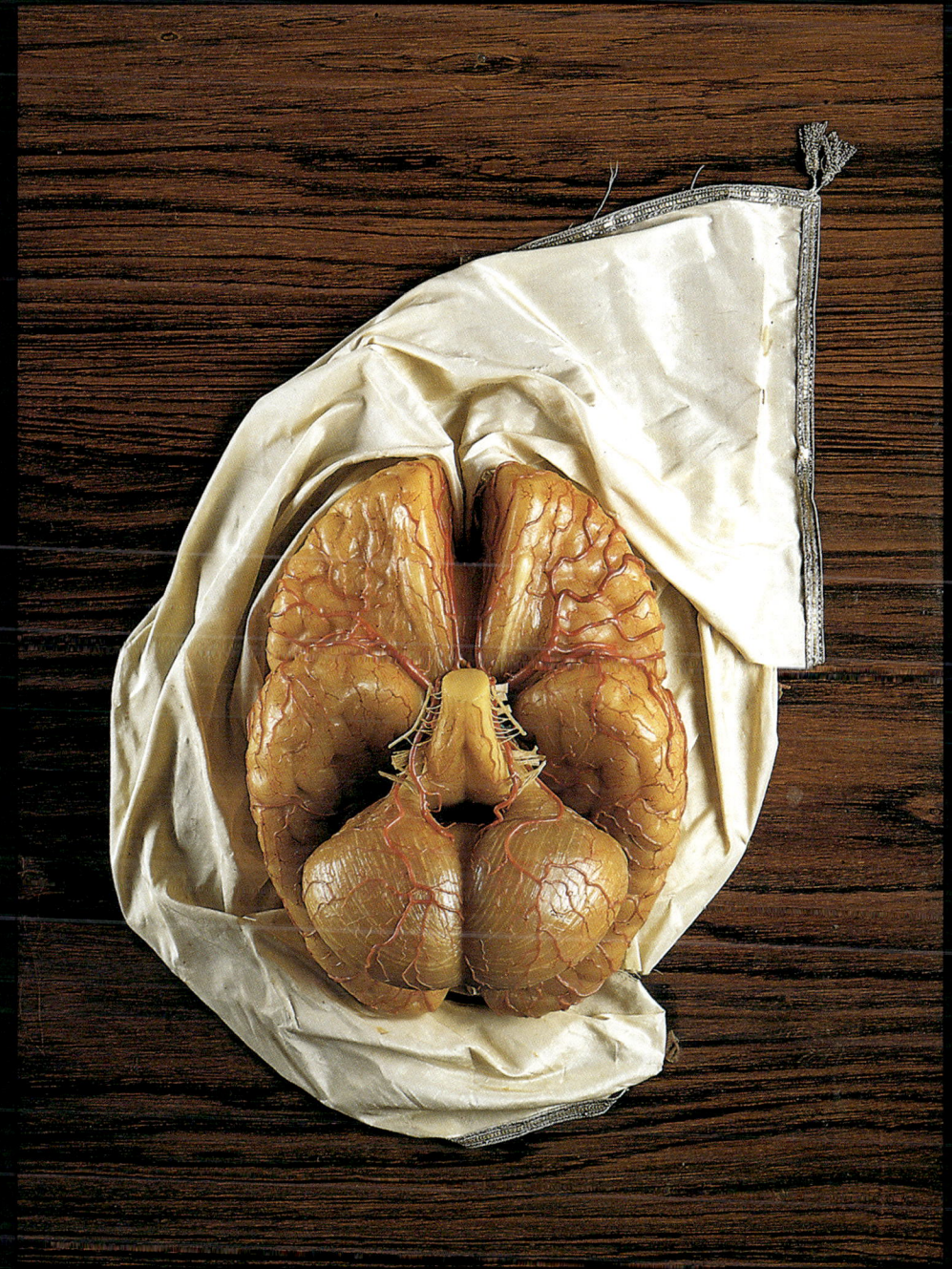

Arteria cerebri anterior et media, Arteria cerebelli posterior inferior
View of the arteries at the base of the brain

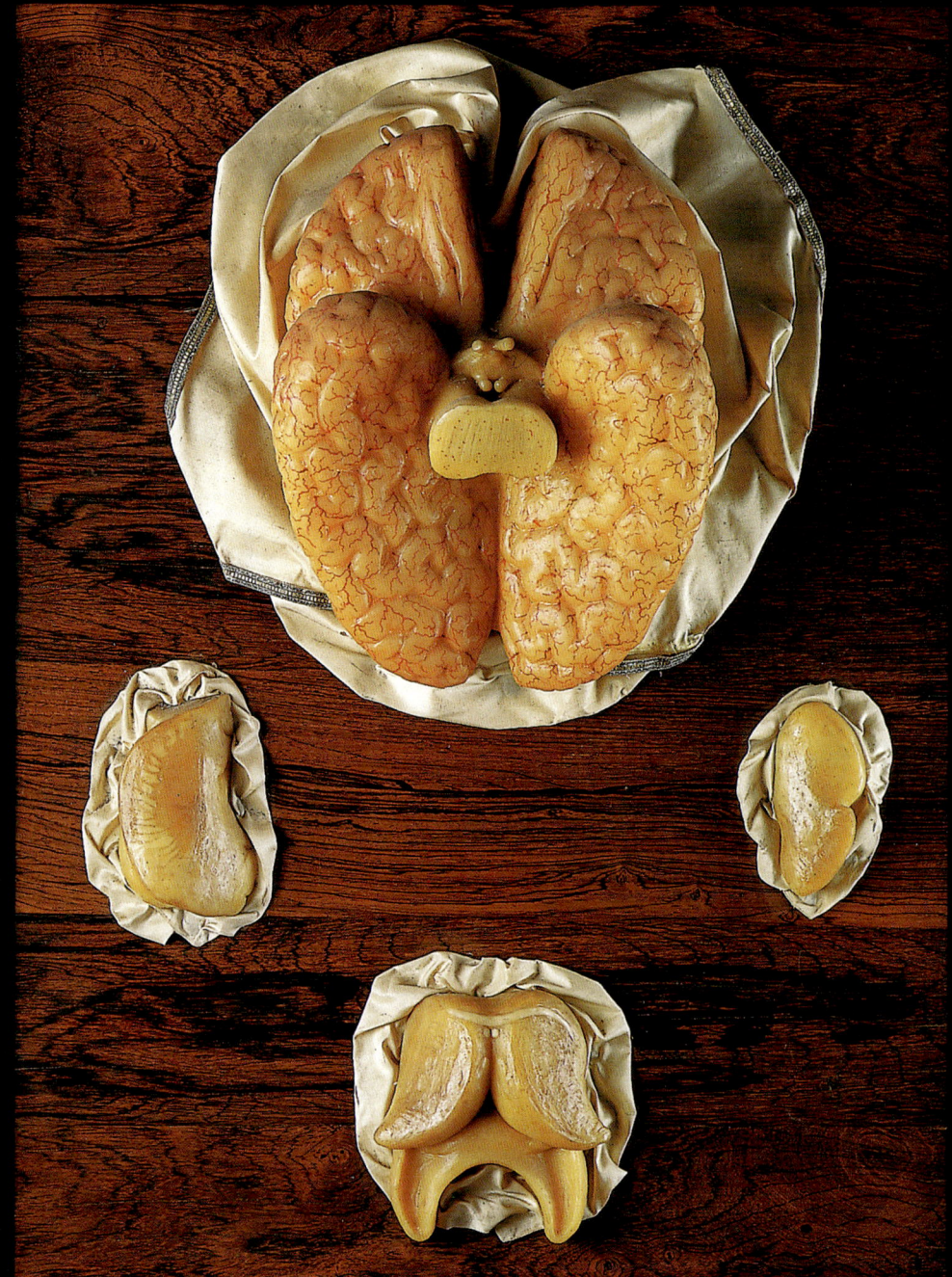

Thalamus

View of the base of the brain. The cerebellum and part of the brainstem have been removed.
The small specimen in the middle shows the thalamus

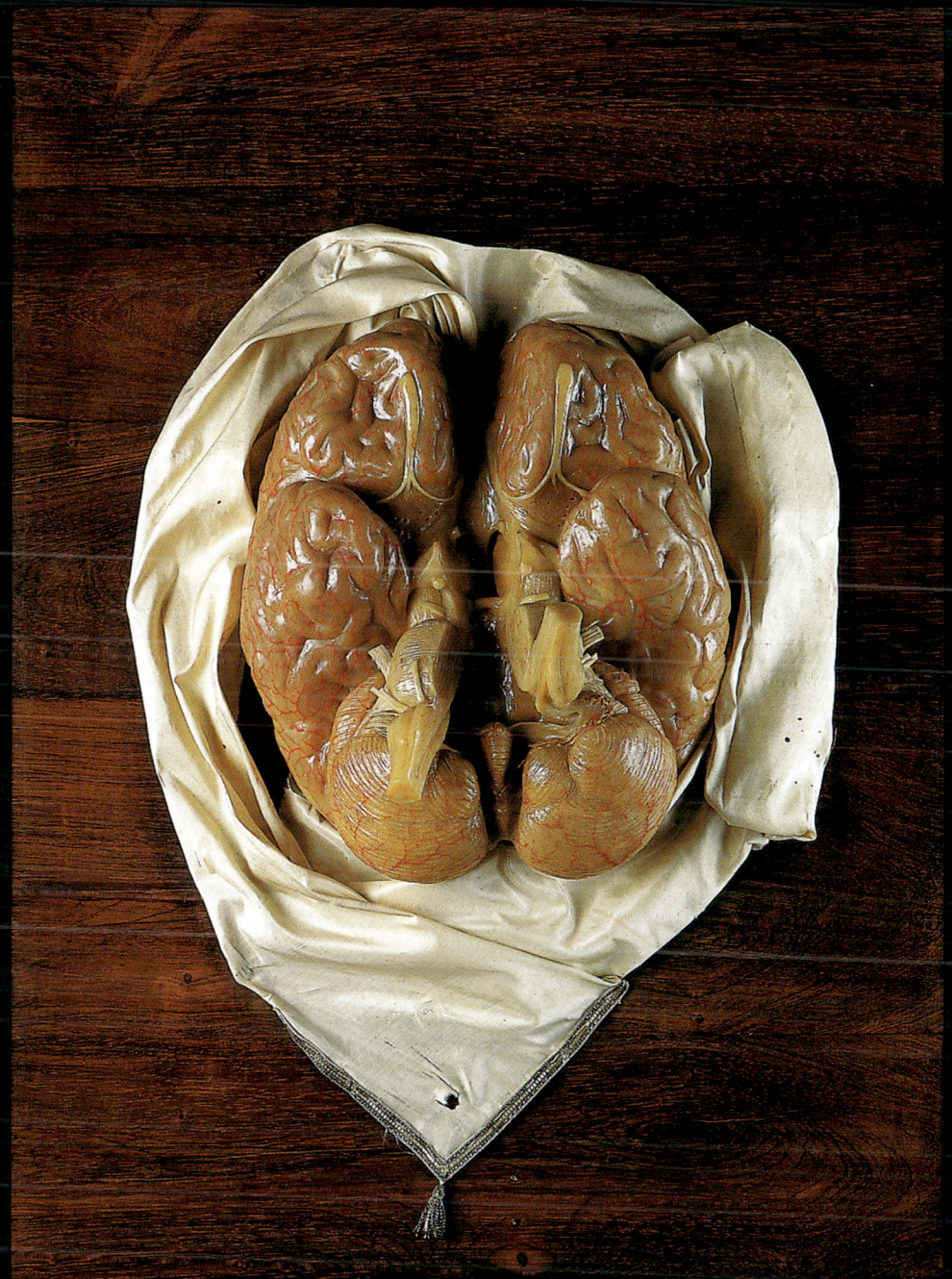

Bulhus olfactorius, Tractus olfactorius medialis et lateralis, Tractus diagonalis Broca

View of the base of the brain showing parts of the olfactory bulb and tract.
The brainstem has been divided in the midline

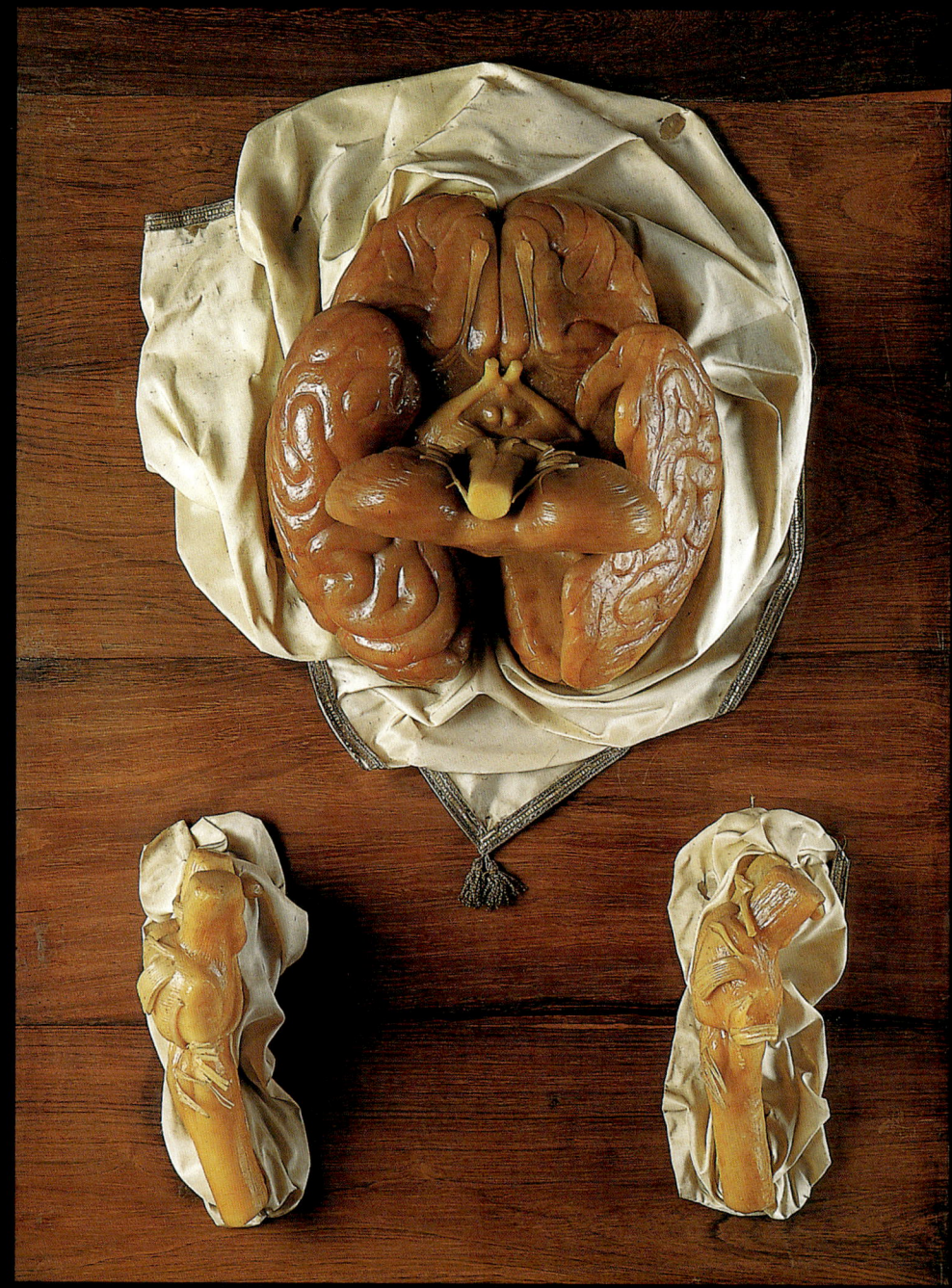

Nervi cranii

View of the base of the brain. The cerebellum has been lifted up and pushed out of shape.
The lower specimen shows a side view of the brainstem with the emerging cranial nerves

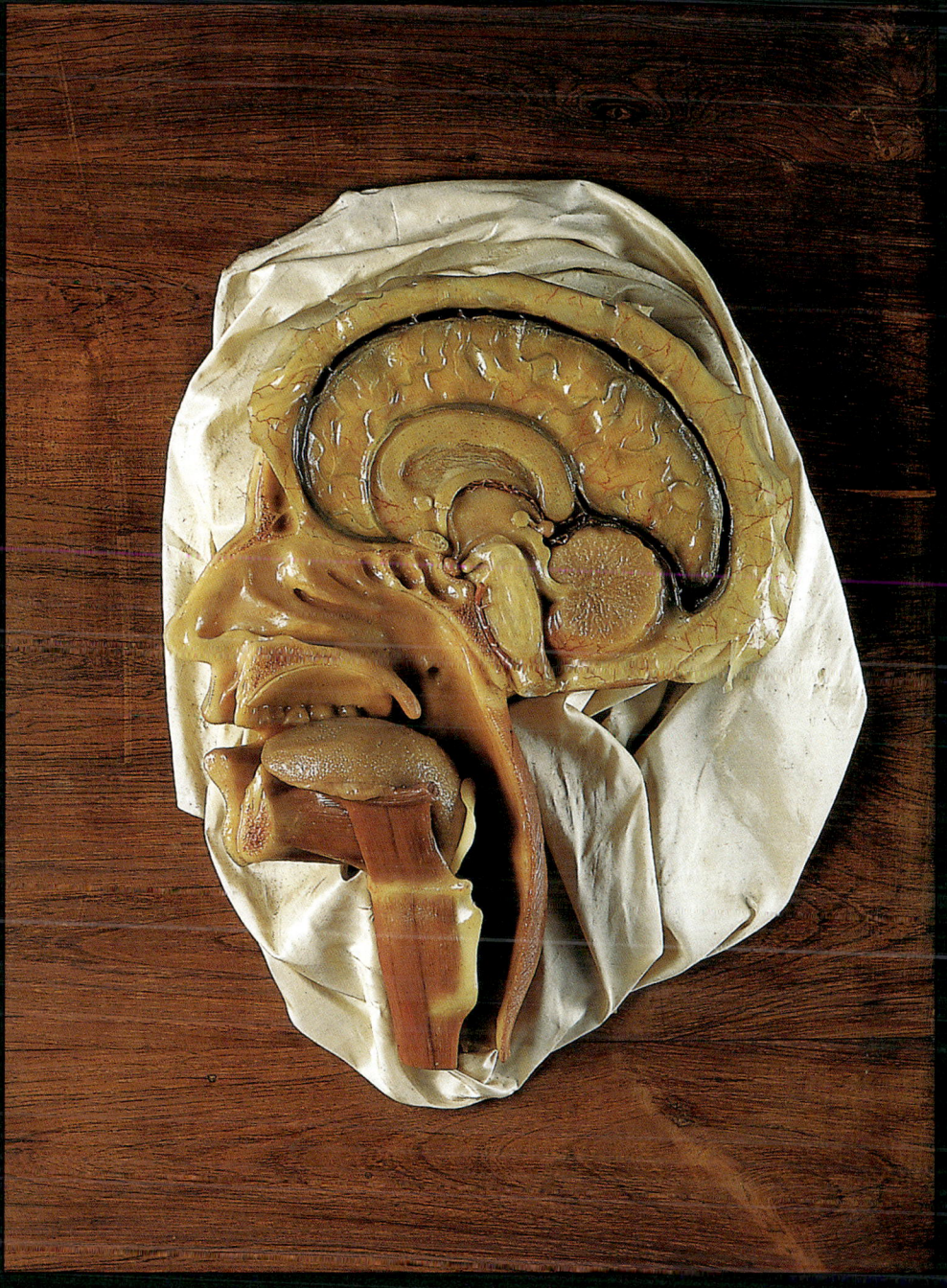

Visccrocranium, Neurocranium, Sinus sphenoidalis, Glandula pituitaria, Ventriculus tertius, Aquaeductus mesencephali, Ventriculus quartus

Specimen of a head which has been divided in the midline to show the relationship between cranial cavity and facial skeleton

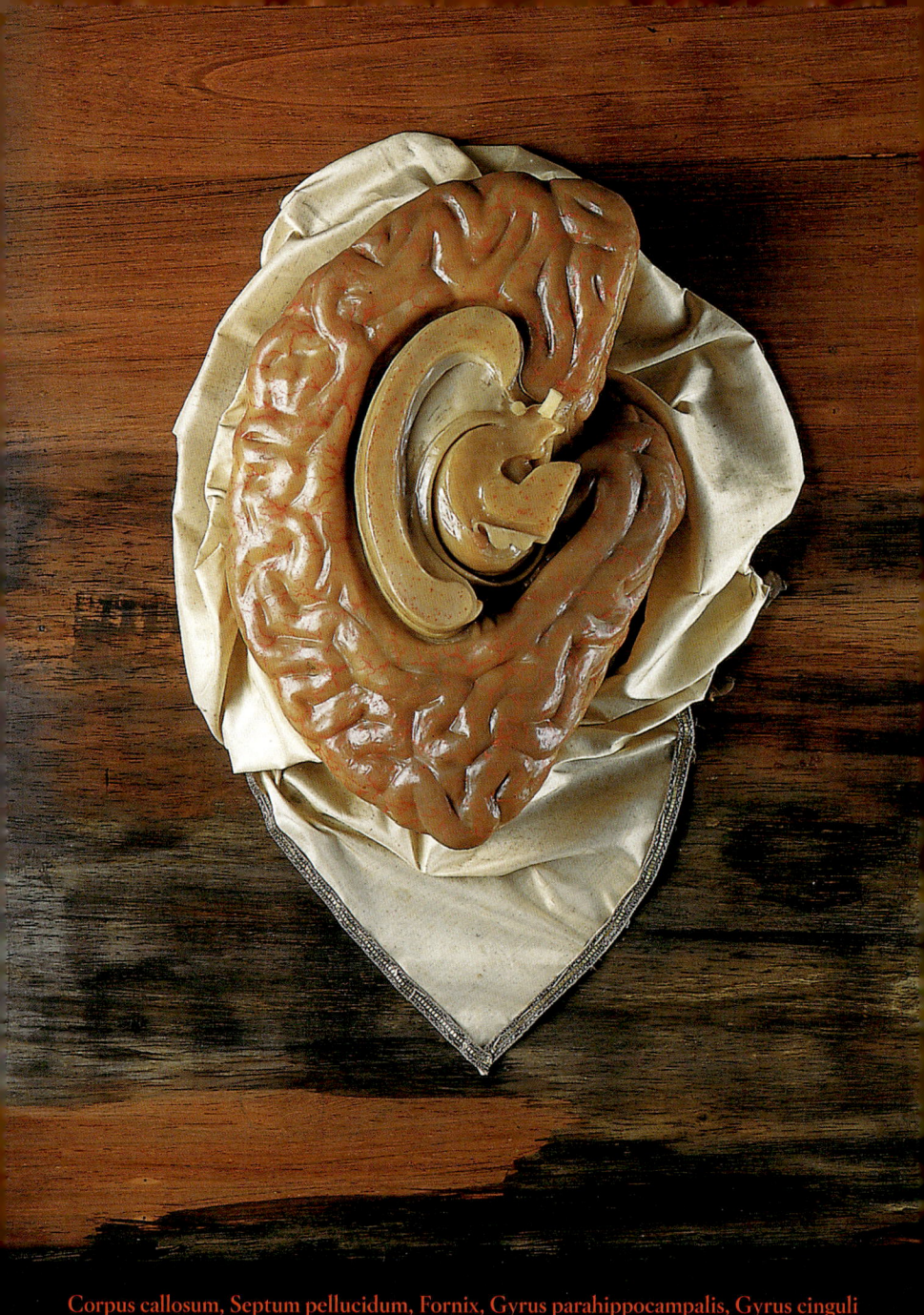

Corpus callosum, Septum pellucidum, Fornix, Gyrus parahippocampalis, Gyrus cinguli

Specimen showing the left half of the brain. The C-shaped structure below the gyri is the corpus callosum, which is the main collection of fibres connecting the two halves of the cerebrum.

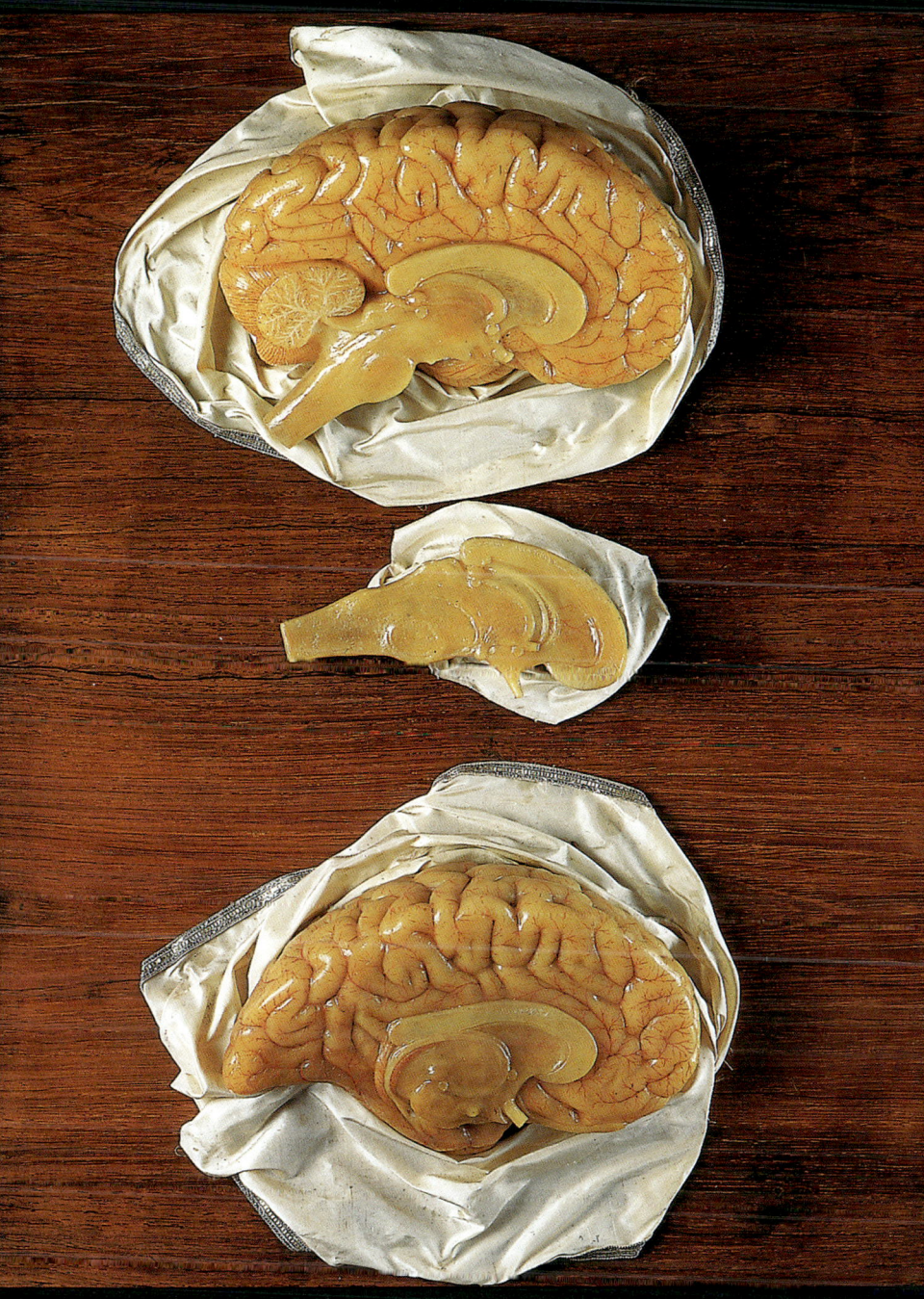

Telencephalon, Diencephalon, Mesencephalon, Rhombencephalon

View of the left half of a brain which has been divided in the midline. In the lower specimen
the brainstem and cerebellum have been removed

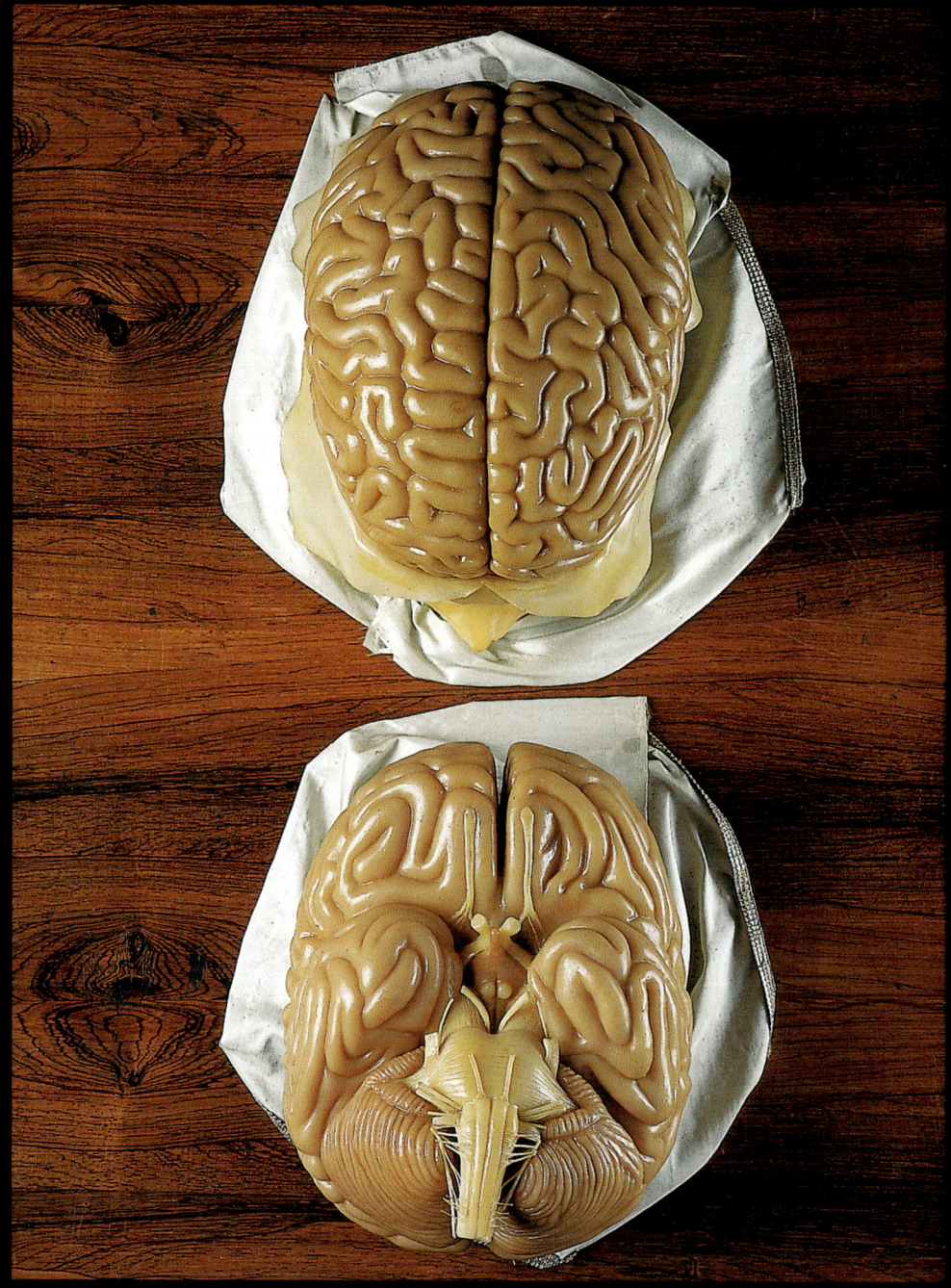

Facies superior et inferior cerebri

View of the brain from above to show the right and left halves,
and from below to show the brainstem and cerebellum

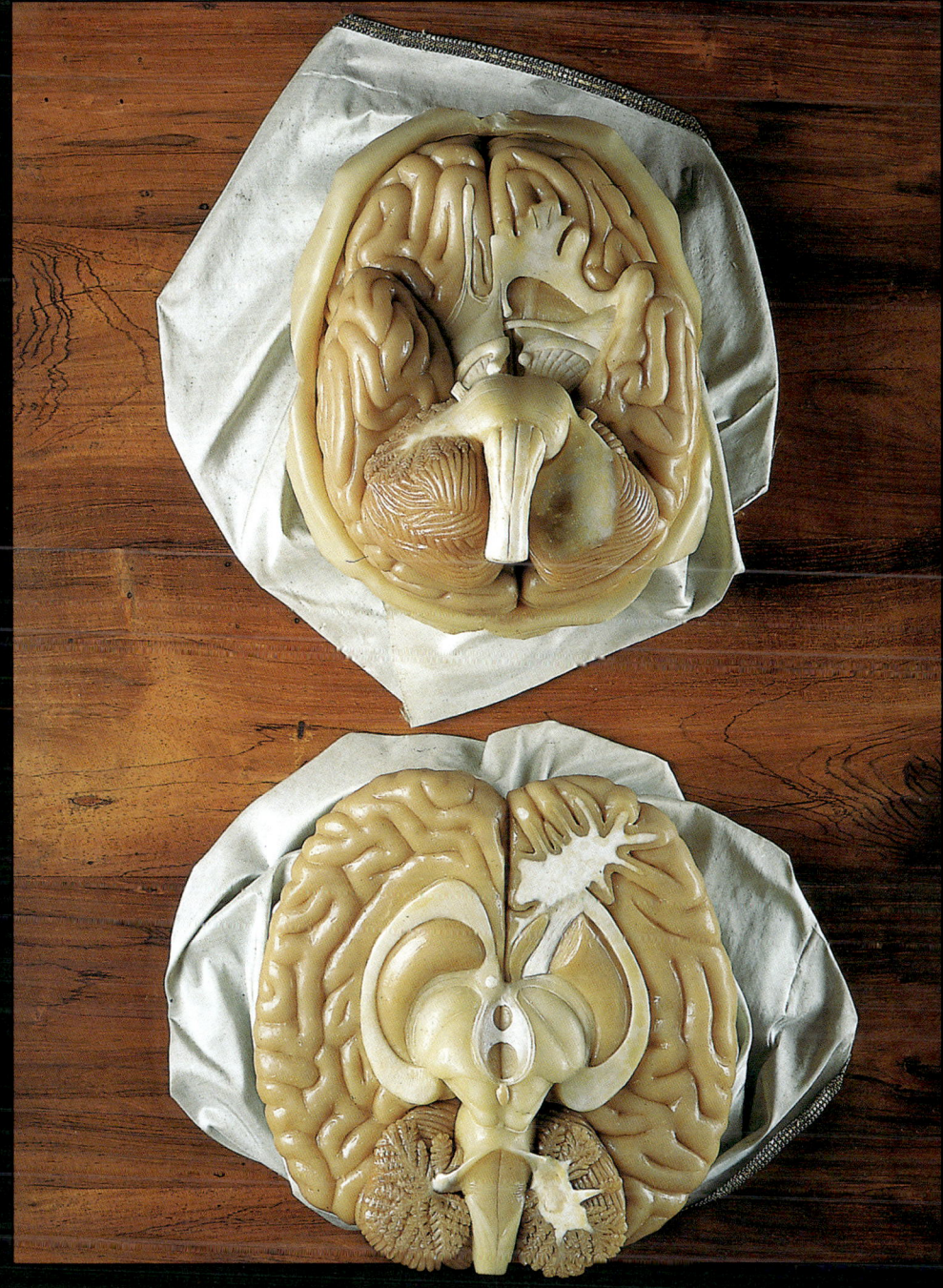

Commissura anterior, Corpus callosum, Nucleus caudatus, Ventriculus tertius cerebri, Thalamus, Glandula pinealis, Lamina quadrigemina, Ventriculus quartus

Specimens showing the white substance of the brain. In the lower specimen the halves of the brain have been separated to show the topography of the commissures in relation to the thalamus and caudate-putamen

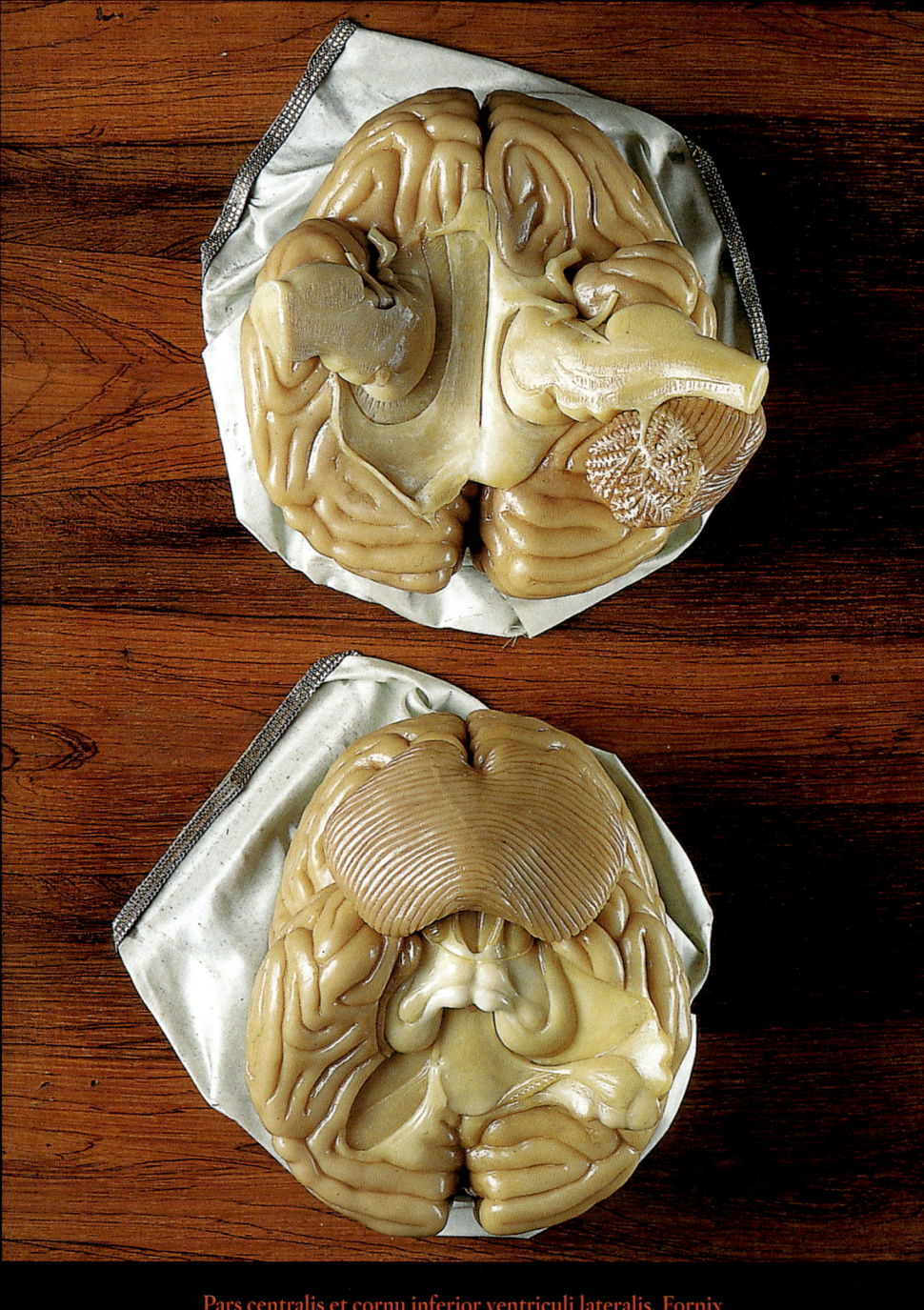

Pars centralis et cornu inferior ventriculi lateralis, Fornix

Specimens to show the cavities of the brain (ventricles)

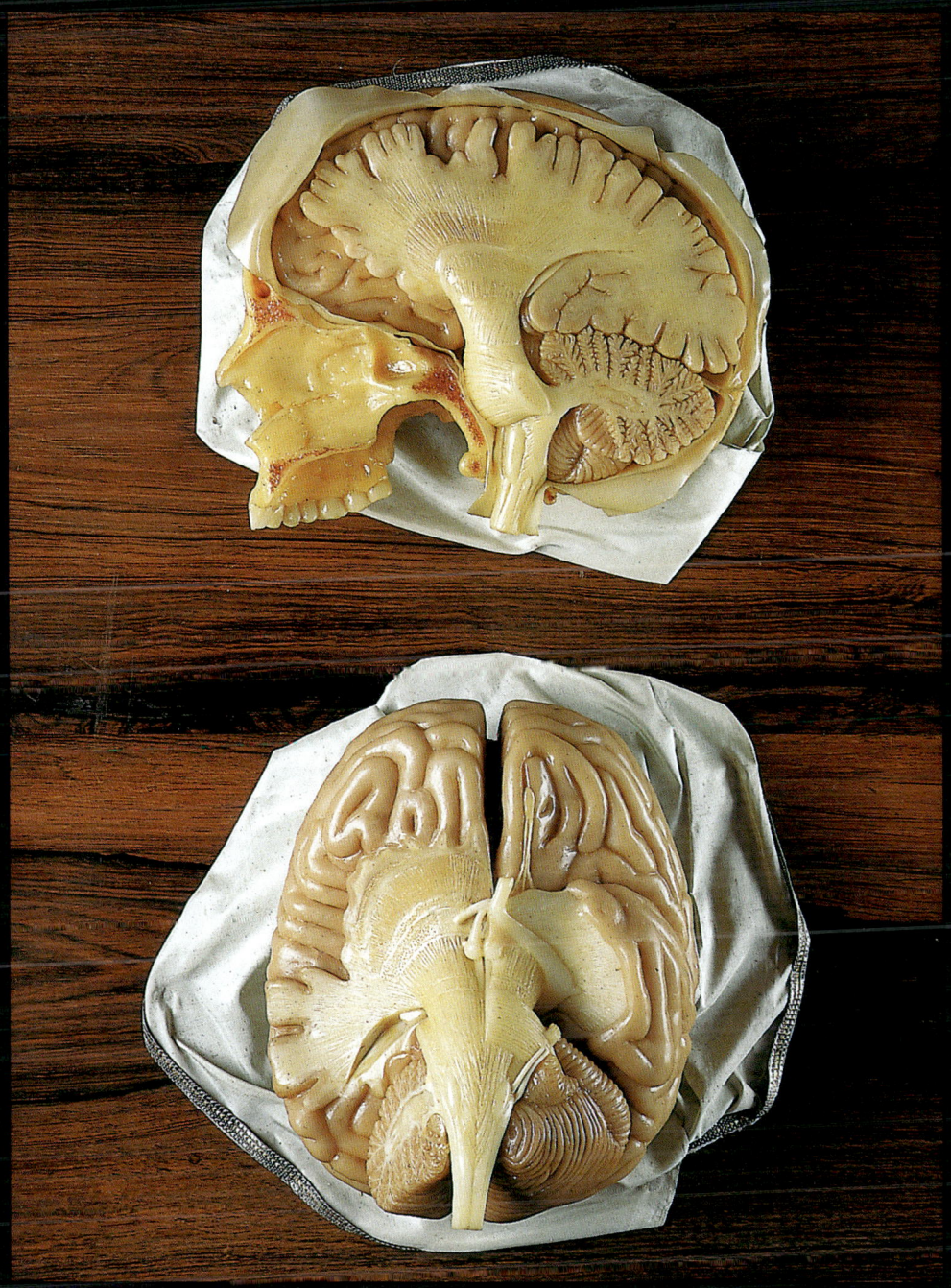

Capsula interna, Pedunculi cerebri, Tractus opticus, Radiatio thalami, Fornix
Display of the nerve fibres between the cerebrum, brainstem and spinal cord

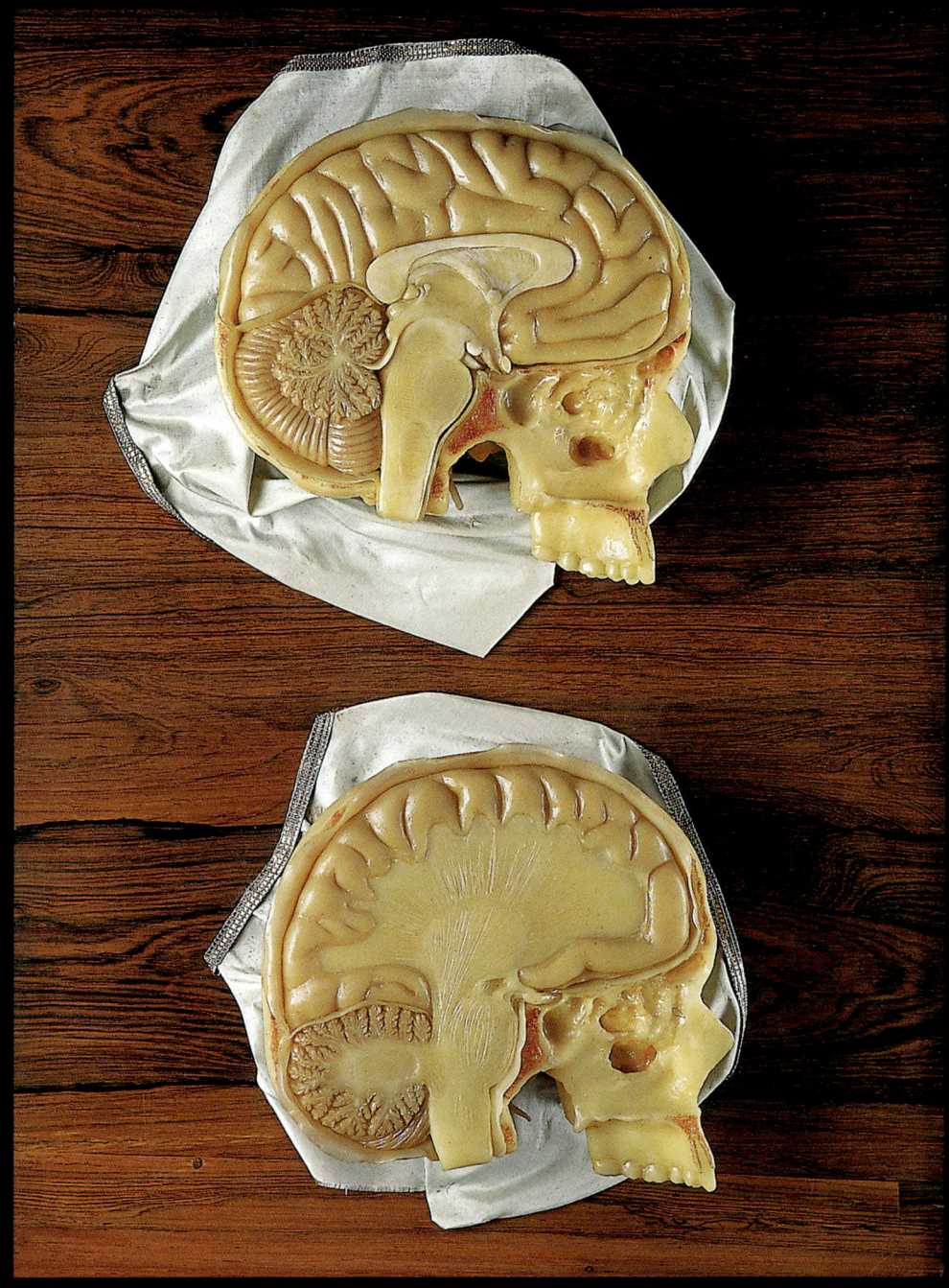

Corpus callosum, Fornix, Septum pellucidum,
Commissura anterior et posterior, Tentorium cerebelli

Median (above) and paramedian (below) sections through the head

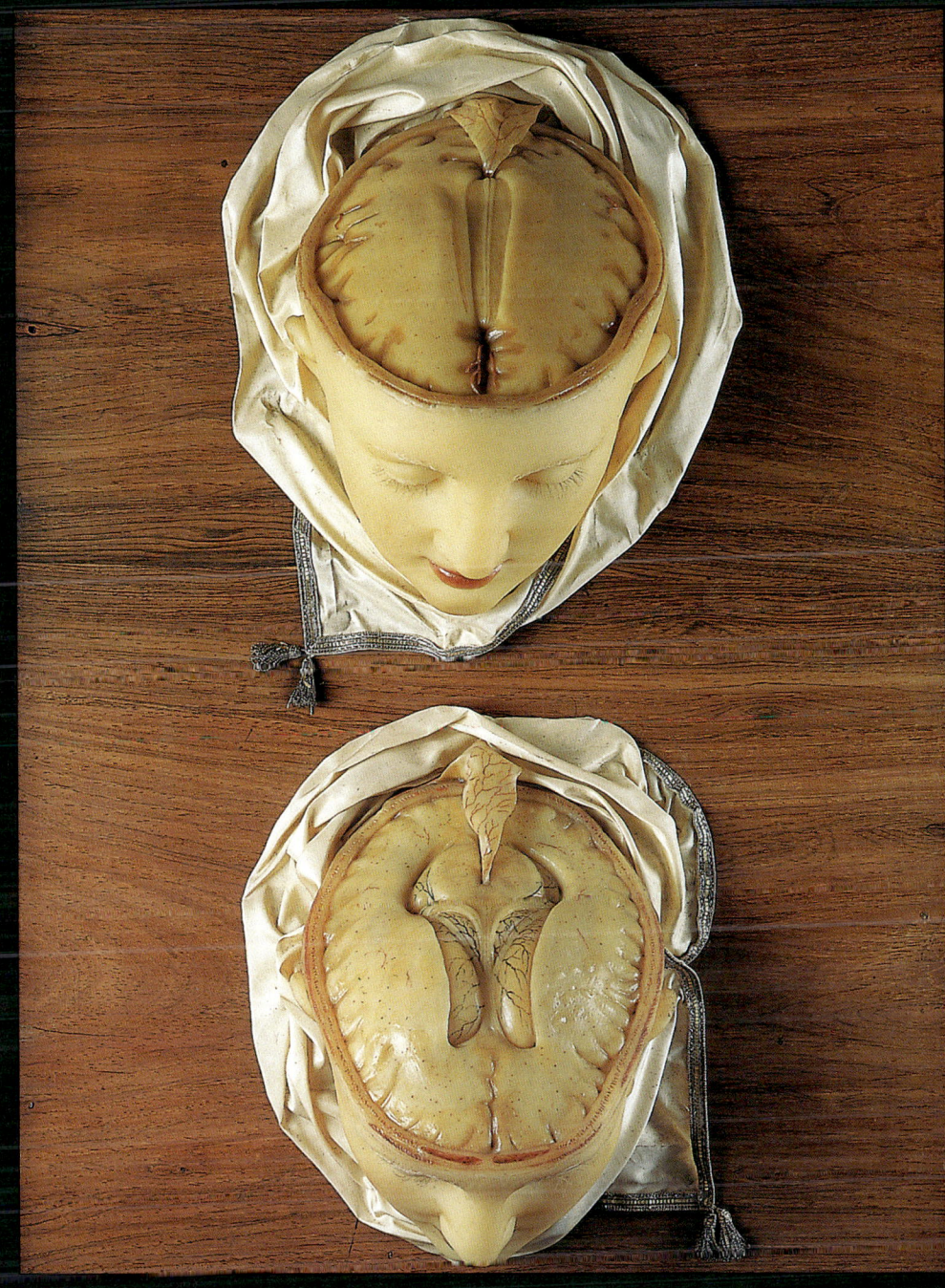

Nucleus caudatus, Striae longitudinales, Centrum ovale, Ventriculi laterales

Horizontal sections through the skull a little above the corpus callosum (upper specimen) and
a little below it (lower specimen). The lower section passes through the ventricles and the caudate nucleus

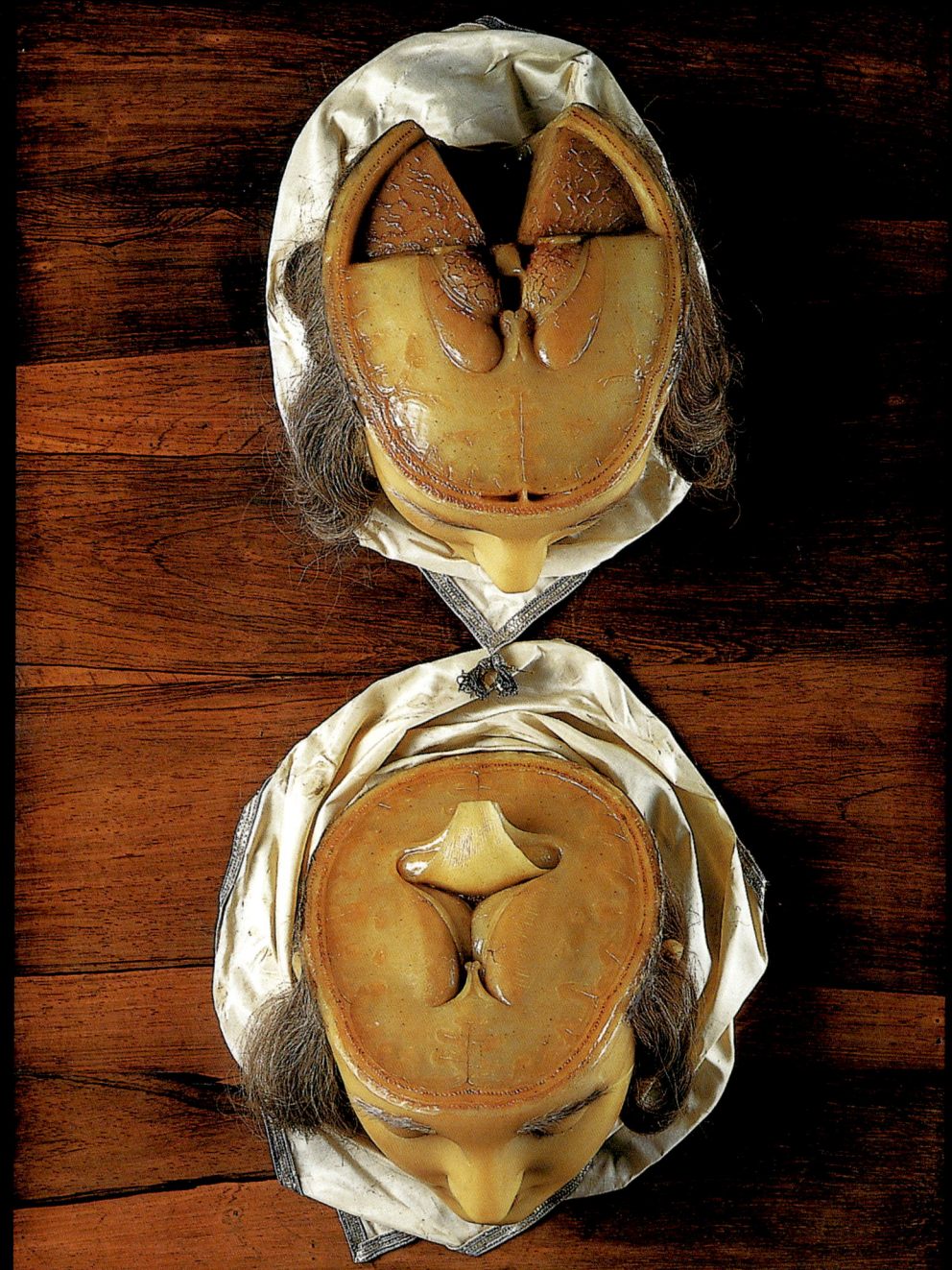

Thalamus, Nucleus caudatus, Ventriculus tertius, Ventriculi laterales

Horizontal sections at various levels through the skull. In the upper specimen
one can see the caudate nucleus, and adjacent to and directly behind it, the thalamus

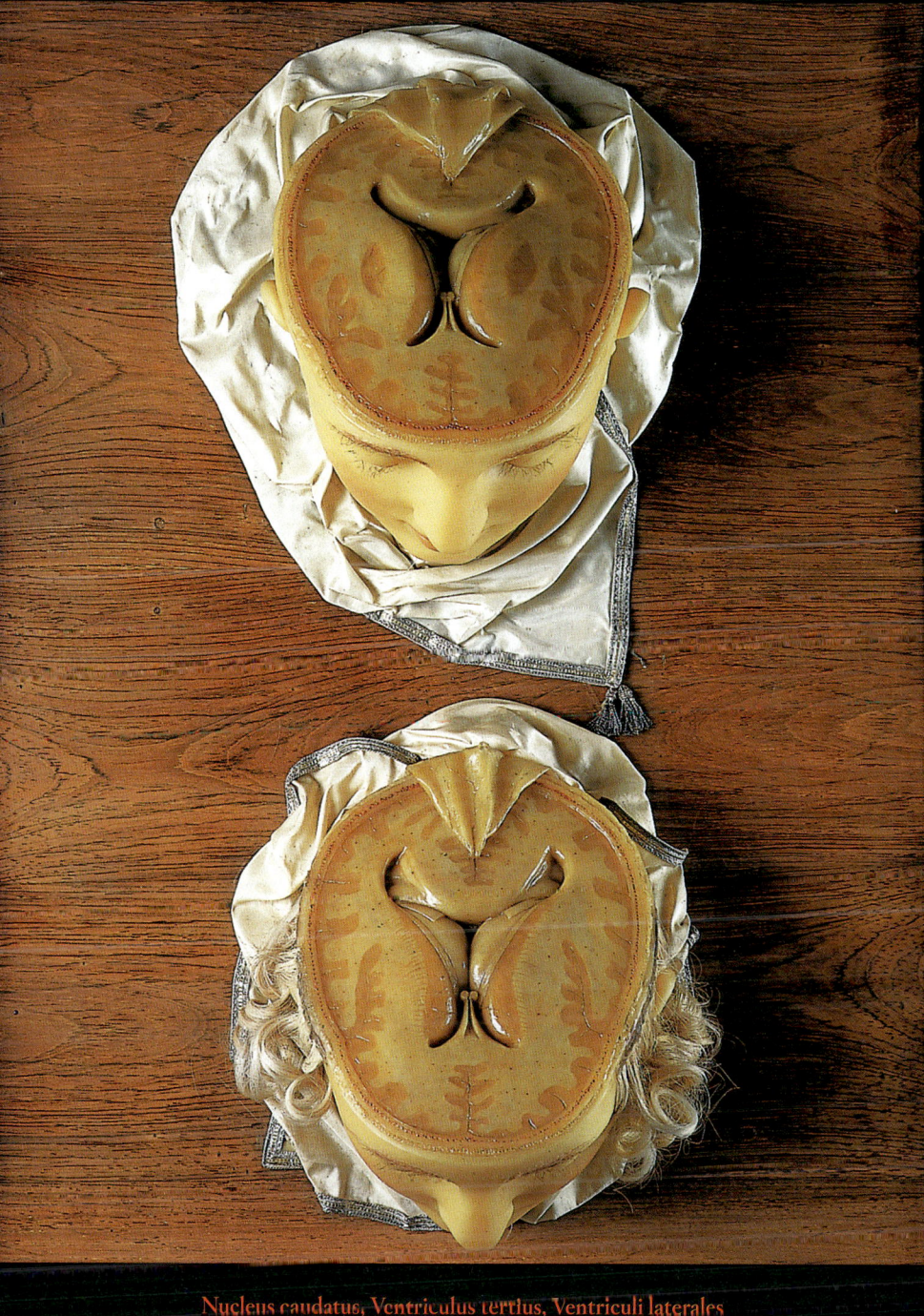

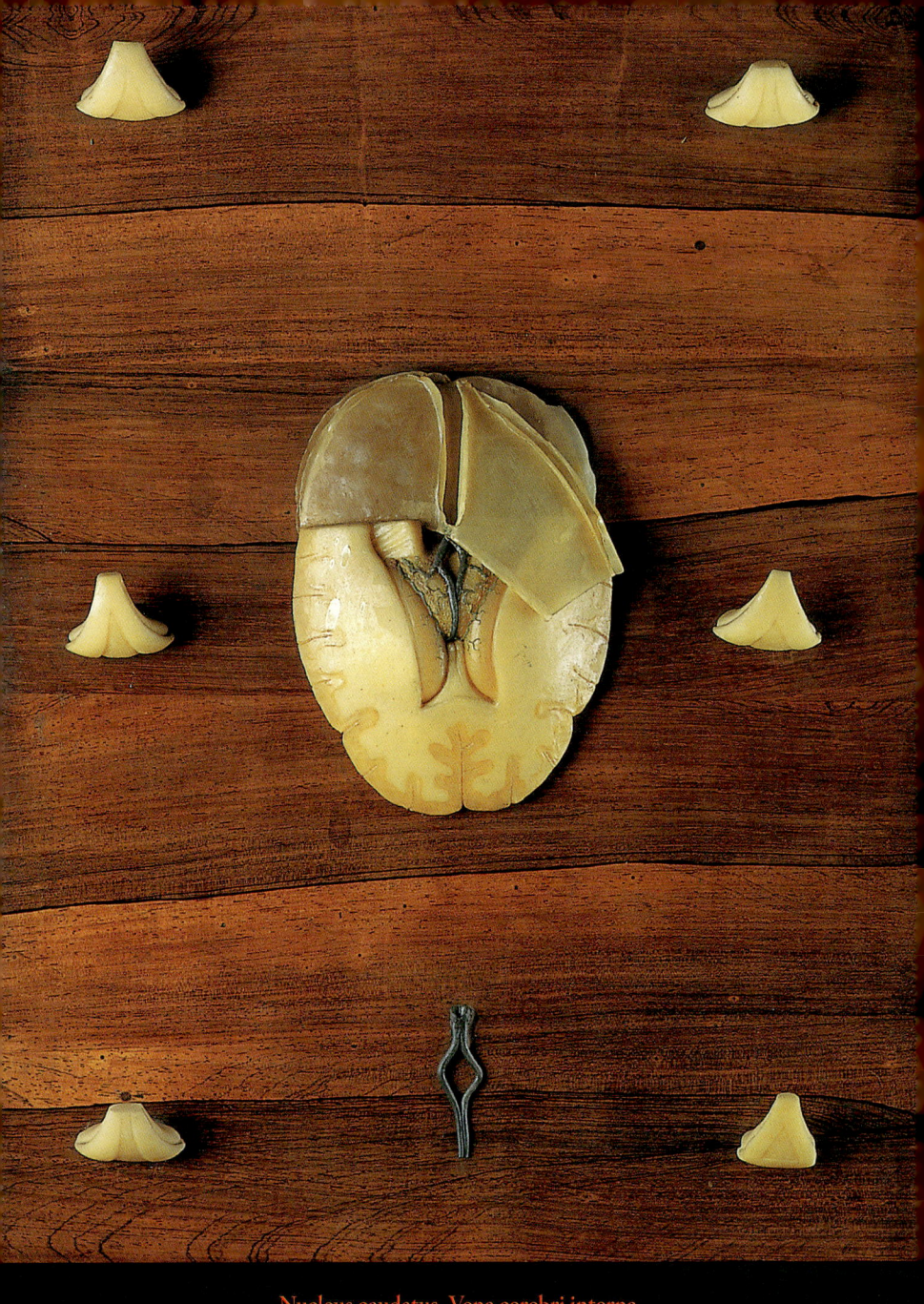

Nucleus caudatus, Vena cerebri interna

Horizontal section through the brain (frontal pole directed downwards).
Caudate nucleus and internal cerebral vein seen from above.

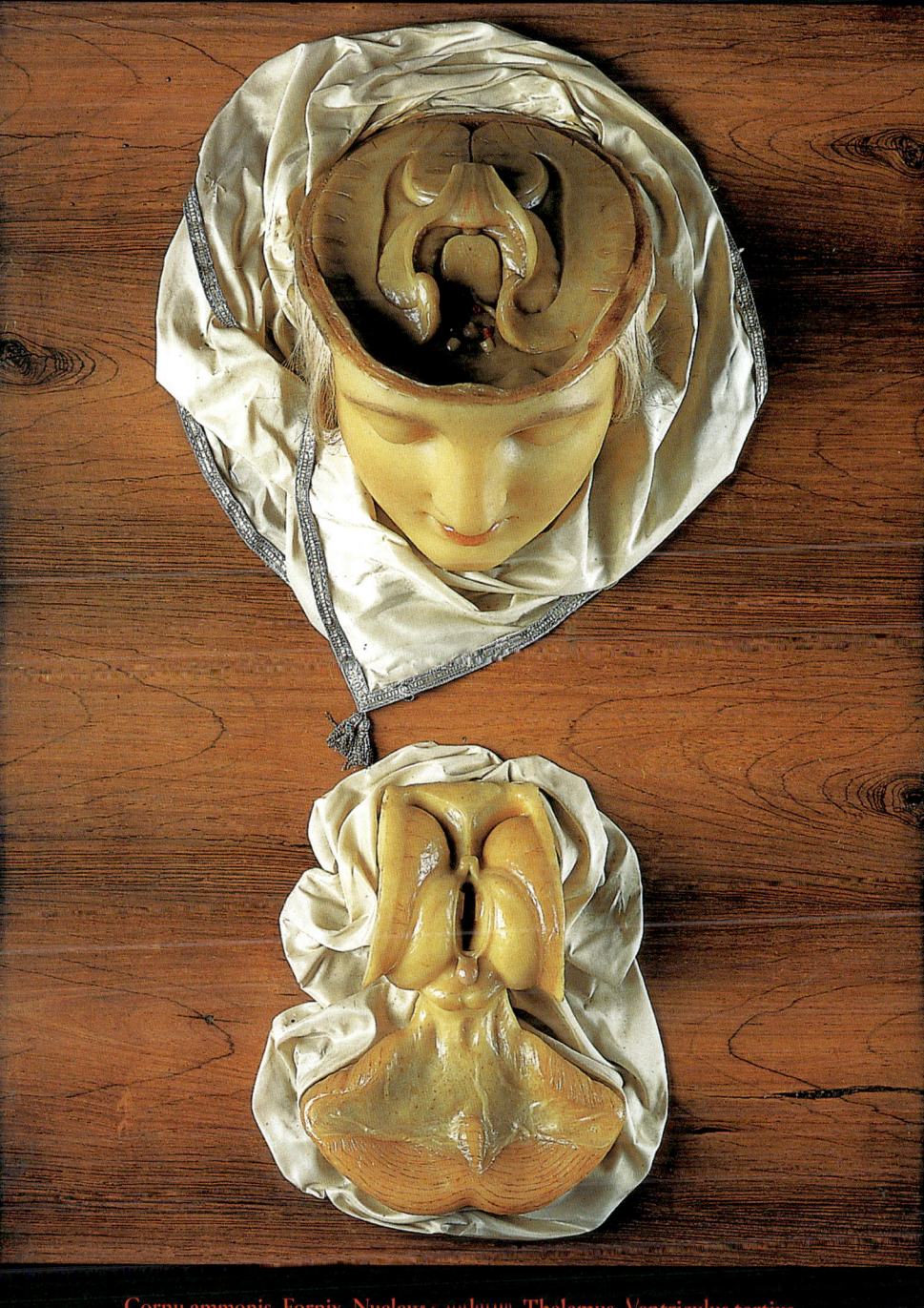

Cornu ammonis, Fornix, Nucleus caudatus, Thalamus, Ventriculus tertius,
Glandula pinealis, Lamina quadrigemina

Upper specimen: hippocampus. The lower specimen shows the centrally placed
cerebral nuclei such as the caudate nucleus and thalamus

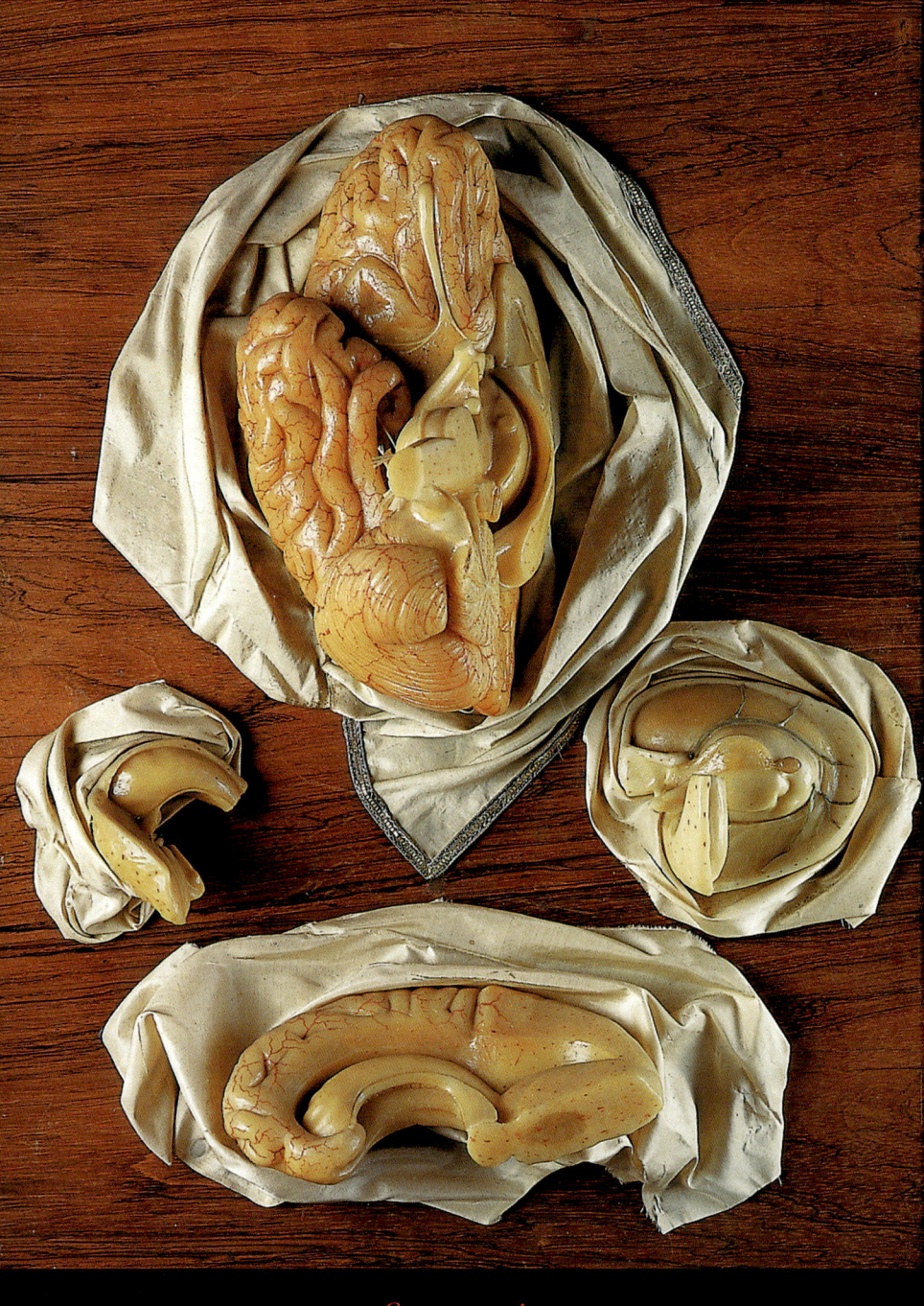

Cornu ammonis

Various views of the hippocampus

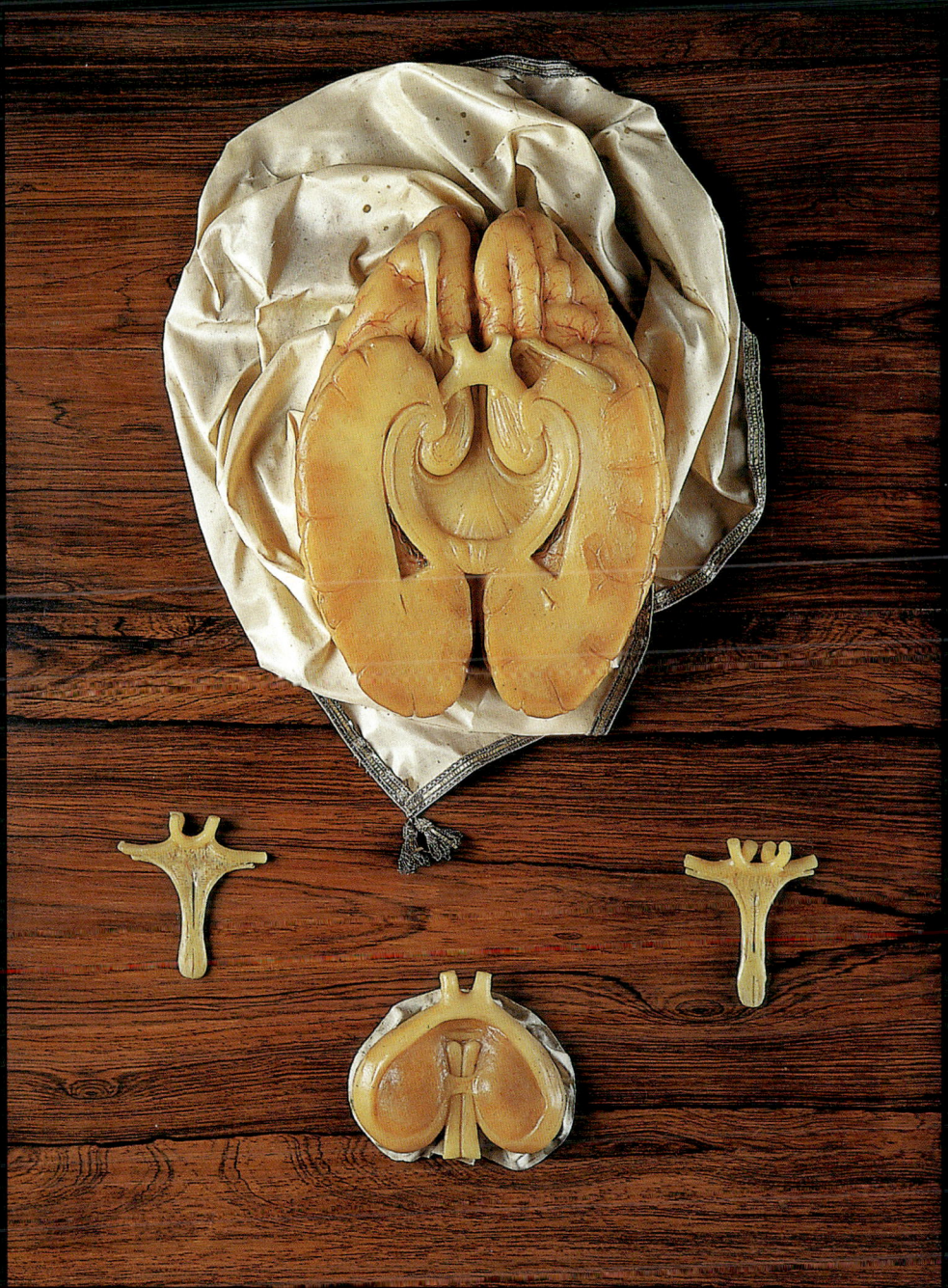

Fornix, Tractus diagonalis Broca, Chiasma opticum, Tractus opticus

Specimens showing the connections (nerve fibres) of the hippocampus

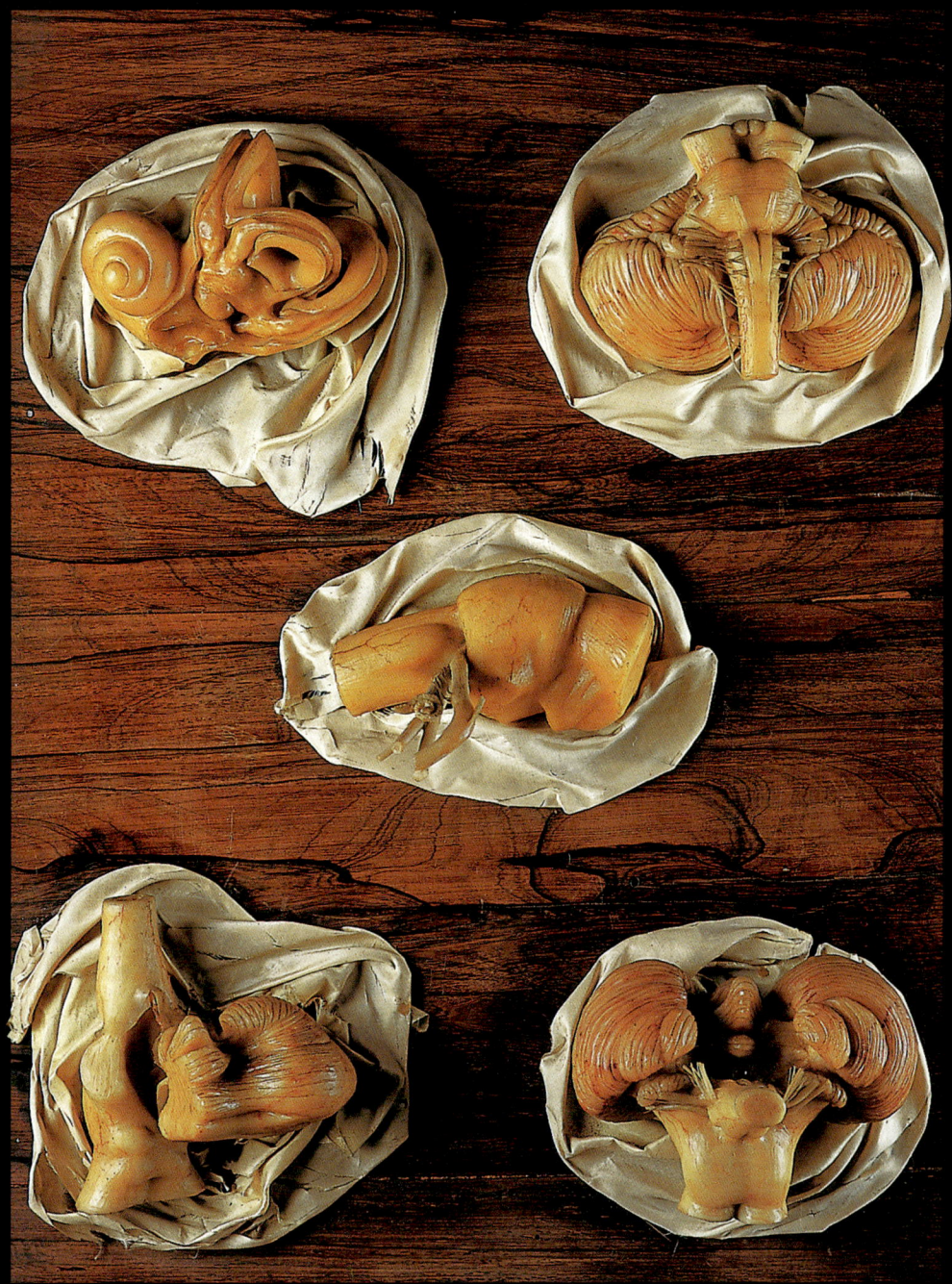

Cochlea, Nervus vestibulo-cochlearis, Nervus facialis

View of the inner ear showing the bony labyrinth (cochlea and three semicircular canals, above left).
The other specimens represent the brainstem with cranial nerves and the cerebellum

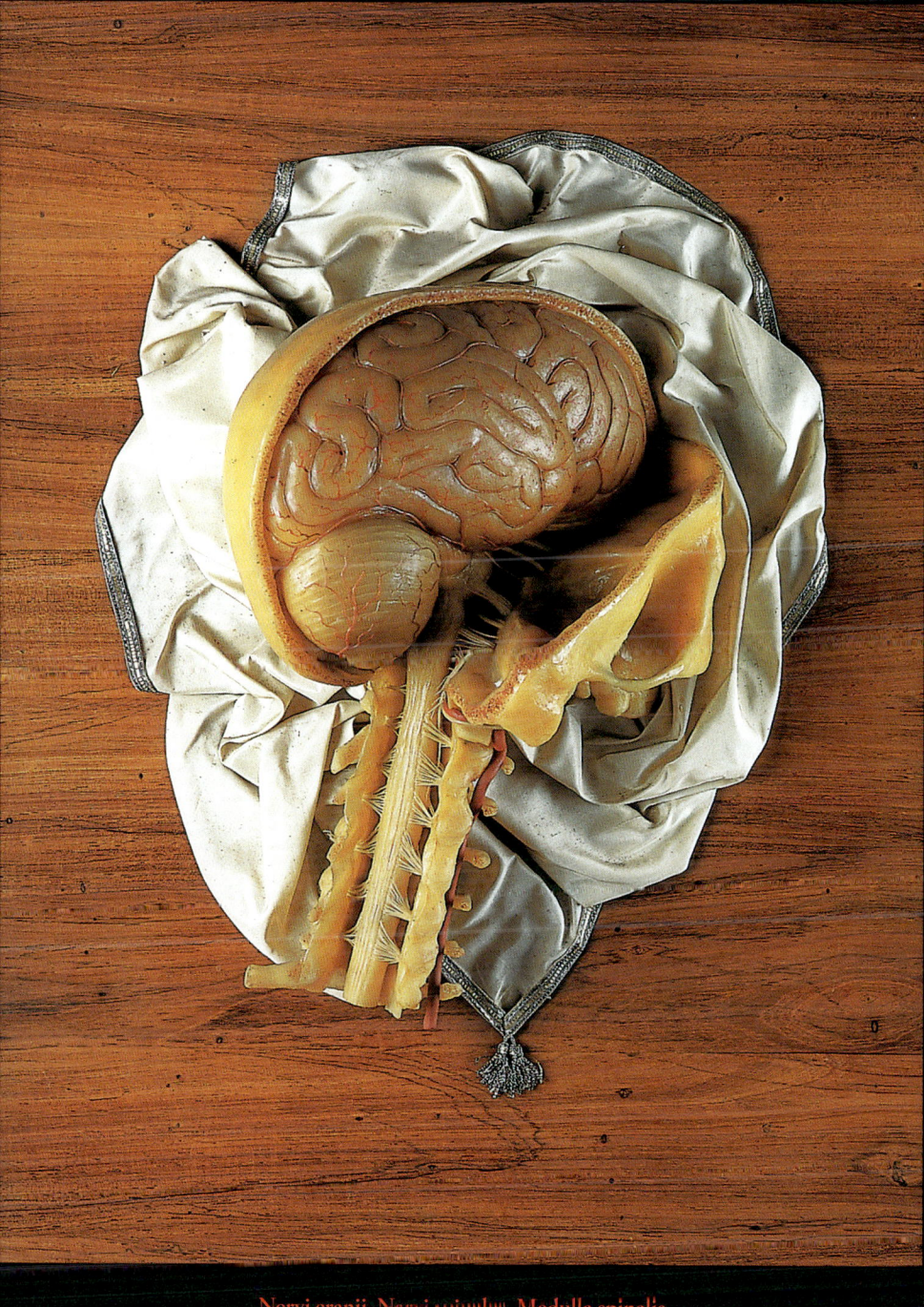

Oblique view from behind (the skull and vertebral canal have been laid open).
The cerebrum and cerebellum have been lifted up to reveal the cranial nerves emerging from the brainstem

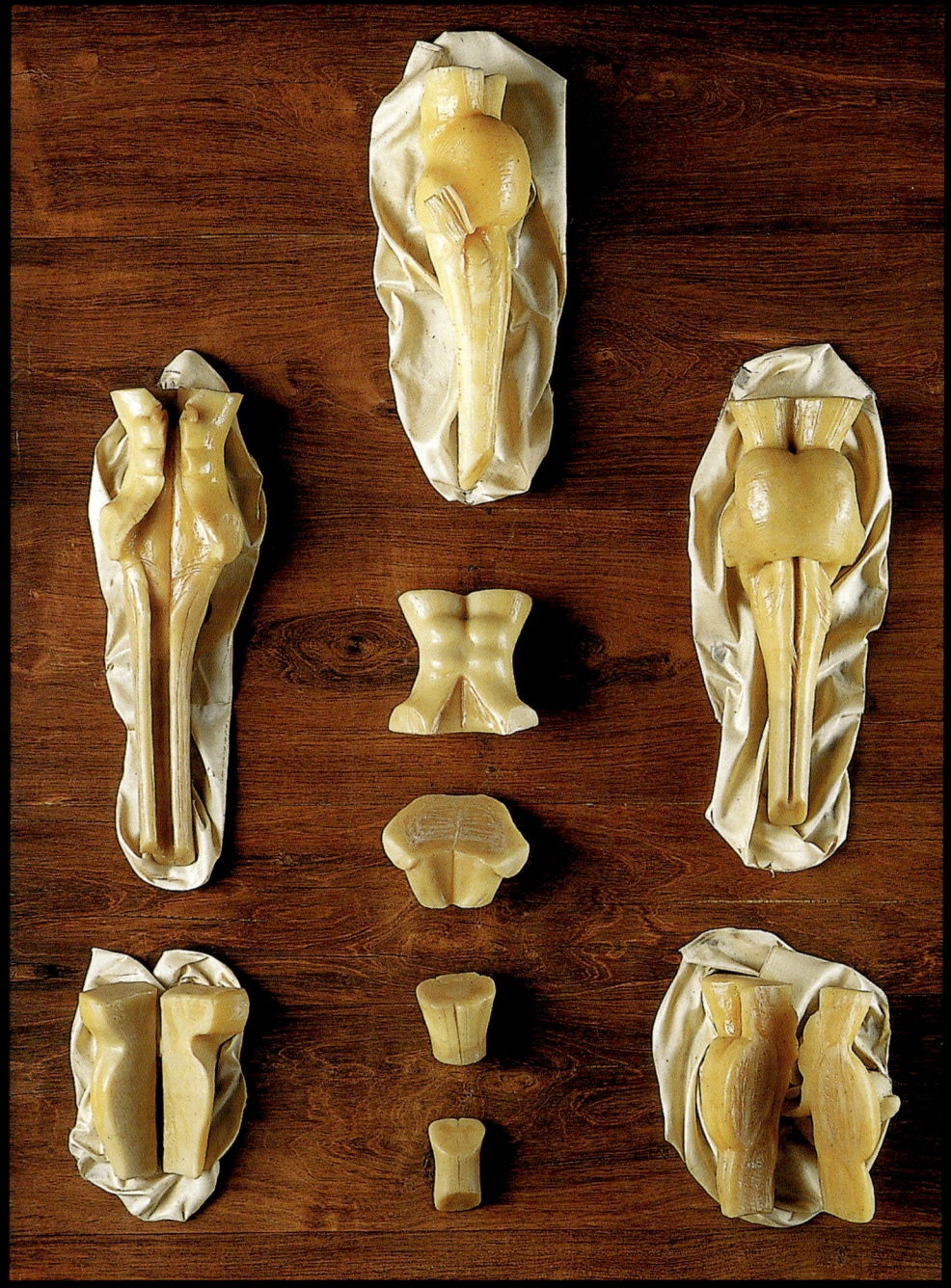

Medulla oblongata, Pons, Pedunculi cerebri
Various displays and views of the brainstem

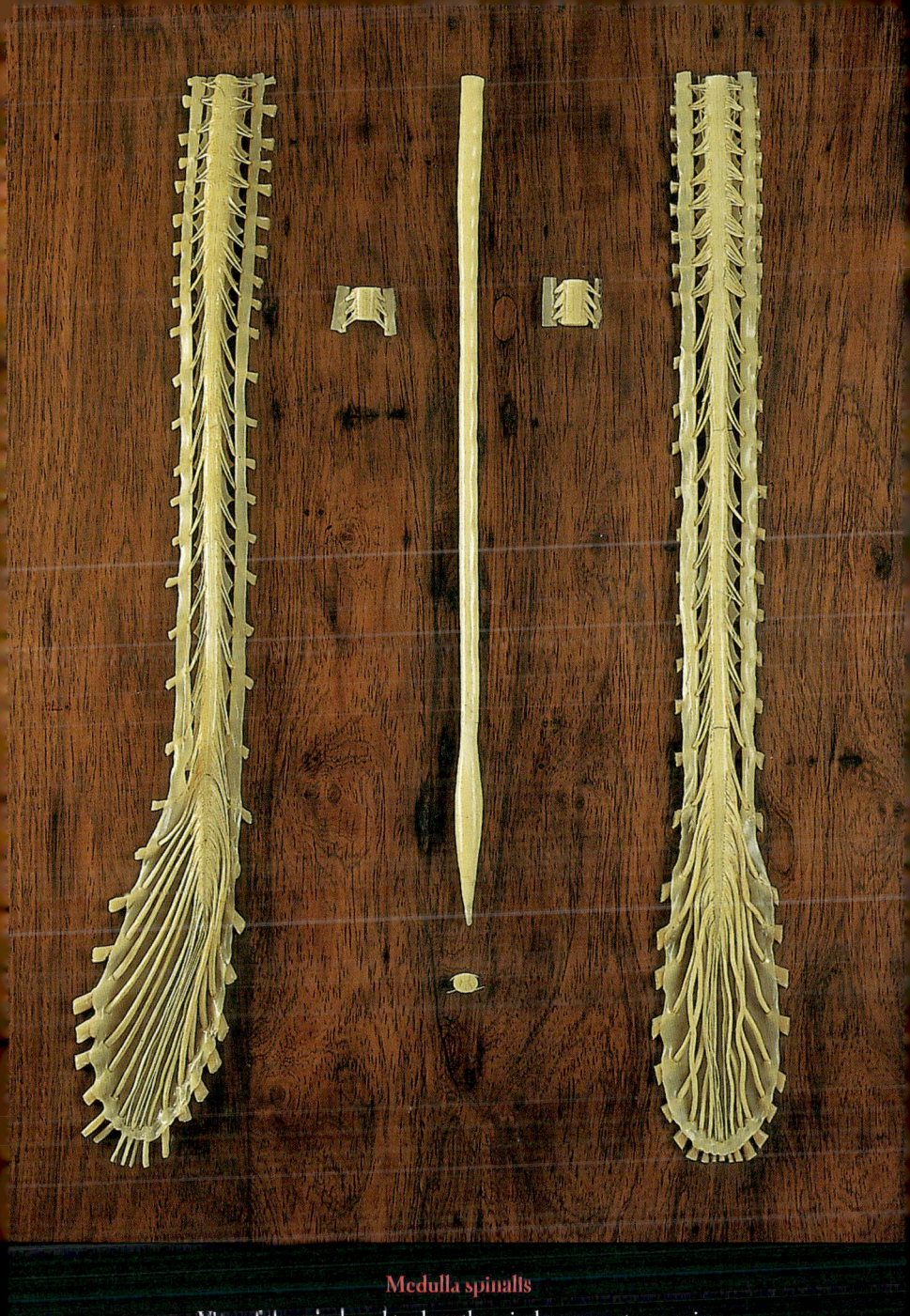

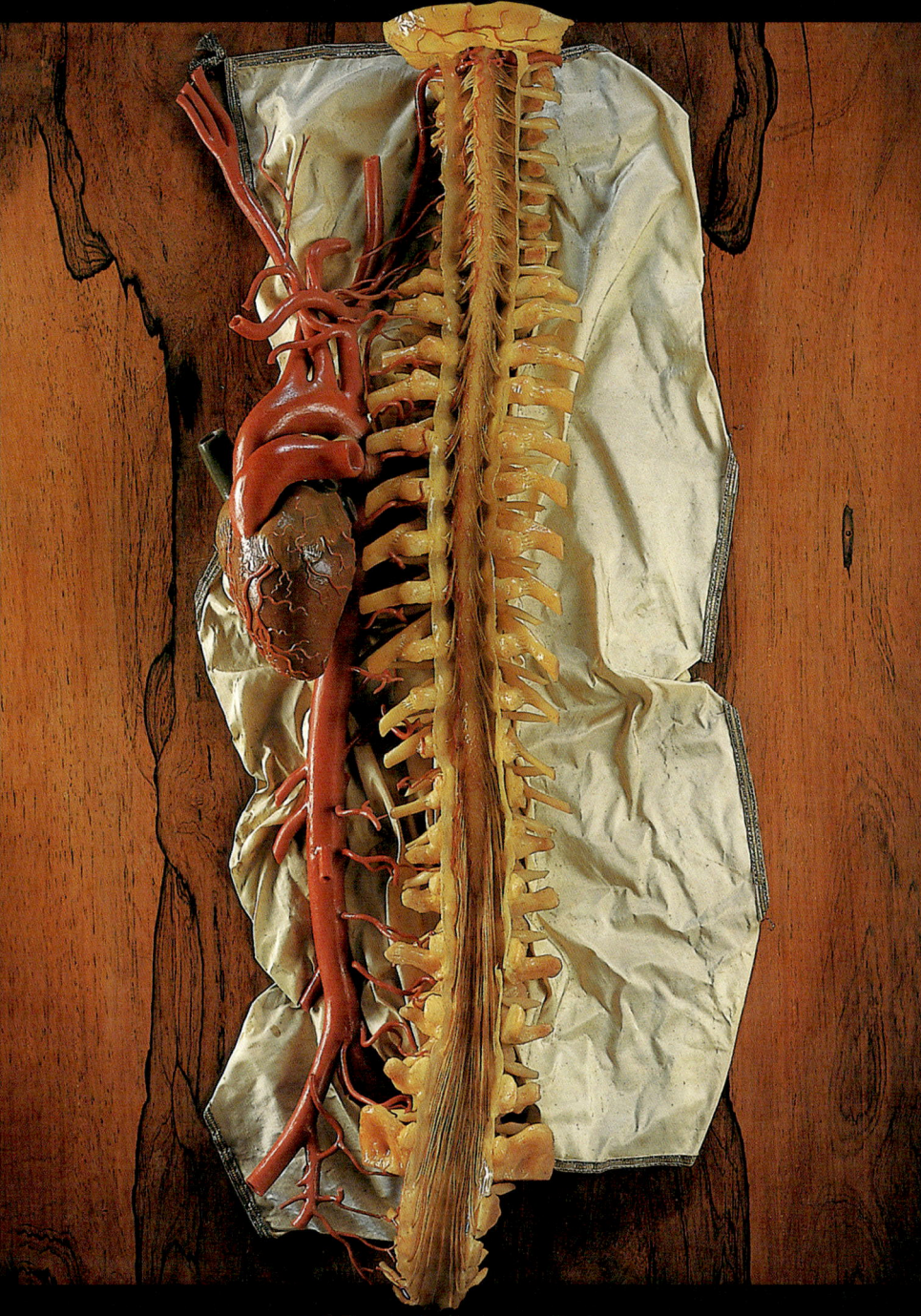

Medulla spinalis, Arteriae radiculares

The vertebral canal has been laid open from behind to show the spinal cord,
the spinal nerve roots and the supplying blood vessels

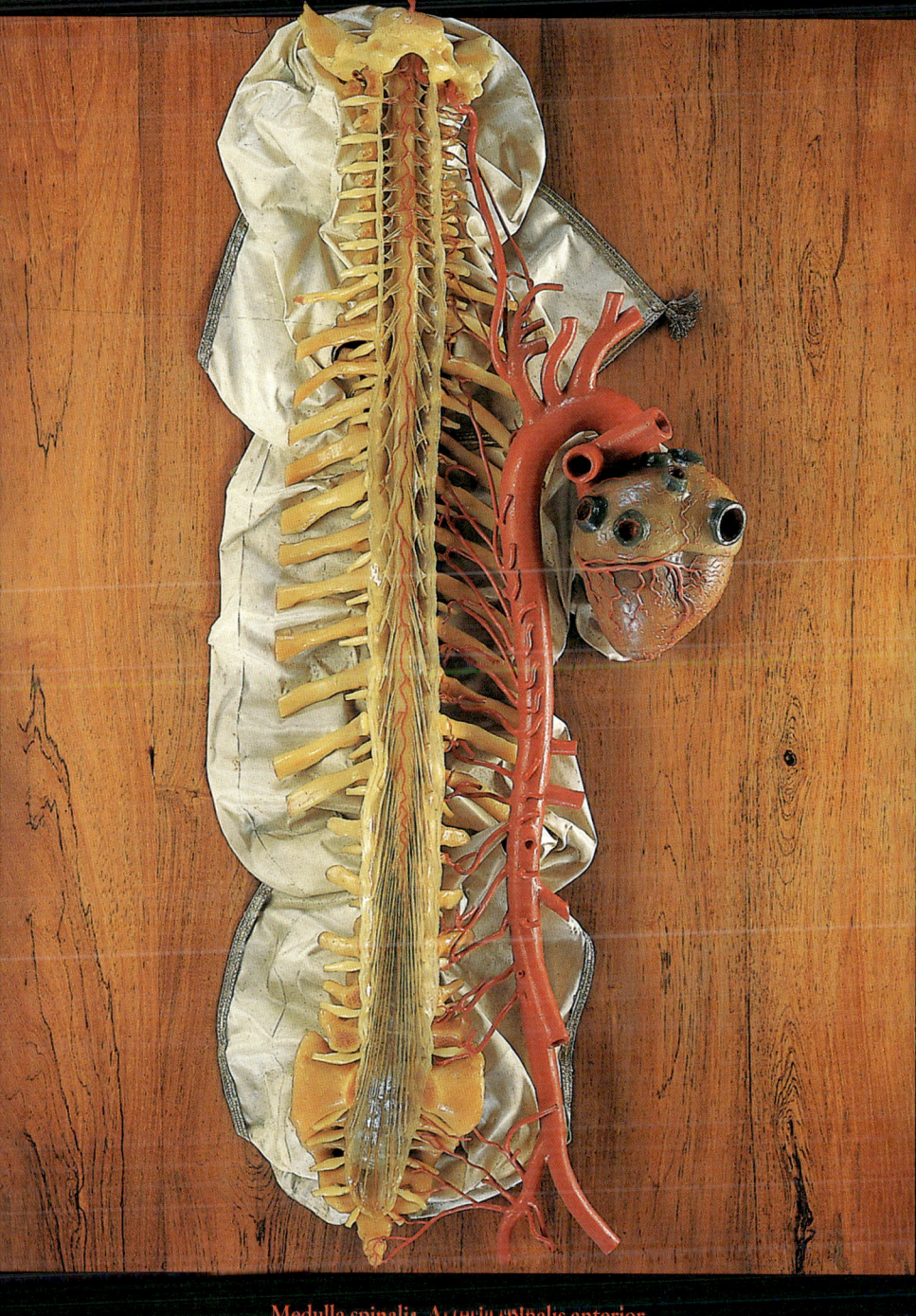

Medulla spinalis, Arteria spinalis anterior

The vertebral canal has been laid open from in front to show the spinal cord,
emerging spinal nerve roots and blood vessels

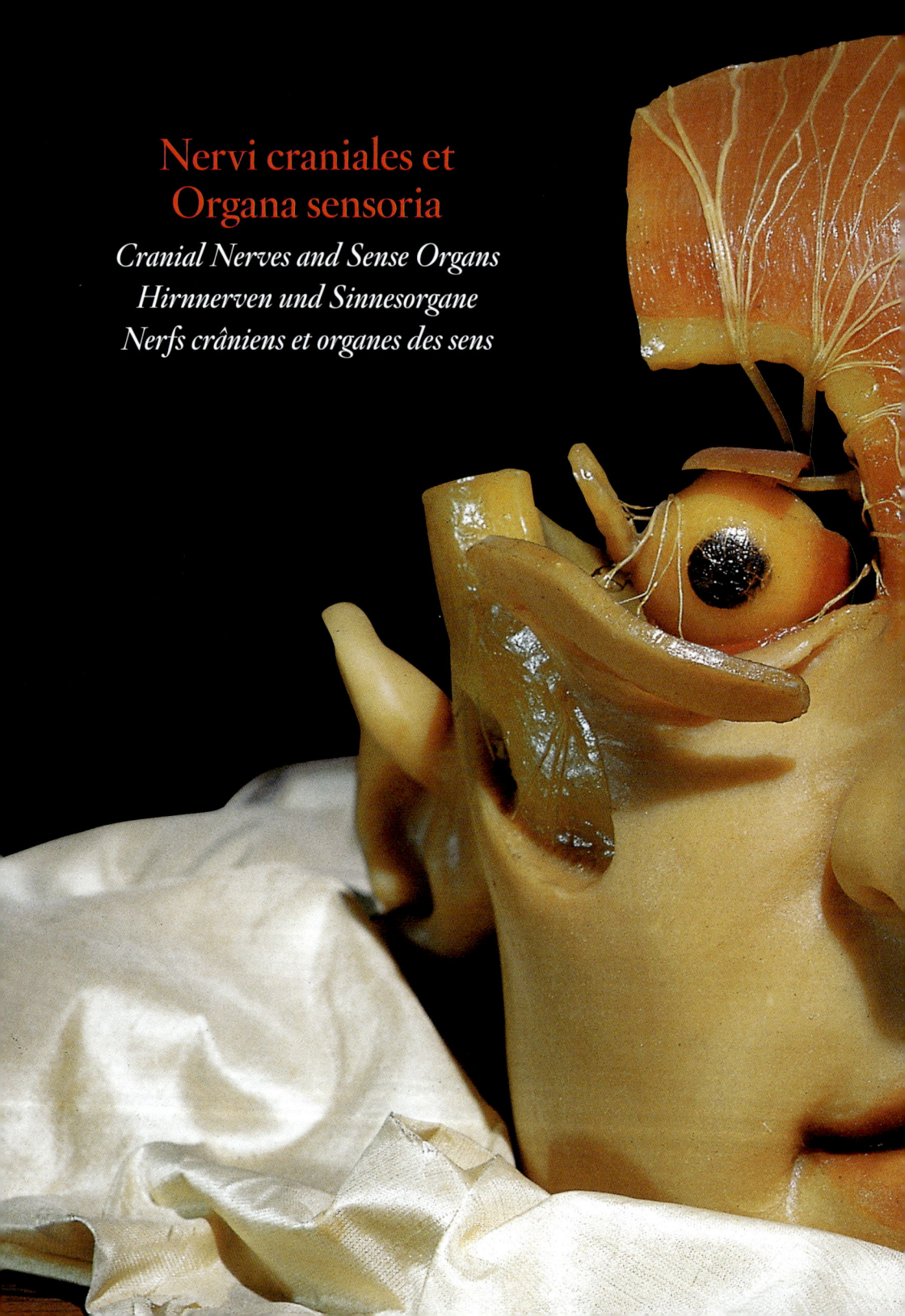

Nervi craniales et
Organa sensoria

Cranial Nerves and Sense Organs
Hirnnerven und Sinnesorgane
Nerfs crâniens et organes des sens

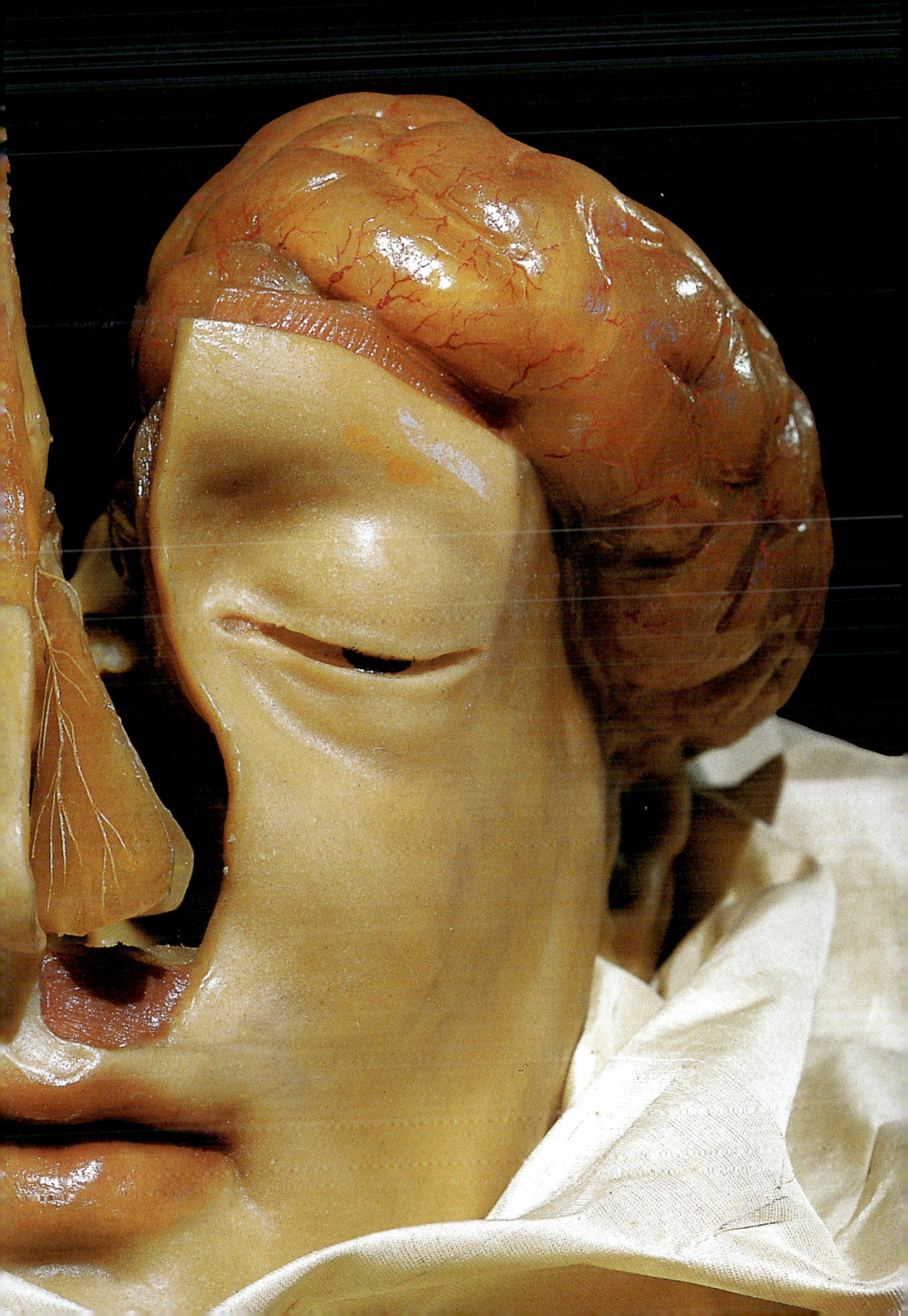

Cranial Nerves and Sense Organs

The sense organs responsible for our perception of stimuli awakened the interest of anatomists very early on, since such special senses as hearing, sight, smell and taste have naturally always been relevant to our conscious experience.

Whereas many of the models show the detailed anatomy of the individual sense organs, other specimens display them in relationship to their central connections with the brain. This type of display is in accordance with the ideas of René Descartes (1596–1650), who interpreted the function of these organs in connection with the brain. It is indeed an essential part of cerebral function to analyse and interpret the particular stimuli which act upon our sense organs.

Hirnnerven und Sinnesorgane

Unsere für die Reizwahrnehmung verantwortlichen Sinnesorgane haben früh das Interesse der Anatomen geweckt, da Sinneswahrnehmungen wie Hören, Sehen, Riechen und Schmecken in unserem Bewusstsein natürlicherweise gegenwärtig sind.

Während zahlreiche Modelle im Detail die Anatomie der einzelnen Sinnesorgane dokumentieren, wird in anderen Präparaten das Sinnesorgan mit seiner nervalen Verbindung zum Gehirn gezeigt. Die Art der Darstellung lehnt sich an das Konzept von René Descartes (1596–1650) an, der die Funktionen der Sinnesorgane im Zusammenhang mit dem Gehirn interpretiert hatte. Dem Gehirn kommen dabei wesentliche Aufgaben bei der Verarbeitung der spezifischen Reize, die auf unsere Sinnesorgane einwirken, zu.

Nerfs crâniens et organes des sens

Les organes des sens, capteurs des stimulations périphériques, ont intéressé précocement les anatomistes car la perception de l'environnement par l'audition, la vue, l'odorat ou le goût est une des bases de la conscience.

L'anatomie des organes des sens isolés est démontrée en détail par de nombreux modèles en cire. Certains modèles montrent les connexions nerveuses des organes des sens avec le cerveau ; cette représentation s'appuie sur les conceptions de René Descartes (1596–1650) intégrant les fonctions des organes des sens à celles du cerveau.

Pages 336–337:
Frontal view of the specimen from page 352

Detail from page 441

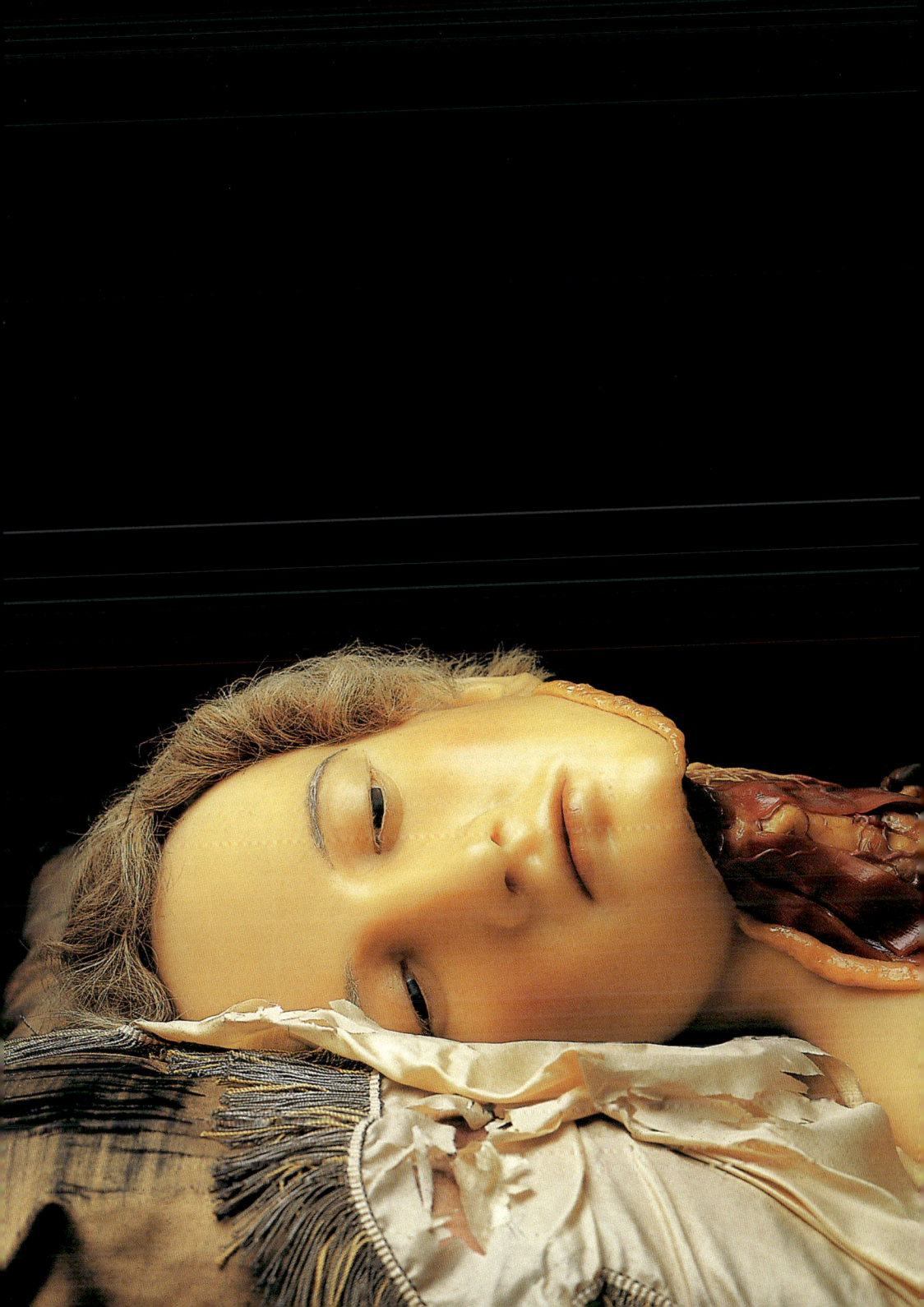

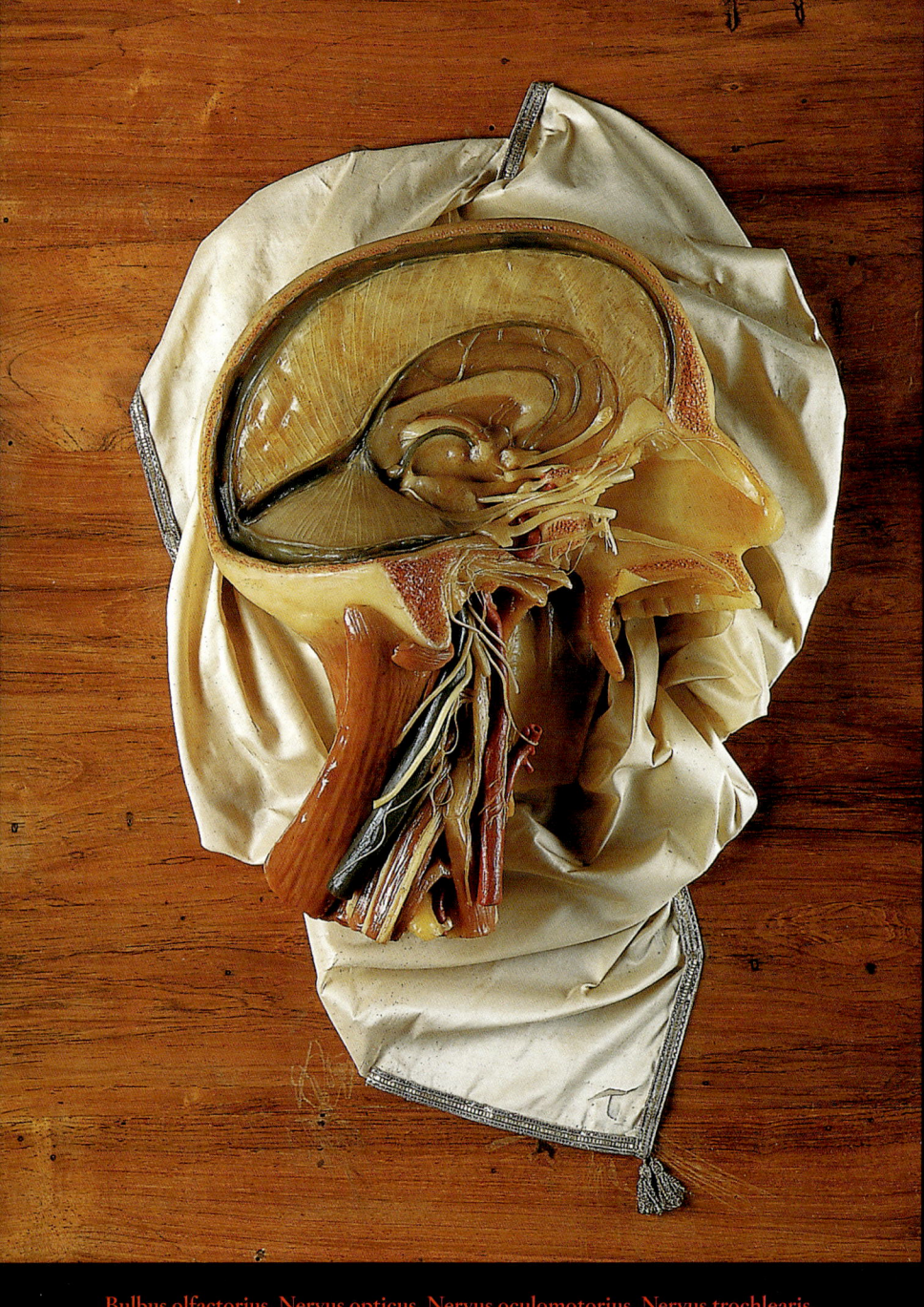

Bulbus olfactorius, Nervus opticus, Nervus oculomotorius, Nervus trochlearis,
Nervus trigeminus, Nervus abducens

Side view of the skull laid open to display the individual cranial nerves. The bulbous thickening right at the top is
part of the olfactory system. Below: the nerves supplying the eye muscles and branches of the trigeminal nerve

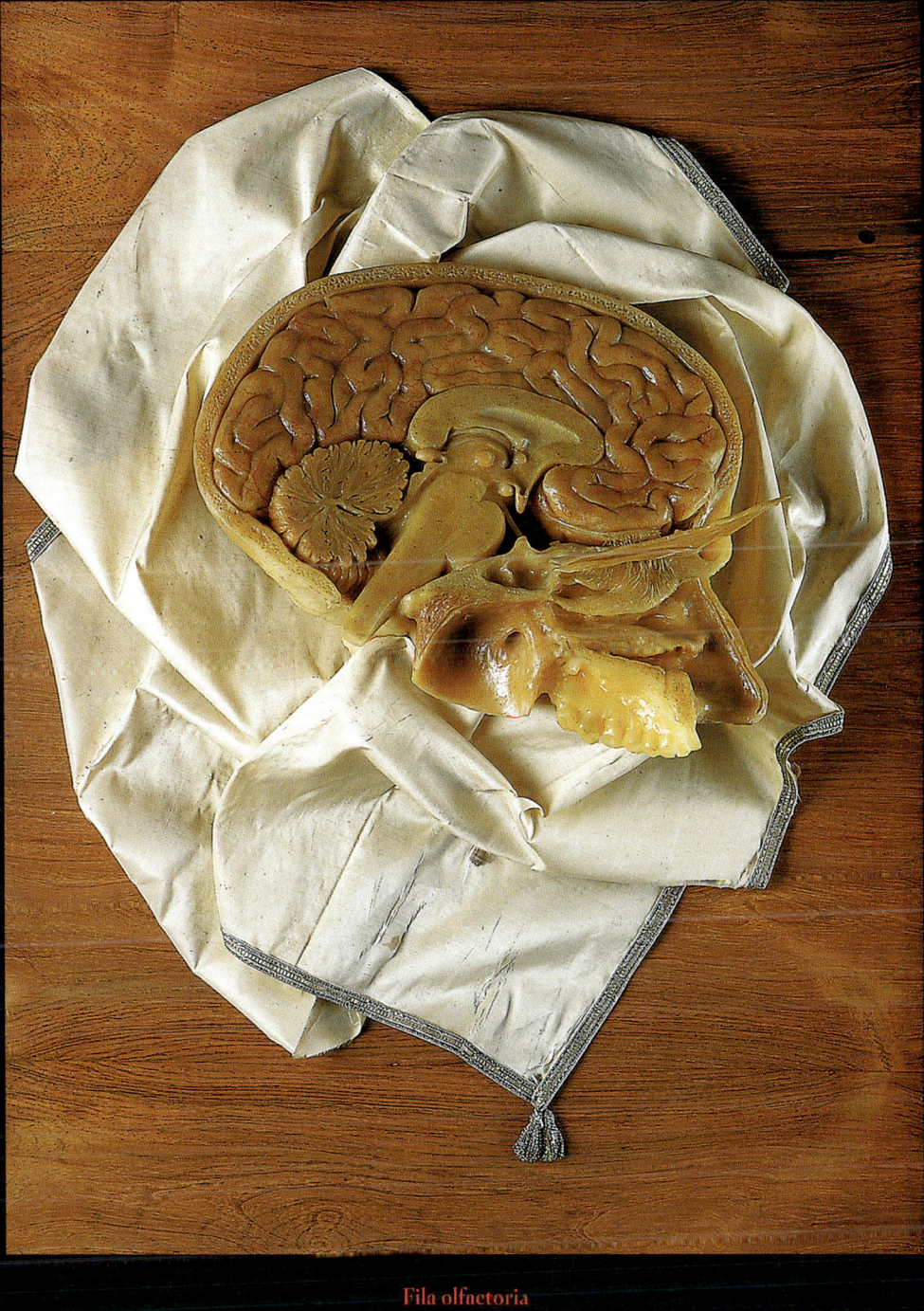

Fila olfactoria

A head bisected in the midline to show the olfactory fibres

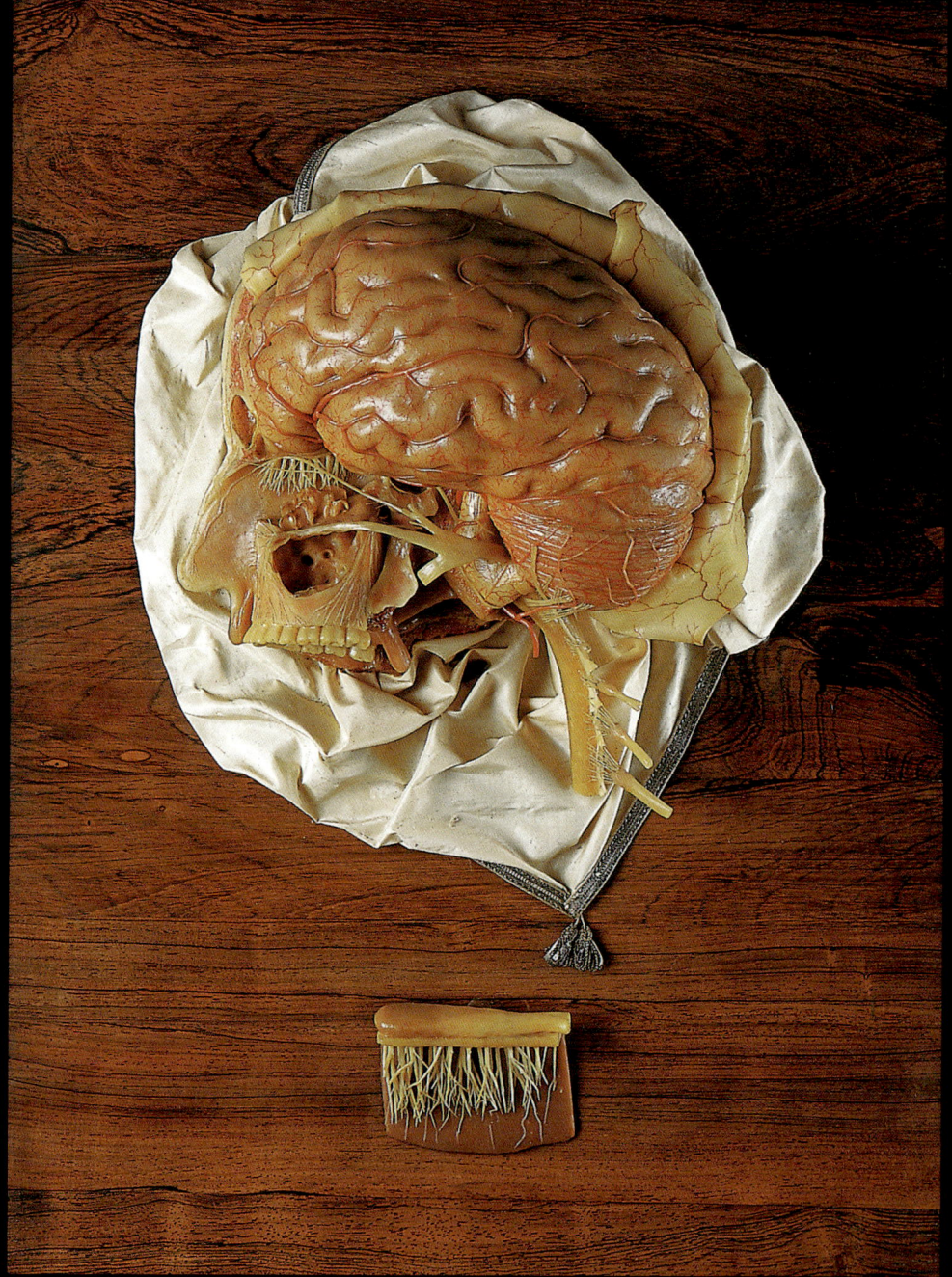

Fila olfactoria, Nervus trigeminus

The olfactory nerve fibres and the trigeminal nerve. The numerous olfactory fibres arise in the
mucous membrane of the nose and enter the cranial cavity. Below: they are shown enlarged

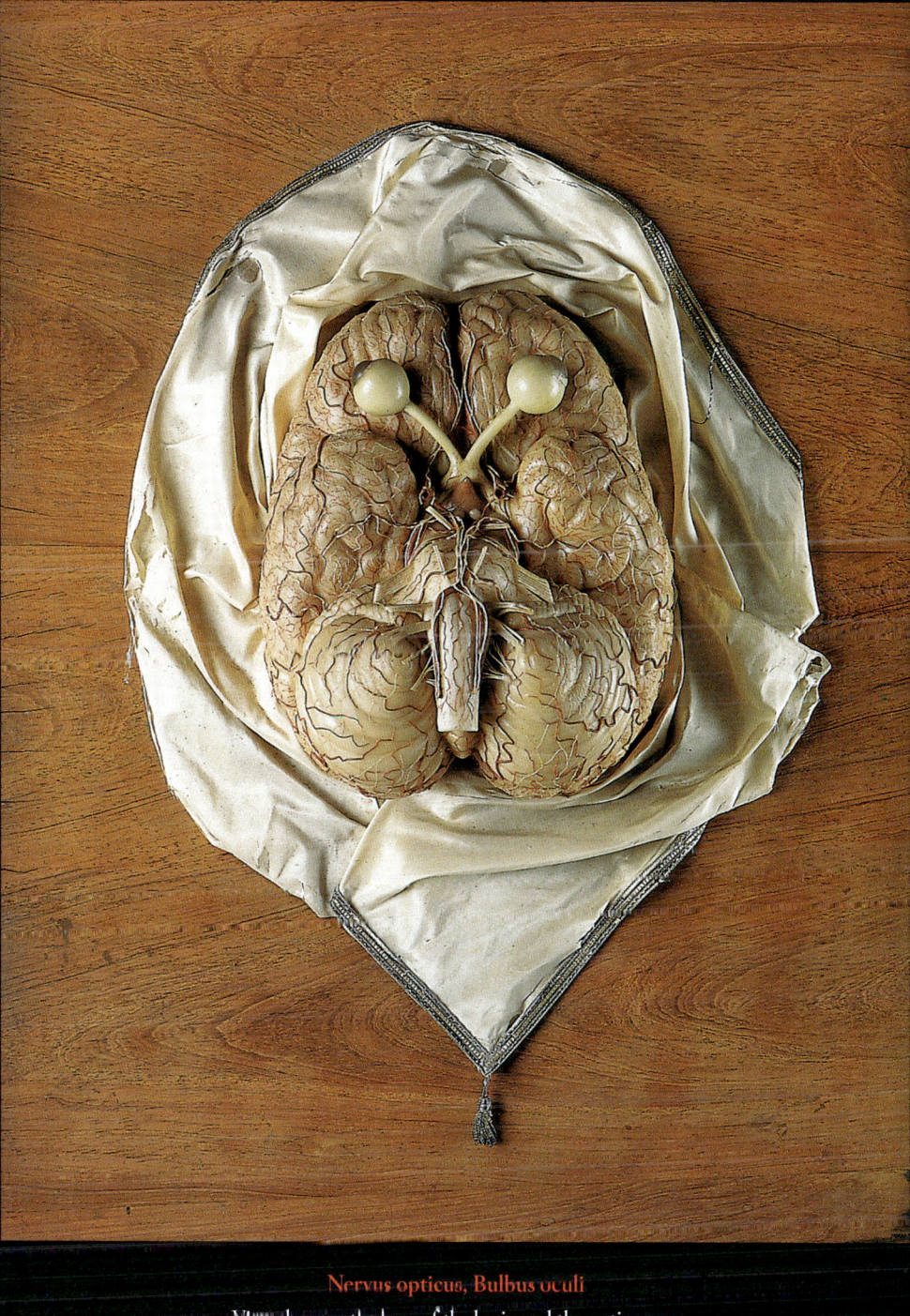

View showing the base of the brain and the optic nerve

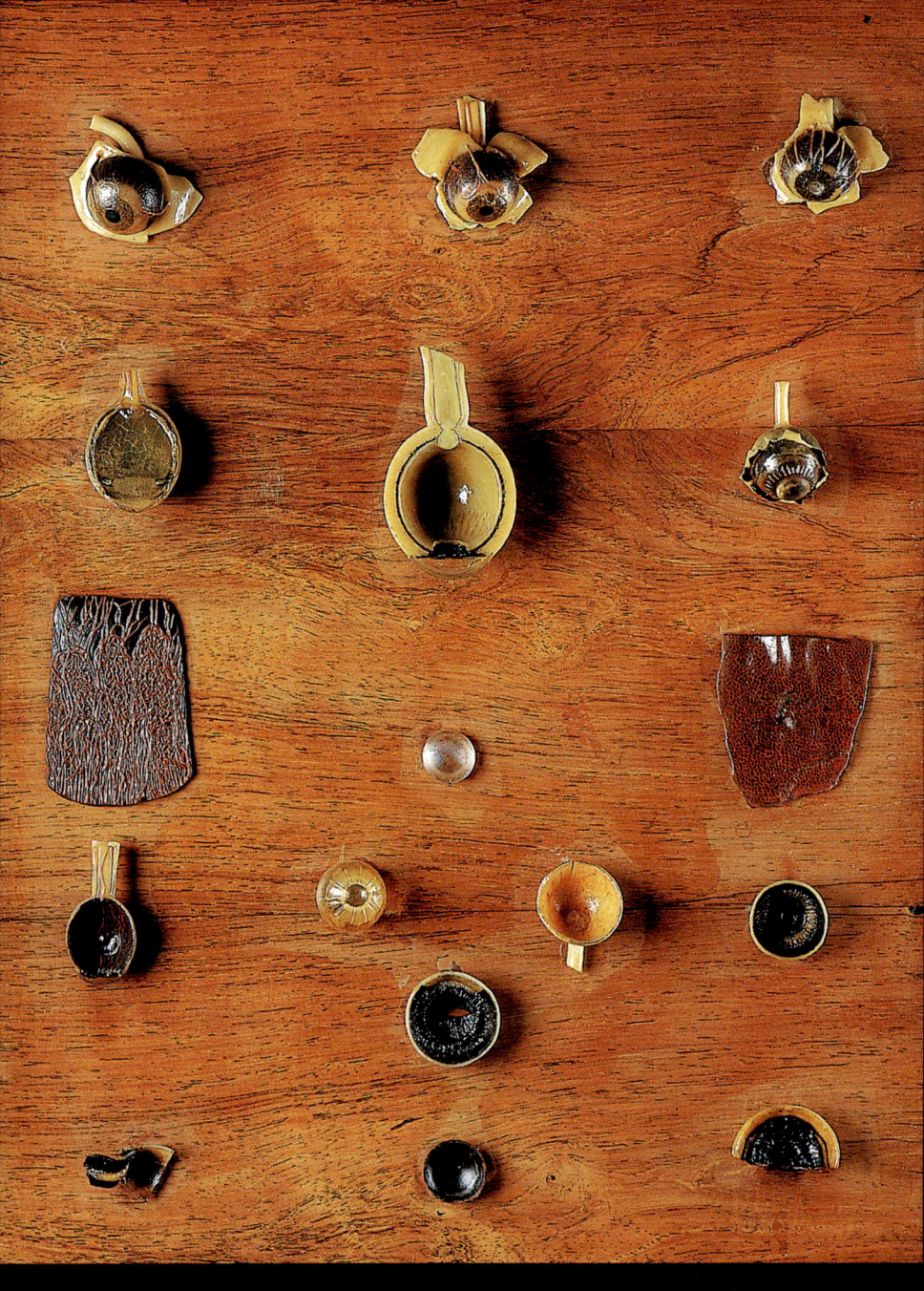

Bulbus oculi

Various dissections of the eyeball

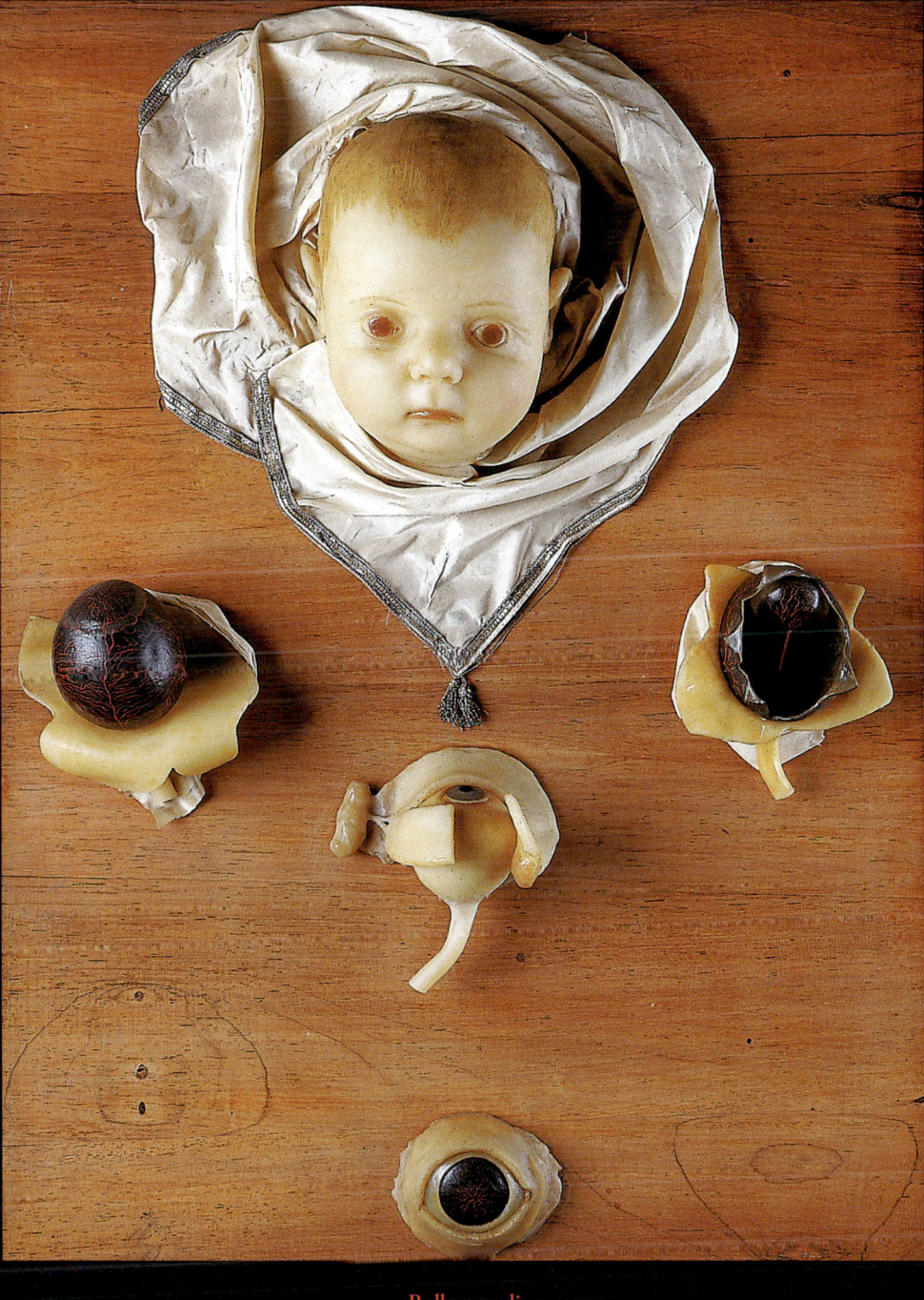

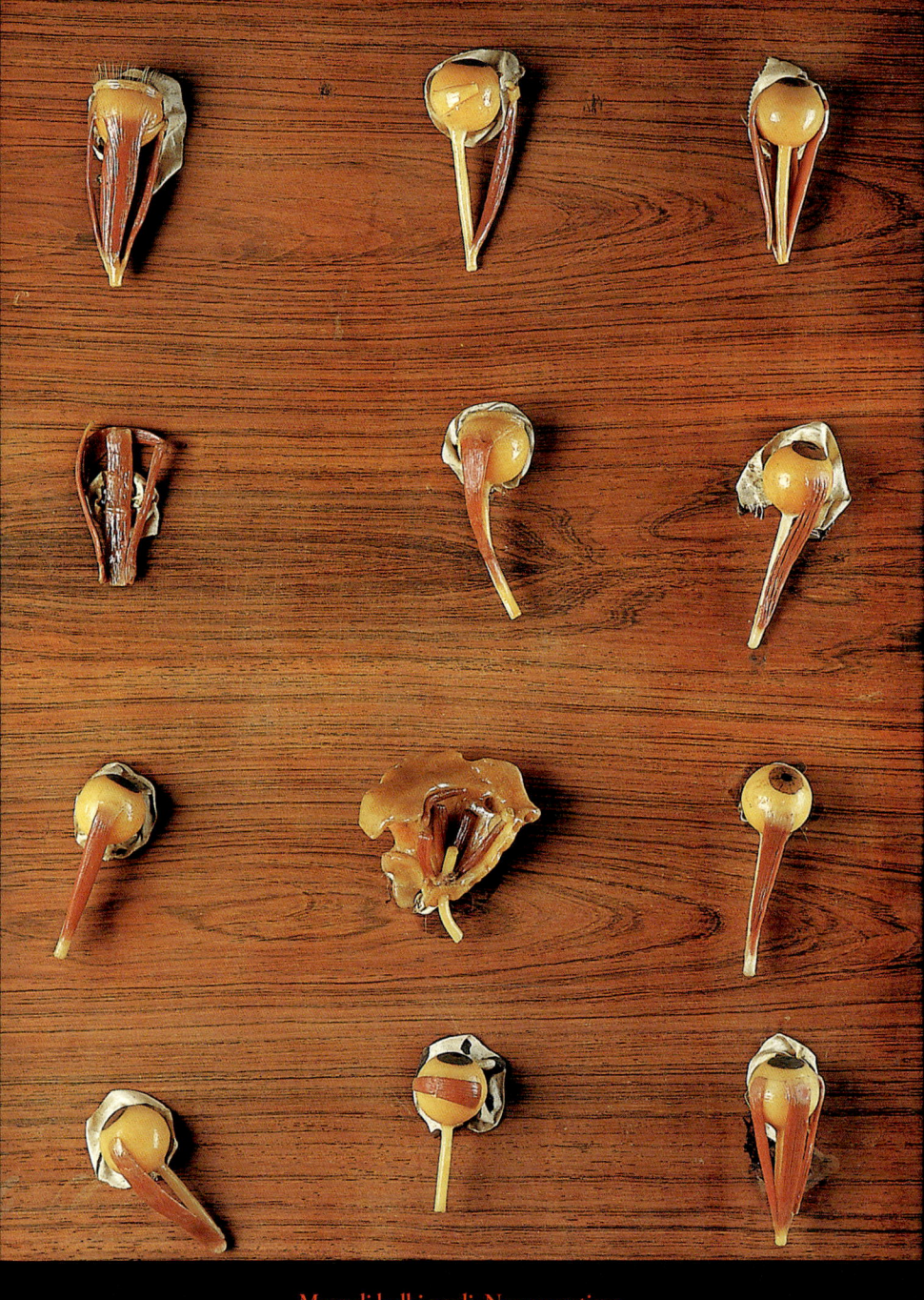

Musculi bulbi oculi, Nervus opticus

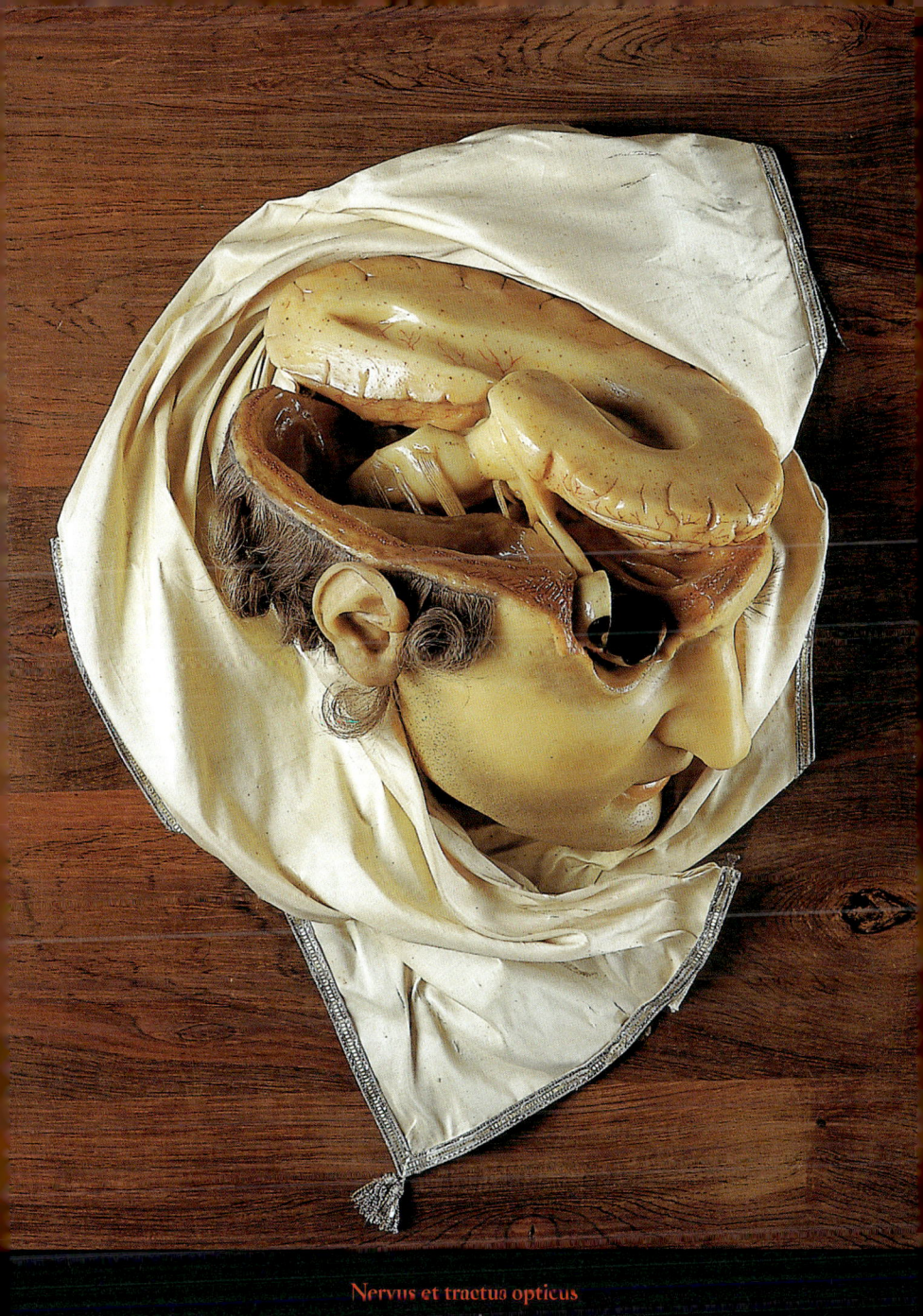

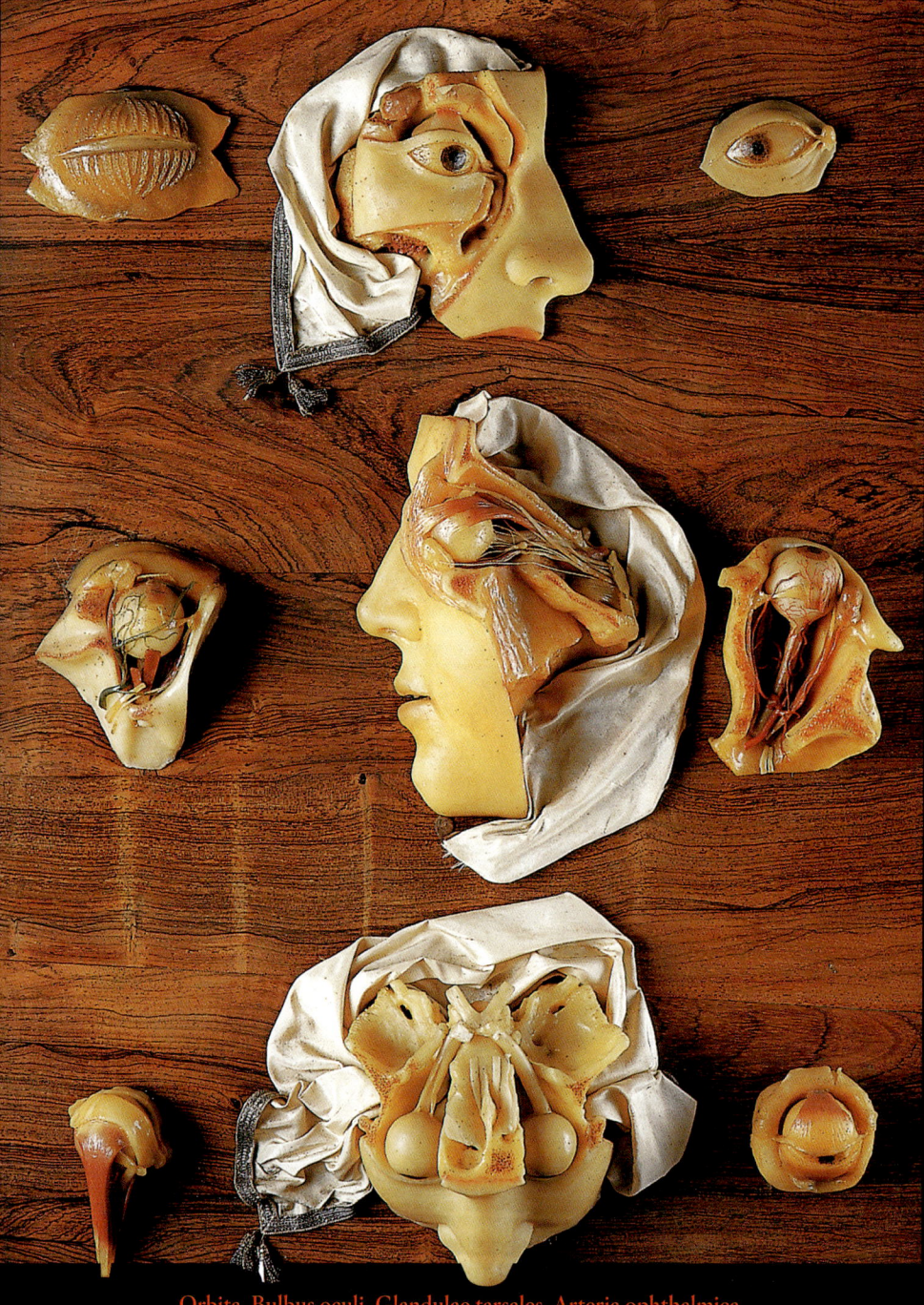

Orbita, Bulbus oculi, Glandulae tarsales, Arteria ophthalmica

Specimens of the orbital cavity, some with the eye muscles, nerves and vessels.
The upper specimen on the left shows the eyelids with the Meibomian glands (sebaceous glands)

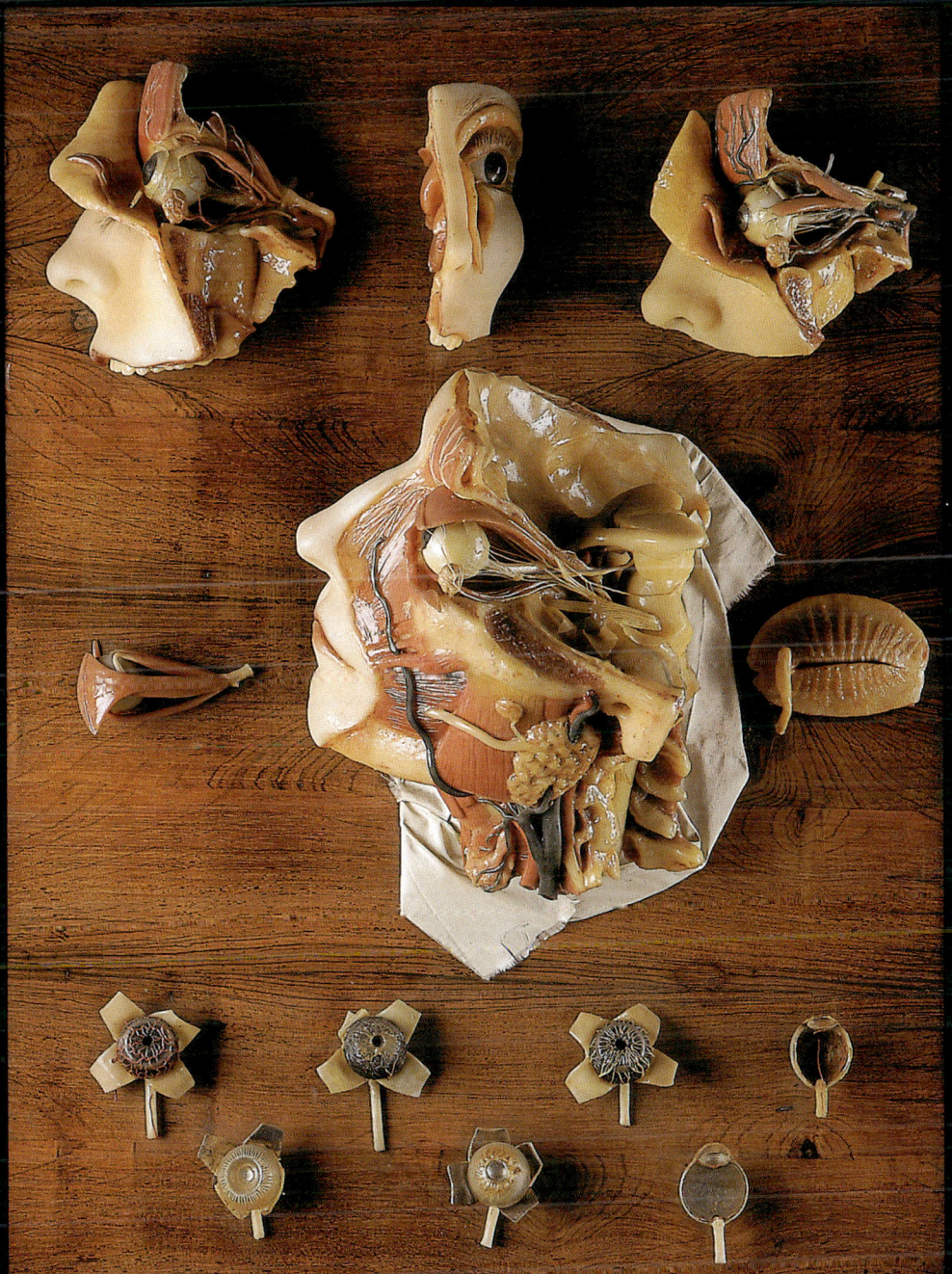

Orbita, Glandula lacrimalis, Ductus nasolacrimalis, Glandulae tarsales

Various presentations of the eyeball and orbit. The lacrimal gland is lying against the eyeball
(middle and upper specimens at the side)

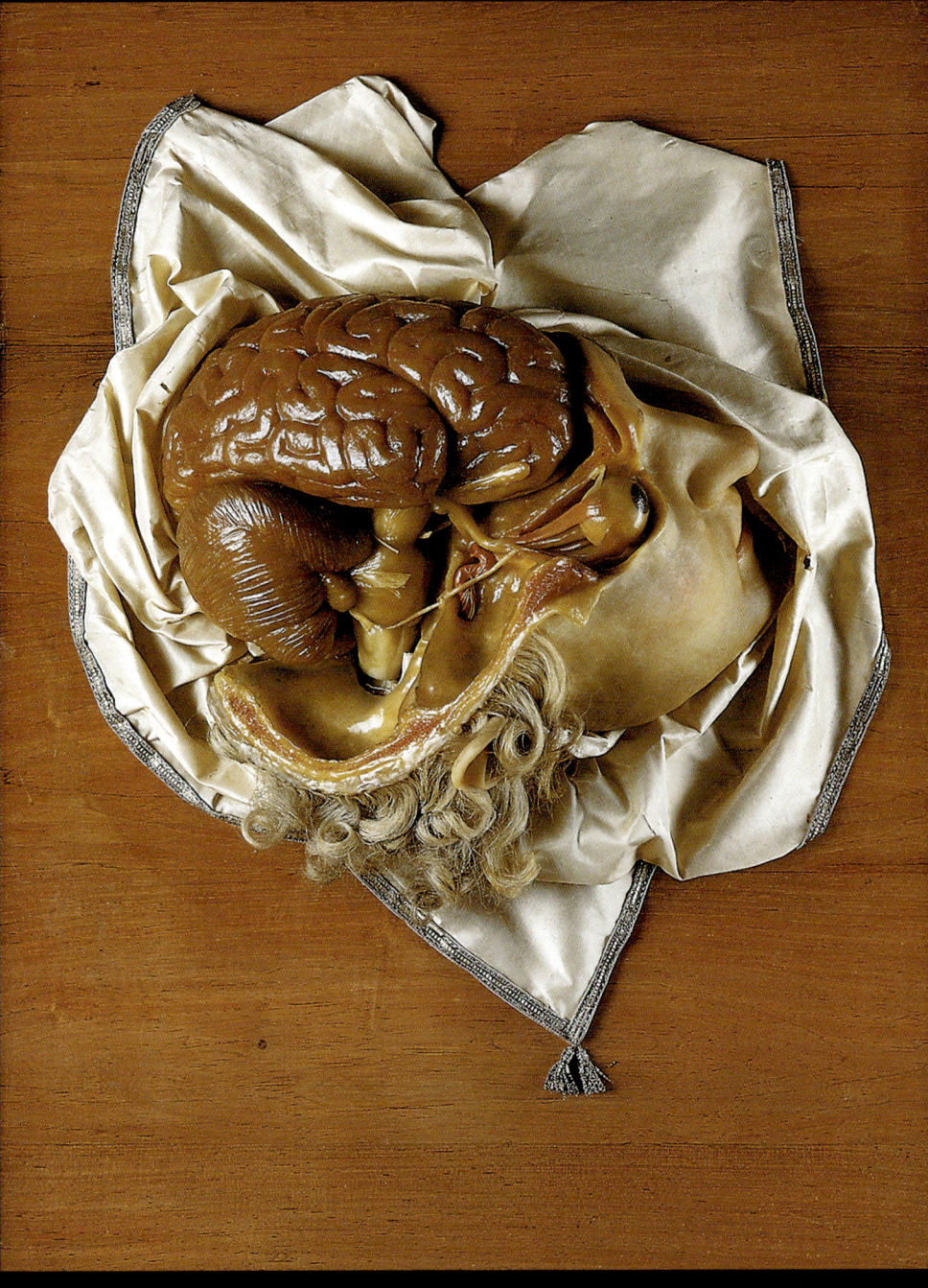

View of the brainstem and orbit. The brain has been lifted up and displaced to one side
to show the course of one of the eye muscle nerves, the abducent nerve

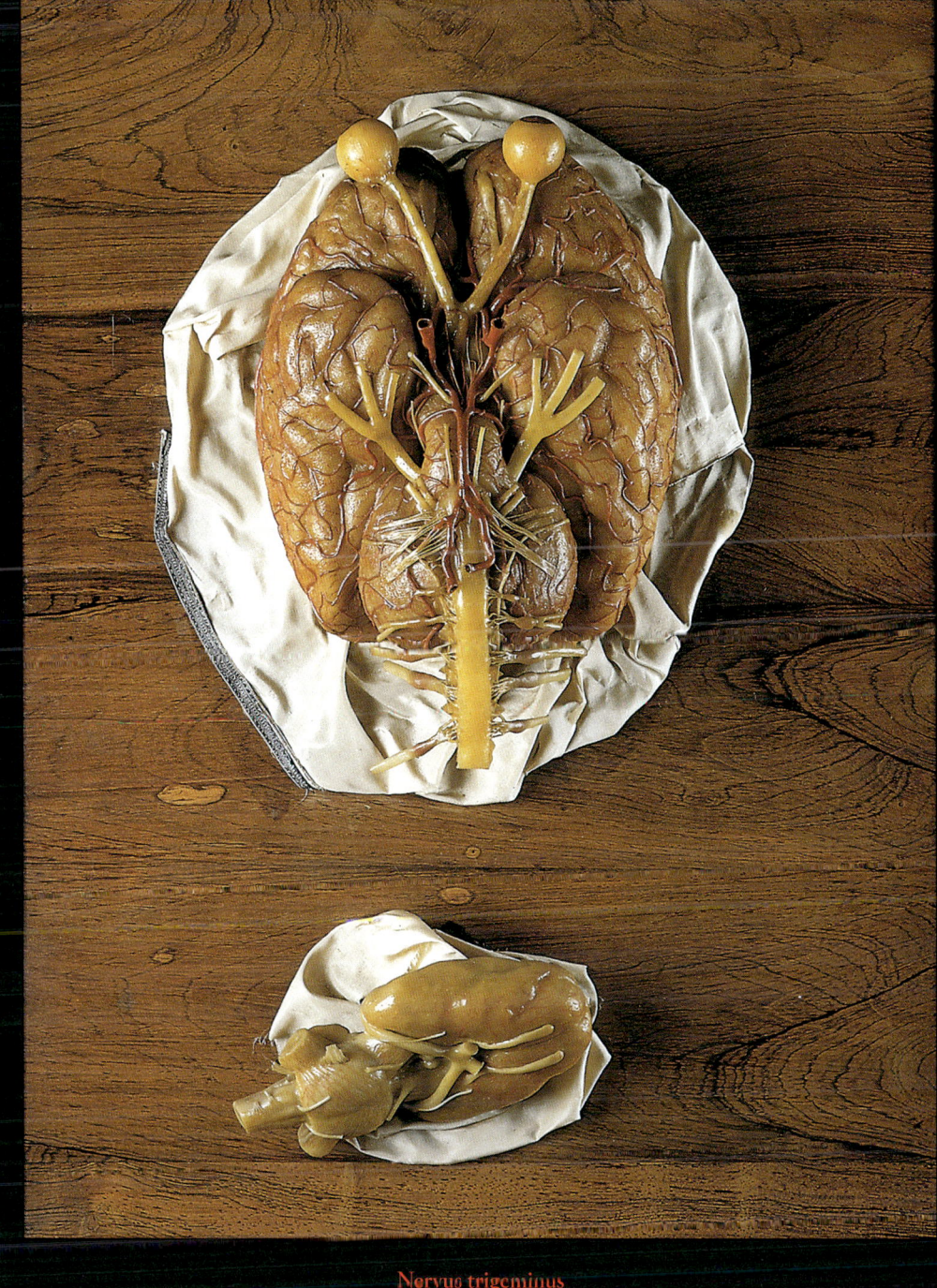

Nervus trigeminus

View of the base of the brain showing the three subdivisions of the trigeminal nerve

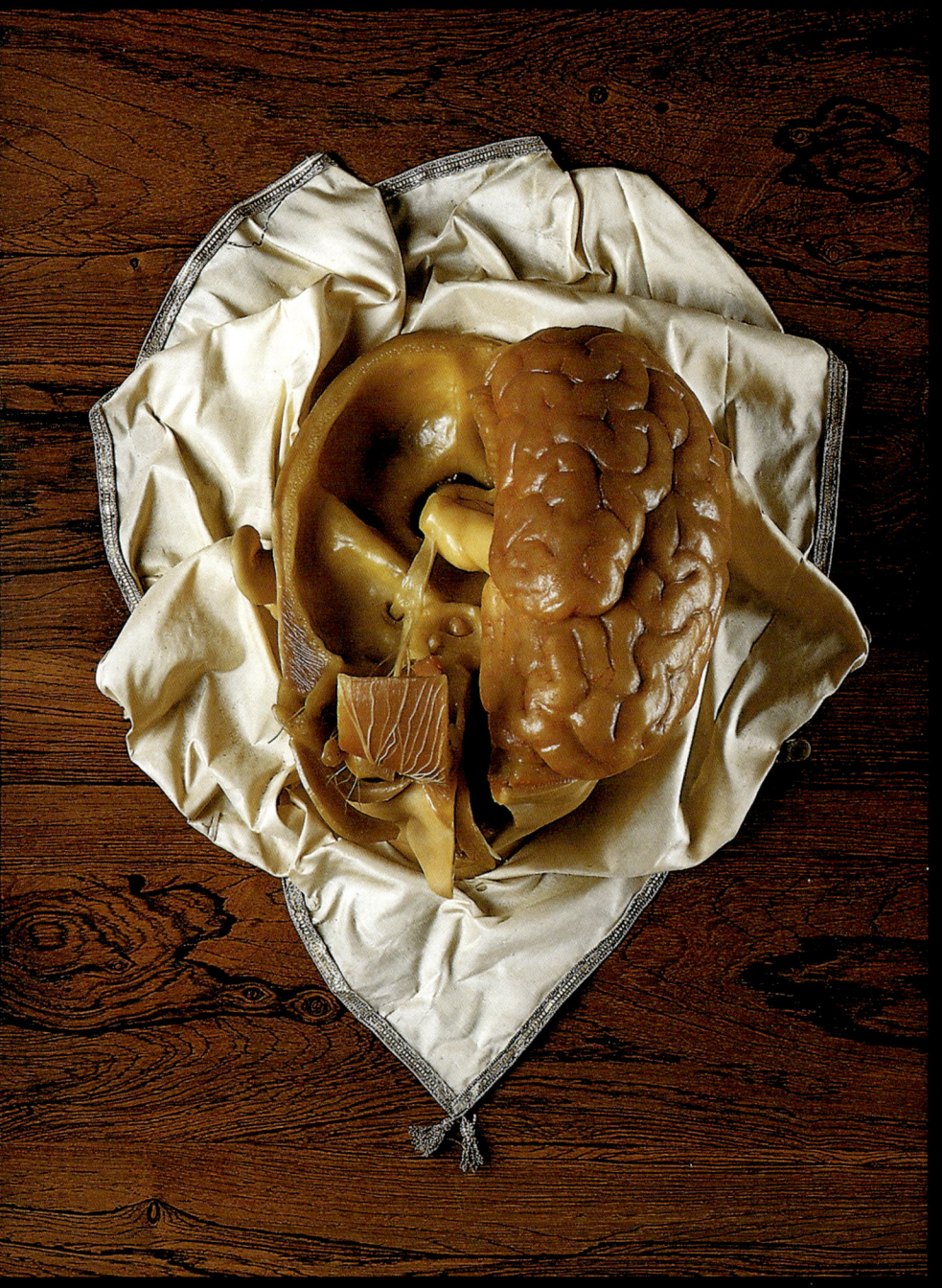

Nervus trigeminus, Nervus ophthalmicus

Display showing the first subdivision of the trigeminal nerve. A part of the frontalis muscle
can be seen on the left, across which single branches of the trigeminal nerve are running

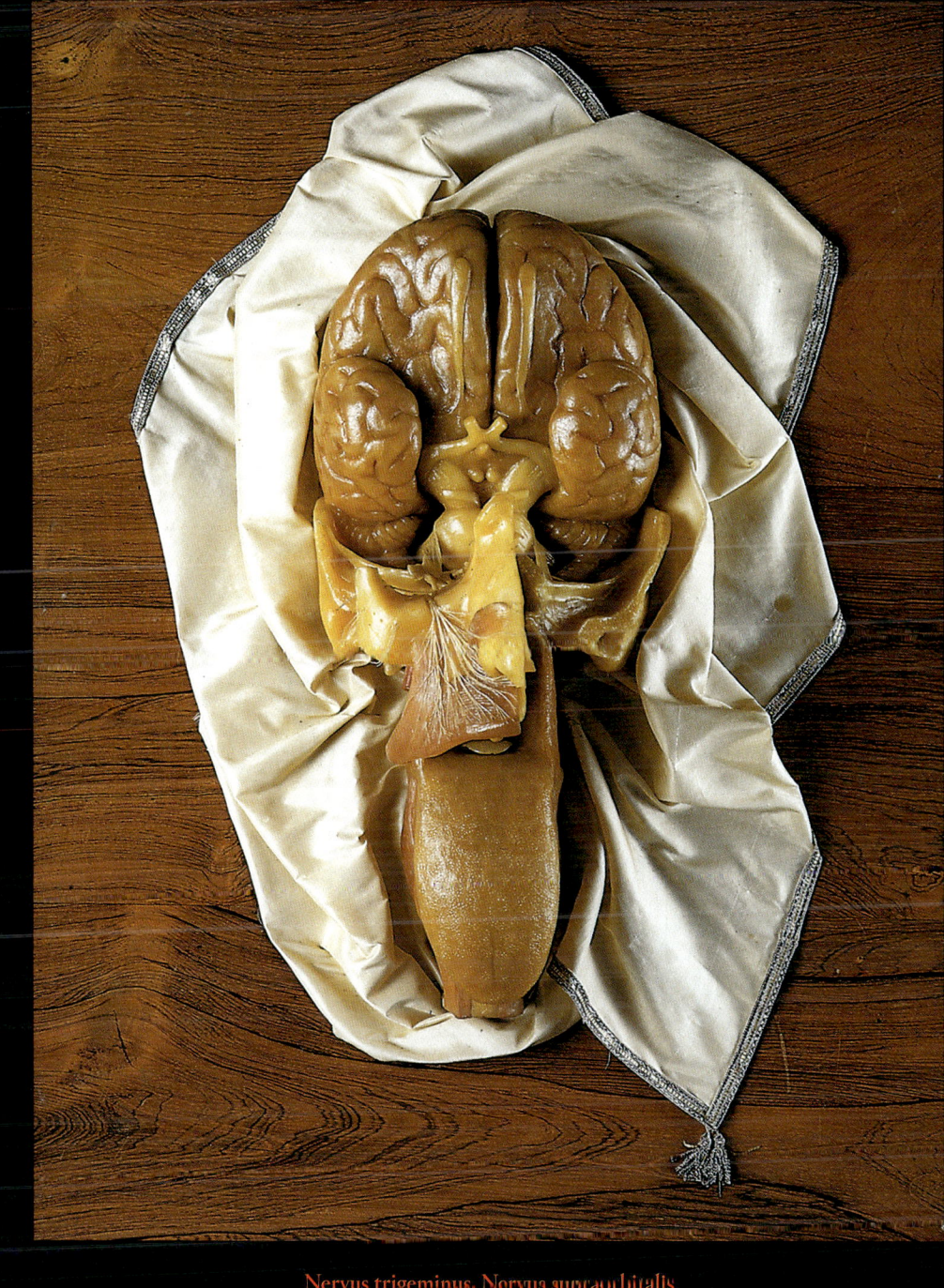

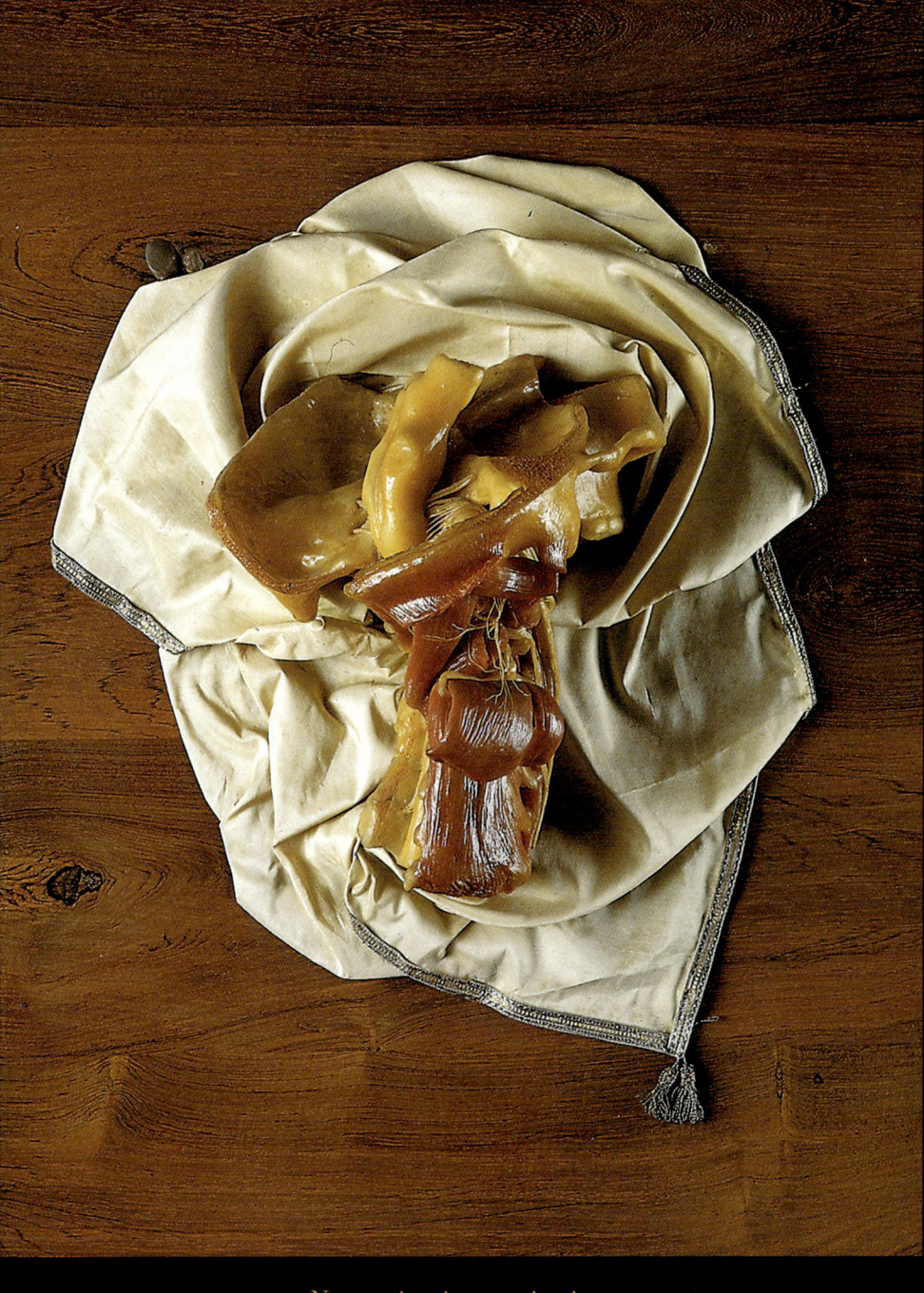

Nervus trigeminus, portio minor

Part of the masticatory muscles with their nerve supply

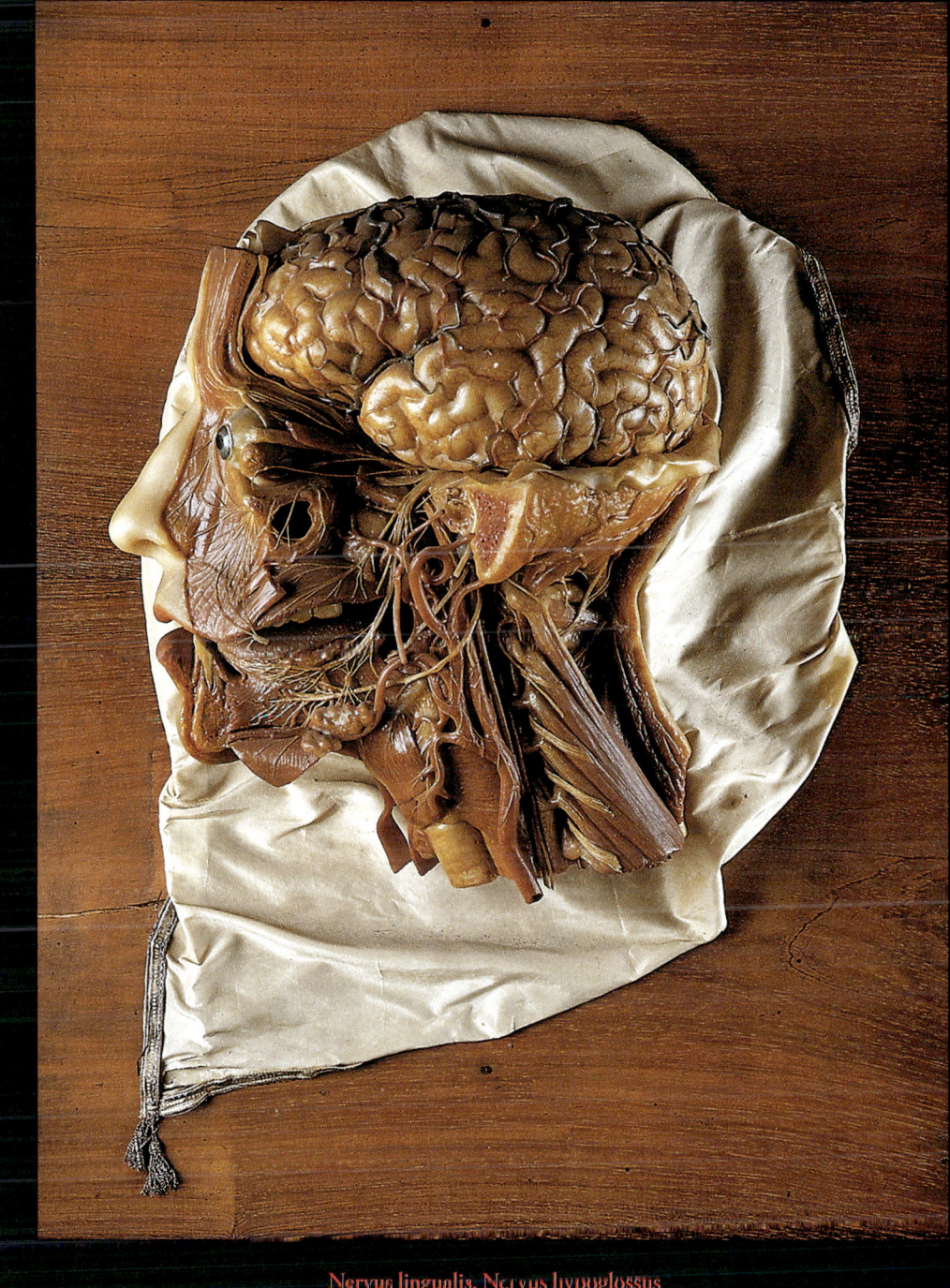

Nervus lingualis, Nervus hypoglossus

Display showing the nerves and arteries of the cavity of the mouth, tongue and throat

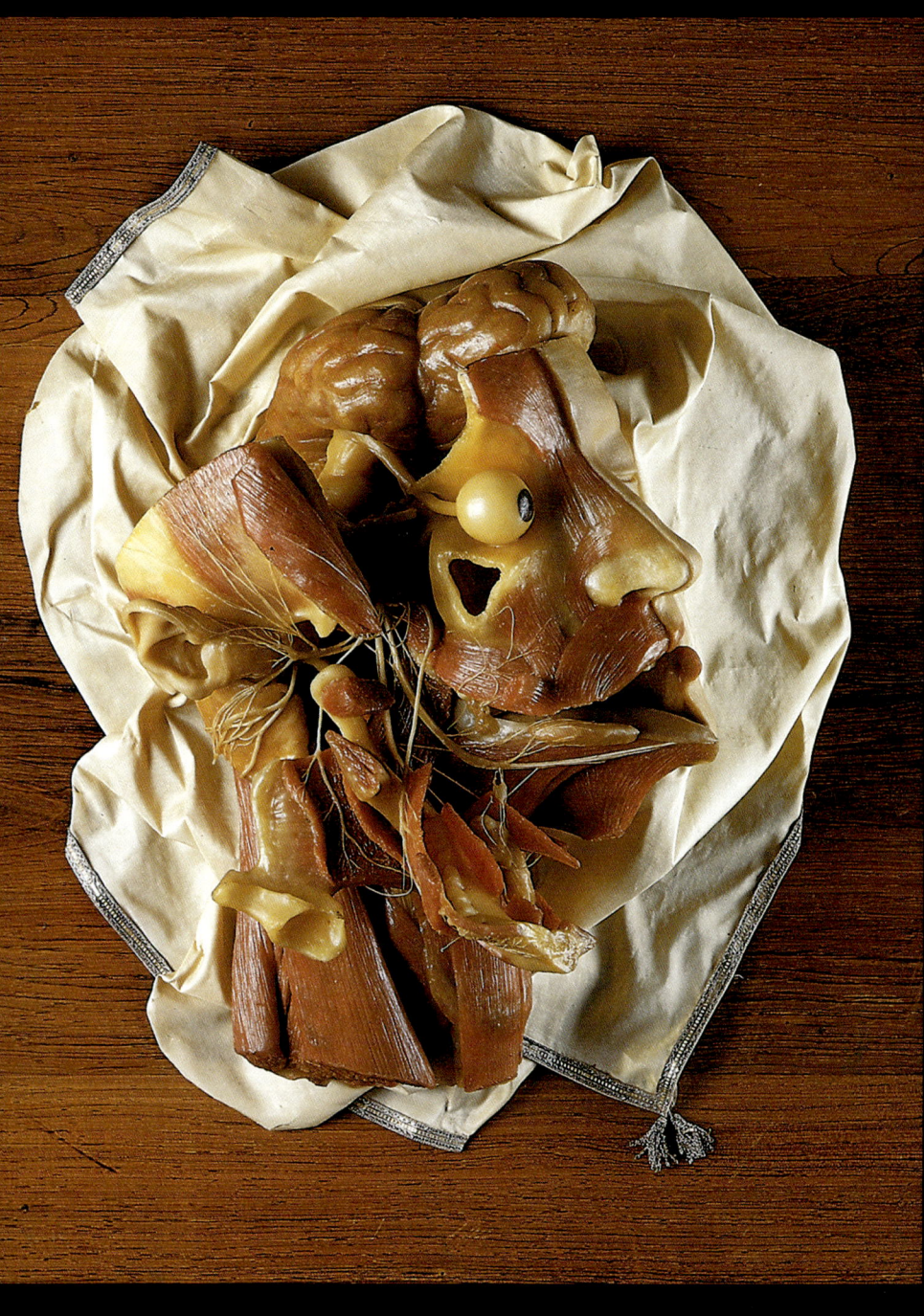

Nervus lingualis

One of the sensory nerves of the tongue

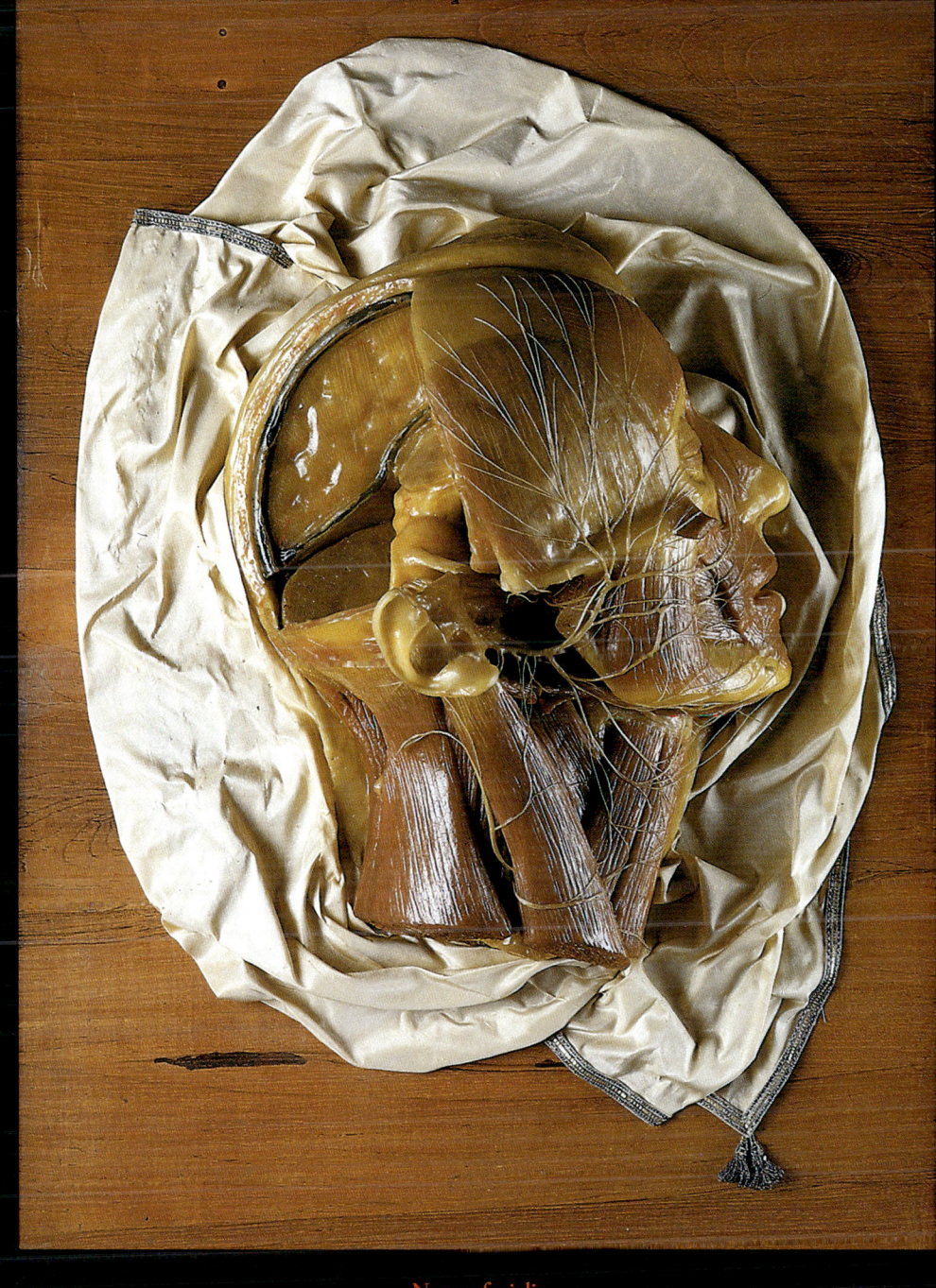

Nervus facialis
Display showing the facial nerve, which supplies the muscles of facial expression

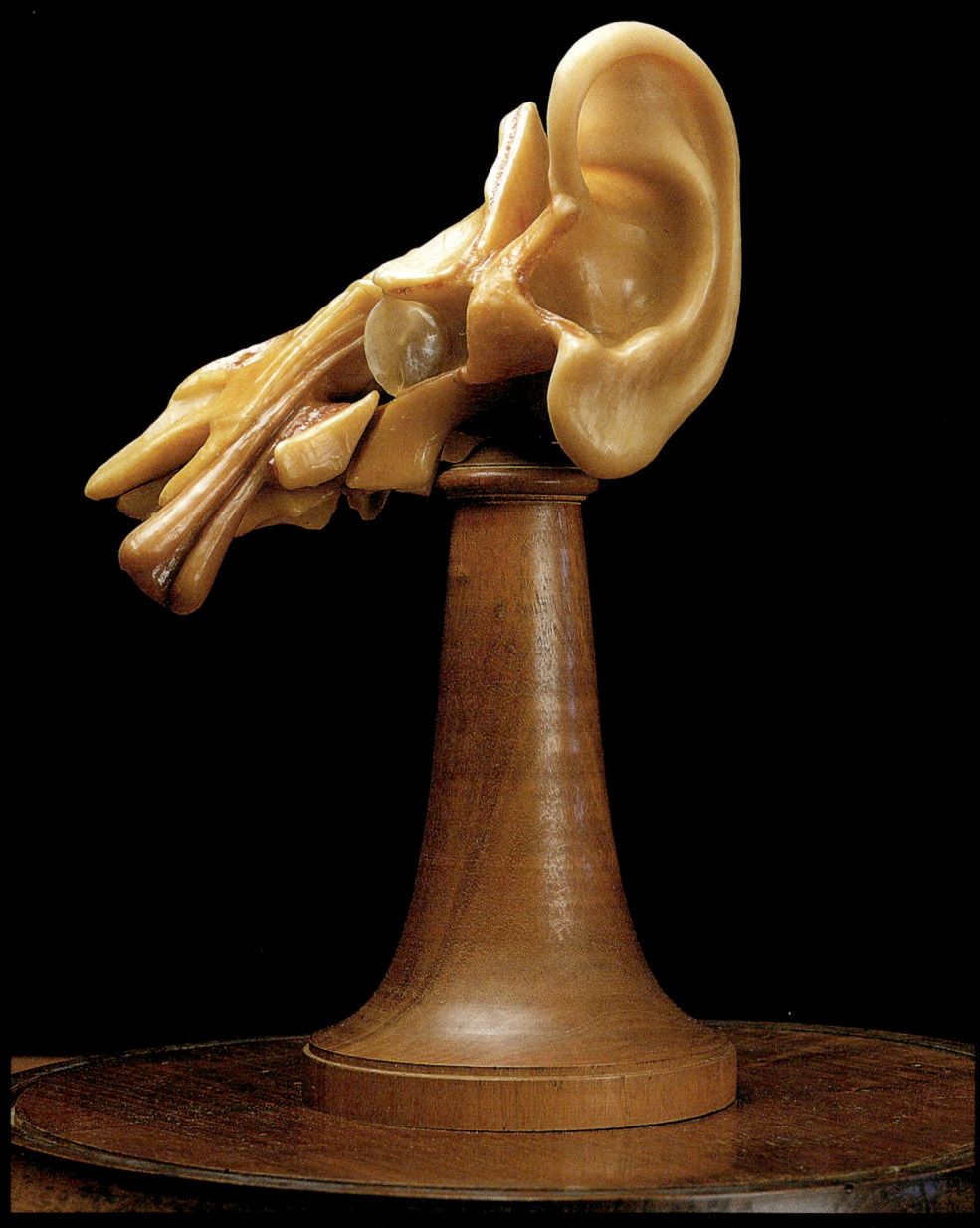

Tuba auditiva, Membrana tympanica
Specimen of the ear with the external auditory meatus, eardrum, middle ear and Eustachian tube

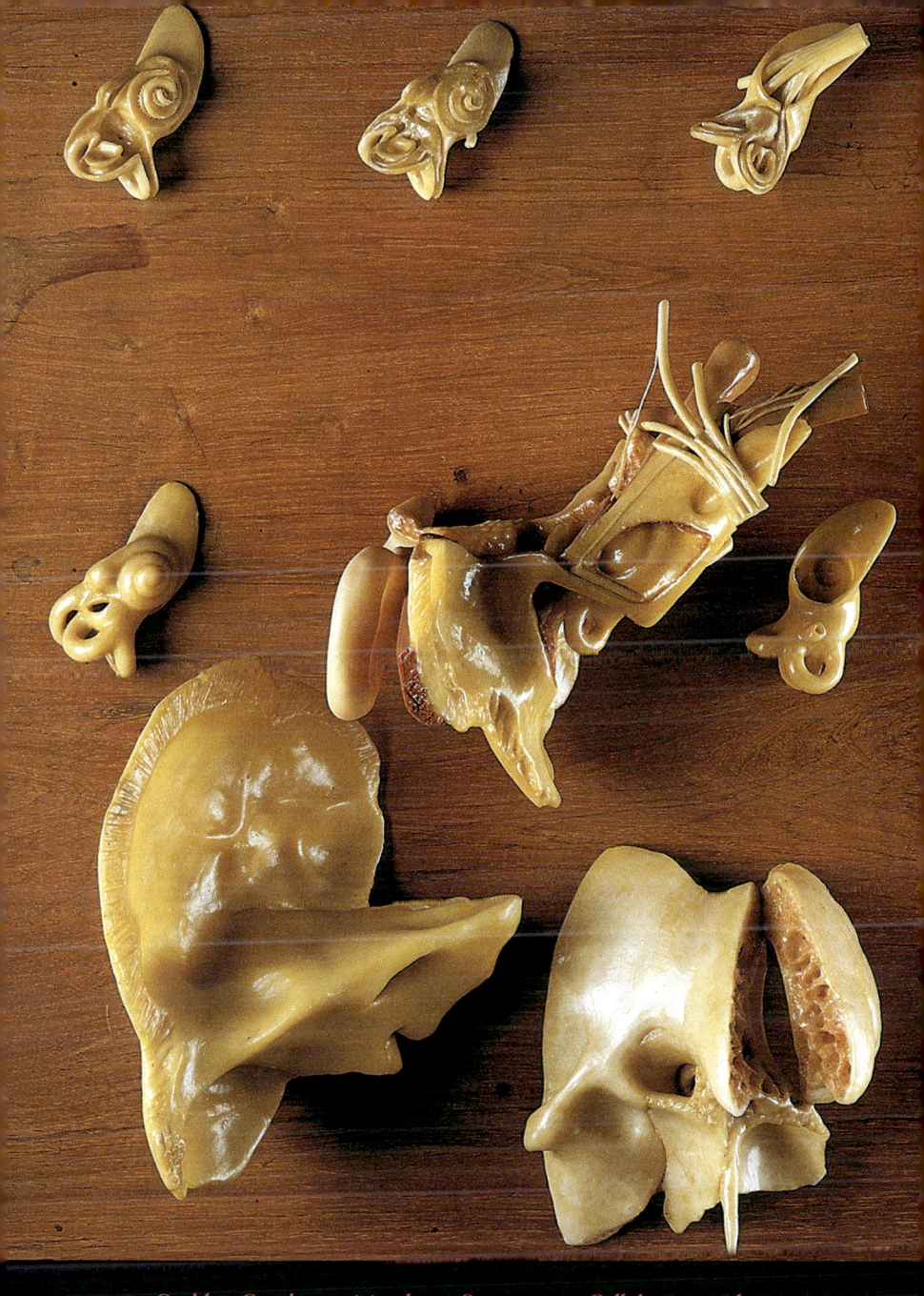

Cochlea, Canales semicirculares, Os petrosum, Cellulae mastoideae

Various preparations of the inner ear with the cochlea and semicircular canals

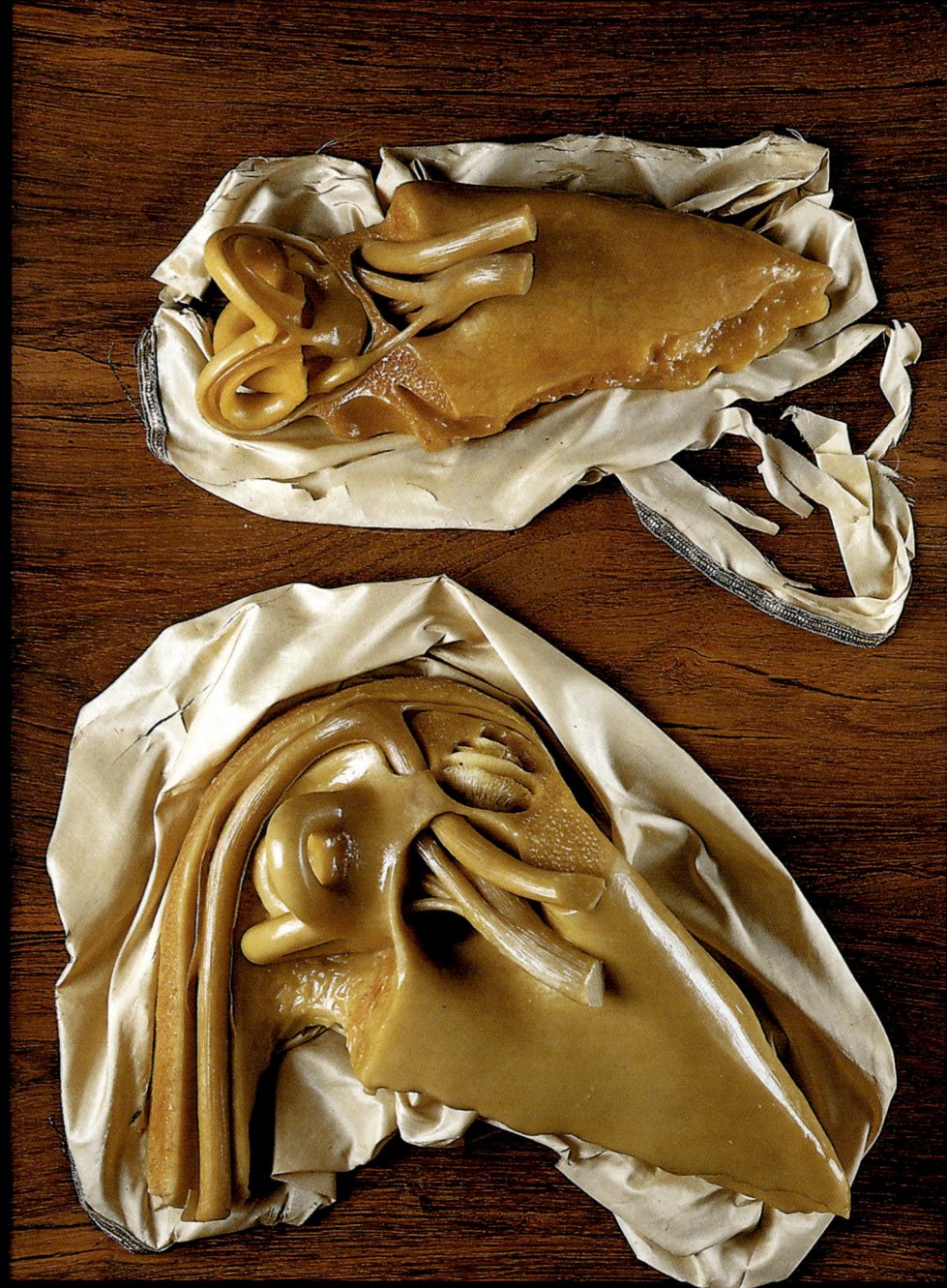

Nervus cochlearis, Genu externus nervi facialis

Display showing the organs of hearing and balance, and the motor nerve of the face
passing through the petrous temporal bone

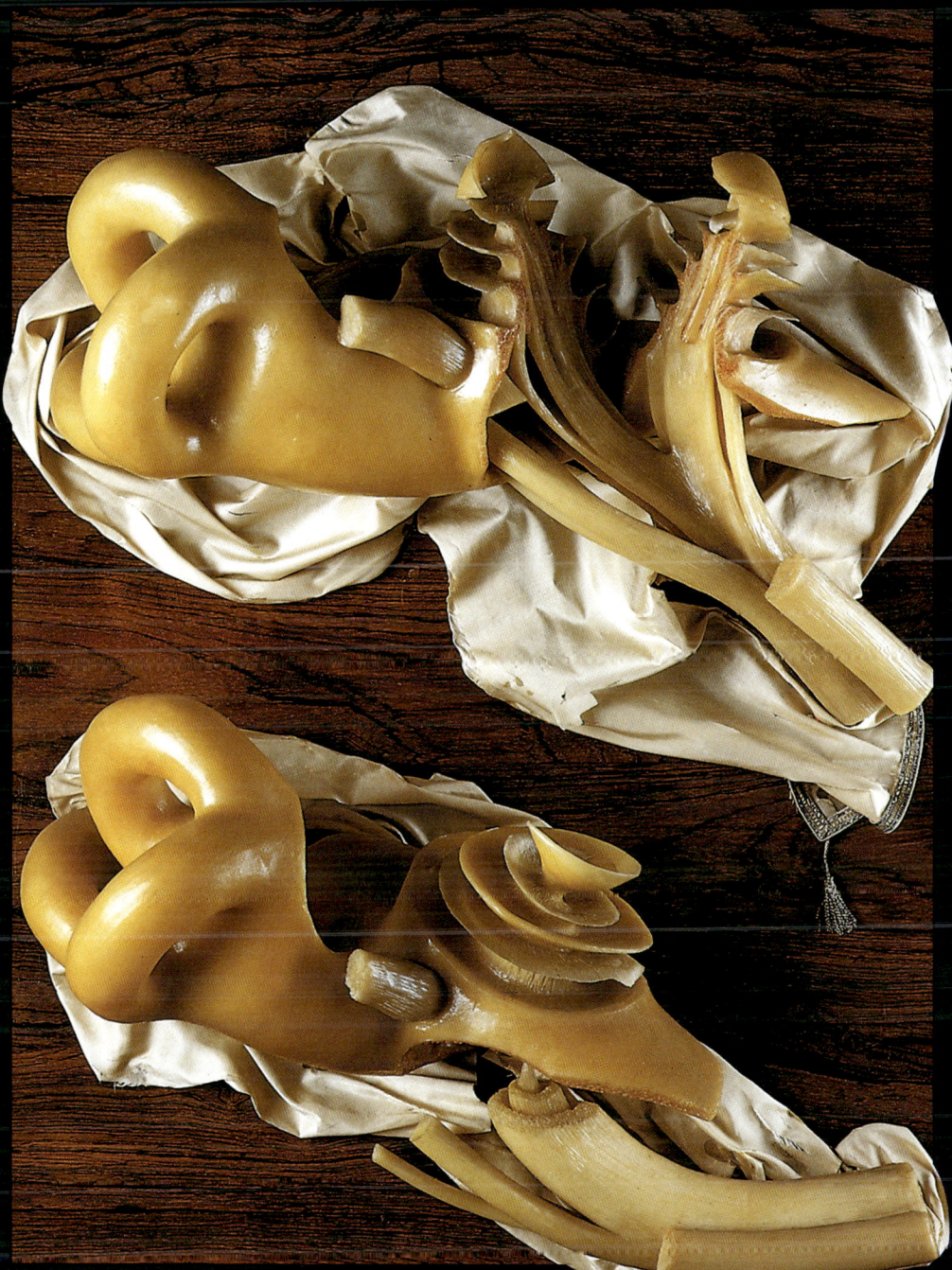

Cochlea, Canales semicirculares

Greatly enlarged specimen of the inner ear showing the semicircular canals
and the convolutions of the cochlea

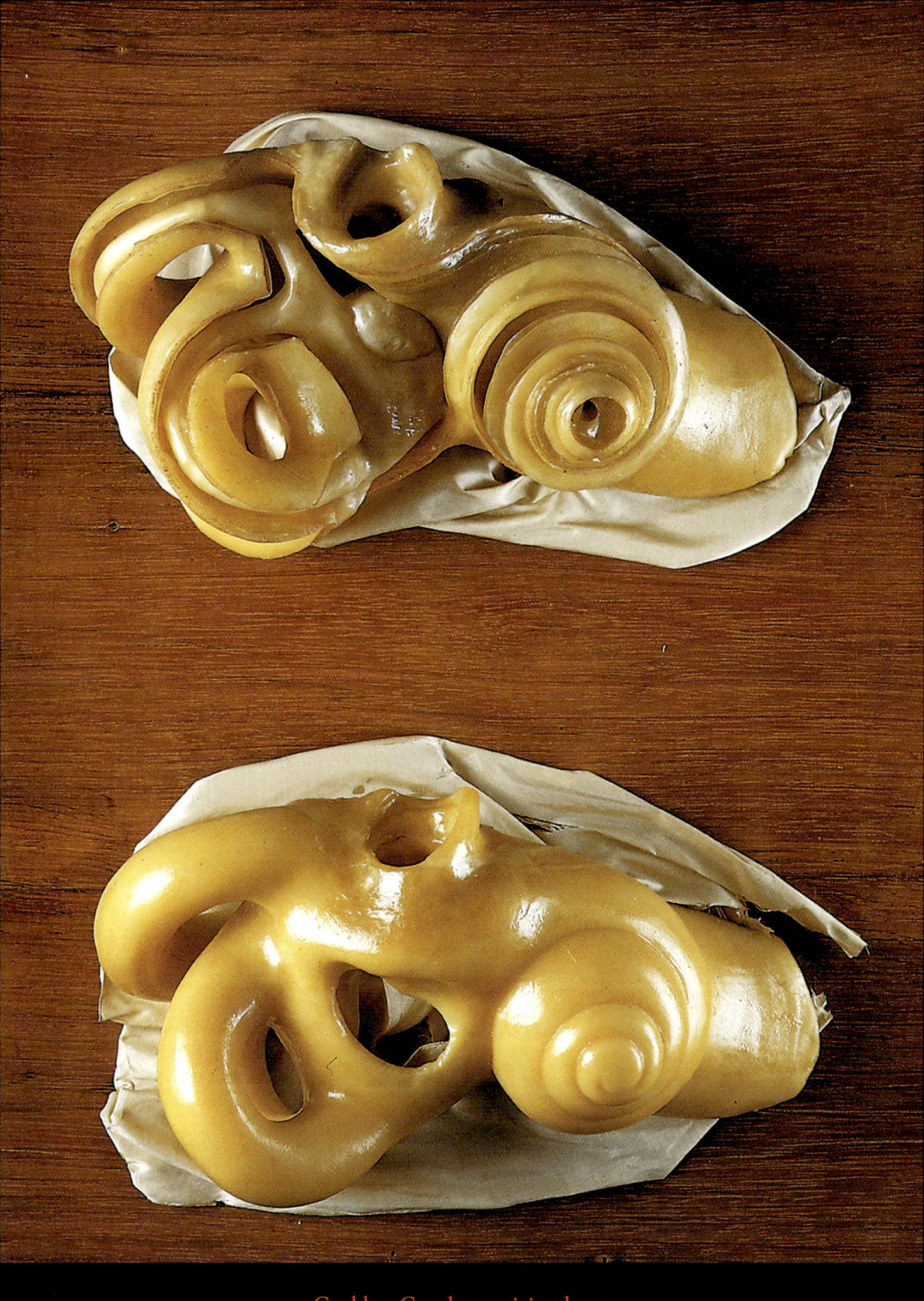

Cochlea, Canales semicirculares

Greatly enlarged specimen of the inner ear showing the semicircular canals
and the convolutions of the cochlea

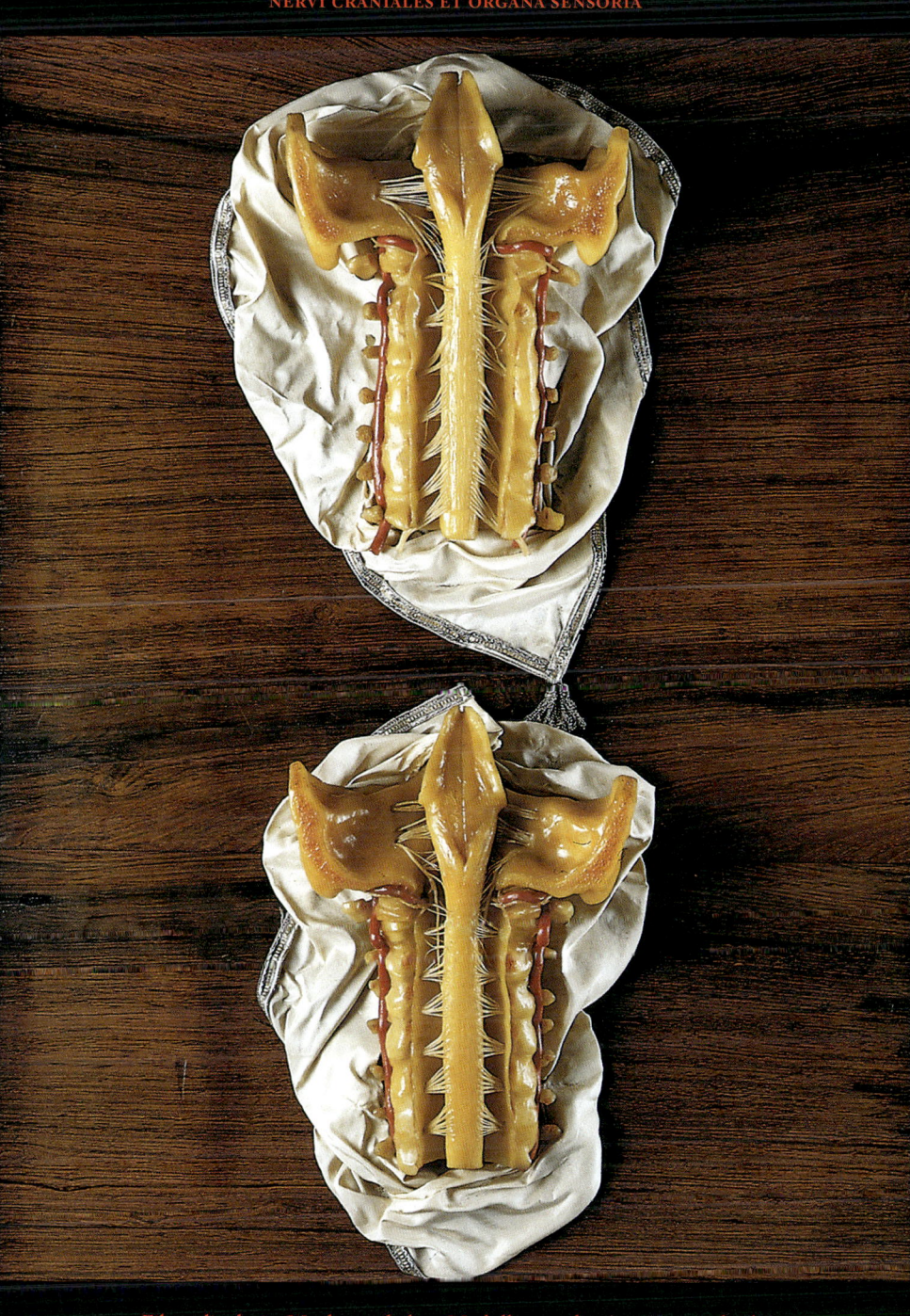

Fila radicularia, Myelencephalon, Medulla spinalis, Arteria vertebralis

Specimen showing the brainstem and upper part of the spinal cord with the attached spinal nerves roots;
seen from behind

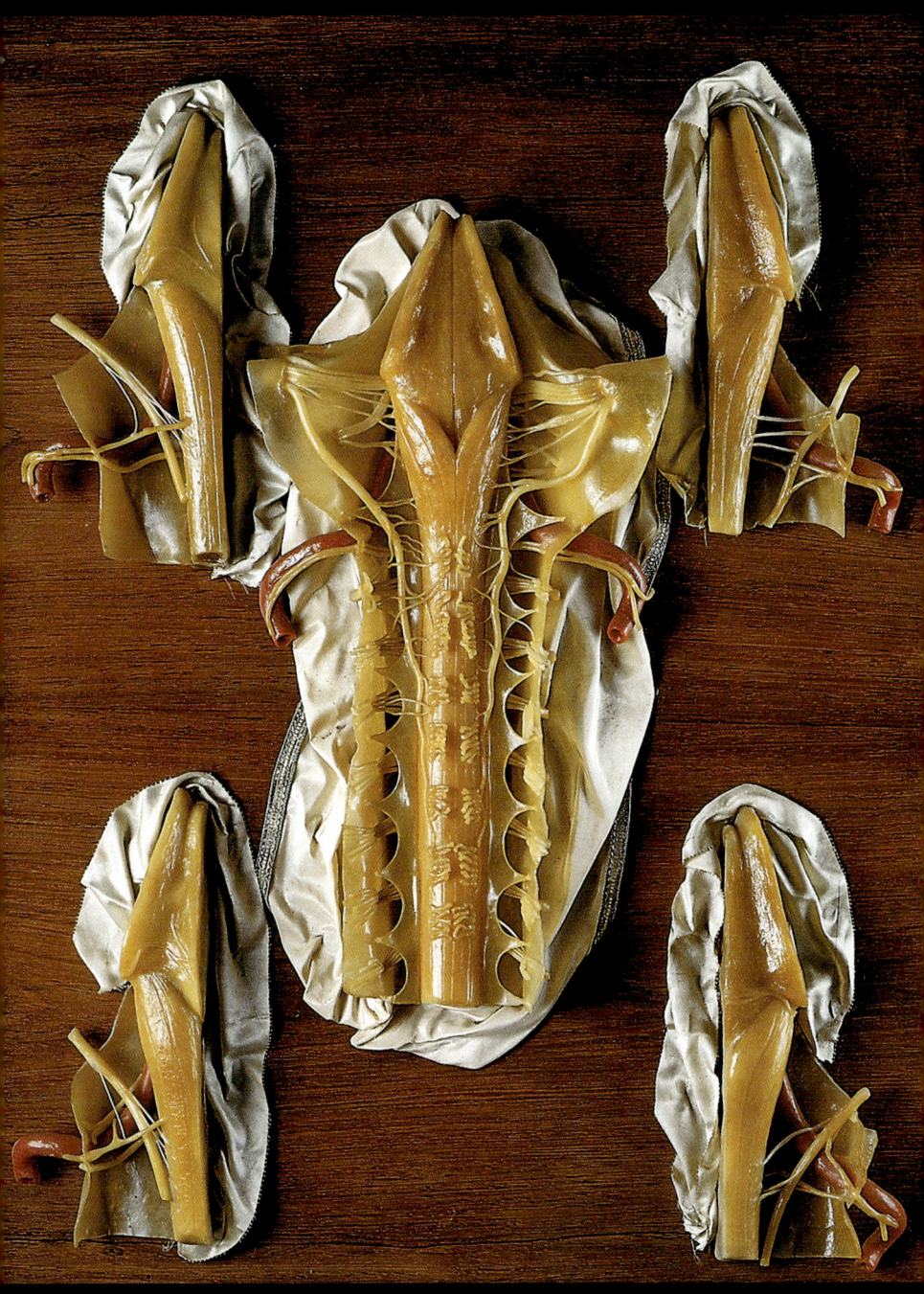

Nervus accessorius, Ligamenta denticulata

The brainstem and upper part of the spinal cord seen from behind.
Variations in the course of the eleventh cranial nerve can be observed.

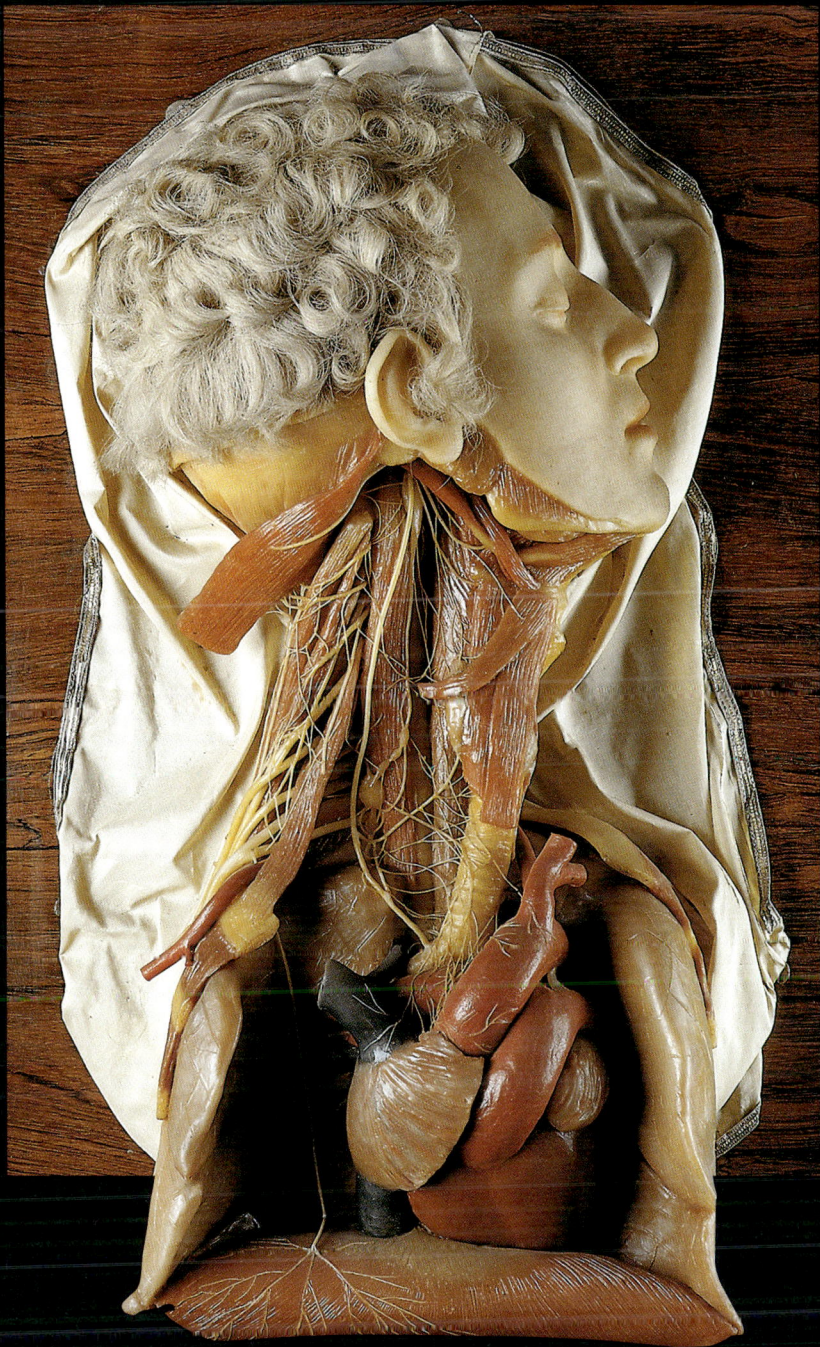

Nervus vagus, Plexus brachialis, Ganglion stellatum
Display showing the tenth cranial nerve, the vagus nerve

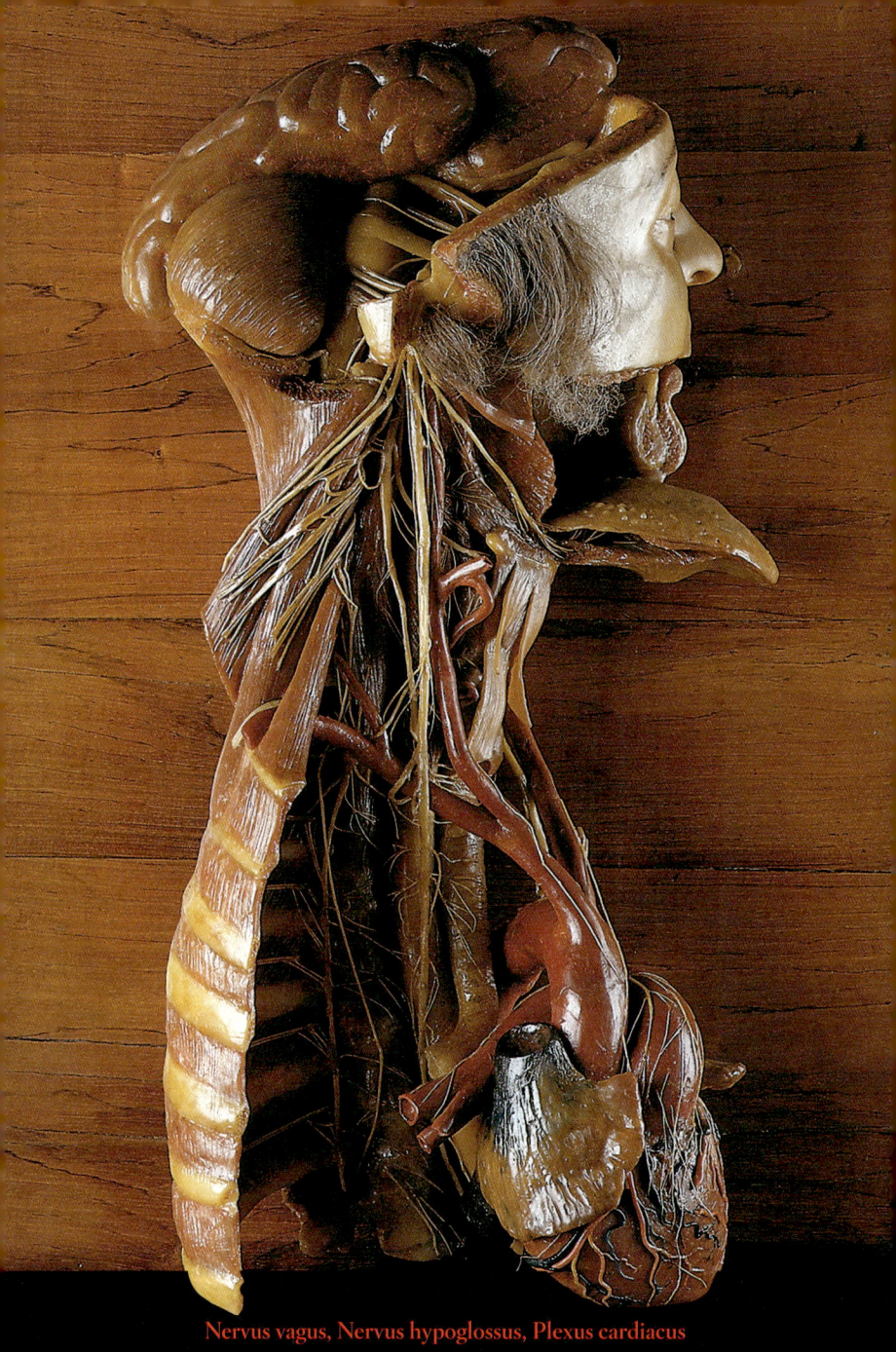

Nervus vagus, Nervus hypoglossus, Plexus cardiacus

Nerves in the region of the neck and thoracic cage.
The vagus nerve follows an almost directly vertical course near the midline.

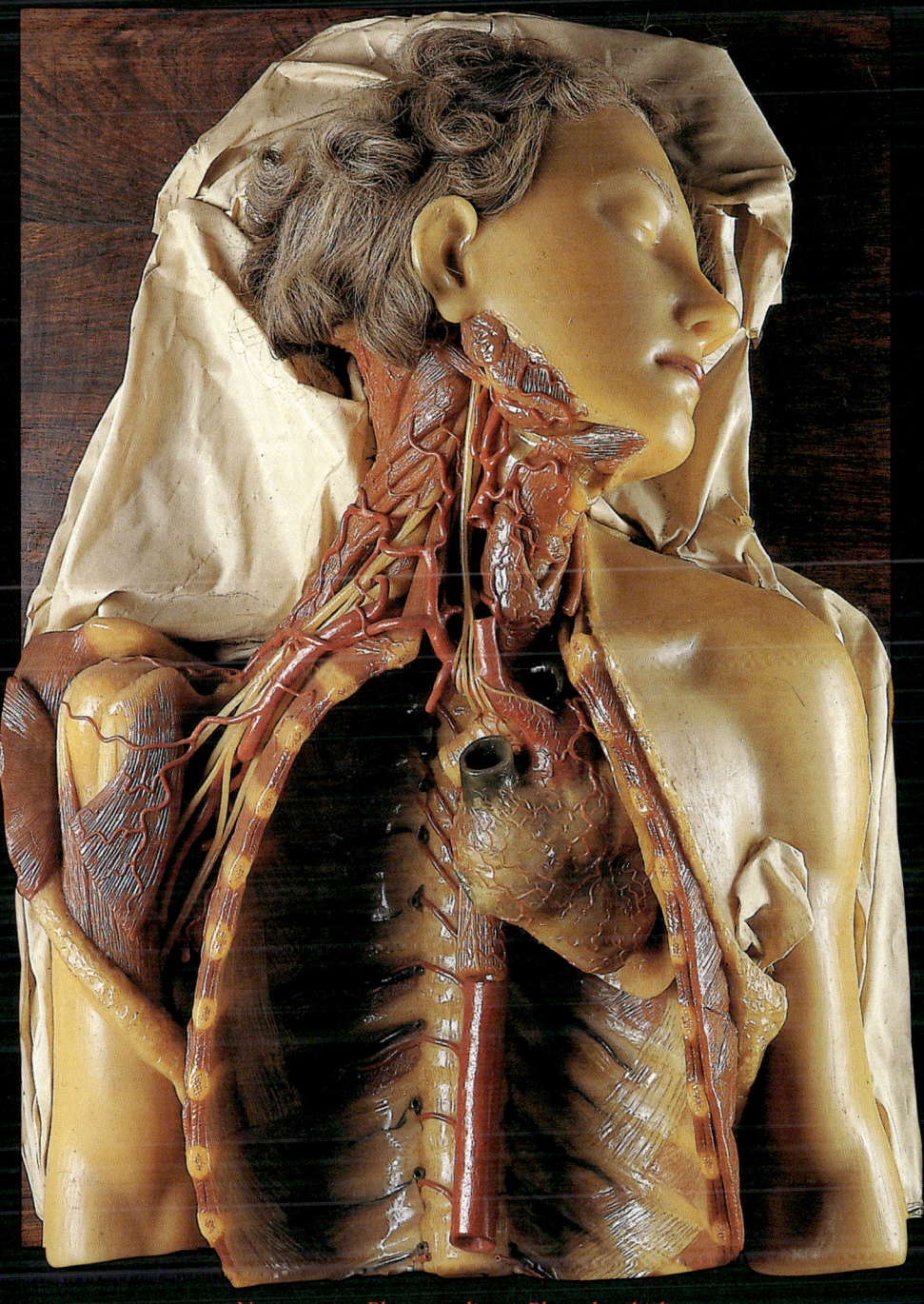

Nervus vagus, Plexus cardiacus, Plexus brachialis
The viscera of neck and thorax showing the course of the vagus nerve

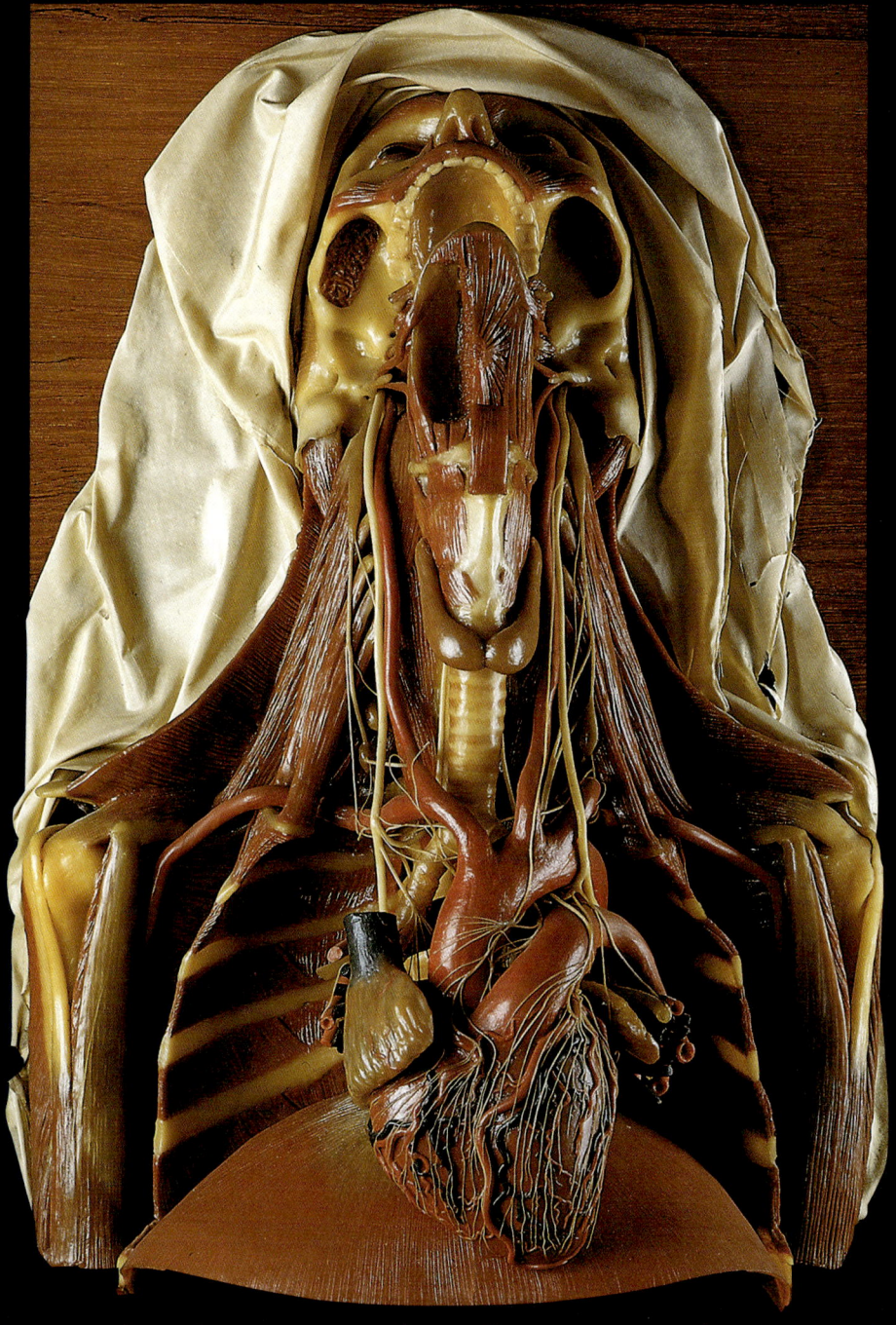

Nervus vagus, Nervus laryngeus recurrens, Plexus cardiacus, Truncus sympathicus, Pars cervicalis
The neurovascular bundle of the neck

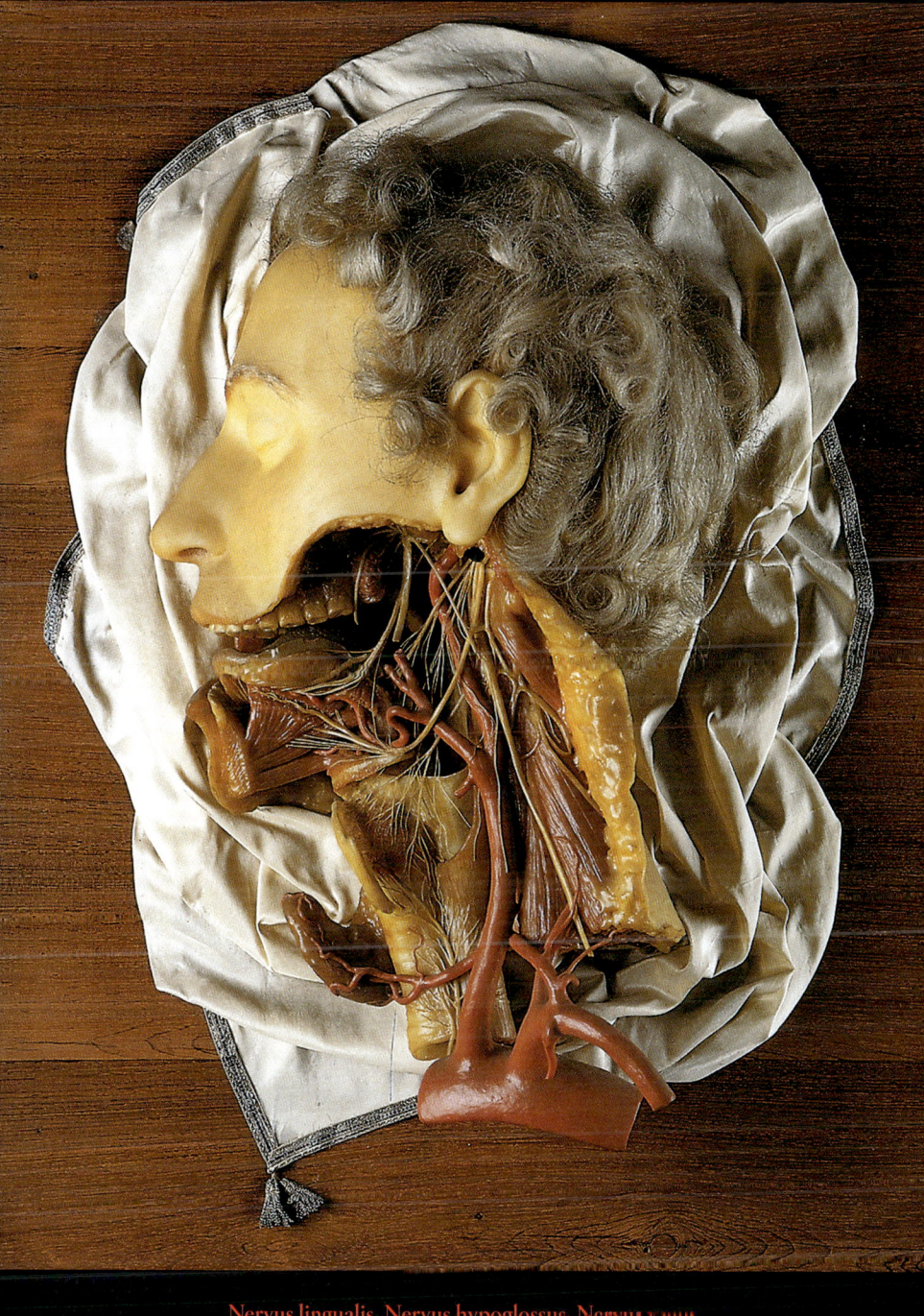

Nervus lingualis, Nervus hypoglossus, Nervus vagus

Neck dissected to show the nerves and vessels. Display showing the innervation of the tongue,
floor of the mouth and larynx

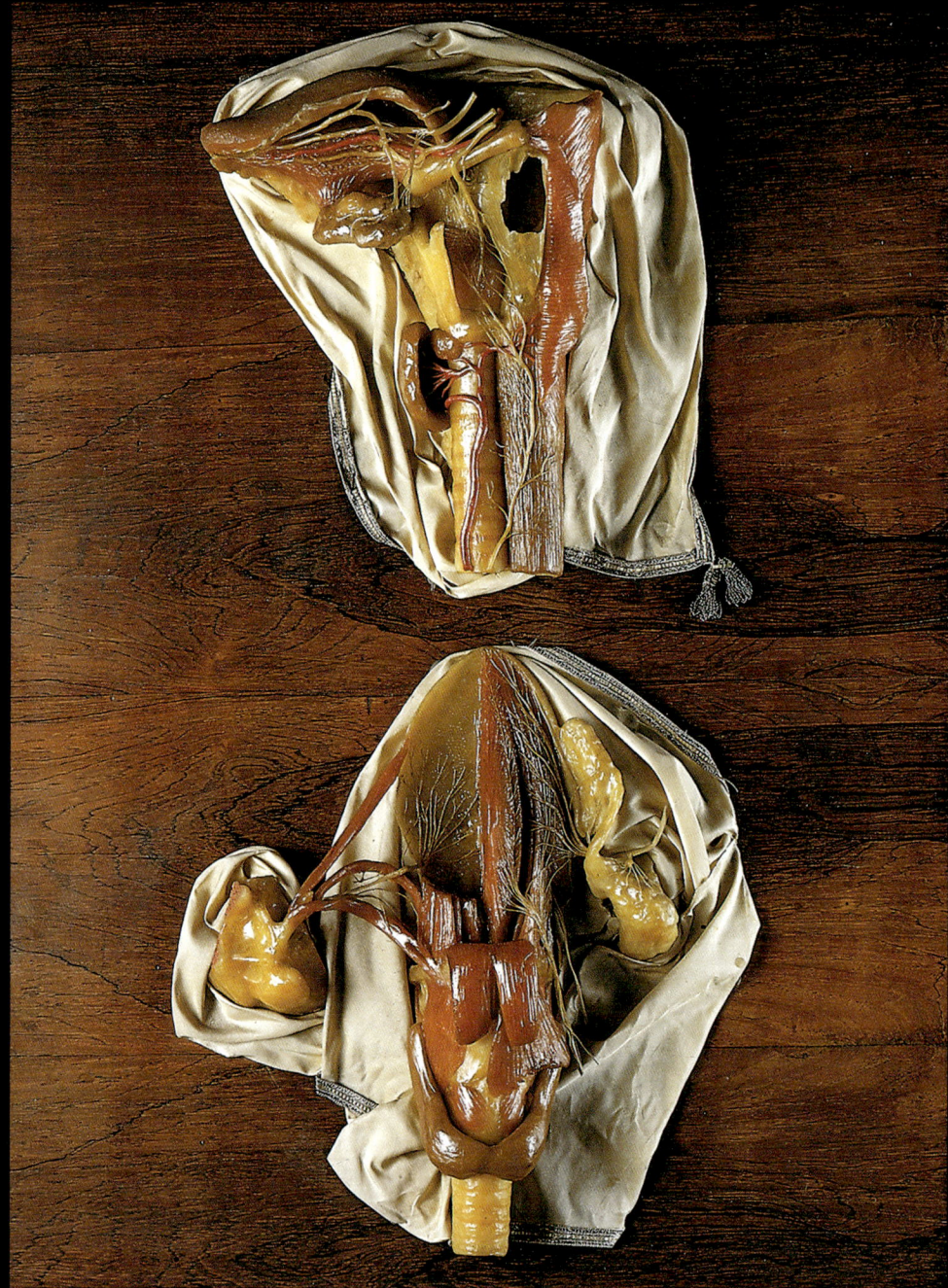

Nervus hypoglossus, Nervus laryngeus recurrens, Nervus vagus, Nervus laryngeus superior

Upper specimen: display showing the innervation of the tongue.
Lower specimen: the innervation of the mucous membrane and muscles of the larynx

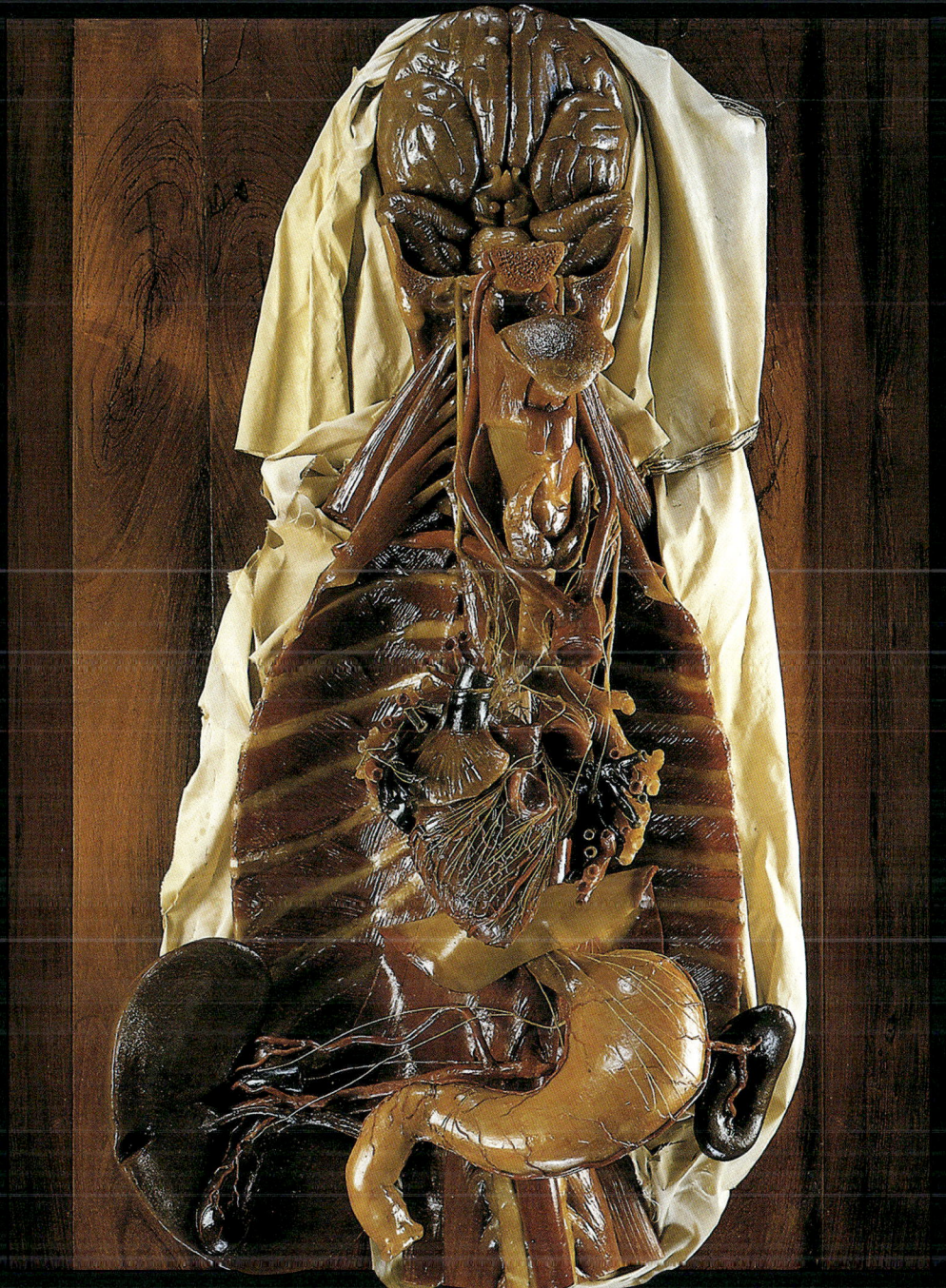

Nervus vagus

Display showing the vagus nerve and the area innervated by it in the thoracic and abdominal cavities

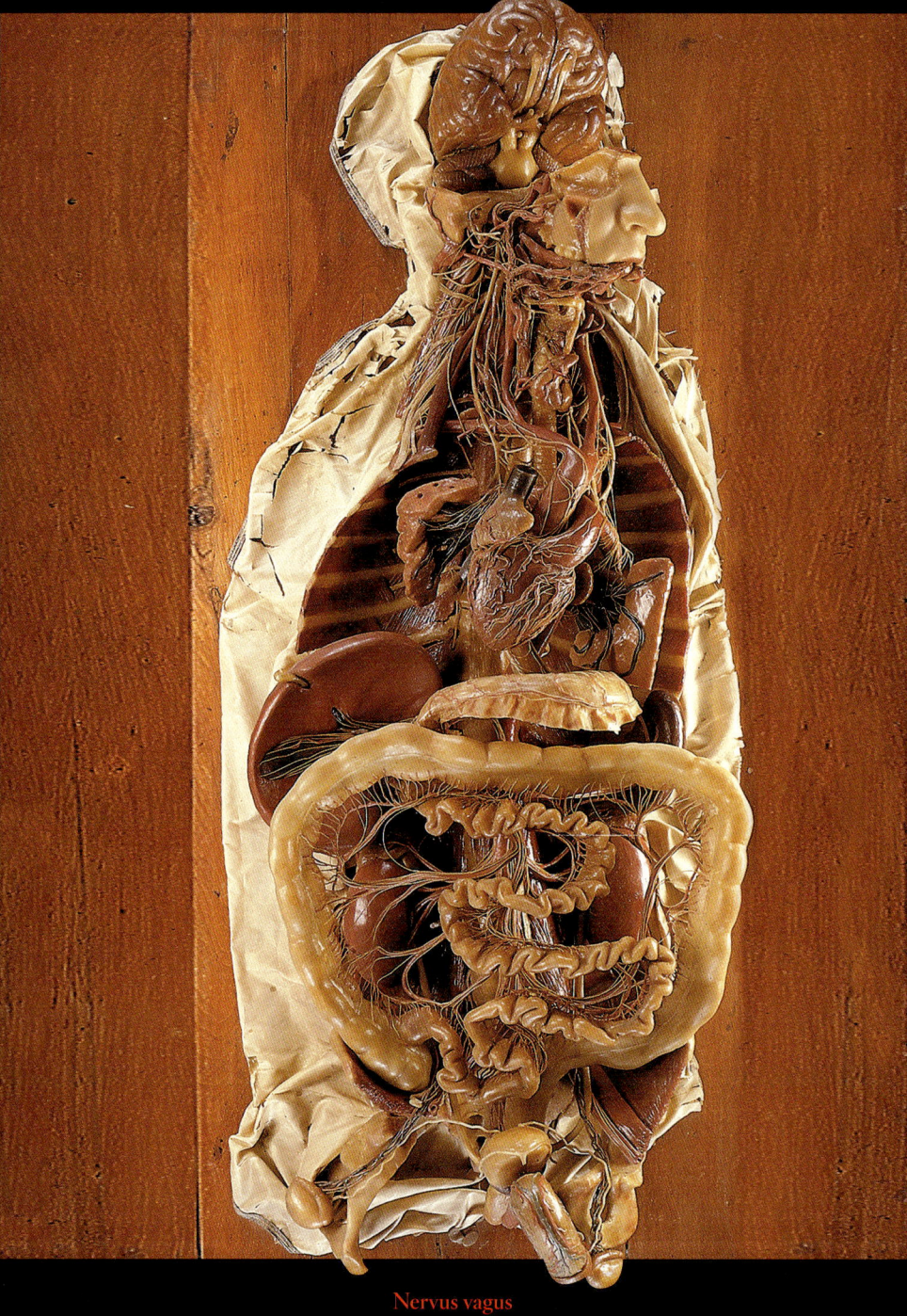

Nervus vagus

Display showing the vagus nerve and its area of innervated thoracic and abdominal cavities

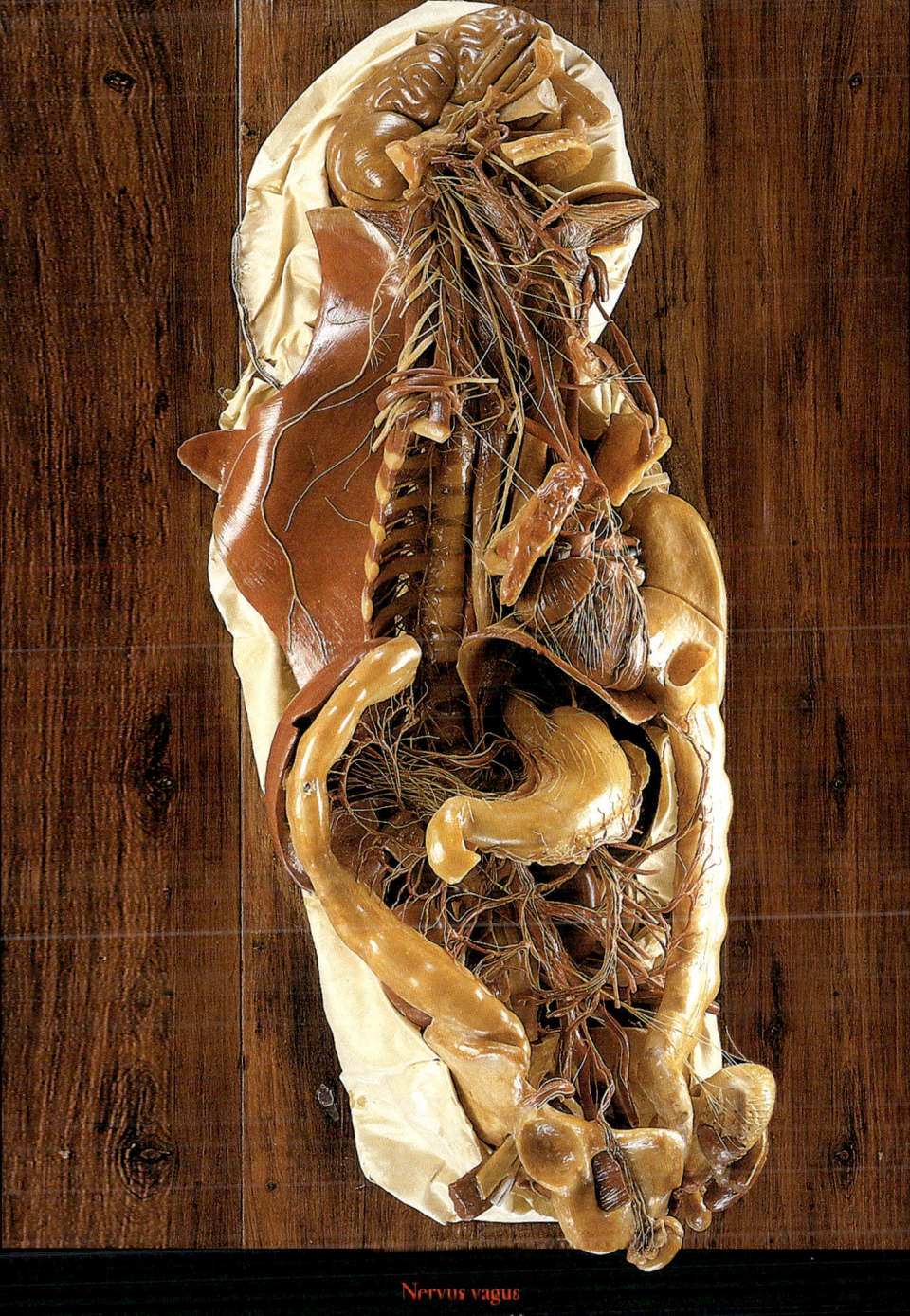

Nervus vagus

Display showing the vagus nerve and the area innervated by it in the thoracic and abdominal cavities

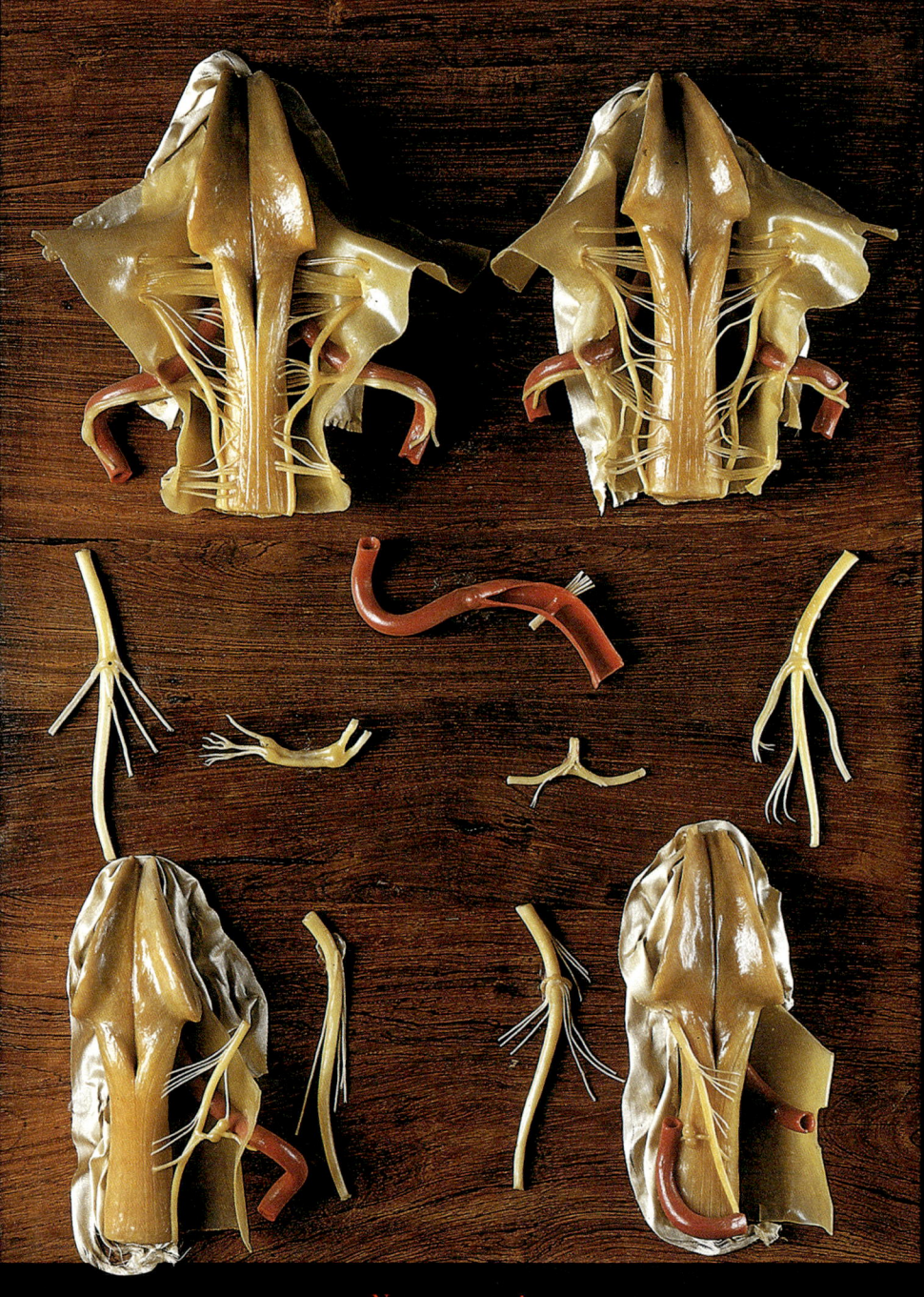

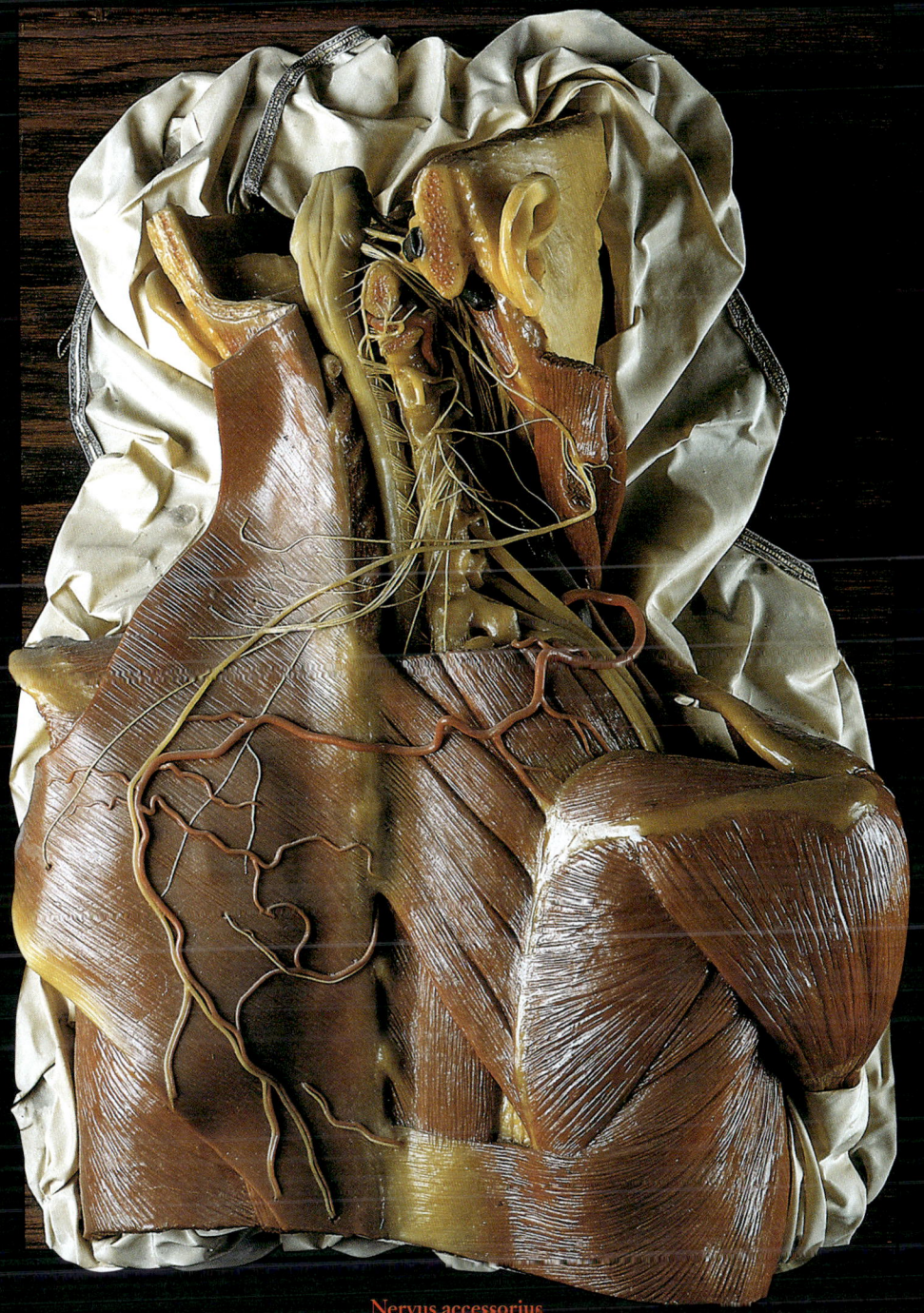

Nervus accessorius

Display showing the course of the eleventh cranial nerve to the trapezius muscle

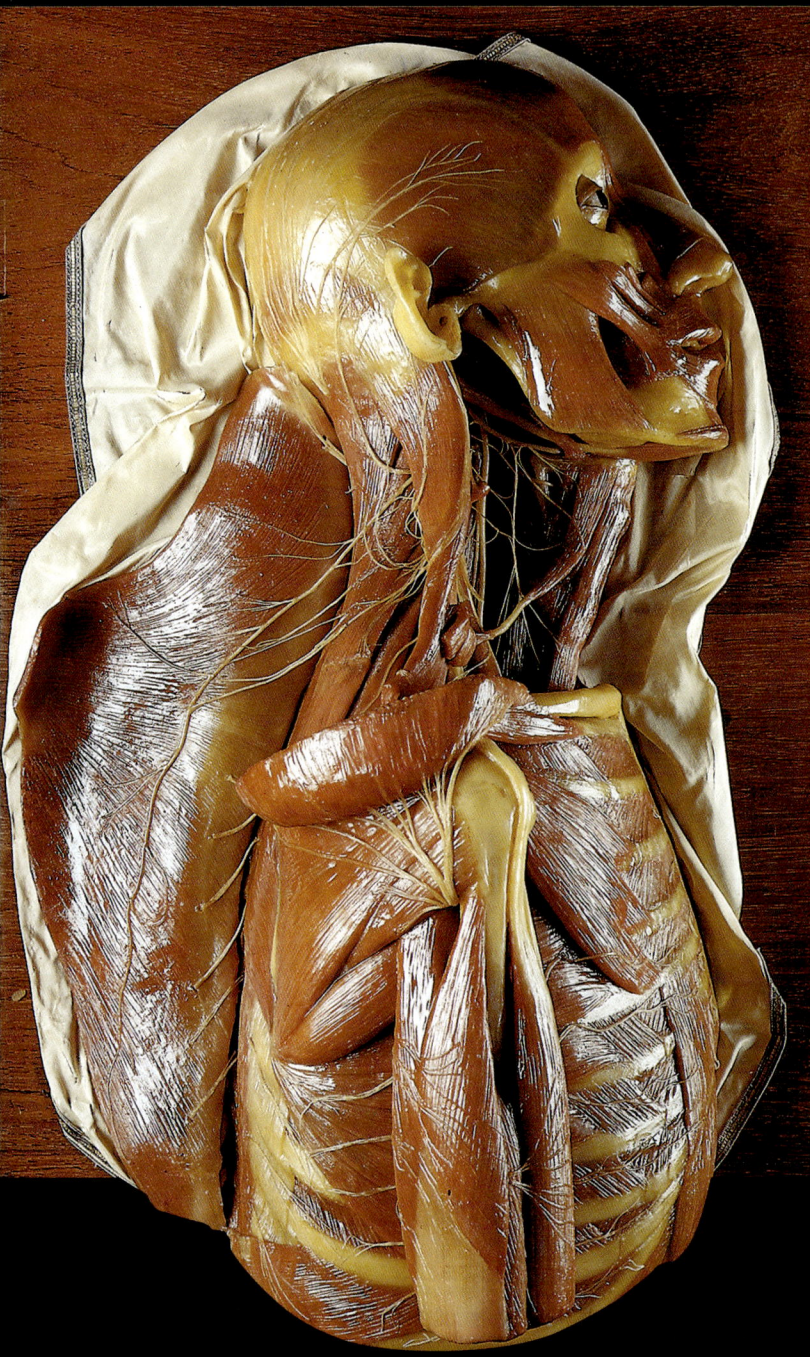

Nervus accessorius, Nervus axillaris

Display showing the course of the eleventh cranial nerve to the trapezius muscle

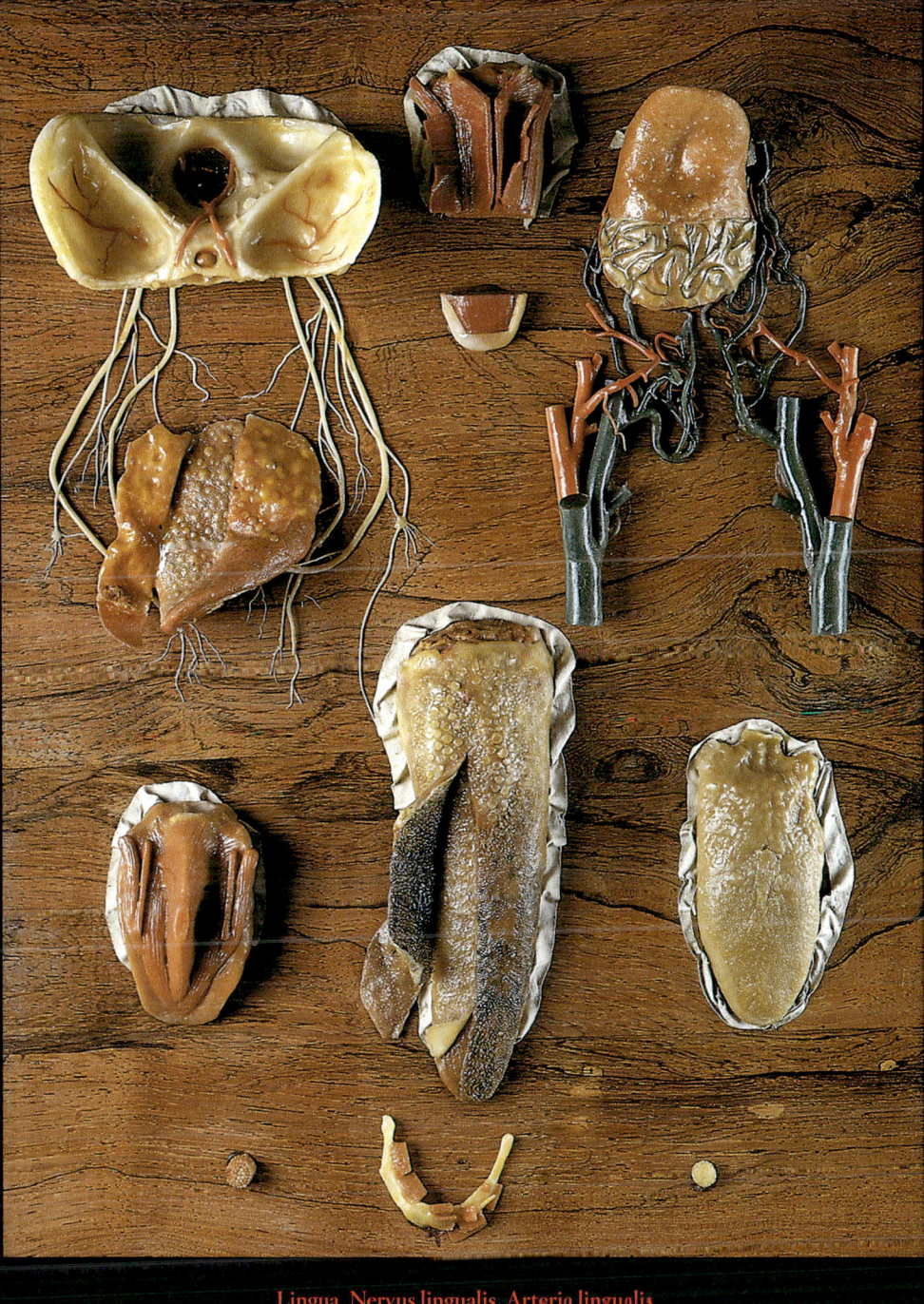

Lingua, Nervus lingualis, Arteria lingualis

Nerves and vessels of the tongue

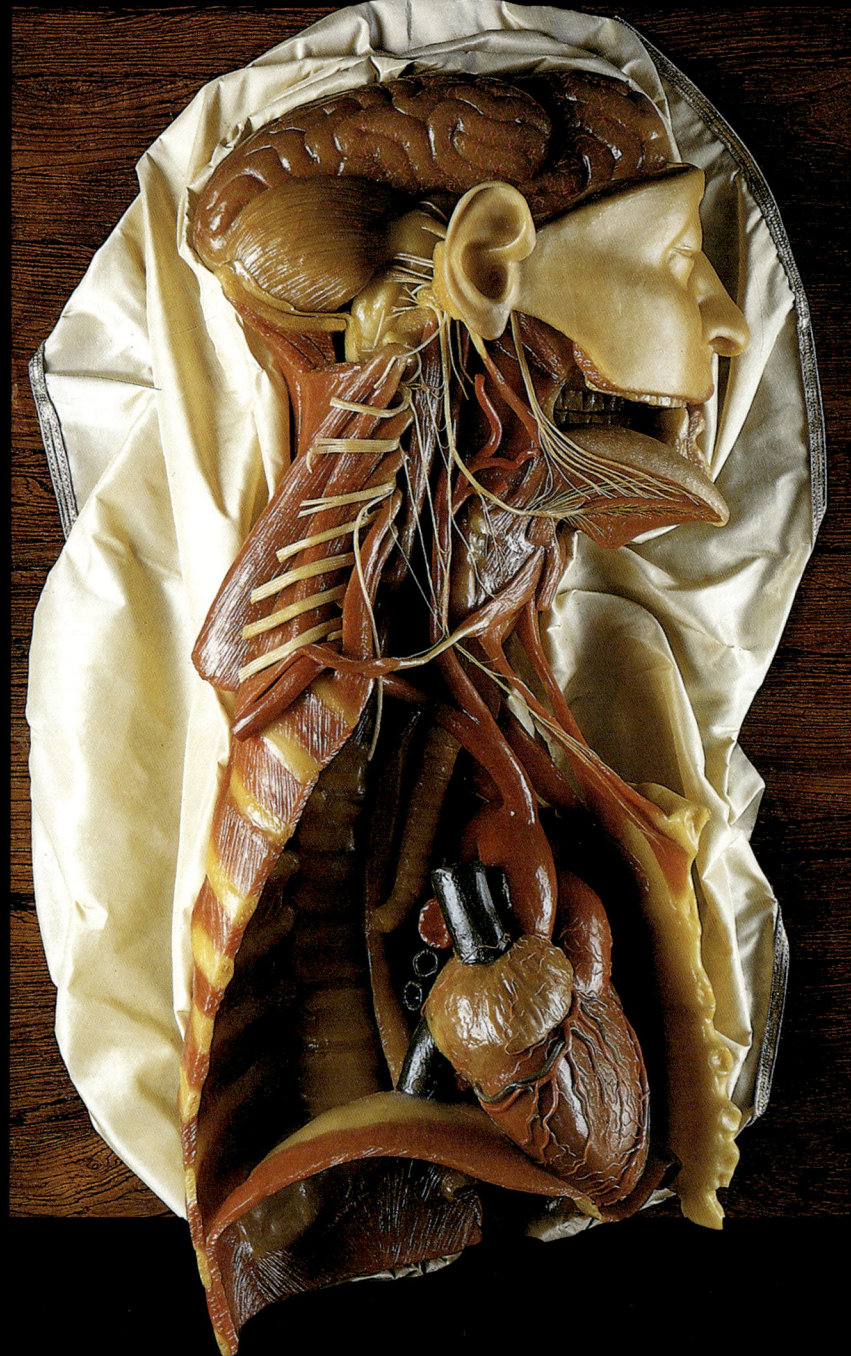

Display showing the innervation of the tongue; the lingual nerve running down in front of,
and the hypoglossal nerve behind, the ear

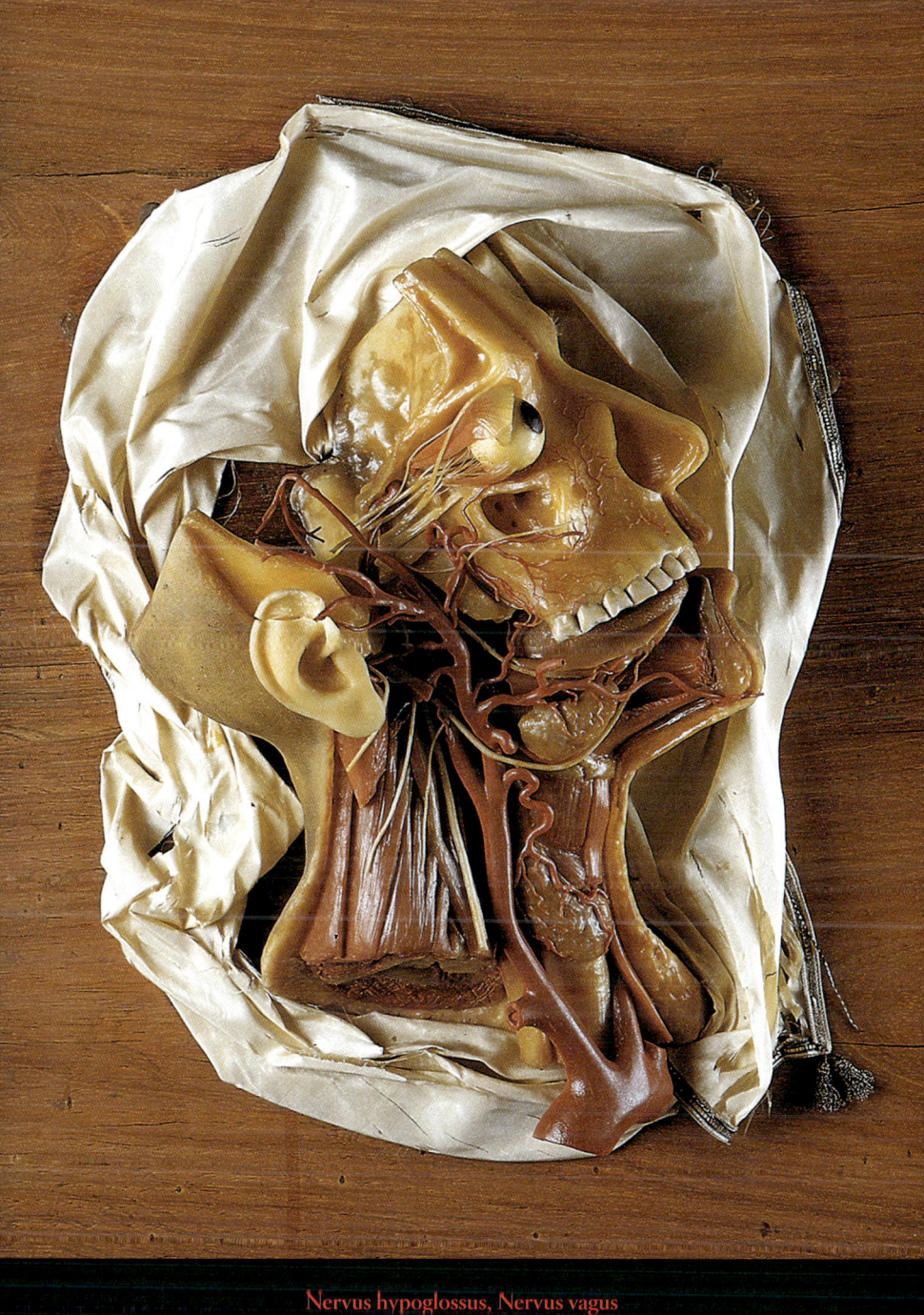

The nerves and arteries of the orbit, tongue and neck

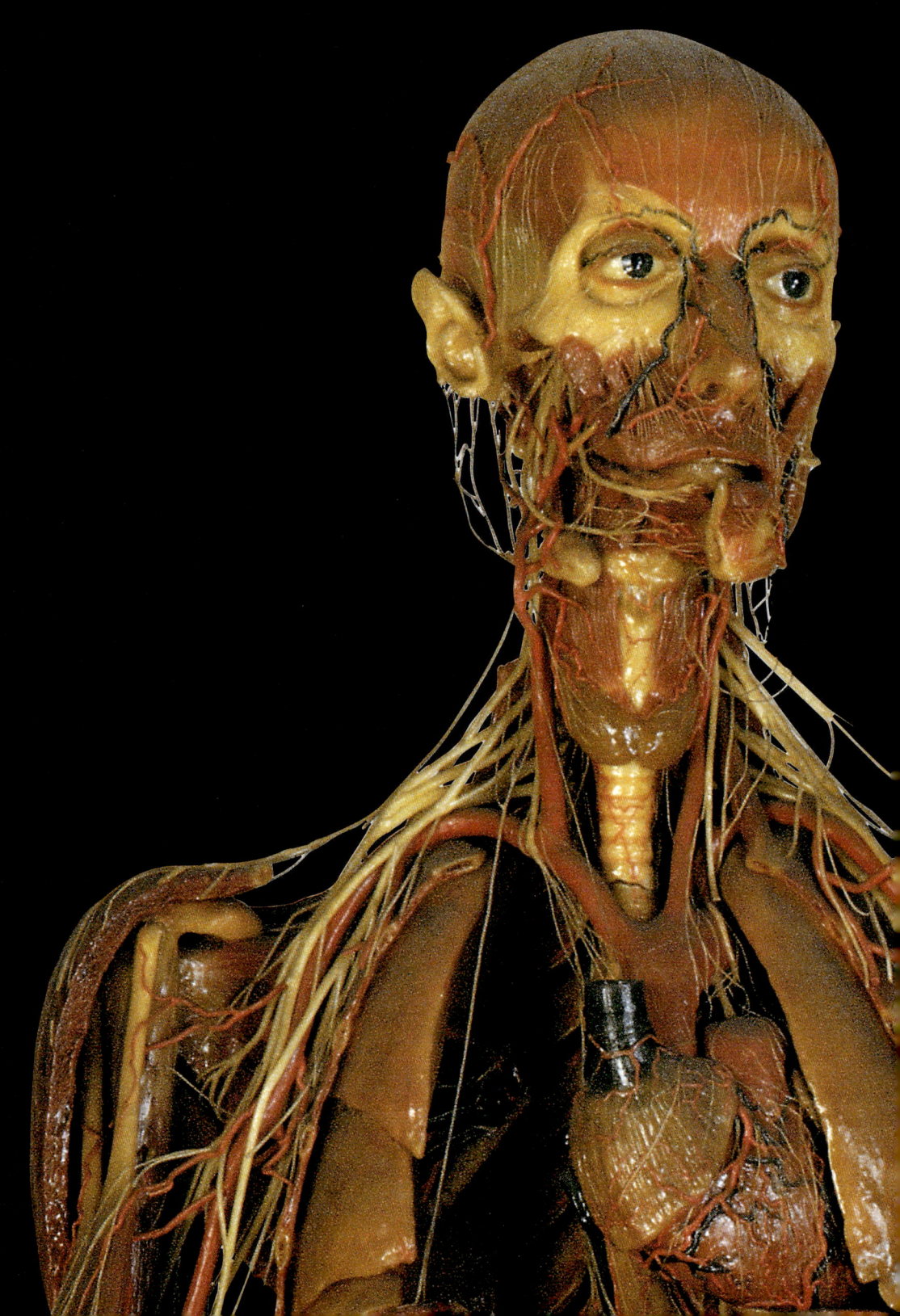

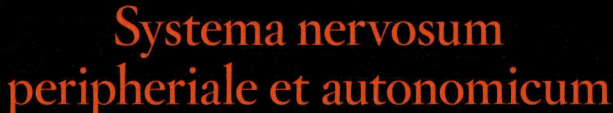

Systema nervosum peripheriale et autonomicum

Spinal Nerves and Autonomic Nervous System

Spinalnerven und autonomes Nervensystem

Nerfs spinaux et système nerveux autonome

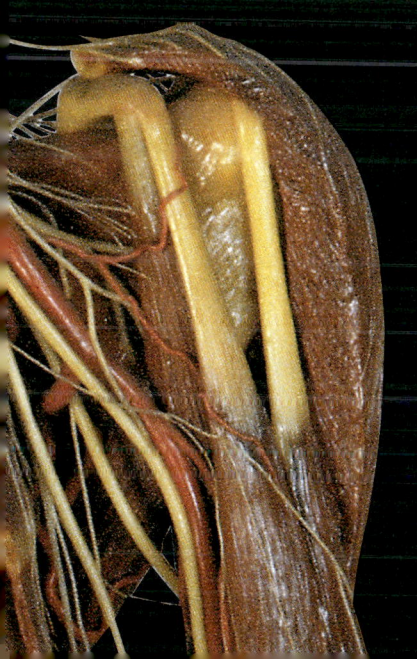

Spinal Nerves and Autonomic Nervous System

The spinal nerves – or nerves of the spinal cord – are shown in their relationship to the central nervous system, similar to the models displaying the course of the cranial nerves. The models show very clearly that the basic principle of segmentation of the spinal cord, together with the emergence of the nerve roots and ganglia and their position within the spinal canal, was already recognised by the end of the 18th century. It is striking to observe how the alternating topographical relationship between the nerves and the adjacent vertebrae is reproduced in the models. Some of the specimens display the large nerve plexuses which lie at the side of the cervical and lumbar vertebrae. It is remarkable how well they reproduce the complexity of the nerve trunks which innervate (that is to say, carry the nerve impulses for) the limbs. In the neck, thoracic and pelvic regions, elements of the sympathetic trunk as branches of the autonomic nervous system can be recognised.

In many of the models of all regions of the body one can fully understand the course and topography of the peripheral nerves. In many cases the complete course of the connective tissue septa and the muscle segmentation are displayed. Again, in other specimens, the most striking thing is the impressive way in which the fine autonomic nerve plexuses of the heart, hilum of the liver and mesentery have been reproduced.

Spinalnerven und autonomes Nervensystem

Die Spinal- oder Rückenmarksnerven werden – ähnlich den Modellen der Hirnnervenverläufe – in ihrem Bezug zum zentralen Nervensystem dargestellt. Diese Modelle zeigen deutlich, dass bereits im ausgehenden 18. Jahrhundert das Grundprinzip der Rückenmarksgliederung, die austretenden Nervenwurzeln und Ganglien sowie die Aufhängung im Wirbelkanal, erkannt war. Die topographische Beziehung der Nervenaustritte zwischen benachbarten Wirbelkörpern ist in bemerkenswerter Weise im Modell umgesetzt. Einige Präparate zeigen die großen Nervengeflechte, die im seitlichen Bereich der Hals- und Lendenwirbelsäule liegen. Sie dokumentieren eindrucksvoll, wie komplex die Nervenstämme zusammengesetzt sind und wie

Pages 380–381:
Detail from page 417

sich schließlich die Nerven für die Innervation (die Reizübertragung durch die Nerven) der Extremitäten formieren. Im Hals-, Brust- und Beckenbereich sind Abschnitte des Truncus sympathicus als einem der Schenkel des autonomen Nervensystems zu erkennen.

In einer großen Anzahl von Modellen aus allen Regionen unseres Körpers kann der Betrachter die Verläufe sowie die Topographie der peripheren Nerven nachvollziehen. Vielfach sind zur Darstellung des gesamten Verlaufes die Bindegewebssepten und Muskellogen aufgeschnitten. In anderen Präparate wiederum liegt das Hauptaugenmerk auf der Wiedergabe der sehr feinen autonomen Nervenfasergeflechte des Herzens, der Leberpforte und des Darmgekröses.

Nerfs spinaux et système nerveux végétatif

Les nerfs spinaux, ou nerfs rachidiens, sont représentés, comme les nerfs crâniens, en connexion avec le système nerveux central. Ces préparations montrent clairement qu'à la fin du XVIIIᵉ siècle étaient connus les principes fondamentaux de la subdivision de la mœlle épinière, la disposition des racines et des ganglions des nerfs spinaux, leurs moyens de suspension dans le canal vertébral. Les rapports topographiques des nerfs spinaux entre deux vertèbres voisines sont représentés de façon remarquable. Quelques préparations montrent les grands plexus nerveux disposés de part et d'autre des segments cervical et lombaire de la colonne vertébrale ; elles montrent la complexité de la formation des troncs nerveux et la constitution des nerfs destinés aux membres. Au niveau du cou, du thorax et du bassin, peuvent être reconnus des éléments du tronc sympathique, composante du système nerveux végétatif.

Le trajet et les rapports topographiques des nerfs périphériques peuvent être suivis et analysés sur de nombreux modèles concernant toutes les régions du corps ; souvent l'ouverture des cloisons fibreuses et des loges musculaires permet de suivre un nerf sur tout son trajet. Sur certaines préparations, au niveau du cœur, du hile du foie, ou du mésentère, l'attention est attirée sur la disposition des réseaux nerveux constitués par les fines fibres nerveuses végétatives.

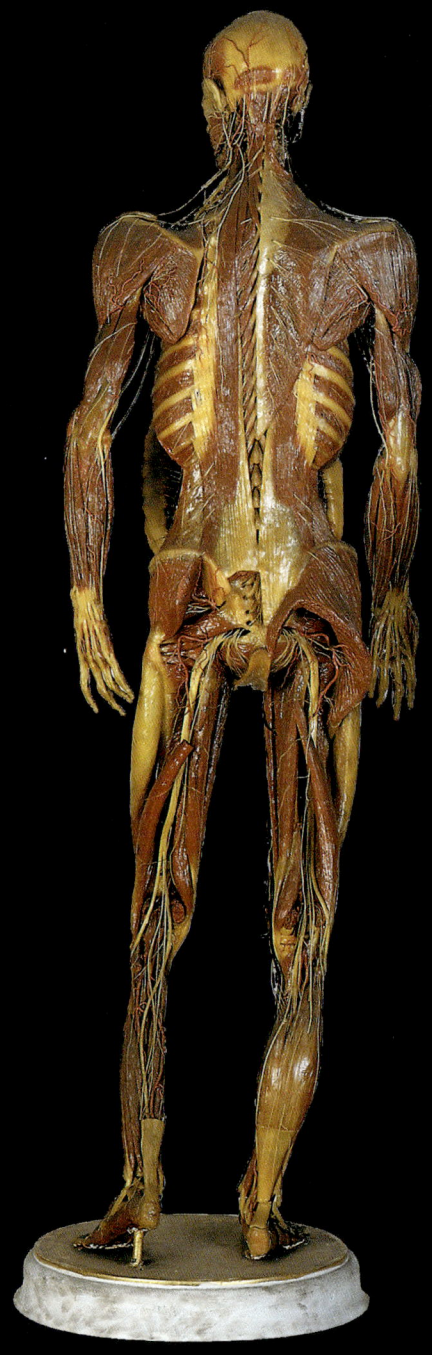

Nervus ischiadicus

Whole body specimen with muscles and nerves from behind

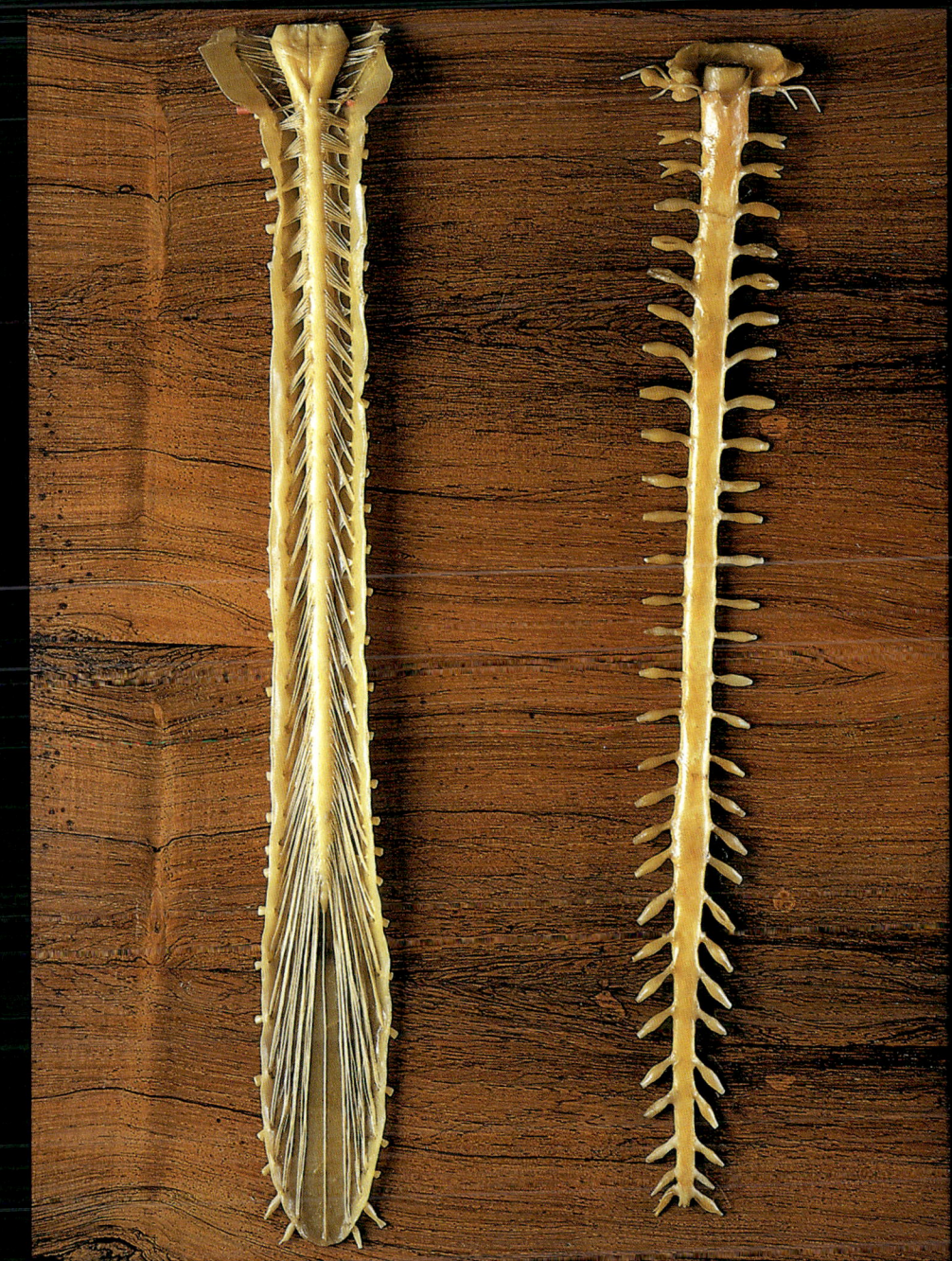

Medulla spinalis, Dura mater spinalis, Ganglia spinalia

View of the spinal cord from behind. Left: with the dura mater. Right: showing the spinal ganglia

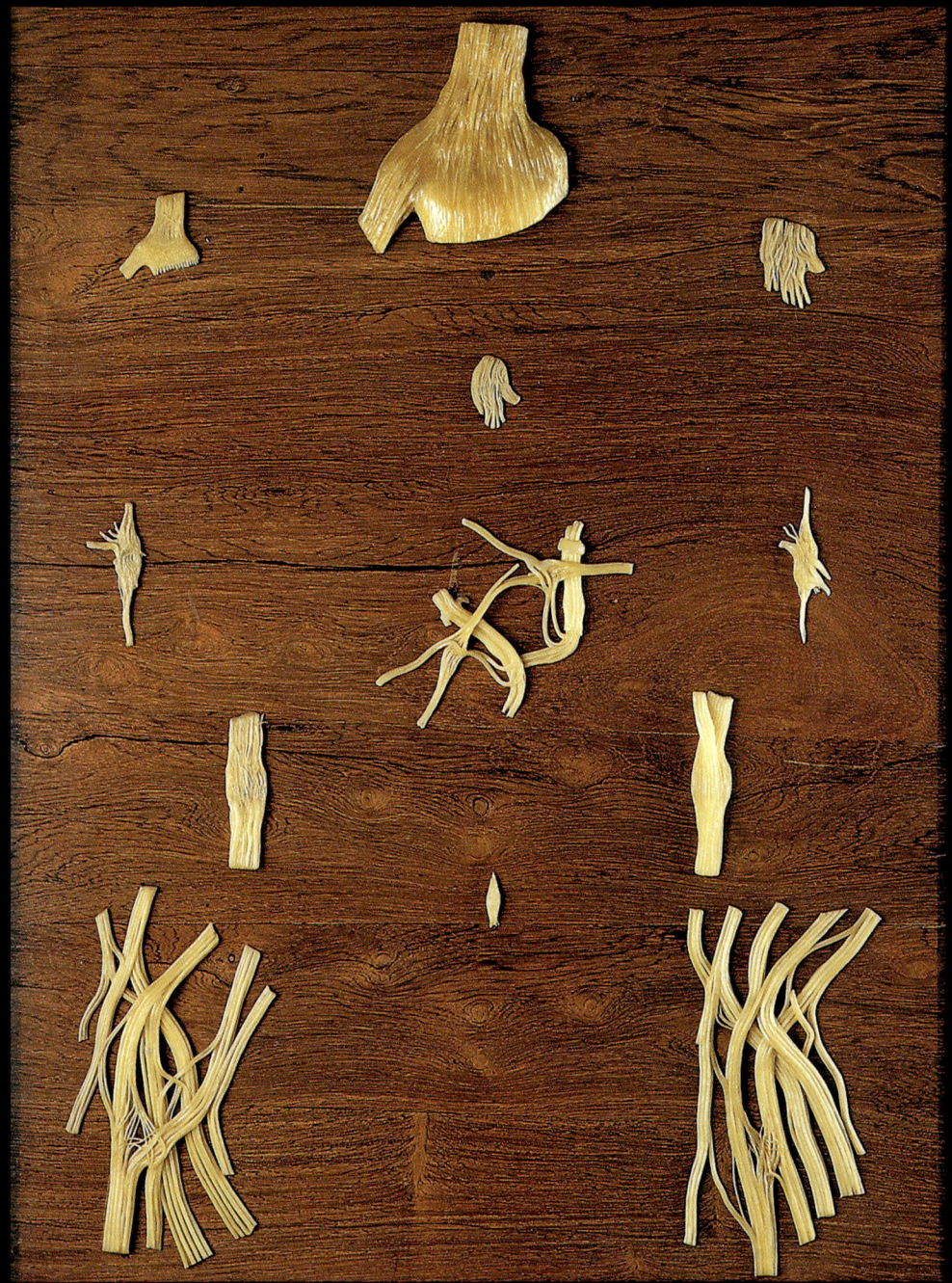

Plexus brachialis, Ganglia autonomia, Ganglion trigeminale
Below: display showing the nerve plexus of the arm. Above: the trigeminal ganglion

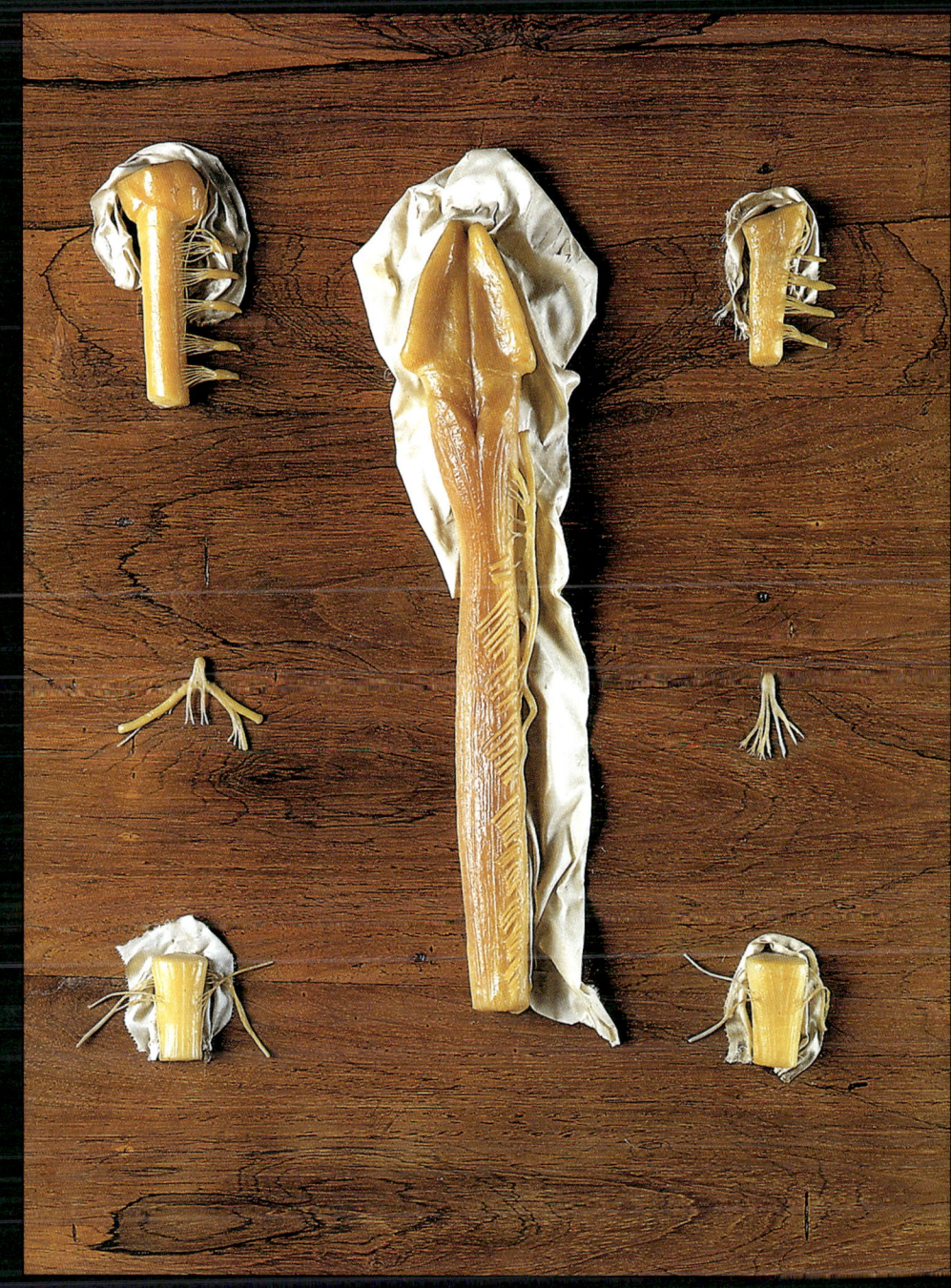

Medulla spinalis, Fila radicularia

Brainstem and upper spinal cord seen from behind. The nerve roots are sometimes only shown on one side

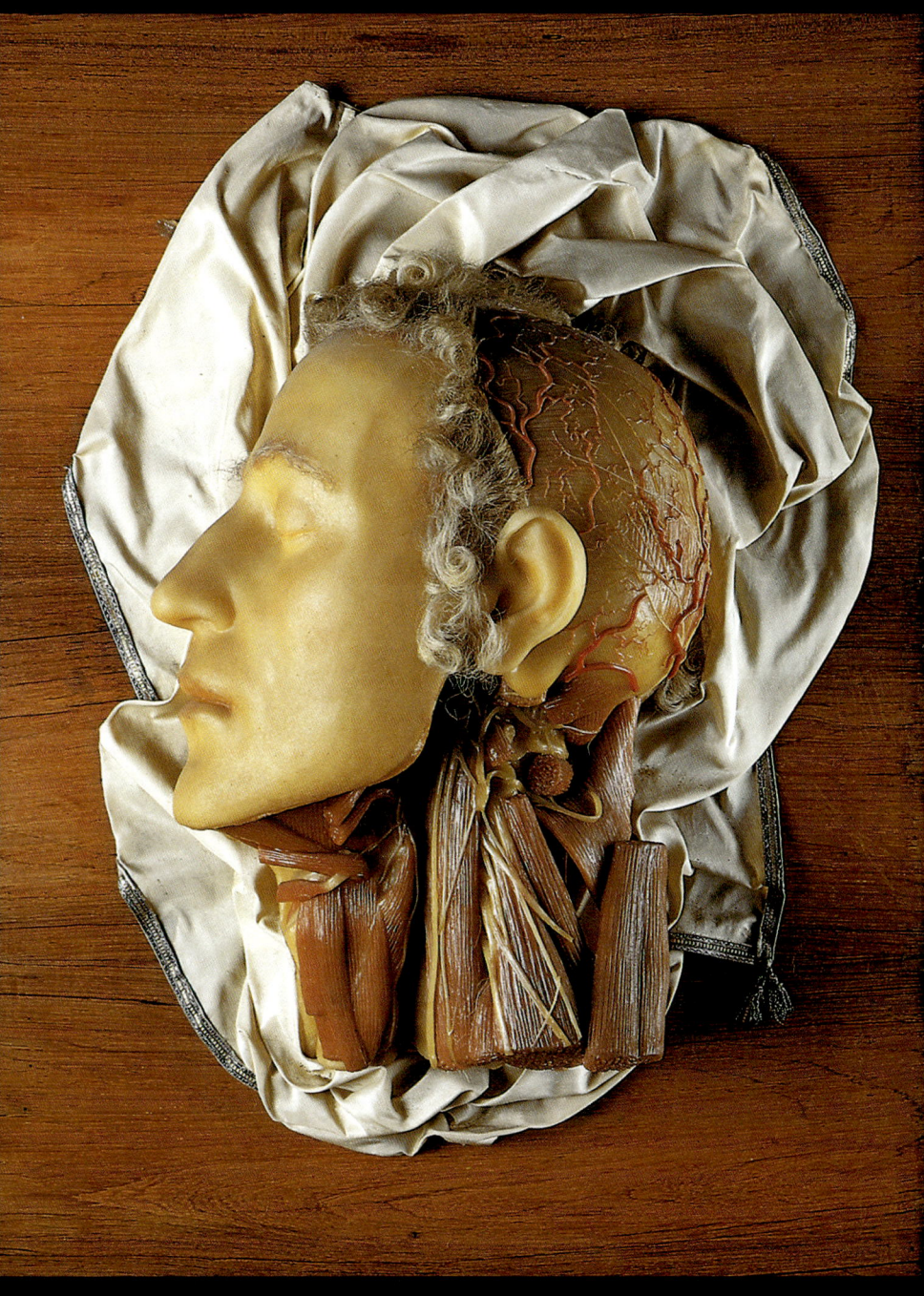

Plexus cervicalis, Ganglion cervicale superius
View of the cervical nerves and of the back of the head after removal of the skin

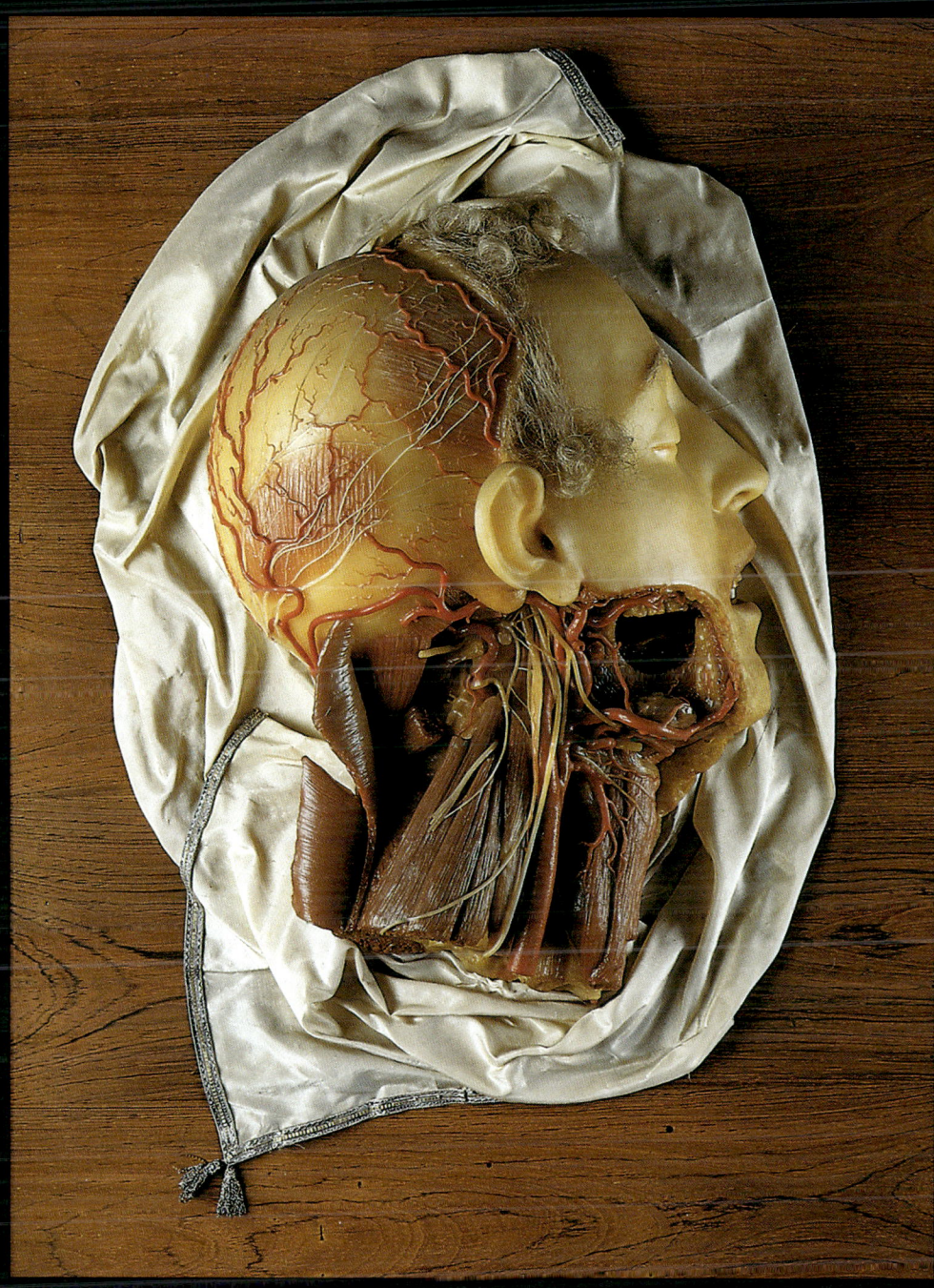

Nervus occipitalis major, Nervus vagus

Display showing the nerves at the back of the head and the neurovascular bundles
in the deep region of the neck

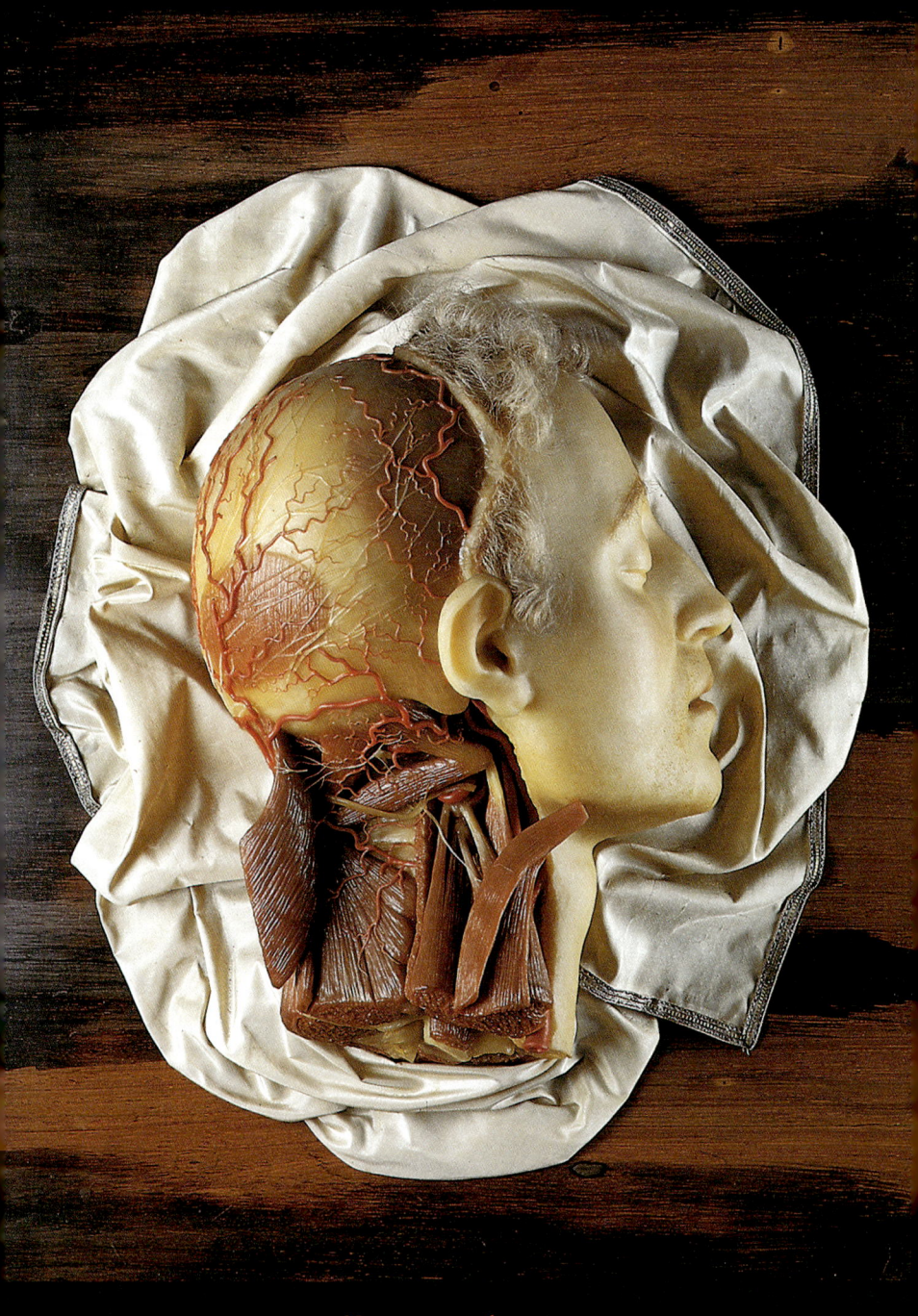

Nervus occipitalis major

Display showing the course of a nerve at the back of the head

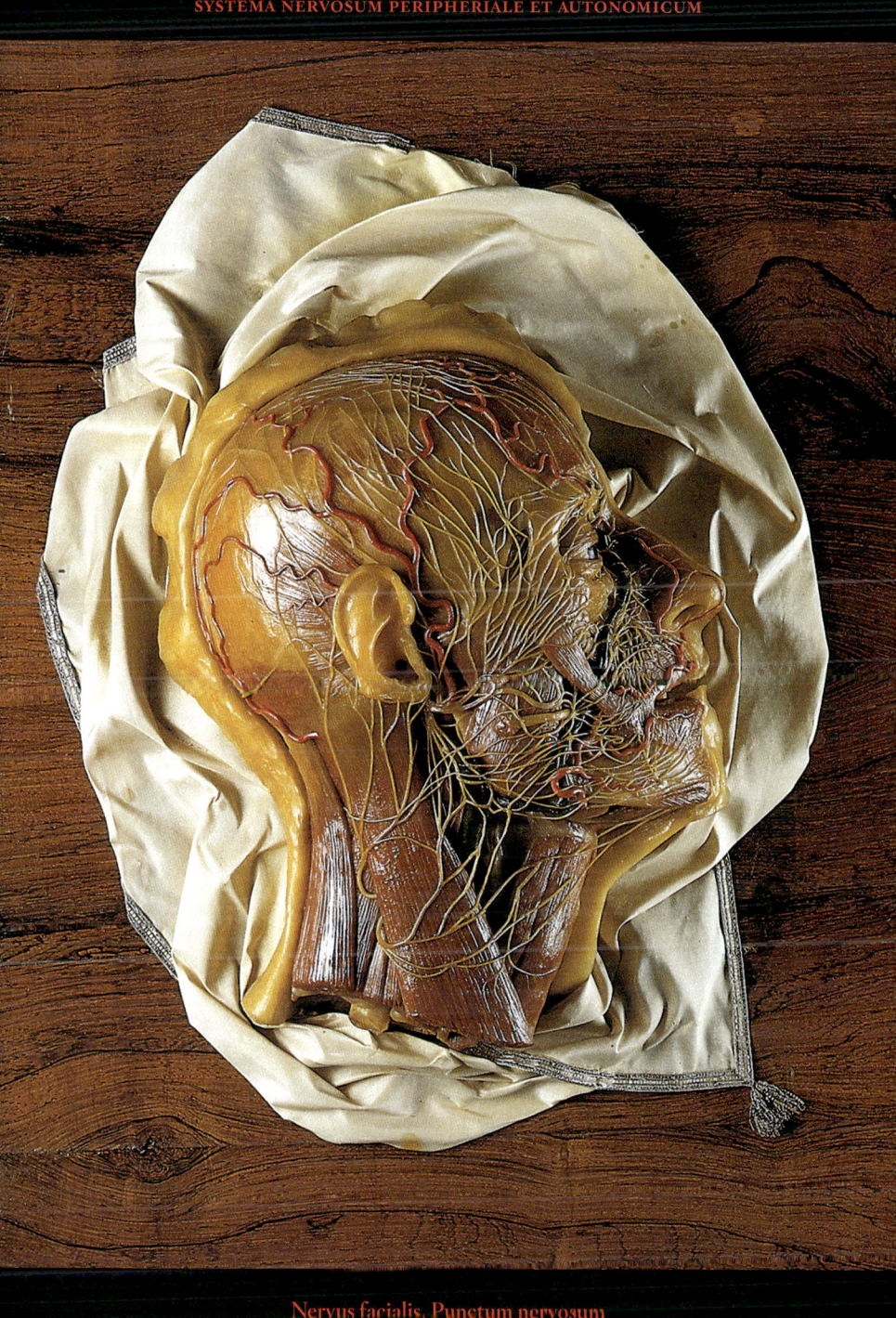

Nervus facialis, Punctum nervosum

Display showing the branches of the facial nerve and the cutaneous branches
of the cervical plexus

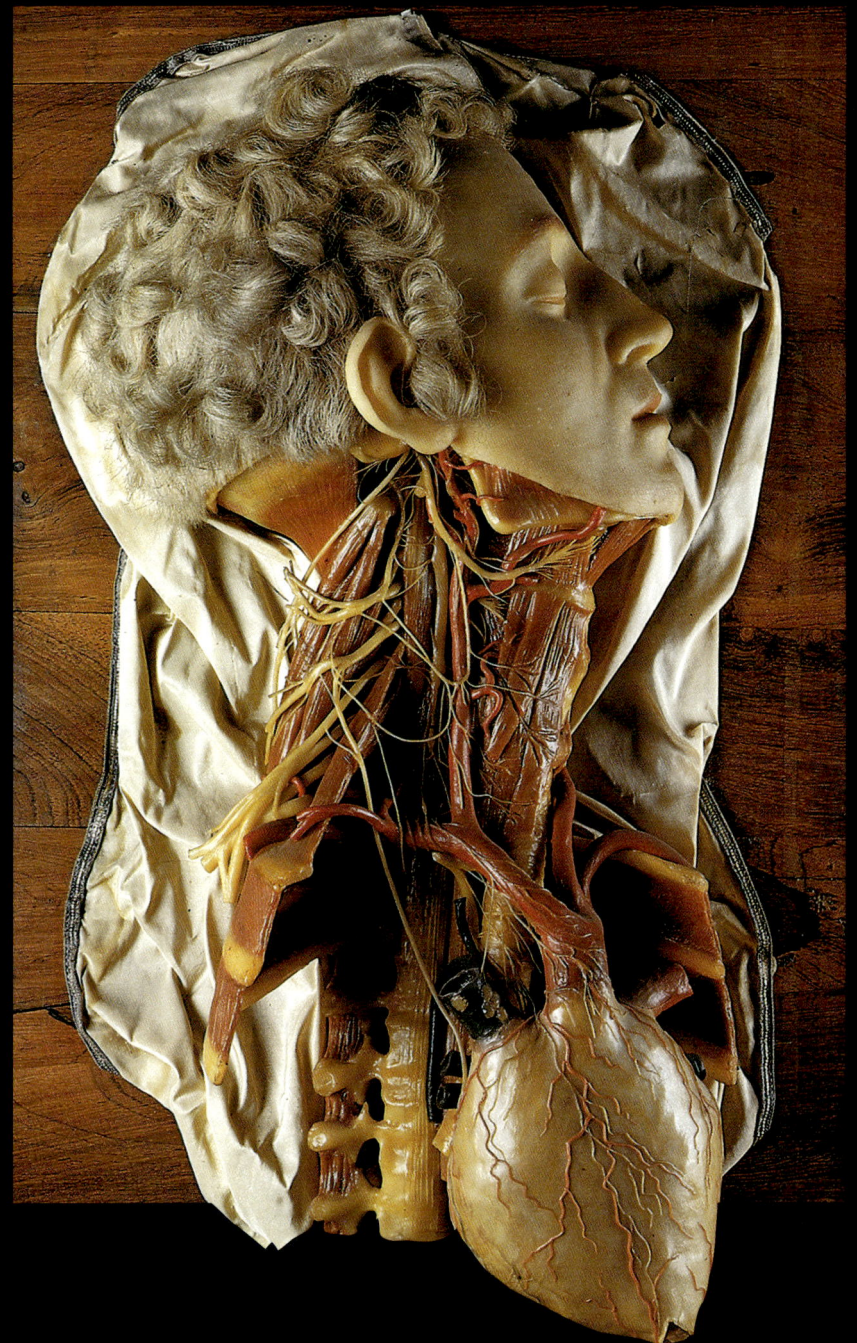

Plexus cervicalis, Nervus phrenicus, Nervus hypoglossus, Ansa cervicalis profunda, Plexus brachialis

View of the deep region of the neck and into the opened thoracic cage. The heart can be seen.
Observe the nerves supplying the diaphragm running along the side of the pericardial wall

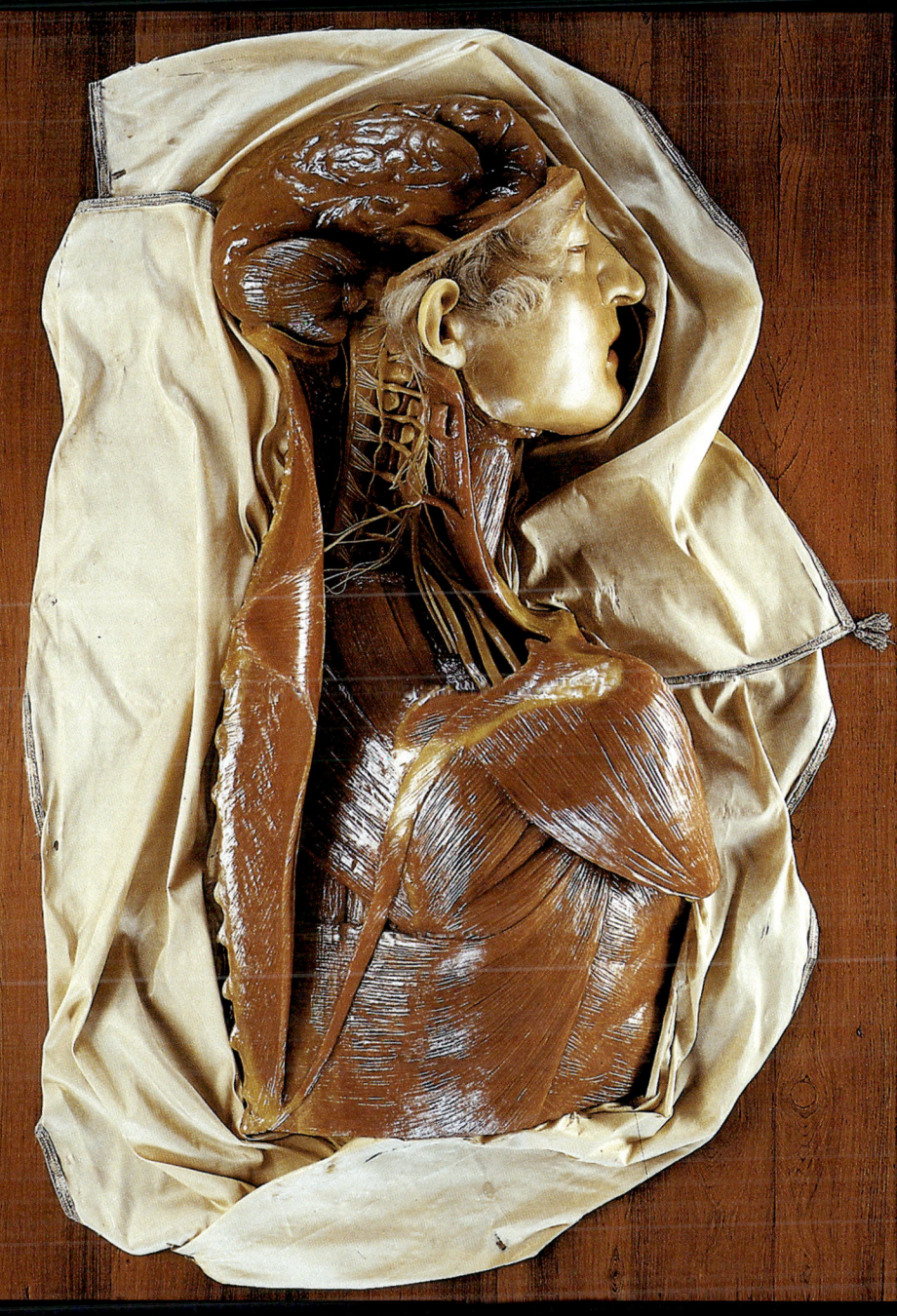

Plexus brachialis

The nerve plexus of the upper limb

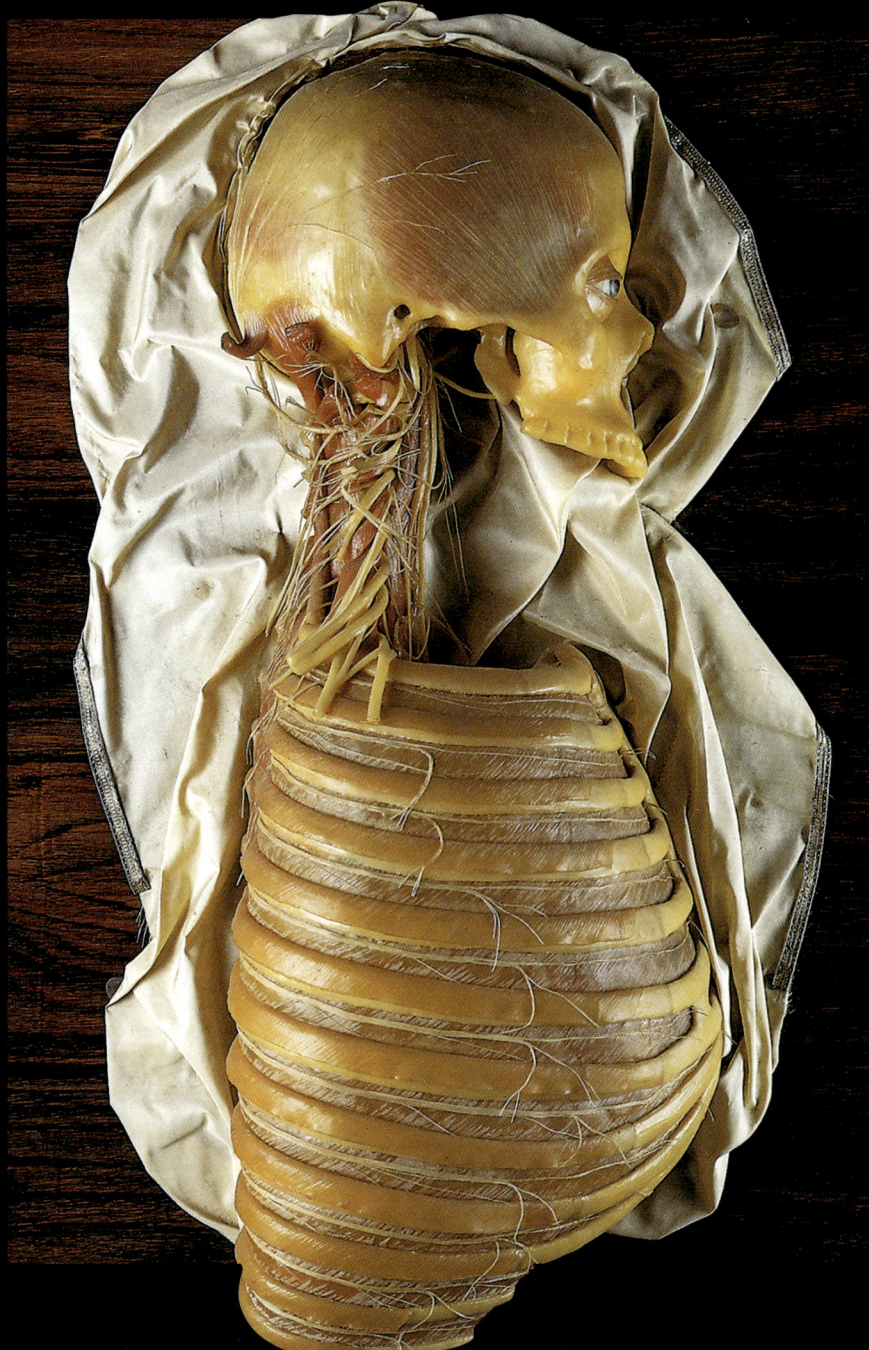

Plexus cervicalis, Plexus brachialis

The nerves in a deeper layer of the neck. Above the thoracic cage
one can see parts of the nerve plexus of the upper limb

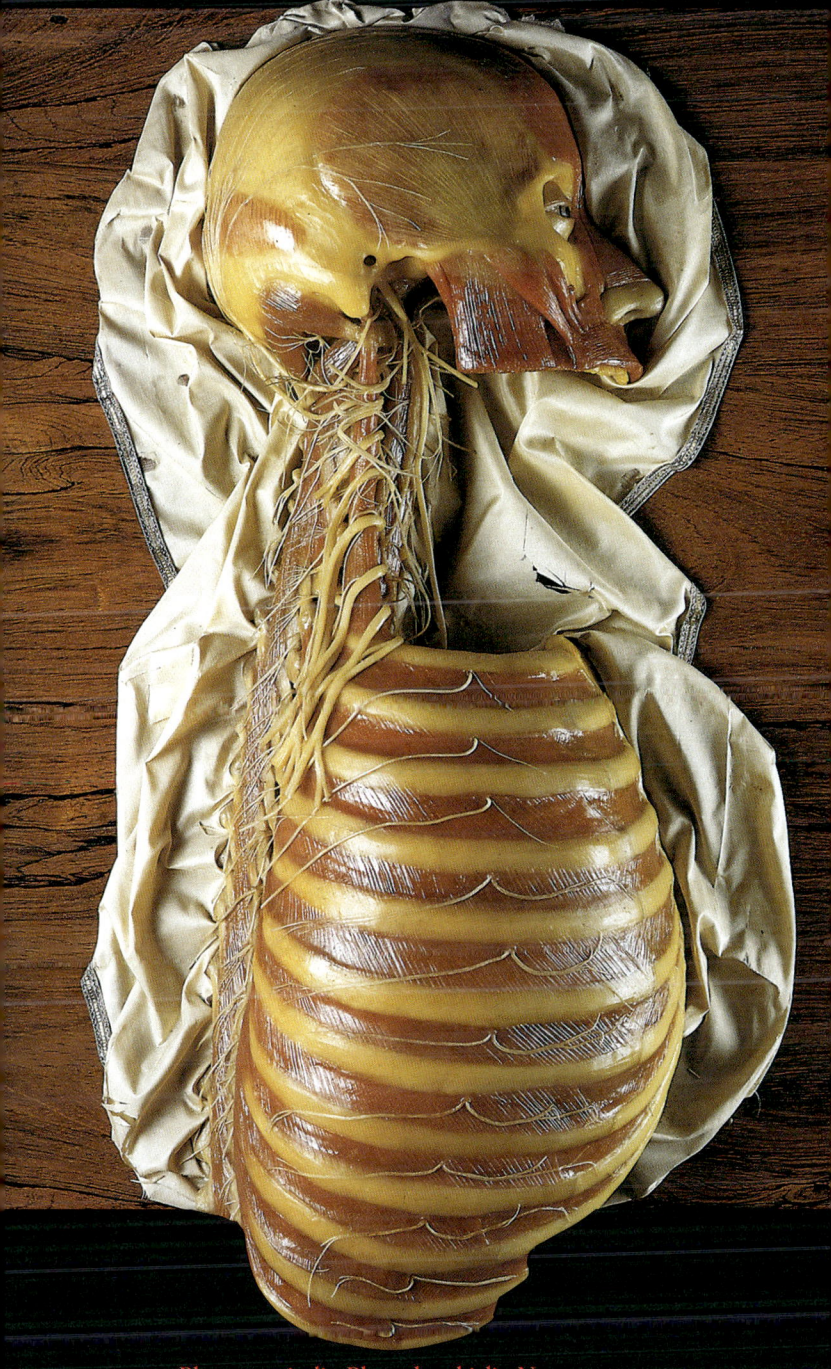

Plexus cervicalis, Plexus brachialis, Nervus vagus

Picture four showing cervical plexus and plexus of the upper limb

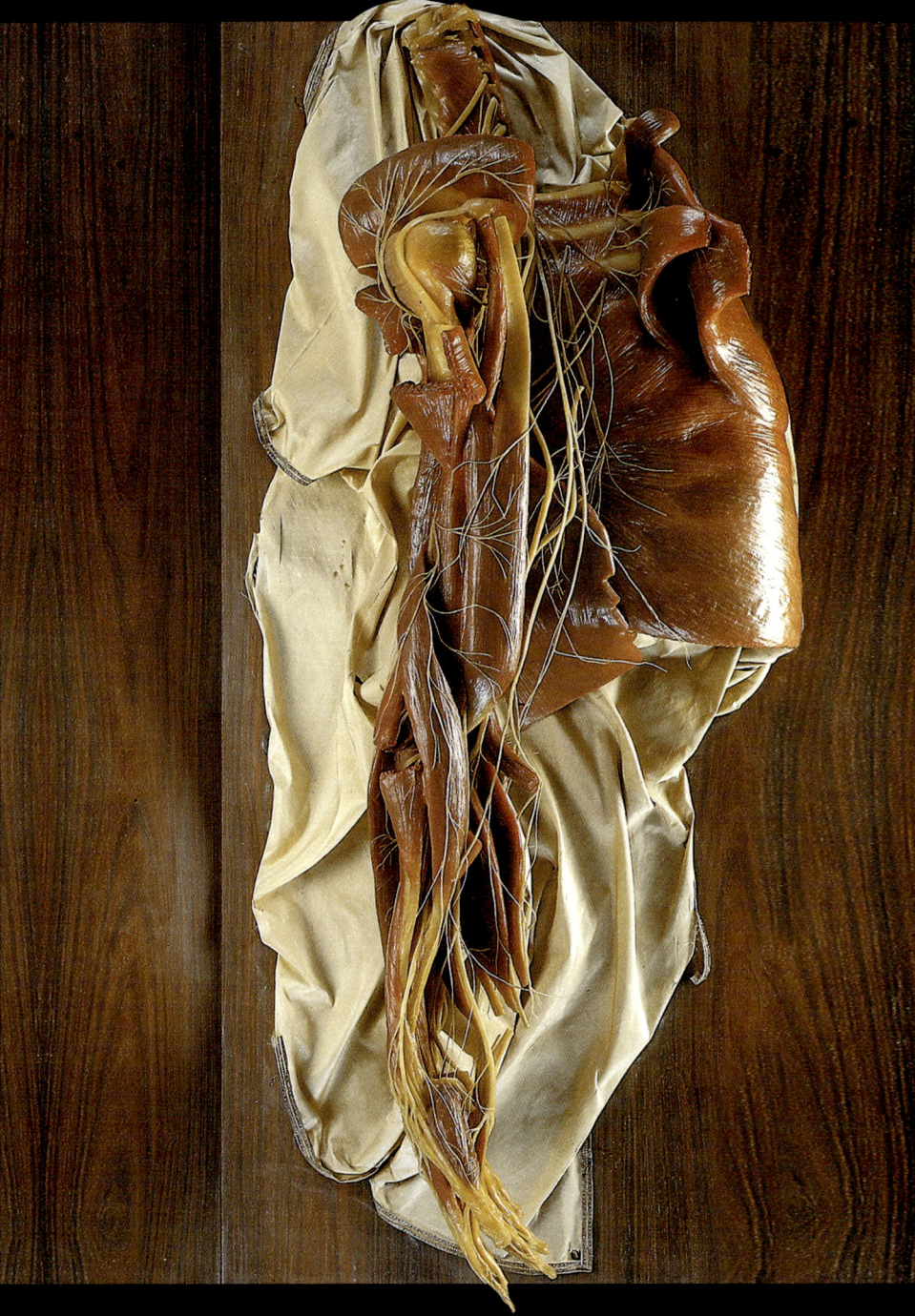

Nervi membri superioris

The nerves of the upper limb and side of the chest wall

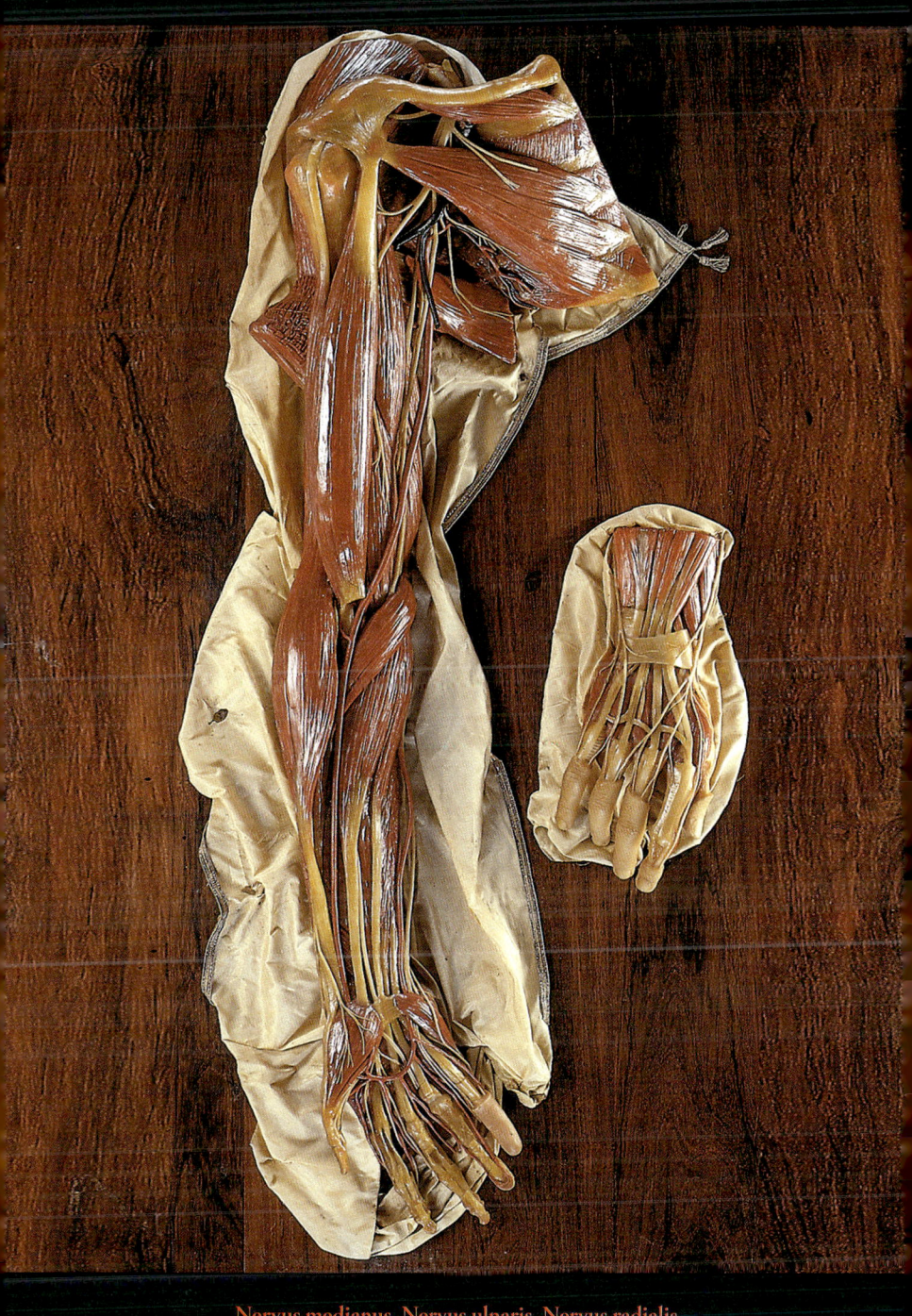

Nervus medianus, Nervus ulnaris, Nervus radialis

The nerves and muscles of the upper limb

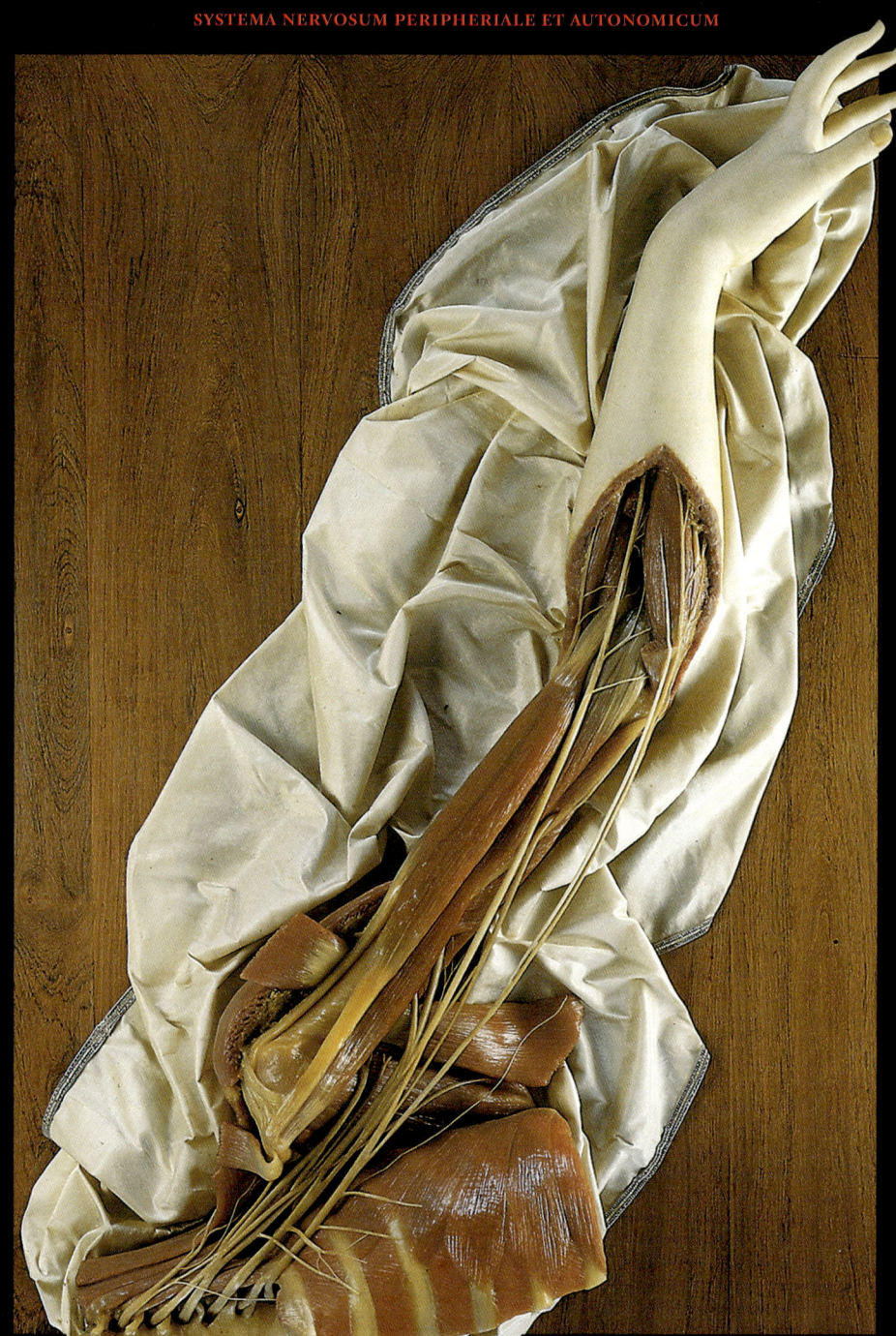

Plexus brachialis

The nerves and plexus of the upper limb

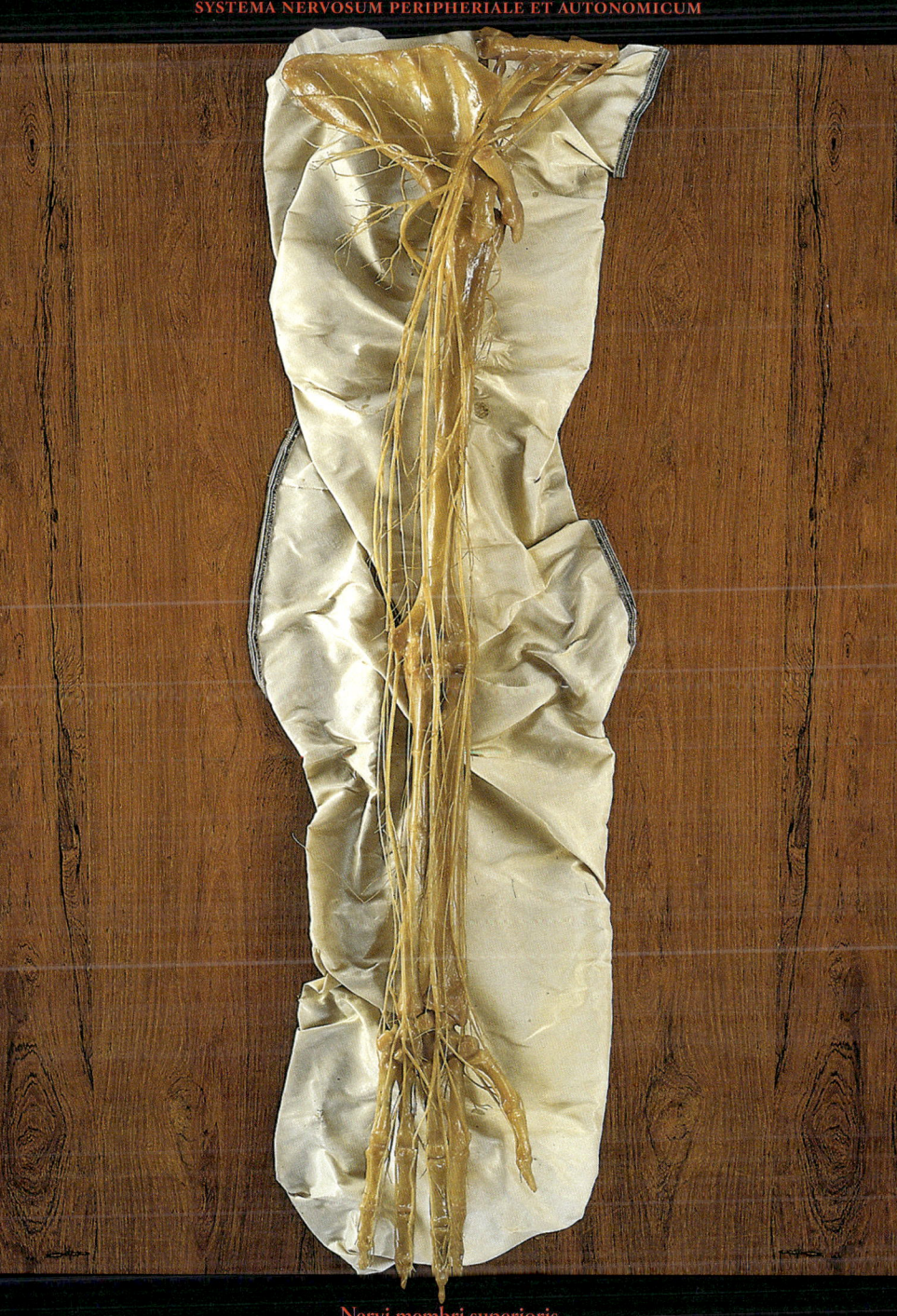

Nervi membri superioris

The skeleton of the left arm and shoulder seen from in front. The nerves and plexus of the arm, together with the associated region of the spinal cord, can also be seen

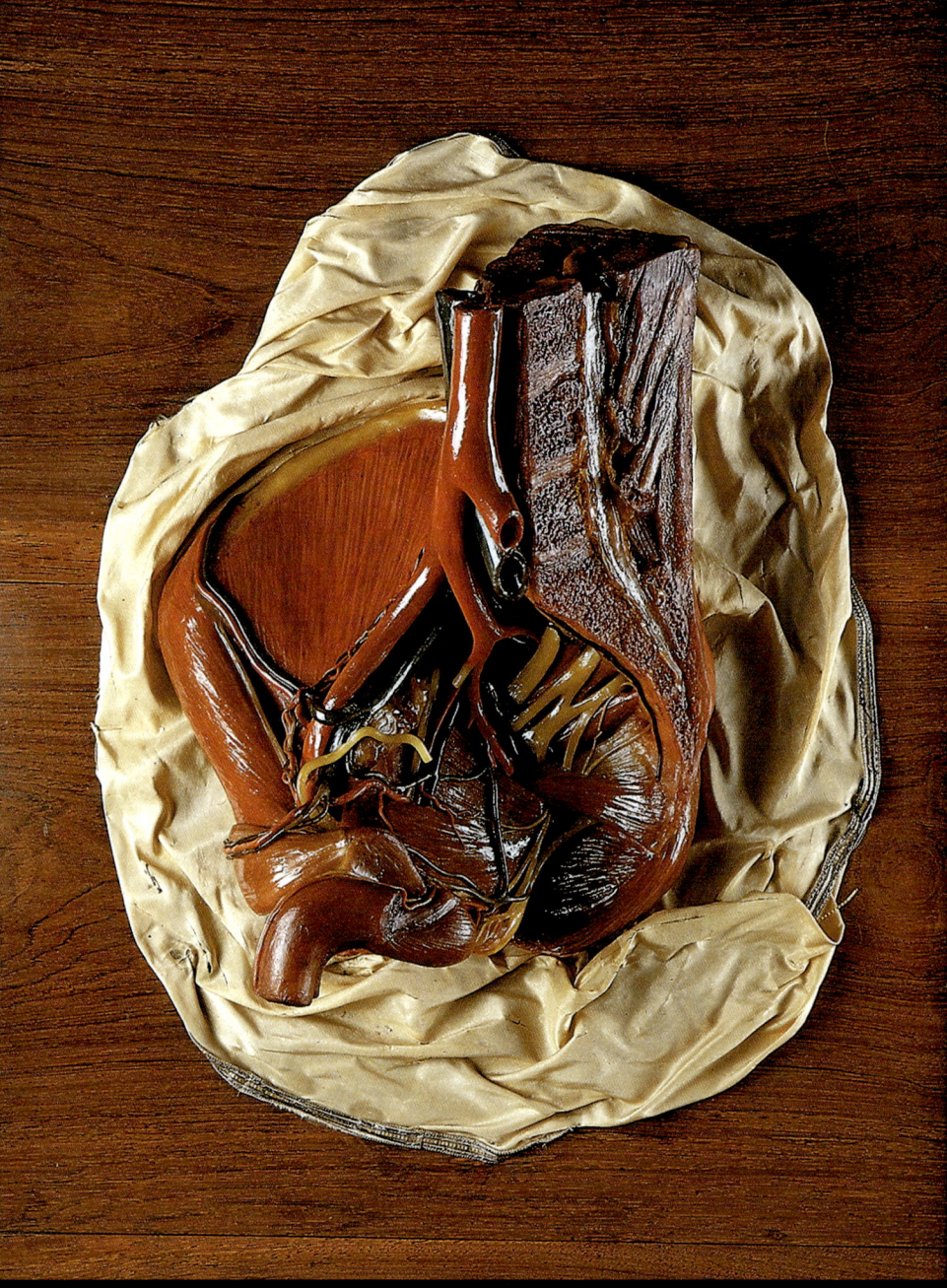

Plexus lumbosacralis

A male pelvis divided in the midline; view of the nerve plexus of the lower limb

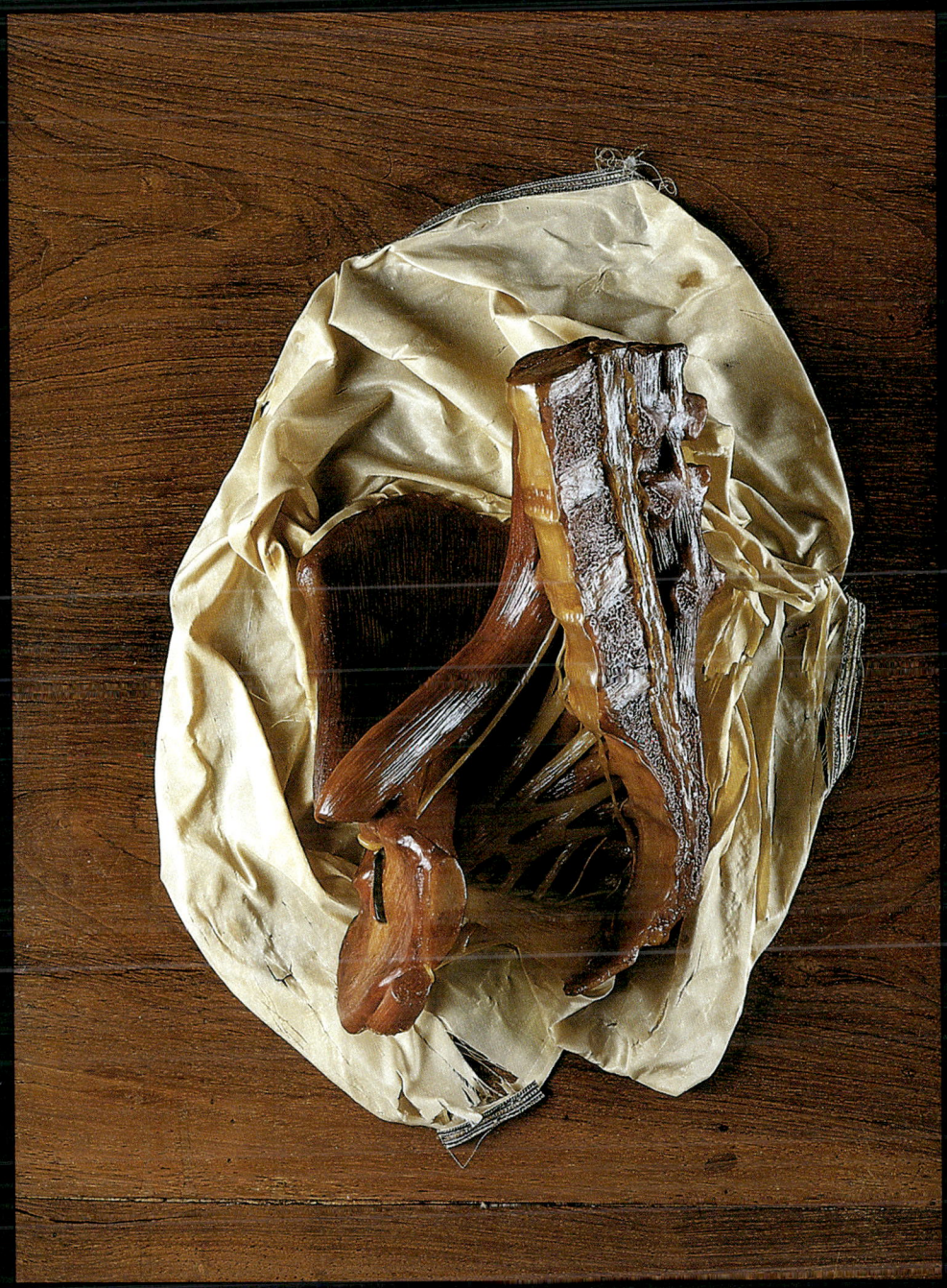

Plexus lumbosacralis
Pelvis divided in the midline; view of the nerve plexus of the lower limb

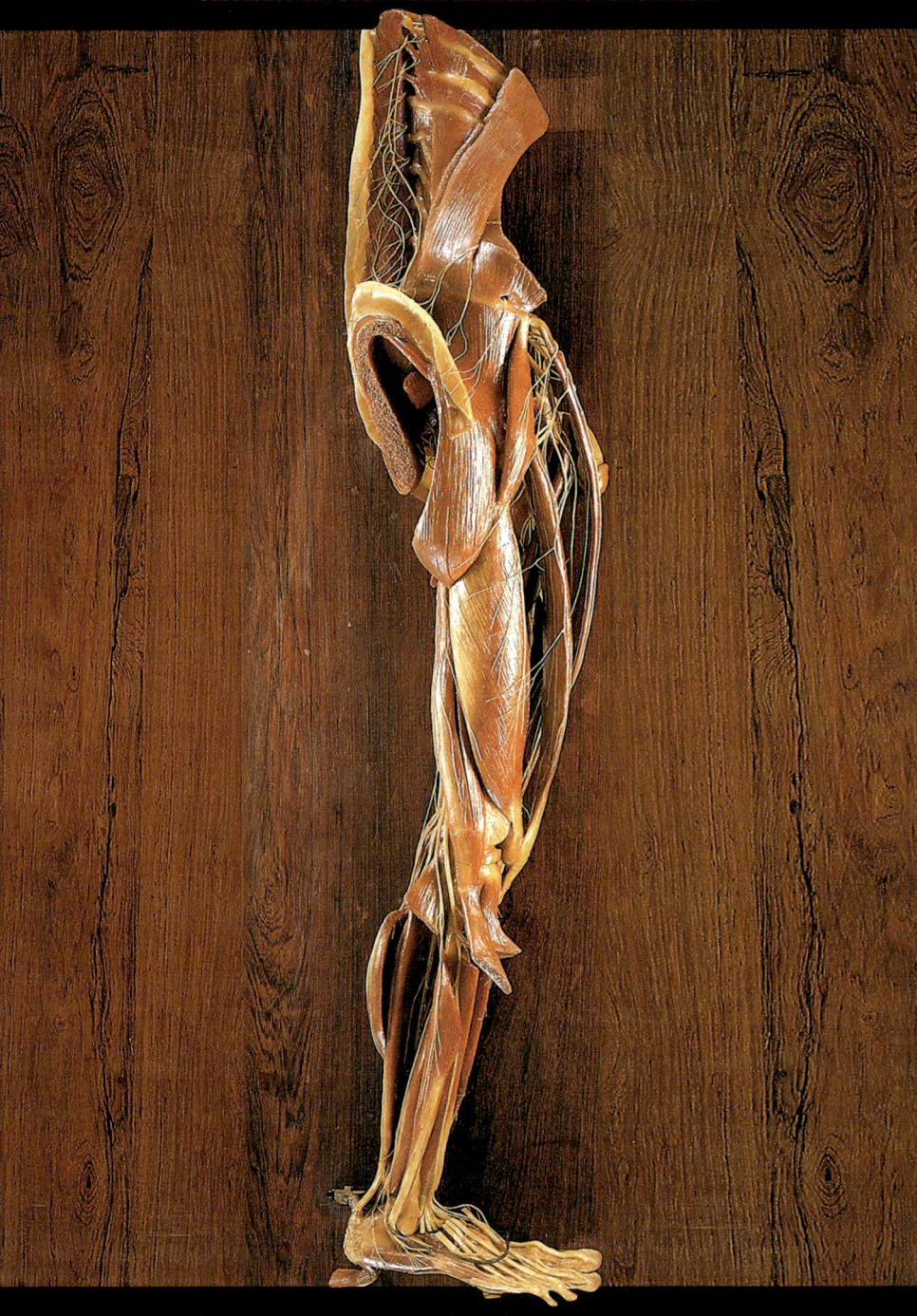

Nervi membri inferiores

The nerves and muscles of the lower limb

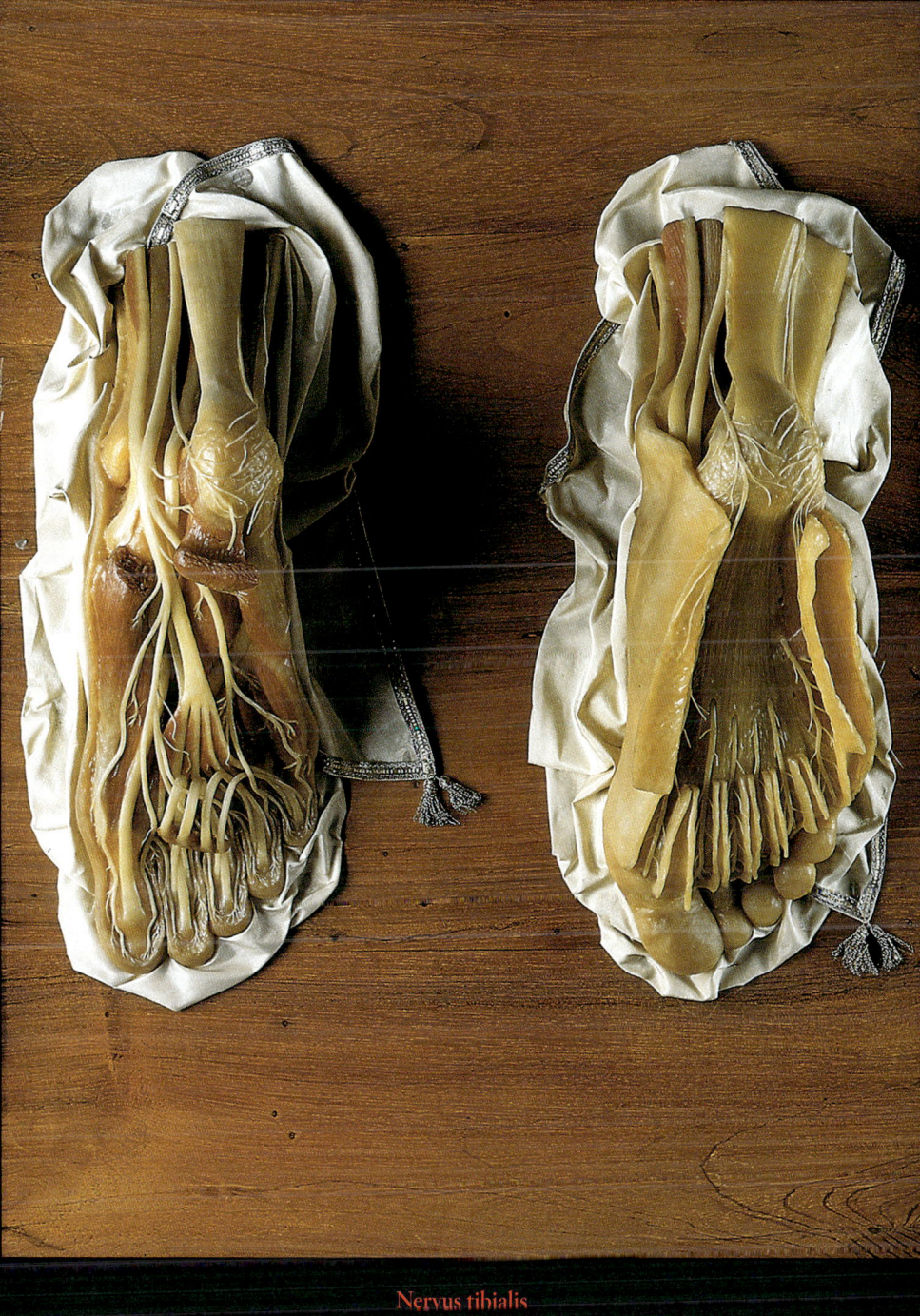

Nervus tibialis

Display showing the nerves supplying the muscles and skin of the sole of the foot

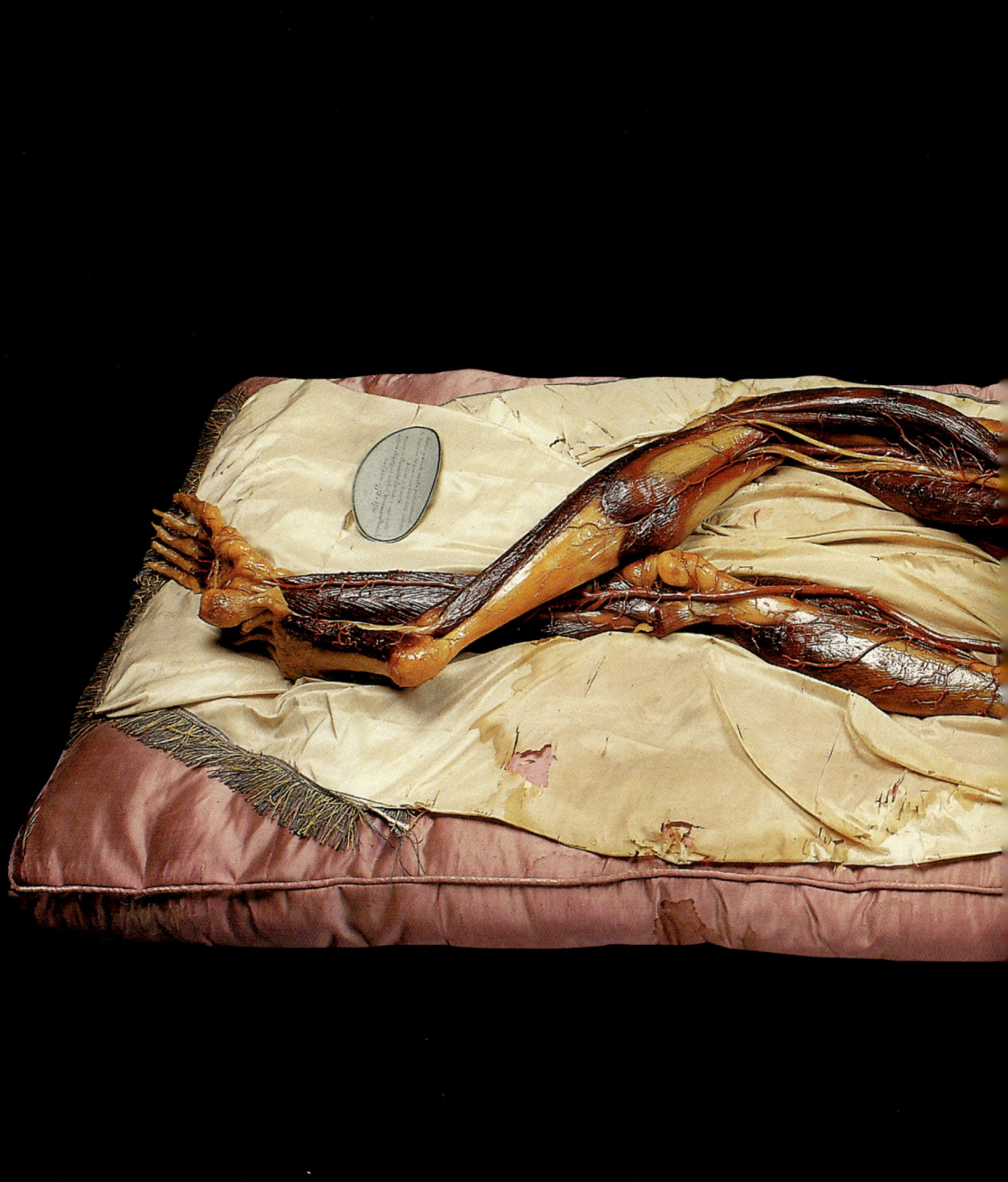

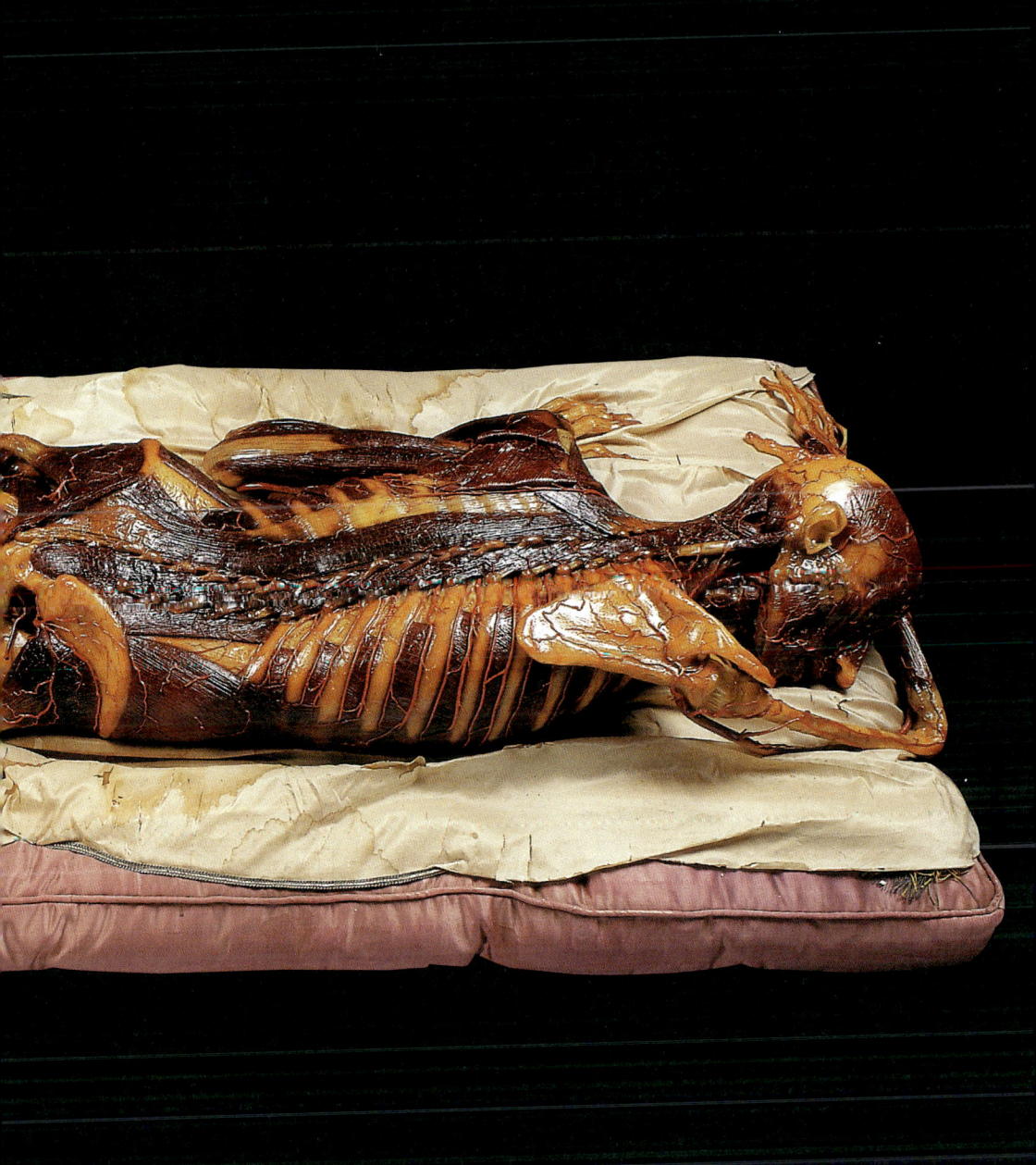

Arteriae
Whole body specimen with the arteries displayed

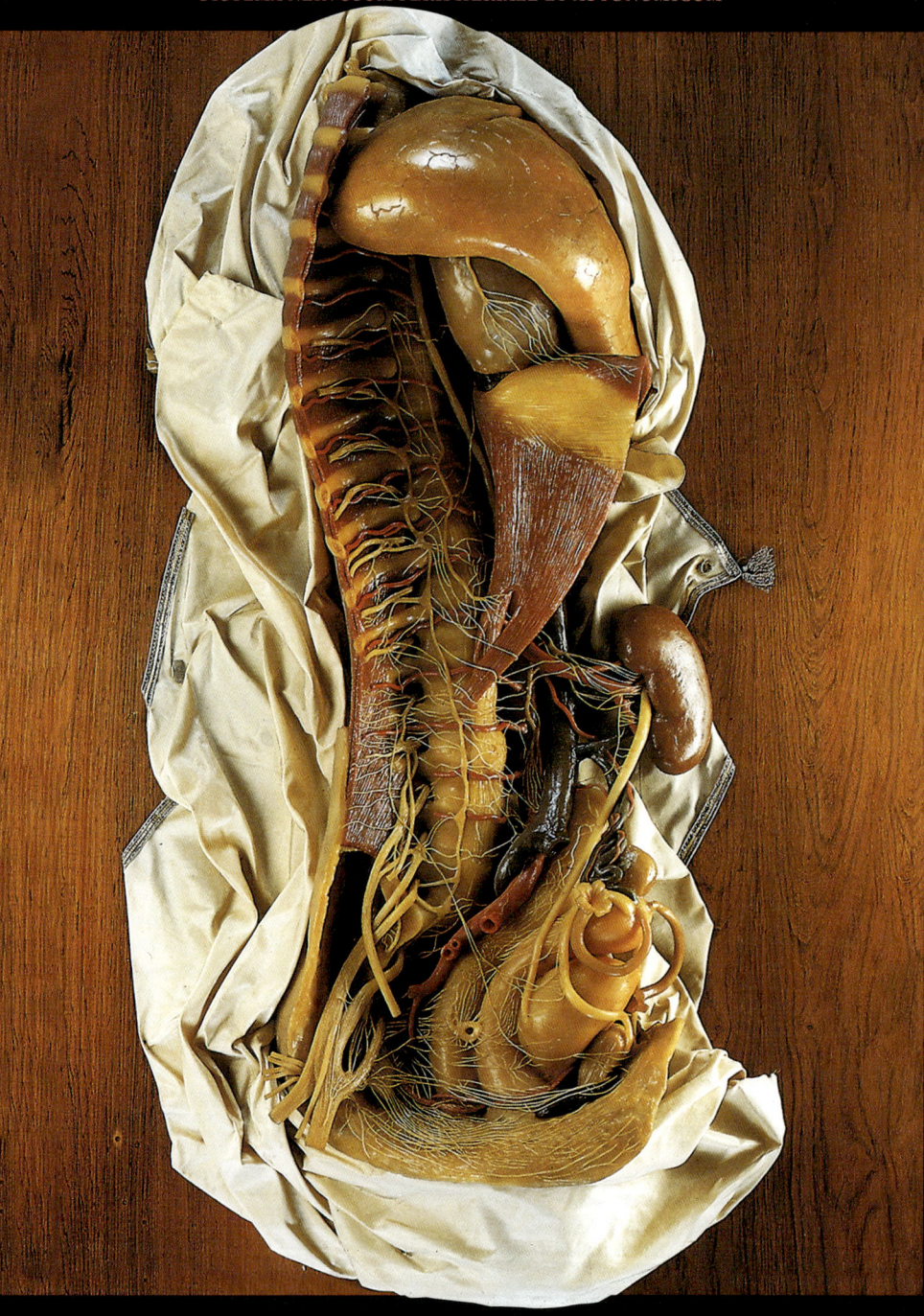

Truncus sympathicus, Plexus lumbosacralis, Plexus pelvinus
The autonomic nerve plexuses of the thoracic and abdominal cavities

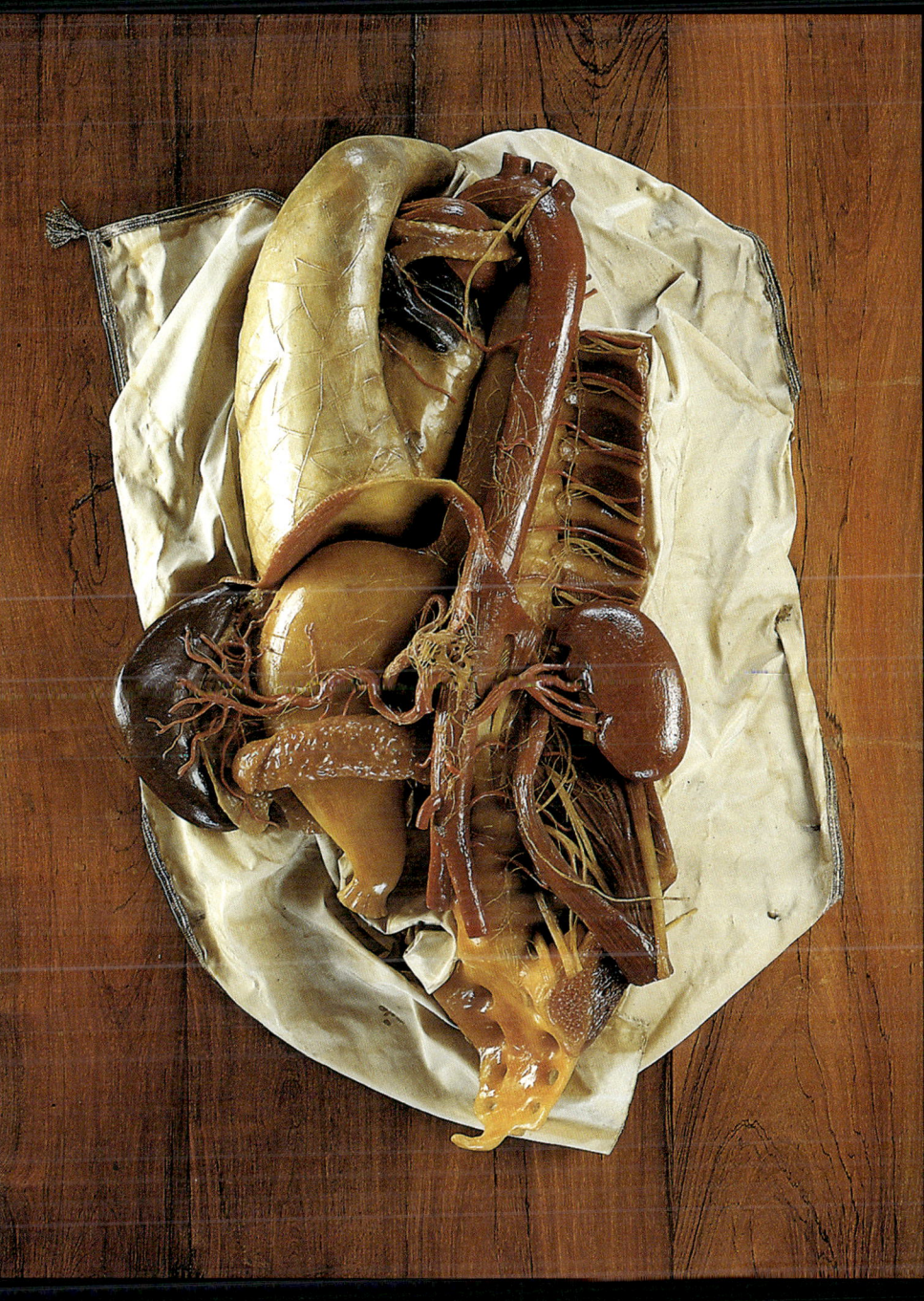

Plexus coeliacus, Plexus lumbalis

Specimen showing the viscera with the arteries and their associated autonomic nerve fibre plexuses

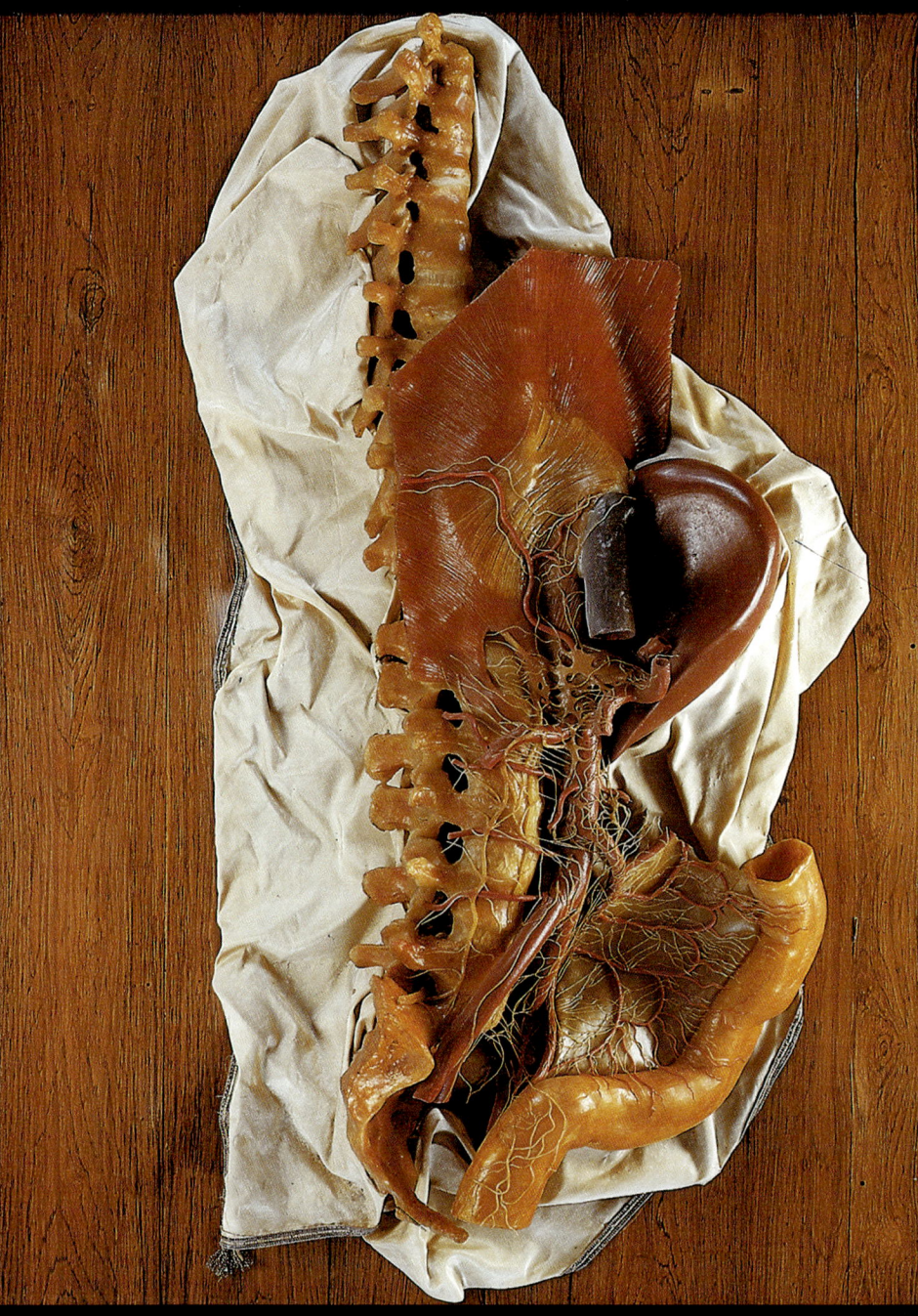

Ganglion coeliacum, Plexus mesentericus

Display showing the autonomic nerve plexuses near the vertebral column and
associated with the abdominal arteries

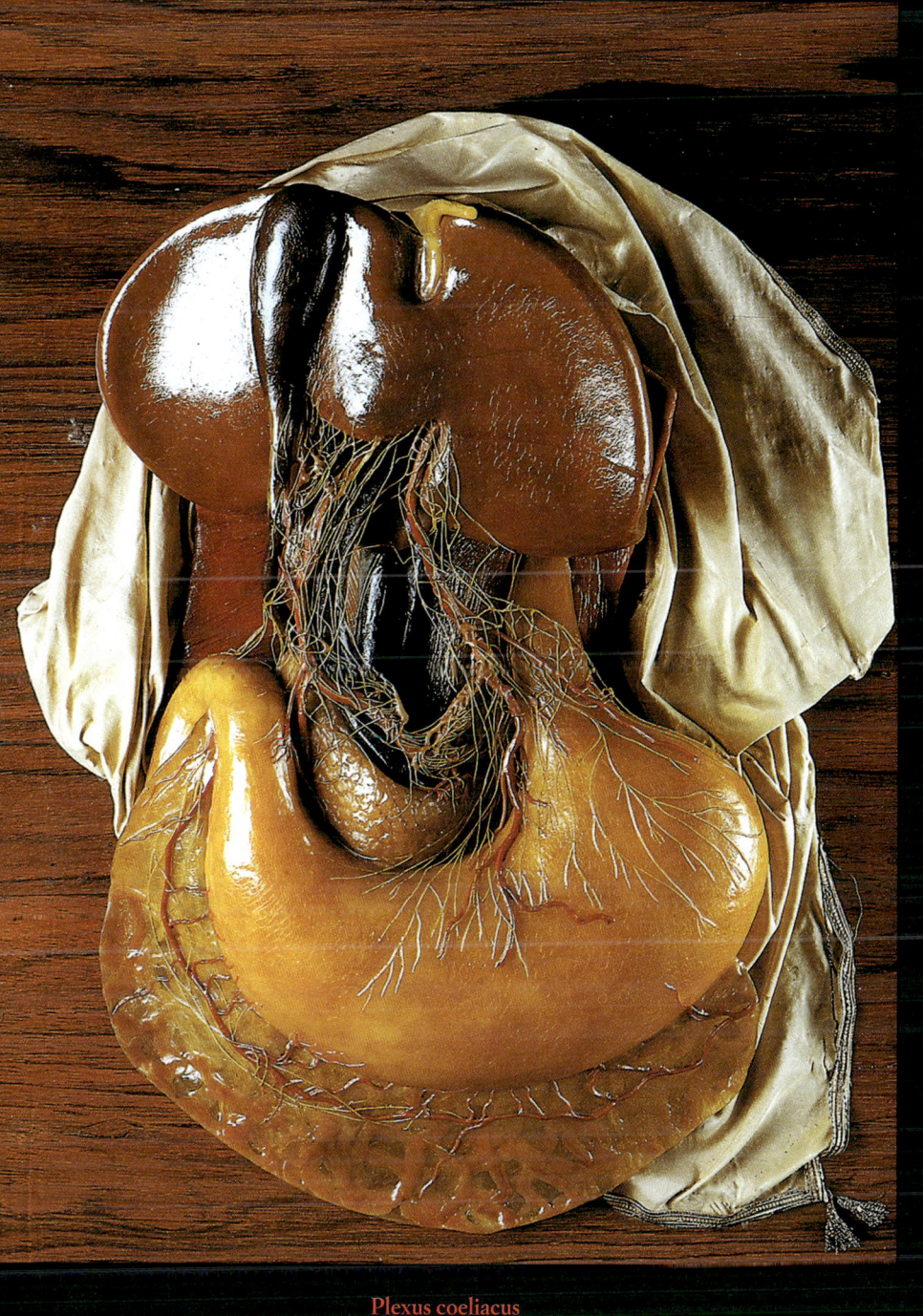

Plexus coeliacus

Display showing the autonomic nerve plexus associated with stomach and liver

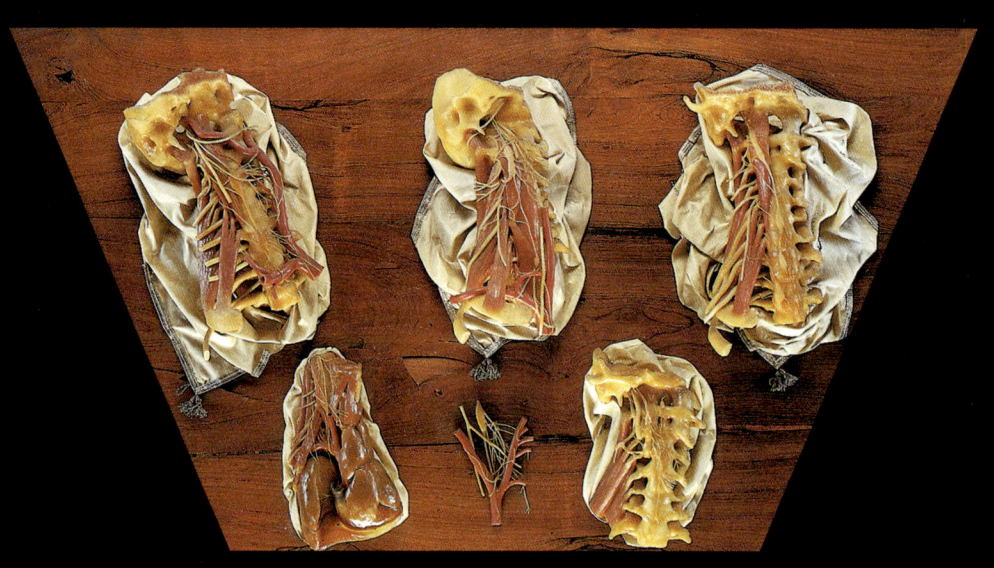

Truncus sympathicus, pars cervicalis, Ganglion cervicale superius et medius, Ganglion stellatum

Display showing the sympathetic nerves and their ganglia in the deep region of the neck

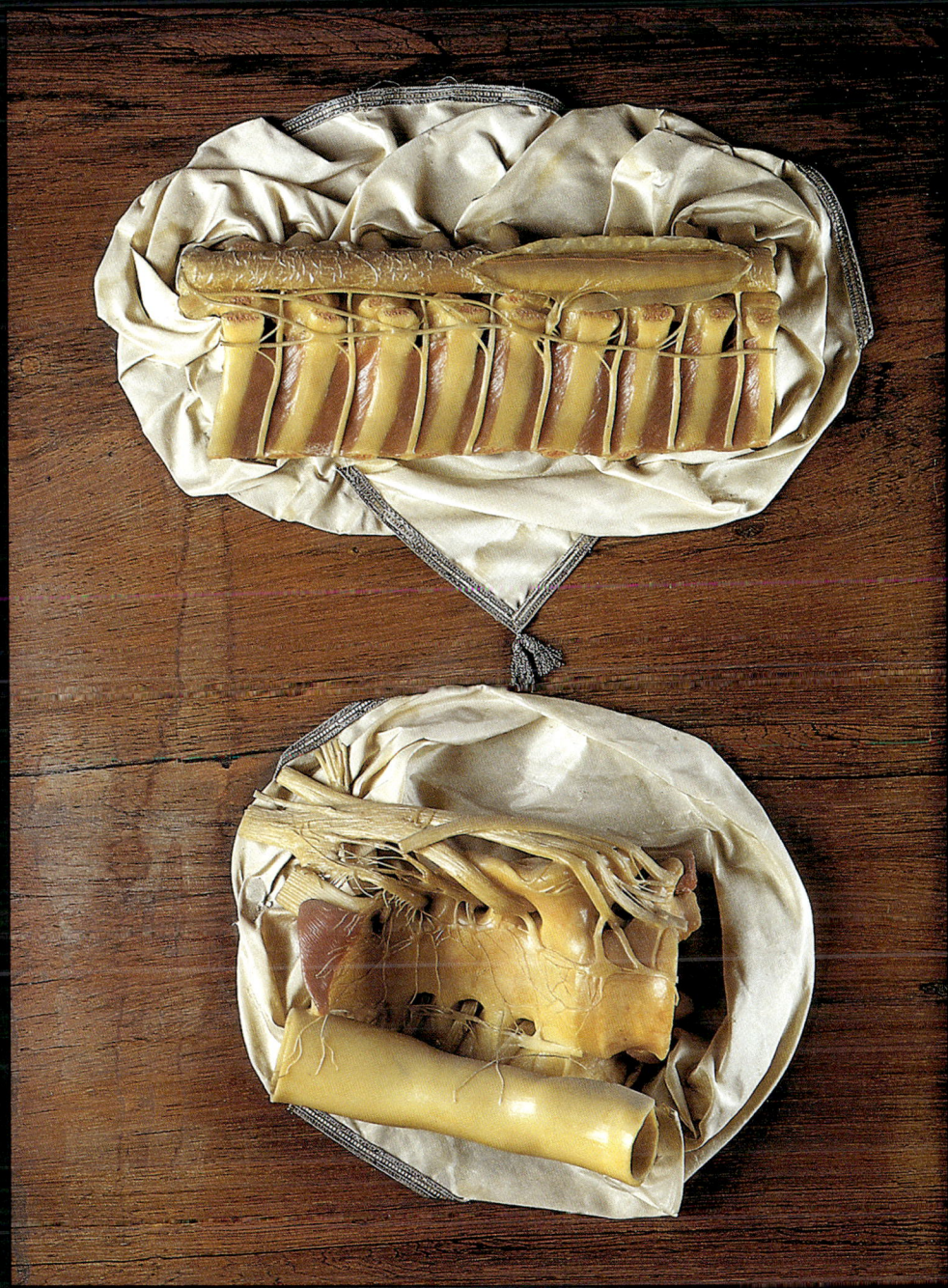

Truncus sympathicus, pars thoracalis, pars lumbosacralis

Display showing the sympathetic nerves and their ganglia in the region of the thoracic cage (upper specimen)
and in the region of the sacrum (lower specimen)

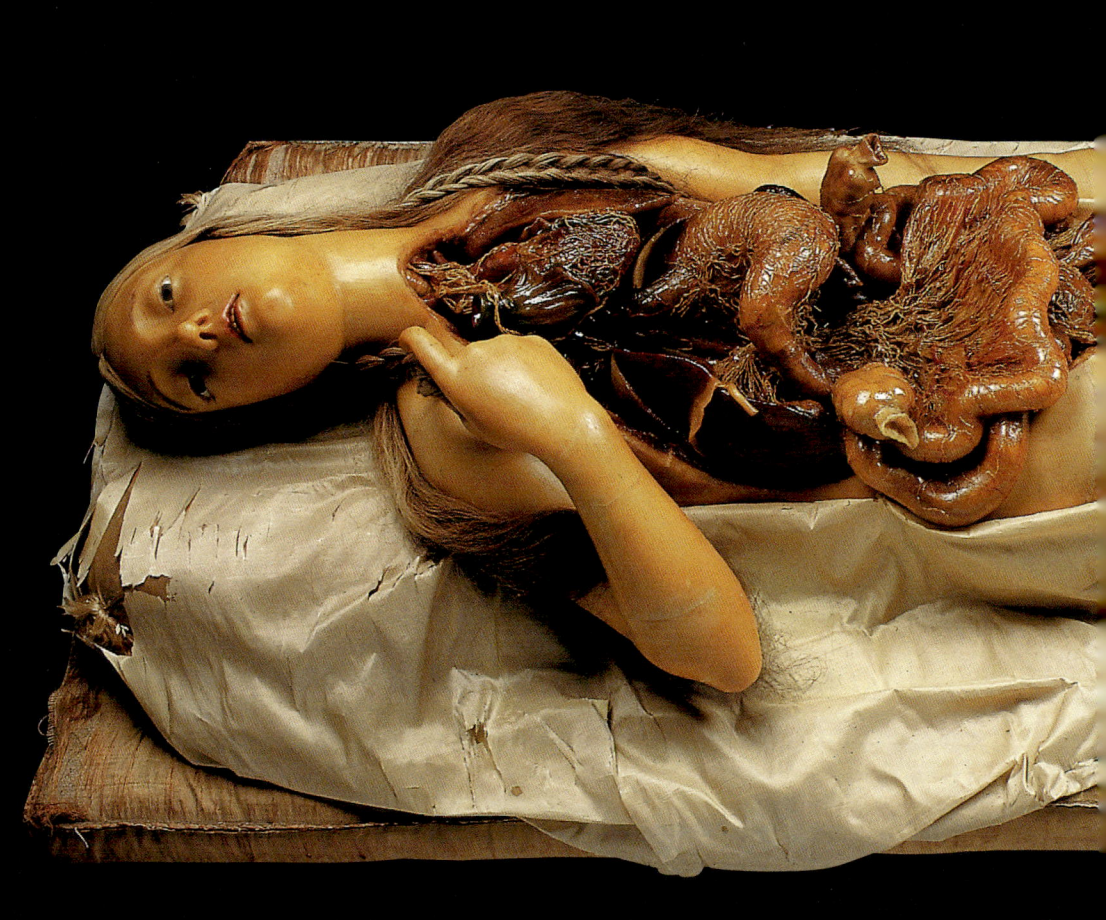

Systema digestorium et respiratorium

*Viscera | Eingeweide | Appareil digestif,
appareil respiratoire*

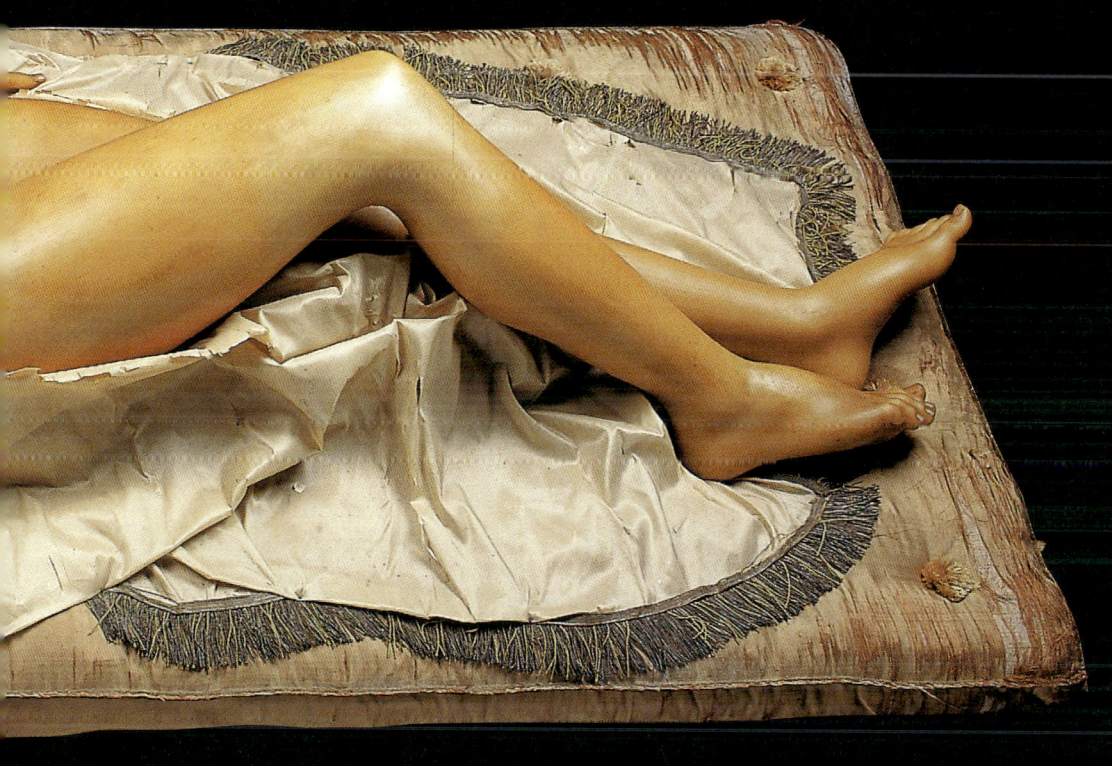

Viscera

The recognition of surgery as a legitimate part of the medical faculty coincided in time with the collecting of these models and had a lasting effect on anatomical teaching and research. As a result of the new status acquired by surgery, accurate anatomical knowledge acquired a new significance. Particular attention was paid to the anatomical dissection of the organs of the thorax, abdomen and true pelvis. Continued developments in dissection techniques led to increasingly detailed specimens representing the topographical (spatial) relationship between individual organs, and also their blood and nerve supply. In contrast to the drawings and wood-cuts, the preparation of wax models of original specimens gave an entirely new impetus to the study of anatomy, allowing the complex structures to be represented in three dimensions.

A large number of the models in this collection show the organs within the thoracic and abdominal cavities in their natural positions. Beginning with these overall displays, many of the detailed models also show the complete extent of the individual organs together with their blood supply, and for this purpose adjacent organs were displaced from their natural relation-ship or completely removed from the body. In these specimens the arteries and veins are clearly distinguished, and coloured either red or blue. Sections of the various organs are arranged to show their internal structure, including the relative internal measurements of hollow organs, as in the case of sections through the digestive tract in which the internal surfaces are also illustrated.

Pages 412–413:
Whole body specimen showing the lymphatic vessels
in the thoracic and abdominal cavities (cf. pp. 470–471)

Eingeweide

Die Anerkennung der Chirurgie als Fach der Medizin fiel zeitlich mit der Herstellung der Modellsammlung zusammen und hatte nachhaltigen Einfluss auf die Lehr- und Forschungsinhalte der Anatomie. In der praktischen Umsetzung durch die Chirurgie erlangten die genauen Kenntnisse der Anatomie einen neuen Stellenwert. Das besondere Interesse galt der anatomischen Sektion der Organsysteme von Brust- und Bauchhöhle sowie des kleinen Beckens. Immer weiter entwickelte Sektionstechniken ermöglichten eine zunehmend detaillierte präparatorische Darstellung der topographischen (räumlichen) Beziehungen der einzelnen Organe sowie ihrer Gefäß- und Nervenversorgung. Im Gegensatz zu Zeichnungen und Holzschnitten eröffneten die vom Originalpräparat hergestellten Wachsmodelle dem anatomischen Studium eine völlig neue Dimension, indem sie die komplexen Strukturen des menschlichen Körpers dreidimensional wiedergaben.

Eine Vielzahl von Modellen dieser Sammlung zeigt die Organe des Brust- und Bauchraums in ihrer natürlichen Lage. Ausgehend von diesen Übersichtsdarstellungen sind in vielen Detailmodellen zusätzlich die Organe in ihrer gesamten Ausdehnung sowie ihre Gefäßversorgung zu sehen. Hierzu wurden benachbarte Organe aus ihrer natürlichen Position verlagert oder ganz herausgenommen. Arterien und Venen sind in diesen Präparaten zur klaren Differenzierung rot bzw. dunkelblau koloriert. Anschnitte der verschiedenen Organe zeigen deren inneren Aufbau, einschließlich der luminalen Größenverhältnisse von Hohlorganen, wie dies beispielsweise der Anschnitt des Darms mit dem Schleimhautrelief verdeutlicht.

Les viscères

La chirurgie a été reconnue comme discipline médicale à part entière à l'époque où a été réalisée la collection de modèles de la Specola ; cette intégration de la chirurgie a influencé l'enseignement et la recherche en anatomie. Les connaissances précises en anatomie ont été valorisées par leur transposition dans la pratique chirurgicale.

La dissection des organes et appareils contenus dans le thorax, l'abdomen et le bassin a été l'objet d'un intérêt particulier. Le perfectionnement continu des techniques de dissection a entraîné une représentation de plus en plus détaillée des rapports topographiques entre les différents organes ainsi que la mise en évidence de leur vascularisation et de leur innervation. Contrairement aux dessins et aux gravures, le modèle en cire réalisé à partir d'une préparation originale offrait à l'étude de l'anatomie une dimension entièrement nouvelle par la représentation tridimensionnelle des structures complexes du corps humain.

Les organes du thorax ou de l'abdomen dans leur situation naturelle sont présentés par un grand nombre de modèles de cette collection. À côté de ces représentations d'ensemble, des détails des organes sont aussi représentés par de nombreux modèles comme les pédicules vasculaires. Pour permettre ces observations, les organes voisins sont alors déplacés de leur position naturelle ou entièrement enlevés. Les artères et les veines sont clairement différenciées dans ces cires par l'emploi des couleurs rouge ou bleu foncé respectivement. Les tranches de section des différents organes permettent de voir leur structure interne, ou le relief de la lumière des viscères creux comme celui de la muqueuse intestinale.

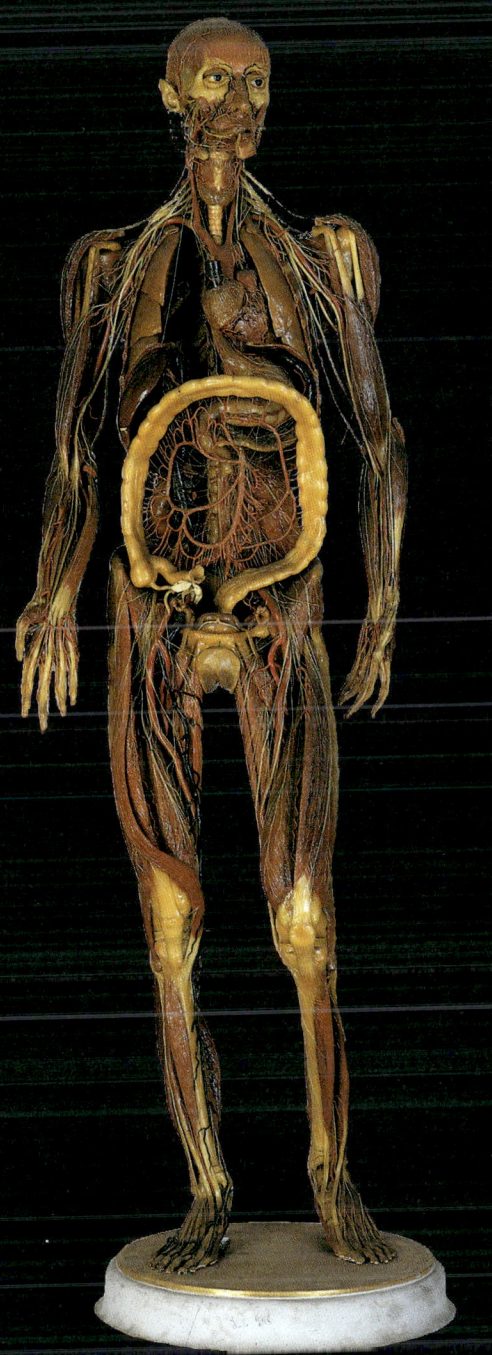

Cavitas thoracis, Cavitas abdominalis

Whole body specimen with thoracic and abdominal cavities laid open.
The loops of the small intestine have been removed to display the arteries, veins and nerves

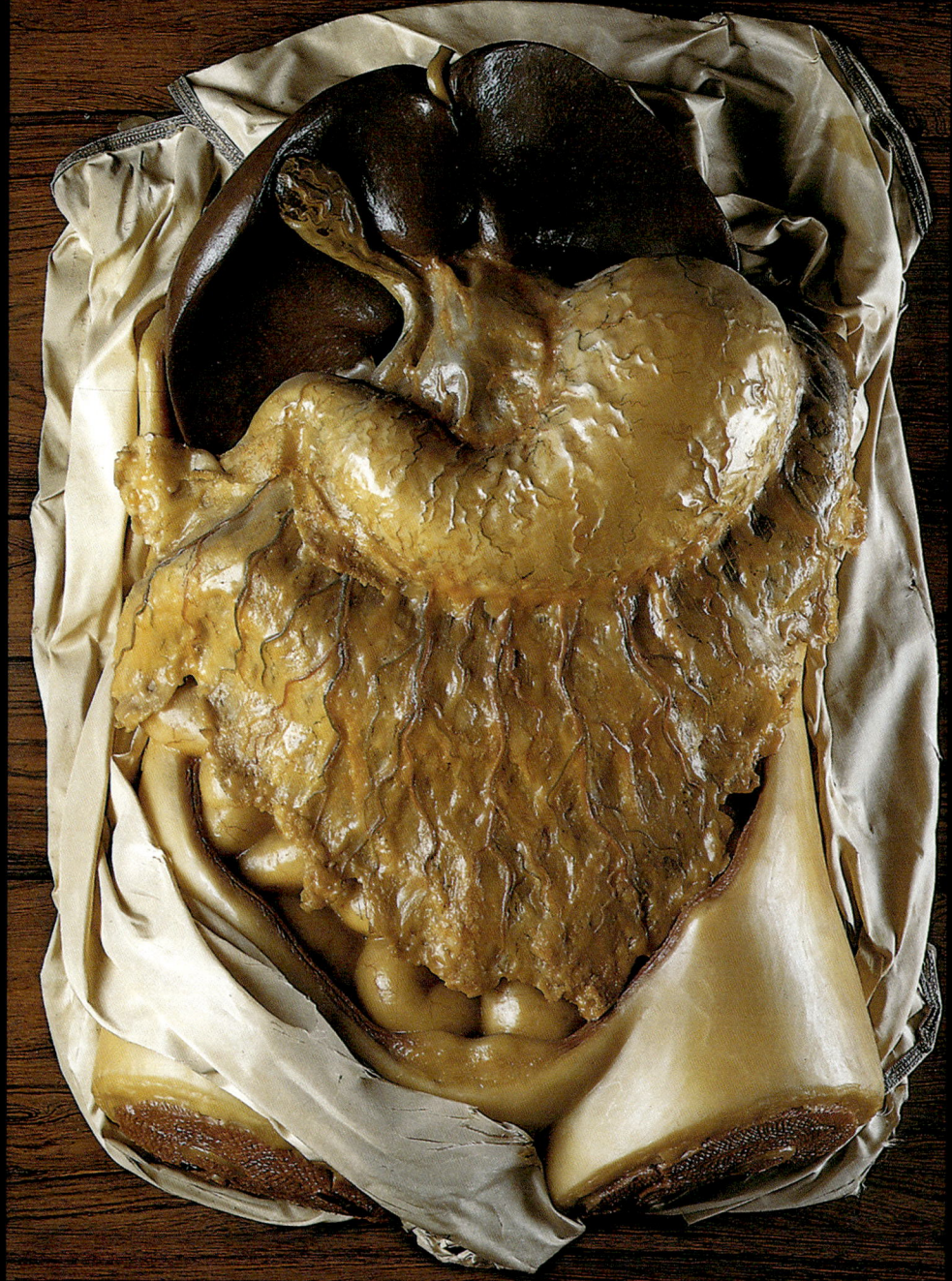

Ventriculus, Hepar, Vesica fellea, Omentum majus et minus

Specimen showing the viscera, including the stomach, liver and lesser and greater omentum.
The liver has been lifted up to show the gallbladder

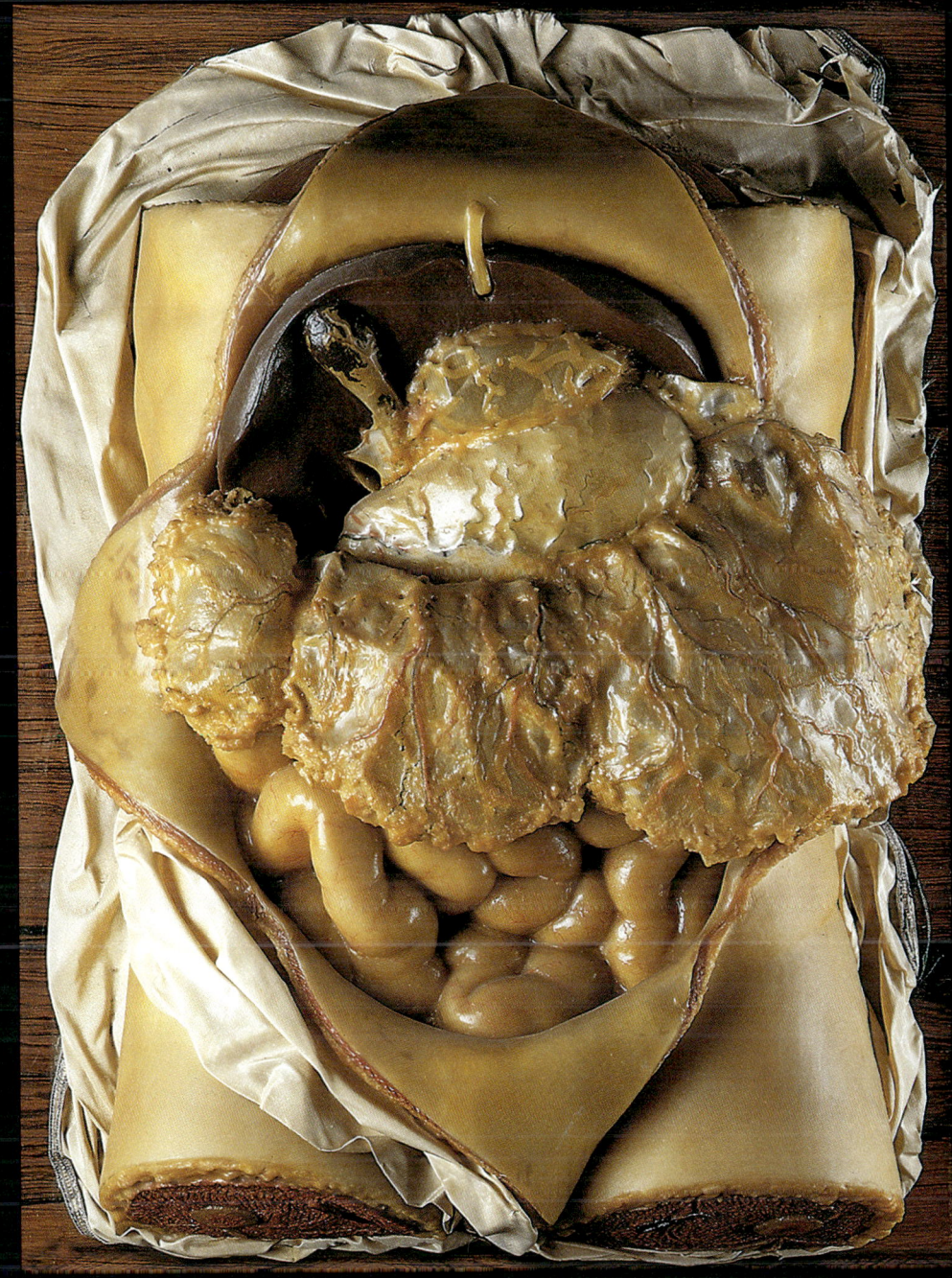

Ventriculus, Hepar, Vesica fellea, Omentum majus et minus, Intestinum tenue

Specimen showing the viscera, including the stomach, liver and greater and lesser omentum,
as well as the loops of the small intestine

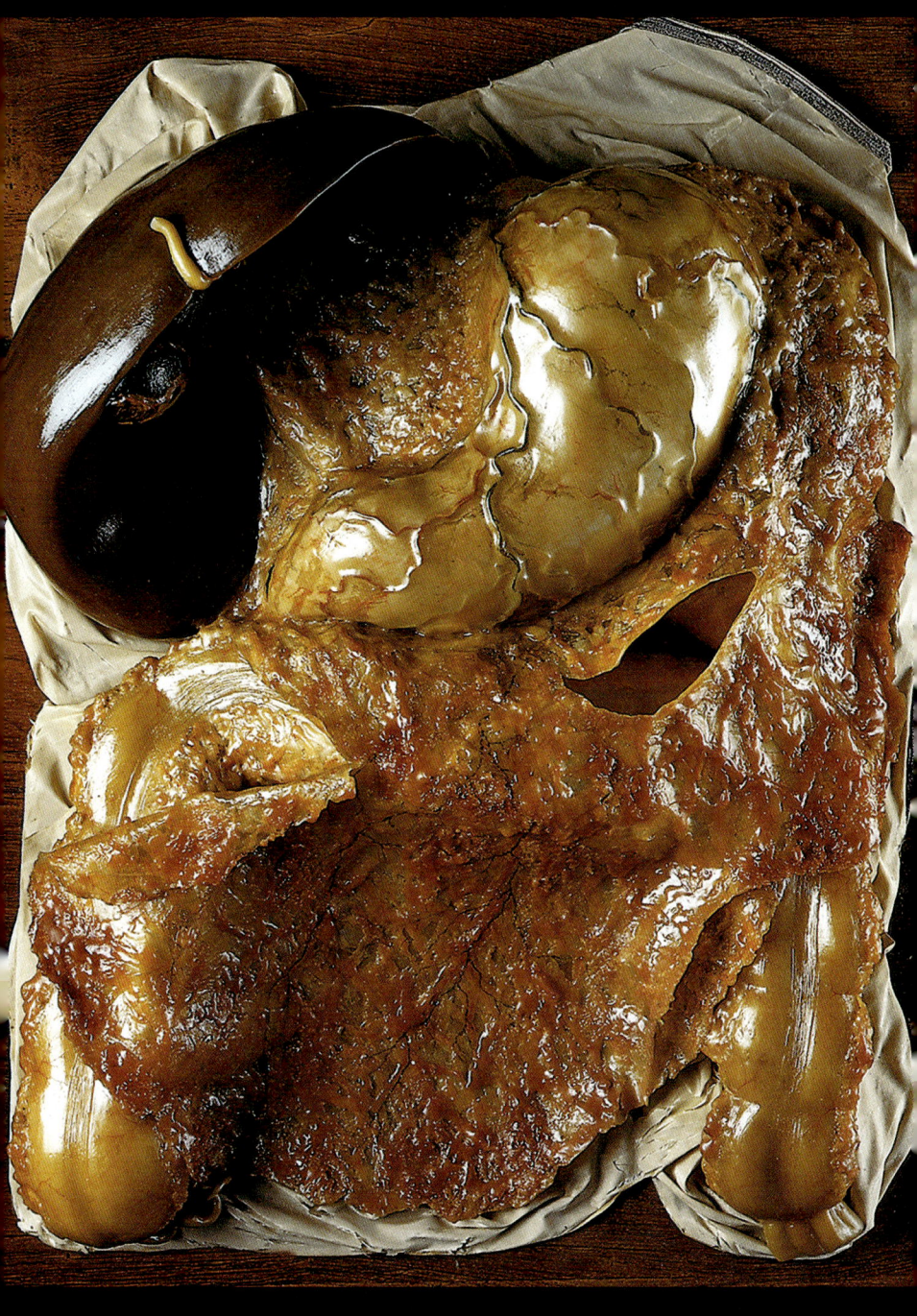

Ventriculus, Hepar, Colon, Omentum majus et minus

Specimen showing the viscera, including the stomach, liver, large intestine
and the lesser and greater omentum

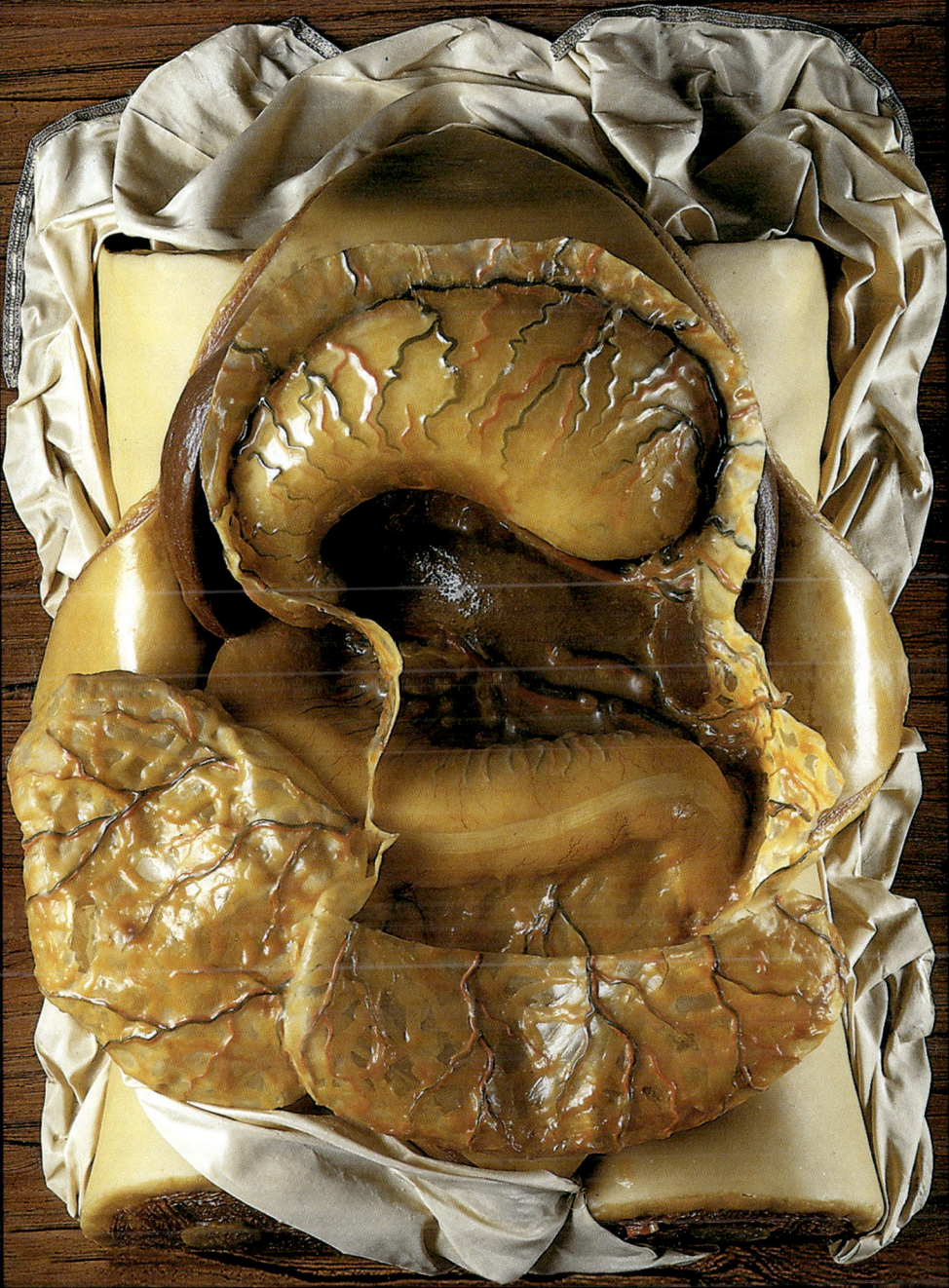

Bursa omentalis

View showing the stomach bed

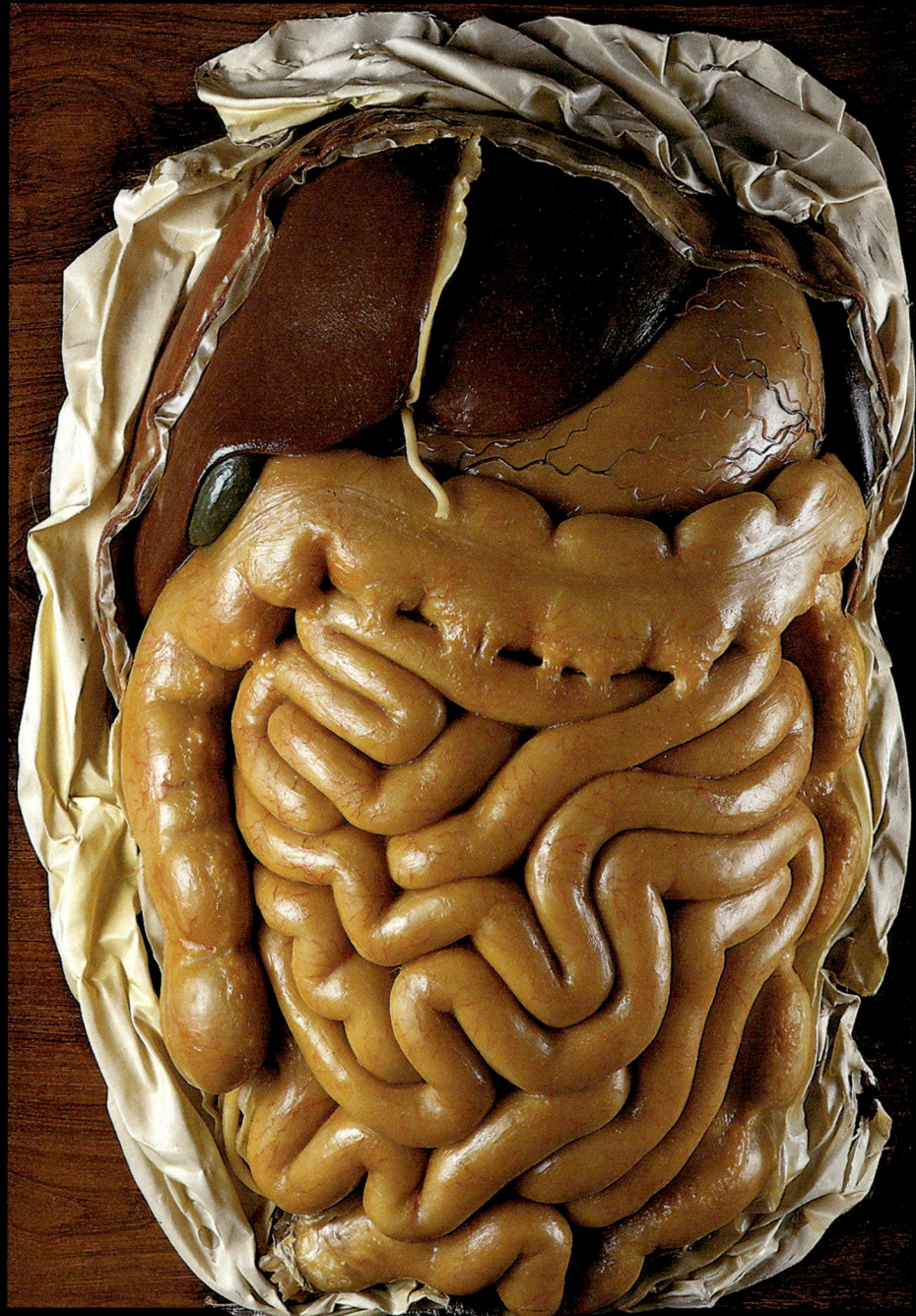

Colon, Ligamentum falciforme, Ligamentum teres hepatis

View showing the loops of the small intestine and parts of the large intestine
after removal of the greater omentum

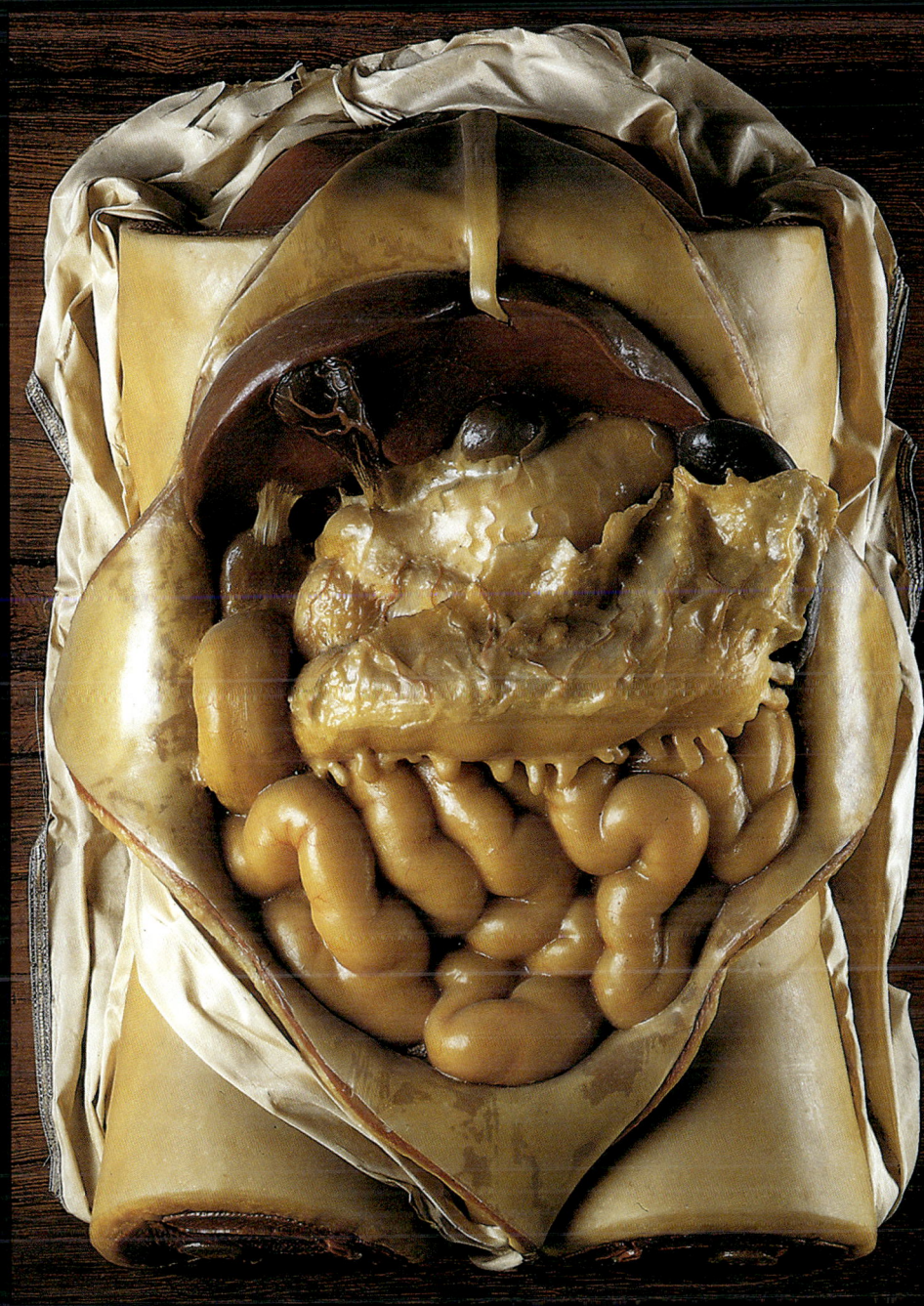

Ventriculus, Hepar, Vesica fellea, Intestinum tenue

View showing the loops of the small intestine and a part of the large intestine
after removal of the greater omentum

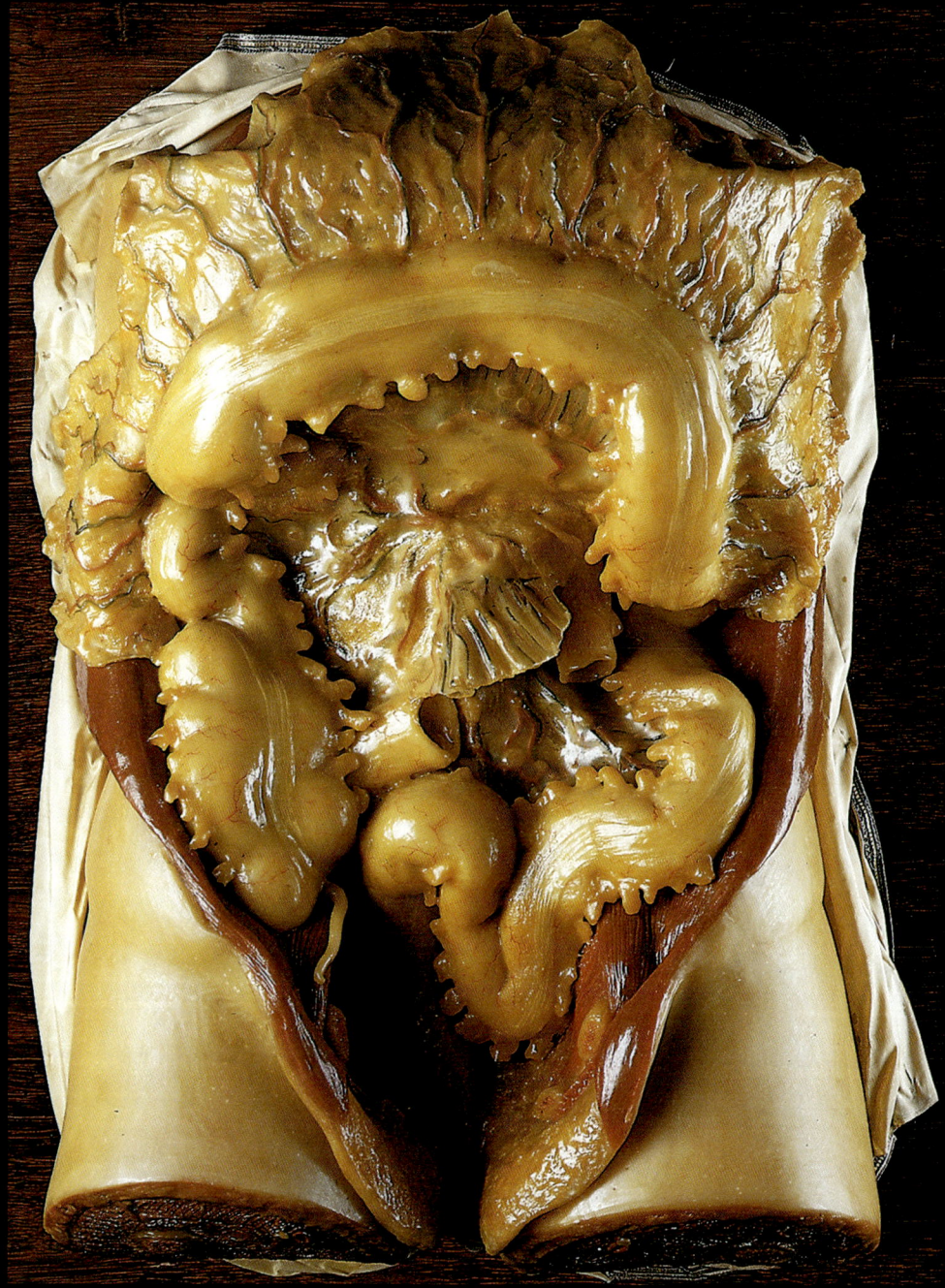

Radix mesenterii, Colon, Appendix vermiformis

View of the large intestine after removal of the small intestine.
The vermiform appendix can be seen on the left

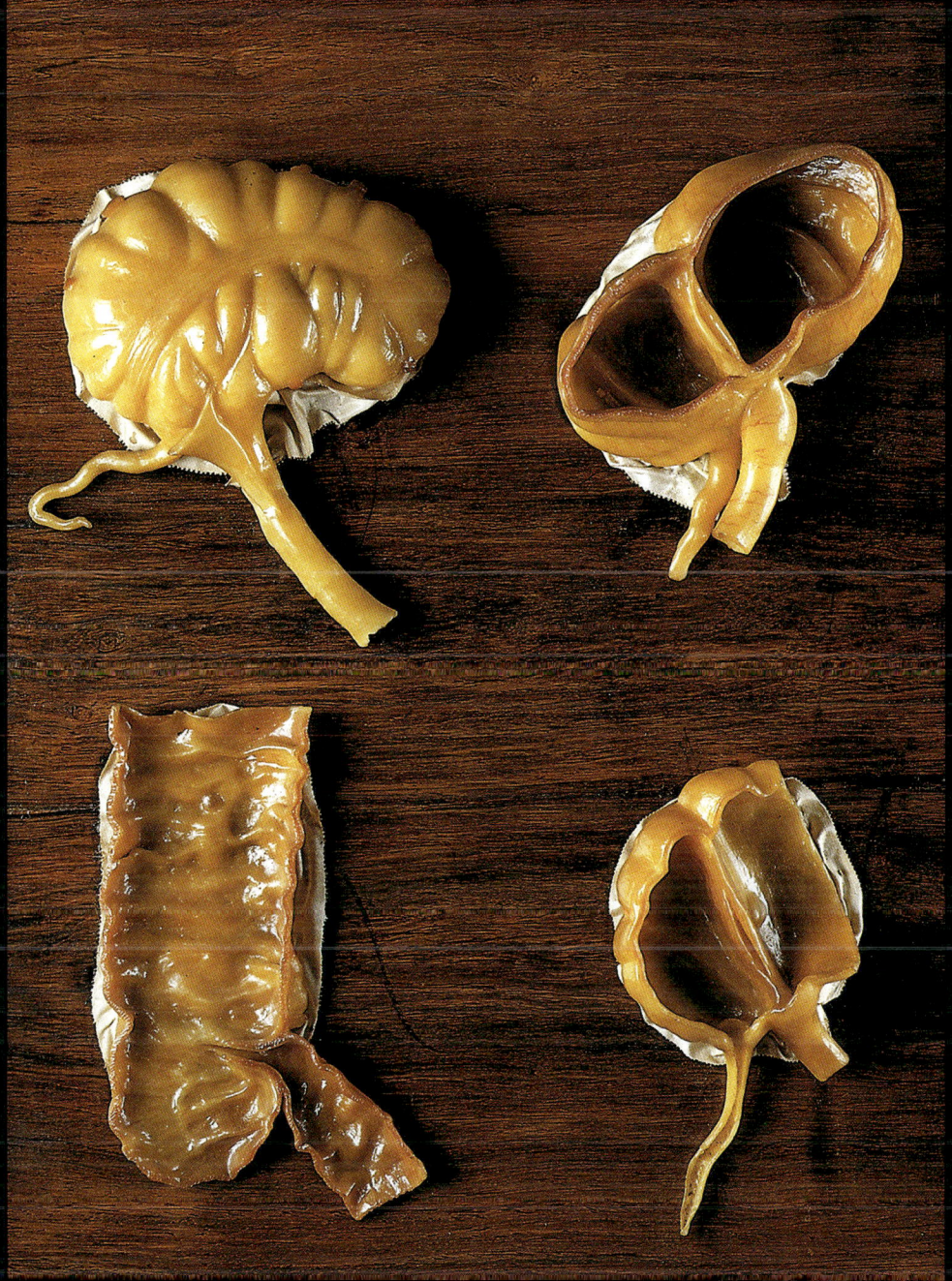

Caecum, Appendix vermiformis, Valvula iliocaecalis

Display showing the vermiform appendix and the entry of the small into the large intestine.
The opened specimens show the valve in the region of the entry

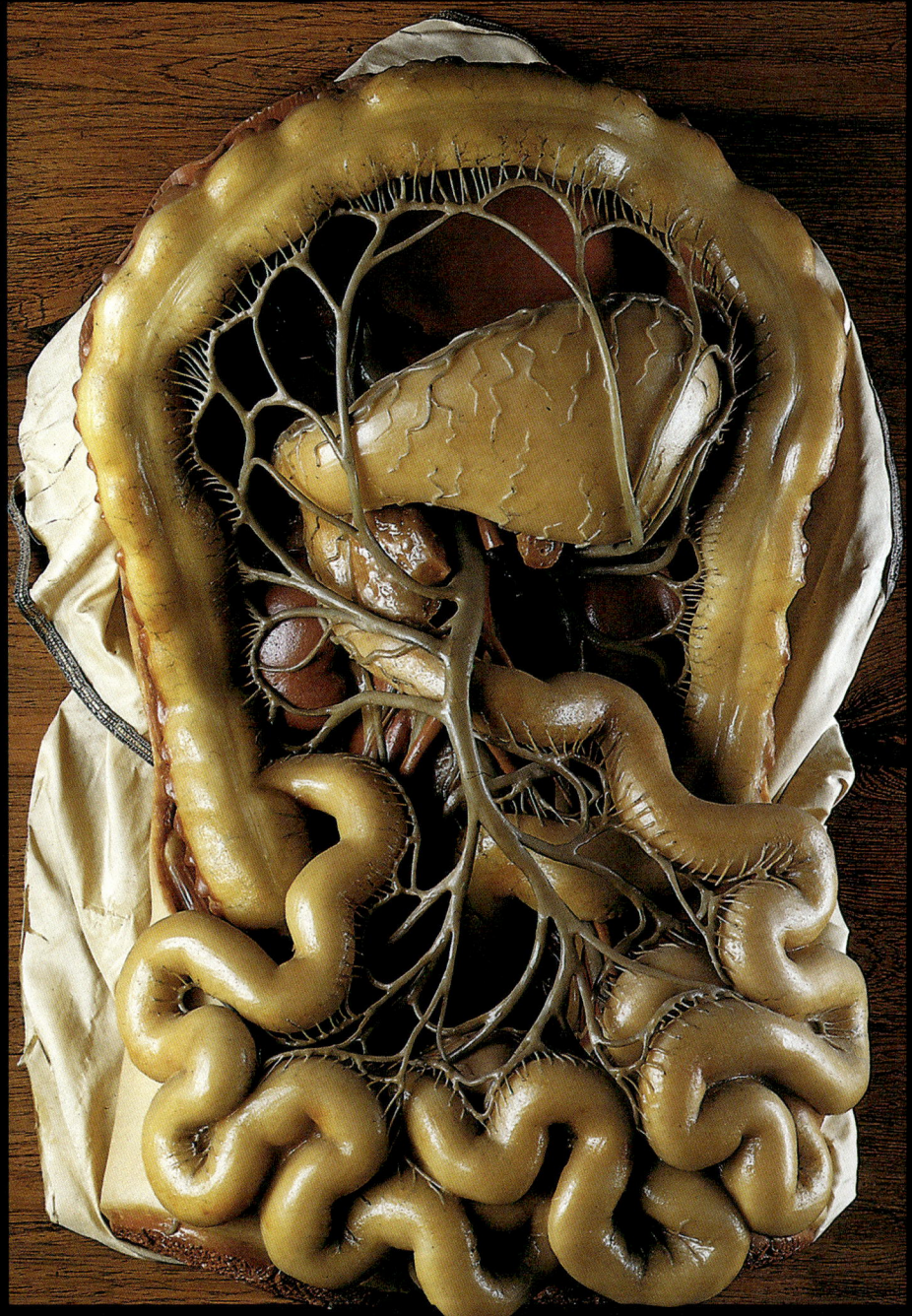

Vena mesenterica superior

Display showing the venous drainage of the various parts of the large and small intestines

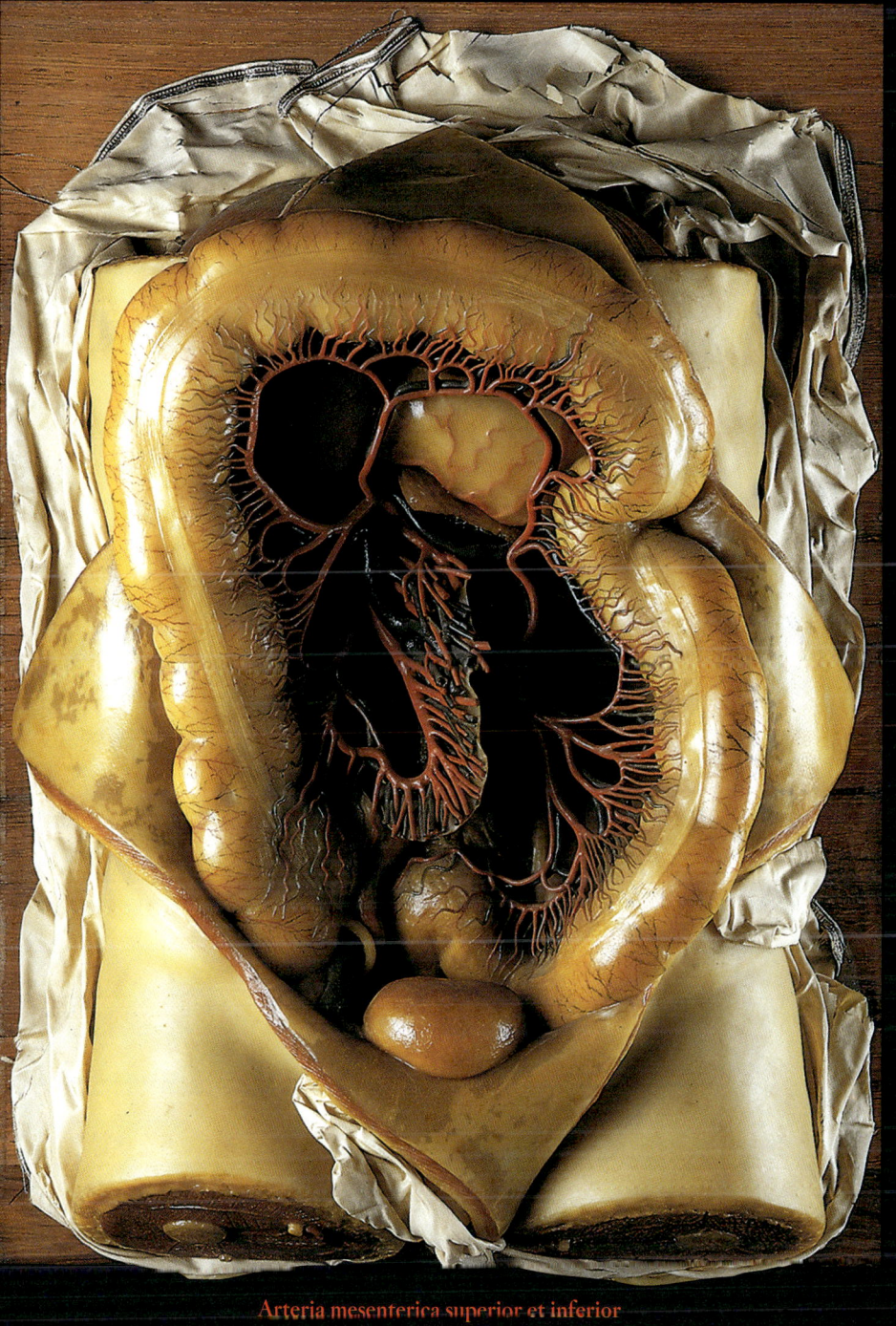

Arteria mesenterica superior et inferior

Display showing the arterial blood supply of the large intestine

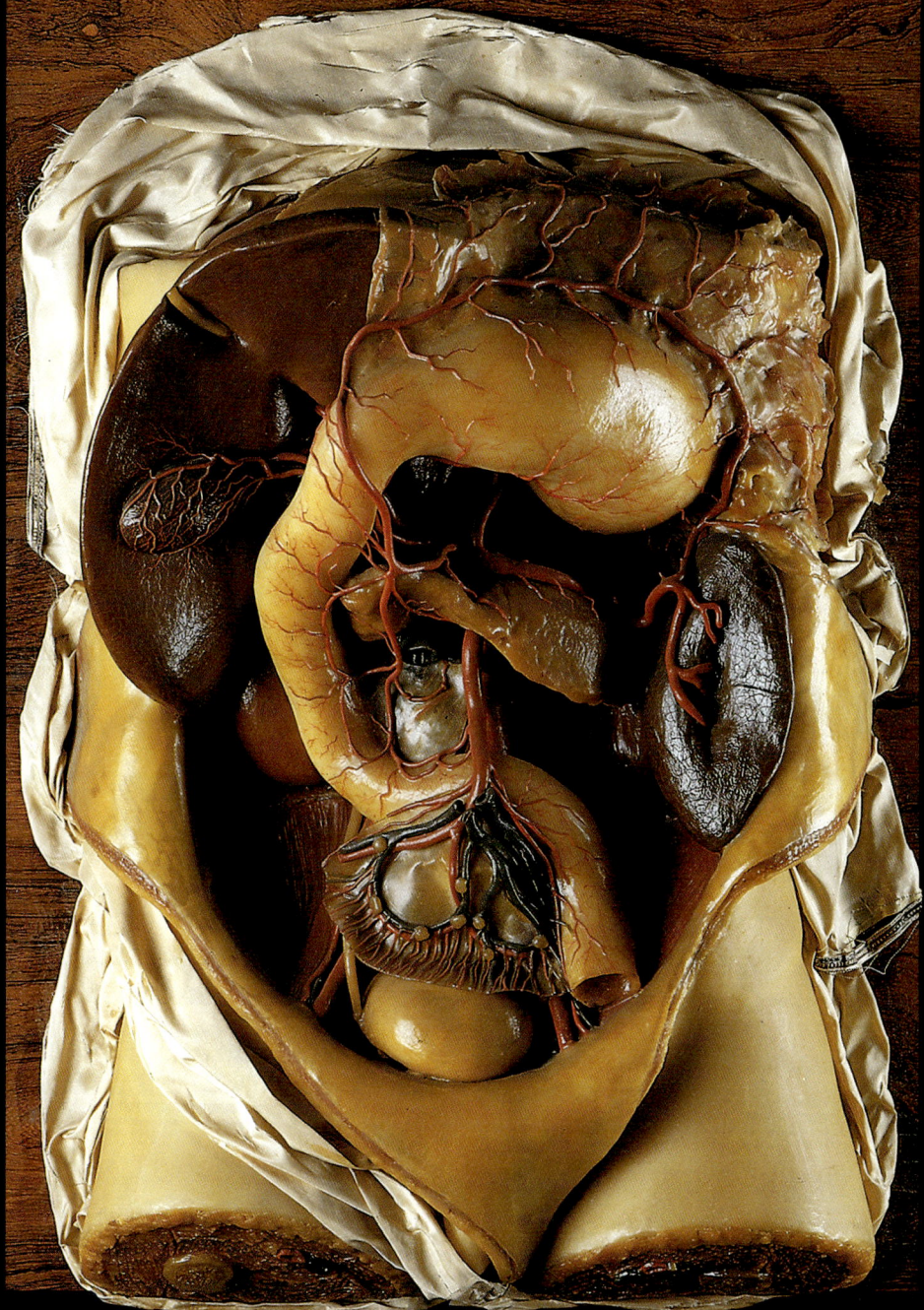

Truncus coeliacus, Arteria mesenterica superior
Display showing the arterial blood supply of the stomach, liver, spleen and pancreas

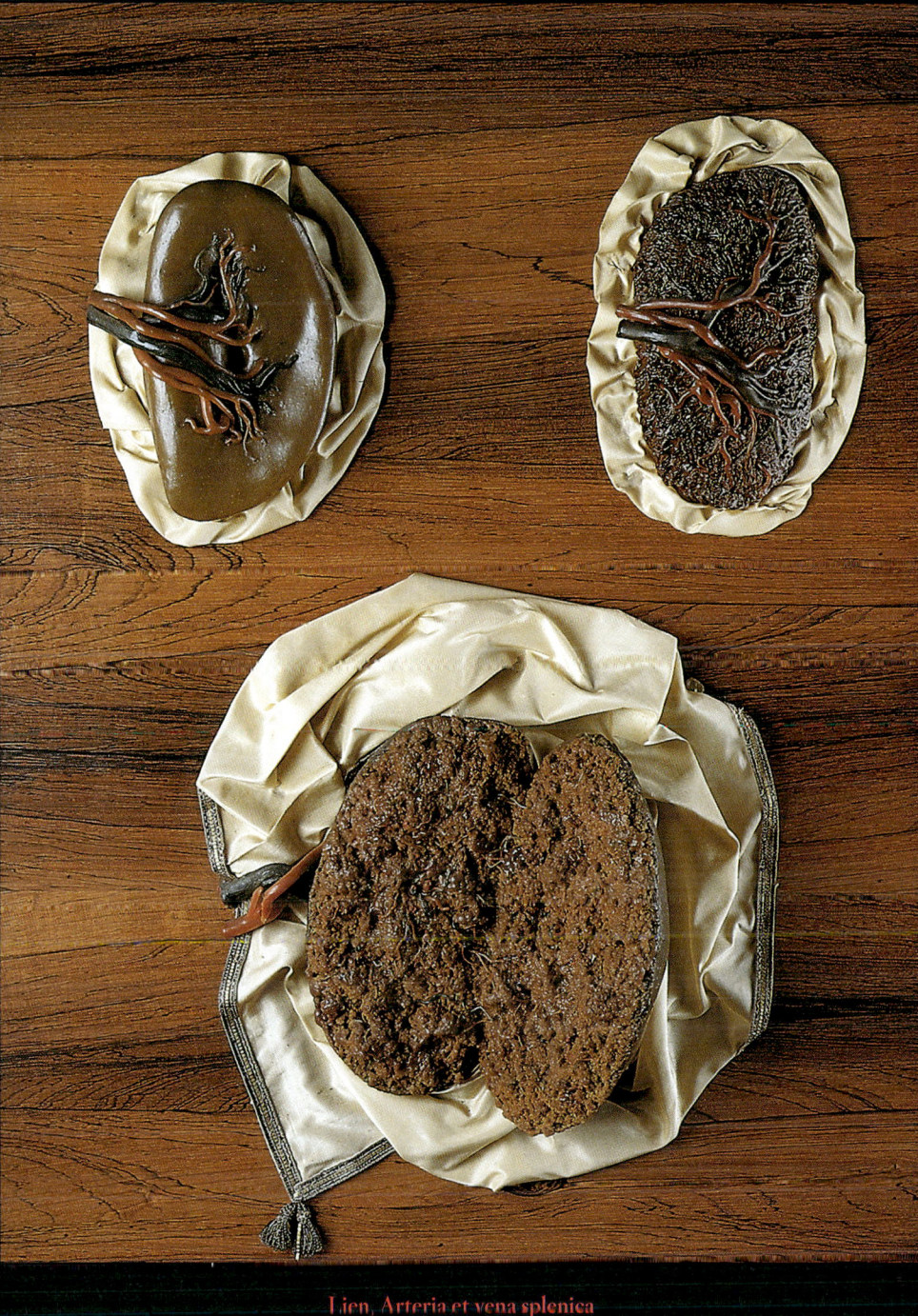

Specimens of the spleen and its vessels. The lower specimen is a bisection of the spleen

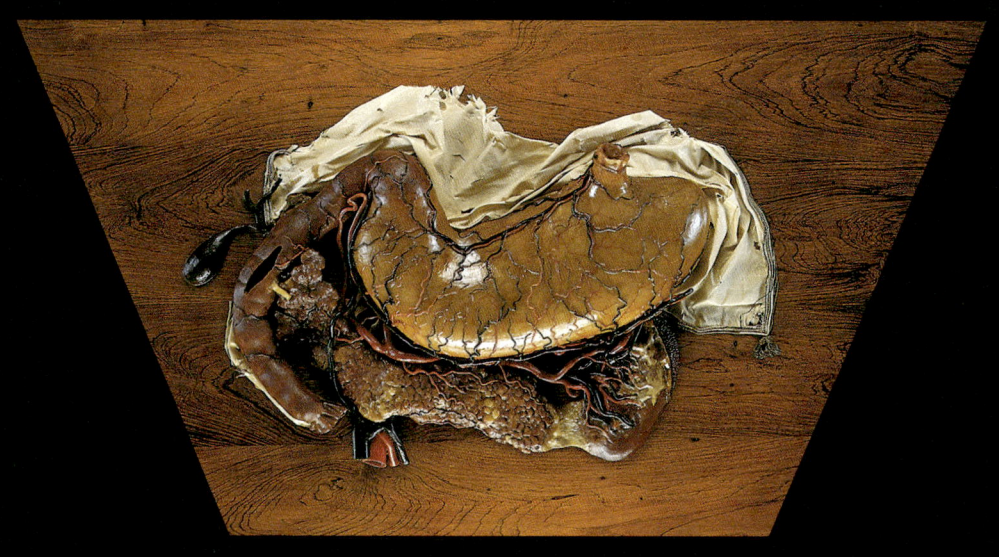

Ventriculus, Duodenum, Pancreas
Display showing the stomach, duodenum, gallbladder and pancreas

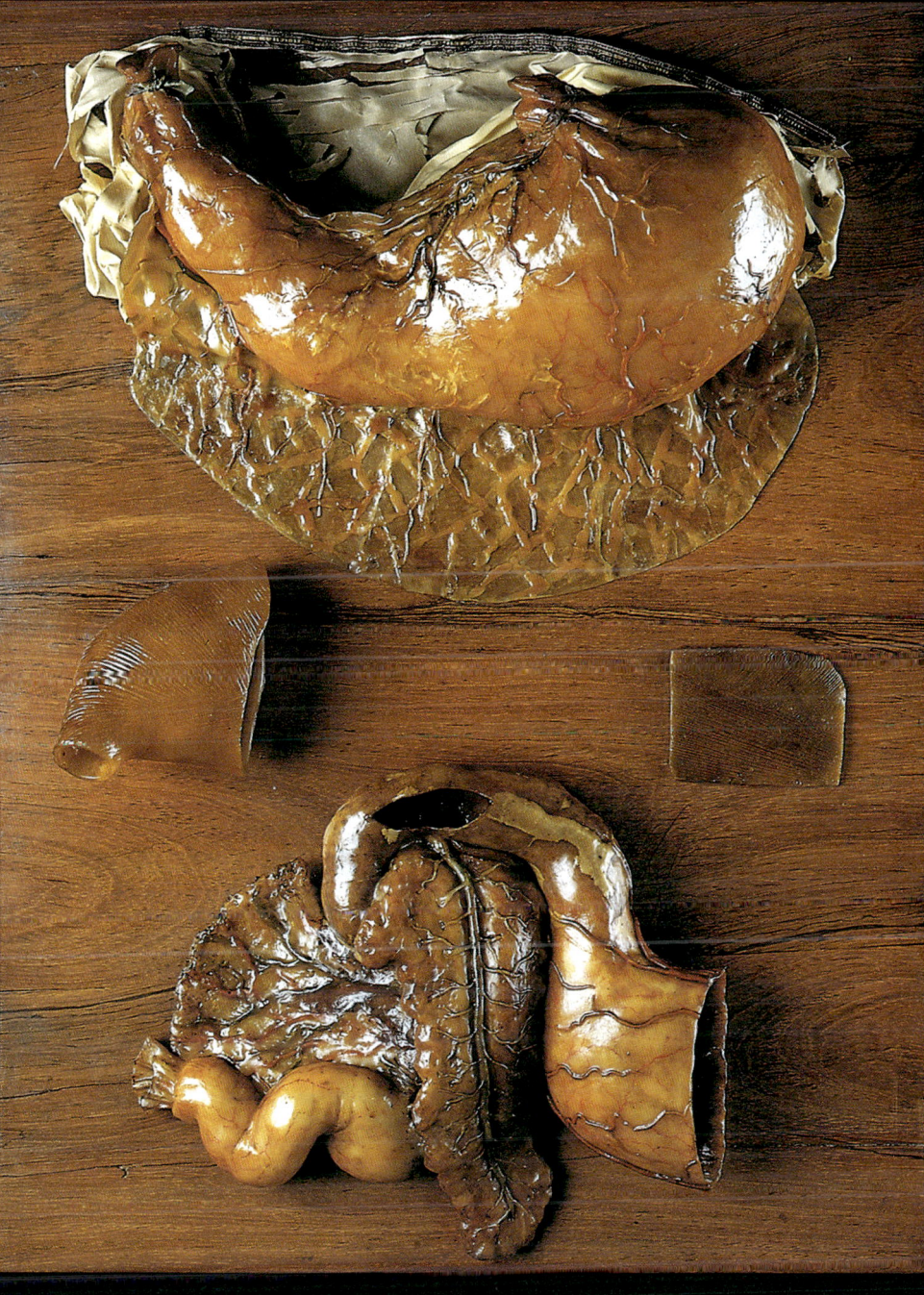

Ductus pancreaticus

Display showing the stomach (upper specimen) and the pancreatic duct
as far as its entry into the duodenum (lower specimen)

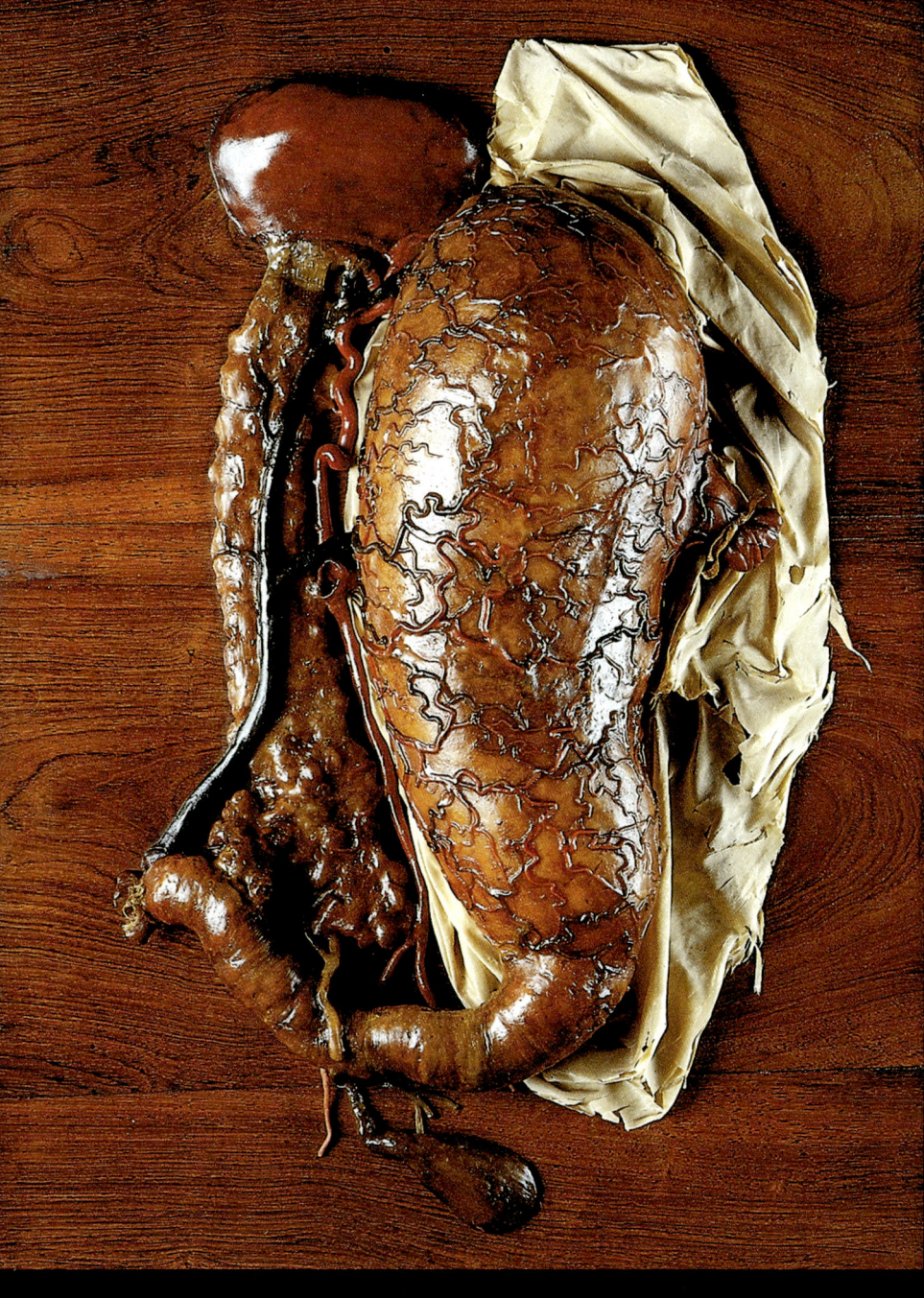

Vena splenica

Stomach, pancreas and splenic vein

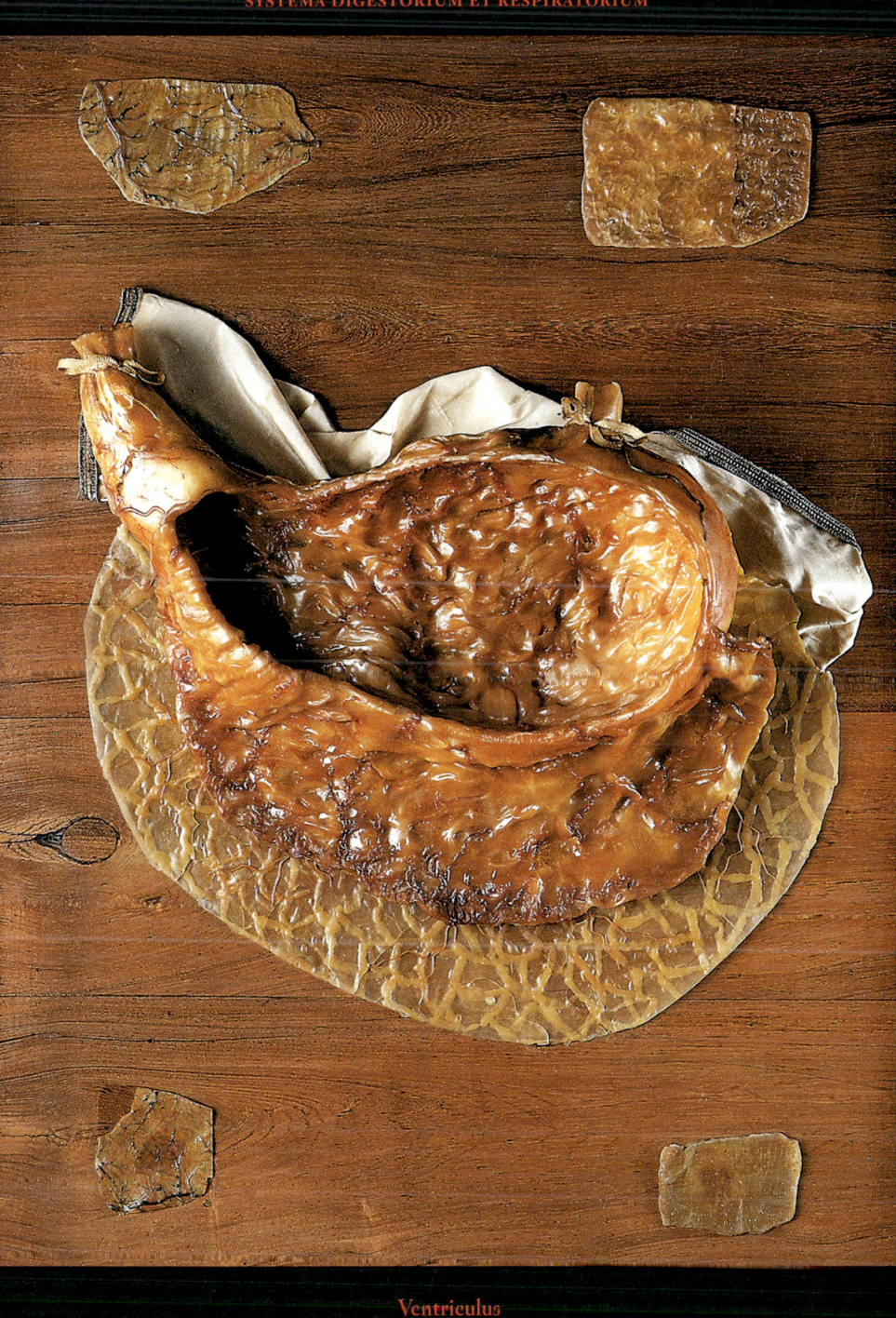

Ventriculus

View into the opened stomach

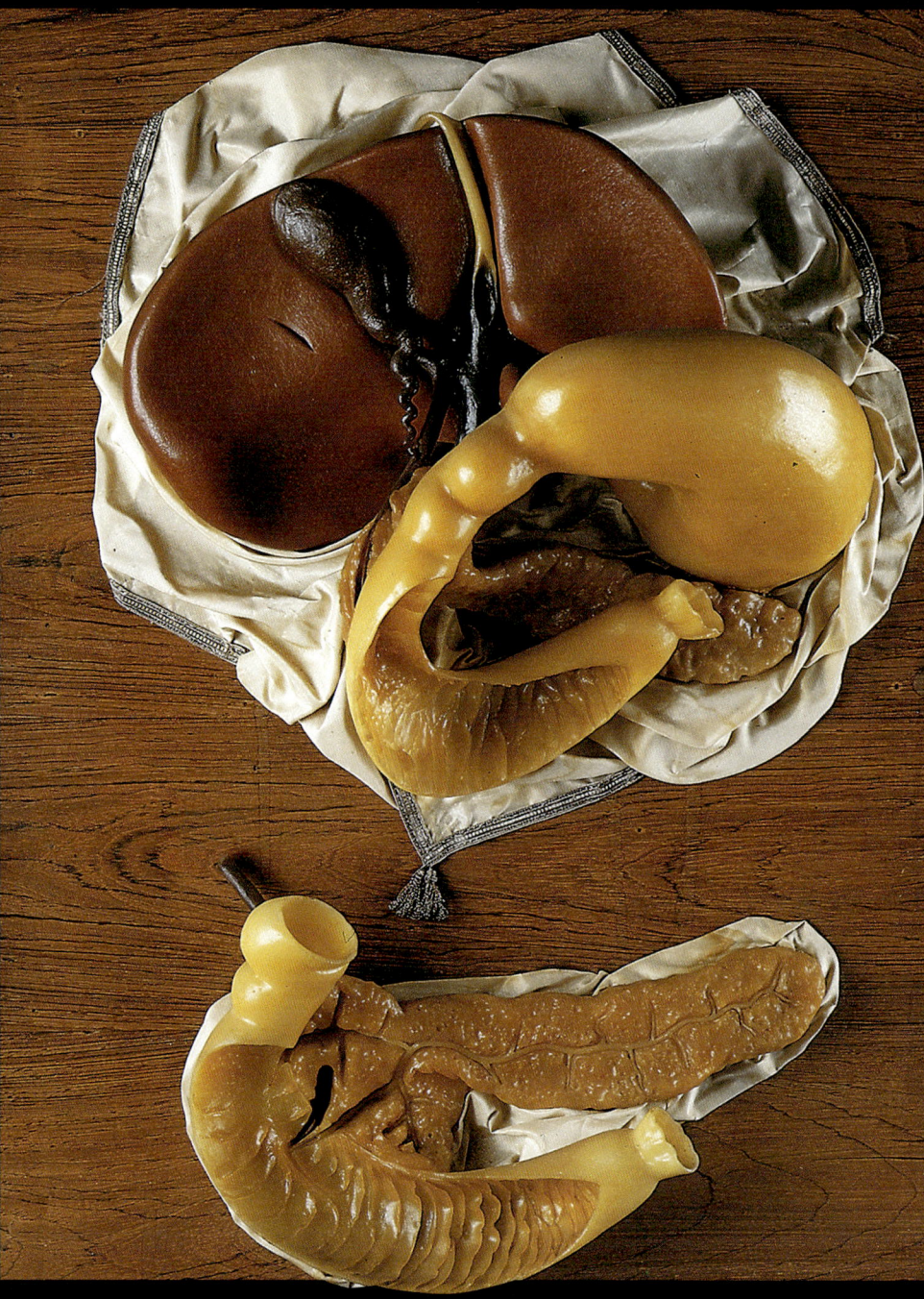

Ligamentum teres hepatis, Vena umbilicalis, Ductus pancreaticus major et minor

Upper specimen: stomach, duodenum, liver and gallbladder.
Lower specimen: the C-shaped form of the duodenum and the pancreatic duct

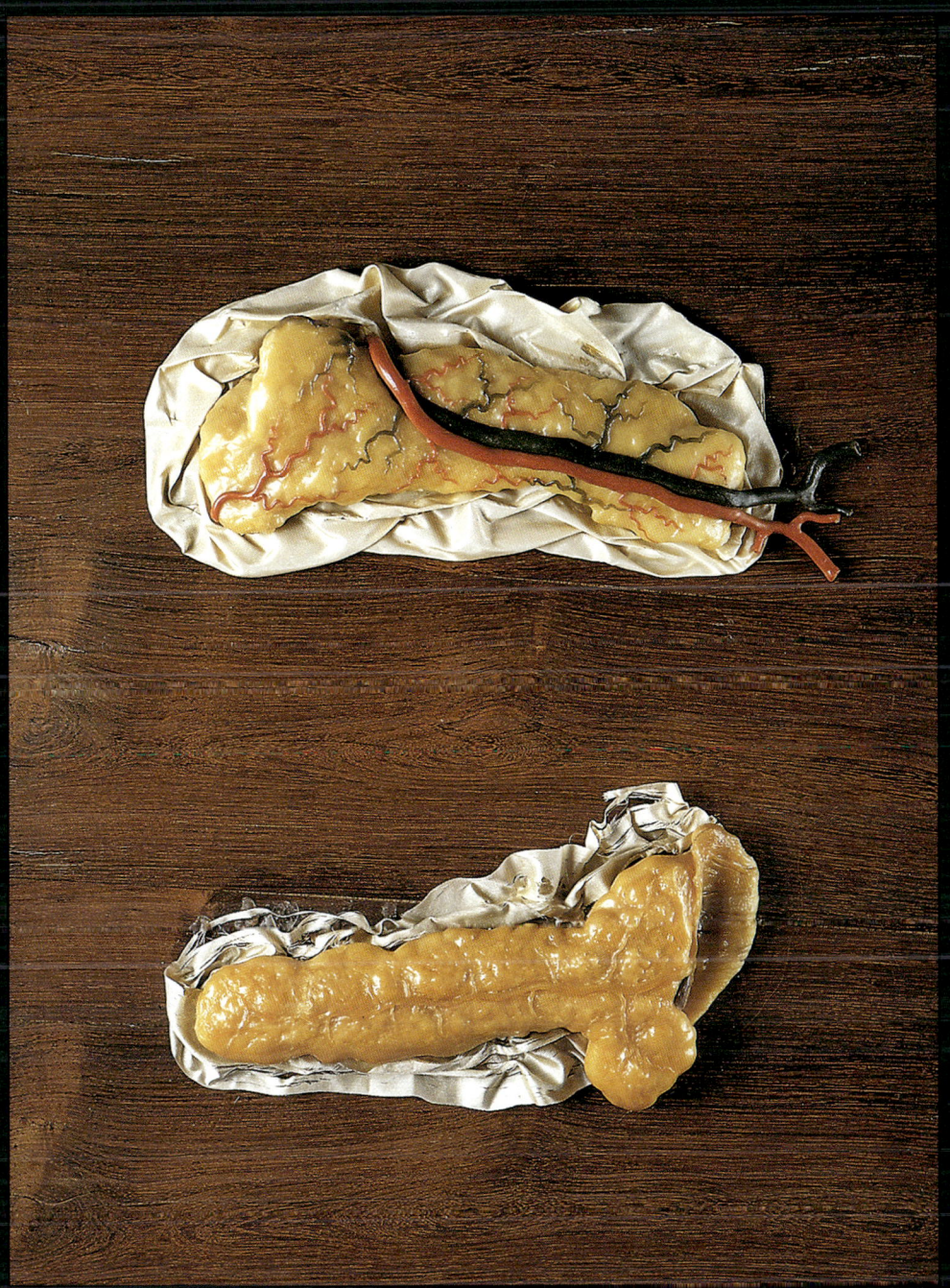

Pancreas, Arteria et vena splenica

The pancreas; in the upper specimen with the splenic vessels

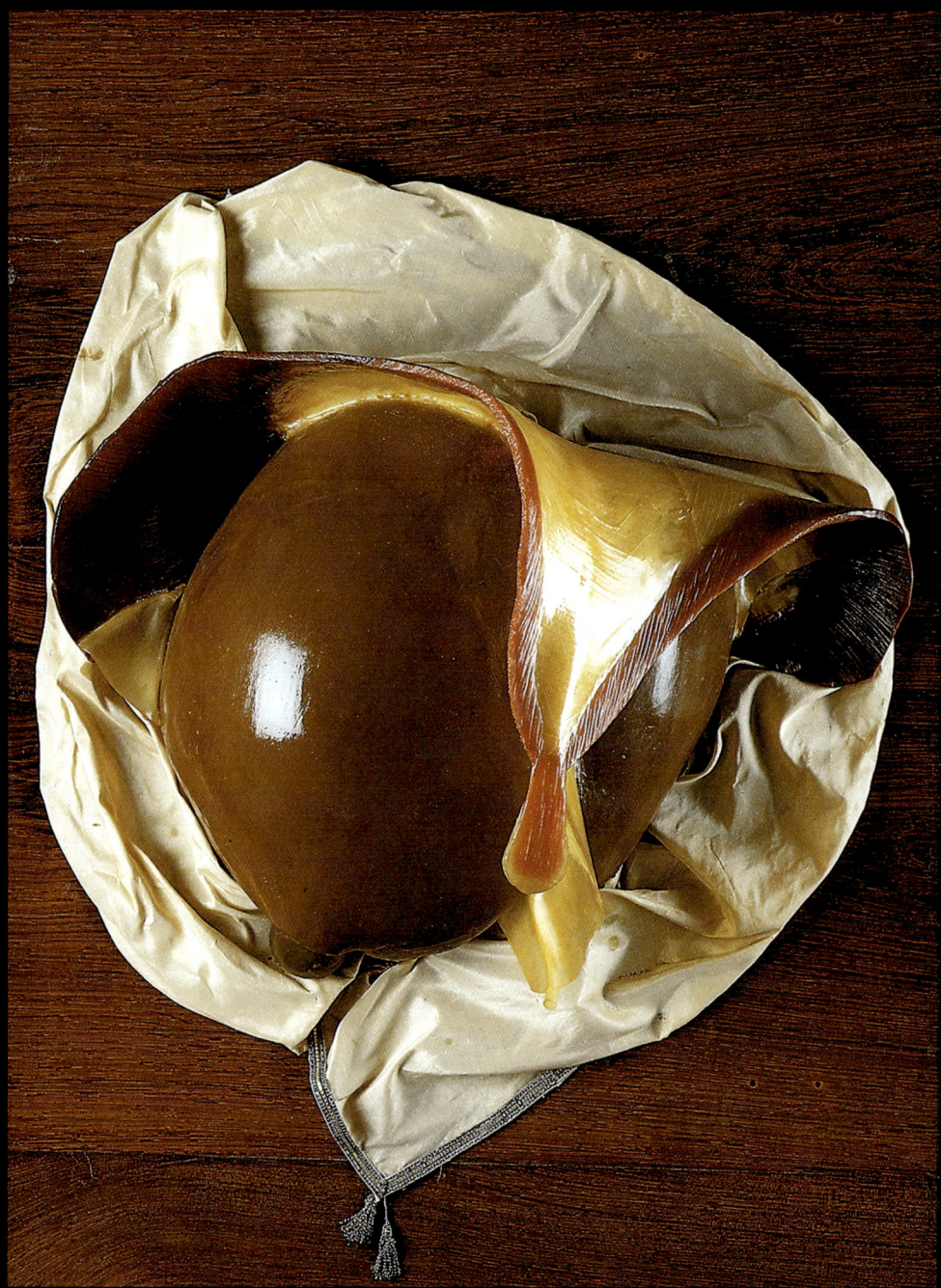

Diaphragma, Hepar, Ligamentum falciforme

Display showing the relationship of the liver to the diaphragm

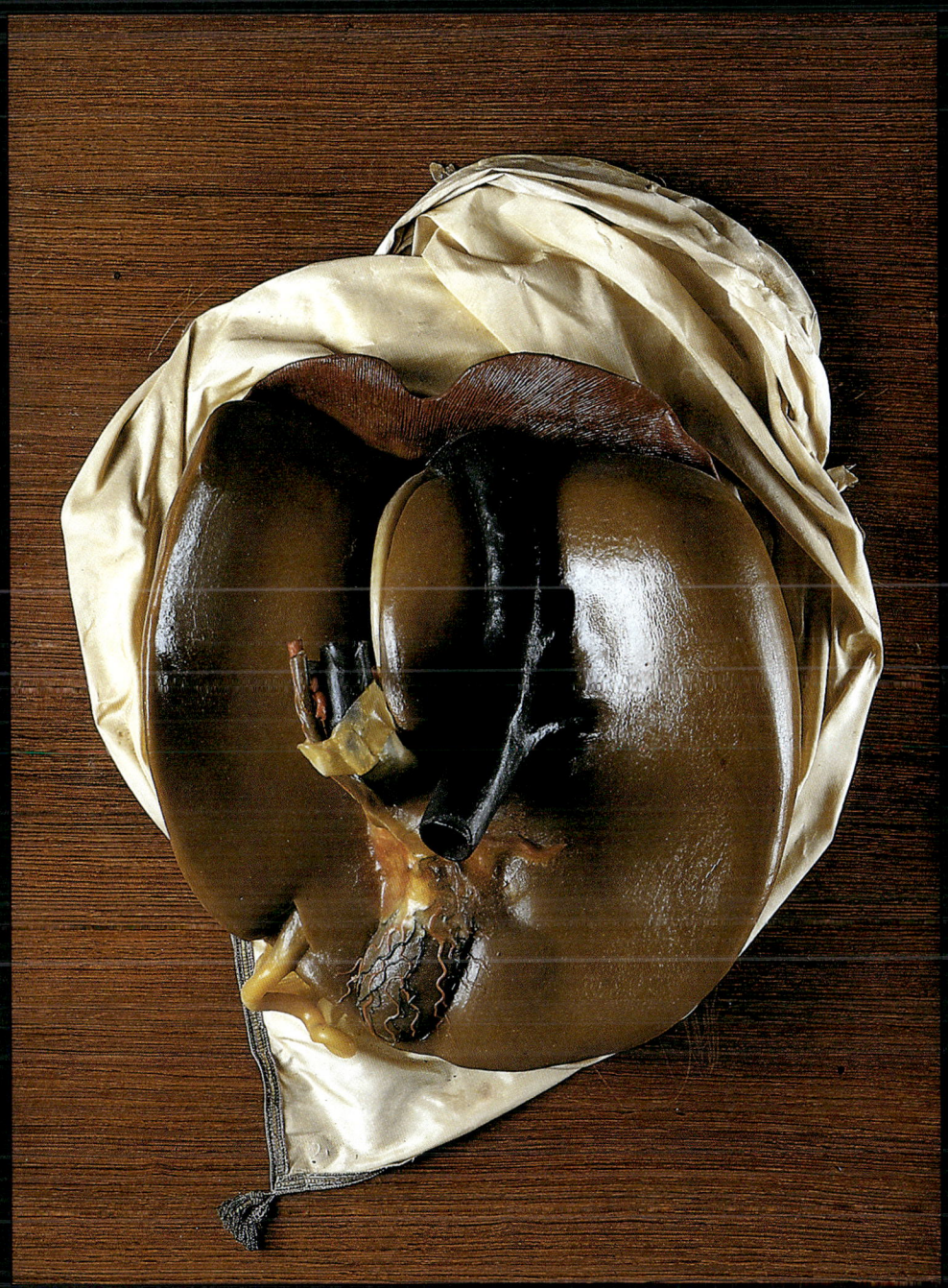

Hepar, Facies visceralis, Vena cava inferior

Posterior surface of the liver showing its various lobes
and the related course of the inferior vena cava

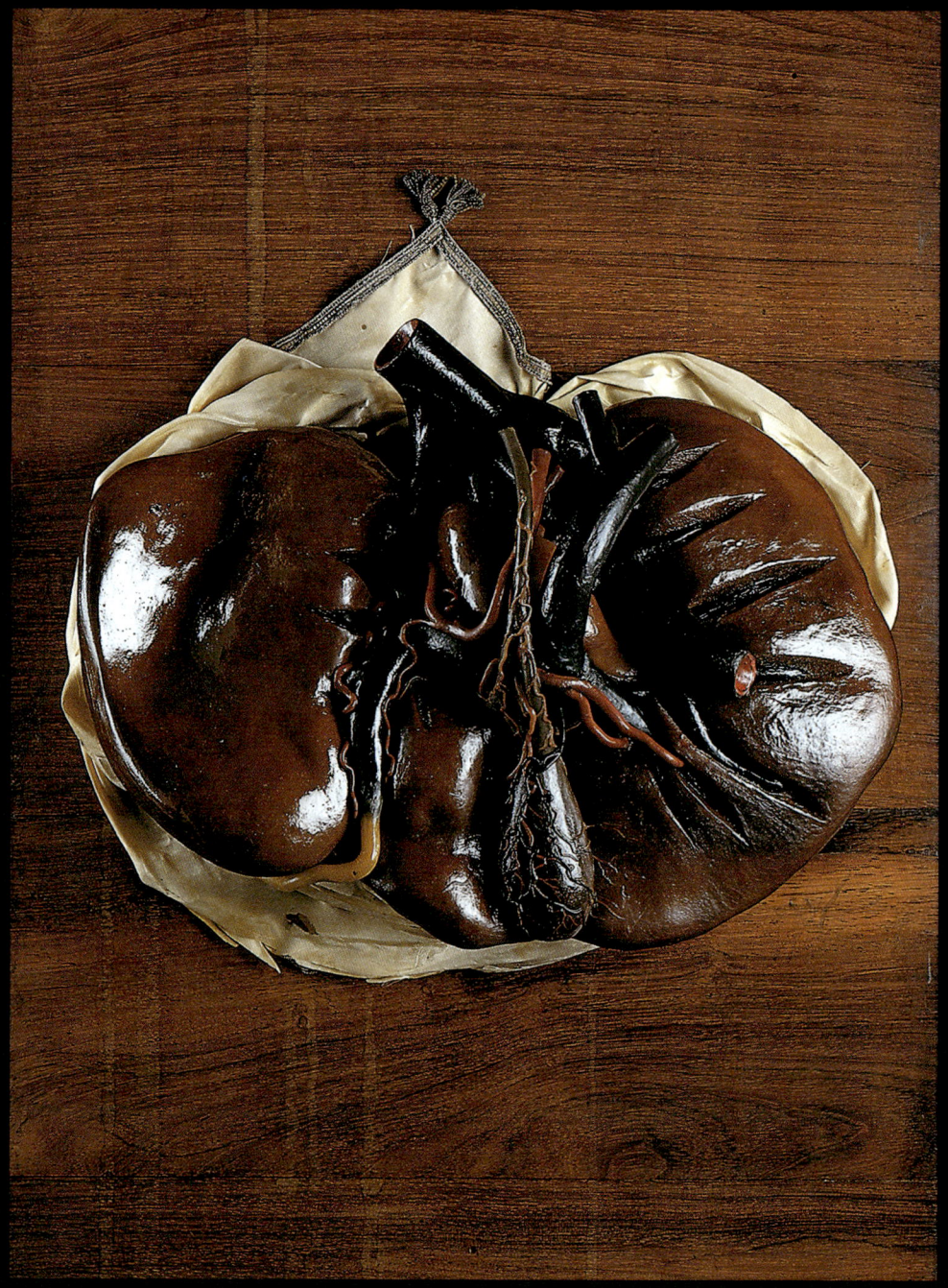

Vena portae

Posterior surface of the liver showing the vessels of the liver

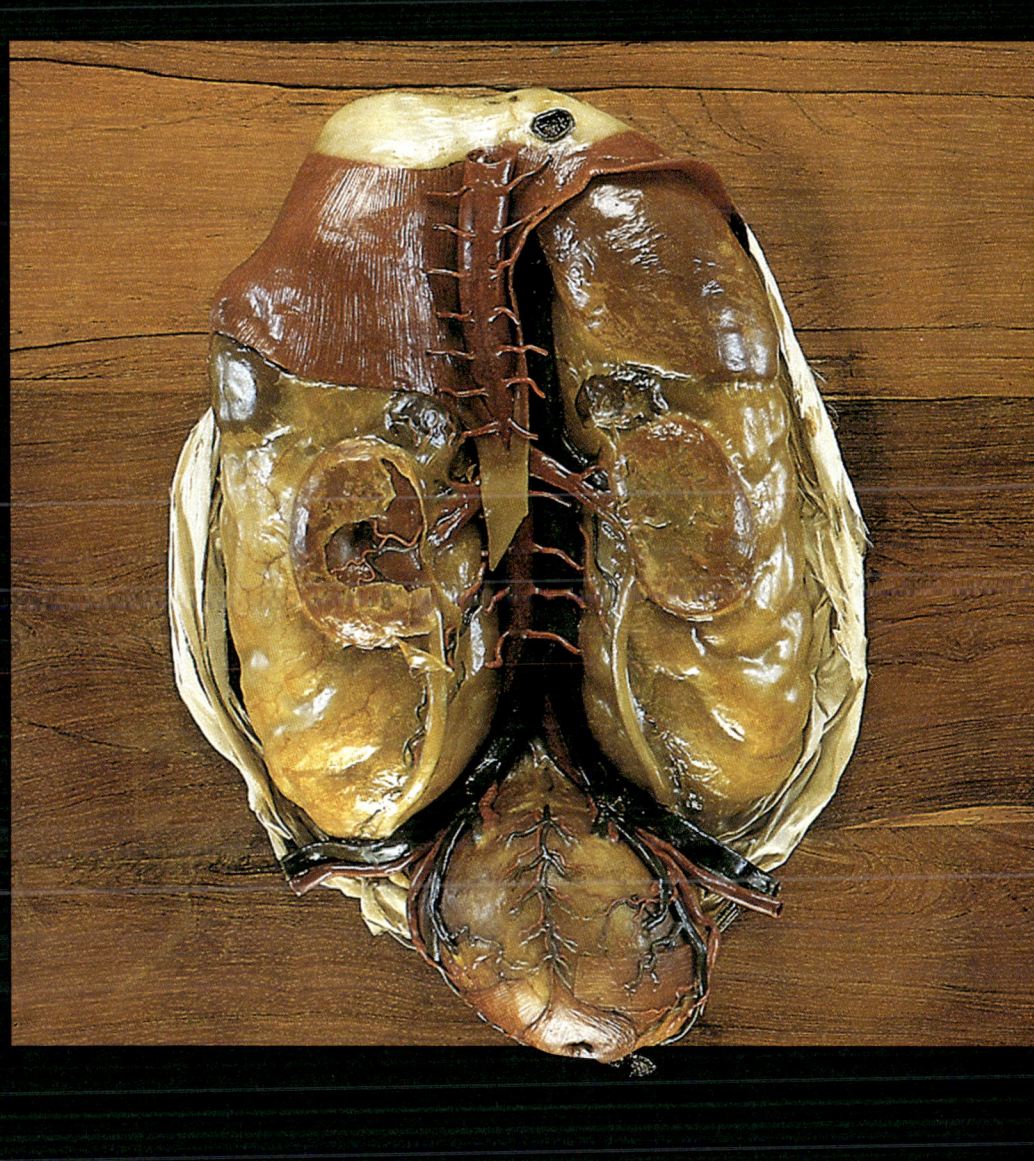

Diaphragma, Peritoneum parietale, Ren, Ureter, Musculus levator ani

Specimen showing the diaphragm, peritoneum, kidneys and ureters, seen from behind

Systema urogenitale et genitale

Urinary and Reproductive System
Harn- und Geschlechtsorgane
Appareil urinaire, appareil génital

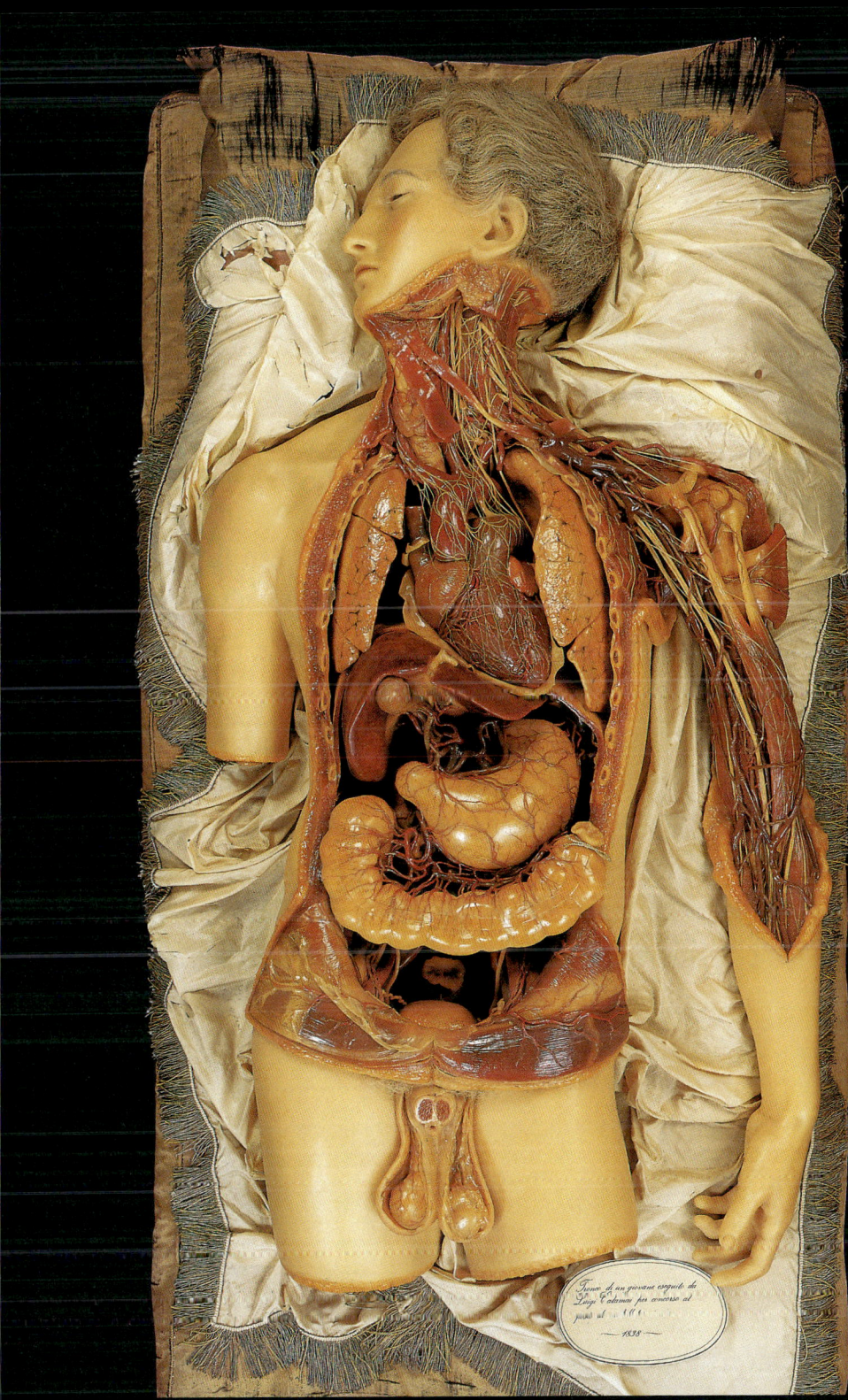

Tronco di un giovane eseguito da
Luigi Calamai per concorso al
premio al ... 1838

1838

Urinary and Reproductive System

The wide range of exhibits in this section reflect very clearly the enlightened interest so characteristic of that time, and which was so far removed from the later prudery of the Victorian era.

The models are anatomically correct representations of the female and male reproductive organs shown in their natural position in the true pelvis. In addition, detailed specimens of functionally related organs are displayed from various points of view. Additional information is given by sections through organs which allow the observer a view into the wide or narrow inside space. Particular emphasis has been laid upon presentations of the uterus taken from the various stages of pregnancy.

The models of embryos are especially informative, as well as those which record the different stages of foetal development up to the time of delivery. Observing the specimens of infants in which the thoracic and abdominal organs are represented in detail, it is the difference in relative size of these organs as compared with those of the adult which is particularly impressive.

Harn- und Geschlechtsorgane

Das breite Spektrum der Exponate in diesem Kapitel spiegelt den aufgeklärten und interessierten Geist der damaligen Zeit eindrucksvoll wider, der sich deutlich von der späteren Prüderie des viktorianischen Zeitalters abhebt.

Die Modelle zeigen anatomisch korrekt die weiblichen und männlichen Geschlechtsorgane in ihrer natürlichen Lage im kleinen Becken. Zudem werden funktionell zueinander in Beziehung stehende Organe als Detailpräparate in verschiedenen Ansichten dargestellt. Zusätzliche Informationen liefern die Schnittpräparate der Organe, die dem Betrachter Einblicke in die verschiedenen weit oder eng gestalteten Innenräume erlauben. Einen thematischen Schwerpunkt bildet die Darstellung der Gebärmutter in den verschiedenen Stadien der Schwangerschaft.

Page 441:
Male torso with the thoracic and abdominal cavities laid open

Besonders informativ sind die Modelle der Embryonen sowie die Präparate, die die fötalen Entwicklungsstadien bis zum geburtsreifen Neugeborenen dokumentieren. Bei der Betrachtung der Säuglingspräparate mit Detaildarstellungen der Brust- und Bauchorgane sind besonders die – im Vergleich zur Organanatomie des Erwachsenen – anderen Größenverhältnisse der sich noch in der Entwicklung befindlichen Organe auffällig.

Appareil urinaire, appareil génital

Le large choix des pièces concernant les organes génitaux et la grossesse reflète l'état d'esprit éclairé et intéressé qui régnait à l'époque de leur réalisation.

Des représentations anatomiques exactes des organes génitaux féminins et masculins en place dans le petit bassin sont données par plusieurs modèles en cire. À ces vues d'ensemble sont adjointes des préparations plus détaillées d'organes ayant des rapports fonctionnels mutuels, vus sous différentes incidences. Des sections et coupes apportent des précisions supplémentaires en donnant à l'observateur une vue dans les diverses cavités, plus difficilement explorables autrement.

La représentation de l'utérus aux différents stades de la grossesse est aussi un thème particulièrement étudié. Les modèles d'embryons ou les préparations démontrant les stades du développement du fœtus jusqu'à la naissance sont particulièrement instructifs. Des préparations montrent avec précision les organes thoraciques et abdominaux de nouveau-nés ; il ressort remarquablement de leur observation les différences de proportions entre les organes du nouveau-né et ceux de l'adulte.

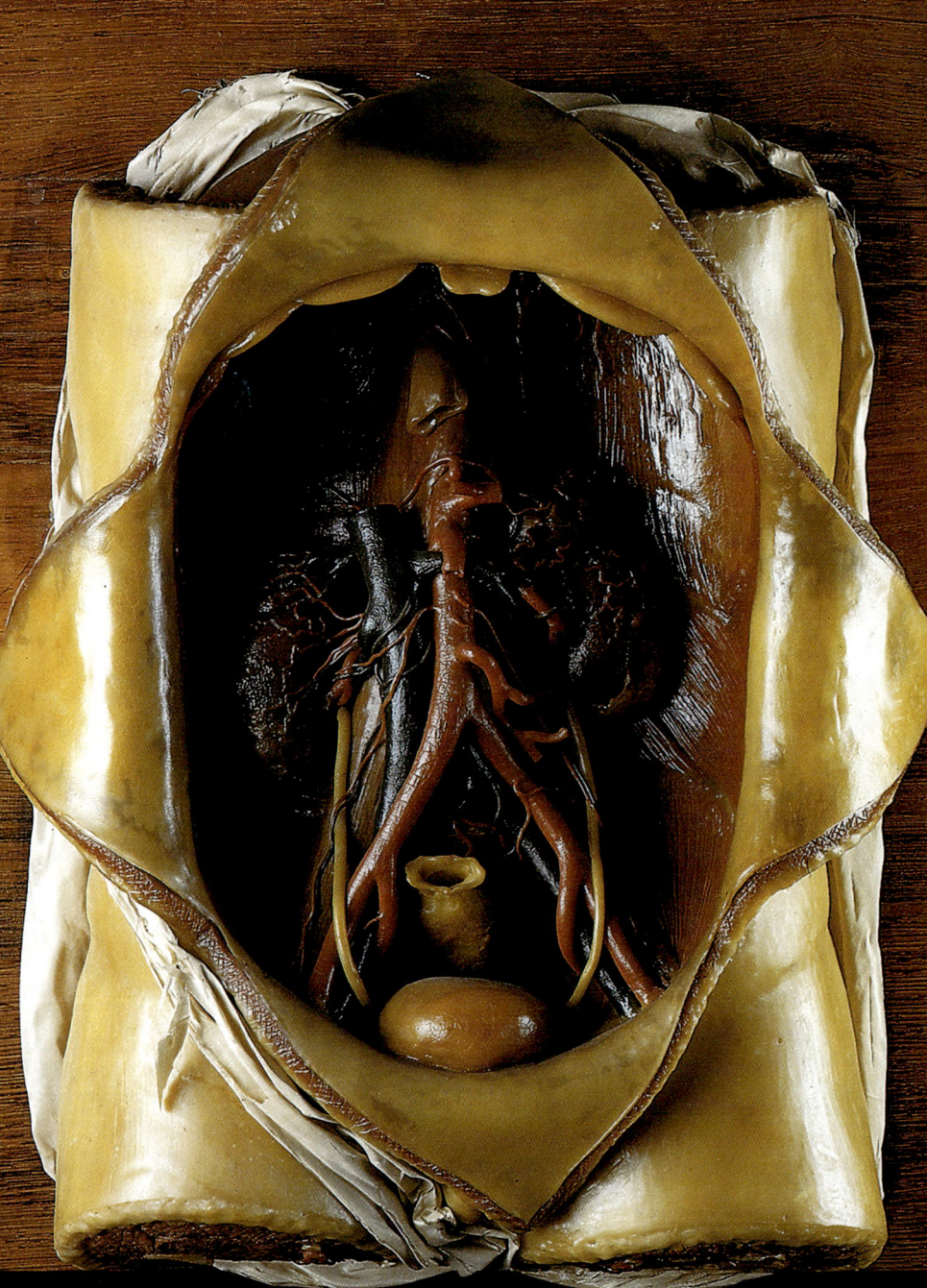

Ren, Ureter, Vesica urinaria, Aorta abdominalis, Vena cava inferior

View of the posterior abdominal wall after removal of the abdominal viscera.
Kidneys, upper urinary tract, urinary bladder, aorta and inferior vena cava

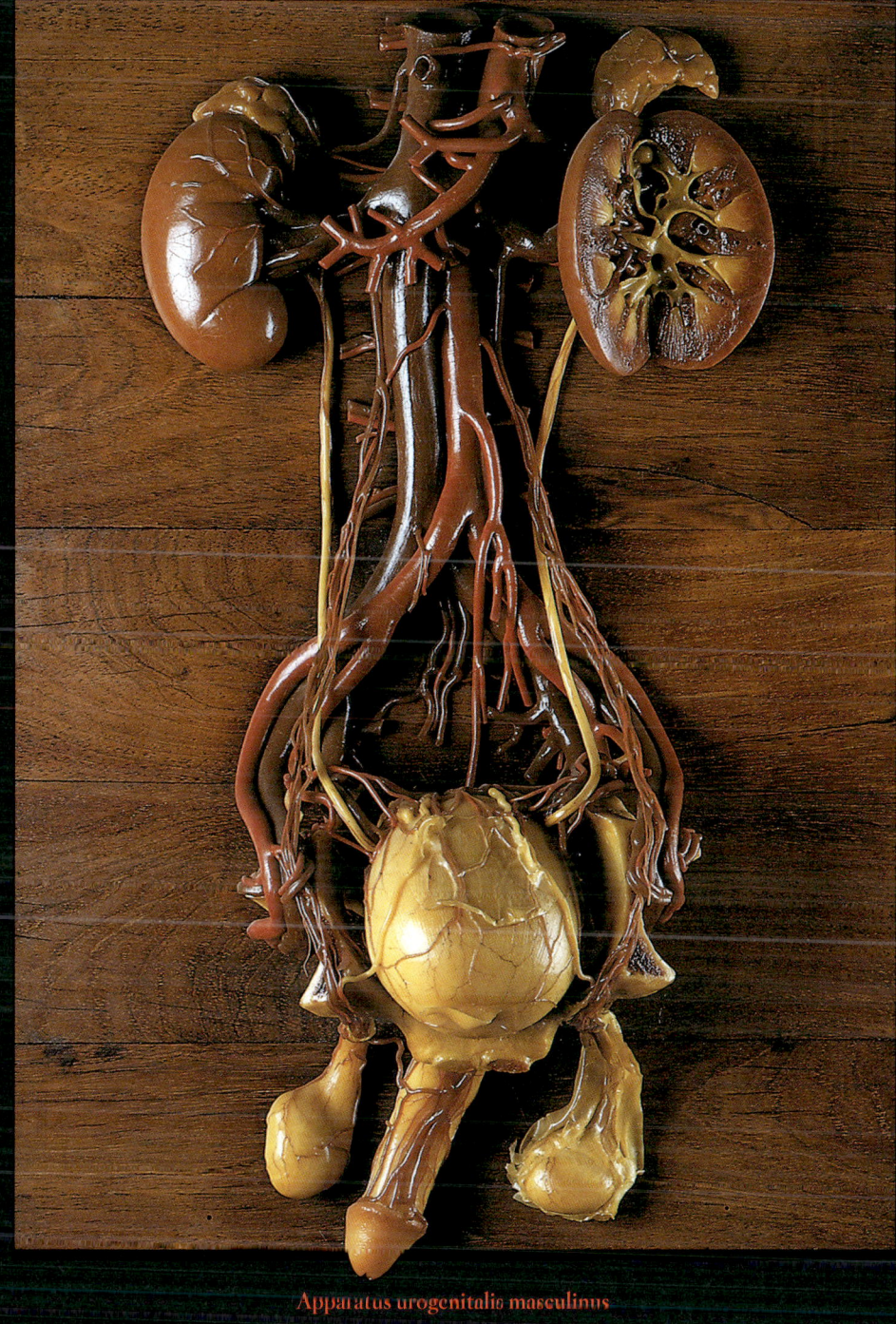

Apparatus urogenitalis masculinus

Male urogenital tract with its blood supply.
The left kidney has been cut open to reveal the calyces and renal pelvis

445

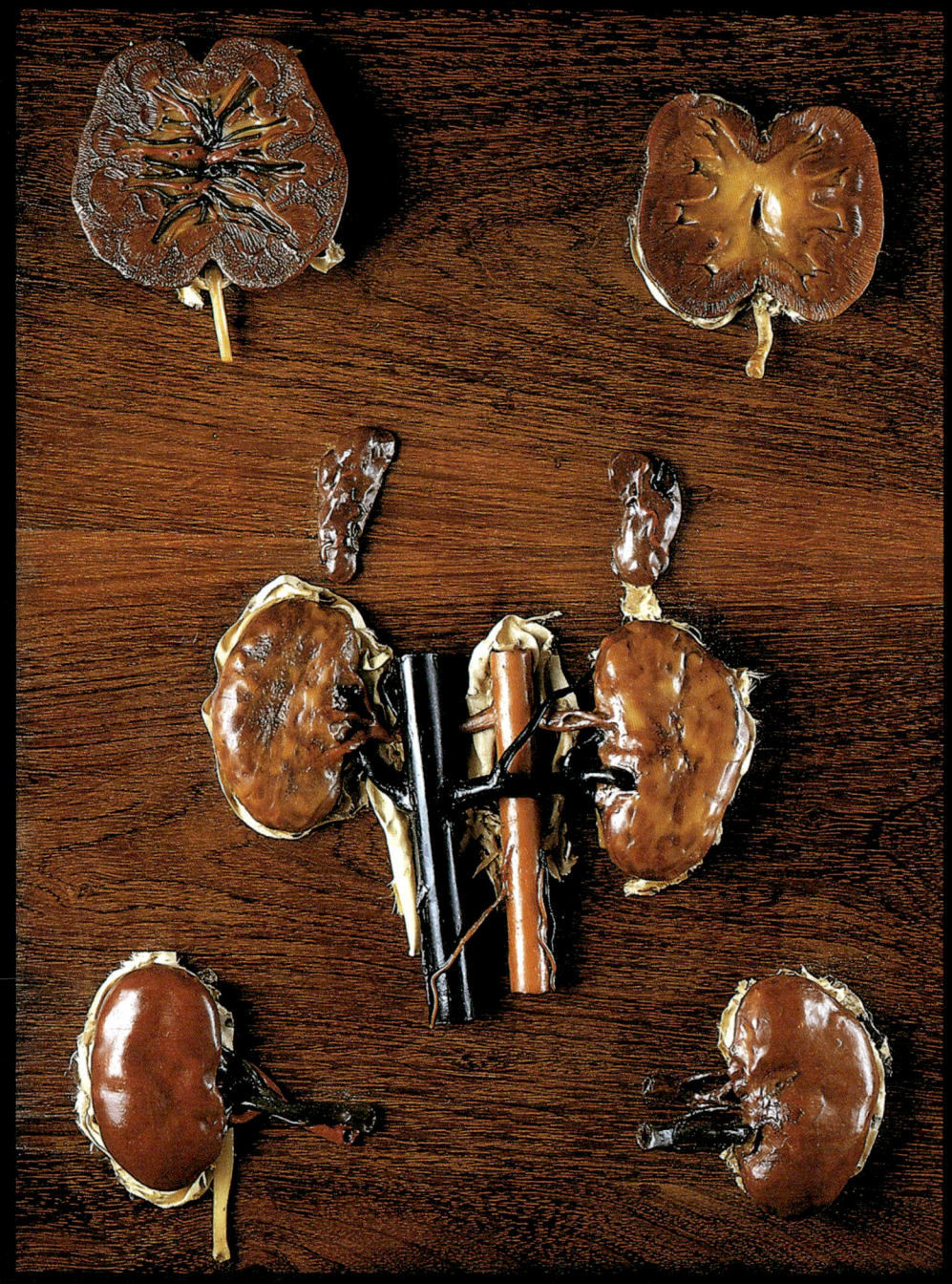

Ren, Glandula suprarenalis
Various aspects of the kidney

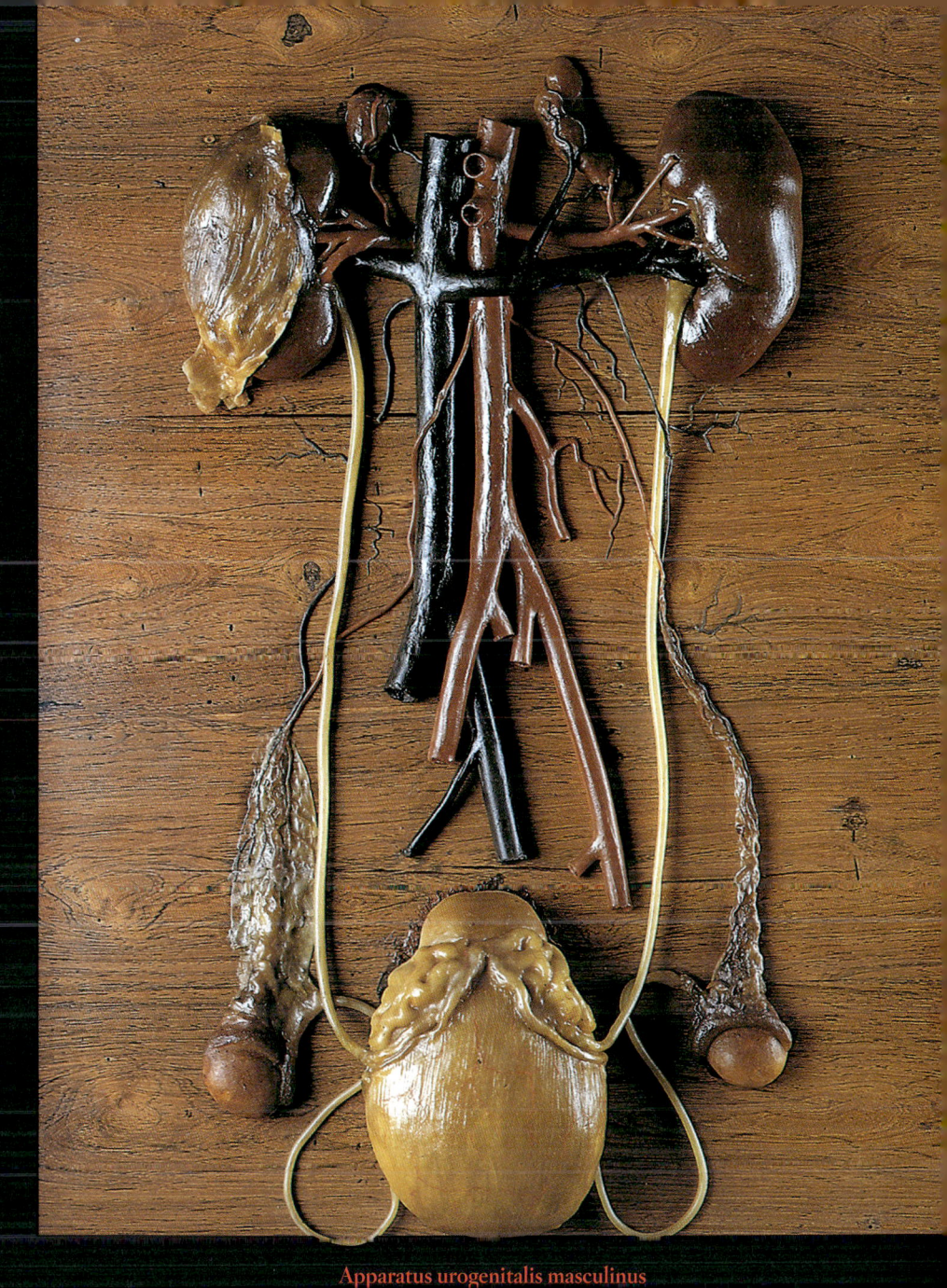

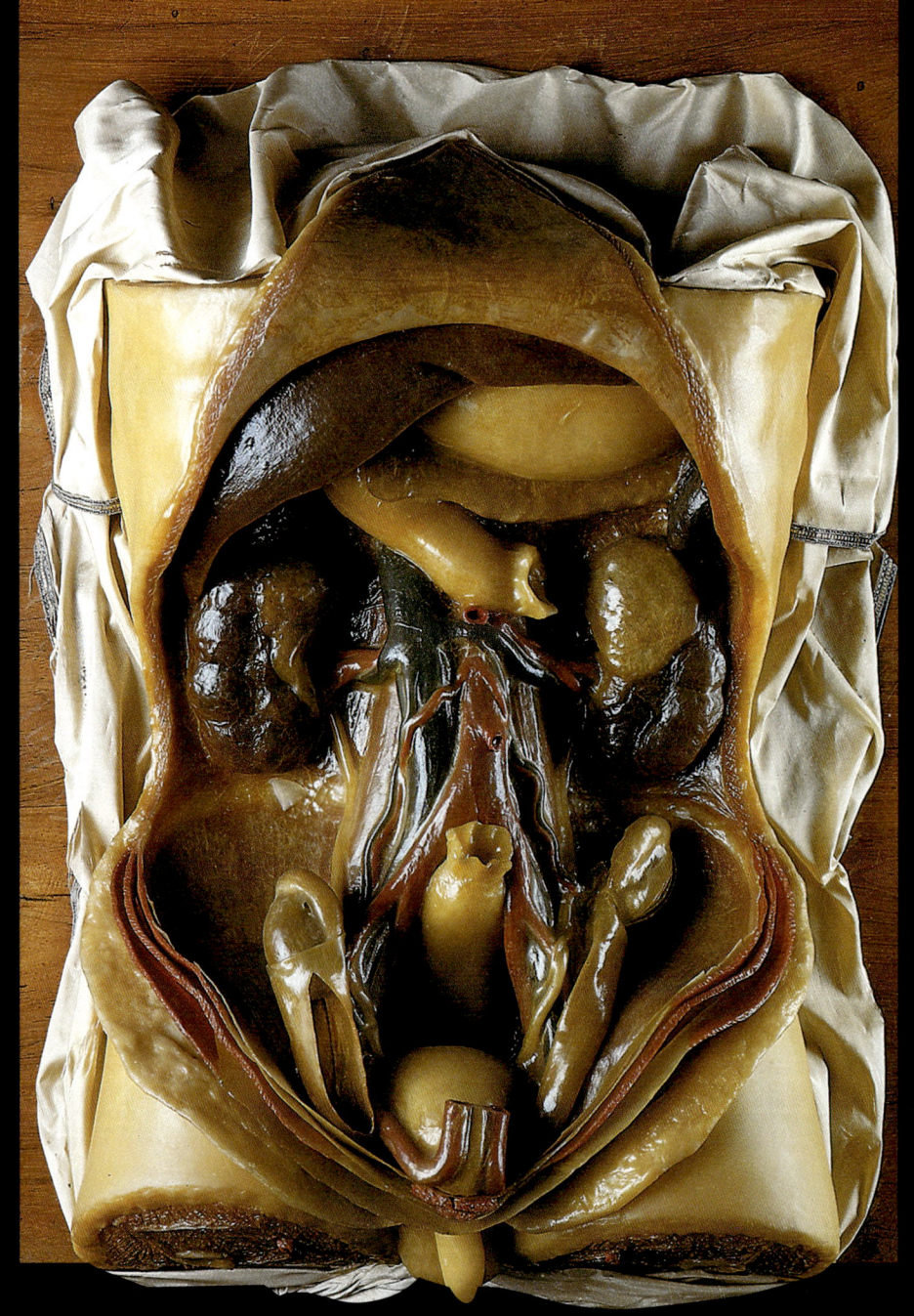

Recessus vaginalis testis

Organs on the posterior abdominal wall of a male foetus. In the true pelvis the right and left testes can be seen.
At this stage of development they have not yet reached the scrotum

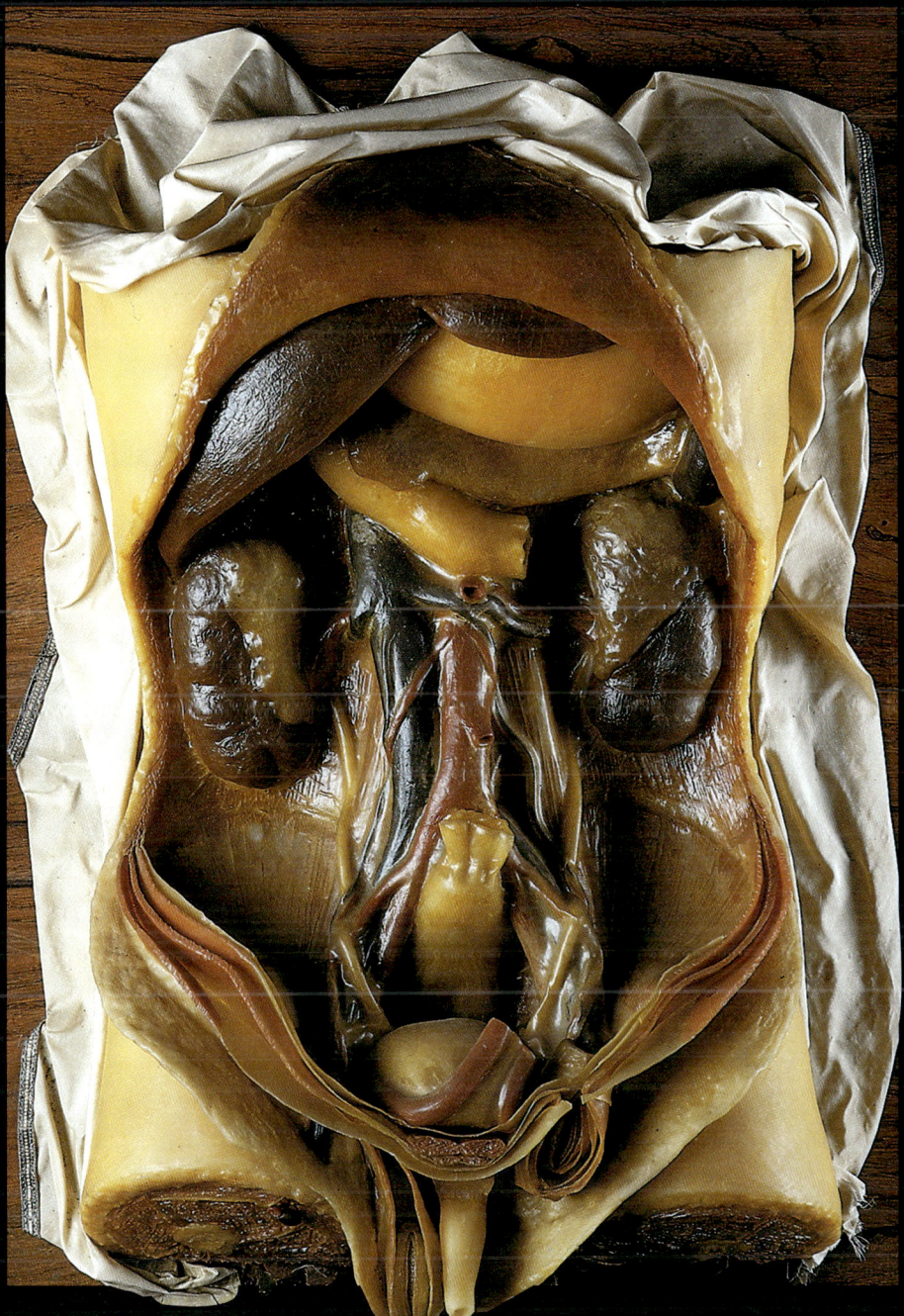

Canalis inguinalis, Tunicae funiculi spermatici, Glandula suprarenalis

Coverings of the testis in a newborn infant

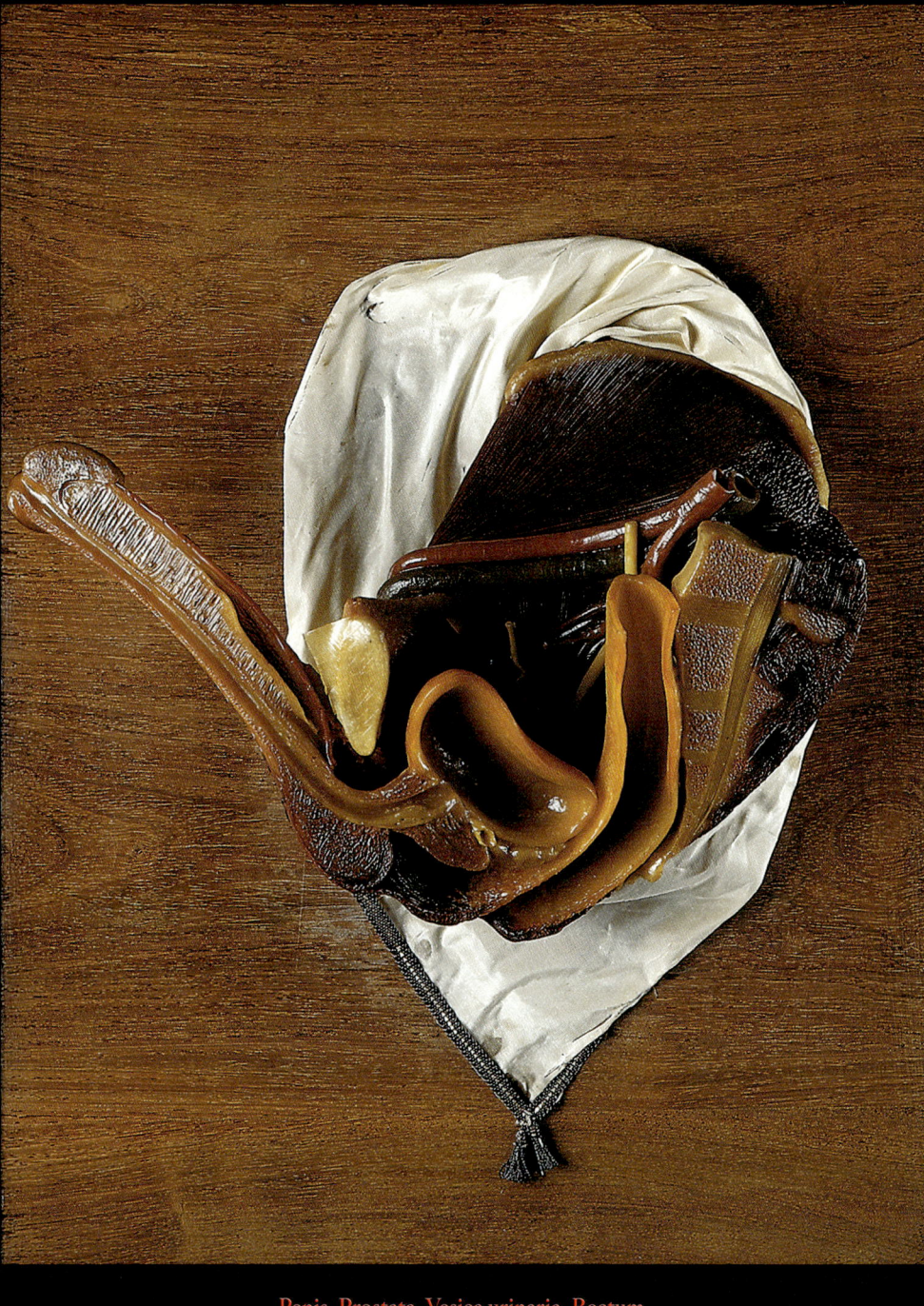

Penis, Prostata, Vesica urinaria, Rectum

Specimen of a male pelvis which has been divided in the midline

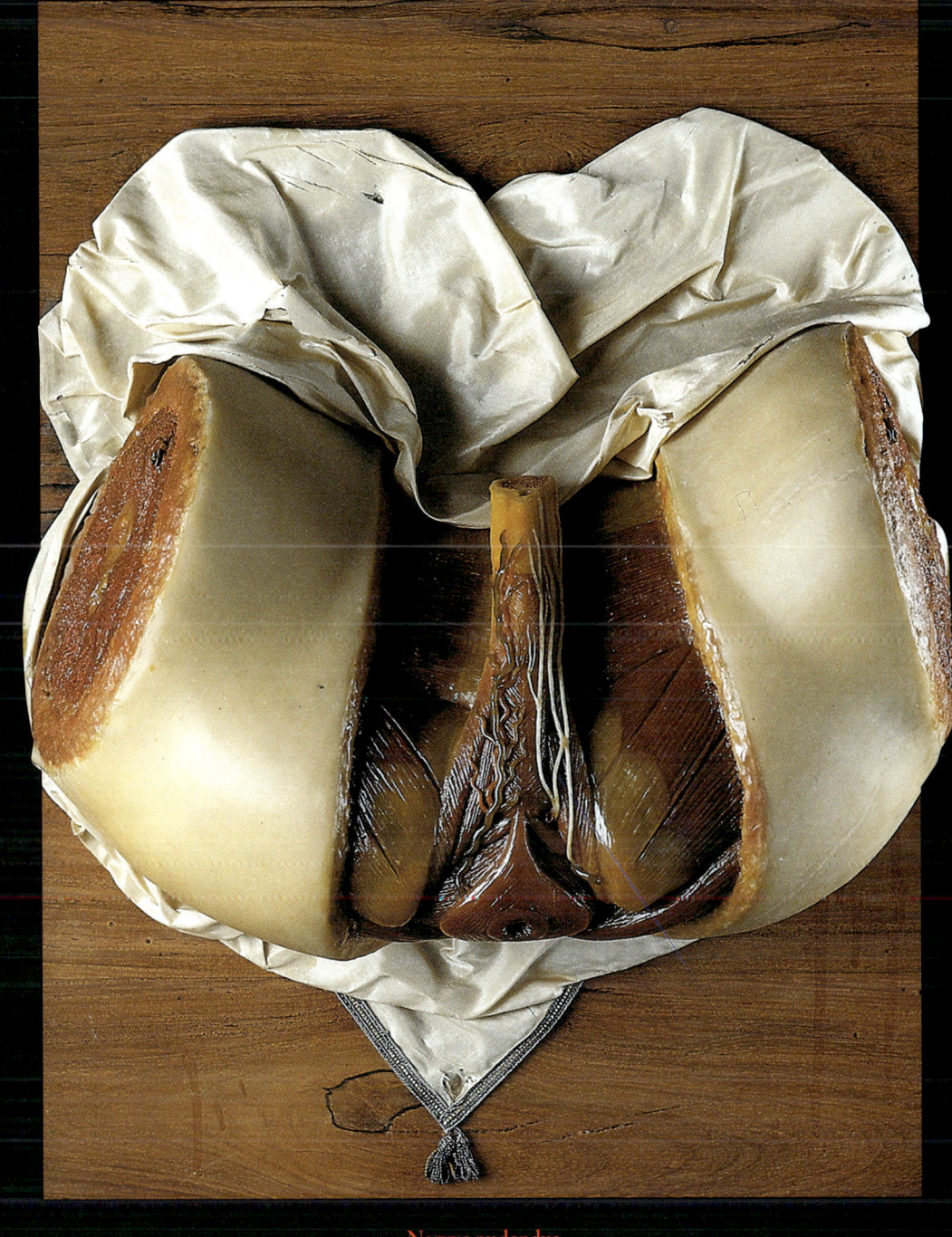

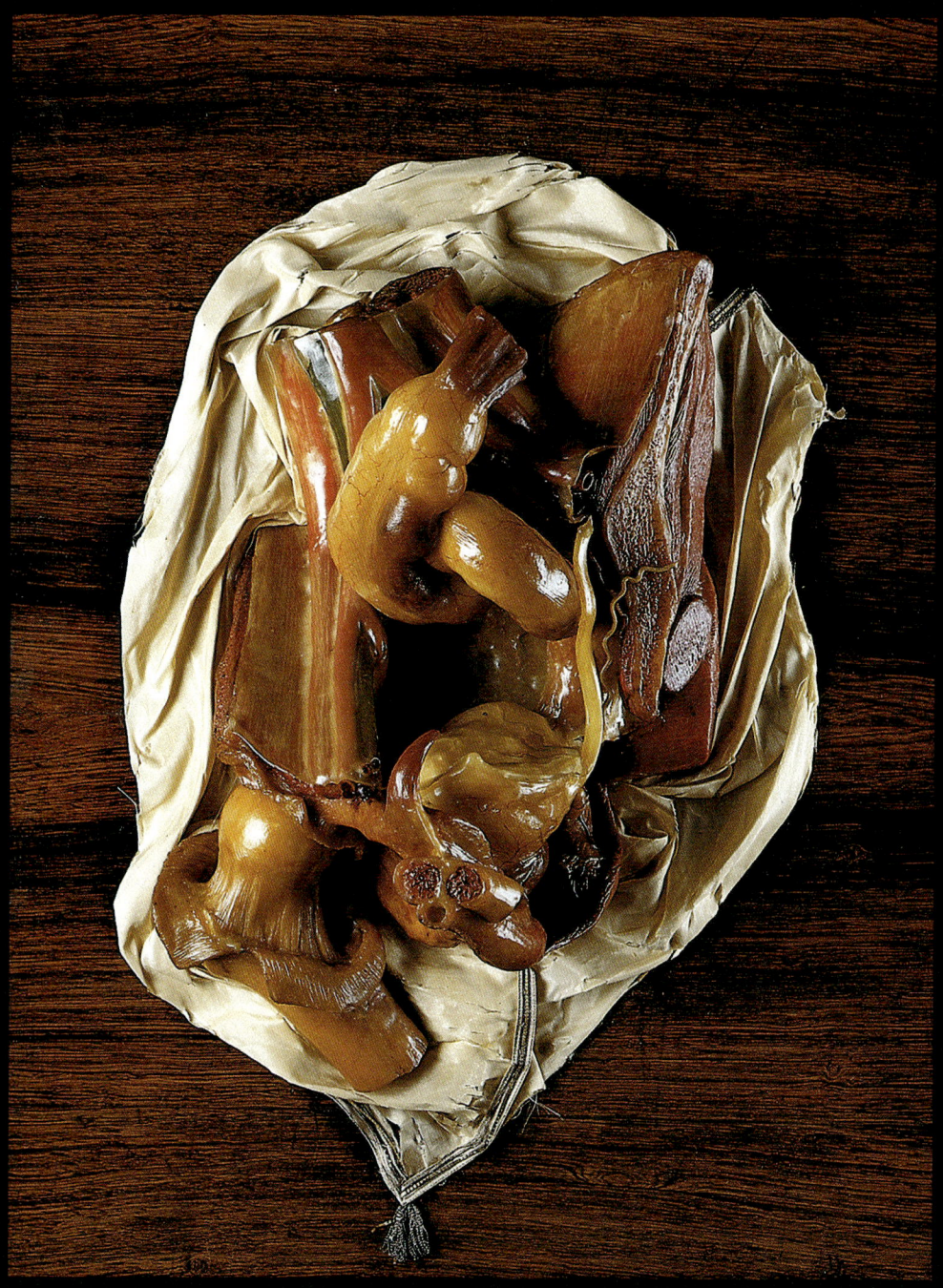

Vesica urinaria, Ureter, Radix penis

View into a hemisected male pelvis

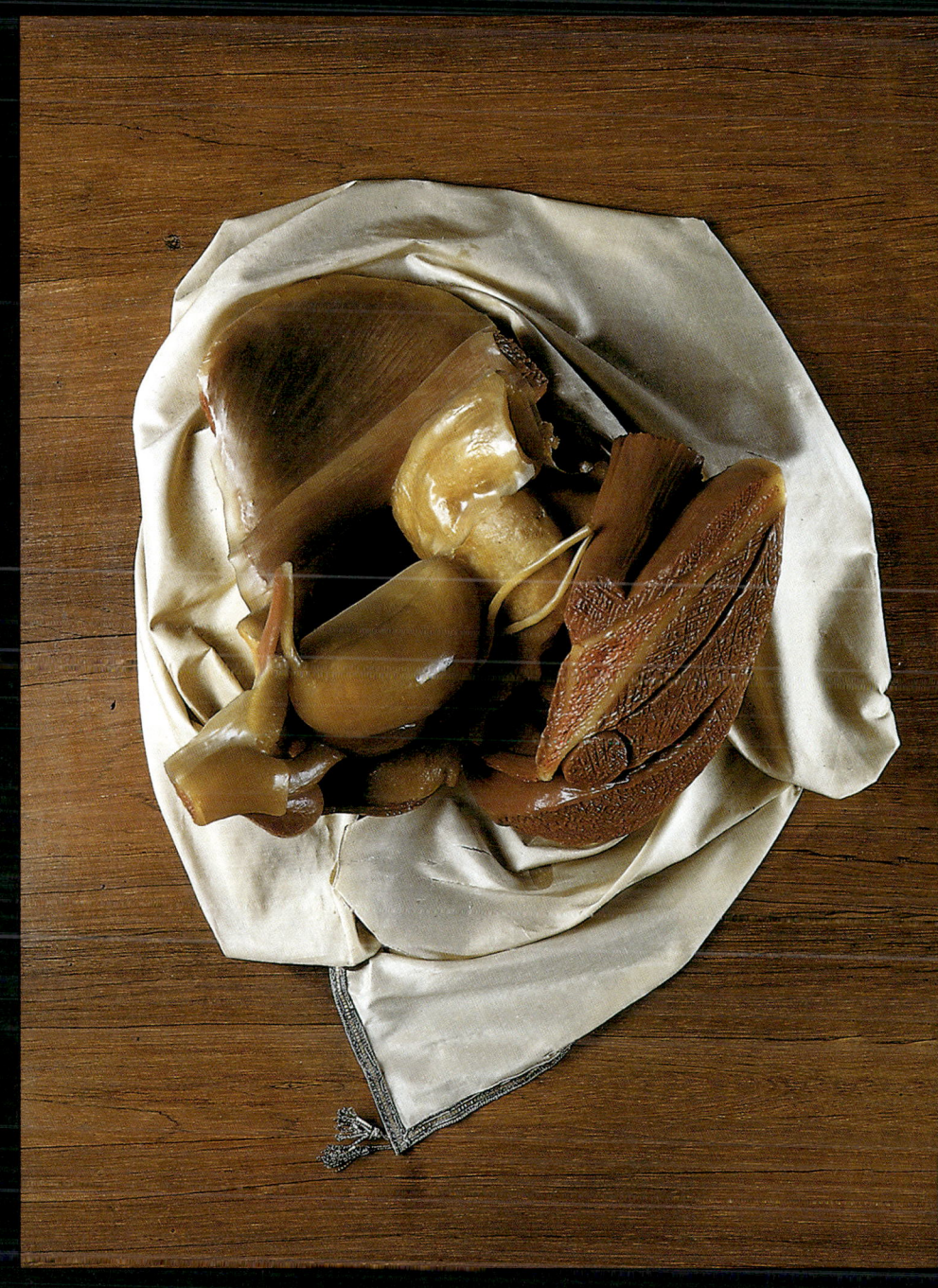

Vesica urinaria

View into the male pelvis showing urinary bladder and rectum

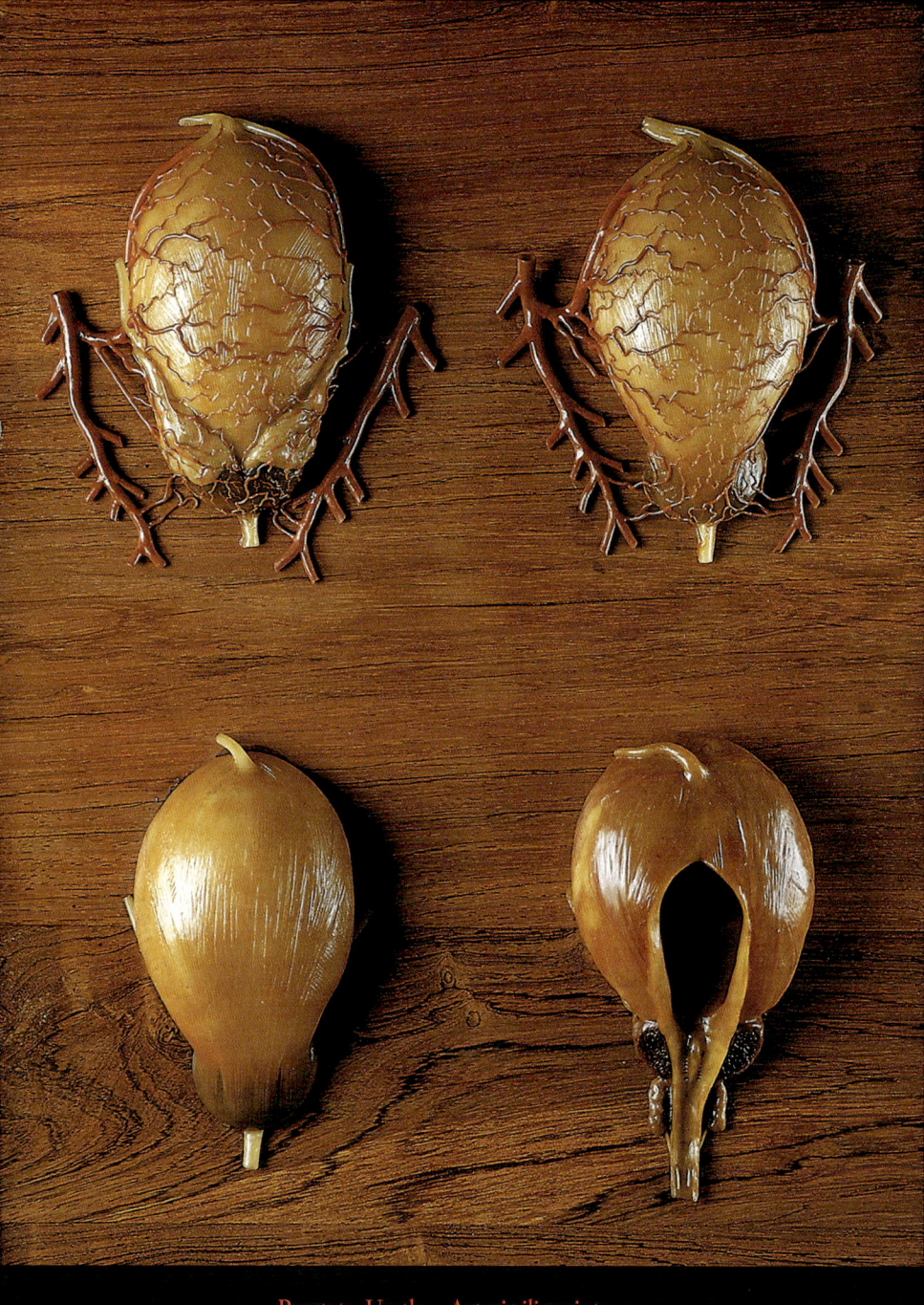

Prostata, Urethra, Arteria iliaca interna

Various aspects of male urinary bladder with prostate gland.
The upper specimen on the left has been opened to show the urinary bladder and urethra

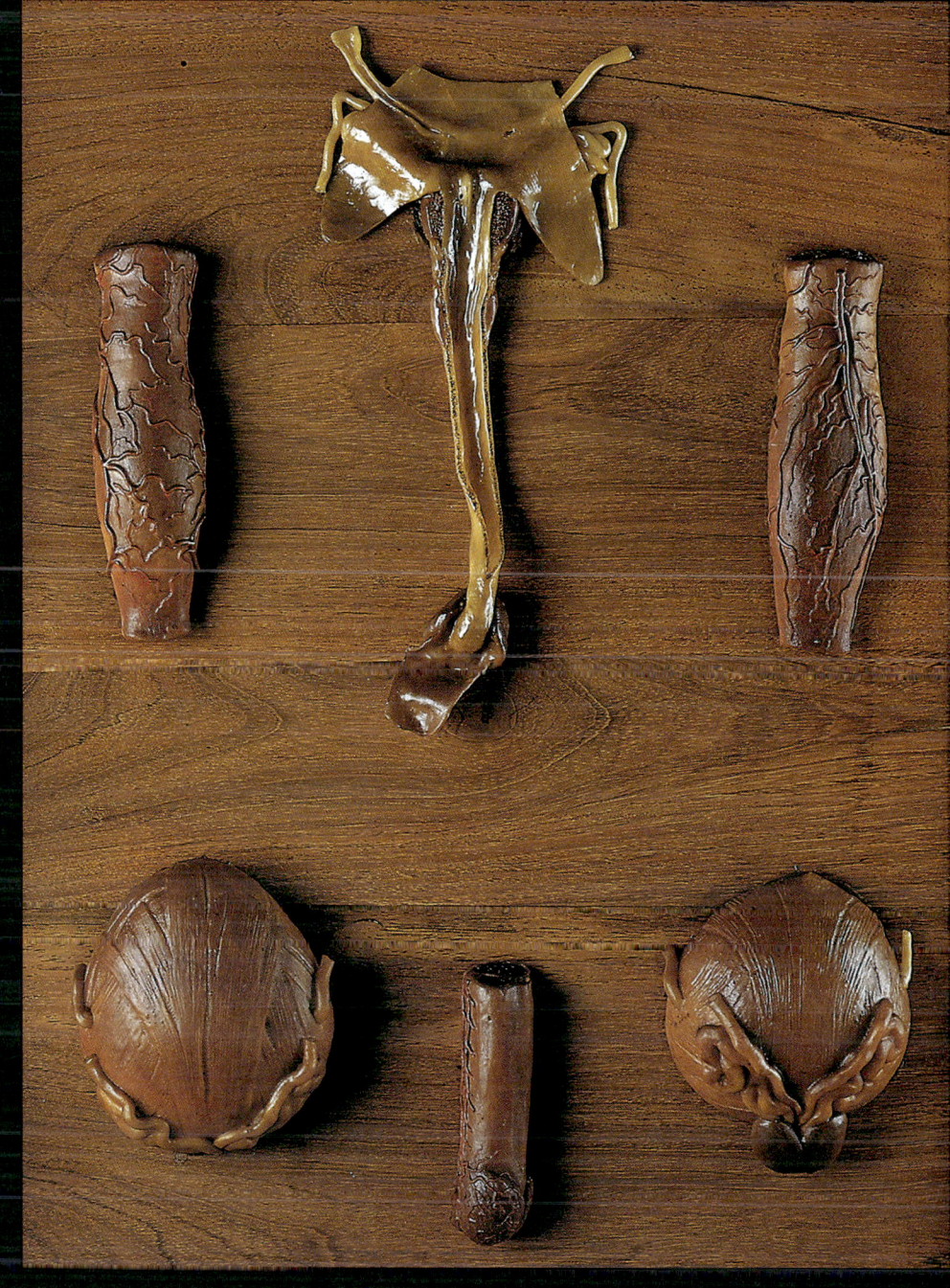

Vesica urinaria, Prostata, Glandula vesiculosa, Penis, Urethra

Specimen of the male genital organs. Above centre: the urethra has been laid open.
Below right: the urinary bladder seen from behind and the chestnut-shaped prostate gland

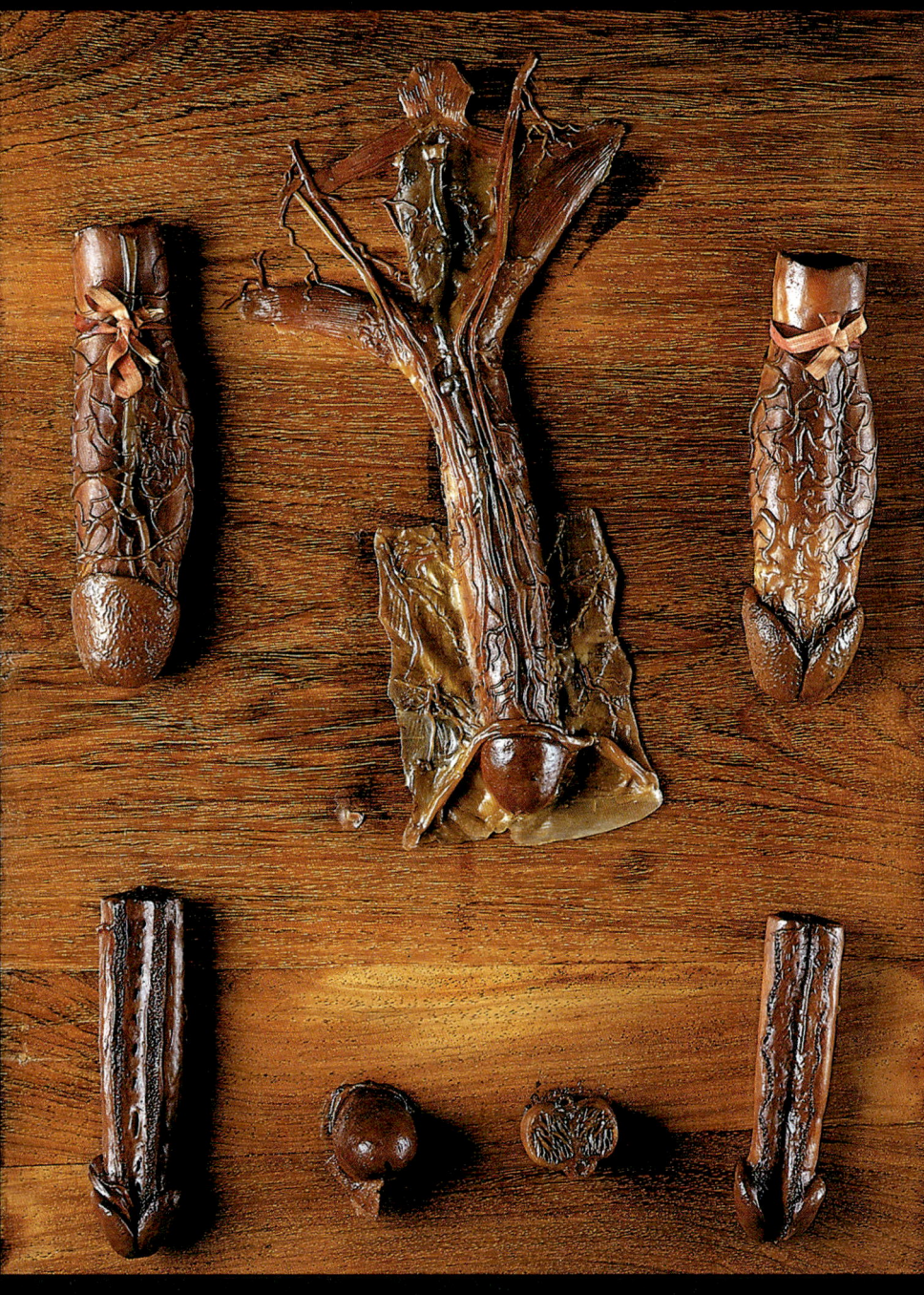

Penis, Arteria dorsalis penis

Various aspects of the blood supply of the penis

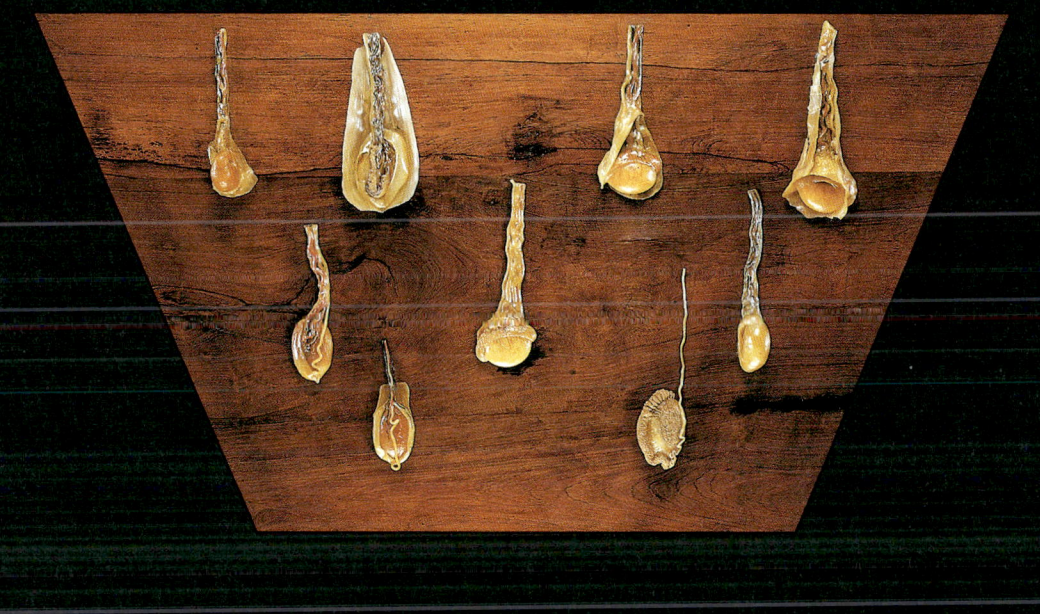

Testis, Funiculus spermaticus, Epididymis
Various aspects of testis, epididymis and their coverings

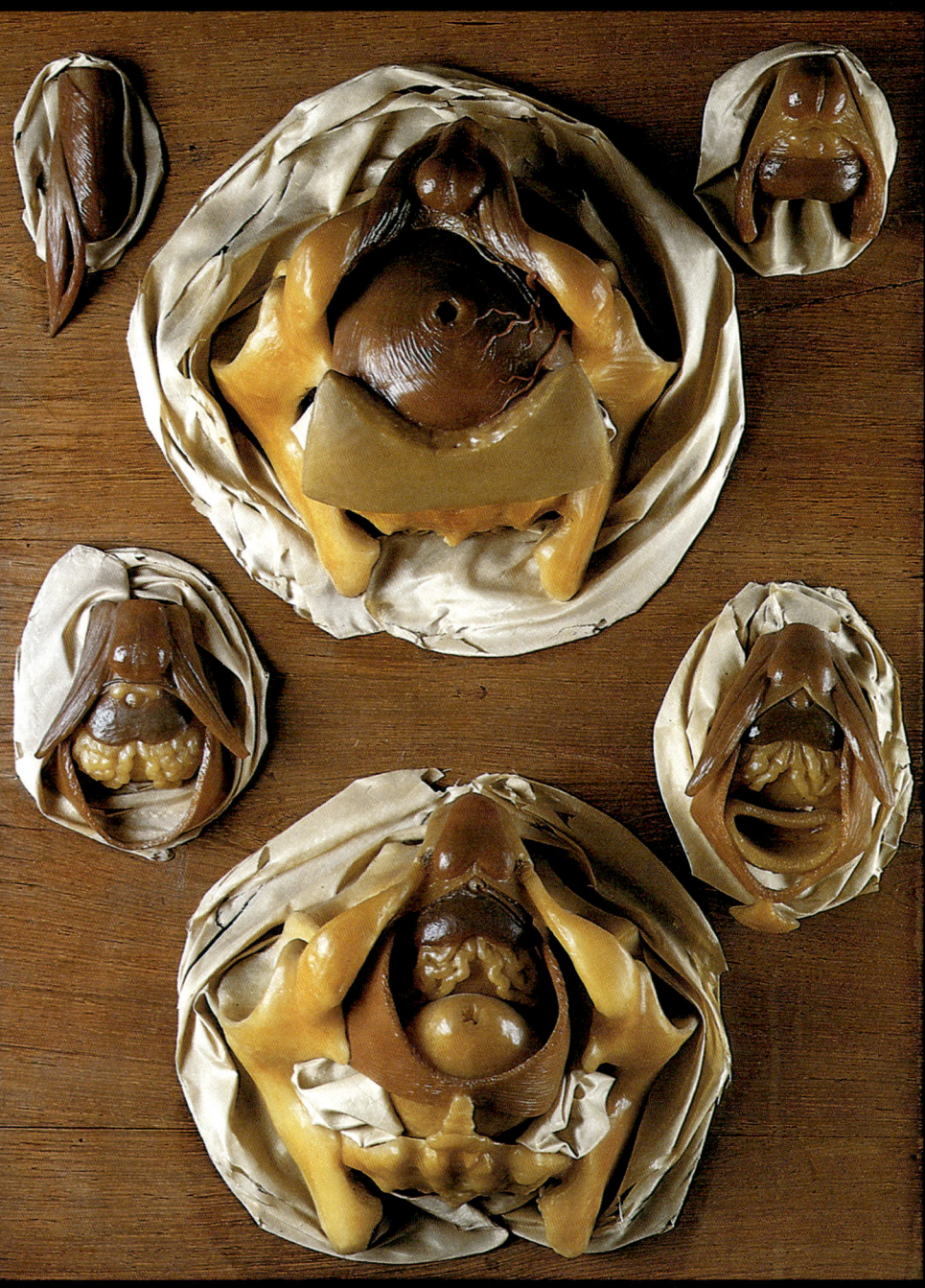

Glandula vesiculosa, Prostata, Musculus levator ani,
Musculus bulbospongiosus, Musculus bulbocavernosus
View of the male pelvic floor from below

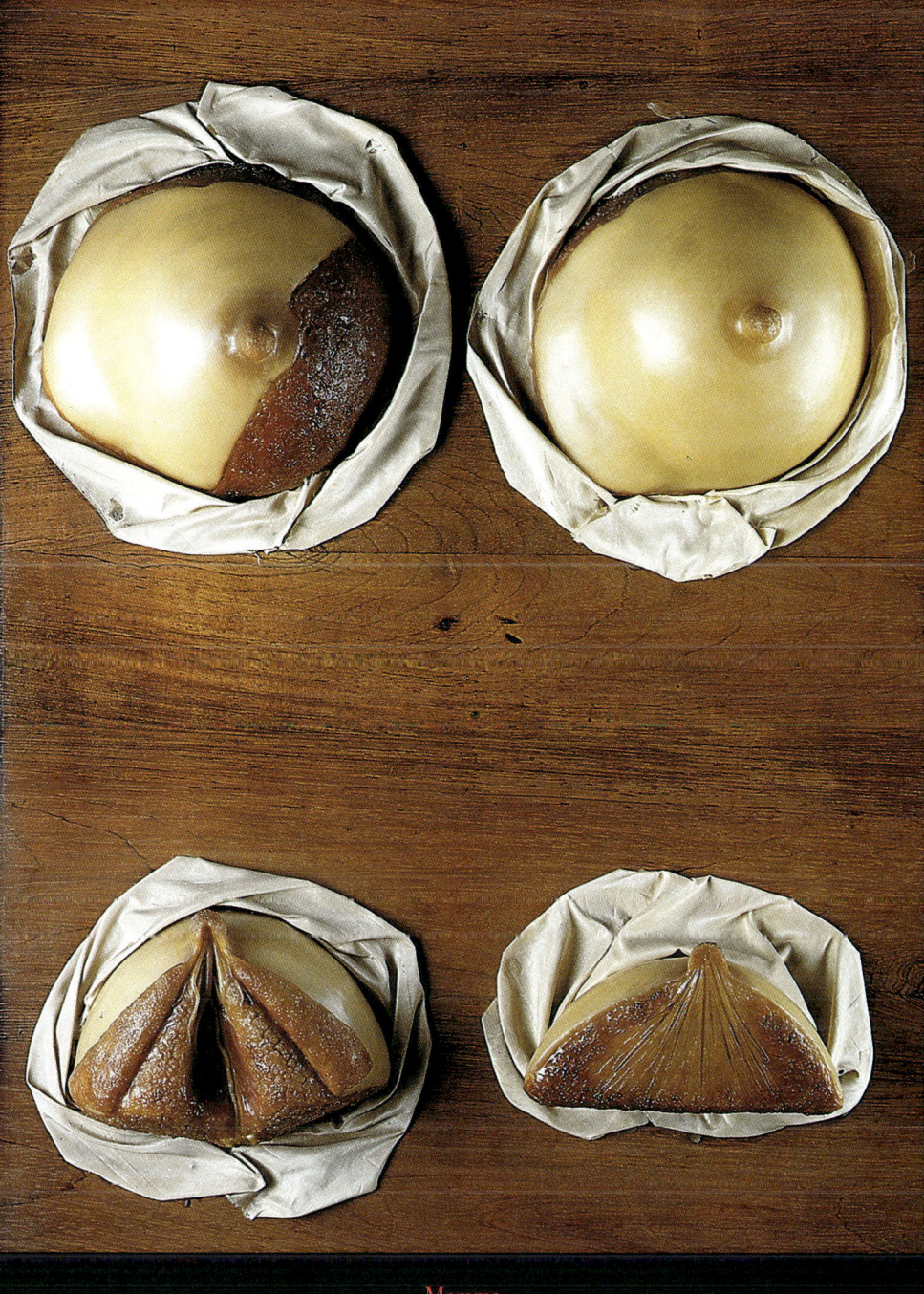

Mamma

Specimens of the female breast

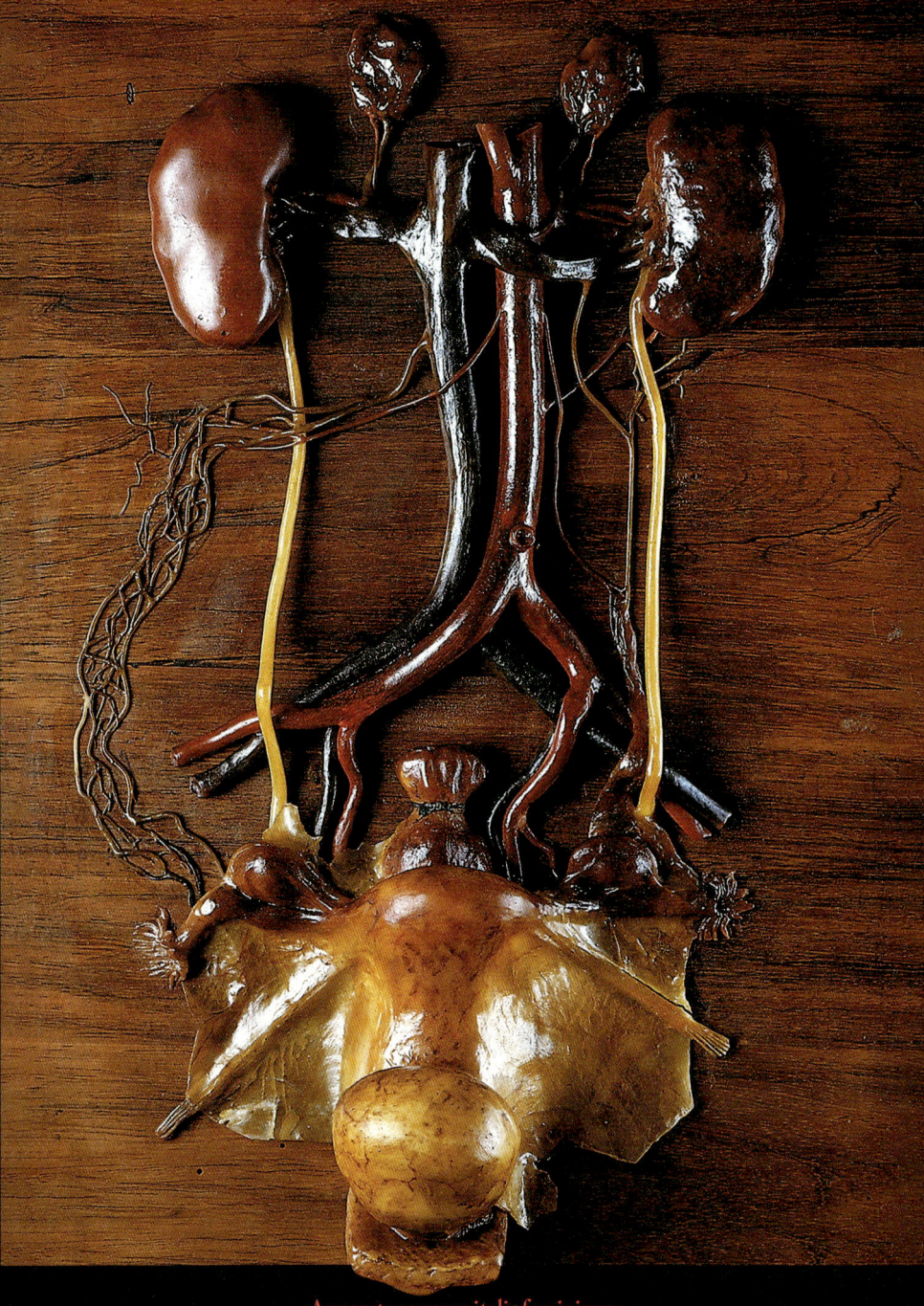

Apparatus urogenitalis femininus

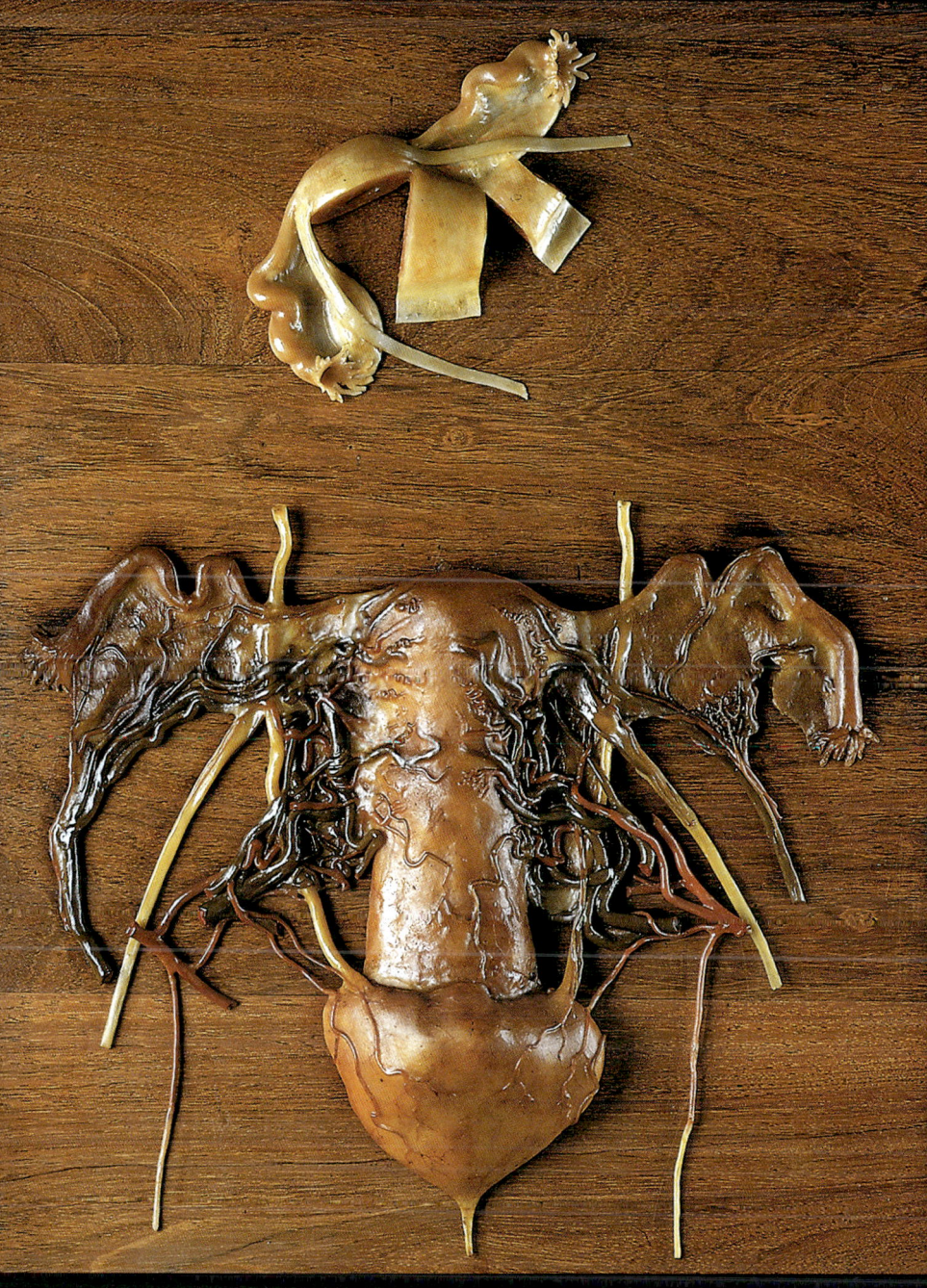

Vesica urinaria, Ureter, Uterus, Tuba uterina, Pars ampullaris tubae uterinae, Arteria et vena uterina, Arteria umbilicalis

View from in front onto the internal genital organs and urinary bladder of a woman.
Above: uterine tube with the infundibulum for reception of the ova

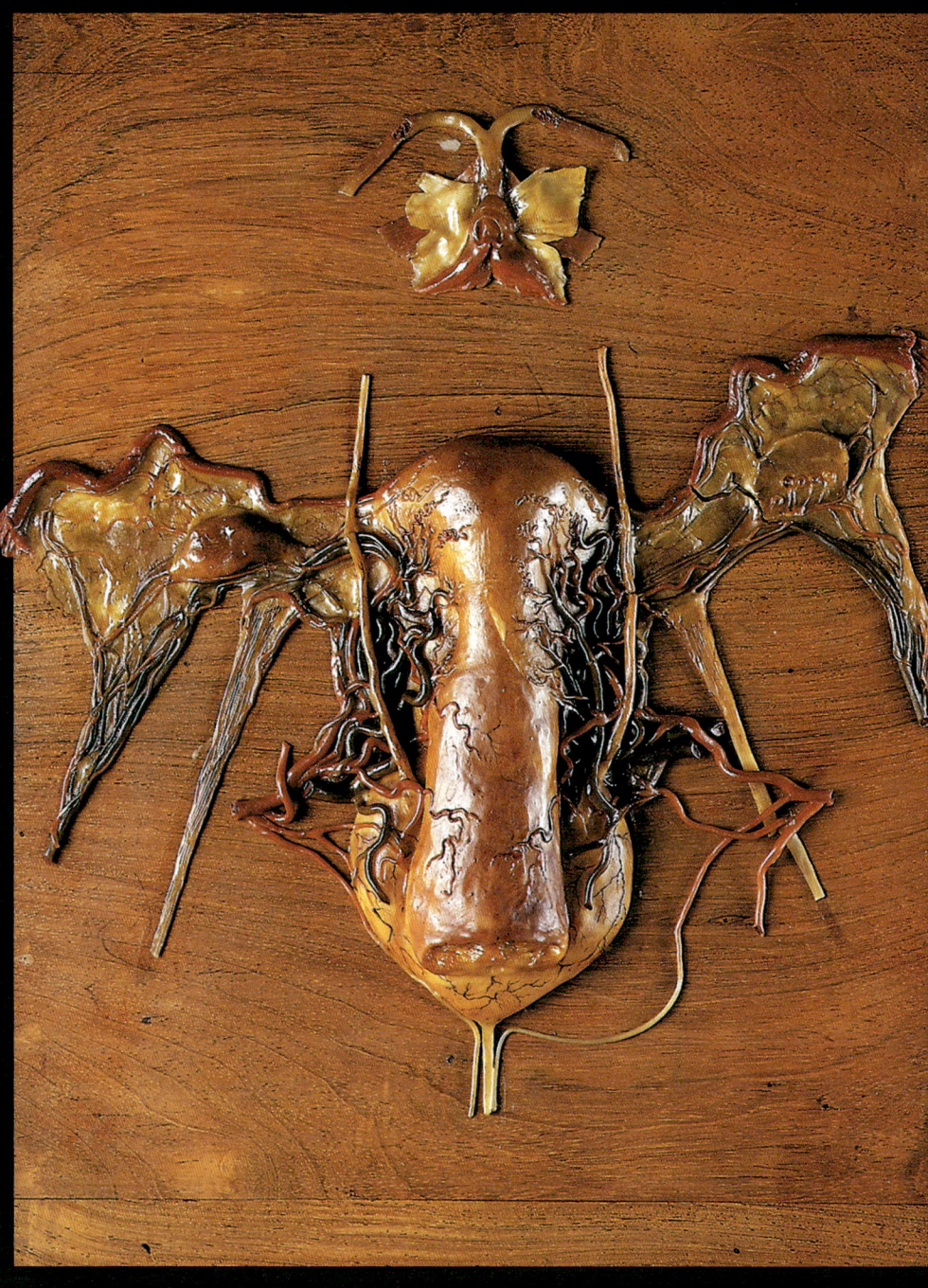

Mesovar, Mesosalpinx, Ovar, Tuba uterina, Arteria et vena uterina, Ureter, Clitoris
View from behind onto the internal genital organs and urinary bladder of a woman (lower specimen).
Above: the clitoris and its muscles

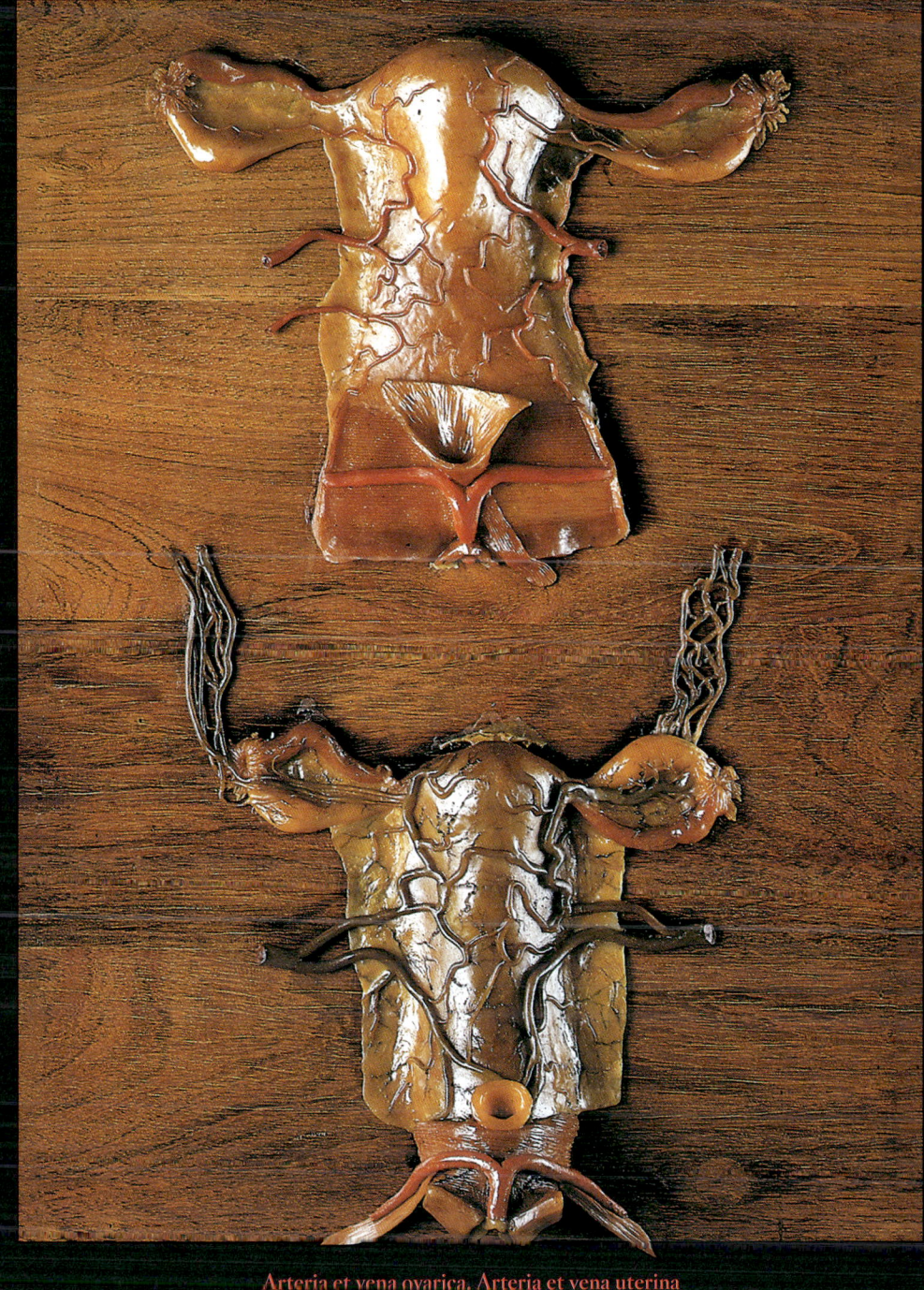

The internal reproductive organs of a woman

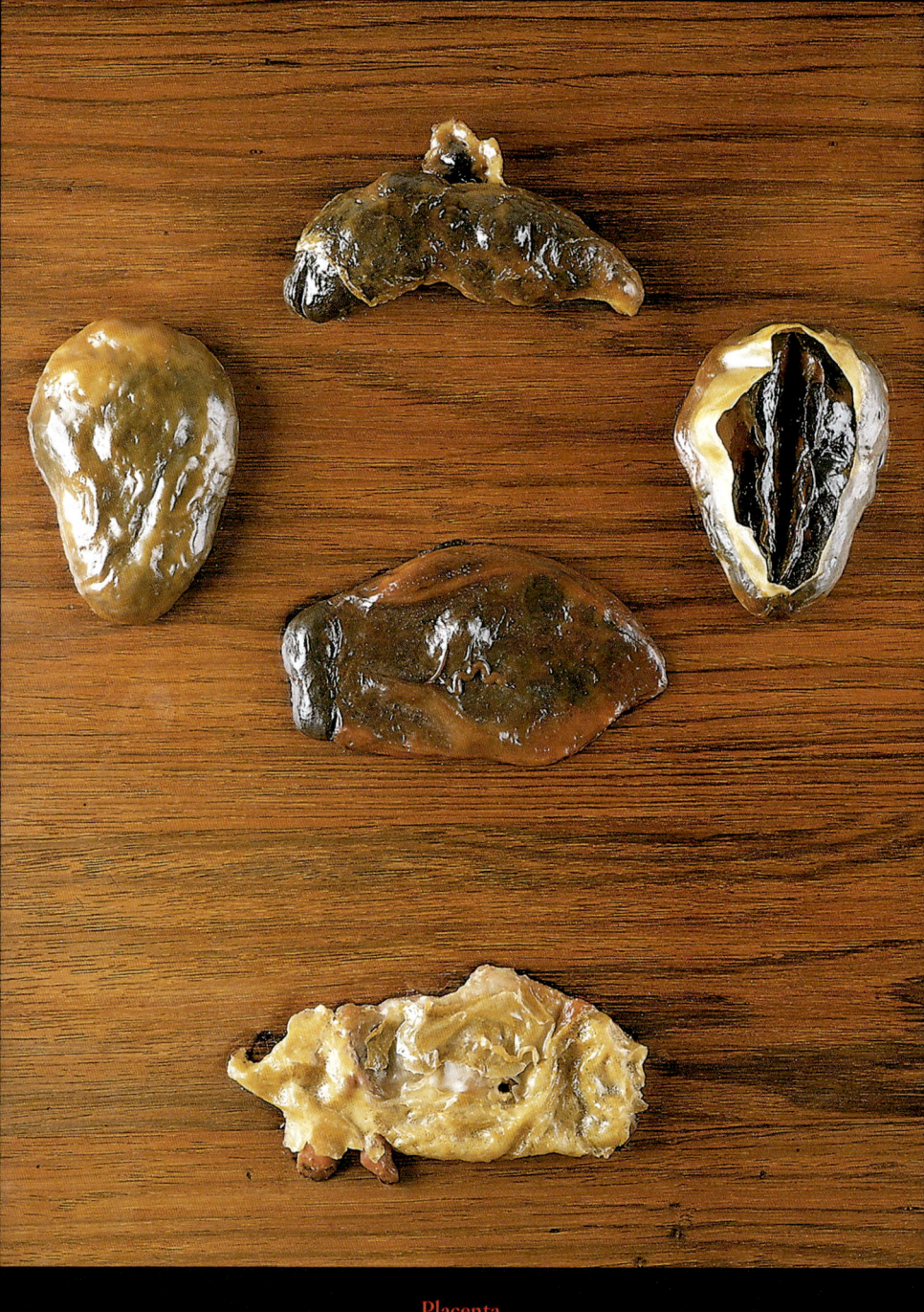

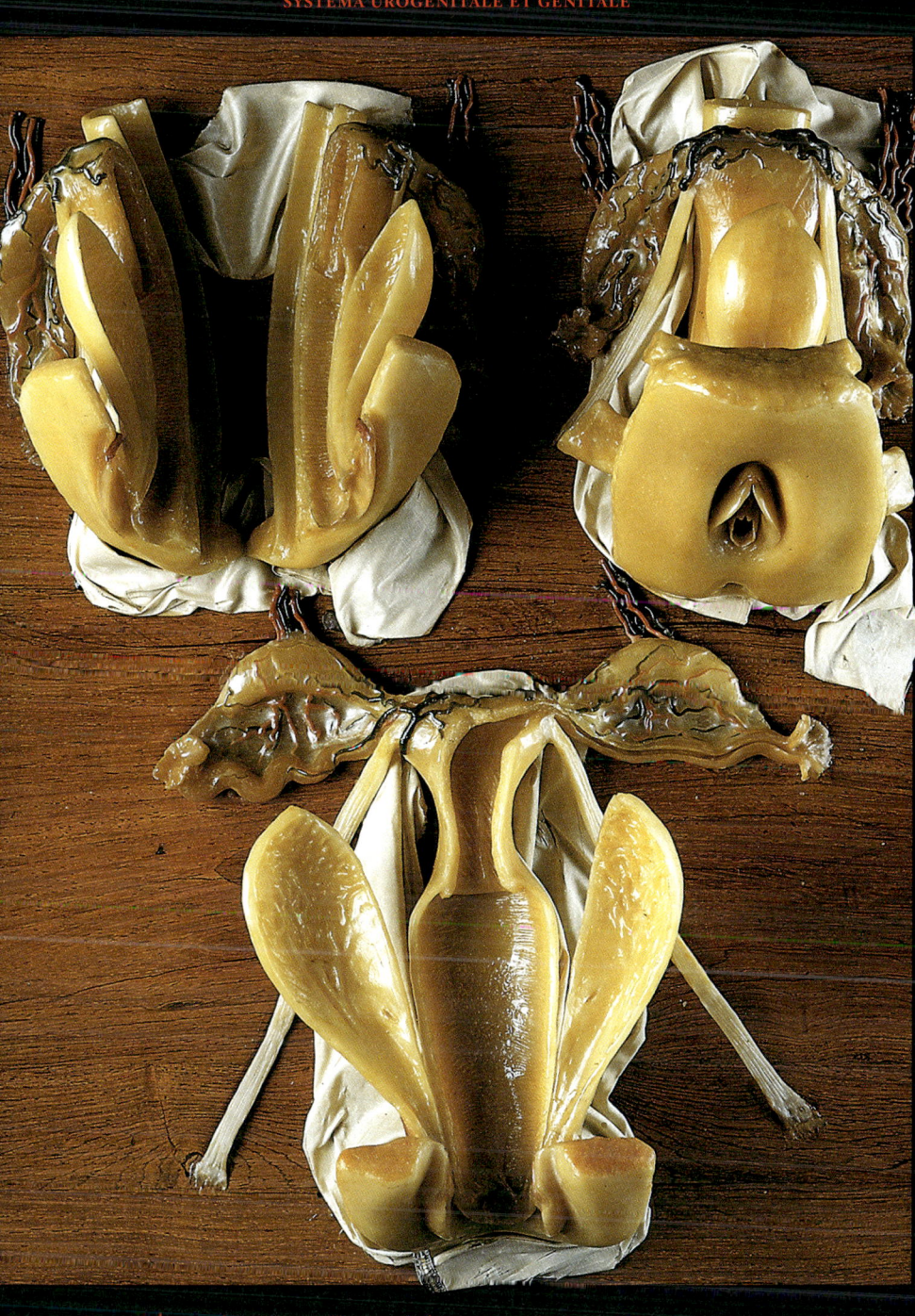

Ovar, Tuba uterina, Uterus, Vagina, Labia majora et minora, Ligamentum teres uteri

Various specimens of the female reproductive organs with ovary, tube, uterus and vagina

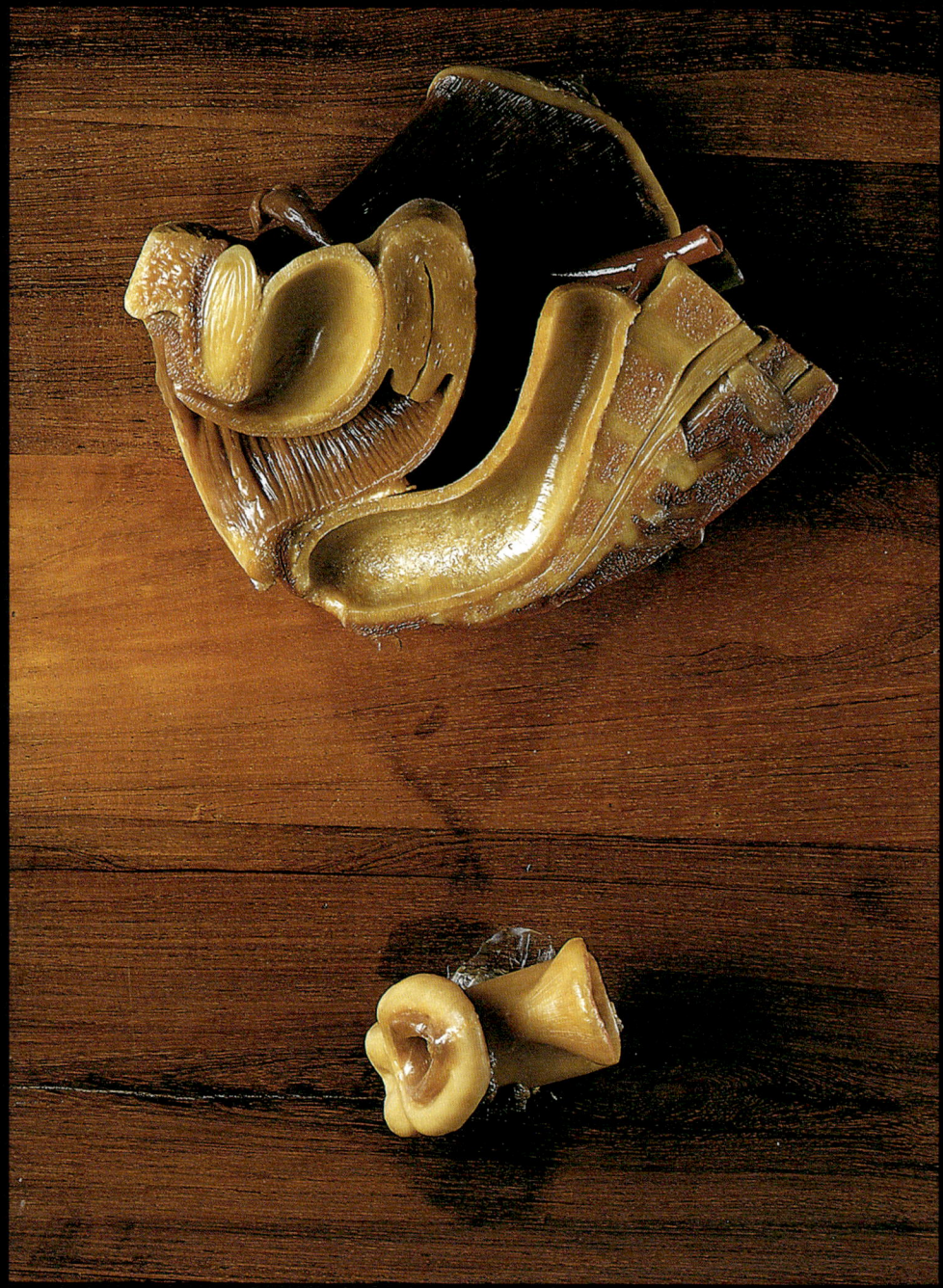

Vagina, Vesica urinaria, Rectum, Excavatio rectouterina

Specimen of a female pelvis divided in the midline

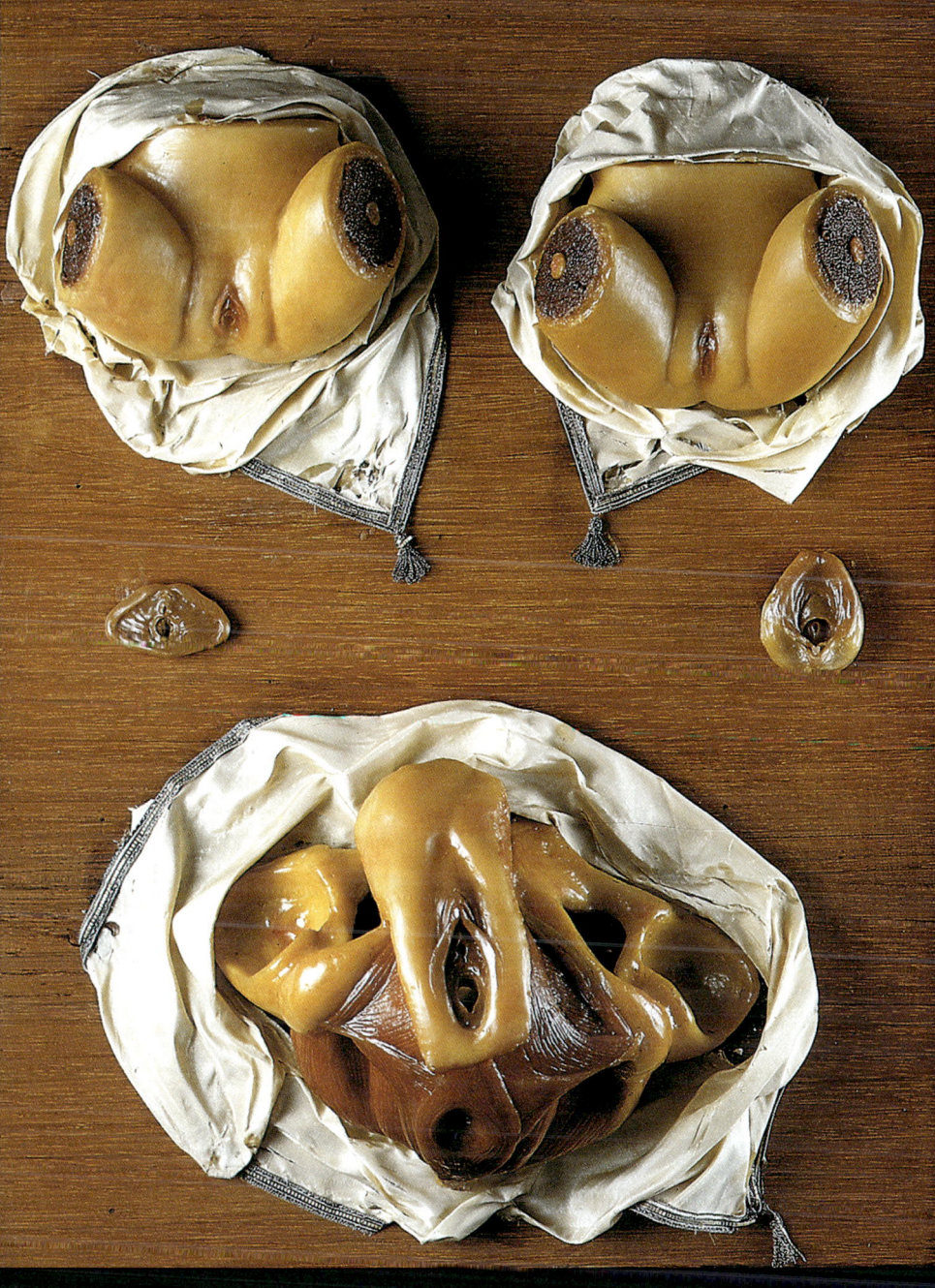

Organa genitalia femina externa

External genital organs of a female infant

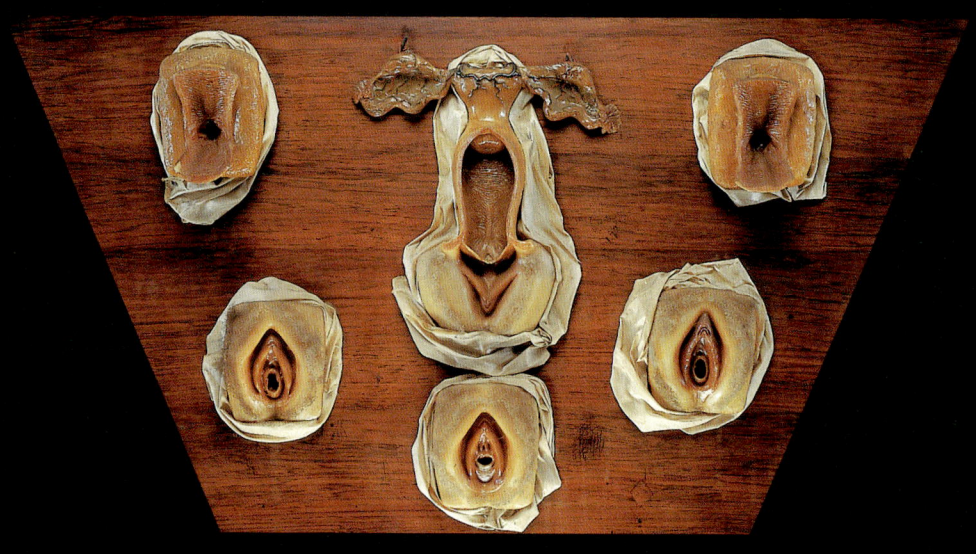

Organa genitalia femina externa
Female external genital organs

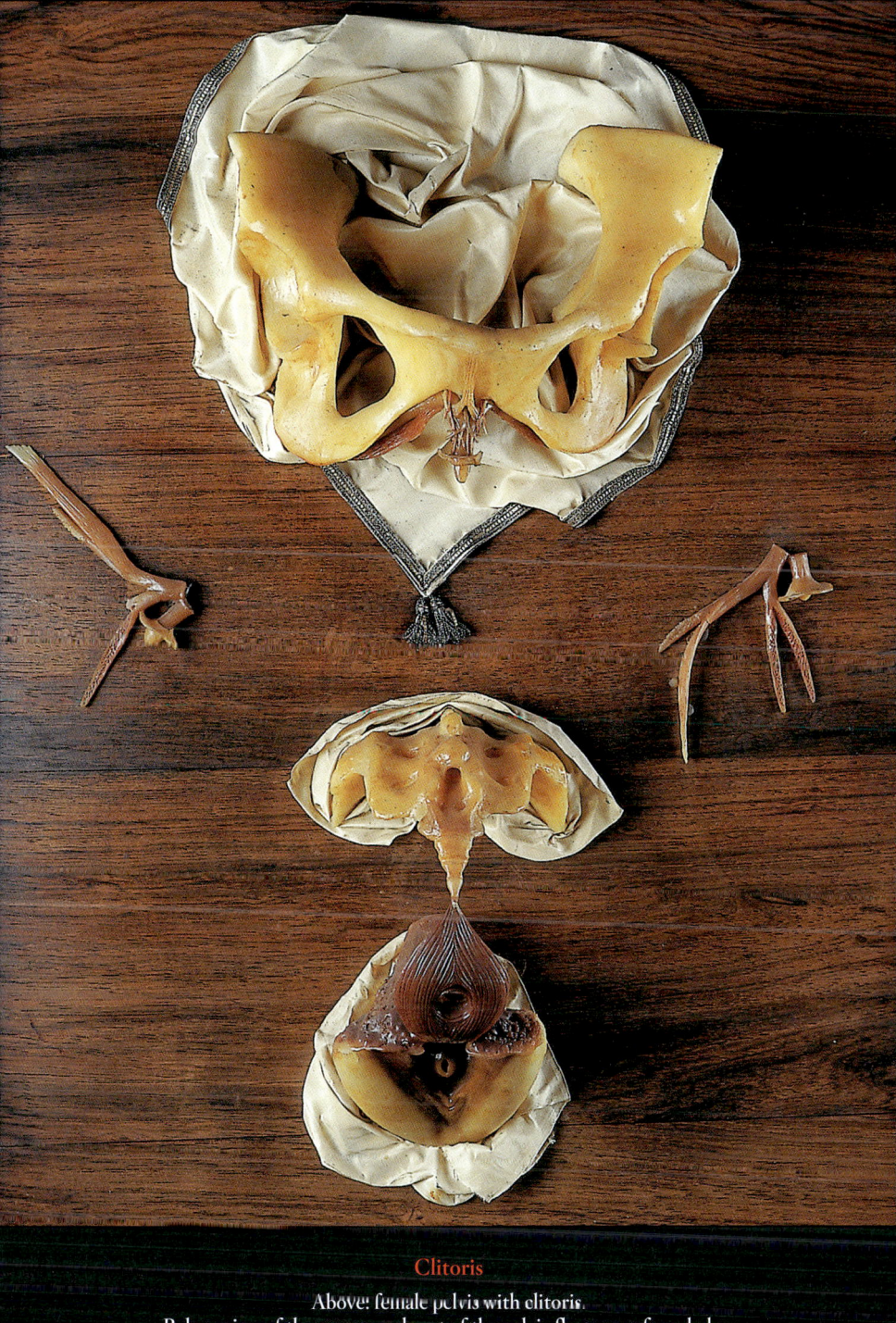

Clitoris

Above: female pelvis with clitoris.
Below: view of the coccyx and part of the pelvic floor, seen from below

Whole body specimen seen from belo

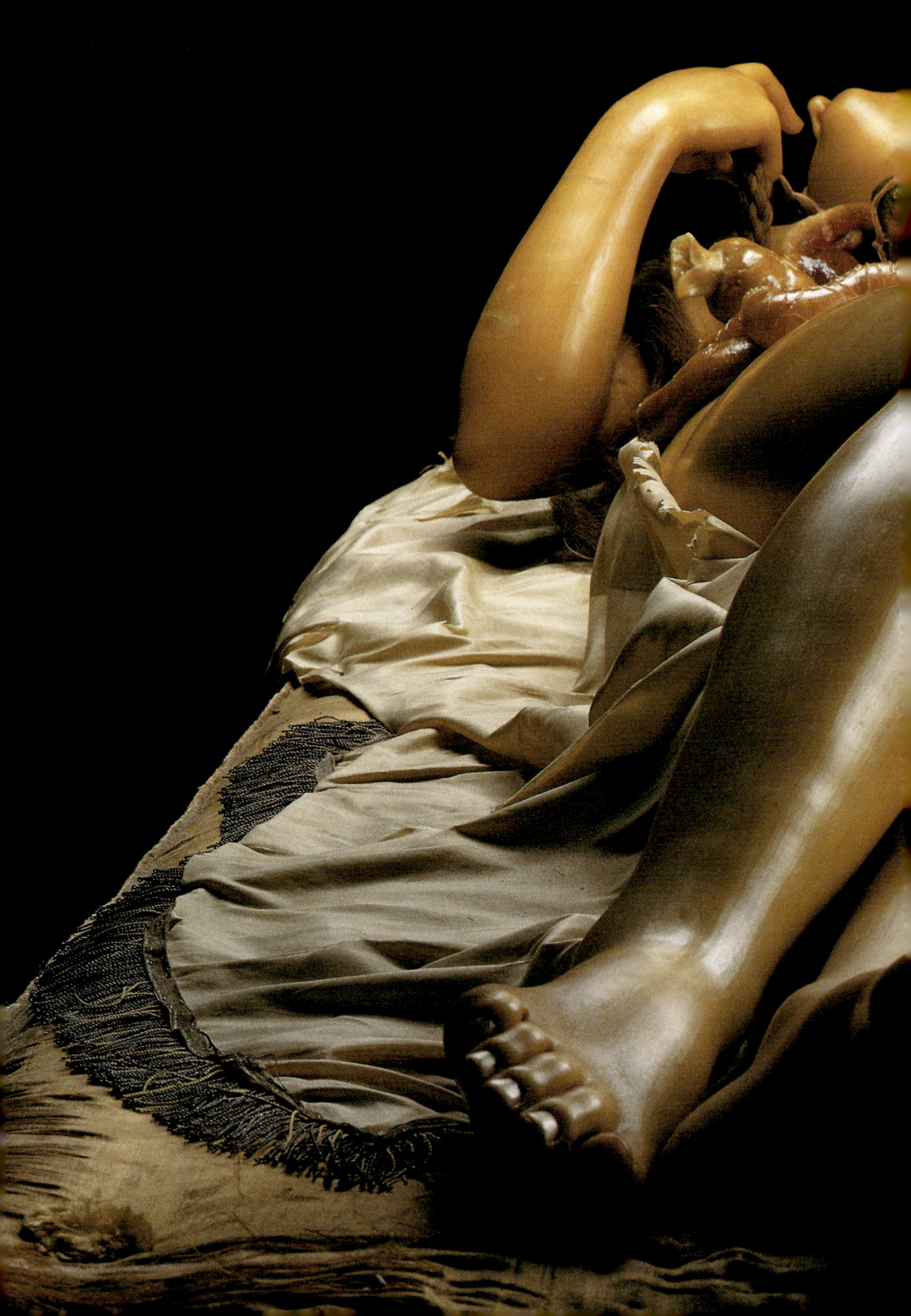

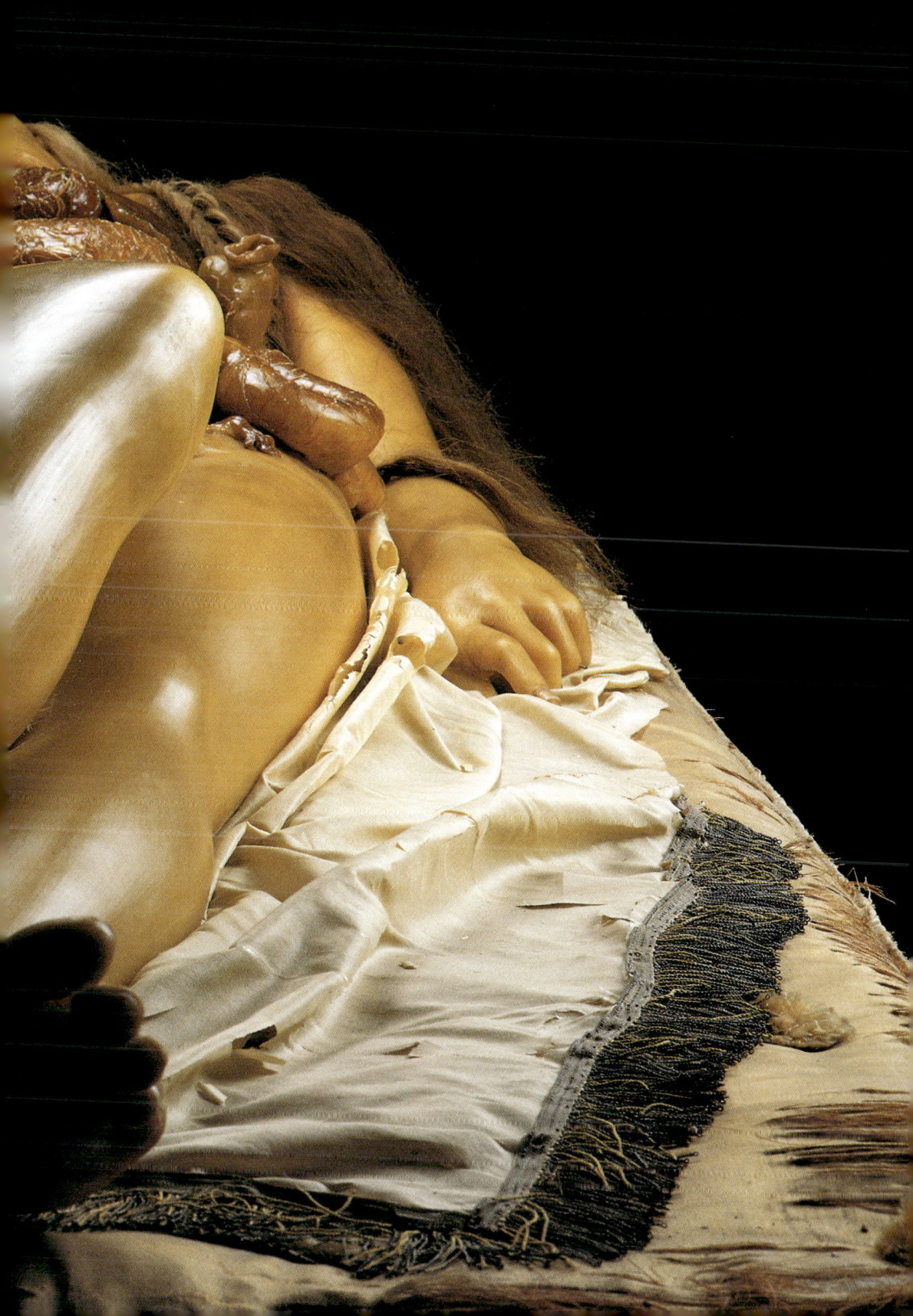

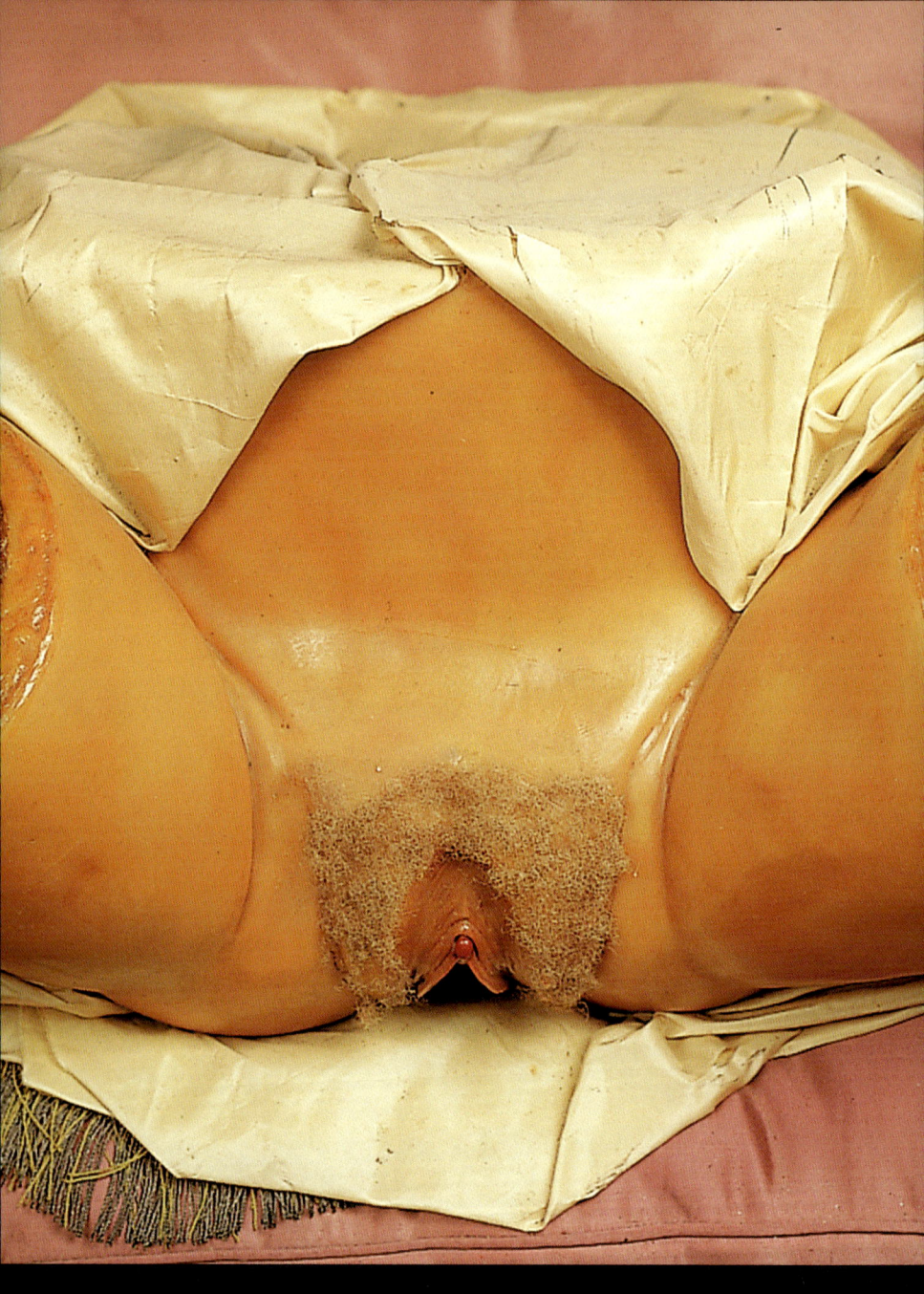

Vulva

Female external genital organs

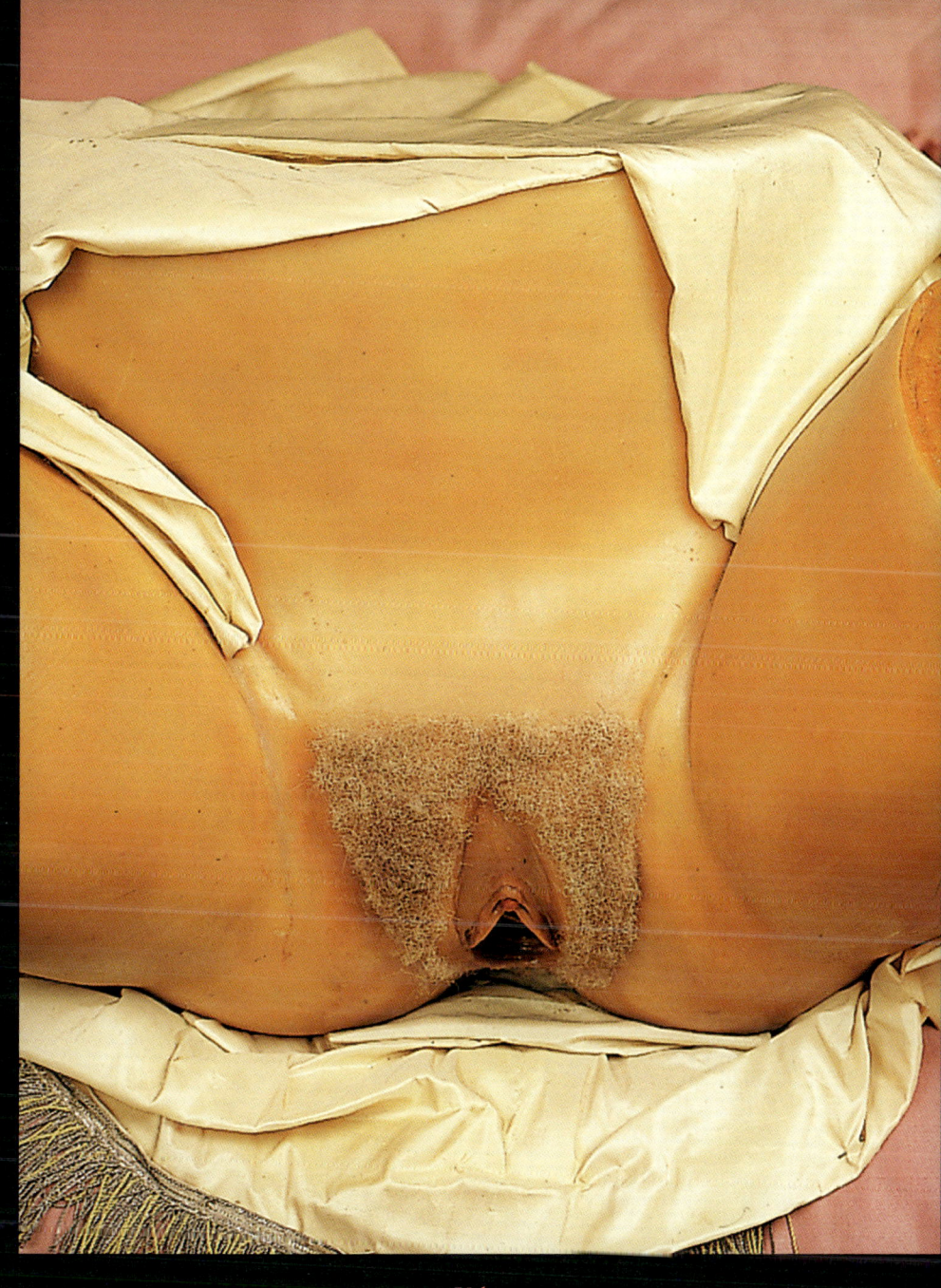

Vulva

Female external genital organs

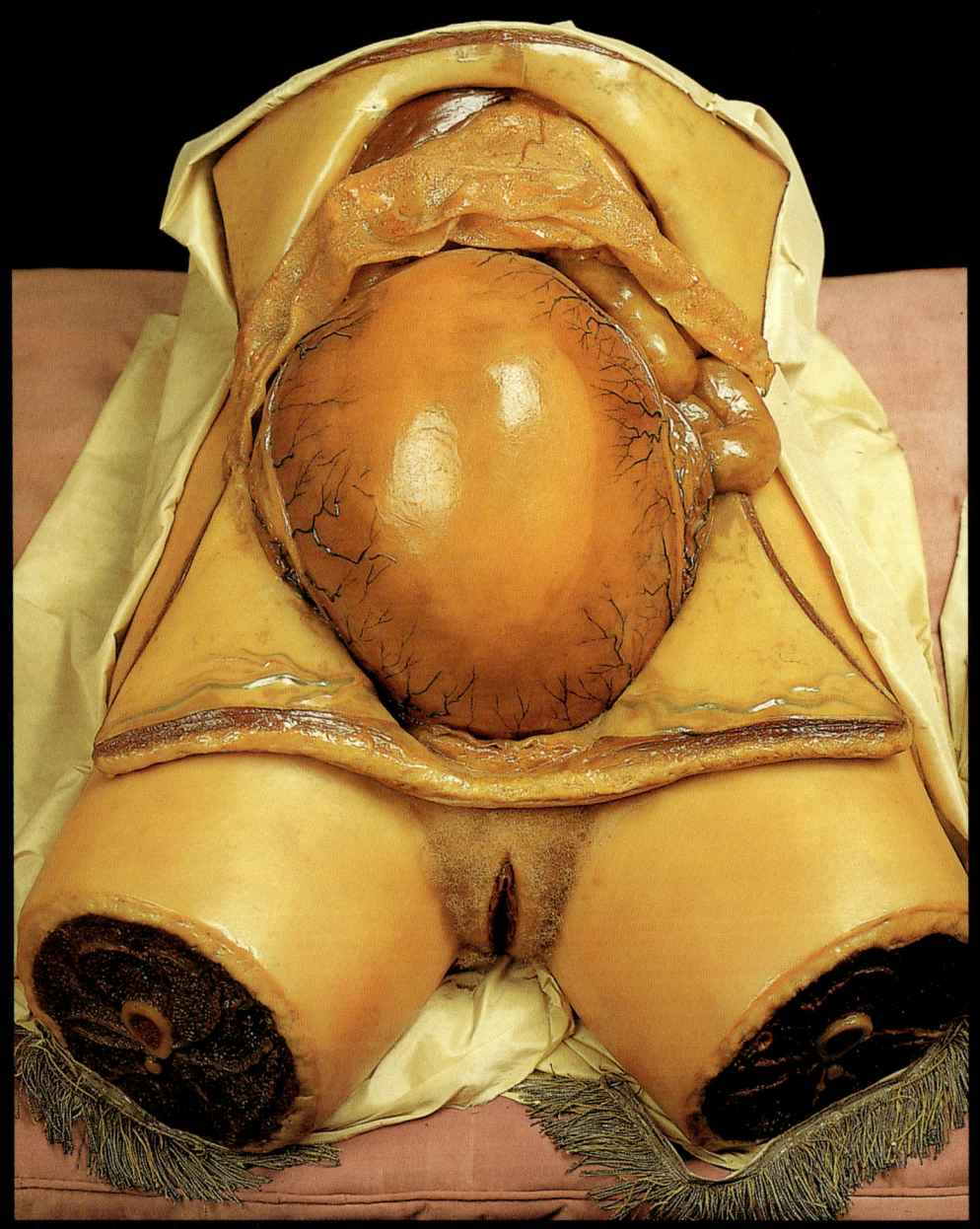

Uterus gravidus

Lower part of the belly of a pregnant woman; abdominal cavity laid open

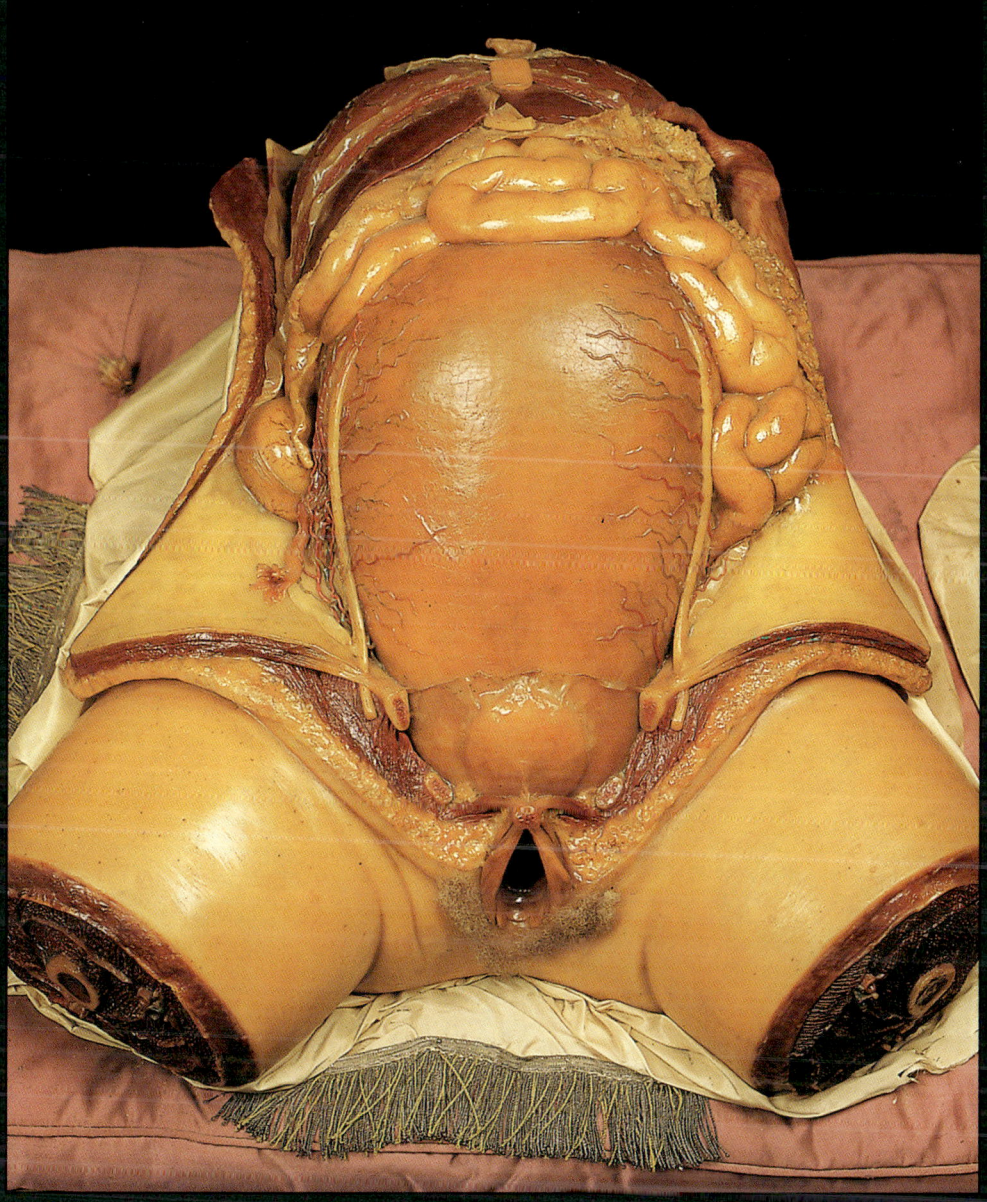

Uterus gravidus

Lower part of the belly of a pregnant woman; abdominal cavity laid open

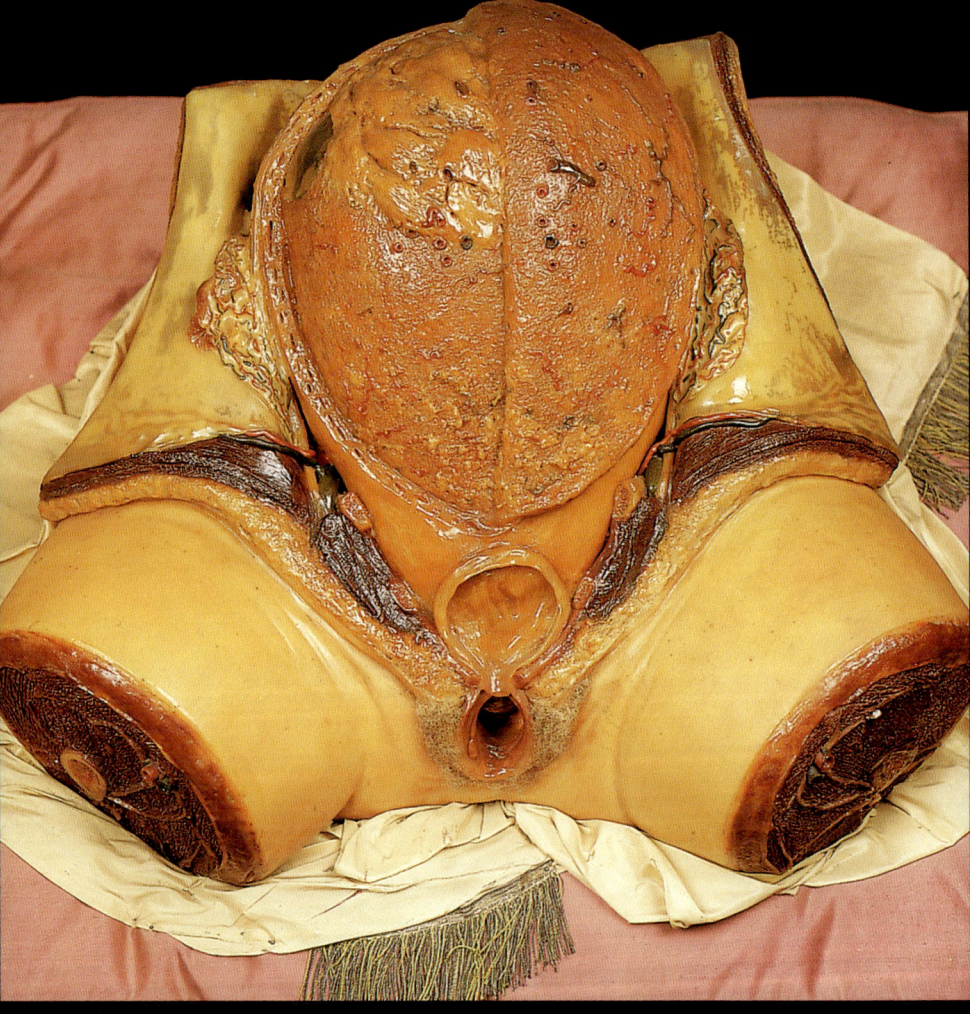

Uterus gravidus

Lower part of the belly of a pregnant woman; abdominal cavity laid open.
The superficial layers of the uterine wall have been removed to show the blood vessels

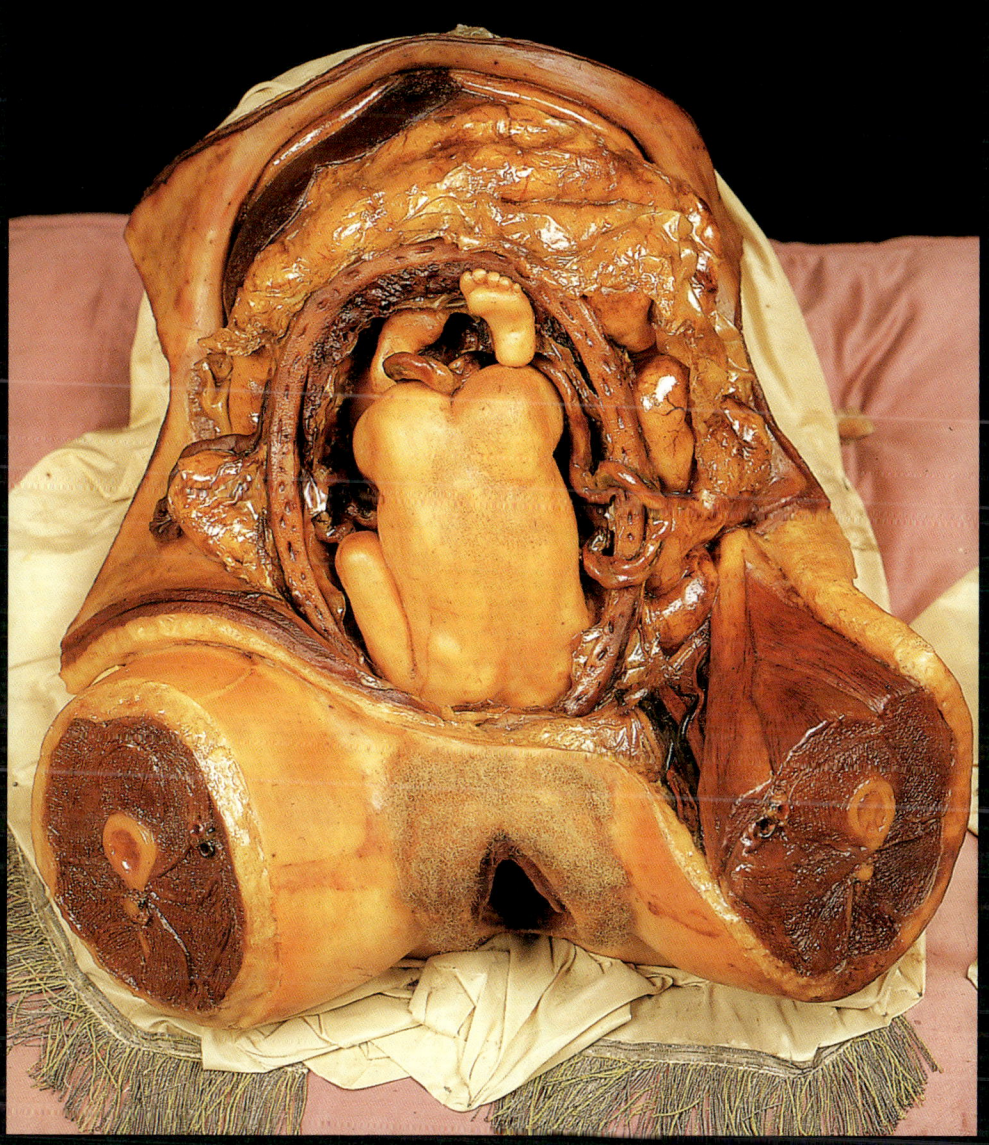

Uterus gravidus, Fetus

View into the opened uterus of a pregnant woman during parturition

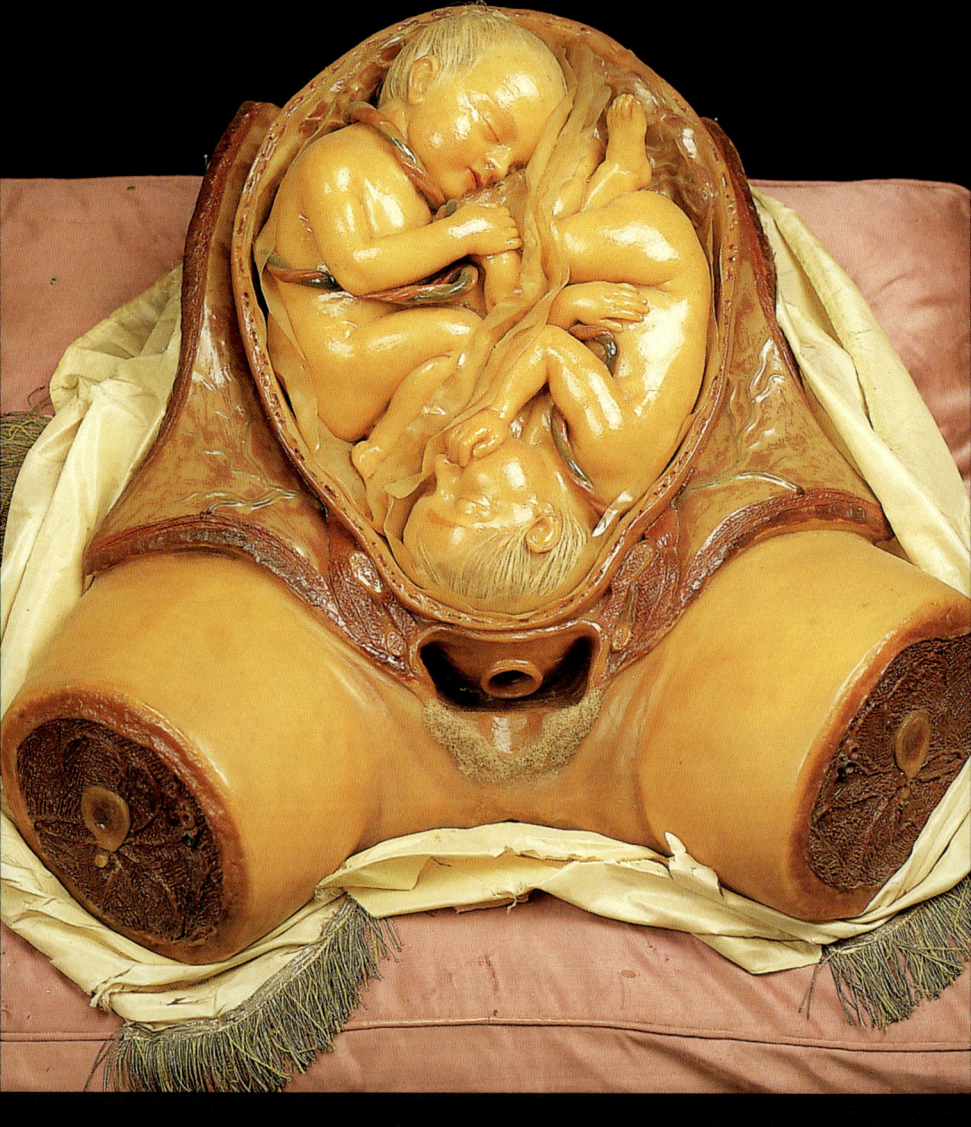

Uterus gravidus, Portio vaginalis uteri, Fetus

Uterus of a pregnant woman laid open; dilatation of external os

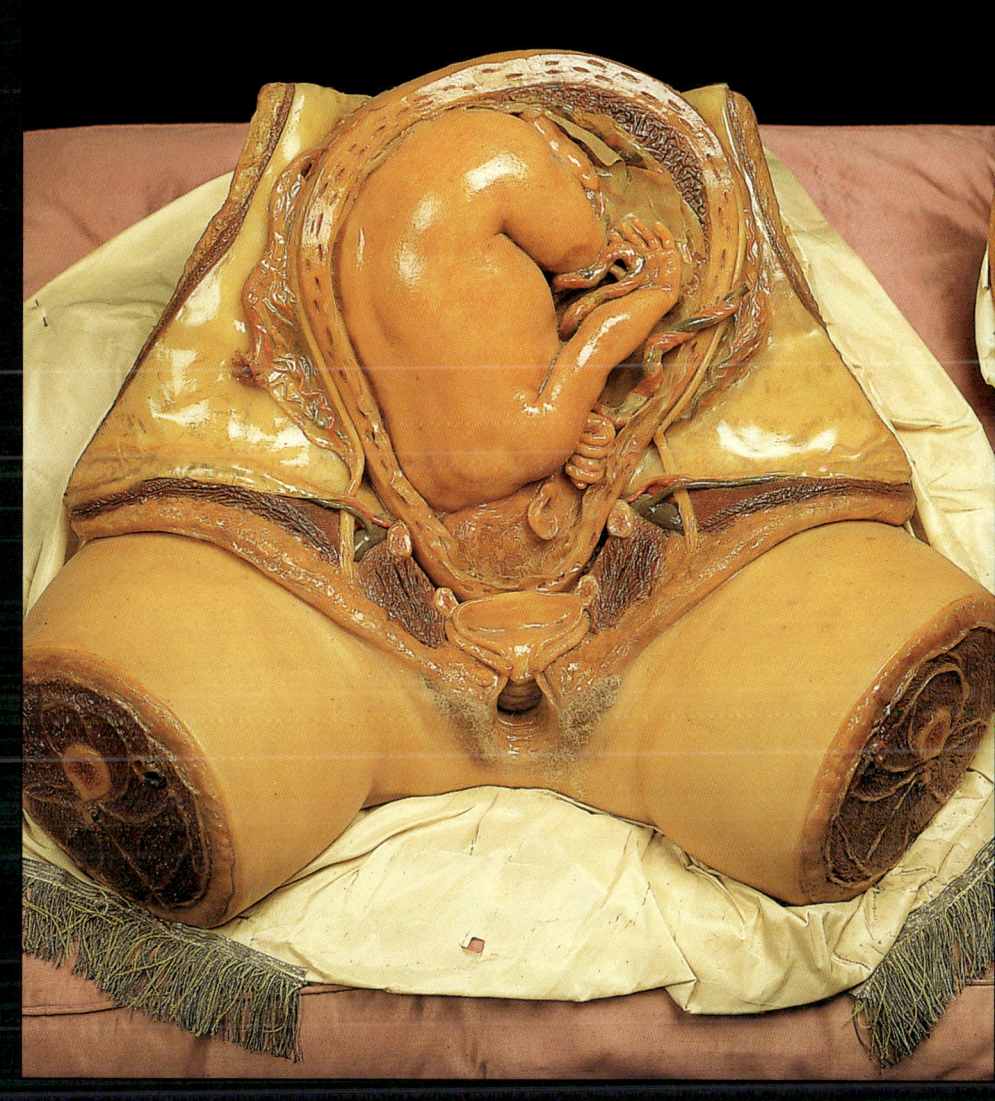

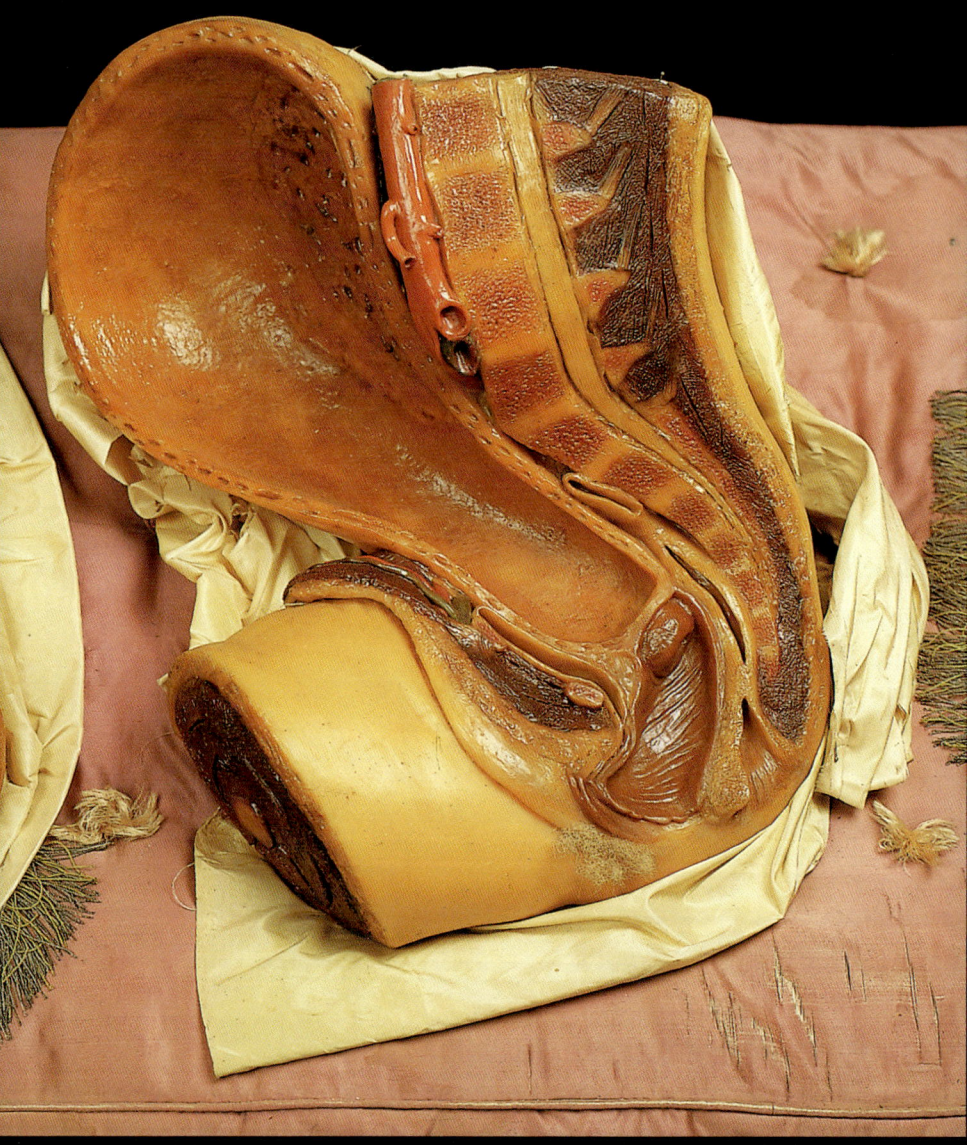

Uterus post partum

Uterus after delivery of child and placenta

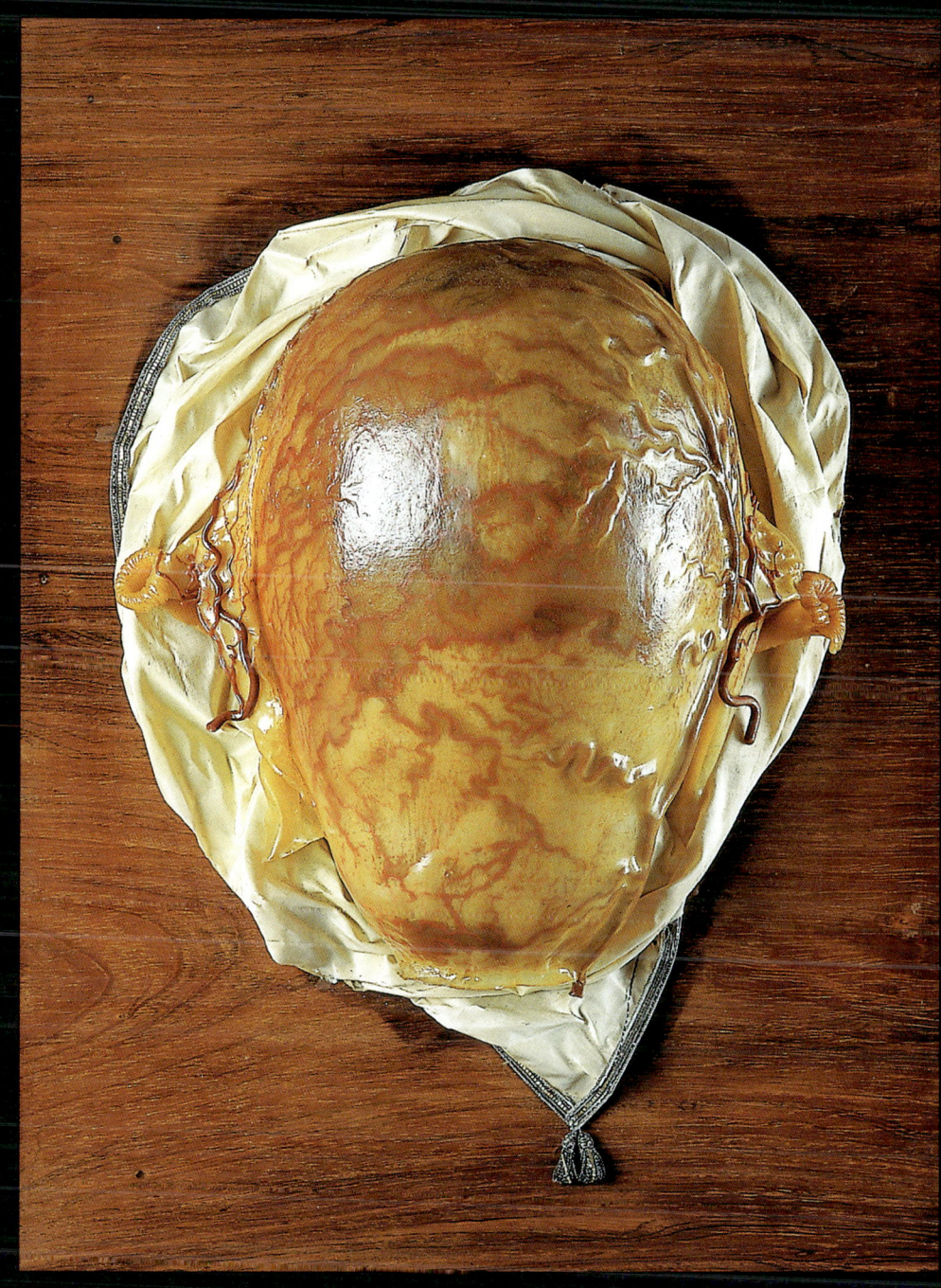

Uterus gravidus

Uterus of a pregnant woman

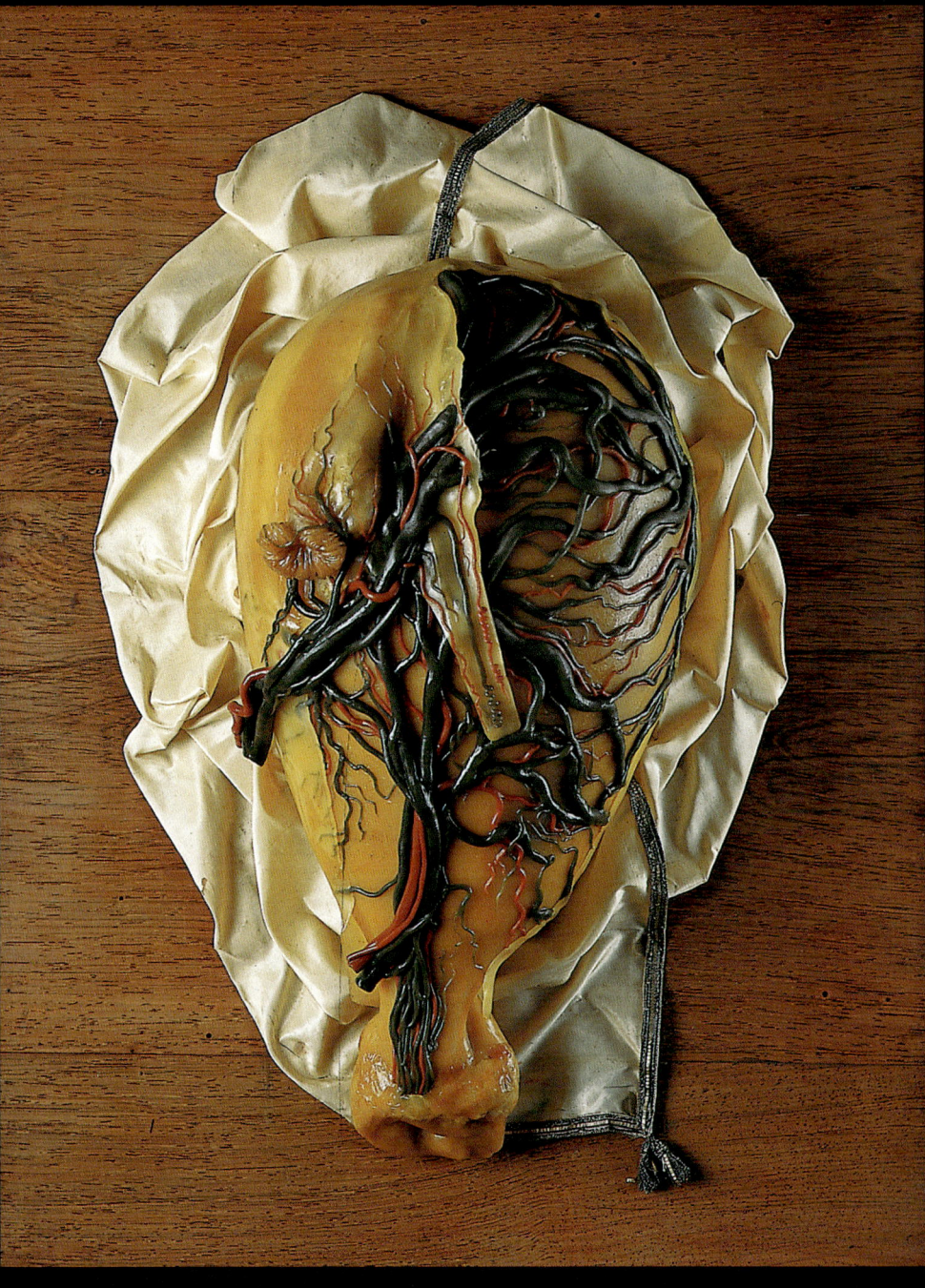

Uterus gravidus

Uterus of a pregnant woman; deeper layer showing blood vessels

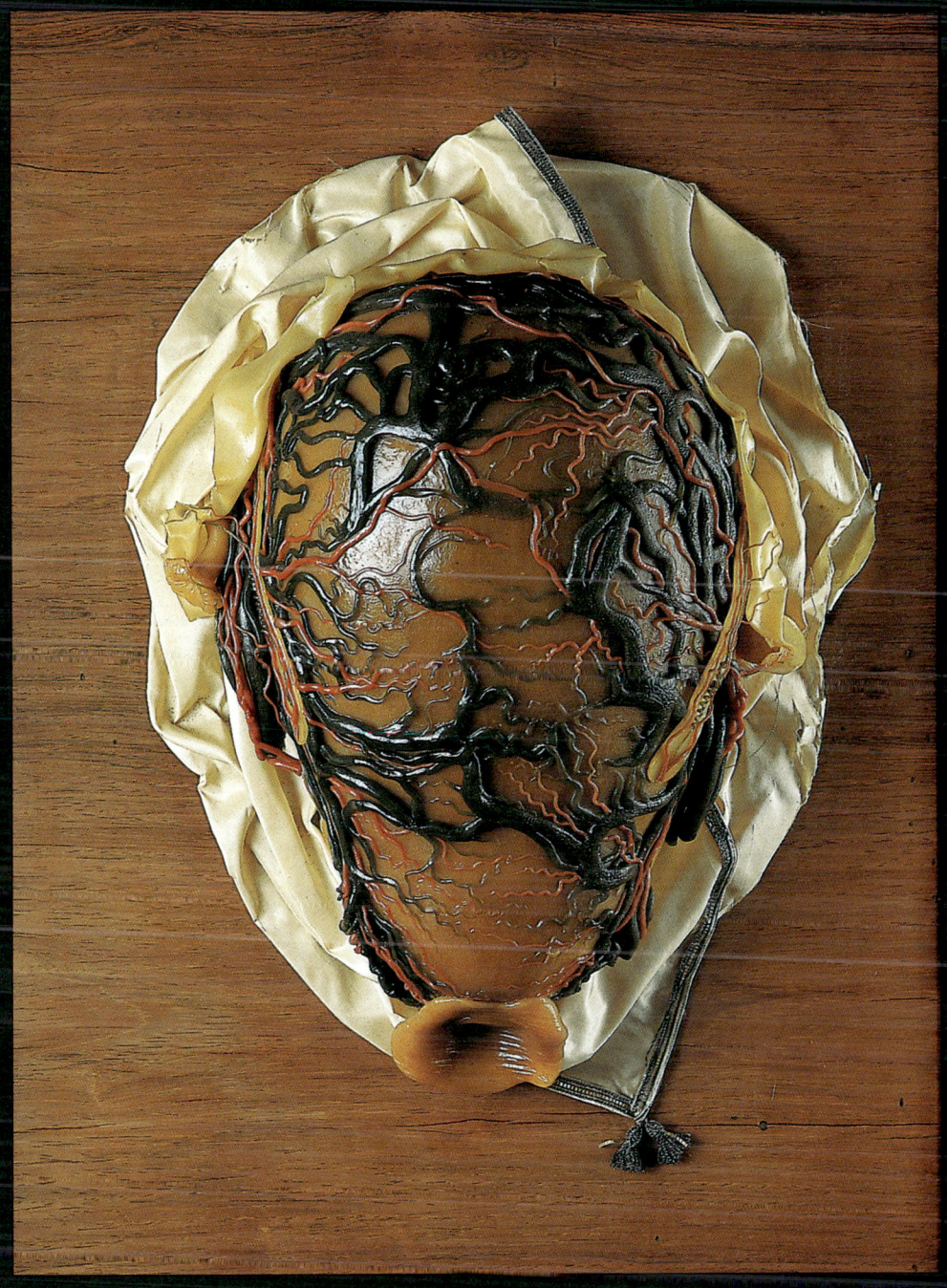

Uterus gravidus
Uterus of a pregnant woman; deeper layer showing blood vessels

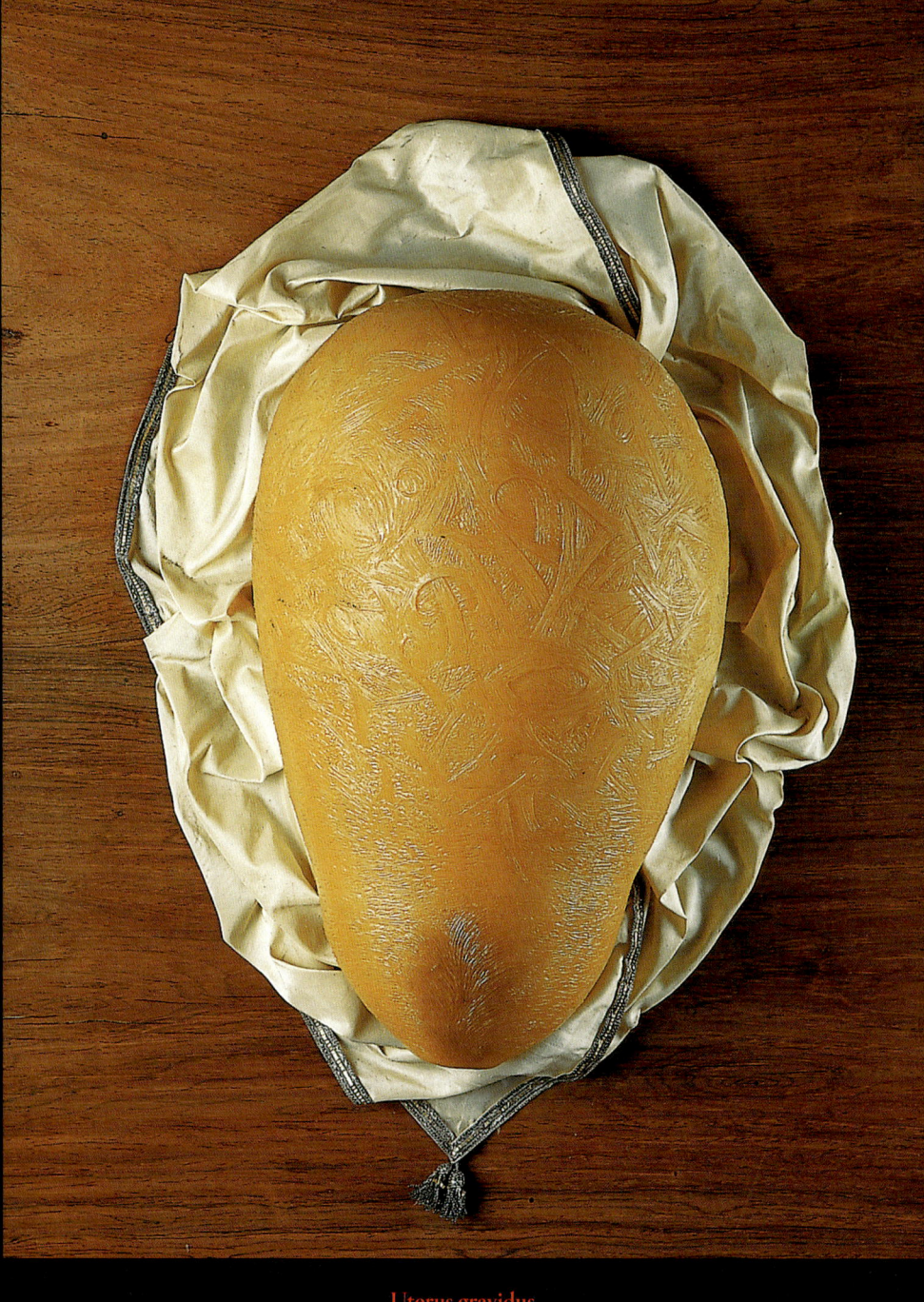

Uterus gravidus

Uterus of a pregnant woman; uterine musculature displayed

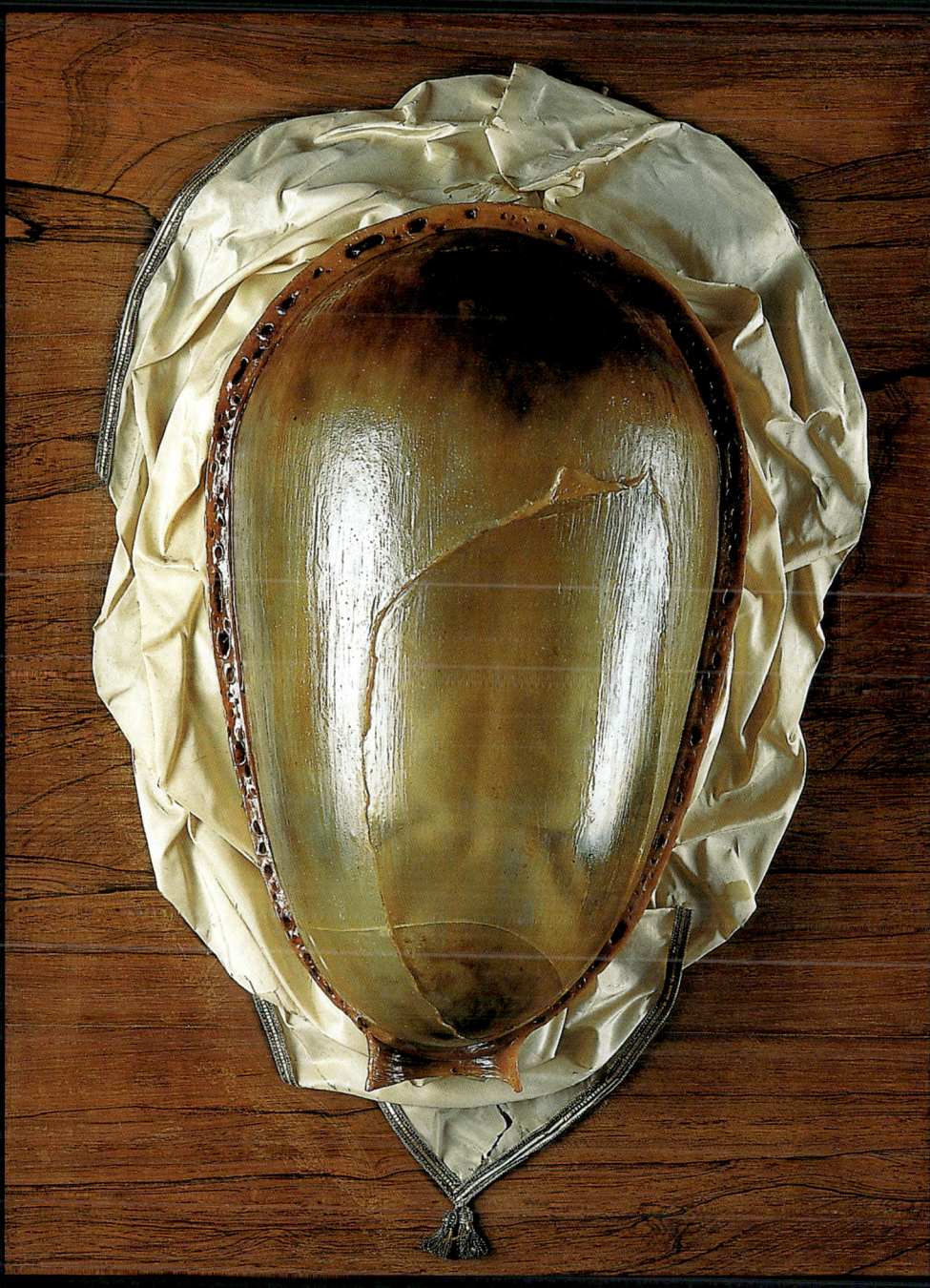

Uterus gravidus, Amnion
Uterus with amniotic sac

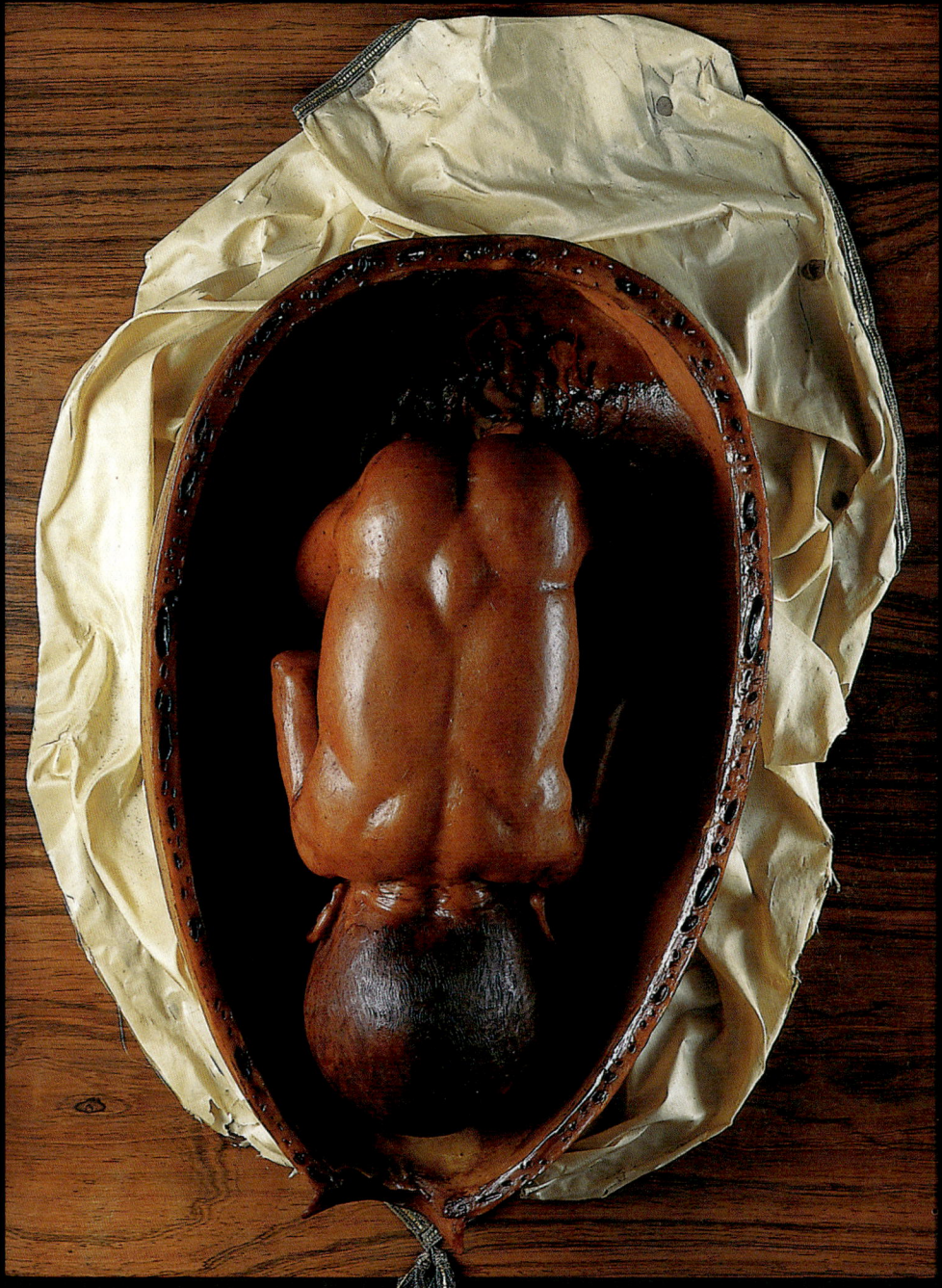

Uterus gravidus, Fetus

Uterus with a foetus

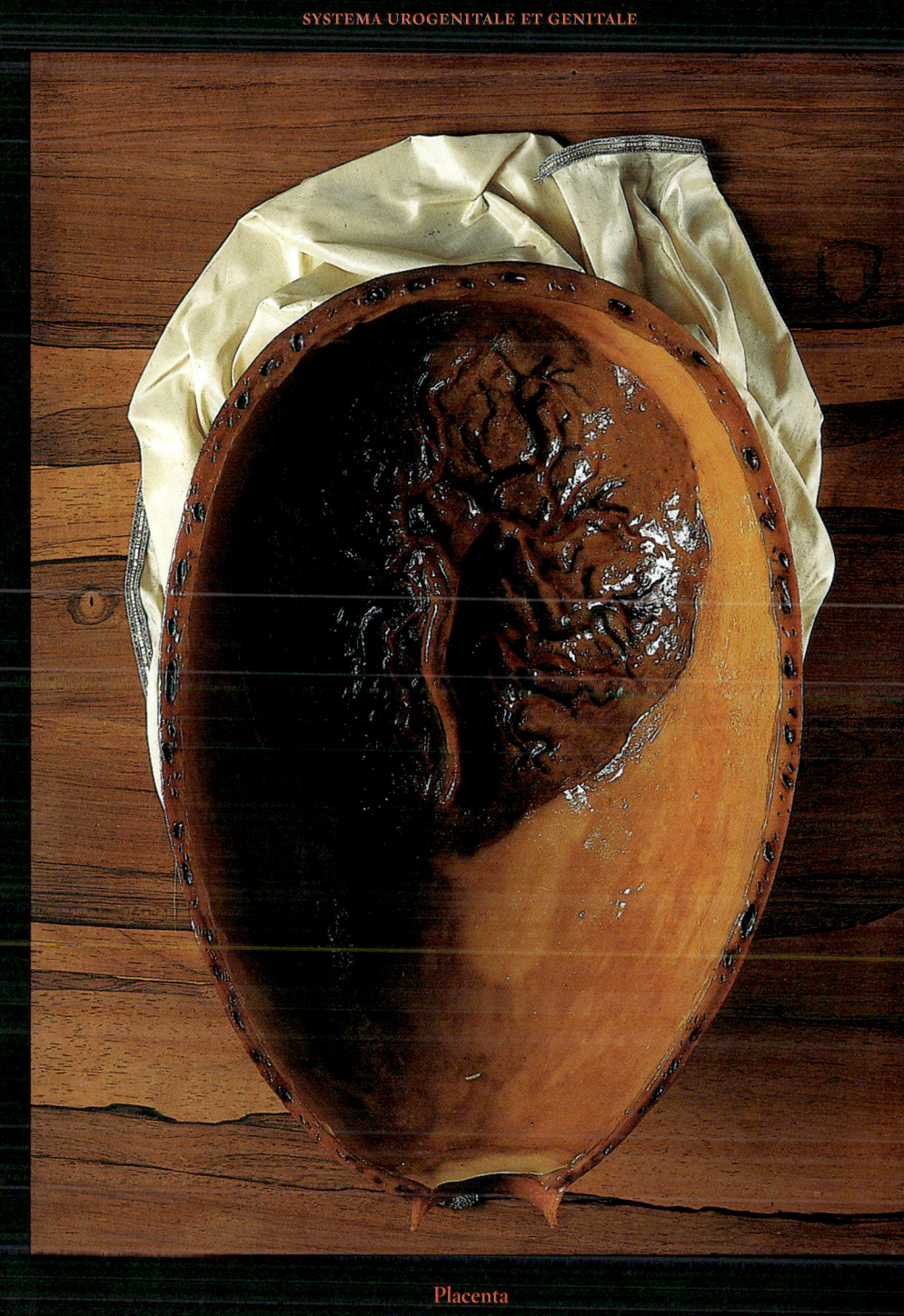

Placenta

Uterus with placenta

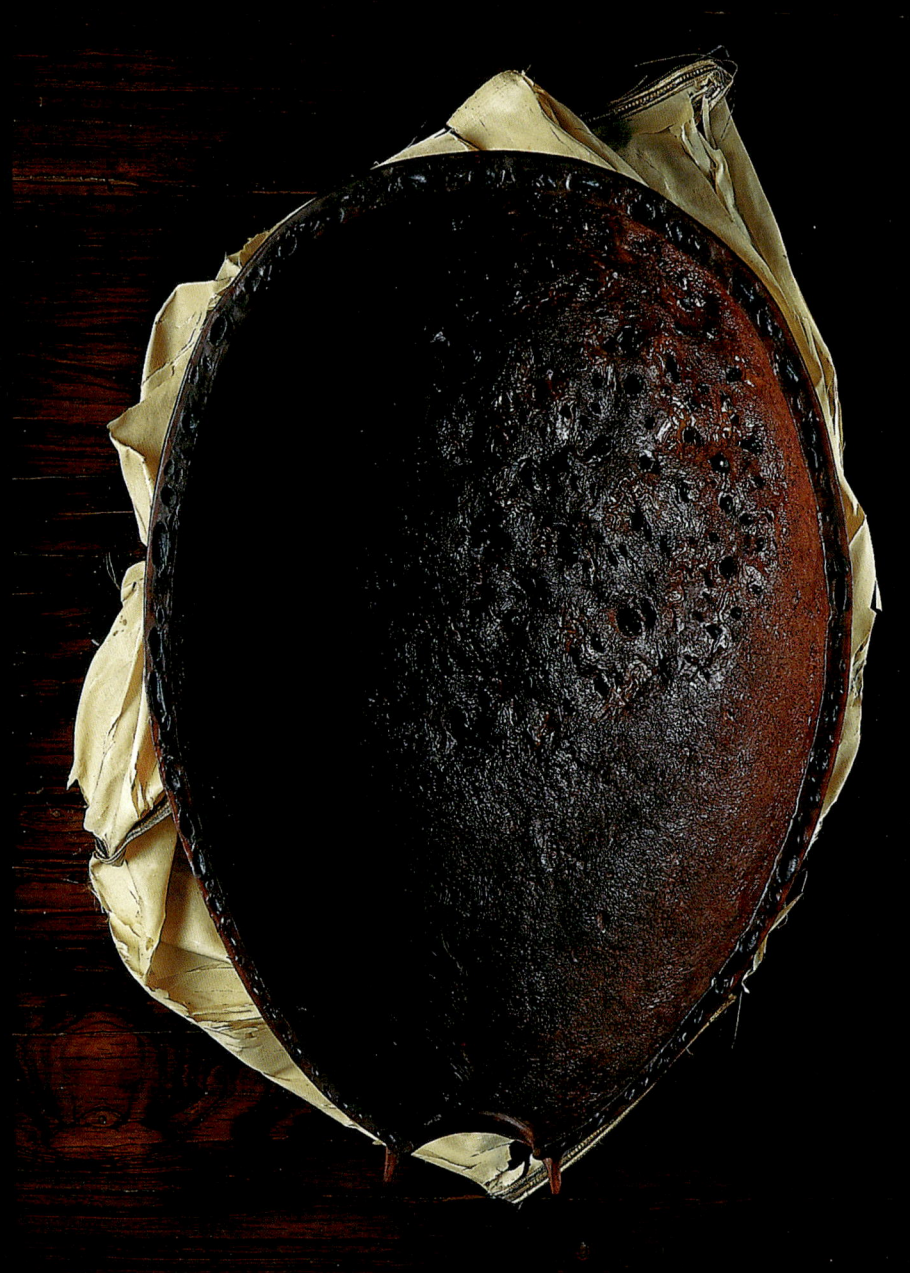

Uterus post partum
Uterus after delivery of the placenta

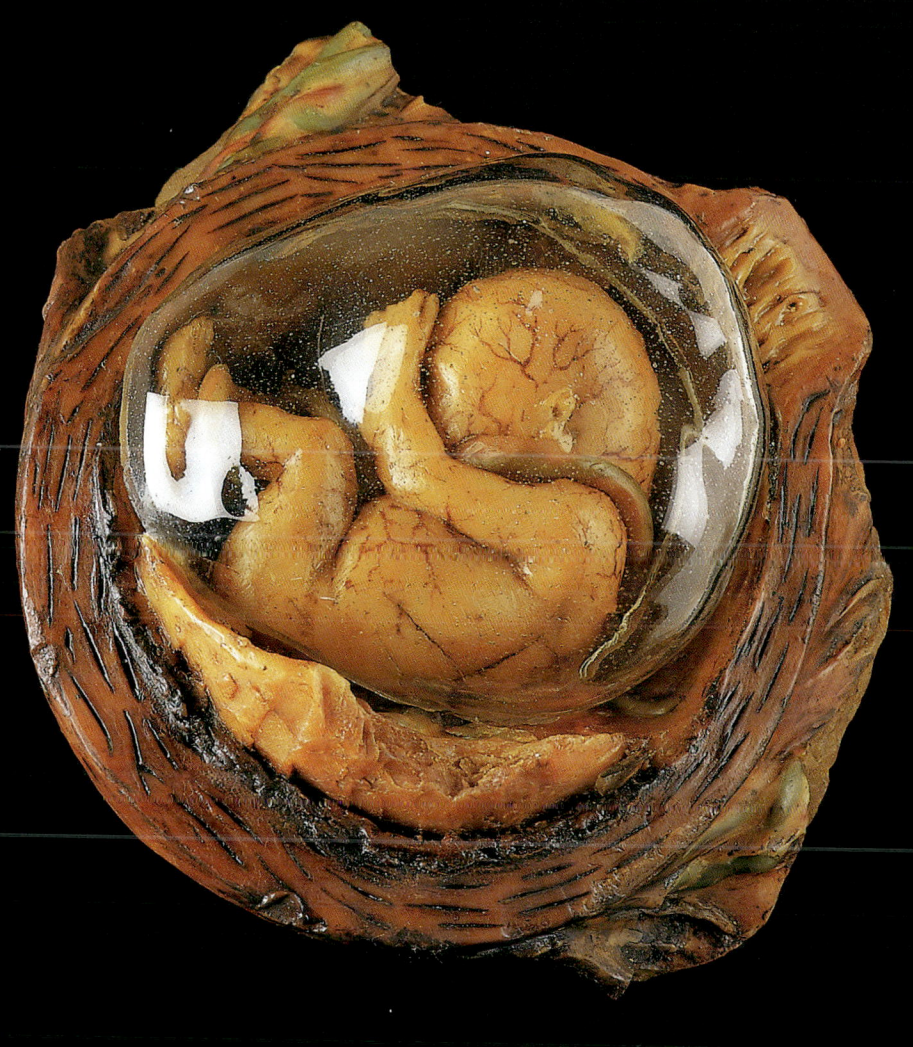

Uterus gravidus, Fetus

Specimen showing a foetus within the amniotic sac

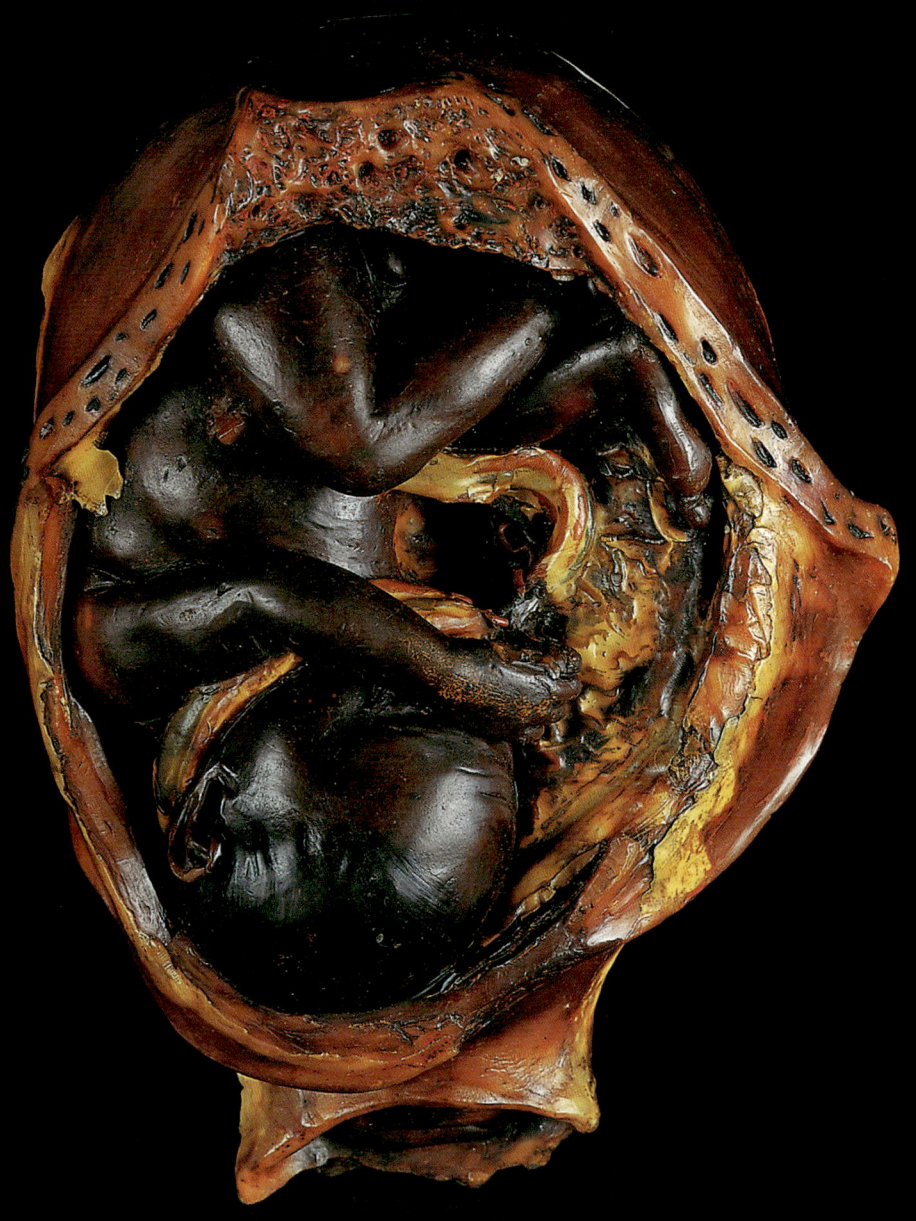

Uterus gravidus, Fetus
View of a foetus inside the uterus

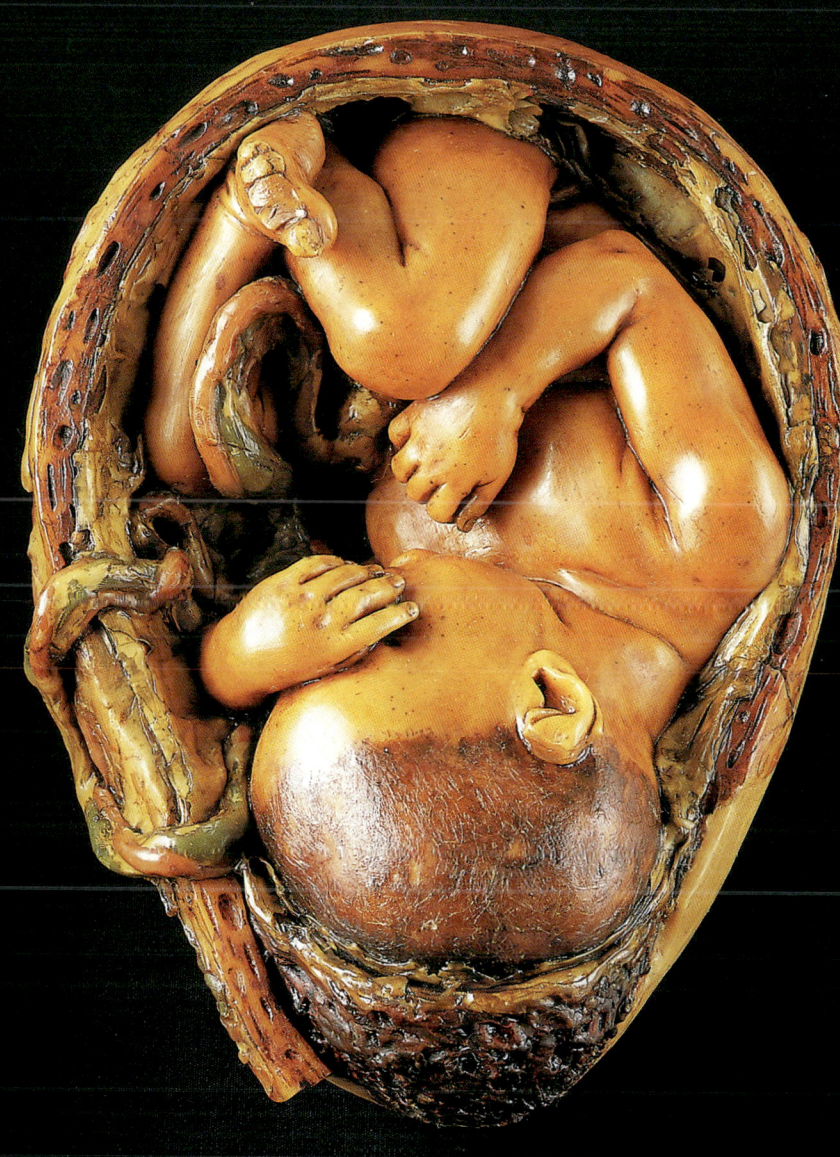

Uterus gravidus, Fetus

View of a foetus inside the uterus

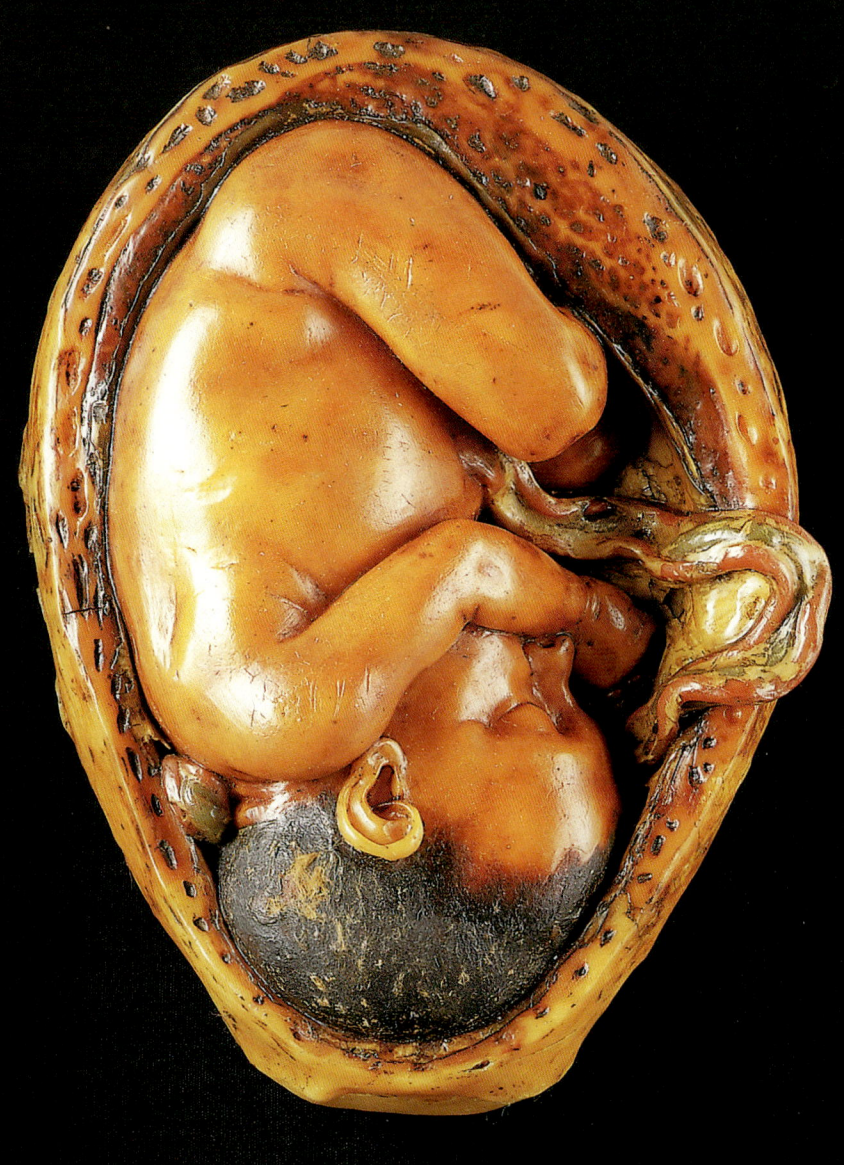

Uterus gravidus, Fetus

View of a foetus inside the uterus

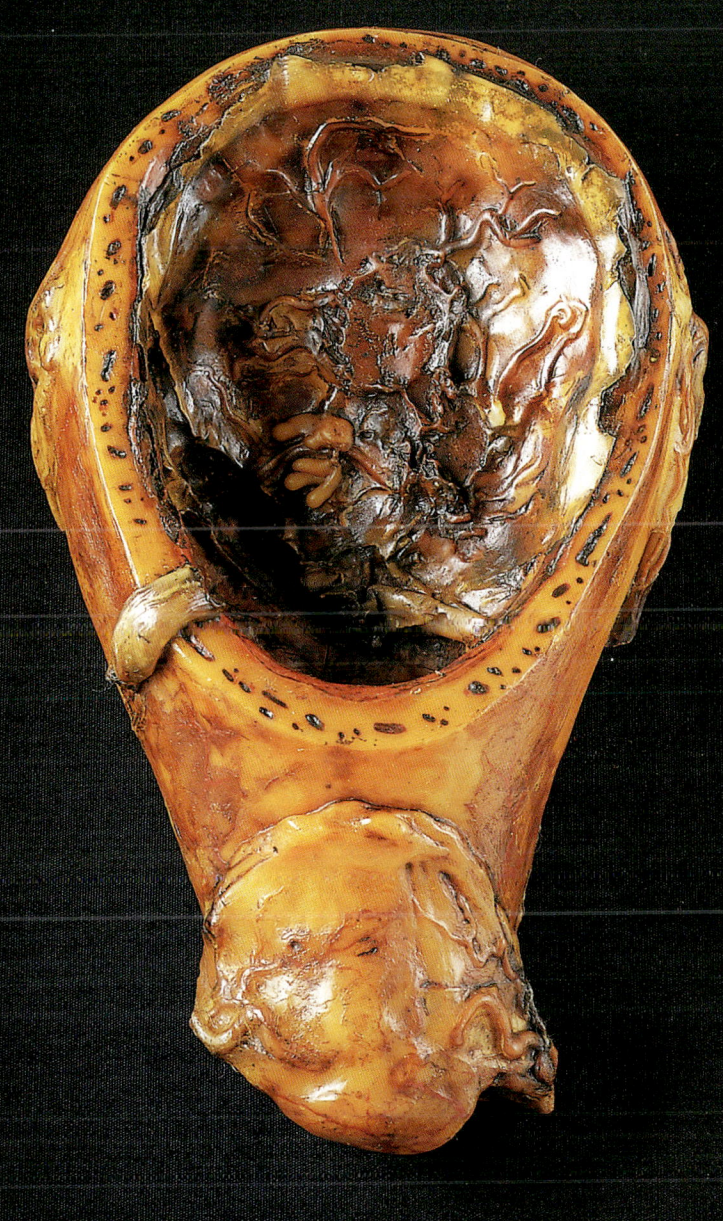

Uterus gravidus, Placenta

The uterus of a pregnant woman containing the placenta

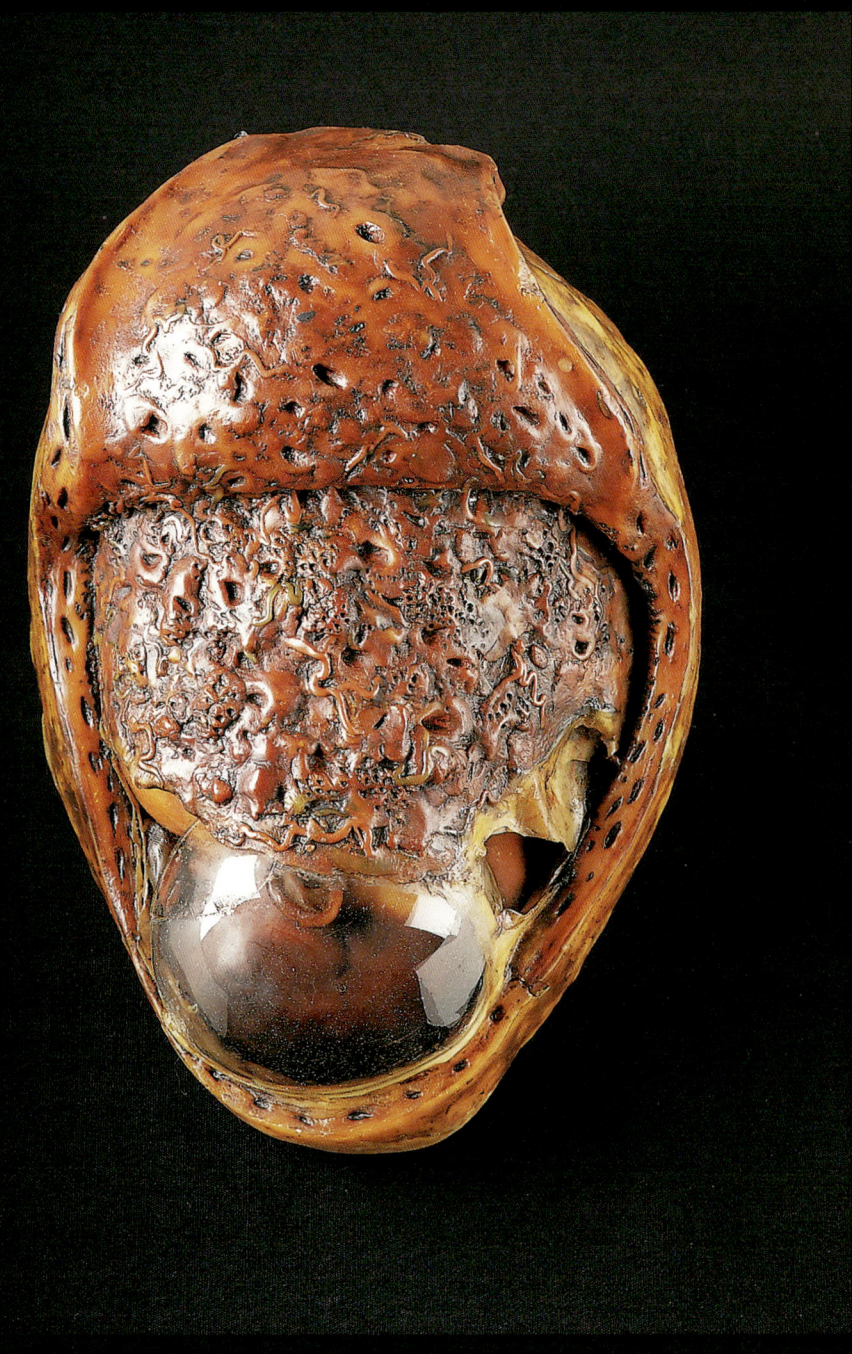

Uterus gravidus, Placenta, Fetus

The uterus of a pregnant woman with amniotic sac, foetus and placenta

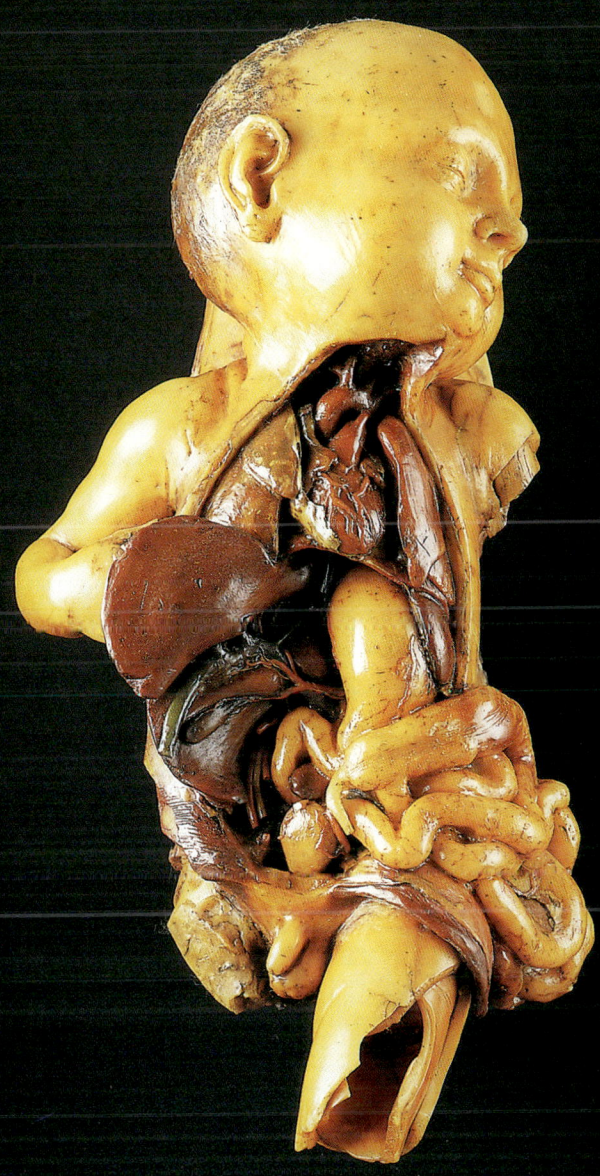

Fetus, Thorax, Abdomen

A newborn infant with the thoracic and abdominal cavities laid open.
The arms and legs of the specimen have been damaged.

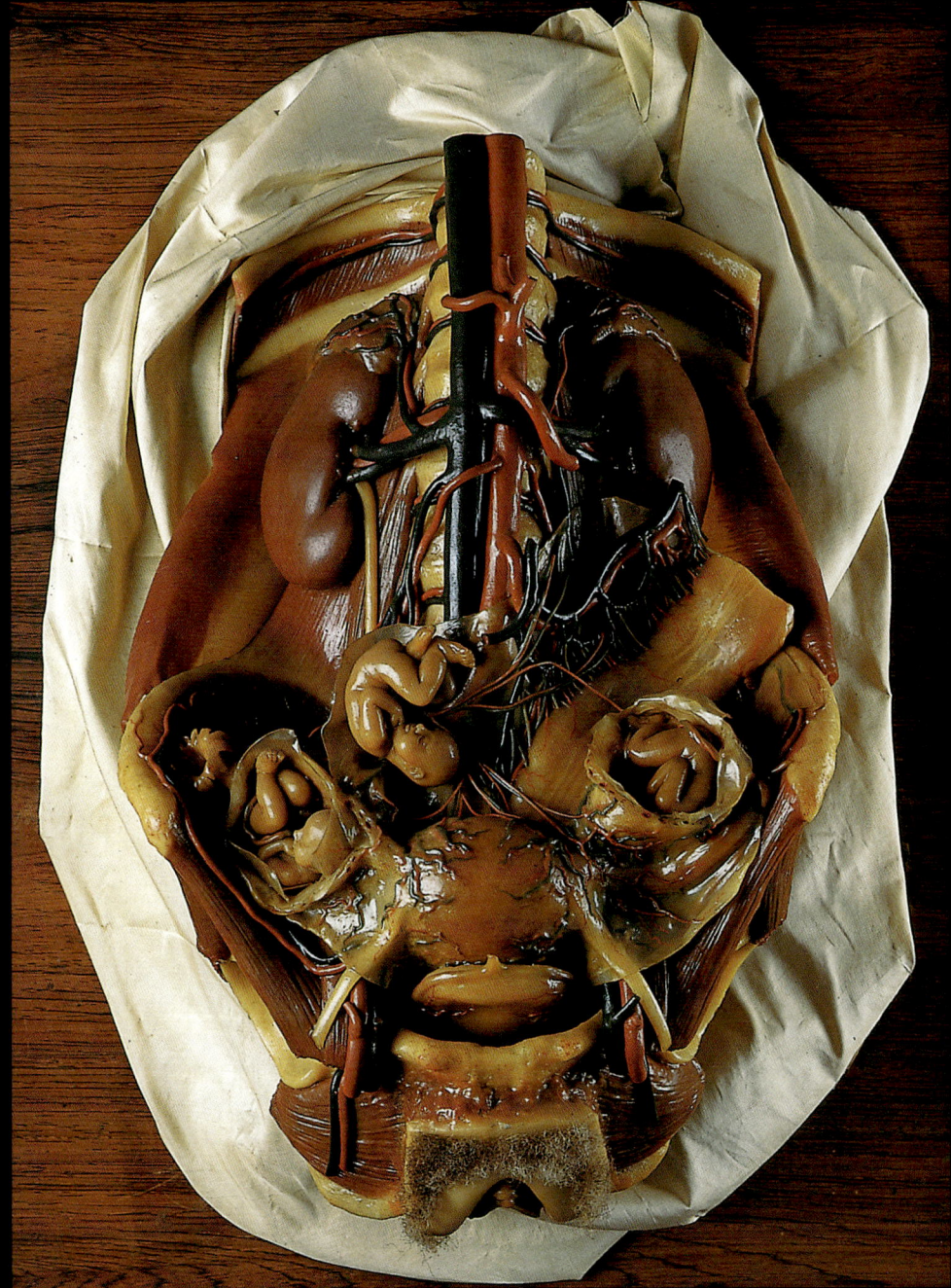

Graviditas extrauterina

Specimen showing abdominal pregnancy

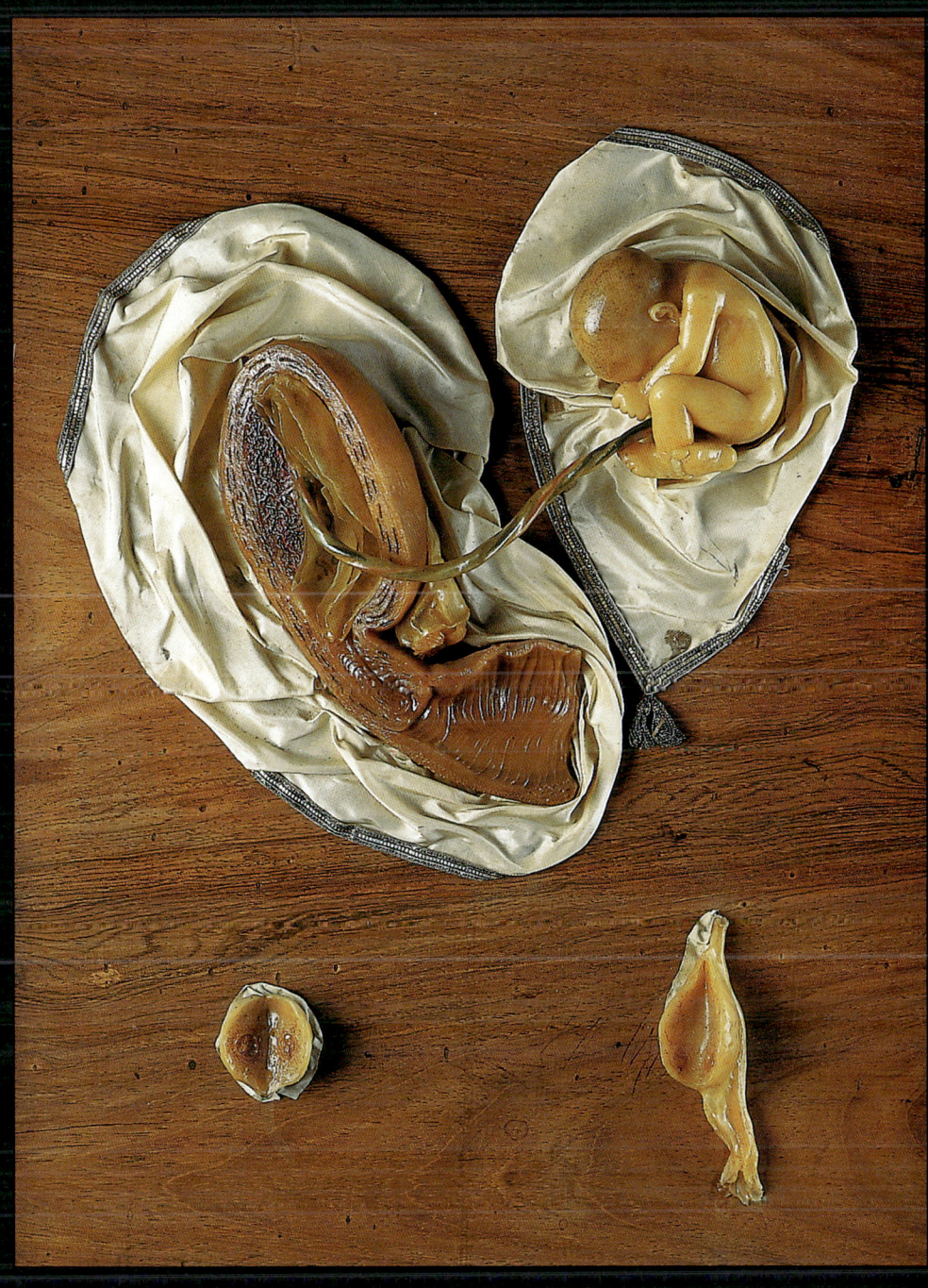

Placenta, Funiculus umbilicalis

Uterus with placenta, umbilical cord and foetus

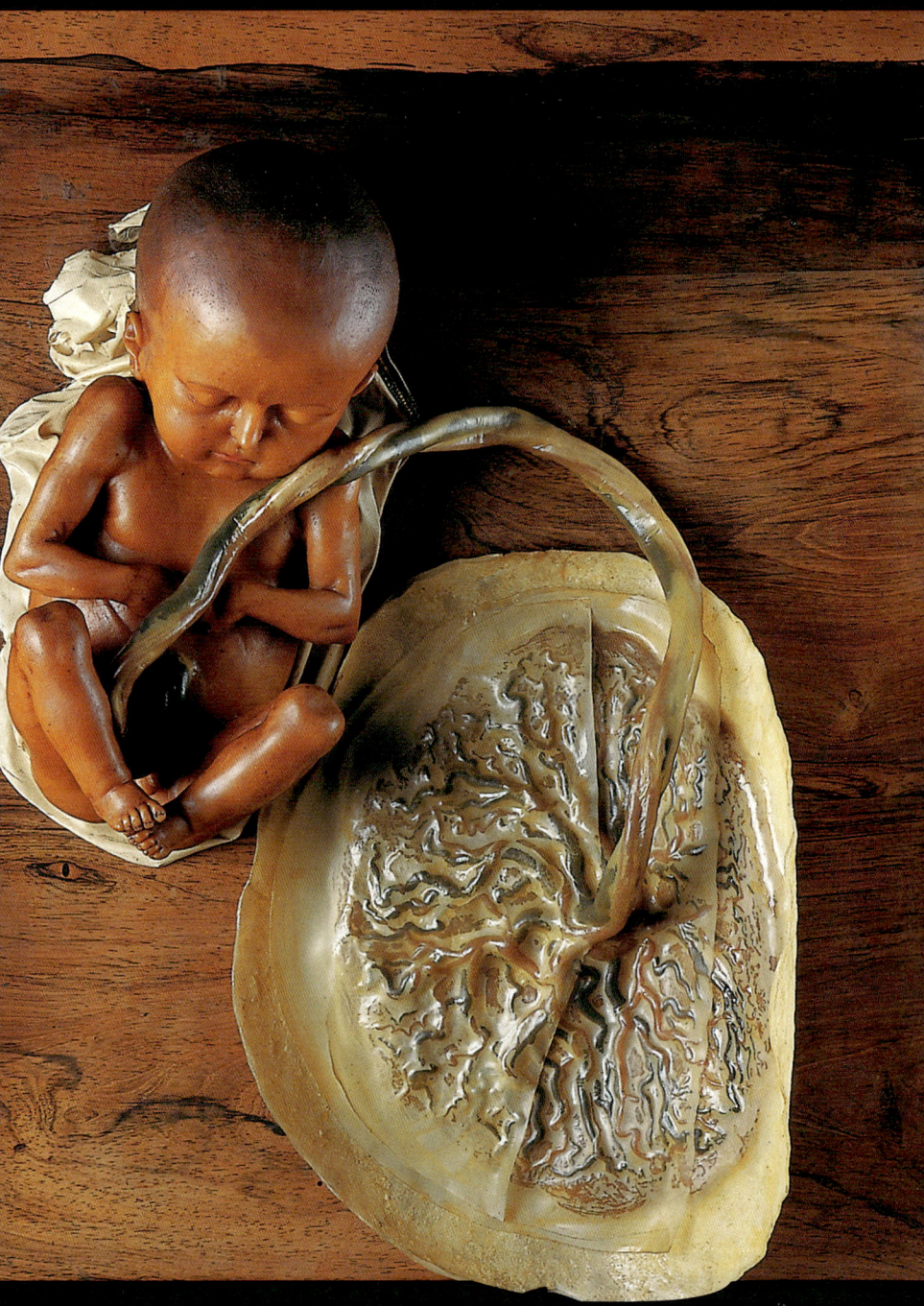

Placenta, Funiculus umbilicalis
Newborn infant with umbilical cord and placenta

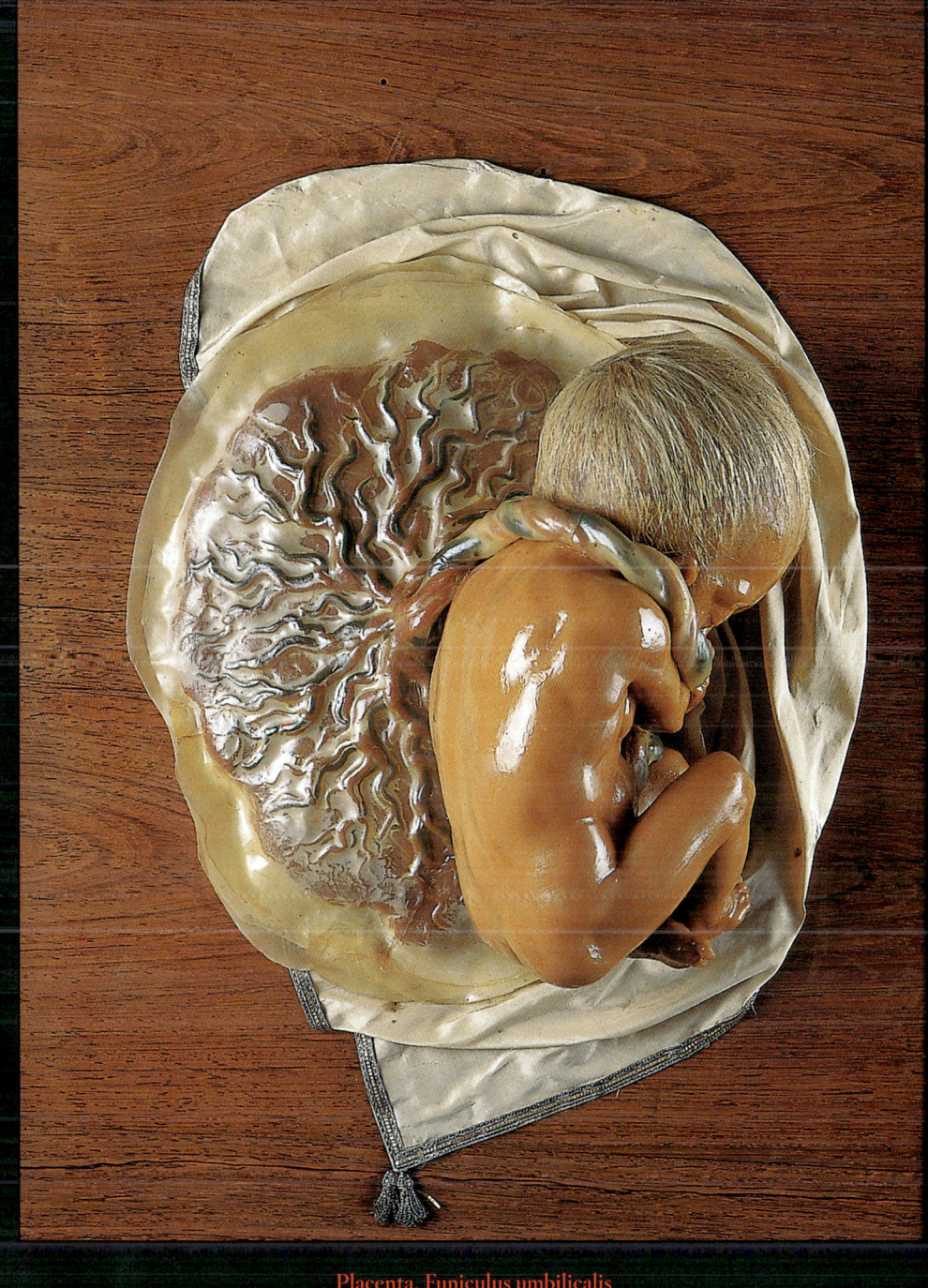

Placenta, Funiculus umbilicalis

Newborn infant with umbilical cord and placenta

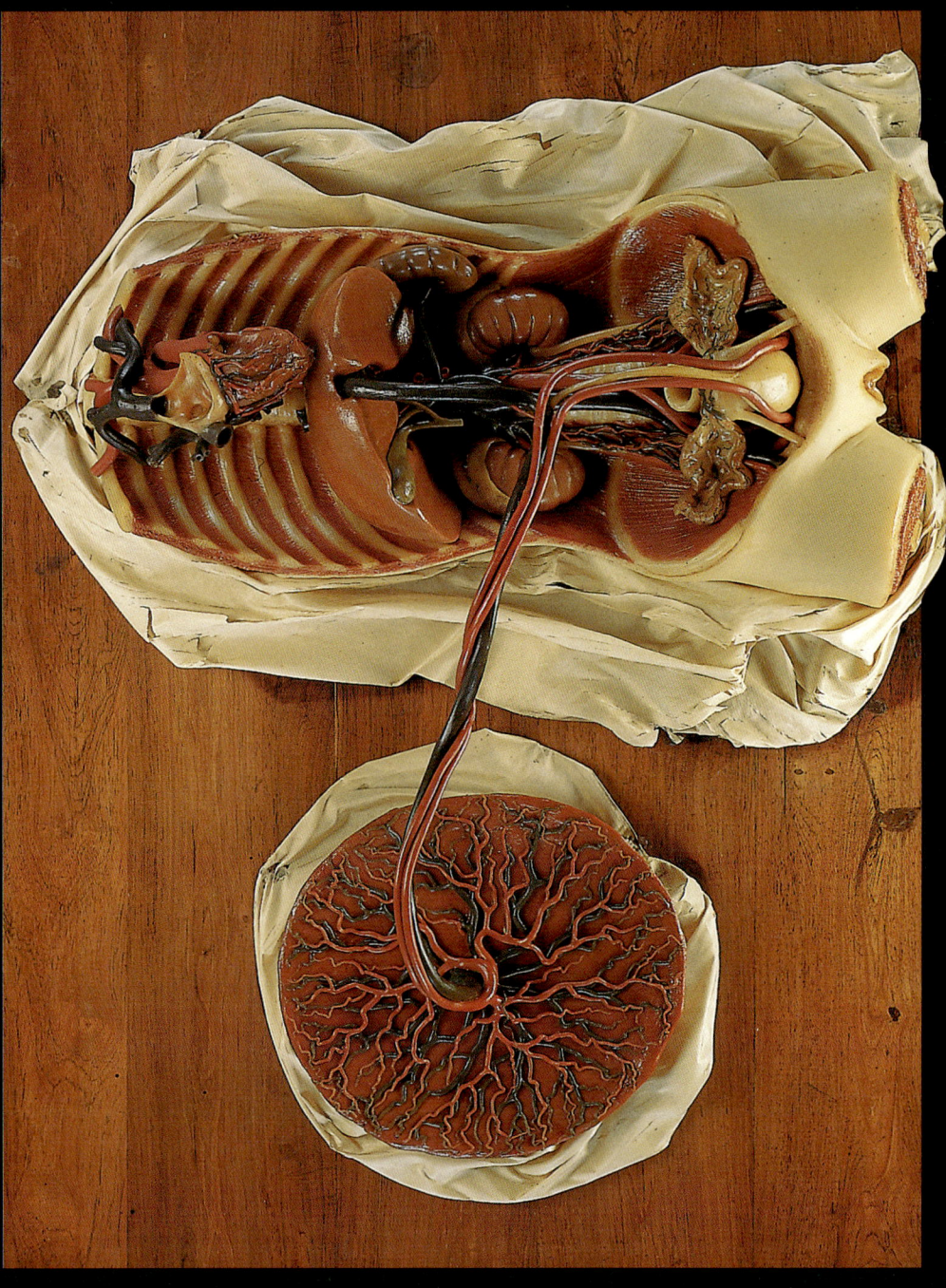

Placenta, Fetus

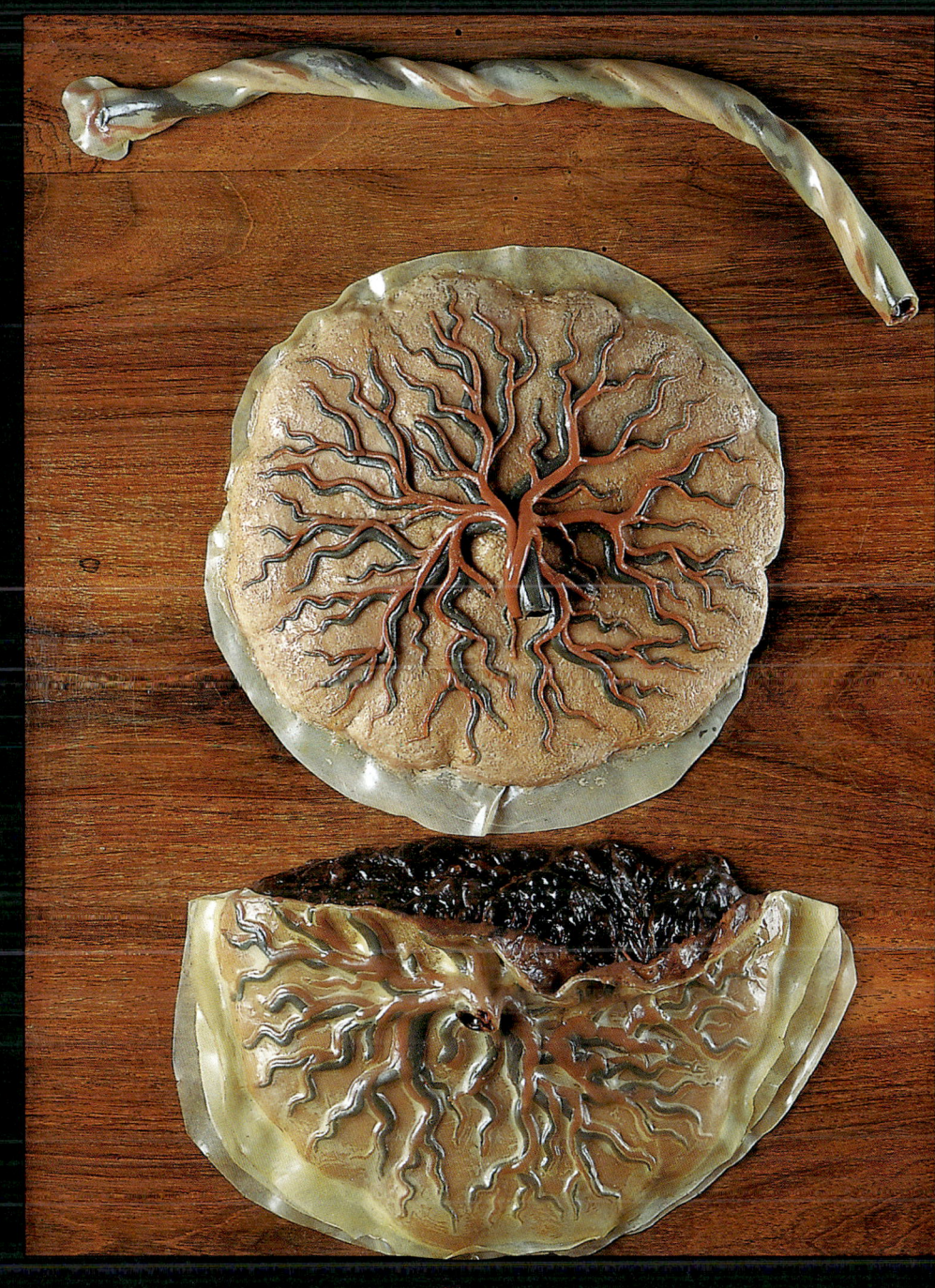

Placenta

Placenta and umbilical cord

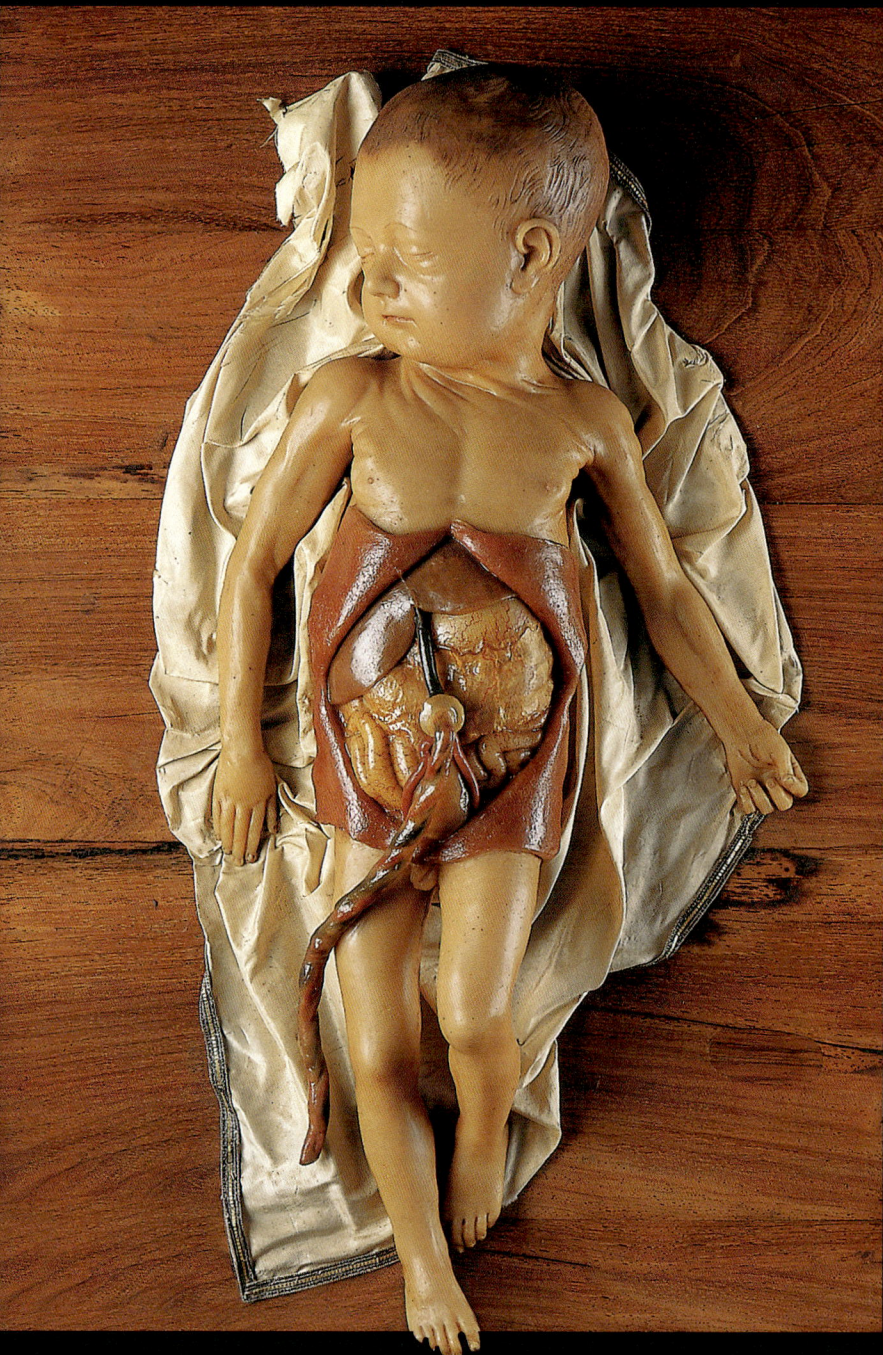

Funiculus umbilicalis, Urachus

Foetus with abdominal cavity laid open to display the umbilical arteries and vein

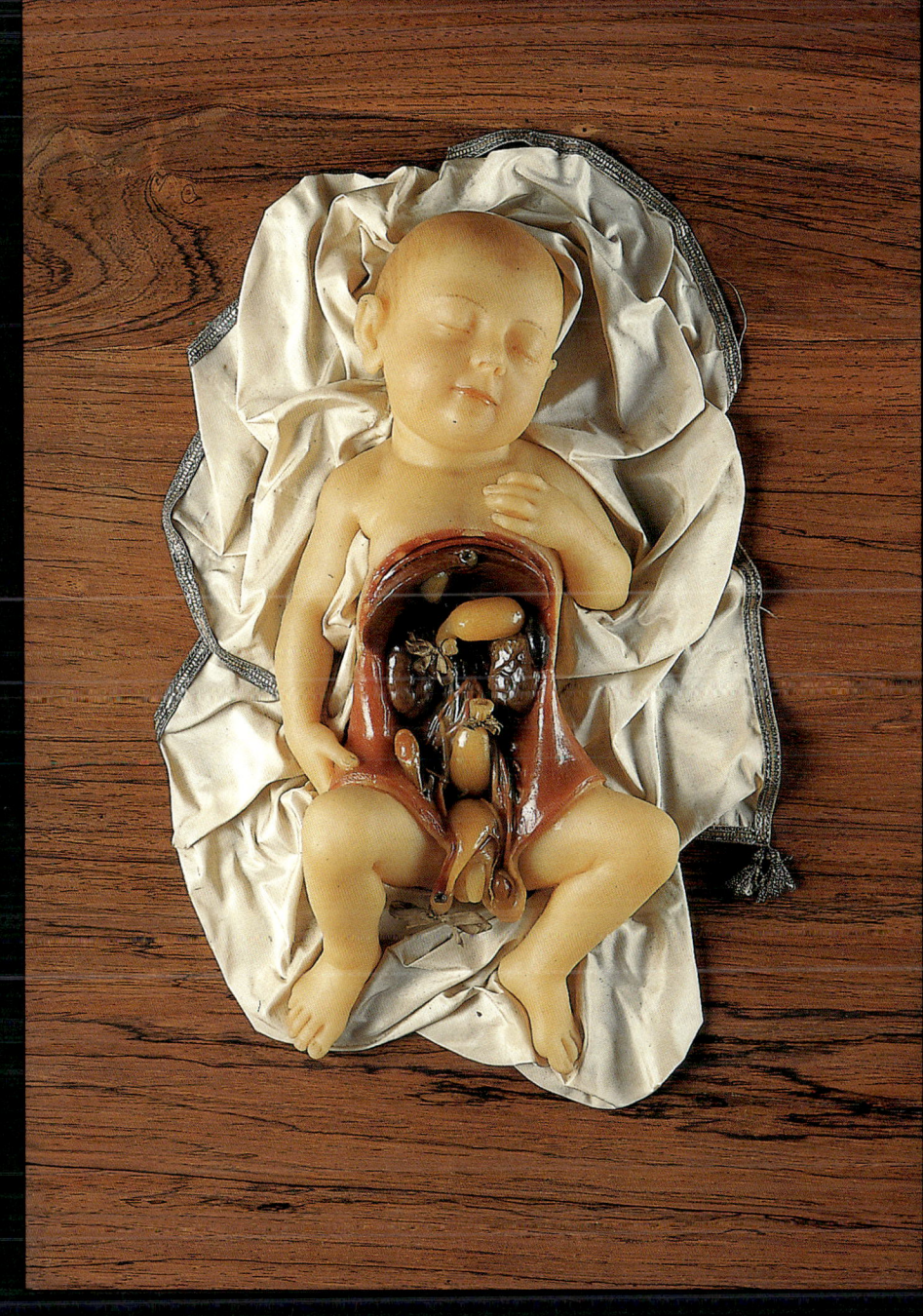

Fetus, Abdomen

Foetus with abdominal cavity laid open and abdominal organs removed to display the urogenital system. The left testis has descended into the scrotum, the right testis is in the abdominal cavity

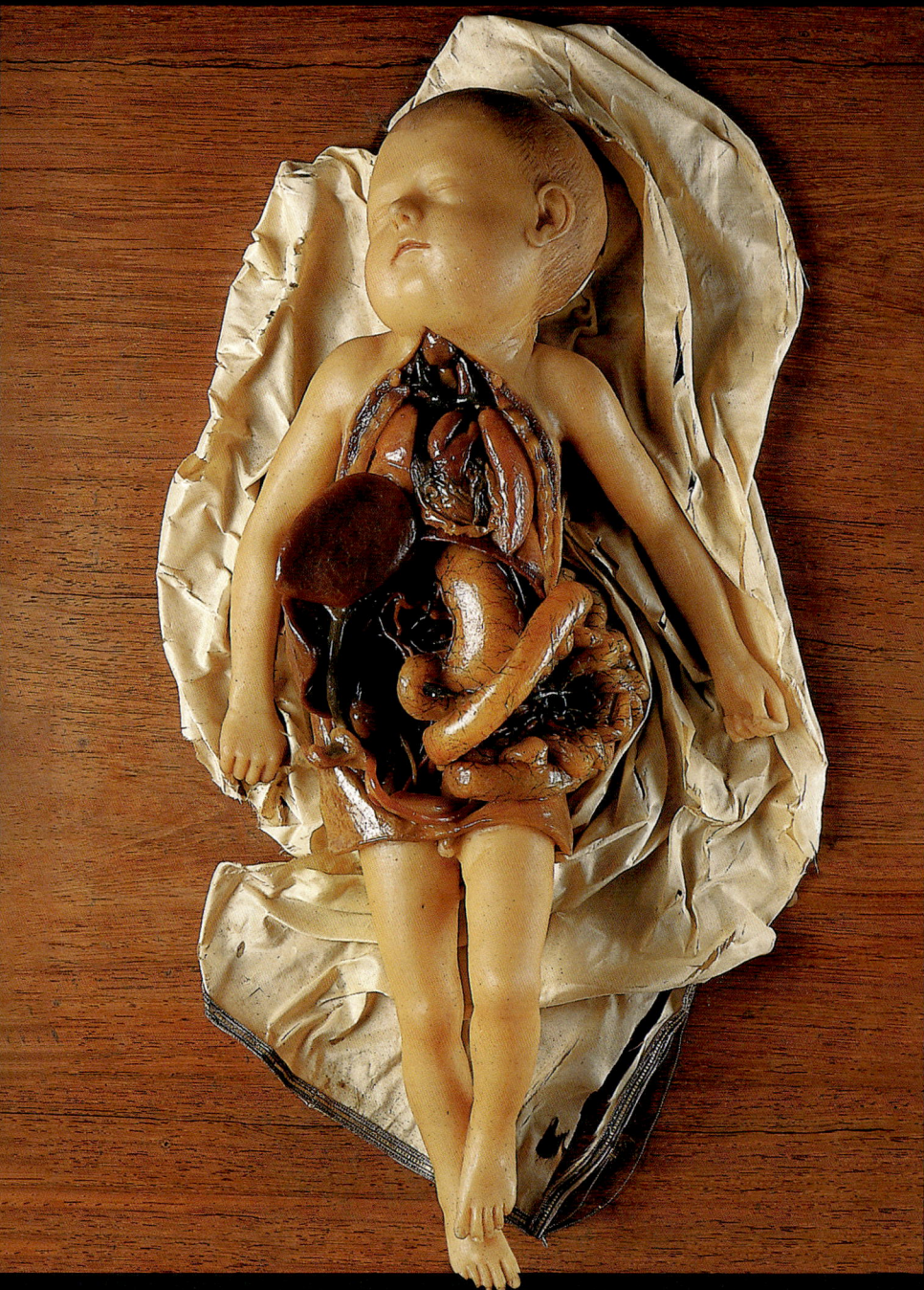

Fetus, Thorax, Abdomen

Foetus with thoracic and abdominal cavities laid open

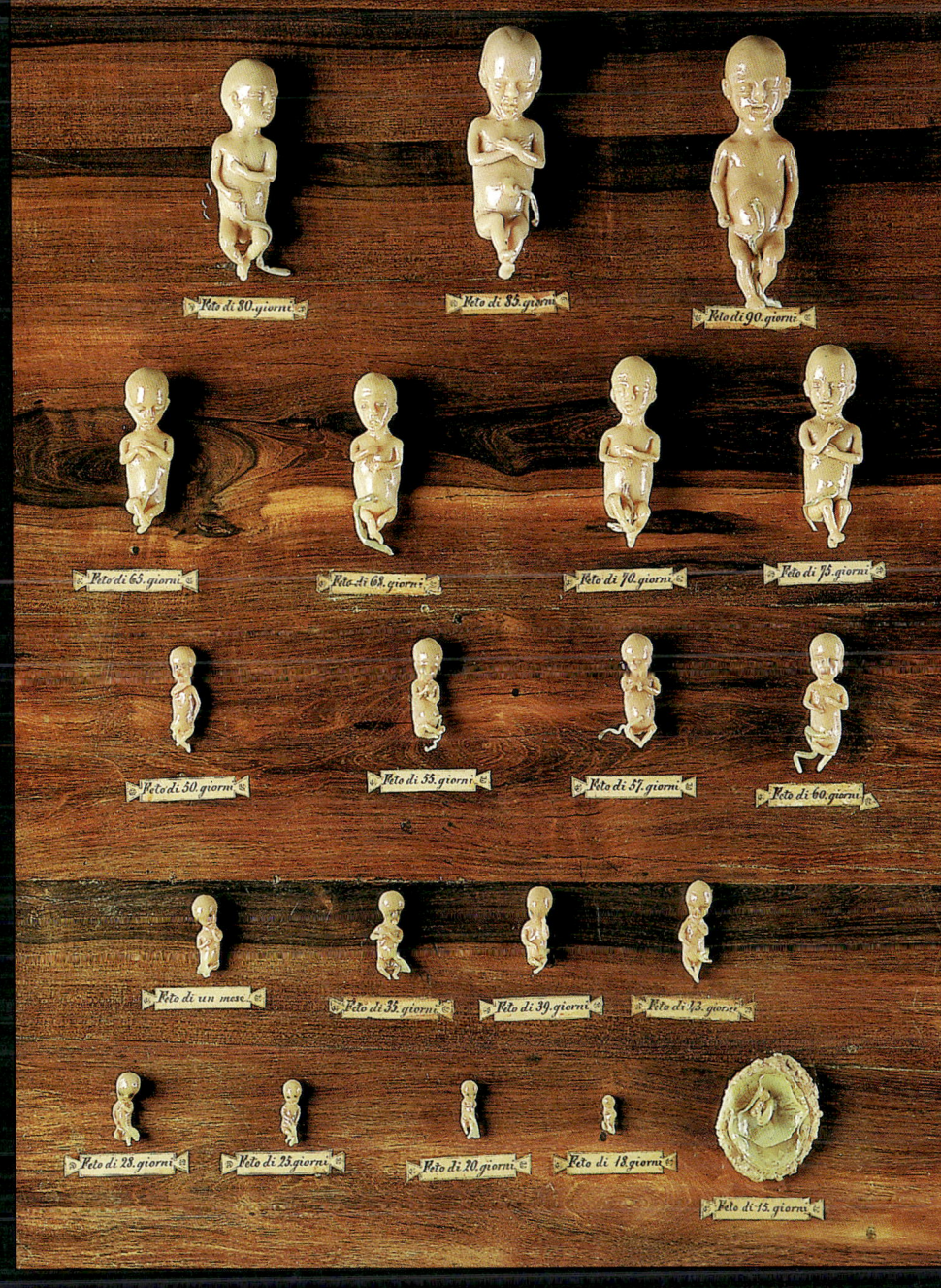

Feto di 80. giorni
Feto di 85. giorni
Feto di 90. giorni

Feto di 65. giorni
Feto di 68. giorni
Feto di 70. giorni
Feto di 75. giorni

Feto di 50. giorni
Feto di 55. giorni
Feto di 57. giorni
Feto di 60. giorni

Feto di un mese
Feto di 35. giorni
Feto di 39. giorni
Feto di 43. giorni

Feto di 28. giorni
Feto di 23. giorni
Feto di 20. giorni
Feto di 18. giorni
Feto di 15. giorni

Fetus, Mens I–III

Display showing the various developmental stages of the embryo up to the third month of pregnancy

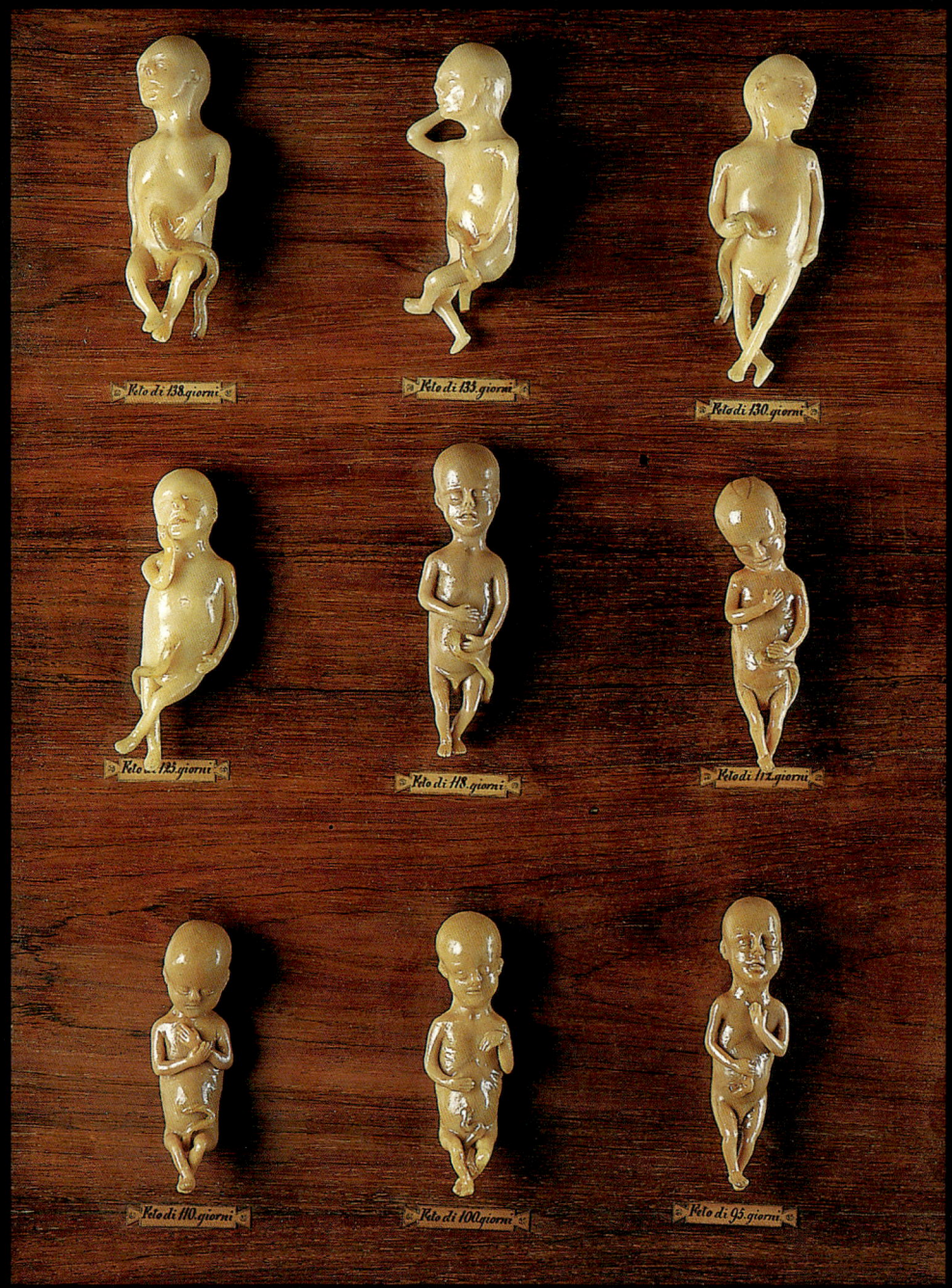

Feto di 138.giorni

Feto di 135.giorni

Feto di 130.giorni

Feto ... 125.giorni

Feto di 118.giorni

Feto di 112.giorni

Feto di 110.giorni

Feto di 100.giorni

Feto di 95.giorni

Fetus, Mens IV

Foetus, fourth month of pregnancy

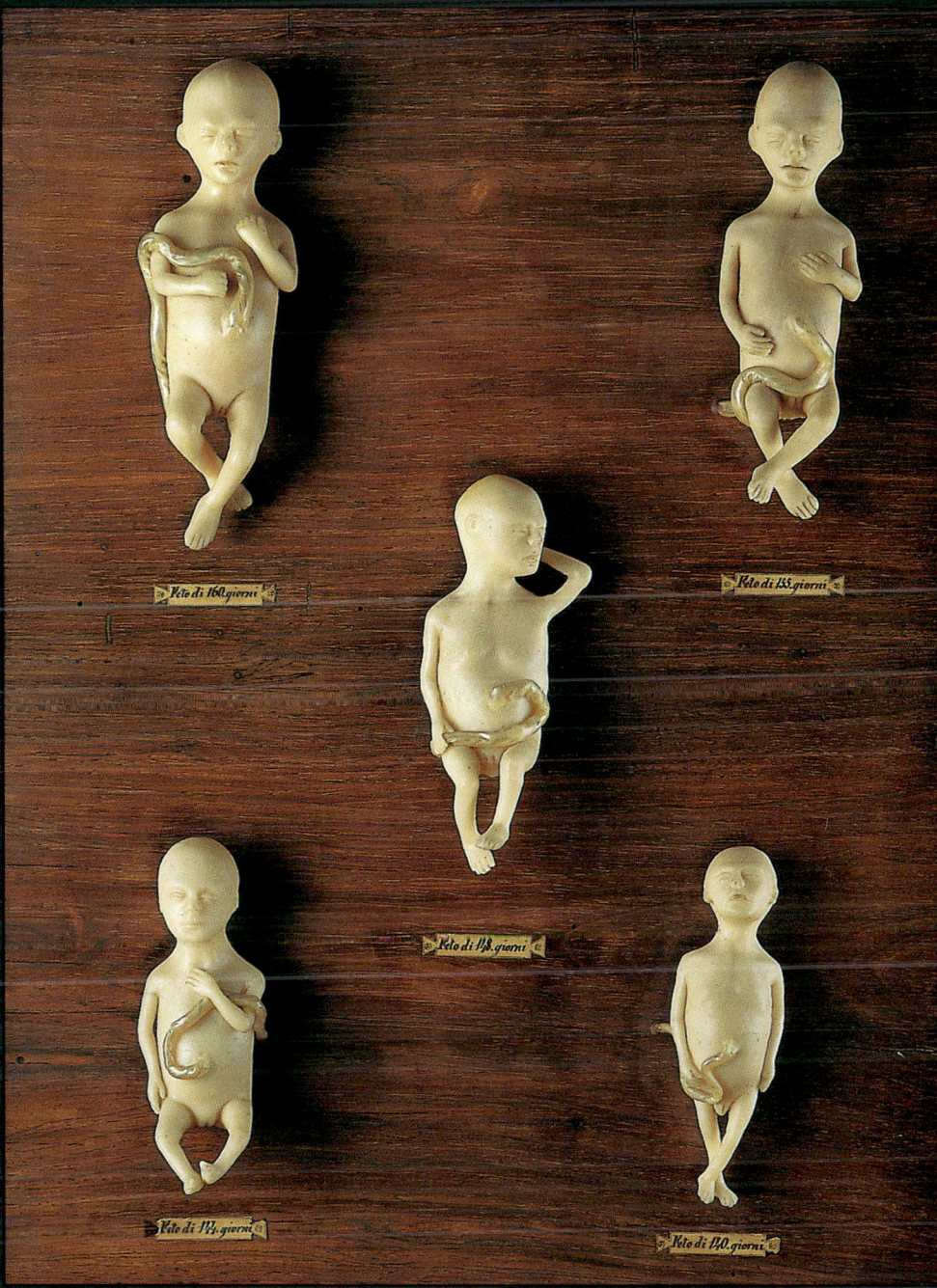

Fetus, Mens V

Foetus, fifth month of pregnancy

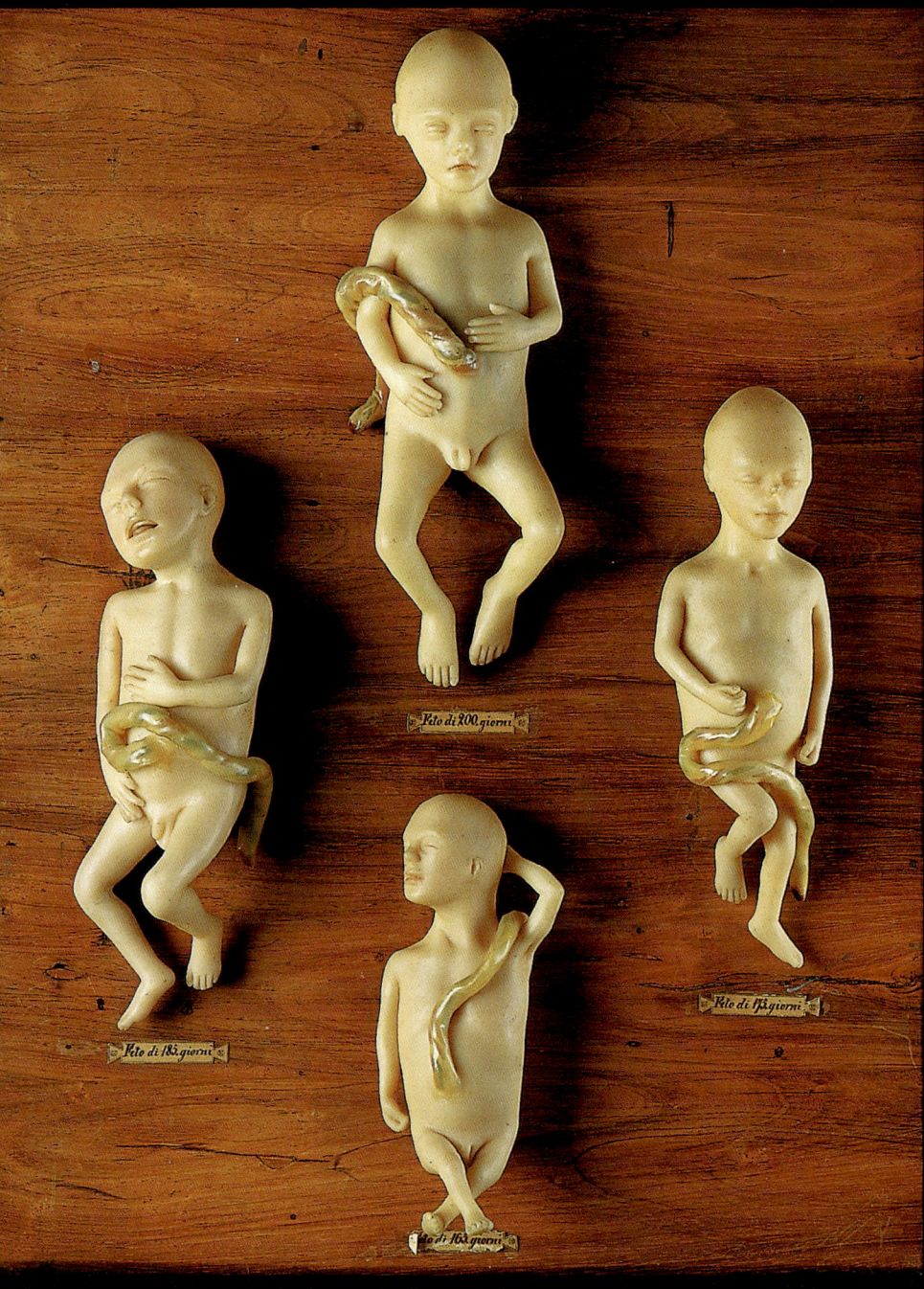

Feto di 200 giorni

Feto di 183 giorni

Feto di 193 giorni

Feto di 163 giorni

Fetus, Mens VI

Foetus, sixth month of pregnancy

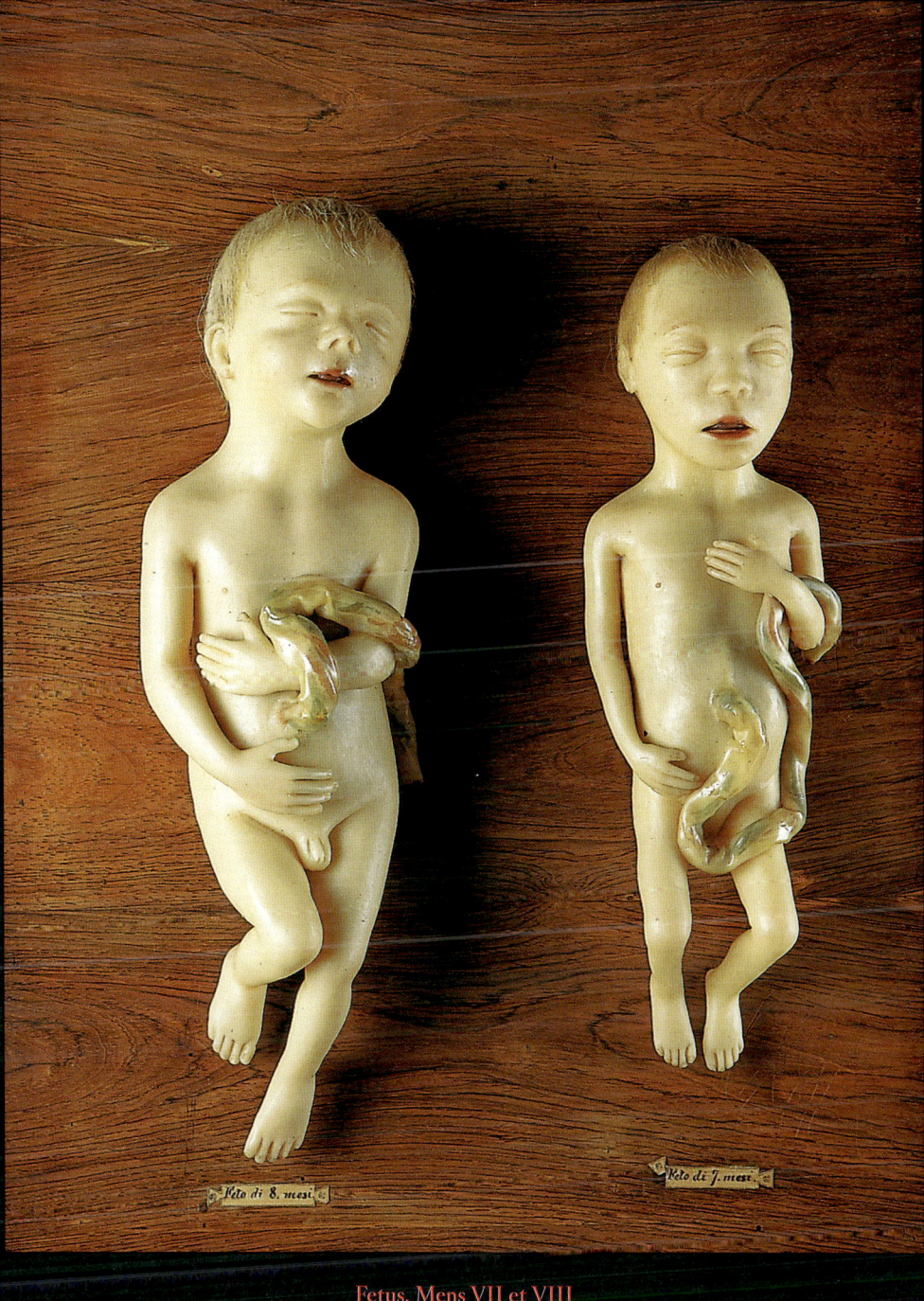

Feto di 8. mesi

Feto di 7. mesi

Fetus, Mens VII et VIII

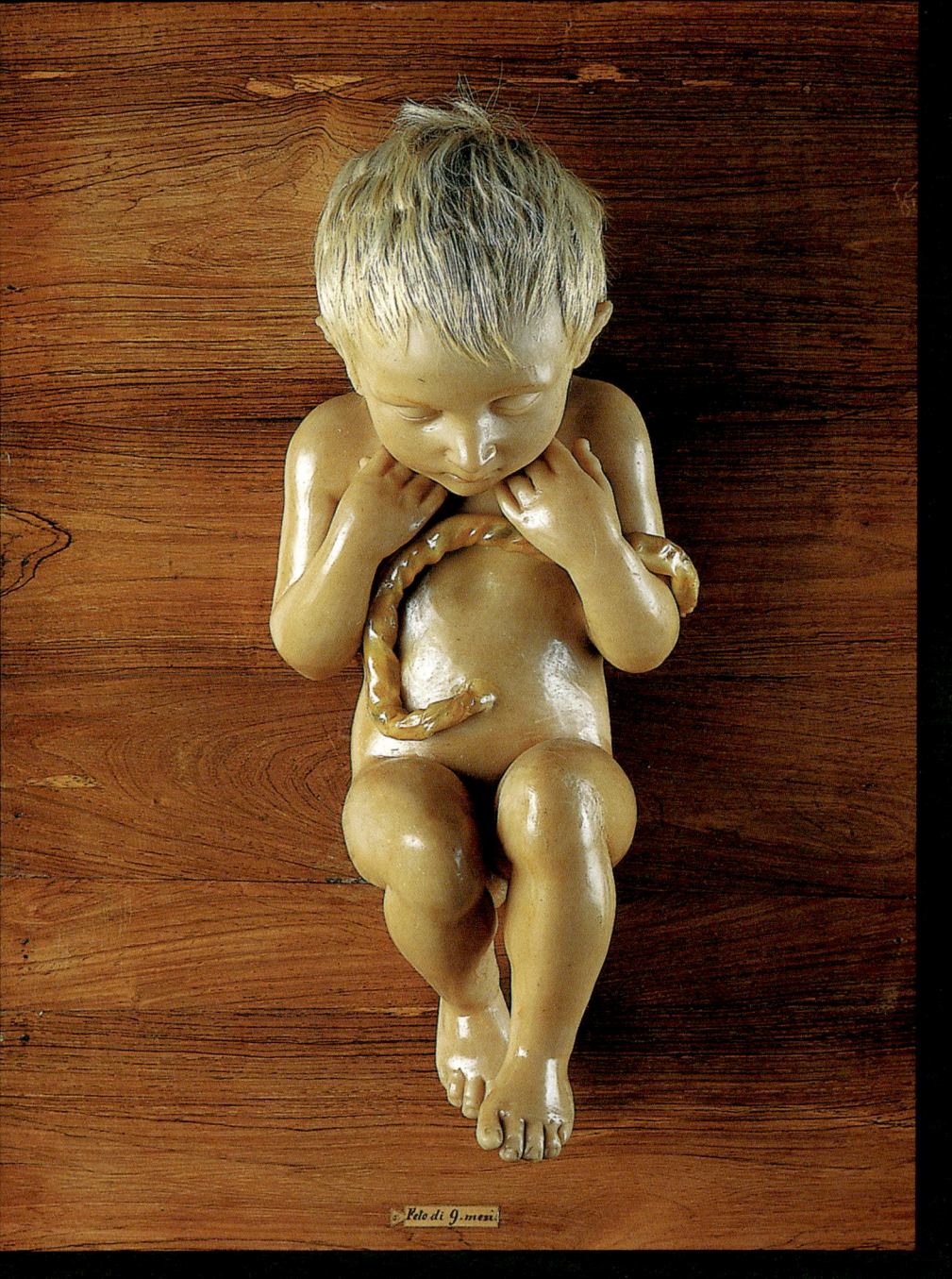

Feto di 9.mesi.

Fetus, Mens IX, Funiculus umbilicalis

Foetus, ninth month of pregnancy